IB BIOLOGY

Student Workbook

BIOZONE

IB Biology

Student Workbook

First edition 2012
Second printing 2013

ISBN 978-1-927173-16-9

Published by **BIOZONE International Ltd**

Printed by REPLIKA PRESS PVT LTD using paper produced from renewable and waste materials

About the Writing Team

Tracey Greenwood joined the staff of Biozone at the beginning of 1993. She has a Ph.D in biology, specialising in lake ecology, and taught undergraduate and graduate biology at the University of Waikato for four years.

Lissa Bainbridge-Smith worked in industry in a research and development capacity for eight years before joining Biozone in 2006. Lissa has an M.Sc from Waikato University.

Kent Pryor has a BSc from Massey University majoring in zoology and ecology. He was a secondary school teacher in biology and chemistry for 9 years before joining Biozone as an author in 2009.

Richard Allan has had 11 years experience teaching senior biology at Hillcrest High School in Hamilton, New Zealand. He attained a Masters degree in biology at Waikato University, New Zealand.

Purchases of this workbook may be made direct from the publisher:

BIOZONE

www.theBIOZONE.com

USA, CANADA & REST OF WORLD:

BIOZONE International Ltd.
P.O. Box 5002, Hamilton 3242, New Zealand
Telephone: +64 7-856 8104
Fax: +64 7-856 9243
Toll FREE phone: 1-866-556-2710 (USA-Canada only)
Toll FREE fax: 1 800 717 8751 (USA-Canada only)
Email: sales@biozone.co.nz
Website: www.the**BIOZONE**.com

UNITED KINGDOM & EUROPE:

BIOZONE Learning Media (UK) Ltd.
Unit 5/6, Greenline Business Park,
Wellington Street, Burton-on-Trent,
DE14 2AS, United Kingdom
Telephone: +44 1283 530 366
Fax: +44 1283 530 961
Email: sales@biozone.co.uk
Website: www.**BIOZONE**.co.uk

AUSTRALIA:

BIOZONE Learning Media Australia
P.O. Box 2841, Burleigh BC,
QLD 4220, Australia
Telephone: +61 7 5535 4896
Fax: +61 7 5508 2432
Email: sales@biozone.com.au
Website: www.**BIOZONE**.com.au

Preface to the First Edition

This first edition of IB Biology has been specifically structured and written to meet the content and skills requirements of Biology for the IB Diploma. Content for both SL (core) and HL is integrated but clearly differentiated throughout and a wealth of activities provide both consolidation and extension of prior knowledge. Learning objectives for each chapter provide students with a concise guide to required outcomes and *Theory of Knowledge* is supported throughout. *IB Biology*'s flexible concept-based structure accommodates diverse learning styles and allows for multiple approaches to teaching essential knowledge. The pedagogical approach of the workbook supports the development of the IB learner's skills, encouraging independence, inquiry, and critical thinking. Key features include:

- Learning Objectives and activities for all SL and HL options available on an IB Options CD-ROM (for separate purchase).
- A contextual approach. We encourage students to become thinkers through the application of their knowledge in appropriate contexts. Chapters may include an account examining a 'biological story' related to the theme of the chapter. This approach provides a context for the material to follow and an opportunity to focus on comprehension and the synthesis of ideas. Throughout the workbook, there are many examples of applying knowledge within context.
- Concept maps introduce each of the two main sections of the workbook, integrating the content across chapters to encourage linking of ideas.
- An easy-to-use chapter introduction summarizing essential knowledge in a numbered list to be completed by the student. Chapter introductions also include a list of key terms and a summary of key concepts.
- An emphasis on acquiring skills in scientific literacy. Each chapter includes a comprehension and/or literacy activity, and the appendix includes references for works cited throughout the text.
- *Weblinks* and *Related Activities* support the material provided on each activity page. Students should bookmark BIOZONE's Weblinks address and visit the site regularly as they progress through the material in the workbook. Each weblink provides a link to a supporting animation or video clip relevant to the material on the page where it is cited.

A Note to the Teacher

This workbook is a student-centered resource, and benefits students by facilitating independent learning and critical thinking. This workbook is just that; a place for your answers notes, asides, and corrections. It is **not a textbook** and regular revisions are our commitment to providing a current, flexible, and engaging resource. The low price is a reflection of this commitment. Please **do not photocopy** the activities. If you think it is worth using, then we recommend that the students themselves own this resource and keep it for their own use. I thank you for your support.
Richard Allan

Acknowledgements

We would like to thank those who have contributed to this edition • Sue FitzGerald, Gwen Gilbert, and Mary McDougall for their efficient handling of the office • Denise Fort, Gemma Conn, and Edith Woischin for graphics support • Paolo Curray for IT support • Suzanne Branford for maintaining the database of journals • TechPool Studios, for their clipart collection of human anatomy: Copyright ©1994, TechPool Studios Corp. USA (some of these images have been modified) • Totem Graphics, for clipart • Corel Corporation, for vector art from the Corel MEGAGALLERY collection.

Photo Credits

Royalty free images, purchased by Biozone International Ltd, are used throughout this workbook and have been obtained from the following sources: **Corel** Corporation from various titles in their Professional Photos CD-ROM collection; **IMSI** (International Microcomputer Software Inc.) images from IMSI's MasterClips® and MasterPhotosTM Collection, 1895 Francisco Blvd. East, San Rafael, CA 94901-5506, USA; ©1996 **Digital Stock**, Medicine and Health Care collection; ©**Hemera** Technologies Inc, 1997-2001; © 2005 JupiterImages Corporation www.clipart.com; ©1994., ©**Digital Vision**; Gazelle Technologies Inc.; **PhotoDisc**®, Inc. USA, www.photodisc.com • 3D modeling software, Poser IV (Curious Labs) and Bryce.

The writing team would like to thank the following individuals and institutions who kindly provided photographs: • Alison Roberts for the image of the plasmodesmata • Wadsworth Centre (NYSDH) for the photo of the cell undergoing cytokinesis • Alan Sheldon, Sheldon's Nature Photography, Wisconsin for the photo of the lizard without its tail • Ed Uthman for the image of the nine week human embryo • Louisa Howard Dartmouth College for image of the mitochondrion • Dartmouth College for image of the chloroplast • Cytogenetics Department, Waikato Hospital (NZ) for karyotype photographs • David Wells at Agresearch for photos on cloning • Roslin Institute for their photo of Dolly • Genesis Research and Development Corp for photos of PCR machines • Shirley Owens MSU for the image of the *Agrobacterium* • Rita Willaert, Flickr, for the photograph of the Nuba woman • Aptychus, Flickr for use of the photograph of the Tamil girl • Dan Butler for the photo of the wounded finger • Dr Roger Wagner, Dept of Biological Sciences, University of Delaware, for the LS of a capillary • Karl Mueller for the photograph of the Malawi grandmother • Dr Douglas Cooper and the University of California San Francisco, for the use of the SEM of a podocyte (http://www.sacs.ucsf.edu/home/cooper/Anat118/urinary/urinary98.htm) • John Mahn PLU, for the cross section of a dicot leaf • University of Florida for the photograph of strawberry runners.

Contributors identified by coded credits are as follows: **BF**: Brian Finerran (Uni. of Canterbury), **BH**: Brendan Hicks (Uni. of Waikato), **BOB**: Barry O'Brien (Uni. of Waikato), **CDC**: Centers for Disease Control and Prevention, Atlanta, USA, **DH**: Don Horne **EII**: Education Interactive Imaging, **HGSI**: Dena Borchardt at Human Genome Sciences Inc.,**NASA**: National Aeronautics and Space Administration, **NIH**: National Institutes of Health, **NYSDEC**: New York State Dept of Environmental Conservation, **RA**: Richard Allan, **RCN**: Ralph Cocklin, **TG**: Tracey Greenwood, **USDA**: United States Department of Agriculture, **WBS**: Warwick Silvester (Uni. of Waikato), **WMU**: Waikato Microscope Unit.

We also acknowledge the photographers that have made their images available through **Wikimedia Commons** under Creative Commons Licences 2.5. or 3.0: • D. Dibenski • Kaylaya • FontanaCG • Greenpeace • DualFreq • Romain Behar • DEQ • Gina Mikel • Andrew Dunn www.andrewdunnphoto.com • Olaf Leillinger • Ute Frevert • UC Regents David Campus • Phil Camill • LJ Grauke • Kristian Peters • Jjron (John O'Neill) • Ernie • Wmpearl

Special thanks to all the partners of the Biozone staff

Cover Photograph

The snowy owl (*Bubo scandiacus*) is a migratory bird inhabiting the Arctic tundra in warmer months, and migrating south to North America, Europe, and Asia in winter. It is diurnal, hunting its primary food source (lemmings) day and night. Snowy owls breed on the Arctic tundra and are highly territorial, defending the nest vigorously against much larger animals (including wolves). Its magnificent plumage ranges from snowy white (in older males), to white with dark bars and spots (in females and juveniles).

PHOTO: iStock © Adam Bennie

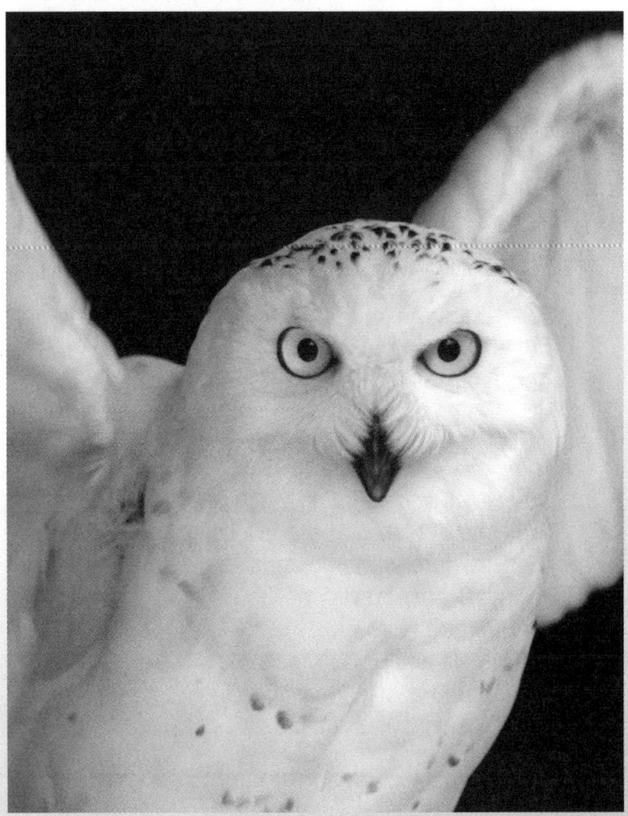

Contents

Note to the Teacher & Acknowledgments ... iii
Getting The Most From This Resource 1
Using the Activities 2
Resources Information 3
Using BIOZONE's Website 4
Command Words 5
IB Biology Guide 6
Concept Map 8

Concept Map: Cell Biology, Genetics, Ecology, and Evolution

Science Practices and Statistical Analysis

Objectives and Key Concepts 9
The Scientific Method 10
Hypotheses and Predictions 11
Accuracy and Precision 13
Variables and Data 14
Manipulating Raw Data 15
Planning a Quantitative Investigation 17
A Simple Investigation 19
Further Data Transformations 22
Constructing Tables 23
Constructing Graphs 24
Evaluating Your Results 25
Descriptive Statistics 27
Interpreting Sample Variability 29
Spearman Rank Correlation 31
The Student's *t* Test 32
Student's *t* Test Exercise 33
KEY TERMS: Mix and Match 35

Cell Biology

Objectives and Key Concepts 36
Cell Theory 37
Cell Sizes 38
Unicellular Eukaryotes 39
Calculating Linear Magnification 40
Surface Area and Volume 41
Prokaryotic Cells 43
Multicellularity 44
Stem Cells and Differentiation 45
Plant Cells 47
Animal Cells 49
Cell Structures and Organelles 51
Identifying Structures in an Animal Cell 54
Identifying Structures in a Plant Cell 55
Interpreting Electron Micrographs 56

Cell Processes

Objectives and Key Concepts 57
The Role of Membranes in Cells 58
The Structure of Membranes 59
Passive Transport Processes 61
Ion Pumps 64
Exocytosis and Endocytosis 65
Active and Passive Transport Summary 66
Cell Division 67
Cancer: Cells out of Control 68
Mitosis and the Cell Cycle 69
Apoptosis: Programmed Cell Death 71
KEY TERMS: Mix and Match 72

The Chemistry of Cells

Objectives and Key Concepts 73
The Biochemical Nature of the Cell 74
Organic Molecules 75
The Role of Water 76
Monosaccharides and Disaccharides 77
Condensation and Hydrolysis of Sugars 78
Polysaccharides 79
Lipids 81
Amino Acids 83
The Properties of Amino Acids 84
Proteins 85
Protein Structure and Function 86
KEY TERMS: Mix and Match 87

The Structure and Function of DNA

Objectives and Key Concepts 88
Nucleotides and Nucleic Acids 89
Packaging DNA in the Nucleus 91
DNA Molecules 93
The Genetic Code 94
Creating a DNA Molecule 95
DNA Replication 99
Enzyme Control of DNA Replication 101
Review of DNA Replication 102
Genes to Proteins 103
Transcription 104
Translation 105
Protein Synthesis Summary 106
KEY TERMS: Word Find 107

Enzymes and Metabolism

Objectives and Key Concepts 108
Enzymes 109
How Enzymes Work 110
Enzyme Reaction Rates 111
Enzyme Cofactors and Inhibitors 112
Control of Metabolic Pathways 114
Applications of Enzymes 115
ATP and Metabolism 116
The Role of ATP in Cells 117
Cell Respiration 118
The Biochemistry of Respiration 119
Chemiosmosis 121
Anaerobic Pathways 122
Investigating Yeast Fermentation 123
Photosynthesis 125
Pigments and Light Absorption 126
Factors Affecting Photosynthetic Rate 127
Light Dependent Reactions 128
Light Independent Reactions 130
KEY TERMS: Mix and Match 131

Activity is marked: ☐ to be done; ☑ when completed

Contents

Chromosomes and Meiosis

Objectives and Key Concepts 132
- Genomes 133
- Alleles 134
- Mitosis vs Meiosis 135
- Non-Disjunction in Meiosis 136
- Stages in Meiosis 137
- Crossing Over 138
- Modelling Meiosis 139
- Changes to the DNA Sequence 141
- Sickle Cell Mutation 142
- Karyotypes 143
- Human Karyotype Exercise 145
- KEY TERMS: Word Find 148

Heredity

Objectives and Key Concepts 149
- A Gene That Can Tell Your Future? 150
- Variation 151
- Mendel's Pea Plant Experiments 153
- Mendel's Laws of Inheritance 154
- Basic Genetic Crosses 155
- The Test Cross 156
- Monohybrid Cross 157
- Codominance of Alleles 158
- Codominance in Multiple Allele Systems 159
- Problems Involving Monohybrid Inheritance 161
- Dihybrid Cross 162
- Inheritance of Linked Genes 163
- Recombination and Dihybrid Inheritance. 165
- Detecting Linkage in Dihybrid Inheritance 167
- Sex Determination 168
- Sex Linkage 169
- Inheritance Patterns 171
- Problems Involving Dihybrid Inheritance . 172
- Pedigree Analysis 174
- Polygenes 177
- Genetic Counseling 179
- KEY TERMS: Crossword 180

Genetic Engineering and Biotechnology

Objectives and Key Concepts 181
- Amazing Organisms, Amazing Enzymes 182
- Polymerase Chain Reaction 183
- Gel Electrophoresis 185
- DNA Profiling using PCR 186
- Forensic Applications of DNA Profiling 188
- Finding the Connection 189
- The Human Genome Project 190
- What is Genetic Modification? 192
- Restriction Enzymes 193
- Ligation 195
- Applications of GMOs 196
- *In vivo* Gene Cloning 197
- Using Recombinant Bacteria 199
- Golden Rice 201
- Production of Insulin 203
- Food for the Masses 205
- The Ethics of GMO Technology 207
- Cloning by Somatic Cell Nuclear Transfer 209
- Therapeutic Cloning 211
- KEY TERMS: Mix and Match 212

Ecology

Objectives and Key Concepts 213
- Components of an Ecosystem 214
- Food Chains 215
- Energy Inputs and Outputs 216
- Food Webs 217
- Energy Flow in an Ecosystem 219
- Ecological Pyramids 221
- Nutrient Cycles 223
- The Carbon Cycle 224
- Global Warming 226
- Global Warming and Effects on Biodiversity 228
- Global Warming and Effects on the Arctic 230
- Applying the Precautionary Principle 231
- Features of Populations 232
- Population Regulation 233
- Population Growth 234
- Population Growth Curves 235
- KEY TERMS: Mix and Match 236

Evolution

Objectives and Key Concepts 237
- Genes, Inheritance, and Selection 238
- The Fossil Record 239
- Selection and Population Change 240
- Homologous Structures 241
- Darwin's Theory 242
- Natural Selection 243
- Directional Selection in Moths 244
- Selection for Skin Color in Humans 245
- Disruptive Selection in Darwin's Finches. 247
- Insecticide Resistance 248
- Evolution in Response to Nutrient Levels 249
- The Evolution of Antibiotic Resistance 250
- Antigenic Variability in Pathogens 251

Classification

Objectives and Key Concepts 252
- Classification System 253
- Features of Taxonomic Groups 255
- Features of Fungi and Plants 260
- Features of Animal Taxa 261
- Classification Keys 263
- Keying Out Plant Species 265
- KEY TERMS: Crossword 266

Concept Map: Human Health and Physiology, and Plant Science

Digestion

Objectives and Key Concepts 268
- The Role of the Digestive System 269
- The Human Digestive Tract 270

Activity is marked: ☐ to be done; ☑ when completed

Contents

- The Digestive Role of the Liver 273
- Digestion, Absorption, and Transport....... 274
- KEY TERMS: Word Find 276

The Transport System

Objectives and Key Concepts 277
- The Human Circulatory System 278
- The Human Heart 279
- The Cardiac Cycle 281
- Control of Heart Activity 282
- Review of the Human Heart..................... 283
- Blood Vessels 284
- Capillary Networks.................................. 286
- Blood 287
- KEY TERMS: Mix and Match.................... 289

Defense Against Disease

Objectives and Key Concepts 290
- Pathogens and Disease 291
- Viral Diseases in Humans....................... 292
- The Body's Defenses.............................. 293
- Blood Clotting and Defense 295
- The Action of Phagocytes........................ 296
- Inflammation 297
- The Immune System............................... 298
- Antibodies 299
- Clonal Selection 300
- Acquired Immunity 301
- Vaccines and Vaccination 303
- Monoclonal Antibodies............................ 305
- HIV and AIDS 307
- The Impact of HIV/AIDS in Africa 309
- KEY TERMS: Word Find 310

Gas Exchange

Objectives and Key Concepts 311
- Introduction to Gas Exchange 312
- The Human Ventilation System................ 313
- Breathing in Humans 315
- Measuring Lung Function 316
- Gas Transport in Humans 318
- CONTEXT: The Effects of High Altitude... 319

Nerves, Hormones, and Homeostasis

Objectives and Key Concepts 320
- Nervous Regulatory Systems 321
- The Human Nervous System................... 322
- Neuron Structure and Function 323
- Reflexes 325
- Transmission of Nerve Impulses 326
- Chemical Synapses 327
- Hormonal Regulatory Systems 328
- Principles of Homeostasis 329
- Thermoregulation in Humans 330
- Acid-Base Balance 331
- Control of Blood Glucose 332
- Diabetes Mellitus 333
- Waste Products in Humans 334
- Water Budget in Humans......................... 335
- The Kidney... 336
- The Physiology of the Kidney 337
- Control of Urine Output........................... 339
- KEY TERMS: Word Find 340

Reproduction

Objectives and Key Concepts 341
- The Male Reproductive System 342
- The Female Reproductive System.......... 343
- The Menstrual Cycle 344
- Spermatogenesis 345
- Oogenesis.. 346
- Gametes 347
- Fertilization and Early Growth................. 348
- The Placenta 350
- The Hormones of Pregnancy.................. 351
- Birth 352
- *In Vitro* Fertilization 353
- KEY TERMS: Mix and Match 355

Muscles and Movement

Objectives and Key Concepts 356
- The Basis of Human Movement 357
- The Mechanics of Movement.................. 359
- Skeletal Muscle Structure and Function . 361
- The Sliding Filament Theory................... 363
- KEY TERMS: Word Find 364

Plant Science

Objectives and Key Concepts 365
- The General Structure of Plants 366
- Monocots vs Dicots 367
- Leaf Structure and Gas Exchange 368
- Dicot Stems and Roots........................... 369
- Modifications in Plants 371
- Qualitative Practical Work....................... 373
- Plant Meristems 374
- Support in Plants 376
- Xylem... 377
- Phloem 378
- Tropisms and Growth Responses 379
- Transport and Effects of Auxins 380
- Investigating Phototropism 381
- Uptake at the Root.................................. 382
- Transpiration 383
- Investigating Plant Transpiration............. 385
- Xerophytes... 387
- Translocation 389
- Insect Pollinated Flowers........................ 390
- Pollination and Fertilization..................... 391
- Seed Dispersal 392
- A Most Accomplished Traveller............... 393
- Seed Structure and Germination 394
- Events in Germination 395
- Photoperiodism in Plants........................ 396

APPENDIX 398
INDEX.. 401

Activity is marked: ☐ to be done; ☑ when completed

Getting The Most From This Resource

This workbook is designed as a resource that will help to increase your understanding of the content and skills requirements of **IB Biology**, and reinforce and extend the ideas developed by your teacher. This workbook includes many useful features to help you locate activities and information relating each topic.

BIOZONE encourages the development the of IB learner profile

Thinkers - applying thinking skills critically

Relating to others - open minded communicators

Using language, symbols, and text

Managing self - independent inquirers

Principled - act with integrity and honesty

Features of the Section Concept Map

The chapter panels identify and summarize the content of each chapter.

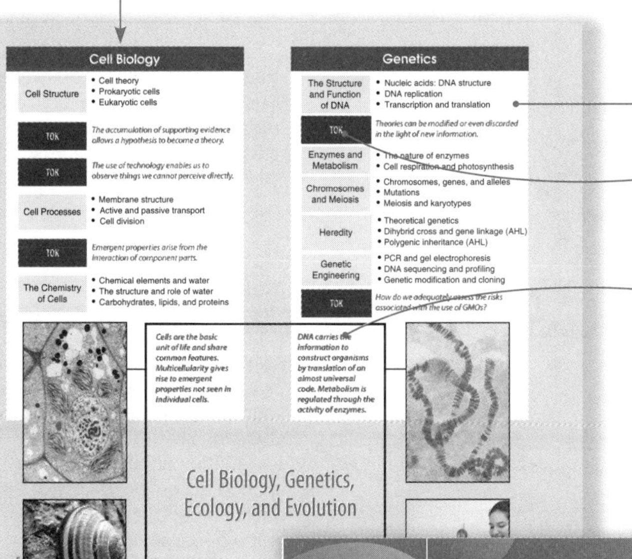

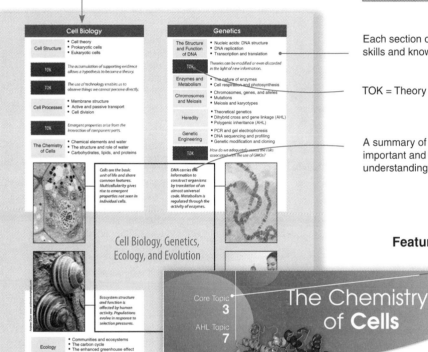

Each section of the workbook emphasizes skills and knowledge to be gained.

TOK = Theory of Knowledge connection.

A summary of why this material is important and where it fits into your understanding of your course content.

Features of the Chapter Topic Page

The topic code or codes to which this chapter applies. Divided into Core and AHL.

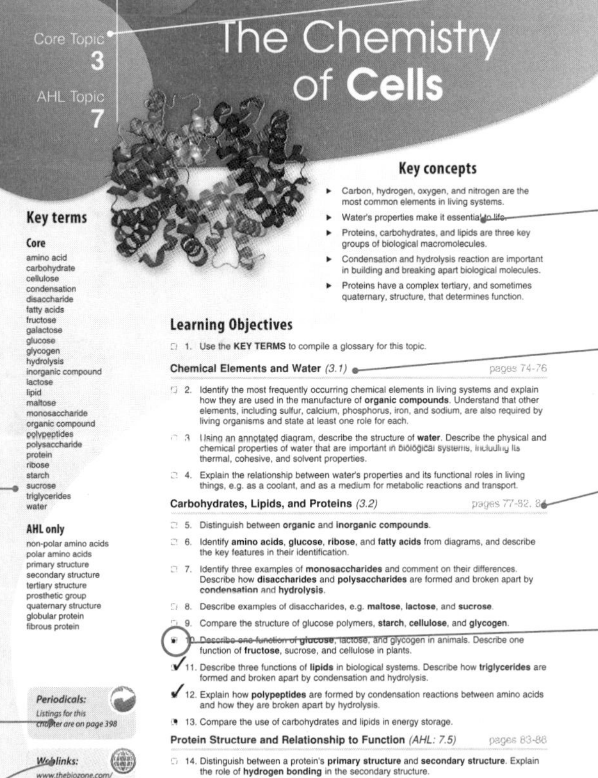

The important key ideas in this chapter. You should have a thorough understanding of the concepts summarized here.

The code from the IB syllabus is indicated for each subsection.

The page numbers direct you to material related to this subsection of work.

You can use the check boxes to mark objectives to be completed (a **dot** to be done; a **tick** when completed).

The learning objectives provide a point by point summary of what you should have achieved by the end of the chapter.

A list of key terms used in the chapter. Where applicable, the key terms are divided into core and AHL only lists. These terms appear in the chapter's vocabulary activity and can be used to create a glossary for revision purposes. The list represents the minimum literacy requirement for the chapter.

Periodicals of interest are identified by title on a tab on the activity page to which they are relevant. The full citation appears in the **Appendix** on the page indicated.

The Weblinks cited on many of the activity pages can be accessed through the web links page at: *www.thebiozone.com/weblink/IB-3169.html*

See page 4 for more details.

Student Review Series provide color review slides for purchase. Download via the free BIOZONE App, available on the App Store.

Using the Activities

The activities and exercises make up most of the content of this workbook. They are designed to reinforce the concepts you have learned about in the topic. Your teacher may use the activity pages to introduce a topic for the first time, or you may use them to revise ideas already covered. They are excellent for use in the classroom, and as homework exercises and revision. In most cases, the activities should not be attempted until you have carried out the necessary background reading from your textbook.

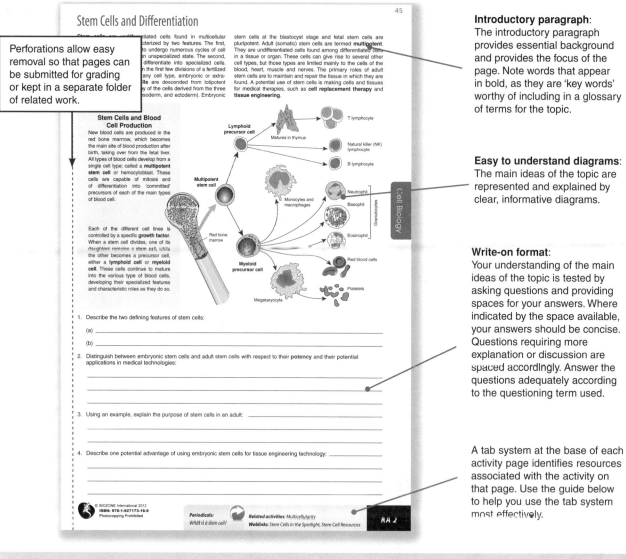

Perforations allow easy removal so that pages can be submitted for grading or kept in a separate folder of related work.

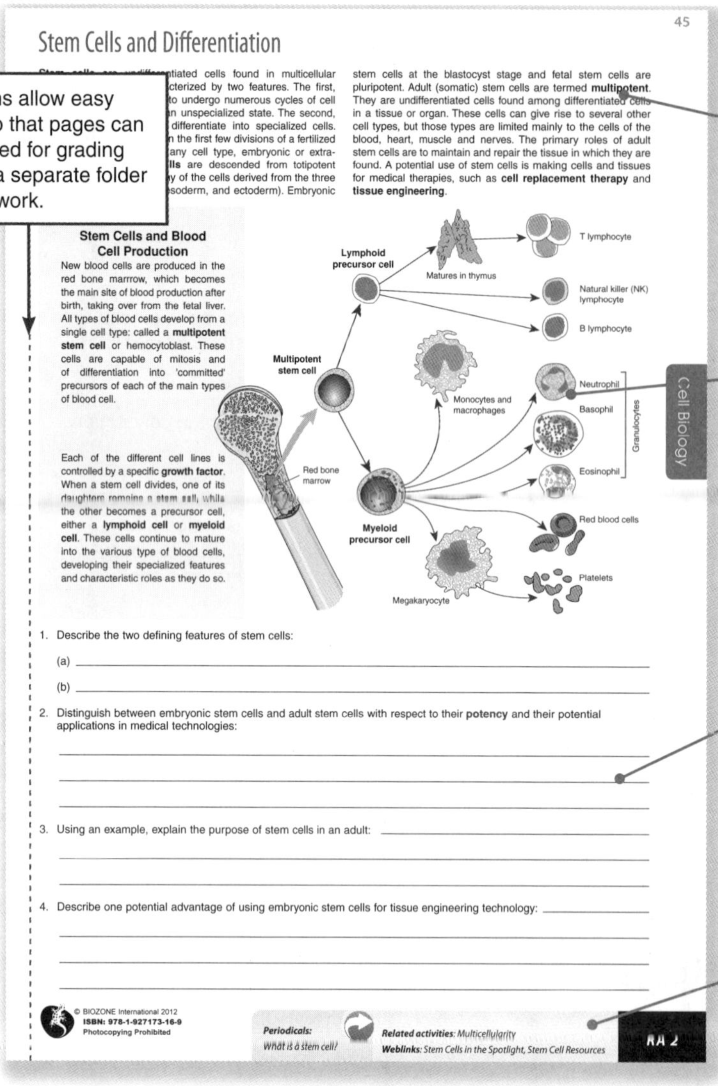

45

Stem Cells and Differentiation

...tiated cells found in multicellular ...cterized by two features. The first, ...to undergo numerous cycles of cell ...n unspecialized state. The second, ...differentiate into specialized cells. ...n the first few divisions of a fertilized ...any cell type, embryonic or extra- ...**lls** are descended from totipotent ...y of the cells derived from the three ...soderm, and ectoderm). Embryonic stem cells at the blastocyst stage and fetal stem cells are pluripotent. Adult (somatic) stem cells are termed **multipotent**. They are undifferentiated cells found among differentiated cells in a tissue or organ. These cells can give rise to several other cell types, but those types are limited mainly to the cells of the blood, heart, muscle and nerves. The primary roles of adult stem cells are to maintain and repair the tissue in which they are found. A potential use of stem cells is making cells and tissues for medical therapies, such as **cell replacement therapy** and **tissue engineering**.

Stem Cells and Blood Cell Production

New blood cells are produced in the red bone marrrow, which becomes the main site of blood production after birth, taking over from the fetal liver. All types of blood cells develop from a single cell type: called a **multipotent stem cell** or hemocytoblast. These cells are capable of mitosis and of differentiation into 'committed' precursors of each of the main types of blood cell.

Each of the different cell lines is controlled by a specific **growth factor**. When a stem cell divides, one of its daughters remains a stem cell, while the other becomes a precursor cell, either a **lymphoid cell** or **myeloid cell**. These cells continue to mature into the various type of blood cells, developing their specialized features and characteristic roles as they do so.

Lymphoid precursor cell
Matures in thymus
T lymphocyte
Natural killer (NK) lymphocyte
B lymphocyte
Multipotent stem cell
Monocytes and macrophages
Neutrophil
Basophil
Eosinophil
Granulocytes
Red bone marrow
Myeloid precursor cell
Red blood cells
Megakaryocyte
Platelets

Cell Biology

1. Describe the two defining features of stem cells:

(a) ______

(b) ______

2. Distinguish between embryonic stem cells and adult stem cells with respect to their **potency** and their potential applications in medical technologies:

3. Using an example, explain the purpose of stem cells in an adult: ______

4. Describe one potential advantage of using embryonic stem cells for tissue engineering technology: ______

© BIOZONE International 2012
ISBN: 978-1-927173-16-9
Photocopying Prohibited

Periodicals: What is a stem cell? *Related activities:* Multicellularity *Weblinks:* Stem Cells in the Spotlight, Stem Cell Resources RA 2

Introductory paragraph: The introductory paragraph provides essential background and provides the focus of the page. Note words that appear in bold, as they are 'key words' worthy of including in a glossary of terms for the topic.

Easy to understand diagrams: The main ideas of the topic are represented and explained by clear, informative diagrams.

Write-on format: Your understanding of the main ideas of the topic is tested by asking questions and providing spaces for your answers. Where indicated by the space available, your answers should be concise. Questions requiring more explanation or discussion are spaced accordingly. Answer the questions adequately according to the questioning term used.

A tab system at the base of each activity page identifies resources associated with the activity on that page. Use the guide below to help you use the tab system most effectively.

Using page tabs more effectively

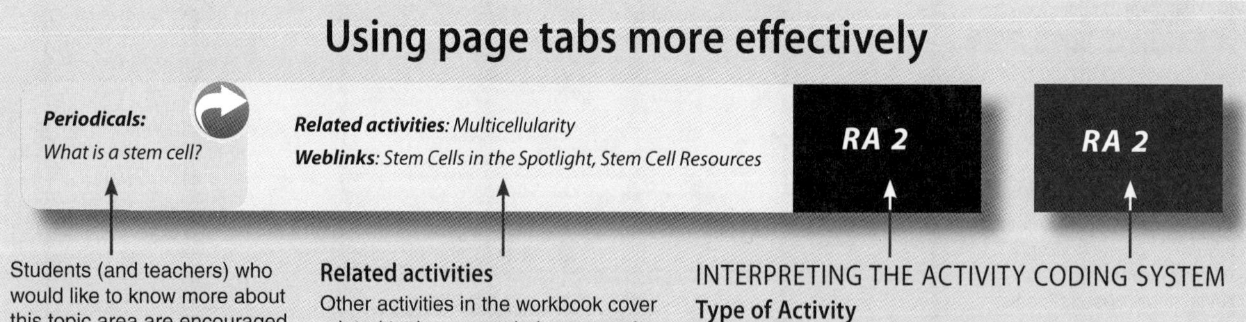

Periodicals:
What is a stem cell?
Related activities: Multicellularity
Weblinks: Stem Cells in the Spotlight, Stem Cell Resources
RA 2
RA 2

Students (and teachers) who would like to know more about this topic area are encouraged to locate the periodical cited on the Periodicals tab. Articles of interest directly relevant to the topic content are cited. The full citation appears in the Appendix as indicated at the beginning of the topic chapter.

Related activities

Other activities in the workbook cover related topics or may help answer the questions on the page. In most cases, extra information for activities that are coded R can be found on the pages indicated here.

Weblinks

This citation indicates a valuable video clip or animation that can be accessed from the Weblinks page specifically for this workbook. *www.thebiozone.com/weblink/IB-3169.html*

INTERPRETING THE ACTIVITY CODING SYSTEM

Type of Activity

D = includes some data handling or interpretation
P = includes a paper practical
R = *may* require extra reading (e.g. text or other activity)
A = includes application of knowledge to solve a problem
E = extension material

Level of Activity

1 = generally simpler, including mostly describe questions
2 = more challenging, including explain questions
3 = challenging content and/or questions, including discuss

Tab Color

Black = Core material
Dark Blue = AHL material

Resources Information

Your set (comprehensive or course) textbook should be a starting point for information about the content of your course. There are also many other resources available, including journals, magazines, supplementary texts, dictionaries, computer software, and the internet. Your teacher will have some prescribed resources for your use, but a few of the readily available periodicals are listed here for quick reference. The titles of relevant articles are listed with the activity to which they relate and are cited in the appendix. For further details or to make purchases or subscriptions, link to the publisher via BIOZONE's website: **www.thebiozone.com** Please note that listing any product in this workbook does not, in any way, denote BIOZONE's endorsement of that product and BIOZONE does not have any business affiliation with the publishers listed herein.

Supplementary Texts

ASKING QUESTIONS IN BIOLOGY
3rd Edition

Barnard, C., F. Gilbert, & P. McGregor, 2007
Asking Questions in Biology: Key Skills for Practical Assessments & Project Work, 256 pp.
Publisher: Benjamin Cummings
ISBN: 978-0132224352
Comments: *Covers many aspects of design, analysis and presentation of practical work in high school biology.*

Comprehensive Textbooks

IB STUDY GUIDES
Biology
OXFORD

A. Allott, 2007
Biology for the IB Diploma - Standard and Higher Level 2e
Publisher: Oxford University Press
Pages: 192
ISBN: 978-0199151431
Comments: *Updated edition to meet the IB Diploma programme from 2007. Includes both core (SL/HL) and option material.*

IB DIPLOMA PROGRAMME
Biology
Course Companion
OXFORD

A. Allott and D. Mindorff, 2007
IB Diploma Programme: Biology Course Companion
Publisher: Oxford
Pages: 400
ISBN: 978-0-19-915145-5
Comments: *Support for students of the IB Diploma Programme and written for the 2007 syllabus including material for all core and option units at both standard and higher level.*

Biology
C J Clegg

C.J. Clegg, 2007
Biology for the IB Diploma
Publisher: Hodder Murray
Pages: 448
ISBN: 978-0340926529
Comments: *This new text has been specifically written to cater for the International Baccalaureate diploma course. The content covers the core topics, and options are provided on the accompanying CD-ROM.*

Biology
IB DIPLOMA

A. Damon *et. al.*
Pearson Baccalaureate, Standard Level Biology, 2e, 2009
Publisher: Pearson
ISBN: 978-04359943-96
Pages: 536
Comments: *Comprehensive coverage of the latest syllabus requirements and all options for SL. Links to TOK and opportunity to make exam-style assessments using past questions.*

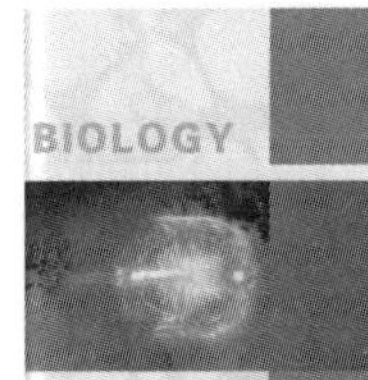
BIOLOGY

M. Peeters, C. Talbot, & A. Mayrhofer, 2008
Biology (3e)
Publisher: IBID Press
Pages: 432
ISBN: 978-18766590-28
Comments: *Rewritten with contributions for IB examiners. Includes references to TOK throughout, a glossary of all key terms, and student's exercises for each chapter.*

Biology
for the IB Diploma
Brenda Walpole,
Ashby Merson-Davies
and Leighton Dann

B. Walpole, A. Merson-Davis, & L. Dann, 2011
Biology for the IB Dipoma Coursebook
Publisher: Cambridge
Pages: 608
ISBN: 978-05211717-86
Comments: *Full coverage of the IB Biology syllabus including all eight options. Links to TOK throughout to stimulate thought and discussion. Includes key terms definitions and exam style questions for each chapter.*

HIGHER LEVEL
Biology
IB DIPLOMA

W. Ward, P. Tosto, R. McGonegal, A. Damon
Higher Level Biology for the IB Diploma, 2e, 2008
Publisher: Pearson
ISBN: 978-04359944-57
Pages: 720
Comments: *Comprehensive coverage of he latest syllabus requirements and all options for HL. Links to TOK and opportunity to make exam-style assessments using past questions.*

Periodicals, Magazines and Journals

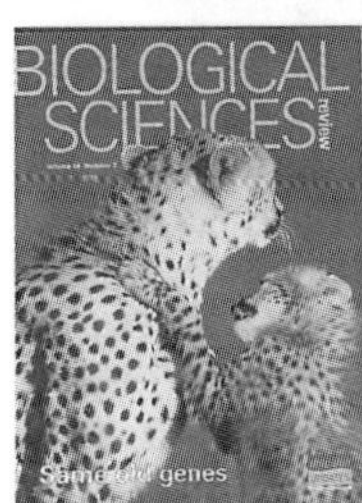
BIOLOGICAL SCIENCES
review

Biological Sciences Review (Biol. Sci. Rev.) *An excellent quarterly publication for teachers and students of biology. The content is current and the language is accessible.* Subscriptions available from Philip Allan Publishers, Market Place, Deddington, Oxfordshire OX 15 OSE.
Tel. 01869 338652
Fax: 01869 338803
E-mail: sales@philipallan.co.uk

NewScientist
THE FIRST HUMAN CLONES
DOES THE FUTURE START HERE?
GRAVITY

New Scientist: *Published weekly and found in many libraries. It often summarizes the findings published in other journals. Articles range from news releases to features.*
Subscription enquiries:
Tel. (UK and international): +44 (0)1444 475636. (US & Canada) 1 888 822 3242.
E-mail: ns.subs@qss-uk.com

SCIENTIFIC AMERICAN
EXCLUSIVE REPORT
THE FIRST Human Clone

Scientific American: *A monthly magazine containing mostly specialist feature articles. Articles range in level of reading difficulty and assumed knowledge.*
Subscription enquiries:
Tel. (US & Canada) 800-333-1199.
Tel. (outside North America): 515-247-7631
Web: www.sciam.com

Biology Dictionaries

HENDERSON'S
Dictionary of
BIOLOGY
Eleanor Lawrence

Henderson, E. Lawrence. **Henderson's Dictionary of Biological Terms**, 2008, 776 pp. Benjamin Cummings. **ISBN**: 978-0321505798
This edition has been updated, rewritten for clarity, and reorganised for ease of use. An essential reference and the dictionary of choice for many.

Using BIOZONE's Website

The current internet address (URL) for the web site is displayed here. You can type a new address directly into this space.

Use Google to search for web sites of interest. The more precise your search words are, the better the list of results. EXAMPLE: If you type in "biotechnology", your search will return an overwhelmingly large number of sites, many of which will not be useful to you. Be more specific, e.g. "biotechnology medicine DNA uses".

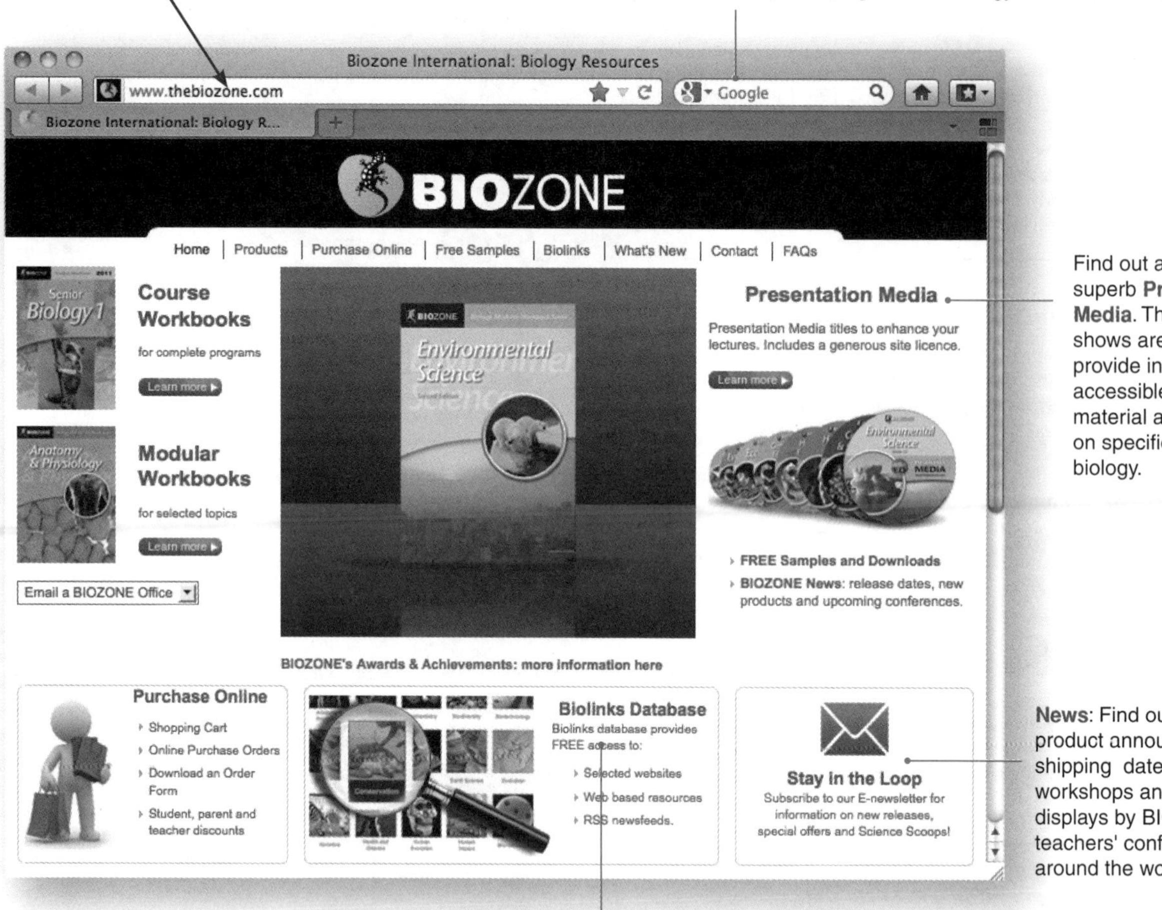

Biozone International: Biology Resources
www.thebiozone.com
Google
Biozone International: Biology R...
BIOZONE
Home | Products | Purchase Online | Free Samples | Biolinks | What's New | Contact | FAQs
Course Workbooks
for complete programs
Learn more
Modular Workbooks
for selected topics
Learn more
Email a BIOZONE Office
Environmental Science
Presentation Media
Presentation Media titles to enhance your lectures. Includes a generous site licence.
Learn more
FREE Samples and Downloads
BIOZONE News: release dates, new products and upcoming conferences.
BIOZONE's Awards & Achievements: more information here
Purchase Online
Shopping Cart
Online Purchase Orders
Download an Order Form
Student, parent and teacher discounts
Biolinks Database
Biolinks database provides FREE access to:
Selected websites
Web based resources
RSS newsfeeds.
Stay in the Loop
Subscribe to our E-newsletter for information on new releases, special offers and Science Scoops!

Find out about our superb **Presentation Media**. These slide shows are designed to provide in-depth, highly accessible illustrative material and notes on specific areas of biology.

News: Find out about product announcements, shipping dates, and workshops and trade displays by BIOZONE at teachers' conferences around the world.

Access the **Biolinks** database of web sites related to each major area of biology. It's a great way to quickly find out more on topics of interest.

Weblinks: www.thebiozone.com/weblink/IB-3169.html

Throughout this workbook, some pages make reference to additional or alternative activities, as well as web sites and periodicals that have particular relevance to the activity. See example of page reference below:

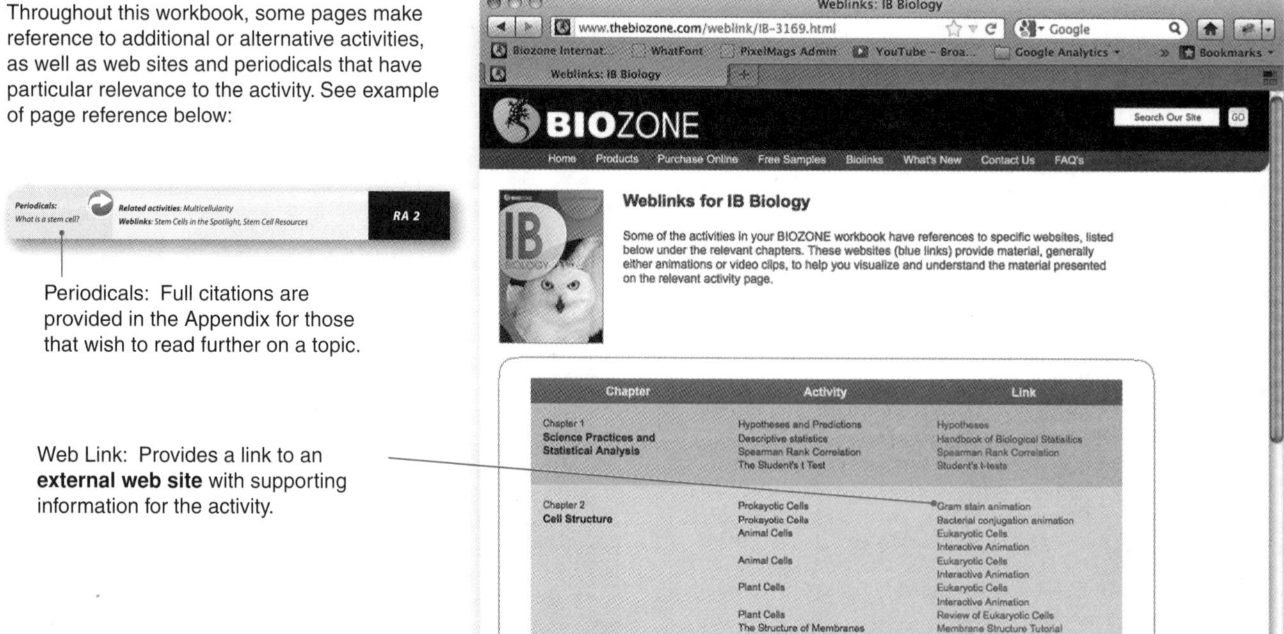

Periodicals:
What is a stem cell?
Related activities: Multicellularity
Weblinks: Stem Cells in the Spotlight, Stem Cell Resources
RA 2
Weblinks: IB Biology
www.thebiozone.com/weblink/IB-3169.html
BIOZONE
Search Our Site
Home Products Purchase Online Free Samples Biolinks What's New Contact Us FAQ's
Weblinks for IB Biology
Some of the activities in your BIOZONE workbook have references to specific websites, listed below under the relevant chapters. These websites (blue links) provide more material, generally either animations or video clips, to help you visualize and understand the material presented on the relevant activity page.
Chapter
Activity
Link
Chapter 1
Science Practices and Statistical Analysis
Hypotheses and Predictions
Descriptive statistics
Spearman Rank Correlation
The Student's t Test
Hypotheses
Handbook of Biological Statistics
Spearman Rank Correlation
Student's t-tests
Chapter 2
Cell Structure
Prokaryotic Cells
Prokaryotic Cells
Animal Cells
Animal Cells
Plant Cells
Plant Cells
The Structure of Membranes
Gram stain animation
Bacterial conjugation animation
Eukaryotic Cells
Interactive Animation
Eukaryotic Cells
Interactive Animation
Eukaryotic Cells
Interactive Animation
Review of Eukaryotic Cells
Membrane Structure Tutorial

Periodicals: Full citations are provided in the Appendix for those that wish to read further on a topic.

Web Link: Provides a link to an **external web site** with supporting information for the activity.

Command Terms

Questions come in a variety of forms. Whether you are studying for an exam, or writing an essay, it is important to understand exactly what the question is asking. A question has two parts to it: one part of the question will provide you with information, the second part of the question will provide you with instructions as to how to answer the question. Following these instructions is most important. Often students in examinations know the material but fail to follow instructions and, as a consequence, do not answer the question appropriately. Examiners often use certain key words to introduce questions. Look out for them and be absolutely clear as to what they mean. Below is a list of commonly used terms that you will come across and a brief explanation of each.

Objective 1

Demonstrate an understanding of scientific facts and concepts, scientific methods and techniques, scientific terminology, and methods of presenting scientific information.

Define: Give the precise meaning of a word or phrase as concisely as possible.

Draw: Represent by means of pencil lines.

Label: Add labels to a diagram.

List: Give a sequence of names or other brief answers with no elaboration. Each one should be clearly distinguishable from the others.

Measure: Find a value for a quantity.

State: Give a specific name, value, or other answer. No supporting argument or calculation is necessary.

Objective 2

Apply and use scientific facts and concepts, scientific methods and techniques, scientific terminology to communicate effectively, and appropriate methods to present scientific information.

Annotate: Add **brief** notes to a diagram, drawing or graph.

Apply: Use an idea, equation, principle, theory, or law in a new situation.

Calculate: Find an answer using mathematical methods. Show the working unless instructed not to.

Describe: Give a detailed account, including all the relevant information.

Distinguish: Give the difference(s) between two or more different items.

Estimate: Find an approximate value for an unknown quantity, based on the information provided and application of scientific knowledge.

Identify: Find an answer from a number of possibilities.

Outline: Give a brief account or summary. Include essential information only.

Objective 3

Construct, analyze, and evaluate hypotheses, research questions, and predictions, scientific methods and techniques, and scientific explanations.

Analyze: Interpret data to reach stated conclusions.

Comment: Give a judgement based on a given statement or result of a calculation.

Compare: Give an account of similarities and differences between two or more items, referring to both (or all) of them throughout. Comparisons can be given in a table. Comparisons generally ask for similarities more than differences (see contrast).

Construct: Represent or develop in graphical form.

Contrast: Show differences. Set in opposition.

Deduce: Reach a conclusion from information given.

Derive: Manipulate a mathematical equation to give a new equation or result.

Design: Produce a plan, object, simulation or model.

Determine: Find the only possible answer.

Discuss: Give an account including, where possible, a range of arguments, assessments of the relative importance of various factors, or comparison of alternative hypotheses.

Evaluate: Assess the implications and limitations.

Explain: Give a clear account including causes, reasons, or mechanisms.

Predict: Give an expected result.

Show: Give the steps in a calculation or derivation

Sketch: Represent by way of a graph, showing a line and labeled but unscaled axes but with important features, such as intercept, shown.

Solve: Obtain an answer using algebraic and/or numerical methods.

Suggest: Propose a hypothesis or other possible explanation.

Other commonly used terms

Account for: Provide a satisfactory explanation or reason for an observation.

Appreciate: To understand the meaning or relevance of a particular situation.

Illustrate: Give concrete examples. Explain clearly by using comparisons or examples.

Interpret: Comment upon, give examples, describe relationships. Describe, then evaluate.

Summarize: Give a brief, condensed account. Include conclusions and avoid unnecessary details.

Students should familiarize themselves with this list of terms and, where necessary throughout the course, they should refer back to them when answering questions. The list of terms mentioned above is not exhaustive and students should compare this list with past examination papers and essays and add any new terms (and their meaning) to the list above. The aim is to become familiar with interpreting the question and answering it appropriately.

IB Biology Course Guide

The International Baccalaureate (IB) biology course is divided into three sections: core, additional higher level material, and option material. All **IB candidates** must complete the **core** topics. Higher level students are also required to undertake Additional Higher Level **(AHL)** material as part of the core. Options fall into three categories (see the following page): those specific to standard level students **(OPT-SL)**, one only specific to higher level students **(OPT-HL)** and those offered to both **(OPT-SL/HL)**. All candidates are required to study two options. All candidates must also carry out **practical work** and must participate in **the group 4 project**. In the guide below, we have indicated where the relevant material can be found.

Topic		See workbook
CORE:	*(All students)*	
1	**Statistical analysis**	
1.1	Mean and SD, t-test, correlation.	Science Practices and Statistical Analysis
	For this CORE topic also see the TRC: Spreadsheets and Statistics	
2	**Cells**	
2.1	Cell theory. Cell and organelle sizes. Surface area to volume ratio. Emergent properties. Cell specialization and differentiation. Stem cells.	Cell Structure
2.2	Prokaryotic cells: ultrastructure & function.	Cell Structure
2.3	Eukaryotic cells: ultrastructure & function. Prokaryotic vs eukaryotic cells. Plant vs animal cells. Extracellular components.	Cell Structure
2.4	Membrane structure. Active and passive transport. Diffusion and osmosis.	Cell Processes
2.5	Cell division and the origins of cancer.	Cell Processes
3	**The chemistry of life**	
3.1	Elements of life. The properties and importance of water.	The Chemistry of Cells
3.2	Structure and function of carbohydrates, lipids, and proteins.	The Chemistry of Cells
3.3	Nucleotides and the structure of DNA.	The Structure and Function of DNA
3.4	Semi-conservative DNA replication.	The Structure and Function of DNA
3.5	RNA and DNA structure. The genetic code. Transcription. Translation.	The Structure and Function of DNA
3.6	Enzyme structure and function.	Enzymes and Metabolism
3.7	Cellular respiration and ATP production.	Enzymes and Metabolism
3.8	Biochemistry of photosynthesis. Factors affecting photosynthetic rates.	Enzymes and Metabolism
4	**Genetics**	
4.1	Eukaryote chromosomes. Genomes. Gene mutations and consequences.	Chromosomes and Meiosis
4.2	Meiosis and non-disjunction. Karyotyping and pre-natal diagnosis.	Chromosomes and Meiosis
4.3	Alleles and single gene inheritance, sex linkage, pedigrees.	Heredity
4.4	PCR, gel electrophoresis, DNA profiling. HGP. Transformation. GMOs. Cloning.	Genetic Engineering and Biotechnology
5	**Ecology and evolution**	
5.1	Ecosystems. Food chains and webs. Trophic levels. Ecological pyramids. The role of decomposers in recycling nutrients.	Ecology
5.2	The greenhouse effect. The carbon cycle. Precautionary principle. Global warming.	Ecology
5.3	Factors influencing population size. Population growth.	Ecology
5.4	Genetic variation. Sexual reproduction as a source of variation in species. Evidence for evolution: natural selection. Evolution in response to environmental change.	Evolution
5.5	Classification. Binomial nomenclature. Features of plant & animal phyla. Keys.	Classification
6	**Human health and physiology**	
6.1	Role of enzymes in digestion. Structure and function of the digestive system.	Digestion
6.2	Structure and function of the heart. The control of heart activity. Blood & vessels.	The Transport System

Topic		See workbook
6.3	Pathogens and their transmission. Antibiotics. Role of skin as a barrier to infection. Role of phagocytic leukocytes. Antigens & antibody production. HIV/AIDS.	Defence Against Disease
6.4	Gas exchange. Ventilation systems. Control of breathing.	Gas Exchange
6.5	Principles of homeostasis. Control of body temperature and blood glucose. Diabetes. Role of the nervous and endocrine systems in homeostasis.	Nerves, Hormones, and Homeostasis
6.6	Human reproduction and the role of hormones. Reproductive technologies and ethical issues.	Reproduction
COMPULSORY:	**AHL Topics** *(HL students only)*	
7	**Nucleic acids and proteins**	
7.1	DNA structure, exons & introns (junk DNA)	The Structure and Function of DNA
7.2	DNA replication, including the role of enzymes and Okazaki fragments.	The Structure and Function of DNA
7.3	DNA alignment, transcription. The removal of introns to form mature mRNA.	The Structure and Function of DNA
7.4	The structure of tRNA and ribosomes. The process of translation. Peptide bonds.	The Structure and Function of DNA, The Chemistry of Cells
7.5	Protein structure and function.	The Chemistry of Cells
7.6	Enzymes: induced fit model. Inhibition. Allostery in the control of metabolism.	Enzymes and Metabolism
8	**Cell respiration and photosynthesis**	
8.1	Structure and function of mitochondria. Biochemistry of cellular respiration.	Enzymes and Metabolism
8.2	Chloroplasts, the biochemistry and control of photosynthesis, chemiosmosis.	Enzymes and Metabolism
9	**Plant science**	
9.1	Structure and growth of a dicot plant. Function and distribution of tissues in leaves. Dicots vs monocots. Plant modifications. Auxins.	Plant Science
9.2	Support in terrestrial plants. Transport in angiosperms: ion movement through soil, active ion uptake by roots, transpiration, translocation. Abscisic acid. Xerophytes.	Plant Science
9.3	Dicot flowers. Pollination and fertilization. Seeds: structure, germination, dispersal. Flowering and phytochrome.	Plant Science
10	**Genetics**	
10.1	Meiosis, and the process of crossing over. Mendel's law of independent assortment.	Chromosomes and Meiosis
10.2	Dihybrid crosses. Types of chromosomes.	Heredity
10.3	Polygenic inheritance.	Heredity
	For extension on this topic also see the TRC: Chromosome Mapping	
11	**Human health and physiology**	
11.1	Blood clotting. Clonal selection. Acquired immunity. Antibodies and monoclonal antibodies. Vaccination.	Defence Against Disease
11.2	Nerves, muscles, bones and movement. Joints. Skeletal muscle and contraction.	Muscles & Movement
11.3	Excretion. Structure and function of the human kidney. Urine production. Diabetes.	Nerves, Hormones, and Homeostasis
11.4	Testis and ovarian structure. Spermatogenesis and oogenesis. Fertilization and embryonic development. The placenta. Birth. Role of hormones.	Reproduction

International Baccalaureate Course *continued*

Topic		See workbook
OPTIONS: OPT - SL *(SL students only)*		
A	**Human nutrition and health**	
A1	Diet and malnutrition. Deficiency & supplements. PKU.	
A2	Energy content of food types. BMI. Obesity and anorexia. Appetite control.	
A3	Special diet issues; breastfeeding vs bottle-feeding, type II diabetes, cholesterol.	
	Provided as a separate complete unit on the IB OPTIONS CD-ROM	
B	**Physiology of exercise**	
B.1	Locomotion in animals. Roles of nerves, muscles, and bones in movement. Joints. Skeletal muscle and contraction.	
B.2	Training and the pulmonary system.	
B.3	Training and the cardiovascular system.	
B.4	Respiration and exercise intensity. Roles of myoglobin and adrenaline. Oxygen debt and lactate in muscle fatigue.	
B.5	Exercise induced injuries and treatment.	
	Provided as a separate complete unit on the IB OPTIONS CD-ROM	
C	**Cells and energy**	
C.1	Protein structure and function. Fibrous and globular proteins.	The Chemistry of Life
C.2	Enzymes: induced fit model. Inhibition. Allostery in the control of metabolism.	The Chemistry of Life
C.3	Biochemistry of cellular respiration.	Cellular Energetics
C.4	The biochemistry of photosynthesis including chemiosmosis. Action and absorption spectra. Limiting factors.	Cellular Energetics
	Provided as a separate complete unit on the IB OPTIONS CD-ROM	
OPTIONS: OPT - SL/HL *(SL and HL students)*		
D	**Evolution**	
D.1	Prebiotic experiments. Comets. Protobionts and prokaryotes. Endosymbiotic theory.	The Origin & Evolution of Life
D.2	Species, gene pools, speciation. Types and pace of evolution. Transient vs balanced polymorphism.	Speciation, Patterns of Evolution
D.3	Fossil dating. Primate features. Hominid features. Diet and brain size correlation. Genetic and cultural evolution.	The Evolution of Humans (**TRC**)
	D.4-D.5 is extension for HL only	
D.4	The Hardy-Weinberg principle.	Speciation
D.5	Biochemical evidence for evolution. Biochemical variations indicating phylogenetic relationships.	The Origin and Evolution of Life
	Classification. Cladistics and cladograms	Classification
	Provided as a separate complete unit on the IB OPTIONS CD-ROM	
E	**Neurobiology and behavior**	
E.1	Stimuli, responses and reflexes in the context of animal behavior. Animal responses and natural selection.	Nerves, Muscles & Movement, Animal Behavior
E.2	Sensory receptors. Structure and function of the human eye and ear.	Nerves, Muscles & Movement
E.3	Innate vs learned behavior and its role in survival. Learned behavior and birdsong.	Animal Behavior
E.4	Presynaptic neurons at synapses. Examples of excitatory and inhibitory psychoactive drugs. Effects of drugs on synaptic transmission. Causes of addiction.	Aspects covered in Nerves, Muscles & Movement
	E.5-E.6 is extension for HL only	
E.5	Structure and function of the human brain. ANS control. Pupil reflex and its use in testing for death. Hormones as painkillers.	Aspects covered in Nerves, Muscles & Movement
E.6	Social behavior and organization. The role of altruism in sociality. Foraging behavior. Mate selection. Rhythmical behavior.	Animal Behavior
	Provided as a separate complete unit on the IB OPTIONS CD-ROM	

Topic		See workbook
F	**Microbes and Biotechnology**	
F.1	Classification. Diversity of Archaea and Eubacteria. Diversity of viruses. Diversity of microscopic eukaryotes.	
F.2	Roles of microbes in ecosystems. Details of the nitrogen cycle including the role of bacteria. Sewage treatment. Biofuels.	
F.3	Reverse transcription. Somatic vs germline, gene therapy. Viral vectors.	
F.4	Microbes involved in food production of beer, wine, bread, and soy sauce. Food preservation. Food poisoning.	
	F.5-F.6 is extension for HL only	
F.5	Metabolism of microbes. Modes of nutrition. Cyanobacterium. Bioremediation.	
F.6	Pathogens and disease: influenza virus, malaria, bacterial infections. Controlling microbes. Epidemiology. Prion hypothesis.	
	Provided as a separate complete unit on the IB OPTIONS CD-ROM	
G	**Ecology and conservation**	
G.1	Factors affecting plant and animal distribution. Sampling. Ecological niche and the competitive exclusion principle. Species interactions. Measuring biomass.	
G.2	Trophic levels. Ecological pyramids. Primary vs secondary succession. Biome vs biosphere. Plant productivity (includes calculating gross and net production, and biomass).	
G.3	Conservation of biodiversity. Diversity index. Human impact on ecosystems: alien species. Biological control. Effect of CFCs on ozone layer. UV radiation absorption.	
	G.4-G.5 is extension for HL only	
G.4	Monitoring environmental change. Biodiversity. Endangered species. Conservation Strategies. Extinction.	
G.5	*r*-strategies and K-strategies.Mark-and-recapture sampling. Fisheries conservation.	
	Provided as a separate complete unit on the IB OPTIONS CD-ROM	
OPTIONS: OPT - HL *(HL students only)*		
H	**Further human physiology**	
H.1	Hormones and their modes of action. Hypothalamus and pituitary gland. Control of ADH secretion.	
H.2	Digestion and digestive juices. Stomach ulcers and stomach cancers. Role of bile.	
H.3	Structure of villi. Absorption of nutrients and transport of digested food.	
H.4	The structure and function of the liver (including role in nutrient processing and detoxification). Liver damage from alcohol.	
H.5	The cardiac cycle and control of heart rhythm. Atherosclerosis, coronary thrombosis and coronary heart disease.	
H.6	Gas exchange: oxygen dissociation curves and the Bohr shift. Ventilation rate and exercise. Breathing at high altitude. Causes and effects of asthma.	
	Provided as a separate complete unit on the IB OPTIONS CD-ROM	

Practical Work *(All students)*

Practical work consists of short and long term investigations, and an interdisciplinary project (The Group 4 project). Also see the "Guide to Practical Work" on the last page of this introductory section.

Short and long term investigations

Investigations should reflect the breadth and depth of the subjects taught at each level, and include a spread of content material from the core, options, and AHL material, where relevant.

The Group 4 project

All candidates must participate in the group 4 project. In this project it is intended that students analyze a topic or problem suitable for investigation in each of the science disciplines offered by the school (not just in biology). This project emphasizes the processes involved in scientific investigations rather than the products of an investigation.

Cell Biology

Cell Structure	• Cell theory • Prokaryotic cells • Eukaryotic cells
TOK	*The accumulation of supporting evidence allows a hypothesis to become a theory.*
TOK	*The use of technology enables us to observe things we cannot perceive directly.*
Cell Processes	• Membrane structure • Active and passive transport • Cell division
TOK	*Emergent properties arise from the interaction of component parts.*
The Chemistry of Cells	• Chemical elements and water • The structure and role of water • Carbohydrates, lipids, and proteins

Genetics

The Structure and Function of DNA	• Nucleic acids: DNA structure • DNA replication • Transcription and translation
TOK	*Theories can be modified or even discarded in the light of new information.*
Enzymes and Metabolism	• The nature of enzymes • Cell respiration and photosynthesis
Chromosomes and Meiosis	• Chromosomes, genes, and alleles • Mutations • Meiosis and karyotypes
Heredity	• Theoretical genetics • Dihybrid cross and gene linkage (AHL) • Polygenic inheritance (AHL)
Genetic Engineering	• PCR and gel electrophoresis • DNA sequencing and profiling • Genetic modification and cloning
TOK	*How do we adequately assess the risks associated with the use of GMOs?*

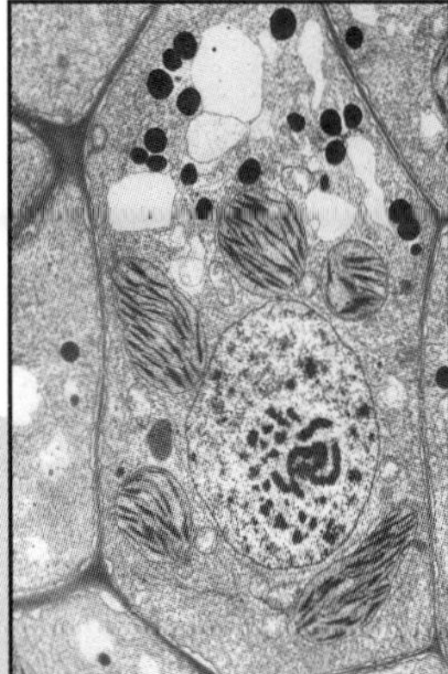

Cells are the basic unit of life and share common features. Multicellularity gives rise to emergent properties not seen in individual cells.

DNA carries the information to construct organisms by translation of an almost universal code. Metabolism is regulated through the activity of enzymes.

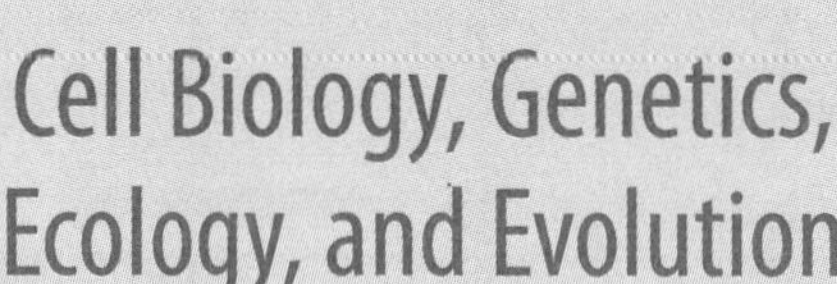
Cell Biology, Genetics,
Ecology, and Evolution

Andrew Dunn www.andrewdunnphoto.com

Ecosystem structure and function is affected by human activity. Populations evolve in response to selection pressures.

Science asks questions and makes predictions about how systems operate. Experimental results can be tested impartially by statistical methods.

Ecology	• Communities and ecosystems • The carbon cycle • The enhanced greenhouse effect • Population ecology
TOK	*Are models useful to our understanding of the functioning of natural systems?*
TOK	*The precautionary principle is justified even if climate changes are less than we expect.*
Evolution	• What is evolution? • Evidence for evolution • Variation and natural selection • Response to environmental change
Classification	• Binomial nomenclature • Classification of plants and animals • Classification keys
TOK	*How do we evaluate the ethical aspects of claims made in a different social climate?*

Ecology and Evolution

Science Practices	• The scientific method • Planning and executing experiments • Graphing and tabulation
Statistical Analysis	• Descriptive statistics • Estimating variability in data • Correlation and Student's t test

Science Practices & Statistical Analysis

Important in this section...

- *Develop understanding of cell structure & function*
- *Understand the basis of inheritance*
- *Understand ecosystem function at all levels*
- *Understand the basis of evolutionary change*
- *Understand how we classify and identify organisms*

Core Topic 1

Science Practices and Statistical Analysis

Key concepts

- Science is based on observation, hypothesis, and investigation. Scientists collect and analyze data to test their hypotheses.
- An experiment should be a **fair test** of the hypothesis.
- The sample mean and standard deviation enable objective analysis of sample data.
- Two variables are correlated when they vary together in some predictable way, but this does not imply cause and effect.

Key terms

Core

accuracy
aim
chi-squared test
control
controlled variable
correlation
dependent variable
descriptive statistics
error bars
fair test
graph
hypothesis
independent variable
median
mode
parameter
precision
qualitative data
quantitative data
raw data
regression
sample mean
sample standard deviation
scientific method
statistic
Student's *t* test
table
trend (of data)
variable

Periodicals:
Listings for this chapter are on page 398

Weblinks:
www.thebiozone.com/weblink/IB-3169.html

Teacher Resource CD-ROM:
Spreadsheets & Statistics

Learning Objectives

☐ 1. Use the **KEY TERMS** to compile a glossary for this topic.

Making Investigations *(Group 4 Project)* — pages 8, 10-14, 17-21, 25-26

☐ 2. Describe and explain the basic principles of the **scientific method**.

☐ 3. Produce an outline of your practical biological investigation, including your **aim** and **hypothesis**, and all information relevant to the study design.

☐ 4. Identify your **dependent** and **independent variables**, their range, and how you will measure them. Identify **controlled variables** and their significance.

☐ 5. Understand the difference between **qualitative** and **quantitative data** and give examples of their appropriate use.

☐ 6. Demonstrate an ability to systematically record data. Evaluate the **accuracy** and **precision** of any recording or measurements you make.

Data Handling and Analysis *(Topic 1)* — pages 15-16, 22-34

☐ 7. Demonstrate an ability to process raw **data**. Calculate percentages, rates, and frequencies for raw data and understand the reason for these manipulations.

☐ 8. Demonstrate an ability to organize different types of data appropriately in a **table**, including any calculated values.

☐ 9. Understand the benefits of graphing data and present different types of data appropriately in **graphs**.

☐ 10. Use **descriptive statistics** (e.g. sample **mean** and **standard deviation**) to help you to summarize and evaluate data. Explain the significance of the standard deviation. Recognize that descriptive statistics for sample data provide estimates of true population **parameters**.

☐ 11. Use **error bars** based on standard deviation (*s*), standard error, or 95% confidence intervals to represent the **variability** in a data set. Use standard deviation to compare the means and spread of data between two or more samples.

☐ 12. Use the **Student's *t* test** to determine the significance of differences between the means of two sets of data. State the criteria for using the *t* test. Recognize that such tests provide an objective standard for statements about the data and the findings they support (*TOK*).

☐ 13. Recognize statistical tests for a difference between data sets (e.g. Student's *t* test, **chi-squared**) and distinguish these from tests for a **trend** or relationship between variables (**correlation** and **regression**).

☐ 14. Understand that data are correlated when there is a relationship between the two variables in question, but neither is assumed to be dependent on the other. Interpret tests for correlation and explain that such tests assess the strength of an association, but can not establish cause and effect.

The Scientific Method

Scientific knowledge grows through a process called the **scientific method**. This process involves observation and measurement, hypothesising and predicting, and planning and executing investigations designed to test formulated **hypotheses**. A scientific hypothesis is a tentative explanation for an observation, which is capable of being tested by experimentation. Hypotheses lead to **predictions** about the system involved and they are accepted or rejected on the basis of the investigation's findings. Acceptance of the hypothesis is not necessarily permanent: explanations may be rejected later in light of new findings.

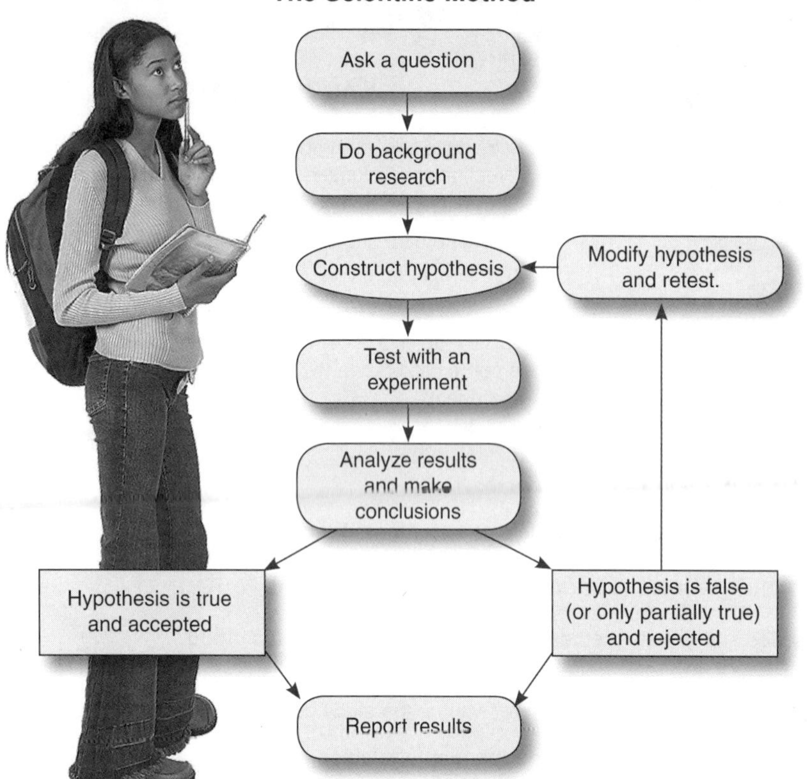
The Scientific Method
Ask a question
Do background research
Construct hypothesis
Modify hypothesis and retest.
Test with an experiment
Analyze results and make conclusions
Hypothesis is true and accepted
Hypothesis is false (or only partially true) and rejected
Report results

Forming a Hypothesis

Features of a sound hypothesis:

- It is based on observations and prior knowledge of the system.
- It offers an explanation for an observation.
- It refers to only one independent variable.
- It is written as a definite statement and not as a question.
- It is testable by experimentation.
- It leads to predictions about the system.

Testing a Hypothesis

Features of a sound method:

- It tests the validity of the hypothesis.
- It is repeatable.
- It includes a control which does not receive treatment.
- All variables are controlled where possible.
- The method includes a dependent and independent variable.
- Only the independent variable is changed (manipulated) between treatment groups.

Hypothesis involving manipulation

Used when the effect of manipulating a variable on a biological entity is being investigated. **Example:** The composition of applied fertilizer influences the rate of growth of plant A.

Hypothesis of choice

Used when investigating species preference, e.g. for a particular habitat type or microclimate. **Example:** Woodpeckers (species A) show a preference for tree type when nesting.

Hypothesis involving observation

Used when organisms are being studied in their natural environment and conditions cannot be changed. **Example:** Fern abundance is influenced by the degree to which the canopy is established.

1. Why might an accepted hypothesis be rejected at a later date? ______________________________

____

2. Explain why a method must be repeatable: ______________________________

____

____

3. In which situation(s) is it difficult, if not impossible, to control all the variables? ______________________________

____

ISBN: 978-1-927173-16-9

Hypotheses and Predictions

A hypothesis offers a tentative explanation to questions generated by observations and leads to one or more **predictions** about the way a biological system will behave. Experiments are constructed to test these predictions. For every hypothesis, there is a corresponding **null hypothesis**; a hypothesis of no difference or no effect. Creating a null hypothesis enables a hypothesis to be tested in a meaningful way using statistical tests. If the results of an experiment are statistically significant, the null hypothesis can be rejected. If a hypothesis is accepted, anyone should be able to test the predictions with the same methods and get a similar result each time. Scientific hypotheses may be modified as more information becomes available.

Observations, Hypotheses, and Predictions

Observation is the basis for formulating hypotheses and making predictions. An observation may generate a number of plausible hypotheses, and each hypothesis will lead to one or more predictions, which can be tested by further investigation.

Observation 1: Some caterpillar species are brightly colored and appear to be conspicuous to predators such as insectivorous birds. Predators appear to avoid these species. These caterpillars are often found in groups, rather than as solitary animals.

Observation 2: Some caterpillar species are cryptic in their appearance or behavior. Their camouflage is so convincing that, when alerted to danger, they are difficult to see against their background. Such caterpillars are usually found alone.

Assumptions

Any biological investigation requires you to make **assumptions** about the biological system you are working with. Assumptions are features of the system (and your investigation) that you assume to be true but do not (or cannot) test. Possible assumptions about the biological system described above include:

- Insectivorous birds have color vision.
- Caterpillars that look bright or cryptic to us, also appear that way to insectivorous birds.
- Insectivorous birds can learn about the palatability of prey by tasting them.

1. Study the example above illustrating the features of cryptic and conspicuous caterpillars, then answer the following:

   (a) Generate a hypothesis to explain the observation that some caterpillars are brightly colored and conspicuous while others are cryptic and blend into their surroundings:

   Hypothesis: ______________________________

   ____

   ____

   (b) State the null form of this hypothesis: ______________________________

   ____

   (c) Describe one of the **assumptions** being made in your hypothesis: ______________________________

   ____

   ____

   (d) Based on your hypothesis, generate a **prediction** about the behavior of insectivorous birds towards caterpillars:

   ____

   ____

ISBN: 978-1-927173-16-9

Related activities: *The Scientific Method*
Weblinks: *Hypothesis*

2. During the course of any investigation, new information may arise as a result of observations unrelated to the original hypothesis. This can lead to the generation of further hypotheses about the system. For each of the incidental observations described below, formulate a prediction, and an outline of an investigation to test it. *The observation described in each case was not related to the hypothesis the experiment was designed to test:*

(a) **Bacterial cultures**

Prediction:

Observation: During an experiment on bacterial growth, these girls noticed that the cultures grew at different rates when the dishes were left overnight in different parts of the laboratory.

Outline of the investigation:

(b) **Plant cloning**

Prediction:

Observation: During an experiment on plant cloning, a scientist noticed that the root length of plant clones varied depending on the concentration of a hormone added to the agar.

Outline of the investigation:

ISBN: 978-1-927173-16-9

Accuracy and Precision

The terms accuracy and precision are often confused, or used interchangeably, but their meanings are different. In any study, **accuracy** refers to how close a measured or derived value is to its true value. Simply put, it is the correctness of the measurement. It can sometimes be a feature of the sampling equipment or its calibration. **Precision** refers to the closeness of repeated measurements to each other, i.e. the ability to be exact. A balance with a fault in it could give very precise (i.e. repeatable) but inaccurate (untrue) results. Using the analogy of a target, repeated measurements are compared to arrows being shot at a target. This analogy can be useful when thinking about the difference between accuracy and precision.

Accurate but imprecise

The measurements are all close to the true value but quite spread apart.

Analogy: The arrows are all close to the bulls-eye.

Precise but inaccurate

The measurements are all clustered close together but not close to the true value.

Analogy: The arrows are all clustered close together but not near the bulls-eye.

Inaccurate and imprecise

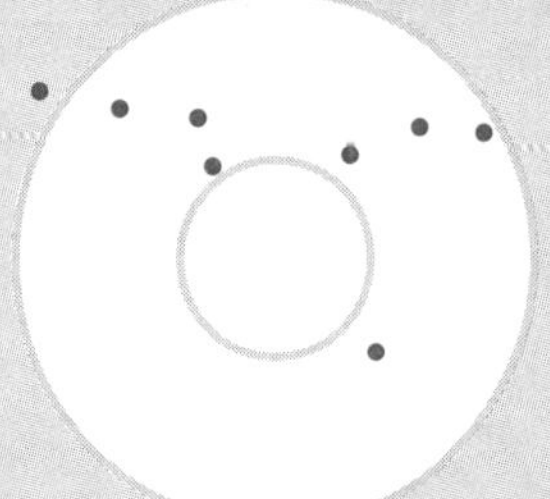

The measurements are all far apart and not close to the true value.

Analogy: The arrows are spread around the target.

Accurate and precise

The measurements are all close to the true value and also clustered close together.

Analogy: The arrows are clustered close together near the bulls-eye.

The accuracy of a measurement refers to how close the measured (or derived) value is to the true value. The precision of a measurement relates to its repeatability. In most laboratory work, we usually have no reason to suspect a piece of equipment is giving inaccurate measurements (is biased), so making precise measures is usually the most important consideration. We can test the precision of our measurements by taking repeated measurements from individual samples.

Population studies present us with an additional problem. When a researcher makes measurements of some variable in a study (e.g. fish length), they are usually trying to obtain an estimate of the true value for a parameter of interest (e.g. the mean size, therefore age, of fish). Populations are variable, so we can more accurately estimate a population parameter if we take a large number of random samples from the population.

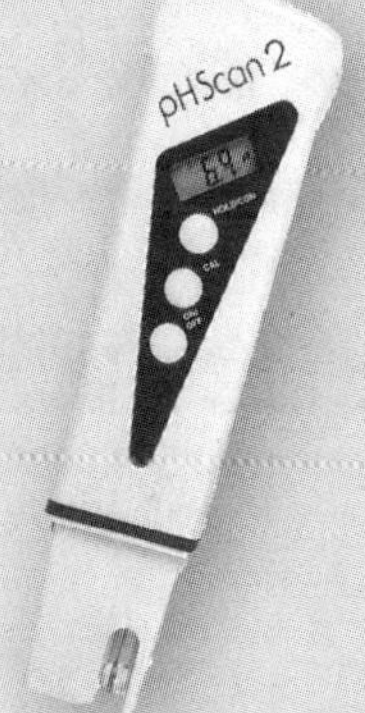
pHScan2

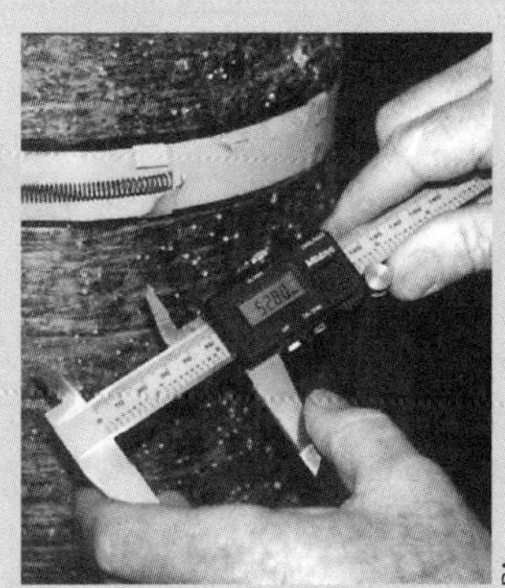
RA

A digital device such as this pH meter (above left) will deliver precise measurements, but its accuracy will depend on correct calibration. The precision of measurements taken with instruments such as callipers (above) will depend on the skill of the operator.

1. Distinguish between accuracy and precision: ______

2. Describe why it is important to take measurements that are both accurate and precise: ______

3. A researcher is trying to determine at what temperature enzyme A becomes denatured. Their temperature probe is incorrectly calibrated. Discuss how this might affect the accuracy and precision of the data collected:

ISBN: 978-1-927173-16-9

Variables and Data

When planning any kind of biological investigation, it is important to consider the type of data that will be collected. It is best, whenever possible, to collect quantitative or numerical data, as these data lend themselves well to analysis and statistical testing. Recording data in a systematic way as you collect it, e.g. using a table or spreadsheet, is important, especially if data manipulation and transformation are required. It is also useful to calculate summary, descriptive statistics (e.g. mean, median) as you proceed. These will help you to recognize important trends and features in your data as they become apparent.

The values for monitored or measured variables, collected during the course of the investigation, are called data. Like their corresponding variables, data may be quantitative, qualitative, or ranked.

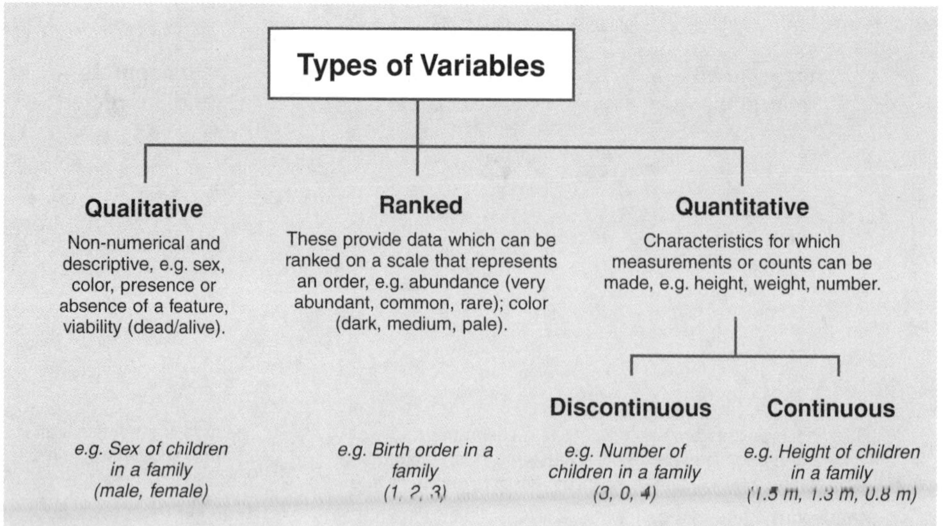

Types of Variables
Qualitative
Non-numerical and descriptive, e.g. sex, color, presence or absence of a feature, viability (dead/alive).
e.g. Sex of children in a family (male, female)
Ranked
These provide data which can be ranked on a scale that represents an order, e.g. abundance (very abundant, common, rare); color (dark, medium, pale).
e.g. Birth order in a family (1, 2, 3)
Quantitative
Characteristics for which measurements or counts can be made, e.g. height, weight, number.
Discontinuous
e.g. Number of children in a family (3, 0, 4)
Continuous
e.g. Height of children in a family (1.5 m, 1.3 m, 0.8 m)

A: Leaf shape

B: Number per litter

C: Fish length

BH

1. For each of the photographic examples (A – C above), classify the variables as quantitative, ranked, or qualitative:

   (a) Leaf shape: __________

   (b) Number per litter: __________

   (c) Fish length: __________

2. Why it is desirable to collect quantitative data where possible in biological studies?

3. How you might measure the color of light (red, blue, green) quantitatively?

4. (a) Give an example of data that could not be collected in a quantitative manner, explaining your answer:

   (b) Sometimes, ranked data are given numerical values, e.g. rare = 1, occasional = 2, frequent = 3, common = 4, abundant = 5. Suggest why these data are sometimes called **semi-quantitative:**

ISBN: 978-1-927173-16-9

Manipulating Raw Data

The data collected by measuring or counting in the field or laboratory are called **raw data.** They often need to be changed (**transformed**) into a form that makes it easier to identify important features of the data (e.g. trends). Some basic calculations, such as totals (the sum of all data values for a variable), are made as a matter of course to compare replicates or as a prelude to other transformations. The calculation of **rate** (amount per unit time) is another example of a commonly performed calculation, and is appropriate for many biological situations (e.g. measuring growth or weight loss or gain). For a line graph, with time as the independent variable plotted against the values of the biological response, the slope of the line is a measure of the rate. Biological investigations often compare the rates of events in different situations (e.g. the rate of photosynthesis in the light and in the dark). Other typical transformations include frequencies (number of times a value occurs) and percentages (fraction of 100).

Tally Chart

Records the number of times a value occurs in a data set

HEIGHT (cm)	TALLY	TOTAL
0-0.99	III	3
1-1.99	卌 I	6
2-2.99	卌 卌	10
3-3.99	卌 卌 II	12
4-4.99	III	3
5-5.99	II	2

- A useful first step in analysis; a neatly constructed tally chart doubles as a simple histogram.
- Cross out each value on the list as you tally it to prevent double entries. Check all values are crossed out at the end and that totals agree.

Example: Height of 6d old seedlings

Percentages

Expressed as a fraction of 100

Women	Body mass (kg)	Lean body mass (kg)	% lean body mass
Athlete	50	38	76.0
Lean	56	41	73.2
Normal weight	65	46	70.8
Overweight	80	48	60.0
Obese	95	52	54.7

- Percentages provide a clear expression of what proportion of data fall into any particular category, e.g. for pie graphs.
- Allows meaningful comparison between different samples.
- Useful to monitor change (e.g. % increase from one year to the next).

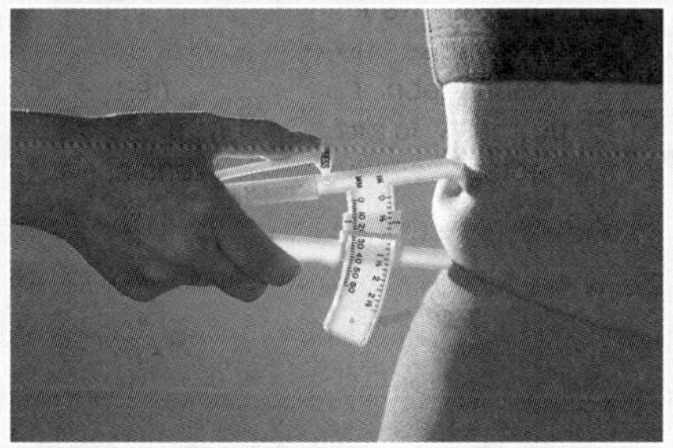

Example: Percentage of lean body mass in women

Rates

Expressed as a measure per unit time

Time (minutes)	Cumulative sweat loss (mL)	Rate of sweat loss (mL min^{-1})
0	0	0
10	50	5
20	130	8
30	220	9
60	560	11.3

- Rates show how a variable changes over a standard time period (e.g. one second, one minute, or one hour).
- Rates allow meaningful comparison of data that may have been recorded over different time periods.

Rate of sweat loss (mLmin^{-1})
0
2
4
6
8
10
12
0
10
20
30
40
50
60
Minutes

Example: Rate of sweat loss in exercise

1. (a) Explain what it means to transform data: ______________________

(b) Briefly explain the general purpose of transforming data: ______________________

2. For each of the following examples, state a suitable transformation, together with a reason for your choice:

(a) Determining relative abundance from counts of four plant species in two different habitat areas:

Suitable transformation: ______________________

Reason: ______________________

(b) Determining the effect of temperature on the production of carbon dioxide by respiring seeds:

Suitable transformation: ______________________

Reason: ______________________

ISBN: 978-1-927173-16-9

Periodicals:
Percentages

Related activities: *Variables and Data*

3. Complete the transformations for each of the tables on the right. The first value is provided in each case.

(a) TABLE: Incidence of cyanogenic clover in different areas:

Working: 124 ÷ 159 = 0.78 = 78%

This is the number of cyanogenic clover out of the total.

Incidence of cyanogenic clover in different areas

Clover plant type	Frost free area		Frost prone area		Totals
	Number	%	Number	%	
Cyanogenic	124	78	26		
Acyanogenic	35		115		
Total	159				

(b) TABLE: Plant water loss using a bubble potometer

Working: (9.0 – 8.0) ÷ 5 min = 0.2

This is the distance the bubble moved over the first 5 minutes. Note that there is no data entry possible for the first reading (0 min) because no difference can be calculated.

Plant water loss using a bubble potometer

Time (min)	Pipette arm reading (cm^3)	Plant water loss ($cm^3\ min^{-1}$)
0	9.0	–
5	8.0	0.2
10	7.2	
15	6.2	
20	4.9	

(c) TABLE: Photosynthetic rate at different light intensities:

Working: 1 ÷ 15 = 0.067

This is time taken for the leaf to float. A reciprocal gives a per minute rate (the variable measured is the time taken for an event to occur).

NOTE: In this experiment, the flotation time is used as a crude measure of photosynthetic rate. As oxygen bubbles are produced as a product of photosynthesis, they stick to the leaf disc and increase its buoyancy. The faster the rate, the sooner they come to the surface. The rates of photosynthesis should be measured over similar time intervals, so the rate is transformed to a 'per minute' basis (the reciprocal of time).

Photosynthetic rate at different light intensities

Light intensity (%)	Average time for leaf disc to float (min)	Reciprocal of time (min^{-1})
100	15	0.067
50	25	
25	50	
11	93	
6	187	

(d) TABLE: Frequency of size classes in a sample of eels:

Working: (7 ÷ 270) x 100 = 2.6 %

This is the number of individuals out of the total that appear in the size class 0-50 mm. The relative frequency is rounded to one decimal place.

Frequency of size classes in a sample of eels

Size class (mm)	Frequency	Relative frequency (%)
0-50	7	2.6
50-99	23	
100-149	59	
150-199	98	
200-249	50	
250-299	30	
300-349	3	
Total	270	

ISBN: 978-1-927173-16-9

Planning a Quantitative Investigation

The middle stage of any investigation (following the planning) is the practical work when the data are collected. Practical work may be laboratory or field based. Typical laboratory based experiments involve investigating how a biological response is affected by manipulating a particular **variable**, e.g. temperature. The data collected for a quantitative practical task should be recorded systematically, with due attention to safe practical techniques, a suitable quantitative method, and accurate measurements to a an appropriate degree of precision. If your quantitative practical task is executed well, and you have taken care throughout, your evaluation of the experimental results will be much more straightforward and less problematic.

Carrying Out Your Practical Work

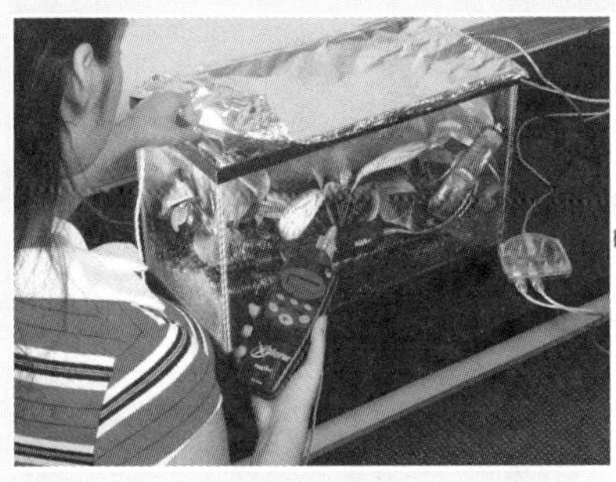

Preparation

Familiarize yourself with the equipment and how to set it up. If necessary, calibrate equipment to give accurate measurements.

Read through the methodology and identify key stages and how long they will take.

Execution

Know how you will take your measurements, how often, and to what degree of precision.

If you are working in a group, assign tasks and make sure everyone knows what they are doing.

Recording

Record your results systematically, in a hand-written table or on a spreadsheet.

Record your results to the appropriate number of significant figures according to the precision of your measurement.

Identifying Variables

A variable is any characteristic or property able to take any one of a range of values. Investigations often look at the effect of changing one variable on another. It is important to identify all variables in an investigation: independent, dependent, and controlled, although there may be nuisance factors of which you are unaware. In all fair tests, only one variable is changed by the investigator.

Dependent variable

- Measured during the investigation.
- Recorded on the y axis of the graph.

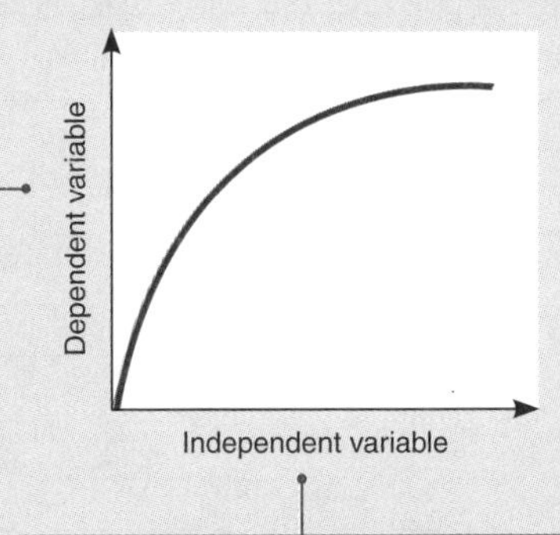
Dependent variable
Independent variable

Controlled variables

- Factors that are kept the same or controlled.
- List these in the method, as appropriate to your own investigation.

Independent variable

- Set by the experimenter.
- Recorded on the graph's x axis.

Experimental Controls

A **control** refers to standard or reference treatment or group in an experiment. It is the same as the experimental (test) group, except that it lacks the one variable being manipulated by the experimenter. Controls are used to demonstrate that the response in the test group is due a specific variable (e.g. temperature). The control undergoes the same preparation, experimental conditions, observations, measurements, and analysis as the test group. This helps to ensure that responses observed in the treatment groups can be reliably interpreted.

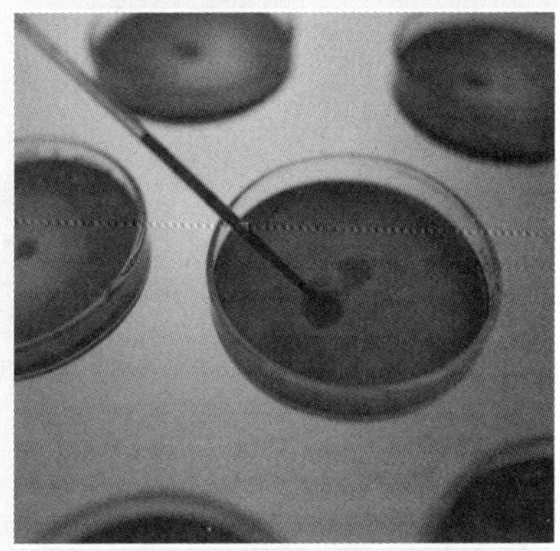

The experiment above tests the effect of a certain nutrient on microbial growth. All the agar plates are prepared in the same way, but the control plate does not have the test nutrient applied. Each plate is inoculated from the same stock solution, incubated under the same conditions, and examined at the same set periods. The control plate sets the baseline; any growth above that seen on the control plate is attributed to the presence of the nutrient.

Examples of Investigations

Aim		Variables	
Investigating the effect of varying...	**on the following...**	**Independent variable**	**Dependent variable**
Temperature	Leaf width	Temperature	Leaf width
Light intensity	Activity of woodlice	Light intensity	Woodlice activity
Soil pH	Plant height at age 6 months	pH	Plant height

ISBN: 978-1-927173-16-9

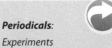

Periodicals: *Experiments*

Related activities: *Variables and Data*

In order to write a sound method for your investigation, you need to determine how the independent, dependent, and controlled variables will be set and measured (or monitored). A good understanding of your methodology is crucial to a successful investigation. You need to be clear about how much data, and what type of data, you will collect. You should also have a good idea about how you plan to analyze those data. Use the example below to practice identifying this type of information.

Case Study: Catalase Activity

Catalase is an enzyme that converts hydrogen peroxide (H_2O_2) to oxygen and water. An experiment investigated the effect of temperature on the rate of the catalase reaction. Small (10 cm^3) test tubes were used for the reactions, each containing 0.5 cm^3 of enzyme and 4 cm^3 of hydrogen peroxide. Reaction rates were assessed at four temperatures (10°C, 20°C, 30°C, and 60°C). For each temperature, there were two reaction tubes (e.g. tubes 1 and 2 were both kept at 10°C). The height of oxygen bubbles present after one minute of reaction was used as a measure of the reaction rate; a faster reaction rate produced more bubbles. The entire experiment, involving eight tubes, was repeated on two separate days.

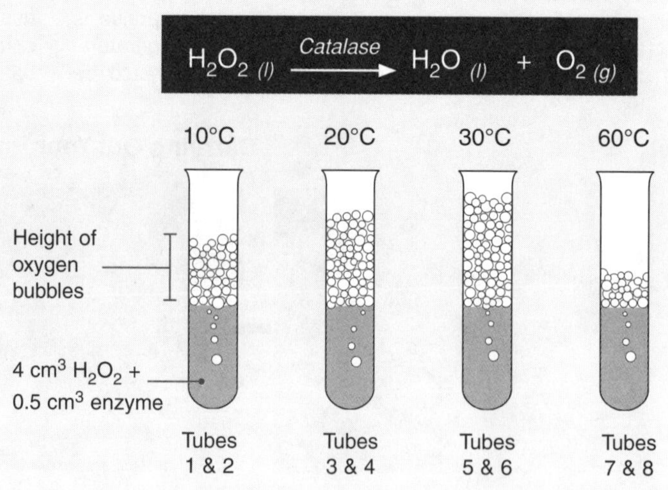
$H_2O_{2\ (l)} \xrightarrow{Catalase} H_2O_{(l)} + O_{2\ (g)}$
10°C
20°C
30°C
60°C
Height of oxygen bubbles
4 cm^3 H_2O_2 + 0.5 cm^3 enzyme
Tubes 1 & 2
Tubes 3 & 4
Tubes 5 & 6
Tubes 7 & 8

1. Write a suitable aim for this experiment: ______________________

2. Write a suitable hypothesis for this experiment: ______________________

3. (a) Identify the **independent variable:** ______________________

   (b) State the range of values for the independent variable: ______________________

   (c) Name the unit for the independent variable: ______________________

   (d) List the equipment needed to set the independent variable, and describe how it was used: ______________________

4. (a) Identify the **dependent variable**: ______________________

   (b) Name the unit for the dependent variable: ______________________

   (c) List the equipment needed to measure the dependent variable, and describe how it was used: ______________________

5. (a) Each temperature represents a treatment/sample/trial (circle one):

   (b) State the number of tubes at each temperature: ______________________

   (c) State the sample size for each treatment: ______________________

   (d) State how many times the whole investigation was repeated: ______________________

6. Explain why it would have been desirable to have included an extra tube containing no enzyme: ______________________

7. Identify three variables that might have been controlled in this experiment, and how they could have been monitored:

   (a) ______________________

   (b) ______________________

   (c) ______________________

8. Explain why controlled variables should be monitored carefully: ______________________

ISBN: 978-1-927173-16-9

A Simple Investigation

Recording data from an experiment is an important skill. Using a table is the preferred way to record your results **systematically**, both during the course of your experiment and in presenting your results. A table can also show calculated values, such as rates or means. An example of a table for recording results is shown below. It relates to a student investigation that followed the observation that plants in a paddock fertilized with a nitrogen fertilizer grew more vigorously than plants in a non-fertilized paddock. The table's first column shows the range of the independent variable. There are spaces for multiple samples, and calculated mean values. The students tested which concentration of a soluble nitrogen fertilizer produced optimal growth.

Radishes

The Aim

To investigate the effect of a nitrogen fertilizer on the growth of plants.

Background

Inorganic fertilizers revolutionized crop farming when they were introduced during the late 19th and early 20th century. Crop yields soared and today it is estimated around 50% of crop yield is attributable to the use of fertilizer. Nitrogen is a very important element for plant growth and several types of purely nitrogen fertilizer are manufactured to supply it, e.g. urea.

Experimental Method

This experiment was designed to test the effect of nitrogen fertilizer on plant growth. Radish seeds were planted in separate identical pots (5 cm x 5 cm wide x 10 cm deep) and grown together in normal room conditions. The radishes were watered every day at 10 am and 3 pm with 1.25 L per treatment. Water soluble fertilizer was mixed and added with the first watering on the 1st, 11th and 21st days. The fertilizer concentrations used were: 0.00, 0.06, 0.12, 0.18, 0.24, and 0.30 g L^{-1} with each treatment receiving a different concentration. The plants were grown for 30 days before being removed, washed, and the root (radish) weighed. Results were tabulated below:

To investigate the effect of nitrogen on plant growth, a group of students set up an experiment using different concentrations of nitrogen fertilizer. Radish seeds were planted into a standard soil mixture and divided into six groups each, with five sample plants (30 plants in total).

Tables should have an accurate, descriptive title. Number tables consecutively through the report.

Heading and subheadings identify each set of data and show units of measurement.

Independent variable in the left column.

Table 1: Mass (g) of radish plant roots under six different fertilizer concentrations (data given to 1dp).

Fertilizer concentration (g L^{-1})	Mass of radish root (g)†						
	Sample (*n*)					Total mass	Mean mass
	1	2	3	4	5		
0	80.1	83.2	82.0	79.1	84.1	408.5	81.7
0.06	109.2	110.3	108.2	107.9	110.7		
0.12	117.9	118.9	118.3	119.1	117.2		
0.18	128.3	127.3	127.7	126.8	DNG*		
0.24	23.6	140.3	139.6	137.9	141.1		
0.30	122.3	121.1	122.6	121.3	123.1		

* DNG: Did not germinate

Control values (if present) should be placed at the beginning of the table.

Values should be shown only to the level of significance allowable by your measuring technique.

Organize the columns so that each category of like numbers or attributes is listed vertically.

Each row should show a different experimental treatment, organism, sampling site etc.

† Based on data from M S Jilani, *et al* Journal Agricultural Research

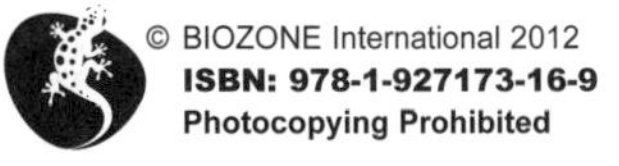

ISBN: 978-1-927173-16-9

Periodicals: *Be confident with calculations*

Related activities: *Descriptive Statistics*
Weblinks: *Introduction to Descriptive Statistics*

1. Identify the independent variable for the experiment and its range: ______

2. What is the sample size for each concentration of fertilizer? ______

3. One of the radishes recorded in Table 1 did not grow as expected and produced an extreme value. Record the **outlying value** here and decide whether or not you should include it in future calculations:

4. Complete the table on the previous page by calculating the **total mass** and **mean mass** of the radish roots:

5. Use the grid below to draw a **line graph** of the experimental results. Use your calculated means and remember to include a title and correctly labelled axes.

6. The students recorded the wet mass of the root (the root still containing water) in their table. What mass should they have actually recorded to get a better representation of the effect of the fertilizer on root mass?

7. Why would measuring just root mass not be a totally accurate way of measuring the effect of fertilizer on radish growth?

8. Describe some other measurements the students could have taken to make their experiment more complete:

ISBN: 978-1-927173-16-9

Calculating Simple Statistics for a Data Set

Statistic	Definition and use	Method of calculation
Mean	• The average of all data entries. • Measure of central tendency for normally distributed data.	• Add up all the data entries. • Divide by the total number of data entries.
Median	• The middle value when data entries are placed in rank order. • A good measure of central tendency for skewed distributions.	• Arrange the data in increasing rank order. • Identify the middle value. • For an even number of entries, find the mid point of the two middle values.
Mode	• The most common data value. • Suitable for bimodal distributions and qualitative data.	• Identify the category with the highest number of data entries using a tally chart or a bar graph.
Range	• The difference between the smallest and largest data values. • Provides a crude indication of data spread.	• Identify the smallest and largest values and find the difference between them.

Data can be simply summarized using **descriptive statistics**. Descriptive statistics, such as mean, median, and mode, can highlight trends or patterns in the data. The mean can be used to compare different groups. You can use more complex statistics to determine if the means of different groups are significantly different.

When NOT to calculate a mean:

In certain situations, calculation of a simple arithmetic mean is inappropriate.

Remember:

- *DO NOT* calculate a mean from values that are already means (averages) themselves.
- *DO NOT* calculate a mean of ratios (e.g. percentages) for several groups of different sizes; go back to the raw values and recalculate.
- *DO NOT* calculate a mean when the measurement scale is not linear, e.g. pH units are not measured on a linear scale.

The students decided to further their experiment by recording the number of leaves on each radish plant:

Table 2: Number of leaves on radish plant under six different fertilizer concentrations.

Fertilizer concentration (g L^{-1})	Number of leaves							
	Sample (*n*)					Mean	Median	Mode
	1	2	3	4	5			
0	9	9	10	8	7			
0.06	15	16	15	16	16			
0.12	16	17	17	17	16			
0.18	18	18	19	18	DNG*			
0.24	6	19	19	18	18			
0.30	18	17	18	19	19			

* DNG: Did not germinate

9. Complete Table 2 by calculating the mean, median and mode for each concentration of fertilizer:

10. Which concentration of fertilizer appeared to produce the best growth results? __________

11. Describe some sources of error for the experiment: __________

12. Write a brief conclusion for the experiment. Include a reference to the aim and results: __________

13. The students decided to replicate the experiment (carry it out again). How might this improve the experiment's results?

ISBN: 978-1-927173-16-9

Further Data Transformations

Raw data usually needs some kind of processing so that trends in the data and relationships between variables can be easily identified and tested statistically. The simplest and most powerful statistical tests generally require data to exhibit a normal distribution, yet many biological variables are not distributed in this way. It is possible to get around this apparent problem by transforming the data. Data transformation can help to account for differences between sample sizes in different treatments. It is also a perfectly legitimate way to normalize data so that its distribution meets the criteria for analysis. It is not a way to manipulate data to get the result you want. Your choice of transformation is based on the type of data you have and how you propose to analyze it. Some experimental results may be so clear, a complex statistical analysis is unnecessary.

Reciprocals

1 / x is the reciprocal of x.

Enzyme concentration (µg mL^{-1})	Reaction time (min)	Reciprocal value
6	25	0.04
12.5	20	0.05
25	14	0.07
50	5	0.20
100	2.5	0.40
150	1.75	0.57

- Reciprocals of time (1/data value) can provide a crude measure of rate in situations where the variable measured is the total time taken to complete a task.

Problem: Responses are measured over different time scales.

Example: Time taken for color change in an enzyme reaction.

Square Root

A square root is a value that when multiplied by itself gives the original number.

Sampling site	No. of Woodlice	Square root
1	10	3.16
2	7	2.65
3	5	2.24
4	3	1.73
5	1	1
6	0	0
7	1	1
8	1	1

- Applied to data that counts something.
- The square root of a negative number cannot be taken. Negative numbers are made positive by the addition of a constant value.
- Helps to normalize skewed data.

Problem: Skewed data.

Example: The number of woodlice distributed across a transect.

Log_{10}

A log transformation has the effect of normalizing data.

- Log transformations are useful for data where there is an exponential increase in numbers (e.g. cell growth).
- Log transformed data will plot as a straight line and the numbers are more manageable.
- To find the log_{10} of a number, e.g. 32, using a calculator, key in log 32 = ___.

The answer should be 1.51.

Problem: Exponential increases.

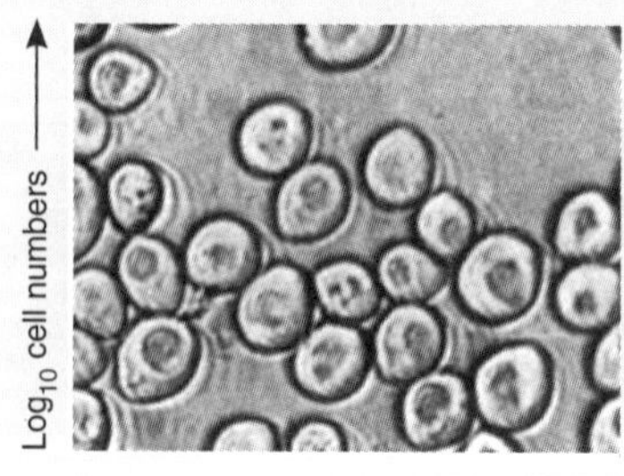
Log_{10} cell numbers
Time

Example: Cell growth in a yeast culture

1. Why might a researcher transform skewed or non-normal data prior to statistical analysis? ____________

2. For each of the following examples, state a suitable transformation, together with a reason for your choice:

(a) Comparing the time taken for chemical precipitation to occur in a flask at different pH values:

Suitable transformation: ____________

Reason: ____________

(b) Analyzing the effect of growth environment on the number of bacterial colonies developing in 50 agar plates:

Suitable transformation: ____________

Reason: ____________

Related activities: Manipulating Raw Data

ISBN: 978-1-927173-16-9

Constructing Tables

Tables provide a convenient way to systematically record and condense a large amount of information for later presentation and analysis. The protocol for creating tables for recording data during the course of an investigation is provided elsewhere, but tables can also provide a useful summary in the results section of a finished report. They provide an accurate record of numerical values and allow you to organise your data in a way that allows you to clarify the relationships and trends that are apparent. Columns can be provided to display the results of any data transformations such as rates. Some basic descriptive statistics (such as mean or standard deviation) may also be included prior to the data being plotted. For complex data sets, graphs tend to be used in preference to tables, although the latter may be provided as an appendix.

Presenting Data in Tables

Tables should have an accurate, descriptive title. Number tables consecutively through the report.

Heading and subheadings identify each set of data and show units of measurement.

Independent variable in the left column.

Table 1: Length and growth of the third internode of bean plants receiving three different hormone treatments (data are given ± standard deviation).

Treatment	Sample size	Mean rate of internode growth (mm day^{-1})	Mean internode length (mm)	Mean mass of tissue added (g day^{-1})
Control	50	0.60 ± 0.04	32.3 ± 3.4	0.36 ± 0.025
Hormone 1	46	1.52 ± 0.08	41.6 ± 3.1	0.51 ± 0.030
Hormone 2	98	0.82 ± 0.05	38.4 ± 2.9	0.56 ± 0.028
Hormone 3	85	2.06 ± 0.19	50.2 ± 1.8	0.68 ± 0.020

Control values (if present) should be placed at the beginning of the table.

Each row should show a different experimental treatment, organism, sampling site etc.

Columns for comparison should be placed alongside each other. Show values only to the level of significance allowable by your measuring technique.

Organize the columns so that each category of like numbers or attributes is listed vertically.

Tables can be used to show a calculated measure of spread of the values about the mean.

1. Describe two advantages of using a table format for data presentation:

   (a) ______________________

   (b) ______________________

2. Why might you tabulate data before you presented it in a graph? ______________________

3. (a) What is the benefit of tabulating basic descriptive statistics rather than raw data? ______________________

   (b) Why would you include a measure of spread (dispersion) for a calculated statistic in a table?

   ____

4. Why should you place control values at the beginning of a table? ______________________

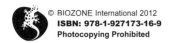

Constructing Graphs

The choice between graphing or tabulation depends on the type and complexity of the data and the information that you are wanting to convey. Graphs are useful for displaying trends or relationships between different variables. Presenting graphs properly requires attention to a few basic details, including correct orientation and labeling of the axes, and accurate plotting of points. Before representing data graphically, it is important to identify the kind of data you have. Common graphs include scatter plots and line graphs (for continuous data), and **bar charts** and histograms (for categorical data). For continuous data with calculated means, points can be connected. On scatter plots, a line of best fit is often drawn.

Guidelines for Line Graphs

Line graphs are used when one variable (the **independent variable**) affects another, the **dependent variable**. If there is a relationship between variables but no implied dependence, then a scatter graph is appropriate. Important features of line graphs include:

- The data must be continuous for both variables. The independent variable is often time or experimental treatment.The dependent variable is usually the biological response.
- Where there is an implied trend, a line of best fit is usually plotted through the data points to show the relationship.
- If fluctuations are important (e.g. with environmental data) the data points are usually connected directly (point to point).
- Line graphs may be drawn with measure of error (right). The data are presented as points (which are calculated means), with **error bars** above and below, indicating the variability or spread in the data (e.g. standard deviation or 95% confidence intervals).

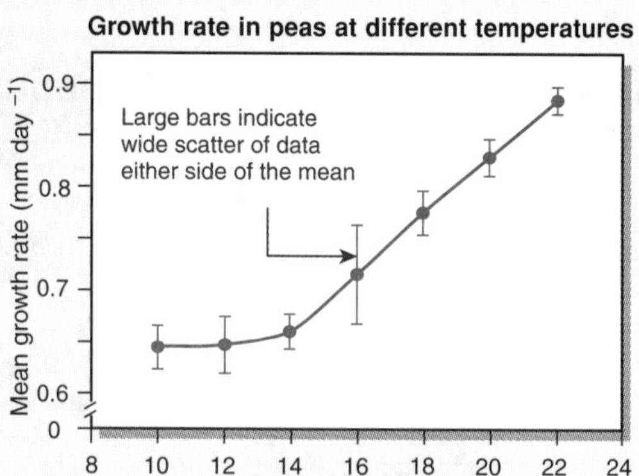
Growth rate in peas at different temperatures
Mean growth rate (mm day^{-1})
0.9
0.8
0.7
0.6
0
Large bars indicate wide scatter of data either side of the mean
8
10
12
14
16
18
20
22
24
Temperature (°C)

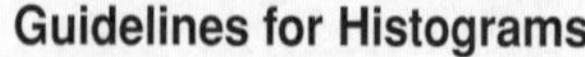

Guidelines for Histograms

Histograms are plots of **continuous** data and are often used to represent frequency distributions, where the y-axis shows the number of times a measurement or value was obtained. For this reason, they are often called frequency histograms. Important features of histograms include:

- The data are numerical and continuous (e.g. height or weight), so the bars touch.
- The x-axis usually records the class interval. The y-axis usually records the number of individuals in each class interval.

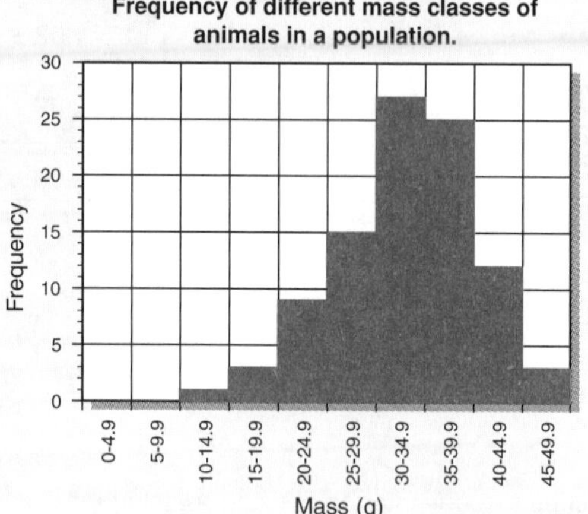
Frequency of different mass classes of animals in a population.
Frequency
30
25
20
15
10
5
0
0-4.9
5-9.9
10-14.9
15-19.9
20-24.9
25-29.9
30-34.9
35-39.9
40-44.9
45-49.9
Mass (g)

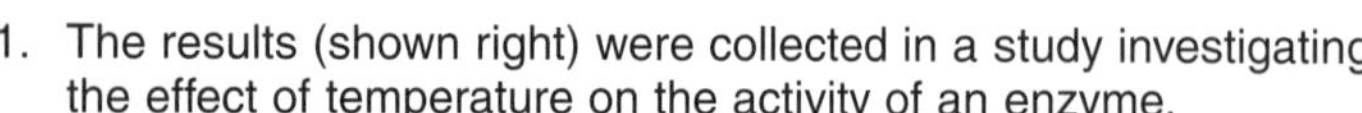

1. The results (shown right) were collected in a study investigating the effect of temperature on the activity of an enzyme.

   (a) Using the results provided in the table (right), plot a line graph on the grid below:

   (b) Estimate the rate of reaction at 15°C: ______________________

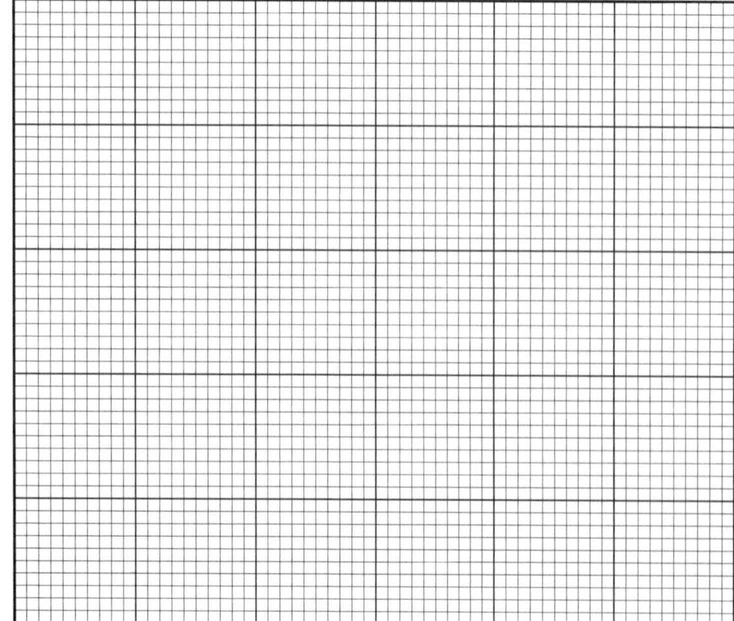

Lab Notebook

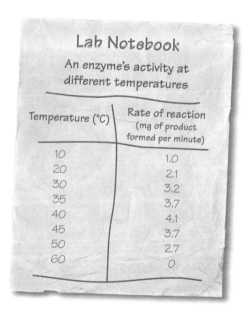

An enzyme's activity at different temperatures

Temperature (°C)	Rate of reaction (mg of product formed per minute)
10	1.0
20	2.1
30	3.2
35	3.7
40	4.1
45	3.7
50	2.7
60	0

ISBN: 978-1-927173-16-9

Evaluating Your Results

Once you have completed the practical part of an experiment, the next task is to evaluate your results in the light of your own hypothesis and your current biological knowledge. A critical evaluation of any study involves analyzing, presenting, and discussing the results, as well as accounting for any deficiencies in your procedures and erroneous results. This activity describes an experiment in which germinating seeds of different ages were tested for their level of catalase activity using hydrogen peroxide solution as the substrate and a simple apparatus to measure oxygen production (see background). Completing this activity, which involves a critical evaluation of the second-hand data provided will help to prepare you for your own evaluative task.

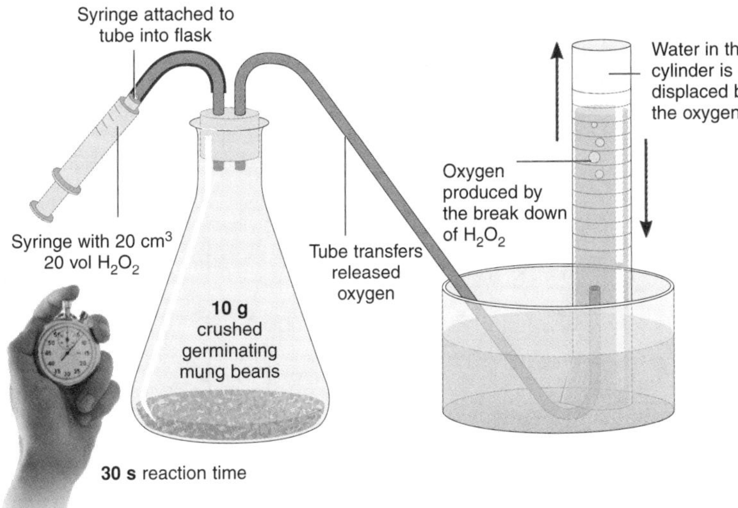
Syringe attached to tube into flask
Water in the cylinder is displaced by the oxygen
Oxygen produced by the break down of H_2O_2
Syringe with 20 cm^3 20 vol H_2O_2
Tube transfers released oxygen
10 g crushed germinating mung beans
30 s reaction time

The Apparatus

In this experiment, 10 g germinating mung bean seeds (0.5, 2, 4, 6, or 10 days old) were ground by hand with a mortar and pestle and placed in a conical flask as above. There were six trials at each of the five seedling ages. With each trial, 20 cm^3 of 20 vol H_2O_2 was added to the flask at time 0 and the reaction was allowed to run for 30 seconds. The oxygen released by the decomposition of the H_2O_2 by catalase in the seedlings was collected via a tube into an inverted measuring cylinder. The volume of oxygen produced is measured by the amount of water displaced from the cylinder. The results from all trials are tabulated below:

The Aim

To investigate the effect of germination age on the level of catalase activity in mung beans.

Background

Germinating seeds are metabolically very active and this metabolism inevitably produces reactive oxygen species, including hydrogen peroxide (H_2O_2). H_2O_2 is helps germination by breaking dormancy, but it is also toxic. To counter the toxic effects of H_2O_2 and prevent cellular damage, germinating seeds also produce **catalase**, an enzyme that catalyzes the breakdown of H_2O_2 to water and oxygen.

A class was divided into six groups with each group testing the seedlings of each age. Each group's set of results (for 0.5, 2, 4, 6, and 10 days) therefore represents one trial.

Stage of germination / days \ Trial #	Volume of oxygen collected after 30s (cm^3)								Mean rate (cm^3 s^{-1} g^{-1})
	1	2	3	4	5	6	Mean	Standard deviation	
0.5	9.5	10	10.7	9.5	10.2	10.5			
2	36.2	30	31.5	37.5	34	40			
4	59	66	69	60.5	66.5	72			
6	39	31.5	32.5	41	40.3	36			
10	20	18.6	24.3	23.2	23.5	25.5			

1. Write the equation for the catalase reaction with hydrogen peroxide: ____________________

2. Complete the table above to summarize the data from the six trials:

   (a) Calculate the mean volume of oxygen for each stage of germination and enter the values in the table.

   (b) Calculate the standard deviation for each mean and enter the values in the table (you may use a spreadsheet).

   (c) Calculate the mean rate of oxygen production in cm^3 per second per gram. For the purposes of this exercise, assume that the weight of germinating seed in every case was 10.0 g.

3. In another scenario, group (trial) #2 obtained the following measurements for volume of oxygen produced: 0.5 d: 4.8 cm^3, 2 d: 29.0 cm^3, 4 d: 70 cm^3, 6 d: 30.0 cm^3, 10 d: 8.8 cm^3 (pencil these values in beside the other group 2 data set).

   (a) Describe how group 2's new data accords with the measurements obtained from the other groups: ____________________

   ____

   (b) Describe how you would approach a reanalysis of the data set incorporating group 2's new data: ____________________

   ____

   ____

ISBN: 978-1-927173-16-9

Periodicals: *Estimating the mean and standard deviation*

Related activities: *Planning A Quantitative Investigation, Manipulating Raw Data*

(c) Explain the rationale for your approach

4. Use the tabulated data to plot an appropriate graph of the results on the grid provided:

5. (a) Describe the trend in the data:

(b) Explain the relationship between stage of germination and catalase activity shown in the data:

6. Describe any potential sources of errors in the apparatus or the procedure:

7. Describe two things that might affect the validity of findings in this experimental design:

8. Describe one improvement you could make to the experiment in order to generate more reliable data:

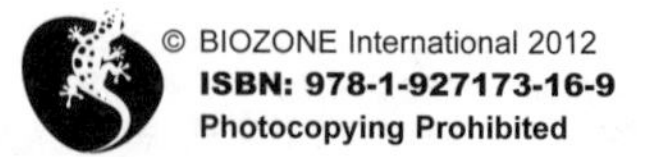

ISBN: 978-1-927173-16-9

Descriptive Statistics

For most investigations, measures of the biological response are made from more than one sampling unit. The sample size (the number of sampling units) will vary depending on the resources available. In lab based investigations, the sample size may be as small as two or three (e.g. two test-tubes in each treatment). In field studies, each individual may be a sampling unit, and the sample size can be very large (e.g. 100 individuals). It is useful to summarize the data collected using **descriptive statistics.**

Descriptive statistics, such as mean, median, and mode, can help to highlight trends or patterns in the data. Each of these statistics is appropriate to certain types of data or distributions, e.g. a mean is not appropriate for data with a skewed distribution (see below). Frequency graphs are useful for indicating the distribution of data. Standard deviation and standard error are statistics used to quantify the amount of spread in the data and evaluate the reliability of estimates of the true (population) mean.

Variation in Data

Whether they are obtained from observation or experiments, most biological data show variability. In a set of data values, it is useful to know the value about which most of the data are grouped; the centre value. This value can be the mean, median, or mode depending on the type of variable involved (see schematic below). The main purpose of these statistics is to summarize important trends in your data and to provide the basis for statistical analyses.

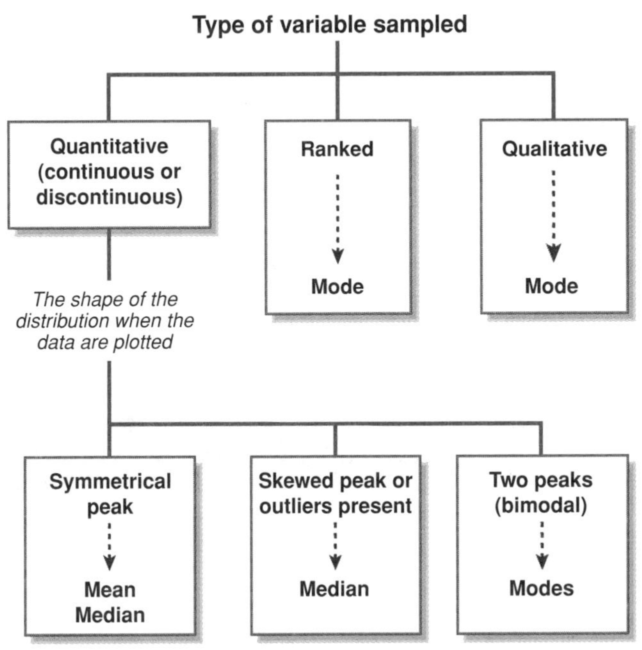
Type of variable sampled
Quantitative (continuous or discontinuous)
Ranked
Qualitative
Mode
Mode
The shape of the distribution when the data are plotted
Symmetrical peak
Mean
Median
Skewed peak or outliers present
Median
Two peaks (bimodal)
Modes

The shape of the distribution will determine which statistic (mean, median, or mode) best describes the central tendency of the sample data.

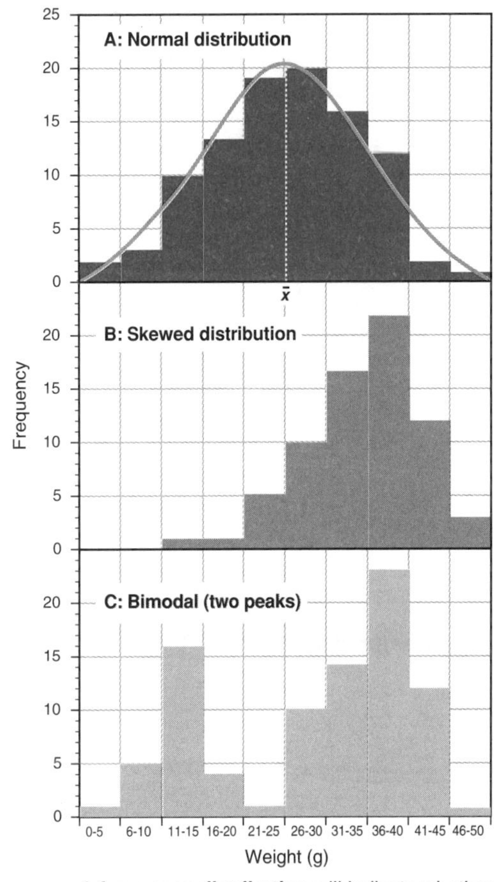
A: Normal distribution
B: Skewed distribution
C: Bimodal (two peaks)
Frequency
Weight (g)
0-5
6-10
11-15
16-20
21-25
26-30
31-35
36-40
41-45
46-50

A **frequency distribution** will indicate whether the data are normal, skewed, or bimodal.

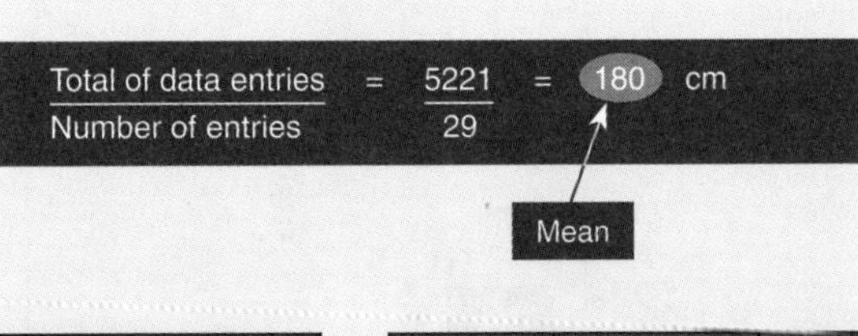
Total of data entries / Number of entries = 5221 / 29 = 180 cm
Mean

Height of swimmers (in rank order)		
174	177	185
175	177	185
175	178	185
175	178	186
176	178	186
176	178	186
176	180	188
176	180	188
176	180	189
177	181	

Mode

Median

Height (cm)	Tally	Total
174	✔	1
175	✔✔✔	3
176	✔✔✔✔✔	5
177	✔✔✔	3
178	✔✔✔✔	4
179		0
180	✔✔✔	3
181	✔	1
182		0
183		0
184		0
185	✔✔✔	3
186	✔✔✔	3
187		0
188	✔✔	2
189	✔	1

Case Study: Height of Swimmers

Data (below) and descriptive statistics (left) from a survey of the height of 29 members of a male swim squad.

Raw data: Height (cm)					
178	177	188	176	186	175
180	181	178	178	176	175
180	185	185	175	189	174
178	186	176	185	177	176
176	188	180	186	177	

1. Give a reason for the difference between the mean, median, and mode for the swimmers' height data:

____

____

____

____

ISBN: 978-1-927173-16-9

Periodicals: *Describing the normal distribution*

Related activities: *Interpreting Sample Variability*
Weblnks: *Handbook of Biological Statistics*

Measuring Spread

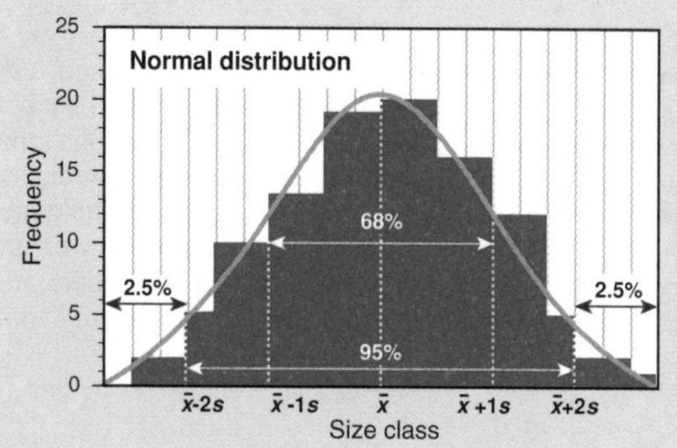
Normal distribution
Frequency
25
20
15
10
5
0
68%
95%
2.5%
2.5%
x̄-2s
x̄-1s
x̄
x̄+1s
x̄+2s
Size class

The **standard deviation** is a frequently used measure of the variability (spread) in a set of data. It is usually presented in the form $\bar{x} \pm s$. In a normally distributed set of data, 68% of all data values will lie within one standard deviation (s) of the mean ($\bar{x}$) and 95% of all data values will lie within two standard deviations of the mean (left).

Two different sets of data can have the same mean and range, yet the distribution of data within the range can be quite different. In both the data sets pictured in the histograms below, 68% of the values lie within the range $\bar{x} \pm 1s$ and 95% of the values lie within $\bar{x} \pm 2s$. However, in B, the data values are more tightly clustered around the mean.

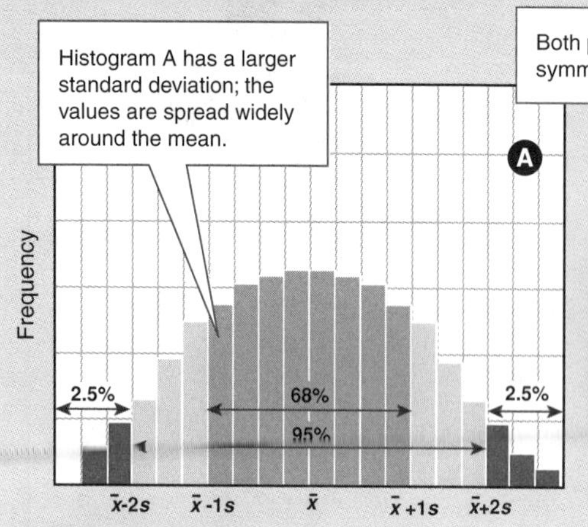
Histogram A has a larger standard deviation; the values are spread widely around the mean.
A
Frequency
2.5%
68%
95%
2.5%
x̄-2s
x̄-1s
x̄
x̄+1s
x̄+2s

Both plots show a normal distribution with a symmetrical spread of values about the mean.

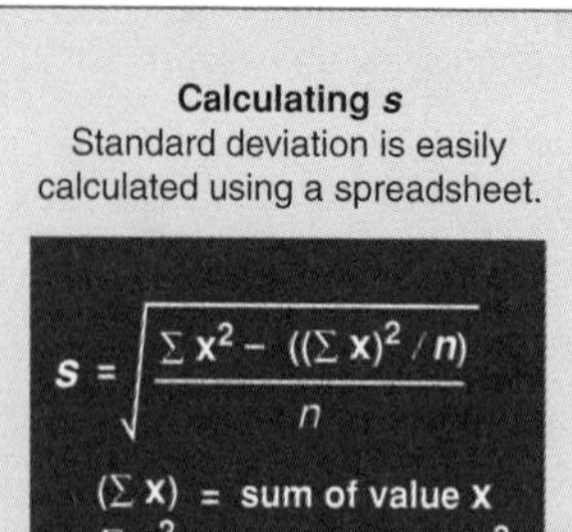

Calculating s
Standard deviation is easily calculated using a spreadsheet.

$$s = \sqrt{\frac{\sum x^2 - ((\sum x)^2 / n)}{n}}$$

$(\sum x)$ = sum of value x
$\sum x^2$ = sum of value x^2
n = sample size

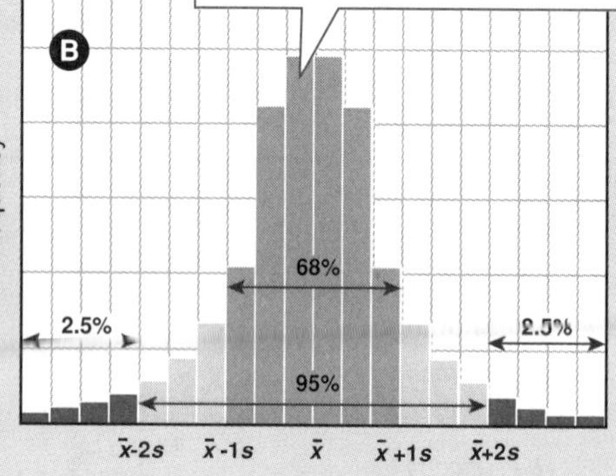
Histogram B has a smaller standard deviation; the values are clustered more tightly around the mean.
B
Frequency
68%
2.5%
2.5%
95%
x̄-2s
x̄-1s
x̄
x̄+1s
x̄+2s

Case Study: Fern Reproduction

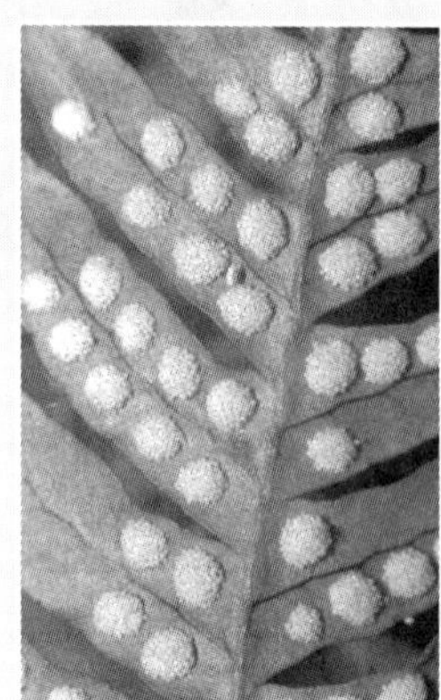
Fern spores

Raw data (below) and descriptive statistics (right) from a survey of the number of sori found on the fronds of a fern plant.

Raw data: Number of sori per frond							
64	60	64	62	68	66	66	63
69	70	63	70	70	63	63	62
71	69	59	70	66	61	61	70
67	64	63	64				

$$\frac{\text{Total of data entries}}{\text{Number of entries}} = \frac{1641}{25} = 66 \text{ sori}$$

Median

Number of sori per frond (in rank order)	
59	66
60	66
61	67
62	68
62	69
63	69
63	70
63	70
63	70
64	70
64	70
64	71
64	

Median

Sori per frond	Tally	Total
59	✓	1
60	✓	1
61	✓	1
62	✓✓	2
63	✓✓✓✓	4
64	✓✓✓✓	4
65		0
66	✓✓	2
67	✓	1
68	✓	1
69	✓✓	2
70	✓✓✓✓✓	5
71	✓	1

Mode

2. Give a reason for the difference between the mean, median, and mode for the fern sori data:

3. Calculate the mean, median, and mode for the data on ladybird masses below. Draw up a tally chart and show all calculations:

Ladybird mass (mg)		
10.1	8.2	7.7
8.0	8.8	7.8
6.7	7.7	8.8
9.8	8.8	8.9
6.2	8.8	8.4

ISBN: 978-1-927173-16-9

Interpreting Sample Variability

Measures of central tendency, such as mean, attempt to identify the most representative value in a set of data, but the description of a data set also requires that we know something about how far the data values are spread around that central measure. As we have seen in the previous activity, the **standard deviation** (s) gives a simple measure of the spread or **dispersion** in data. The **variance** (s^2) is also a measure of dispersion, but the standard deviation is usually preferred because it is expressed in the original units. Two data sets could have exactly the same mean values, but very different values of dispersion. If we were simply to use the central tendency to compare these data sets, the results would (incorrectly) suggest that they were alike. The assumptions we make about a population will be affected by what the sample data tell us. This is why it is important that sample data are unbiased (e.g. collected by **random sampling**) and that the sample set is as large as practicable. This exercise will help to illustrate how our assumptions about a population are influenced by the information provided by the sample data.

Random Sampling, Sample Size, and Dispersion in Data

Sample size and sampling bias can both affect the information we obtain when we sample a population. In this exercise you will calculate some descriptive statistics for some sample data.

The complete set of sample data we are working with comprises 689 length measurements of year zero (young of the year) perch (column left). Basic descriptive statistics for the data have bee calculated for you below and the frequency histogram has also been plotted.

Look at this data set and then complete the exercise to calculate the same statistics from each of two smaller data sets (tabulated right) drawn from the same population. This exercise shows how random sampling, large sample size, and sampling bias affect our statistical assessment of variation in a population.

Complete sample set
n = 689 (random)

Length in mm	*Freq*
25	1
26	0
27	0
28	0
29	0
30	0
31	0
32	2
33	3
34	3
35	4
36	5
37	10
38	23
39	22
40	33
41	39
42	41
43	41
44	36
45	49
46	32
47	14
48	32
49	27
50	25
51	24
52	17
53	18
54	27
55	21
56	20
57	11
58	18
59	16
60	22
61	13
62	8
63	10
64	5
65	7
66	2
67	3
68	3
69	1
70	0
71	1

Length of year zero perch

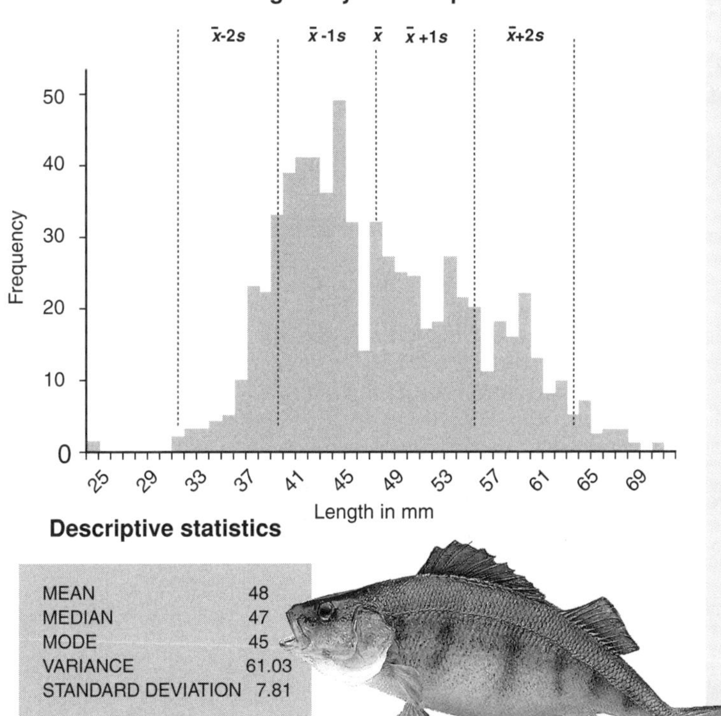

x̄-2s
x̄-1s
x̄
x̄+1s
x̄+2s
50
40
30
20
10
0
Frequency
25
29
33
37
41
45
49
53
57
61
65
69
Length in mm

Descriptive statistics

MEAN	48
MEDIAN	47
MODE	45
VARIANCE	61.03
STANDARD DEVIATION	7.81

Small sample set
n = 30 (random)

Length in mm	*Freq*
25	1
26	0
27	0
28	0
29	0
30	0
31	0
32	0
33	0
34	0
35	2
36	0
37	0
38	3
39	2
40	1
41	3
42	0
43	0
44	0
45	0
46	1
47	0
48	2
49	0
50	0
51	1
52	3
53	0
54	0
55	0
56	0
57	1
58	0
59	3
60	2
61	2
62	0
63	0
64	0
65	0
66	0
67	2
68	1
	30

This population was sampled randomly to obtain this data set

This column records the number of fish of each size

Number of fish in the sample

Small sample set
n = 50 (bias)

Length in mm	*Freq*
46	1
47	0
48	0
49	1
50	0
51	0
52	1
53	1
54	1
55	1
56	0
57	2
58	2
59	4
60	1
61	0
62	8
63	10
64	13
65	2
66	0
67	2
	50

The person gathering this set of data was biased towards selecting larger fish because the mesh size on the net was too large to retain small fish

1. For the complete data set (n = 689) calculate the percentage of data falling within:

   (a) ± one standard deviation of the mean: ______

   (b) ± two standard deviations of the mean: ______

   (c) Explain what this information tells you about the distribution of year zero perch from this site: ______

2. Give another reason why you might reach the same conclusion about the distribution: ______

ISBN: 978-1-927173-16-9

Periodicals:
Estimating mean and standard deviation

Related activities: *Descriptive Statistics*

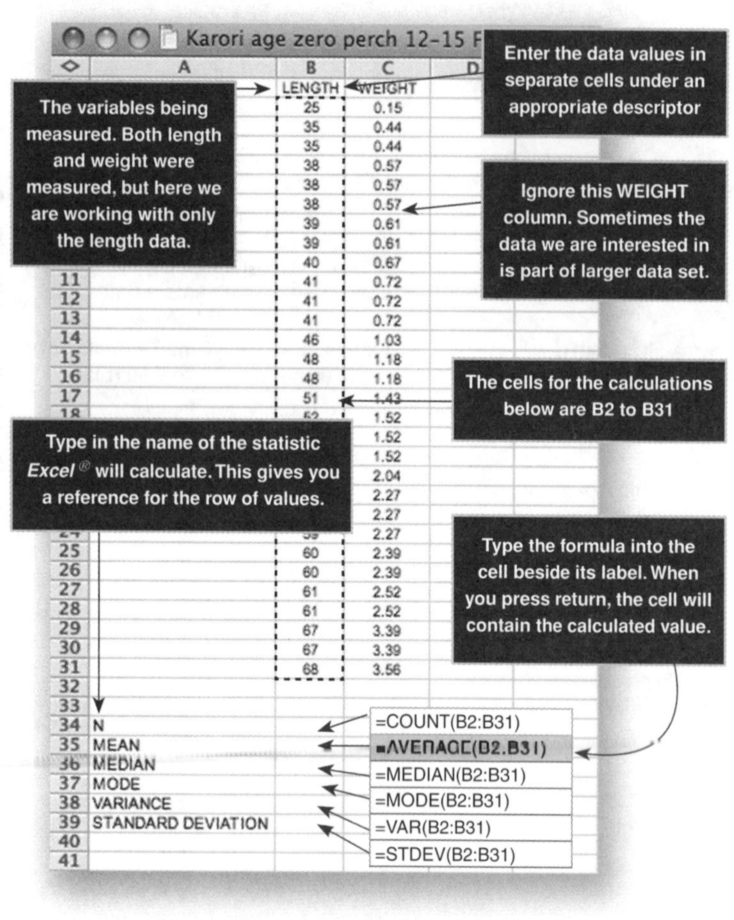
Karori age zero perch 12-15 F
LENGTH
WEIGHT
Enter the data values in separate cells under an appropriate descriptor
The variables being measured. Both length and weight were measured, but here we are working with only the length data.
Ignore this WEIGHT column. Sometimes the data we are interested in is part of larger data set.
The cells for the calculations below are B2 to B31
Type in the name of the statistic Excel® will calculate. This gives you a reference for the row of values.
Type the formula into the cell beside its label. When you press return, the cell will contain the calculated value.
N
MEAN
MEDIAN
MODE
VARIANCE
STANDARD DEVIATION
=COUNT(B2:B31)
=AVERAGE(B2:B31)
=MEDIAN(B2:B31)
=MODE(B2:B31)
=VAR(B2:B31)
=STDEV(B2:B31)

Calculating Descriptive Statistics Using *Excel*®

You can use *Microsoft Excel*® or other similar spreadsheet program to easily calculate descriptive statistics for sample data.

In this first example, the smaller data set (n = 30) is shown as it would appear on an *Excel*® spreadsheet, ready for the calculations to be made. Use this guide to enter your data into a spreadsheet and calculate the descriptive statistics as described.

When using formulae in *Excel*®, = indicates that a formula follows. The cursor will become active and you will be able to select the cells containing the data you are interested in, or you can type the location of the data using the format shown. The data in this case are located in the cells B2 through to B31 (B2:B31).

3. For this set of data, use a spreadsheet to calculate:

   (a) Mean: ______

   (b) Median: ______

   (c) Mode: ______

   (d) Sample variance: ______

   (e) Standard deviation: ______

   Staple the spreadsheet into your workbook.

4. Repeat the calculations for the second small set of sample data (n = 50) on the previous page. Again, calculate the statistics as indicated below and staple the spreadsheet into your workbook:

   (a) Mean: ______ (b) Median: ______ (c) Mode:

   (d) Variance: ______ (e) Standard deviation: ______

5. On a separate sheet, plot **frequency histograms** for each of the two small data sets. Label them n = 30 and n = 50. Staple them into your workbook. If you are proficient in *Excel*® and you have the "Data Analysis" plug in loaded, you can use *Excel*® to plot the histograms for you once you have entered the data.

6. Compare the descriptive statistics you calculated for each data set with reference to the following:

   (a) How close the median and mean to each other in each sample set: ______

   (b) The size of the standard deviation in each case: ______

   (c) How close each small of the sample sets resembles the large sample set of 689 values: ______

7. (a) Compare the two frequency histograms you have plotted for the two smaller sample data sets: ______

   (b) Why do you think two histograms look so different? ______

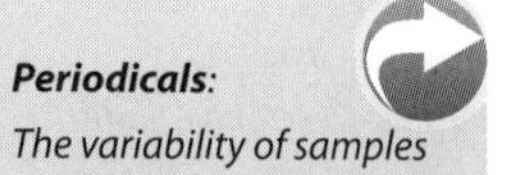

Periodicals:
The variability of samples

ISBN: 978-1-927173-16-9

Spearman Rank Correlation

The Spearman rank correlation is a test used to determine if there is a statistical dependence (correlation) between two variables. The test is appropriate for data that have a non-normal distribution (or where the distribution is not known) and assesses the degree of association between the X and Y variables. For the test to work, the values used must be **monotonic** i.e. the values must increase or decrease together or one increases while the other decreases. A value of 1 indicates a perfect correlation; a value of 0 indicates no correlation between the variables. The example below examines the relationship between the frequency of the drumming sound made by male frigatebirds (Y) and the volume of their throat pouch (X).

Spearman's Rank Data for Frigate Bird Pouch Volume and Drumming Frequency

Bird	Volume of pouch (cm^3)	Rank (R_1)	Frequency of drumming sound (Hz)	Rank (R_2)	Difference (D) (R_1-R_2)	D^2
1	2550		461			
2	2440	1	473	6	-5	25
3	2740		532			
4	2730		465			
5	3010		485			
6	3370		488			
7	3080		527			
8	4910		478			
9	3740		485			
10	5090		434			
11	5090		468			
12	5380		449			
Based on Madsen et al 2004				Σ(Sum)		

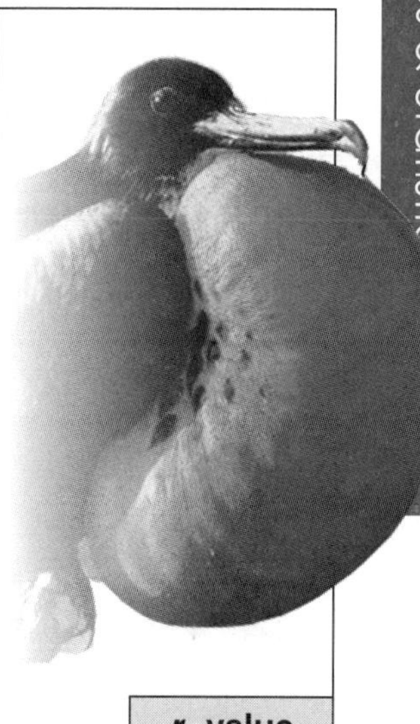

r_s value

Analyzing the Data

Step one: Rank the data for each variable. For each variable, the numbers are ranked in descending order, e.g. for the variable, volume, the highest value 5380 cm^3 is given the rank of 12 while its corresponding frequency value is given the rank of 2. Fill in the rank columns in the table above in the same way. If two numbers have the same rank value, then use the mean rank of the two values (e.g. 1+2 = 3. 3/2= 1.5).

Step two: Calculate the difference (D) between each pair of ranks (R_1-R_2) and enter the value in the table (as a check, the sum of all differences should be 0).

Step three: Square the differences and enter them into the table above (this removes any negative values).

Step four: Sum all the D^2 values and enter the total into the table.

Step five: Use the formula below to calculate the Spearman Rank Correlation Coefficient (r_s). Enter the r_s value in the box above.

$$r_s = 1 - \left(\frac{6\Sigma D^2}{n(n^2-1)}\right)$$

Spearman Rank Correlation Coefficient

Step six: Compare the r_s value to the table of critical values (right) for the appropriate number of pairs. If the r_s value (ignoring sign) is greater than or equal to the critical value then there is a significant correlation. If r_s is positive then there is a positive correlation. If r_s is negative then there is a negative value correlation.

Number of pairs of measurements	Critical value
5	1.00
6	0.89
7	0.79
8	0.74
9	0.68
10	0.65
12	0.59
14	0.54
16	0.51
18	0.48
20	0.45

1. State the null hypothesis for the data set. ____________________

____

____

2 (a) Identify the critical value for the frigate bird data: ____________________

(b) State is the correlation is positive of negative: ____________________

(c) State whether the correlation is significant: ____________________

3. Explain why the data collected must be monotonic if a Spearman rank correlation is to be used: ____________________

____

____

ISBN: 978-1-927173-16-9

The Student's *t* Test

The Student's *t* test is a commonly used test when comparing two sample means, e.g. means for a treatment and a control in an experiment, or the means of some measured characteristic between two animal or plant populations. The test is a powerful one, i.e. it is a good test for distinguishing real but marginal differences between samples. The *t* test is a simple test to apply, but it is only valid for certain situations. It is a two-group test and is not appropriate for multiple use i.e. sample 1 vs 2, then sample 1 vs 3. *You must have only two sample means to compare.* You are also assuming that the data have a normal (not skewed) distribution, and the scatter (standard deviations) of the data points is similar for both samples. You may wish to exclude obvious outliers from your data set for this reason. Below is a simple example outlining the general steps involved in the Student's *t* test. The following is a simple example using a set of data from a fictitious experiment involving a treatment and a control (the units are not relevant in this case, only the values). A portion of the Student's *t* table is provided, sufficient to carry out the test. Follow the example through, making sure that you understand what is being done at each step.

Steps in performing a Student's *t* test

Explanatory notes

Step 1 ***Calculate basic summary statistics for your two data sets***

Control (A): 6.6, 5.5, 6.8, 5.8, 6.1, 5.9 $n_A = 6$, $\bar{x}_A = 6.12$, $s_A = 0.496$

Treatment (B): 6.3, 7.2, 6.5, 7.1, 7.5, 7.3 $n_B = 6$, $\bar{x}_B = 6.98$, $s_B = 0.475$

n_A and n_B are the number of values in the first and second data sets respectively (these need not be the same). $\bar{x}$ is the mean. ***s*** is the standard deviation (a measure of scatter in the data).

Step 2 ***Set up and state your null hypothesis (H_0)***

H_0: there is no treatment effect. The differences in the data sets are the result of chance variation only and they are not really different

The alternative hypothesis is that there is a treatment effect and the two sets of data are truly different.

Step 3 ***Decide if your test is one or two tailed***

This tells you what section of the t table to consult. Most biological tests are two-tailed. Very few are one-tailed.

A one-tailed test looks for a difference only in one particular direction. A two-tailed test looks for any difference (+ or –).

Step 4 ***Calculate the t statistic***

For our sample data above the calculated value of t is –3.09. The degrees of freedom (df) are $n_1 + n_2 - 2 = 10$.

Calculation of the *t* value uses the variance which is simply the square of the standard deviation (s^2). You may compute the *t* value by entering your data onto a computer and using a simple statistical program.

It does not matter if your calculated *t* value is a positive or negative (the sign is irrelevant). If you do not have access to a statistical program, computation of *t* is not difficult. Step 4 (calculating *t*) is described in the following *t* test exercises.

Step 5 ***Consult the t table of critical values***

Selected critical values for Student's t statistic (two-tailed test)

Degrees of freedom	$P = 0.05$	$P = 0.01$	$P = 0.001$
5	2.57	4.03	6.87
10	2.23	3.17	4.59
15	2.13	2.95	4.07
20	2.09	2.85	3.85

Critical value of *t* for 10 degrees of freedom. The calculated *t* value must exceed this

The absolute value of the *t* statistic (3.09) well exceeds the critical value for $P = 0.05$ at 10 degrees of freedom.

We can reject H_0 and conclude that the means are different at the 5% level of significance.

If the calculated absolute value of *t* had been less than 2.23, we could not have rejected H_0.

1. (a) In an experiment, data values were obtained from four plants in experimental conditions and three plants in control conditions. The mean values for each data set (control and experimental conditions) were calculated. The *t* value was calculated to be 2.16. The null hypothesis was: "The plants in the control and experimental conditions are not different". State whether the calculated *t* value supports the null hypothesis or its alternative (consult *t* table above):

   (b) The experiment was repeated, but this time using 6 control and 6 "experimental" plants. The new *t* value was 2.54. State whether the calculated *t* value supports the null hypothesis or its alternative now:

2. Explain why, in terms of applying Student's *t* test, extreme data values (outliers) are often excluded from the data set(s):

3. Explain what you understand by statistical significance (for any statistical test):

Related activities: *Descriptive Statistics*

Weblinks: *Student's t-tests*

ISBN: 978-1-927173-16-9

Student's *t* Test Exercise

Data from two flour beetle populations are given below. Ten samples were taken from each population and the number of beetles in each sample were counted. The student's *t* test is used to determine if the densities of the two populations were significantly different. The exercise below uses a workbook computation to determine a *t* value. Follow the steps to complete the test. The calculations are also very simple yo do using a spreadsheet programme such as *Excel®* (see the following page).

1. (a) Complete the calculations to perform the *t* test for these two populations. Some calculations are provided for you.

x (counts)		$x - \bar{x}$ (deviation from the mean)		$(x - \bar{x})^2$ (deviation from mean)^2	
Popn A	**Popn B**	**Popn A**	**Popn B**	**Popn A**	**Popn B**
465	310	9.3	-10.6	86.5	112.4
475	310	19.3	-10.6	372.5	112.4
415	290				
480	355				
436	350				
435	335				
445	295				
460	315				
471	316				
475	330				
$n_A = 10$	$n_B = 10$			$\Sigma(x - \bar{x})^2$	$\Sigma(x - \bar{x})^2$

The number of samples in each data set

The sum of each column is called the sum of squares

(b) The variance for population A: $s^2_A =$

The variance for population B: $s^2_B =$

(c) The difference between the population means

$(\bar{x}_A - \bar{x}_B) =$

(d) $t_{\text{(calculated)}} =$

(e) Determine the degrees of freedom (d.f.)

d.f. $(n_A + n_B - 2) =$

(f) $P =$

$t_{\text{(critical value)}} =$

(g) Your decision is:

Step 1: Summary statistics

Tabulate the data as shown in the first 2 columns of the table (left). Calculate the mean and give the n value for each data set. Compute the standard deviation if you wish.

Popn A $\bar{x}_A = 455.7$, $n_A = 10$, $s_A = 21.76$

Popn B $\bar{x}_B = 320.6$, $n_B = 10$, $s_B = 21.64$

Step 2: State your null hypothesis

Step 3: Decide if your test is one or two tailed

Calculating the t value

Step 4a: Calculate sums of squares

Complete the computations outlined in the table left. The sum of each of the final two columns (left) is called the sum of squares.

Step 4b: Calculate the variances

Calculate the variance (s^2) for each set of data. This is the sum of squares divided by $n-1$ (number of samples in each data set – 1). In this case the n values are the same, but they need not be.

$$s^2_A = \frac{\Sigma(x - \bar{x})^2_{(A)}}{n_A - 1} \qquad s^2_B = \frac{\Sigma(x - \bar{x})^2_{(B)}}{n_B - 1}$$

Step 4c: Differences between means

Calculate the *actual* difference between the means

$$(\bar{x}_A - \bar{x}_B)$$

Step 4d: Calculate t

Calculate the *t* value. Ask for assistance if you find interpreting the lower part of the equation difficult

$$t = \frac{(\bar{x}_A - \bar{x}_B)}{\sqrt{\frac{s^2_A}{n_A} + \frac{s^2_B}{n_B}}}$$

Step 4e: Determine the degrees of freedom

Degrees of freedom (d.f.) are defined by the number of samples (e.g. counts) taken: d.f. = $n_A + n_B - 2$ where n_A and n_B are the number of counts in each of populations A and B.

Step 5: Consult the t table

Consult the *t*-tables (opposite page) for the critical *t* value at the appropriate degrees of freedom and the acceptable probability level (e.g. P = 0.05).

Step 5a: Make your decision

Make your decision whether or not to reject H_0. If your *t* value is large enough you may be able to reject H_0 at a lower *P* value (e.g. 0.001), increasing your confidence in the alternative hypothesis.

ISBN: 978-1-927173-16-9

2. The previous example (manual calculation for two beetle populations) is outlined below in a spreadsheet (created in *Microsoft Excel®*). The spreadsheet has been shown in a special mode with the formulae displayed. Normally, when using a spreadsheet, the calculated values will appear as the calculation is completed (entered) and a formula is visible only when you click into an individual cell. When setting up a spreadsheet, you can arrange your calculating cells wherever you wish. What is important is that you accurately identify the cells being used for each calculation. Also provided below is a summary of the spreadsheet notations used and a table of critical values of t at different levels of P. Note that, for brevity, only some probability values have been shown. To be significant at the appropriate level of probability, calculated values must be greater than those in the table for the appropriate degrees of freedom.

   (a) Using the data in question 1, set up a spreadsheet as indicated below to calculate t. Save your spreadsheet. Print it out and staple the print-out into your workbook.

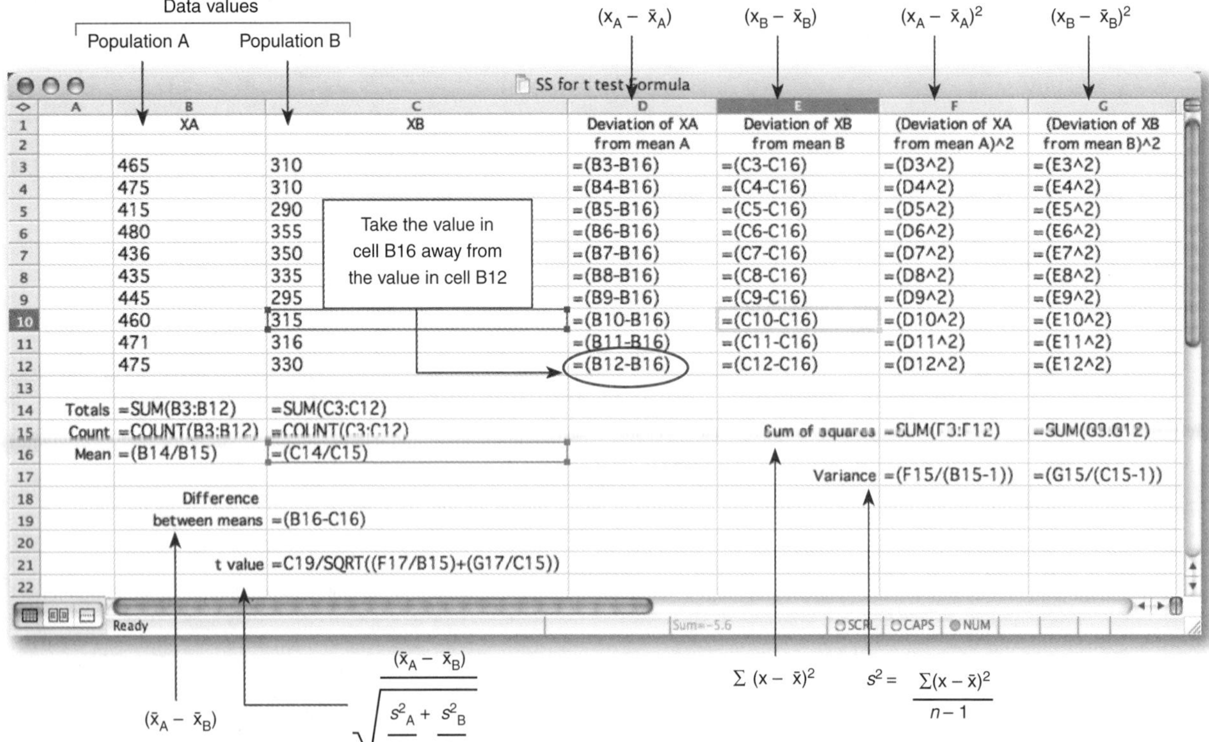
Data values
Population A
Population B
$(x_A - \bar{x}_A)$
$(x_B - \bar{x}_B)$
$(x_A - \bar{x}_A)^2$
$(x_B - \bar{x}_B)^2$
SS for t test formula
Take the value in cell B16 away from the value in cell B12
Ready
Sum=-5.6
SCRL CAPS NUM
$(\bar{x}_A - \bar{x}_B)$
$\frac{(\bar{x}_A - \bar{x}_B)}{\sqrt{\frac{s^2_A}{n_A} + \frac{s^2_B}{n_B}}}$
$\Sigma (x - \bar{x})^2$
$s^2 = \frac{\Sigma(x - \bar{x})^2}{n - 1}$

	A	B	C	D	E	F	G
1		XA	XB	Deviation of XA	Deviation of XB	(Deviation of XA	(Deviation of XB
2				from mean A	from mean B	from mean A)^2	from mean B)^2
3		465	310	=(B3-B16)	=(C3-C16)	=(D3^2)	=(E3^2)
4		475	310	=(B4-B16)	=(C4-C16)	=(D4^2)	=(E4^2)
5		415	290	=(B5-B16)	=(C5-C16)	=(D5^2)	=(E5^2)
6		480	355	=(B6-B16)	=(C6-C16)	=(D6^2)	=(E6^2)
7		436	350	=(B7-B16)	=(C7-C16)	=(D7^2)	=(E7^2)
8		435	335	=(B8-B16)	=(C8-C16)	=(D8^2)	=(E8^2)
9		445	295	=(B9-B16)	=(C9-C16)	=(D9^2)	=(E9^2)
10		460	315	=(B10-B16)	=(C10-C16)	=(D10^2)	=(E10^2)
11		471	316	=(B11-B16)	=(C11-C16)	=(D11^2)	=(E11^2)
12		475	330	=(B12-B16)	=(C12-C16)	=(D12^2)	=(E12^2)
13							
14	Totals	=SUM(B3:B12)	=SUM(C3:C12)				
15	Count	=COUNT(B3:B12)	=COUNT(C3:C12)		Sum of squares	=SUM(F3:F12)	=SUM(G3:G12)
16	Mean	=(B14/B15)	=(C14/C15)				
17					Variance	=(F15/(B15-1))	=(G15/(C15-1))
18		Difference					
19		between means	=(B16-C16)				
20							
21		t value	=C19/SQRT((F17/B15)+(G17/C15))				
22							

Notation	Meaning
Columns and rows	Columns are denoted A, B, C ... at the top of the spreadsheet, rows are 1, 2, 3, ... on the left. Using this notation a cell can be located e.g. C3
=	An "equals" sign before other entries in a cell denotes a formula.
( )	Parentheses are used to group together terms for a single calculation. This is important for larger calculations (see cell C21 above)
C3:C12	Cell locations are separated by a colon. C3:C12 means "every cell between and including C3 and C12"
SUM	Denotes that what follows is added up. =SUM(C3:C12) means "add up the values in cells C3 down to C12"
COUNT	Denotes that the number of values is counted =COUNT(C3:C12) means "count up the number of values in cells C3 down to C12"
SQRT	Denotes "take the square root of what follows"
^2	Denotes an exponent e.g. x^2 means that value x is squared.

Above is a table explaining some of the spreadsheet notations used for the calculation of the t value for the exercise on the previous page. It is not meant to be an exhaustive list for all spreadsheet work, but it should help you to become familiar with some of the terms and how they are used. This list applies to *Microsoft Excel®* . Different spreadsheets may use different notations. These will be described in the spreadsheet manual.

Table of critical values of t at different levels of P.

Degrees of freedom	Level of Probability		
	0.05	0.01	0.001
1	12.71	63.66	636.6
2	4.303	9.925	31.60
3	3.182	5.841	12.92
4	2.776	4.604	8.610
5	2.571	4.032	6.869
6	2.447	3.707	5.959
7	2.365	3.499	5.408
8	2.306	3.355	5.041
9	2.262	3.250	4.781
10	2.228	3.169	4.587
11	2.201	3.106	4.437
12	2.179	3.055	4.318
13	2.160	3.012	4.221
14	2.145	2.977	4.140
15	2.131	2.947	4.073
16	2.120	2.921	4.015
17	2.110	2.898	3.965
18	2.101	2.878	3.922
19	2.093	2.861	3.883
20	2.086	2.845	3.850

   (b) Save your spreadsheet under a different name and enter the following new data values for population B: **425, 478, 428, 465, 439, 475, 469, 445, 421, 438**. Notice that, as you enter the new values, the calculations are updated over the entire spreadsheet. Re-run the t-test using the new t value. State your decision for the two populations now:

   New t value: ________________ Decision on null hypothesis (delete one): Reject / Do not reject

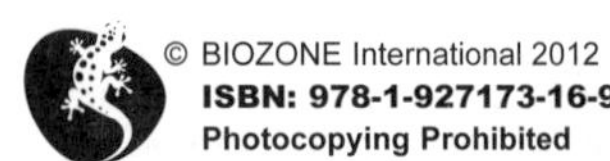

ISBN: 978-1-927173-16-9

KEY TERMS: Mix and Match

INSTRUCTIONS: Test your vocabulary by matching each term to its definition, as identified by its preceding letter code.

- accuracy
- aim
- chi-squared test
- control
- controlled variable
- correlation
- data
- dependent variable
- descriptive statistics
- error bars
- fair test
- graph
- hypothesis
- independent variable
- median
- mode
- parameter
- precision
- qualitative data
- quantitative data
- regression
- sample mean
- sample standard deviation
- scientific method
- statistic
- Student's *t* test
- table
- trend (of data)
- variable

A A relationship in which variables vary together in some predictable way, but cause and effect is not implied.

B The use of an ordered, repeatable method to investigate, manipulate, gather, and record data.

C An experimental situation in which you change only one factor (variable) and keep all other conditions the same.

D A diagram which often displays numerical information in a way that can be used to identify trends in the data.

E Estimate of the true population mean based upon data collected by random sampling. Valid for population data that are normally distributed. The sum of the data divided by the number of data entries (n).

F A short statement describing the purpose or reason for an experiment.

G A calculated statistic expressing the variability of a sample population about its mean.

H The value that occurs most often in a data set.

I Facts collected for analysis.

J A quantity that defines a characteristic of a system

K A variable whose values are set, or systematically altered, by the investigator.

L The central value in a sorted set of data.

M The repeatability of a measurement.

N Data described in descriptors or terms rather than by numbers

O Data able to be expressed in numbers. Numerical values derived from counts or measurements

P The degree of closeness of a measured value to its true amount.

Q A standard (reference) treatment that helps to ensure that the responses to the other treatments can be reliably interpreted.

R A test used to determine if the difference between two sample means is significant.

S Calculated values that summarize the main features of a collection of data.

T A tentative explanation of an observation, capable of being tested by experimentation.

U A calculated measure of some attribute of a sample (e.g. the arithmetic mean).

V A statistical test for determining the significance of departures of observed data from an expected result.

W A relationship between a dependent variable and one or more independent variables.

X Variable that is fixed at a specific amount as part of the design of experiment.

Y A variable whose values are determined by another variable.

Z A factor in an experiment that is subject to change.

AA A relationship between variables in a data set.

BB A representation of data, often summarized, organized in rows and columns.

CC A graphical representation of the variability of data. Used on graphs to indicate the uncertainty in a reported measurement.

ISBN: 978-1-927173-16-9

Core Topic 2

Cell Biology

Key concepts

- Cells are the fundamental units of life. Microscopy can be used to understand cellular structure.
- Surface area to volume ratio limits the size of cells.
- Cell differentiation and multicellularity promote efficiency of function. Stem cells retain their ability to differentiate into cells of different types.
- Prokaryotic cells lack cellular organelles and divide by binary fission.
- Eukaryotic cells have specialized organelles, which localize reactions and promote functional efficiency.

Key terms

Core

binary fission
cell differentiation
cell theory
cell wall
centrioles
chloroplast
cytoplasm
electron microscope
emergent property
endoplasmic reticulum
eukaryotic cell
flagella (*sing.* flagellum)
Golgi
light microscope
linear magnification
lysosomes
magnification
microscopy
mitochondria
multicellularity
nucleolus
nucleus
plasma membrane
prokaryotic cell
resolution
ribosomes
stem cell
surface area to volume ratio
vacuole

Learning Objectives

☐ 1. Use the **KEY TERMS** to compile a glossary for this topic.

Cell Theory *(2.1)*

pages 37-42, 44-46, 56

☐ 2. Outline the **cell theory** and describe the evidence supporting it (*TOK*). Describe the role of **microscopy** to our understanding of cell structure and function.

☐ 3. Describe the criteria for defining life. Comment on the position of viruses with respect to these criteria.

☐ 4. Compare the relative sizes of various molecules, cell structures and organelles, viruses, and cells, using appropriate SI units.

☐ 5. Calculate the **linear magnification** of drawings and the actual size of specimens viewed with a microscope at known magnification.

☐ 6. Distinguish between **magnification** and **resolution**. Identify cell structures and organelles from images produced using light microscopy and electron microscopy.

☐ 7. Explain the significance of **surface area to volume ratio** to cells and relate this to organism size.

☐ 8. Describe the role of **cell differentiation** in the development of tissues and organs in multicellular organisms. Explain how **multicellularity** increases efficiency of function and gives rise to **emergent properties** not exhibited by single cells.

☐ 9. Describe the properties of **stem cells** and explain the role of stem cells in multicellular organisms. Describe an example of the therapeutic use of stem cells.

Prokaryotic Cells *(2.2)*

page 43

☐ 10. Using an annotated diagram, describe the structure and function of a **prokaryotic cell**, e.g. *E. coli*. Identify structures, e.g. **cell wall**, **flagella**, **ribosomes**, and **nucleoid** region in electron micrographs of prokaryotic cells.

☐ 11. Describe **binary fission** in prokaryotes. Recognize prokaryotic binary fission as an asexual form of reproduction in which all the cells are genetically identical.

☐ 12. Compare and contrast the structure of prokaryotic and eukaryotic cells.

Eukaryotic Cells *(2.3)*

pages 47-56

☐ 13. Using an annotated diagram, describe the structure and function of a **eukaryotic cell**, e.g. liver cell. Identify cell structures and organelles in electron micrographs, e.g. **Golgi**, **endoplasmic reticulum**, **mitochondria**, **nucleus**, **nucleolus**, **ribosomes**, **centrioles**, **lysosomes**, **plasma membrane**, and **cytoplasm**.

☐ 14. Compare and contrast the structure of plant and animal cells.

☐ 15. Describe extracellular components in plants and animals and their roles, e.g. cellulose **cell wall** in plants, and **glycoproteins** in animals.

Periodicals:
Listings for this chapter are on page 398

Weblinks:
www.thebiozone.com/weblink/IB-3169.html

BIOZONE APP:
Student Review Series
Cell Structure

Cell Theory

The idea that all living things are composed of cells developed over many years and is strongly linked to the invention and refinement of the microscope. Early microscopes in the 1600s (such as Leeuwenhoek's below) opened up a whole new field of biology; the study of cell biology and microorganisms. The cell theory is a fundamental idea of biology.

Early Microscopes

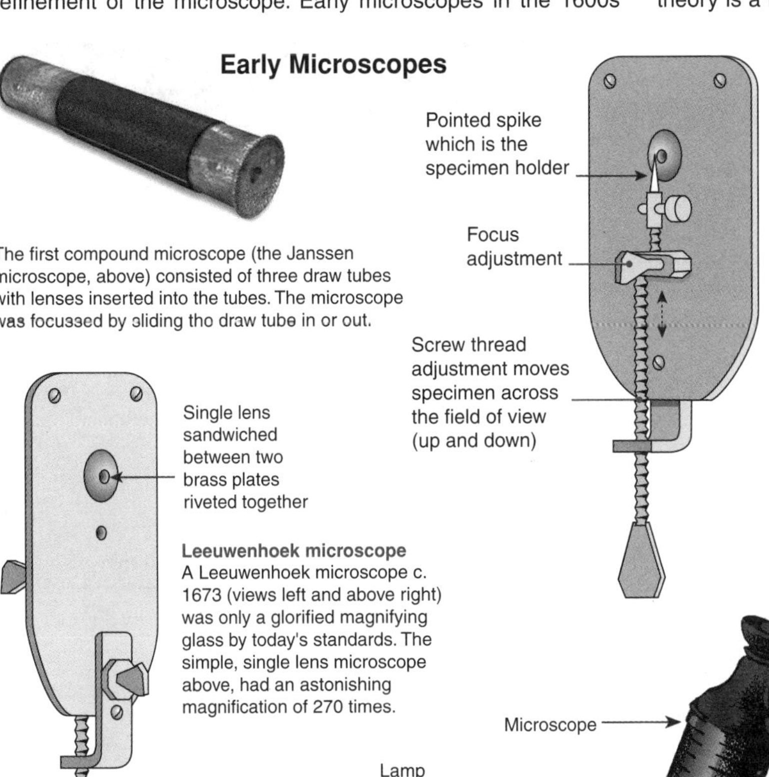
Pointed spike which is the specimen holder
Focus adjustment
Screw thread adjustment moves specimen across the field of view (up and down)
Single lens sandwiched between two brass plates riveted together
Front

The first compound microscope (the Janssen microscope, above) consisted of three draw tubes with lenses inserted into the tubes. The microscope was focussed by sliding the draw tube in or out.

Leeuwenhoek microscope
A Leeuwenhoek microscope c. 1673 (views left and above right) was only a glorified magnifying glass by today's standards. The simple, single lens microscope above, had an astonishing magnification of 270 times.

Microscope
Lamp
Mirror

Robert Hooke c. 1665
Hooke was fascinated by microscopy, and in his book *Micrographia* (1665) he described the use of the compound microscope that he had devised (**right**). He was the first to coin the name cell after he observed the angular spaces that he saw in a thin section of cork.

The Cell Theory

The idea that cells are fundamental units of life is part of the cell theory. The basic principles of the theory (as developed by early biologists) are:

- All living things are composed of cells and cell products.
- New cells are formed only by the division of preexisting cells.
- The cell contains inherited information (genes) that are used as instructions for growth, functioning, and development.
- The cell is the functioning unit of life; the chemical reactions of life take place within cells.

Milestones in Cell Biology

1500s	Convex lenses with a magnification greater than x5 became available.
1595	**Zacharias Janssen** of Holland has been credited with the first compound microscope (more than one lens).
Early 1600s	First compound microscopes used in Europe (used two convex lenses to make objects look larger). Suffered badly from colour distortion; an effect called 'spherical aberration'.
1632 - 1723	**Antoni van Leeuwenhoek** of Holland produced over 500 single lens microscopes, discovering bacteria, human blood cells, spermatozoa, and protozoa. Friend of Robert Hooke.
1661	**Marcello Malpighi** used lenses to study insects. Discovered capillaries and may have described cells in writing of 'globules' and 'saccules'.
1662	**Robert Hooke** of England used the term 'cell' in describing the microscopic structure of cork. He believed that the cell walls were the important part of otherwise empty structures. Published *Micrographia* in 1665.
1672	**Nehemlah Grew** wrote the first of two well-illustrated books on the microscopic anatomy of plants.
1838 - 1839	Botanist **Matthias Schleiden** and zoolozgist **Theodor Schwann** proposed the cell theory based on their observations of plant and animal cells.
1855	**Rudolph Virchow** extended the cell theory by stating that "new cells are formed only by the division of previously existing cells".
1880	**August Weismann** added to Virchow's idea by pointing out that "all the cells living today can trace their ancestry back to ancient times", thus making the link between cell theory and evolution.

1. What impact do you think the invention of microscopes has had on biology? ______________________

____

____

2. Before the development of the cell theory, it was commonly believed that living organisms could arise by spontaneous generation. What does this term mean and why has it been discredited as a theory?

____

____

____

ISBN: 978-1-927173-16-9

Related activities: *Optical Microscopes*

Cell Sizes

Cells are extremely small and can only be seen properly when viewed through the magnifying lenses of a microscope. The diagrams below show a variety of cell types, together with a virus (non-cellular) and a multicellular microscopic animal (as a comparison). For each of these images, note the scale and relate this to the type of microscopy used.

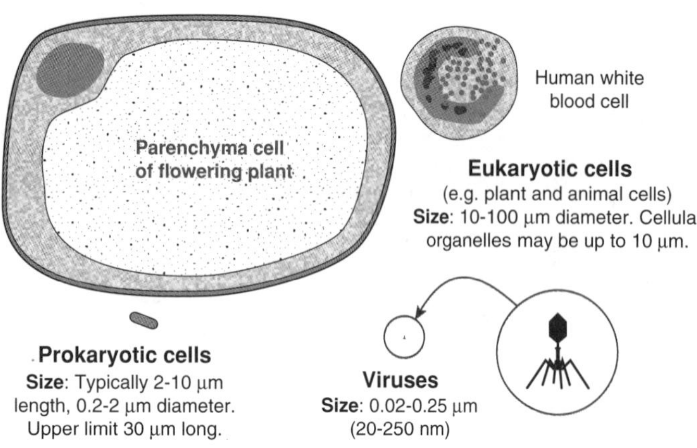

Parenchyma cell of flowering plant
Human white blood cell
Eukaryotic cells
(e.g. plant and animal cells)
Size: 10-100 μm diameter. Cellular organelles may be up to 10 μm.
Prokaryotic cells
Size: Typically 2-10 μm length, 0.2-2 μm diameter. Upper limit 30 μm long.
Viruses
Size: 0.02-0.25 μm (20-250 nm)

Unit of length (International System)		
Unit	**Meters**	**Equivalent**
1 meter (m)	1 m	= 1000 millimeters
1 millimeter (mm)	10^{-3} m	= 1000 micrometers
1 micrometer (μm)	10^{-6} m	= 1000 nanometers
1 nanometer (nm)	10^{-9} m	= 1000 picometers

Micrometers are sometime referred to as microns. Smaller structures are usually measured in nanometers (nm) e.g. molecules (1 nm) and plasma membrane thickness (10 nm).

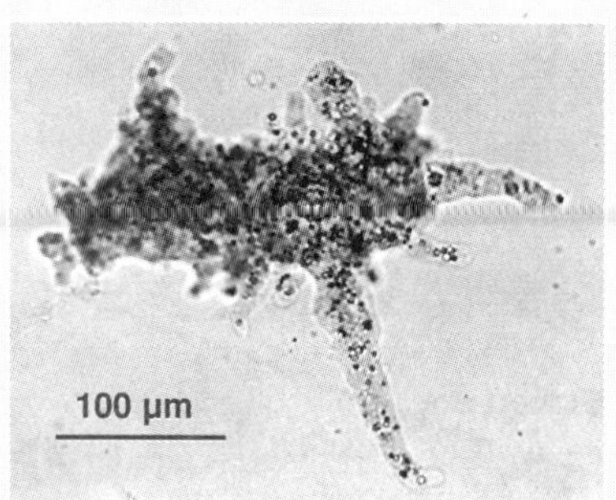

100 μm
RCN

An ***Amoeba*** showing extensions of the cytoplasm called pseudopodia. This protoctist changes its shape, exploring its environment.

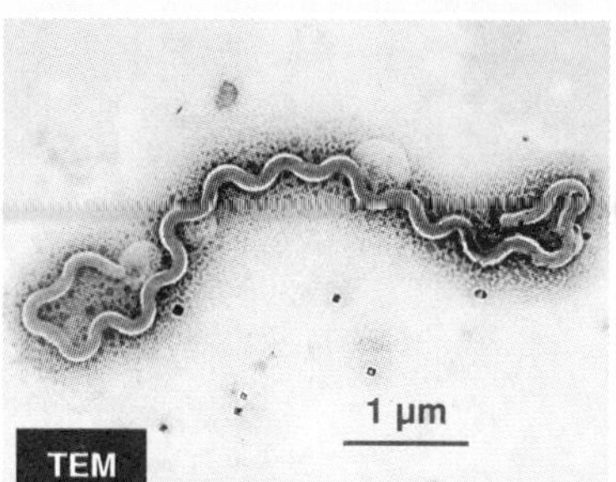

1 μm
TEM
WMU

A long thin cell of the spirochete bacterium ***Leptospira pomona***, which causes the disease leptospirosis.

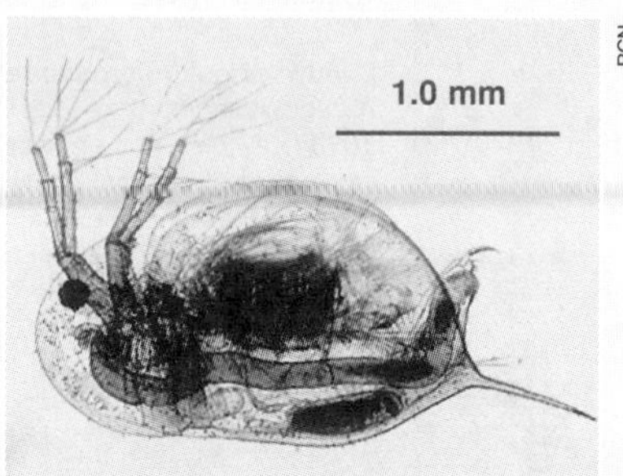

1.0 mm
RCN

Daphnia showing its internal organs. These freshwater microcrustaceans are part of the zooplankton found in lakes and ponds.

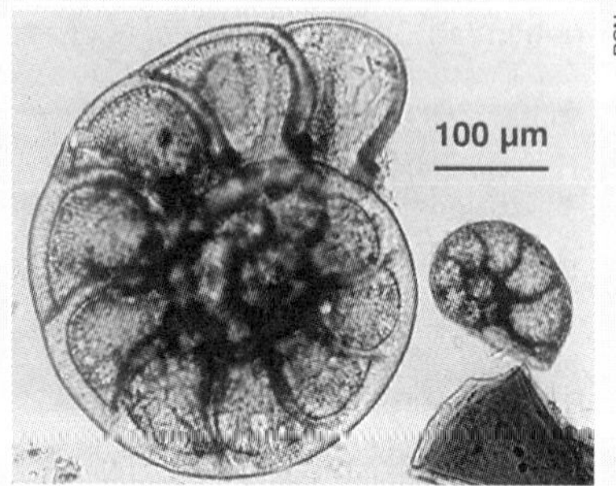

100 μm
RCN

A **foraminiferan** showing its chambered, calcified shell. These single-celled protozoans are marine planktonic amoebae.

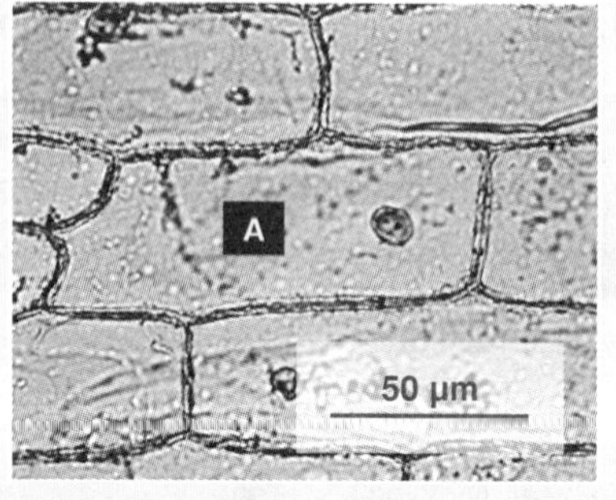

A
50 μm
RCN

Epidermal cells (skin) from an onion bulb showing the nucleus, cell walls and cytoplasm. Most organelles are not visible at this resolution.

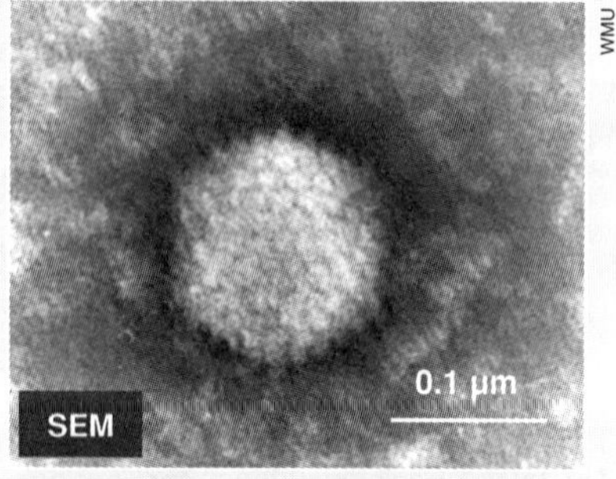

0.1 μm
SEM
WMU

Papillomavirus (human wart virus) showing its polyhedral protein coat (20 triangular faces, 12 corners) made of ball-shaped structures.

1. Using the measurement scales provided on each of the photographs above, determine the longest dimension (length or diameter) of the cell/animal/virus in μm and mm (choose the cell marked '**A**' for epidermal cells):

   (a) *Amoeba*: ________ μm ________ mm (d) Epidermis: ________ μm ________ mm

   (b) Foraminiferan: ________ μm ________ mm (e) *Daphnia*: ________ μm ________ mm

   (c) *Leptospira*: ________ μm ________ mm (f) *Papillomavirus*: ________ μm ________ mm

2. List these six organisms in order of size, from the smallest to the largest: ____________________

   ____

3. Study the scale of your ruler and state which of these six organisms you would be able to see with your unaided eye:

   ____

4. Calculate the equivalent length in millimetres (mm) of the following measurements:

   (a) 0.25 μm: ________ (b) 450 μm: ________ (c) 200 nm: ________

Related activities: *Calculating Linear Magnification*
Weblinks: *Cell Size and Scale*

Periodicals:
Size does matter

ISBN: 978-1-927173-16-9

Unicellular Eukaryotes

Unicellular (single-celled) **eukaryotes** comprise the majority of the diverse kingdom, **Protista**. They are found almost anywhere there is water, including within larger organisms (as parasites or symbionts). The protists are a very diverse group, exhibiting some features typical of generalized eukaryotic cells, as well as specialised features, which may be specific to one genus. Note that even within the genera below there is considerable variation in life functions such as metabolism, reproduction, and nutrition. *Amoeba* and *Paramecium* are both **heterotrophic**, ingesting food particles. *Euglena and Chlamydomonas* are autotrophic, although *Euglena* is heterotrophic when deprived of light. The typical mode of reproduction in most of the major protistan taxa is asexual **binary fission**. Most can also reproduce sexually, most commonly by syngamy (fusion of gametes to produce a zygote).

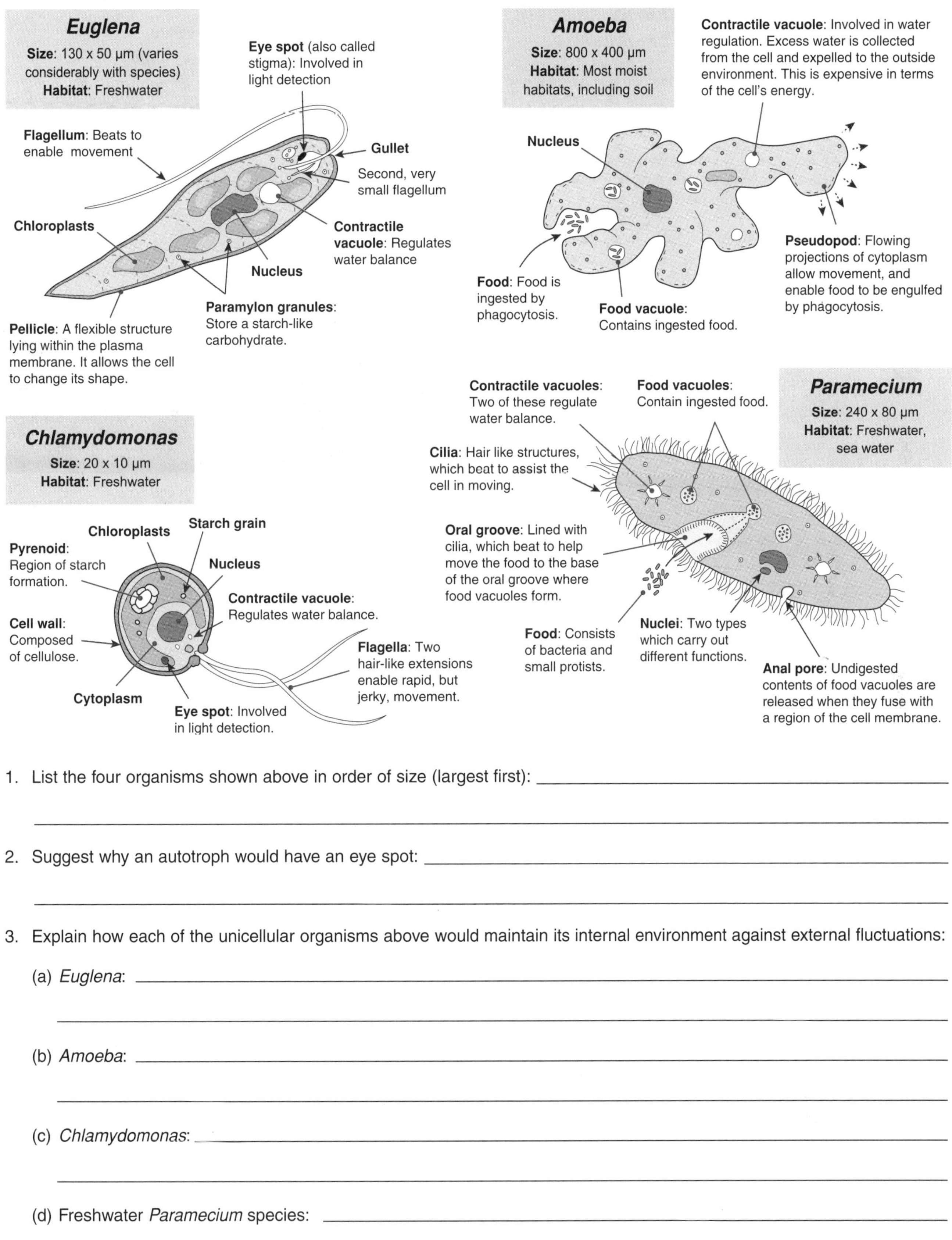
Euglena
Size: 130 x 50 µm (varies considerably with species)
Habitat: Freshwater
Eye spot (also called stigma): Involved in light detection
Flagellum: Beats to enable movement
Gullet
Second, very small flagellum
Chloroplasts
Contractile vacuole: Regulates water balance
Nucleus
Paramylon granules: Store a starch-like carbohydrate.
Pellicle: A flexible structure lying within the plasma membrane. It allows the cell to change its shape.
Amoeba
Size: 800 x 400 µm
Habitat: Most moist habitats, including soil
Contractile vacuole: Involved in water regulation. Excess water is collected from the cell and expelled to the outside environment. This is expensive in terms of the cell's energy.
Nucleus
Pseudopod: Flowing projections of cytoplasm allow movement, and enable food to be engulfed by phagocytosis.
Food: Food is ingested by phagocytosis.
Food vacuole: Contains ingested food.
Chlamydomonas
Size: 20 x 10 µm
Habitat: Freshwater
Chloroplasts
Starch grain
Pyrenoid: Region of starch formation.
Nucleus
Contractile vacuole: Regulates water balance.
Cell wall: Composed of cellulose.
Flagella: Two hair-like extensions enable rapid, but jerky, movement.
Cytoplasm
Eye spot: Involved in light detection.
Contractile vacuoles: Two of these regulate water balance.
Food vacuoles: Contain ingested food.
Paramecium
Size: 240 x 80 µm
Habitat: Freshwater, sea water
Cilia: Hair like structures, which beat to assist the cell in moving.
Oral groove: Lined with cilia, which beat to help move the food to the base of the oral groove where food vacuoles form.
Food: Consists of bacteria and small protists.
Nuclei: Two types which carry out different functions.
Anal pore: Undigested contents of food vacuoles are released when they fuse with a region of the cell membrane.

1. List the four organisms shown above in order of size (largest first): ______

2. Suggest why an autotroph would have an eye spot: ______

3. Explain how each of the unicellular organisms above would maintain its internal environment against external fluctuations:

   (a) *Euglena*: ______

   (b) *Amoeba*: ______

   (c) *Chlamydomonas*: ______

   (d) Freshwater *Paramecium* species: ______

ISBN: 978-1-927173-16-9

Related activities: *Features of Taxonomic Groups*
Weblinks: *Paramecium Animation, Cytoplasmic Streaming in Protists*

Calculating Linear Magnification

Microscopes produce an enlarged (magnified) image of an object allowing it to be observed in greater detail than is possible with the naked eye. **Magnification** refers to the number of times larger an object appears compared to its actual size. The degree of magnification possible depends upon the type of microscopy used. **Linear magnification** is calculated by taking a ratio of the image height to the object's actual height. If this ratio is greater than one, the image is enlarged, if it is less than one, it is reduced. To calculate magnification, all measurements should be converted to the same units. Most often, you will be asked to calculate an object's actual size, in which case you will be told the size of the object and given the magnification.

Calculating Linear Magnification: A Worked Example

1. Measure the body length of the bed bug image (right). Your measurement should be 40 mm (***not*** including the body hairs and antennae).

2. Measure the length of the scale line marked 1.0 mm. You will find it is 10 mm long. The magnification of the scale line can be calculated using equation 1 (below right).

   The magnification of the scale line is **10** (10 mm / 1 mm)

   **NB: The magnification of the bed bug image will also be 10x because the scale line and image are magnified to the same degree.*

3. Calculate the actual (real) size of the bed bug using equation 2 (right):

   The actual size of the bed bug is **4 mm** (40 mm / 10 x magnification)

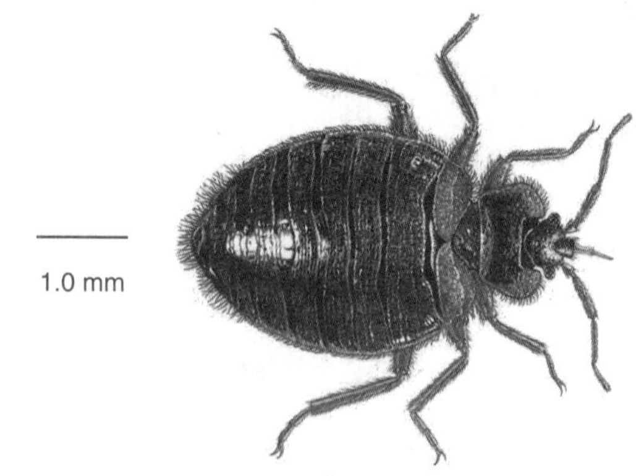
1.0 mm

Microscopy Equations

$$1.\quad \text{Magnification} = \frac{\text{size of the image}}{\text{actual size of object}}$$

$$2.\quad \text{Actual object size} = \frac{\text{size of the image}}{\text{magnification}}$$

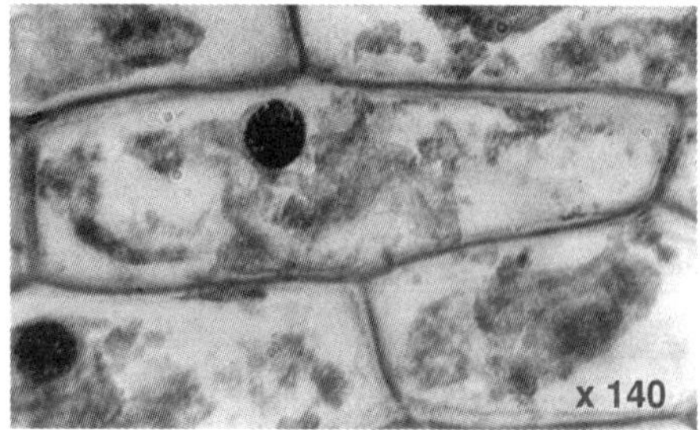
x 140

1. The bright field microscopy image on the left is of onion epidermal cells. The measured length of the onion cell in the centre of the photograph is 52,000 µm (52 mm). The image has been magnified 140 x. Calculate the actual size of the cell:

___

___

___

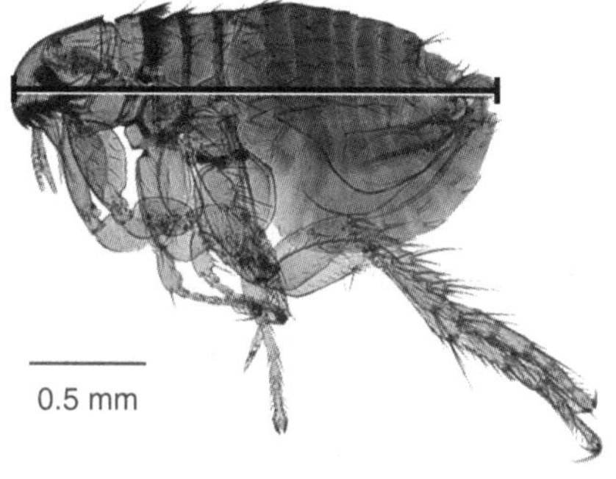
0.5 mm

2. The image of the flea (left) has been captured using light microscopy.

   (a) Calculate the magnification using the scale line on the image:

   ___

   ___

   ___

   (b) The body length of the flea is indicated by a line. Measure along the line and calculate the actual length of the flea:

   ___

   ___

   ___

3. The image size of the *E.coli* cell (left) is 43 mm, and its actual size is 2 µm. Using this information, calculate the magnification of the image:

___

___

___

Related activities: Cell Sizes

Periodicals: Size does matter

ISBN: 978-1-927173-16-9

Surface Area and Volume

When an object (e.g. a cell) is small it has a large surface area in comparison to its volume. In this case diffusion will be an effective way to transport materials (e.g. gases) into the cell. As an object becomes larger, its surface area compared to its volume is smaller. Diffusion is no longer an effective way to transport materials to the inside. For this reason, there is a physical limit for the size of a cell, with the effectiveness of diffusion being the controlling factor.

Diffusion in Organisms of Different Sizes

Respiratory gases and some other substances are exchanged with the surroundings by diffusion or active transport across the plasma membrane.

The **plasma membrane**, which surrounds every cell, functions as a selective barrier that regulates the cell's chemical composition. For each square micrometer of membrane, only so much of a particular substance can cross per second.

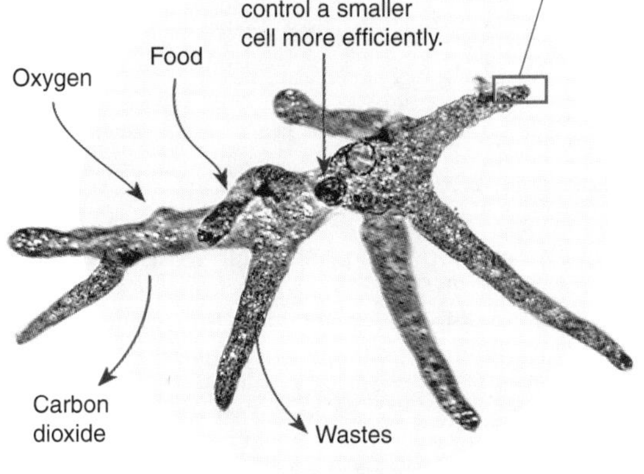
The nucleus can control a smaller cell more efficiently.
Oxygen
Food
Carbon dioxide
Wastes

Amoeba: The small size of single-celled protists, such as *Amoeba*, provides a large surface area relative to the cell's volume. This is adequate for many materials to be moved into and out of the cell by diffusion or active transport.

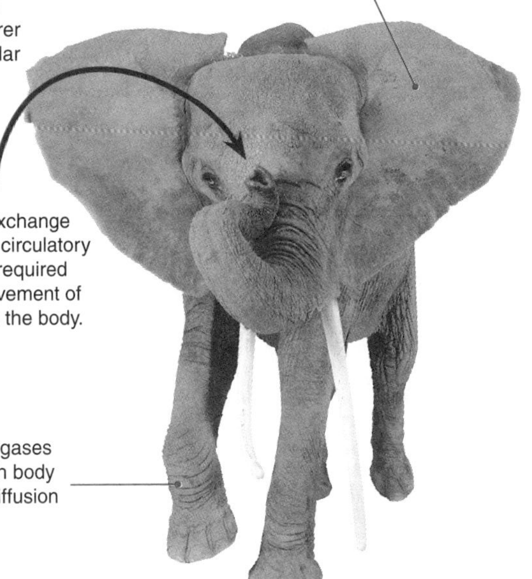
The surface area of an elephant is increased, for radiating body heat, by large flat ears.
A specialized gas exchange surface (lungs) and circulatory (blood) system are required to speed up the movement of substances through the body.
Respiratory gases cannot reach body tissues by diffusion alone.

Multicellular organisms: To overcome the problems of small cell size, plants and animals became multicellular. They provide a small surface area compared to their volume but have evolved various adaptive features to improve their effective surface area.

The diagram below shows four imaginary cells of different sizes (cells do not actually grow to this size, their large size is for the sake of the exercise). They range from a small 2 cm cube to a larger 5 cm cube. This exercise investigates the effect of cell size on the efficiency of diffusion.

2 cm cube

3 cm cube

4 cm cube

5 cm cube

1. Calculate the volume, surface area and the ratio of surface area to volume for each of the four cubes above (the first has been done for you). When completing the table below, show your calculations.

Cube size	Surface area	Volume	Surface area to volume ratio
2 cm cube	2 x 2 x 6 = 24 cm^2 (2 cm x 2 cm x 6 sides)	2 x 2 x 2 = 8 cm^3 (height x width x depth)	24 to 8 = 3:1
3 cm cube			
4 cm cube			
5 cm cube			

ISBN: 978-1-927173-16-9

2. Create a graph, plotting the surface area against the volume of each cube, on the grid on the right. Draw a line connecting the points and label axes and units.

3. Which increases the fastest with increasing size: the **volume** or the **surface area**?

4. Explain what happens to the ratio of surface area to volume with increasing size.

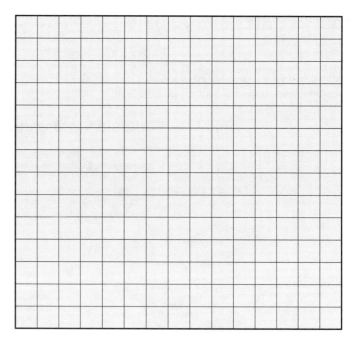

5. The diffusion of molecules into a cell can be modelled by using agar cubes infused with phenolphthalein indicator and soaked in sodium hydroxide (NaOH). Phenolphthalein turns a pink color when in the presence of a base. As the NaOH diffuses into the agar, the phenolphthalein changes to pink and thus indicates how far the NaOH has diffused into the agar. By cutting an agar block into cubes of various sizes, it is possible to show the effect of cell size on diffusion.

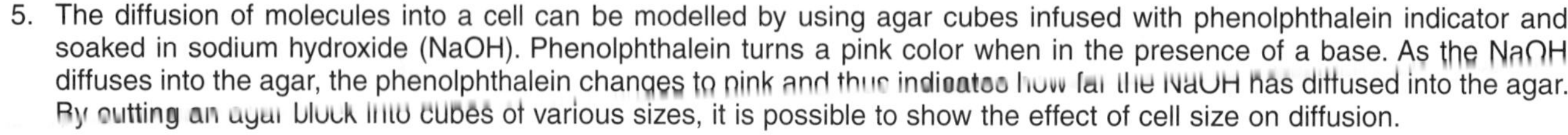

(a) Use the information below to fill in the table on the right:

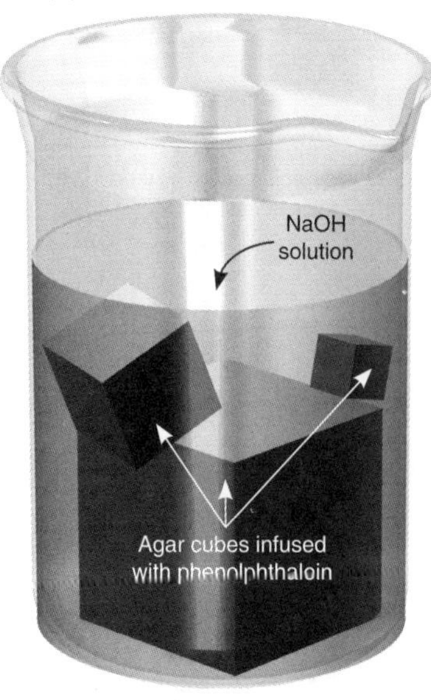
NaOH solution
Agar cubes infused with phenolphthalein

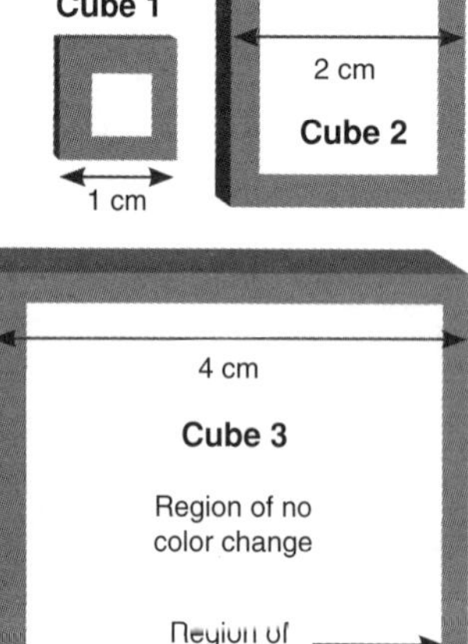
Cube 1
1 cm
2 cm
Cube 2
4 cm
Cube 3
Region of no color change
Region of color change

Cubes shown to same scale

Cube	1	2	3
1. Total volume (cm^3)			
2. Volume not pink (cm^3)			
3. Diffused volume (1. – 2.) (cm^3)			
4. Percentage diffusion			

(b) Diffusion of substances into and out of a cell occurs across the plasma membrane. For a cuboid cell, explain how increasing cell size affects the effective ability of diffusion to provide the materials required by the cell:

6. Explain why a single large cell of 2 cm x 2 cm x 2 cm is less efficient in terms of passively acquiring nutrients than eight cells of 1 cm x 1 cm x 1 cm:

ISBN: 978-1-927173-16-9

Prokaryotic Cells

Bacterial cells are much simpler than the more complex cells of eukaryotes (plant, animal, and fungi). Called prokaryotic cells, they are much smaller than eukaryotic cells and they lack many eukaryotic features (e.g. prokaryotes lack a distinct nucleus and membrane-bound cellular organelles). The bacterial cell wall is an important feature. It is a complex, multi-layered structure and has a role in the virulence of many species. This page illustrates some features of bacterial structure and diversity.

A Generalized Prokaryote: *E. coli*

Plasmids are small, circular DNA molecules that can reproduce independently of the main chromosome. They can be transferred between cells, or species, which allows for the transmission of properties such as antibiotic resistance between bacteria.

A single, circular main chromosome makes bacteria haploid for most genes. It is possible for some genes to be found on both the plasmid and chromosome.

There is no nuclear membrane or nucleus. The chromosomes are in direct contact with the cytoplasm so free ribosomes can attach to mRNA and begin making proteins while the mRNA is still being transcribed from the DNA.

1 μm

Fimbriae are hairlike structures (shorter, straighter, and thinner than flagella) used for attachment, not movement.

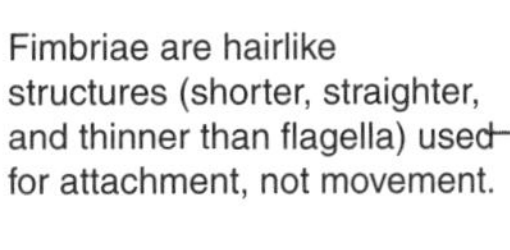

Cytoplasm

Cell surface membrane is similar in composition to eukaryotic membranes, although less rigid.

The glycocalyx is a viscous, gelatinous layer outside the cell wall. In some species, the glycocalyx allows attachment to substrates.

The cell wall gives the cell shape, prevents rupture, and serves as an anchorage point for flagella. It is composed of a carbohydrate macromolecule called peptidoglycan, together with lipopolysaccharides and lipoproteins.

Flagella (*sing.* flagellum) are used for locomotion. There may be one polar flagellum, one or more flagella at each end of the cell, or the flagella may be numerous, as in *E. coli.*

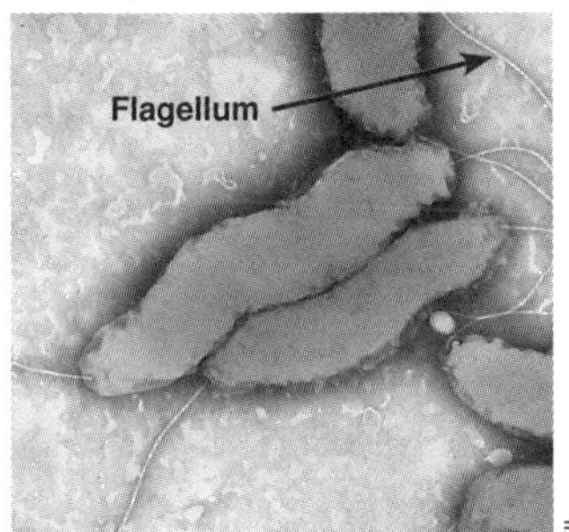
Flagellum

A spiral shape is one of four bacterial morphologies (the others being rods, commas, and cocci). These *Campylobacter* cells are also flagellated.

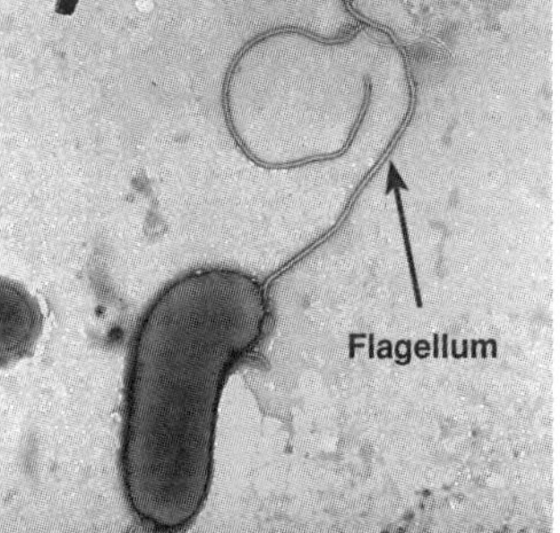
Flagellum
EII

Helicobacter pylori, is a comma-shaped vibrio bacterium that causes stomach ulcers in humans. This bacterium moves by means of polar flagella.

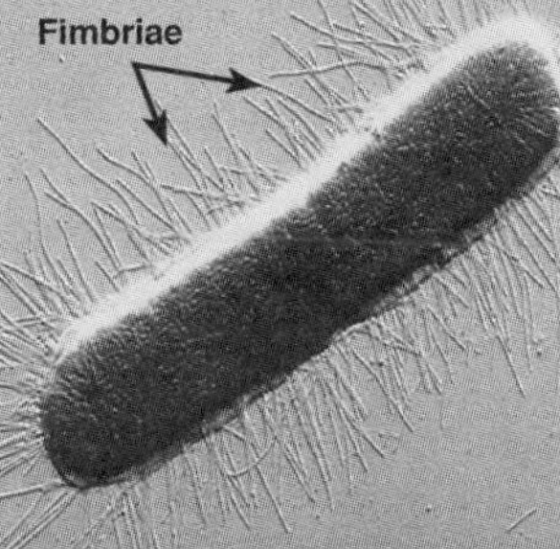
Fimbriae

Escherichia coli is a rod-shaped bacterium, common in the human gut. The fimbriae surrounding the cell are used to adhere to the intestinal wall.

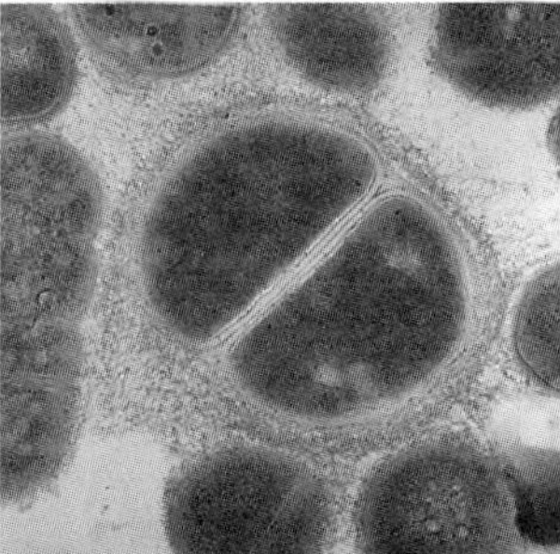

Bacteria usually divide by binary fission. During this process, DNA is copied and the cell splits into two cells, as in these gram positive cocci (round cells).

1. Describe three features distinguishing prokaryotic cells from eukaryotic cells:

   (a) ______________________________

   (b) ______________________________

   (c) ______________________________

2. Describe the function of flagella in bacteria and distinguish them from fimbriae: ______________________________

   ____

   ____

3. Describe the location and general composition of the bacterial cell wall: ______________________________

   ____

   ____

4. Describe the main method by which bacteria reproduce: ______________________________

   ____

ISBN: 978-1-927173-16-9

Multicellularity

With each step in the hierarchy of biological order, new properties emerge that were not present at simpler levels of organization. Life itself is associated with numerous **emergent properties**, including **metabolism** and growth. The cell is the site of life. It is the functioning unit structure from which living organisms are made. Viruses and cells are profoundly different. Viruses are non-cellular, lack the complex structures found in cells, and show only some of the properties we associate with living things. The traditional view of viruses is as a minimal particle, although the identification in 2004 of a new family of viruses, called mimiviruses, is forcing a rethink of this conservative view. By producing specialist cells by cell differentiation, **multicellular** organisms are able to carry out multiple and complex functions. Cell differentiation begins as soon as cells have been formed by cell division. It is achieved via the action of regulatory genes (and, in some cases, hormones) that turn specific genes on or off. Cell differentiation is a serial process. The options of a cell become more restricted as it becomes more specialized. Once its fate is determined, a cell cannot alter its path and change into another cell type.

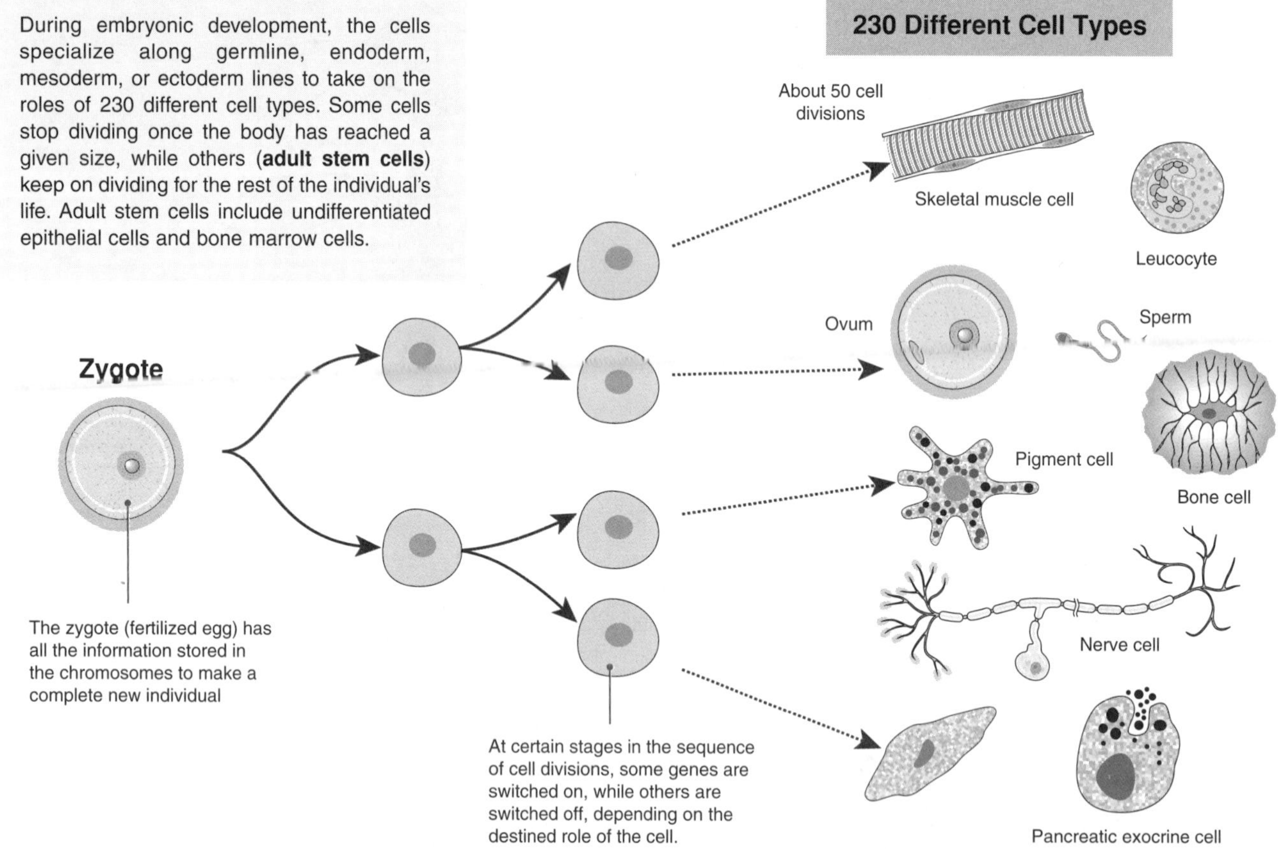
During embryonic development, the cells specialize along germline, endoderm, mesoderm, or ectoderm lines to take on the roles of 230 different cell types. Some cells stop dividing once the body has reached a given size, while others (**adult stem cells**) keep on dividing for the rest of the individual's life. Adult stem cells include undifferentiated epithelial cells and bone marrow cells.
230 Different Cell Types
About 50 cell divisions
Skeletal muscle cell
Leucocyte
Ovum
Sperm
Zygote
Pigment cell
Bone cell
Nerve cell
The zygote (fertilized egg) has all the information stored in the chromosomes to make a complete new individual
At certain stages in the sequence of cell divisions, some genes are switched on, while others are switched off, depending on the destined role of the cell.
Pancreatic exocrine cell

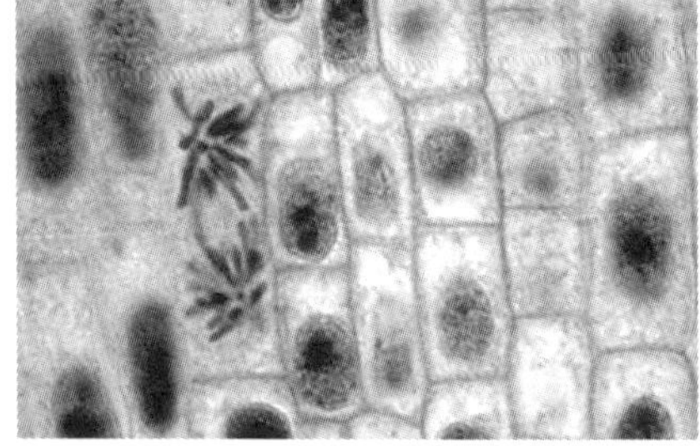
Life is an emergent property of billions of chemical reactions that are driven by the input of energy that produces work and results in decreased entropy (disorder) within the system.

D Dibenski
Swarming is a common behavior in animals. Each individual follows a set of simple rules, but the effect of this causes the swarm to display complex "behaviors" and patterns.

Humans show emergence by spontaneous order. A group of humans, left to self regulate, will form groups and social structures, producing a complex community order.

1. Using examples, explain the concept of emergent properties: ______

2. Explain how cellular differentiation allows a multicellular organism to carry out complex functions: ______

Related activities: *Stem Cells and Differentiation*

ISBN: 978-1-927173-16-9

Stem Cells and Differentiation

Stem cells are undifferentiated cells found in multicellular organisms. They are characterized by two features. The first, **self renewal**, is the ability to undergo numerous cycles of cell division while maintaining an unspecialized state. The second feature, **potency**, is the ability to differentiate into specialized cells. **Totipotent** cells, produced in the first few divisions of a fertilized egg, can differentiate into any cell type, embryonic or extra-embryonic. **Pluripotent cells** are descended from totipotent cells and can give rise to any of the cells derived from the three germ layers (endoderm, mesoderm, and ectoderm). Embryonic stem cells at the blastocyst stage and fetal stem cells are pluripotent. Adult (somatic) stem cells are termed **multipotent**. They are undifferentiated cells found among differentiated cells in a tissue or organ. These cells can give rise to several other cell types, but those types are limited mainly to the cells of the blood, heart, muscle and nerves. The primary roles of adult stem cells are to maintain and repair the tissue in which they are found. A potential use of stem cells is making cells and tissues for medical therapies, such as **cell replacement therapy** and **tissue engineering**.

Stem Cells and Blood Cell Production

New blood cells are produced in the red bone marrow, which becomes the main site of blood production after birth, taking over from the fetal liver. All types of blood cells develop from a single cell type: called a **multipotent stem cell** or hemocytoblast. These cells are capable of mitosis and of differentiation into 'committed' precursors of each of the main types of blood cell.

Each of the different cell lines is controlled by a specific **growth factor**. When a stem cell divides, one of its daughters remains a stem cell, while the other becomes a precursor cell, either a **lymphoid cell** or **myeloid cell**. These cells continue to mature into the various type of blood cells, developing their specialized features and characteristic roles as they do so.

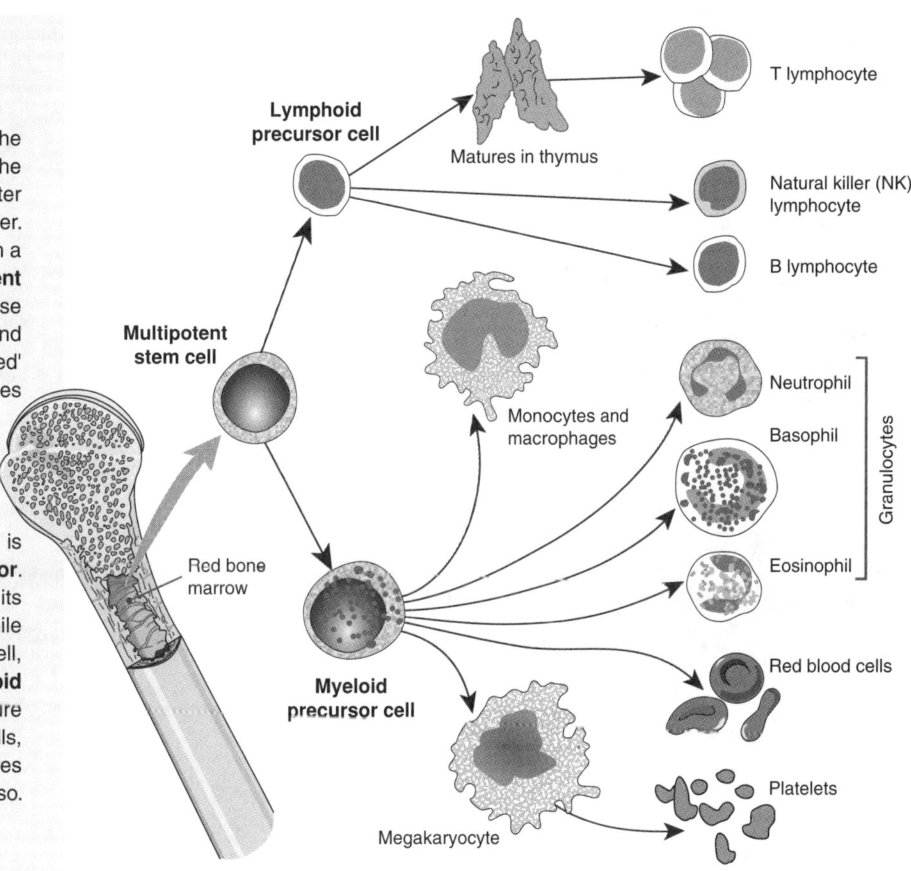

Lymphoid precursor cell
Matures in thymus
T lymphocyte
Natural killer (NK) lymphocyte
B lymphocyte
Multipotent stem cell
Monocytes and macrophages
Neutrophil
Basophil
Eosinophil
Granulocytes
Red bone marrow
Myeloid precursor cell
Red blood cells
Platelets
Megakaryocyte

1. Describe the two defining features of stem cells:

   (a) ______

   (b) ______

2. Distinguish between embryonic stem cells and adult stem cells with respect to their **potency** and their potential applications in medical technologies:

   ______

3. Using an example, explain the purpose of stem cells in an adult: ______

4. Describe one potential advantage of using embryonic stem cells for tissue engineering technology: ______

ISBN: 978-1-927173-16-9

Engineering a Living Skin

New technologies such as cell replacement therapy and tissue engineering require a disease-free and plentiful supply of cells of specific types. Tissue engineering, for example, involves inducing living cells to grow on a scaffold of natural or synthetic material to produce a three-dimensional tissue such as bone or skin.

In 1998, an artificial skin called **Apligraf** became the first product of this type to be approved for use as a biomedical device. It is now widely used in place of skin grafts to treat diabetic ulcers and burns, with the patient's own cells and tissues helping to complete the biological repair. Producing Apligraf is a three stage process (right), which results in a bilayered, living structure capable of stimulating wound repair through its own growth factors and proteins. The cells used to start the culture are usually obtained from discarded neonatal foreskins collected after circumcision. The key to future tissue engineering will be the developments in stem cell research. The best source of stem cells is from very early embryos, but some adult tissues (e.g. bone marrow) also contain stem cells.

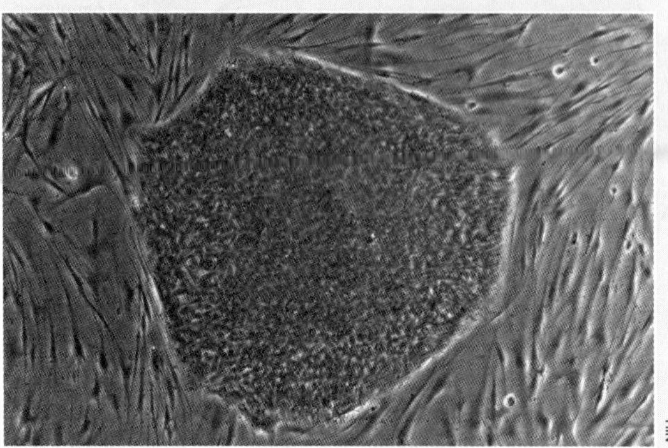
NIH

Human embryonic stem cells (ESCs) growing on mouse embryonic fibroblasts. The mouse fibroblasts act as feeder cells for the culture, releasing nutrients and providing a surface for the ESCs to grow on.

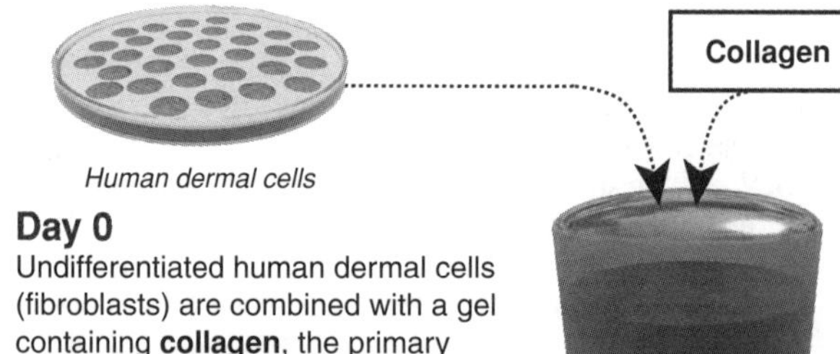
Collagen
Human dermal cells

Day 0

Undifferentiated human dermal cells (fibroblasts) are combined with a gel containing **collagen**, the primary protein of skin. The dermal cells move through the gel, rearranging the collagen and producing a fibrous, living matrix similar to the natural dermis.

Step 1
Form the lower dermal layer

Human epidermal cells

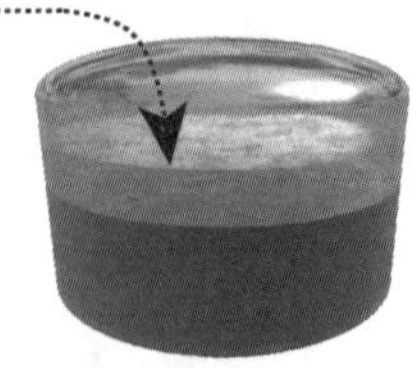

Day 6

Human epidermal cells (called **keratinocytes**) are placed on top of the dermal layer. These cells multiply to cover the dermal layer.

Step 2
Form the upper epidermal layer

Air exposure

Day 10

Exposing the culture to air prompts the epidermal cells to form the outer protective (keratinized) layer of skin. The final size of the Apligraf product is about 75 mm and, from this, tens of thousands of pieces can be made.

Step 3
Form the outer layer

5. Describe the benefits of using a tissue engineered skin product, such as Apligraf, to treat wounds that require grafts:

6. (a) Describe one of the major difficulties with transplantation of cells:

   (b) Explain one way in which this problem could be overcome:

   (c) Describe one of the ethical concerns associated with this solution:

7. Discuss some of the difficulties which must be overcome when growing *in vitro* tissue cultures:

ISBN: 978-1-927173-16-9

Plant Cells

Eukaryotic cells have a similar basic structure, although they may vary tremendously in size, shape, and function. Certain features are common to almost all eukaryotic cells, including their three main regions: a **nucleus** (usually located near the centre of the cell), surrounded by a watery **cytoplasm**, which is itself enclosed by the **plasma membrane**. Plant cells share many structures and organelles in common with animal cells, but also have several unique features. Plant cells are enclosed in a cellulose cell wall, which gives them a regular and uniform appearance. The cell wall protects the cell, maintains its shape, and prevents excessive water uptake. It provides rigidity to plant structures but permits the free passage of materials into and out of the cell.

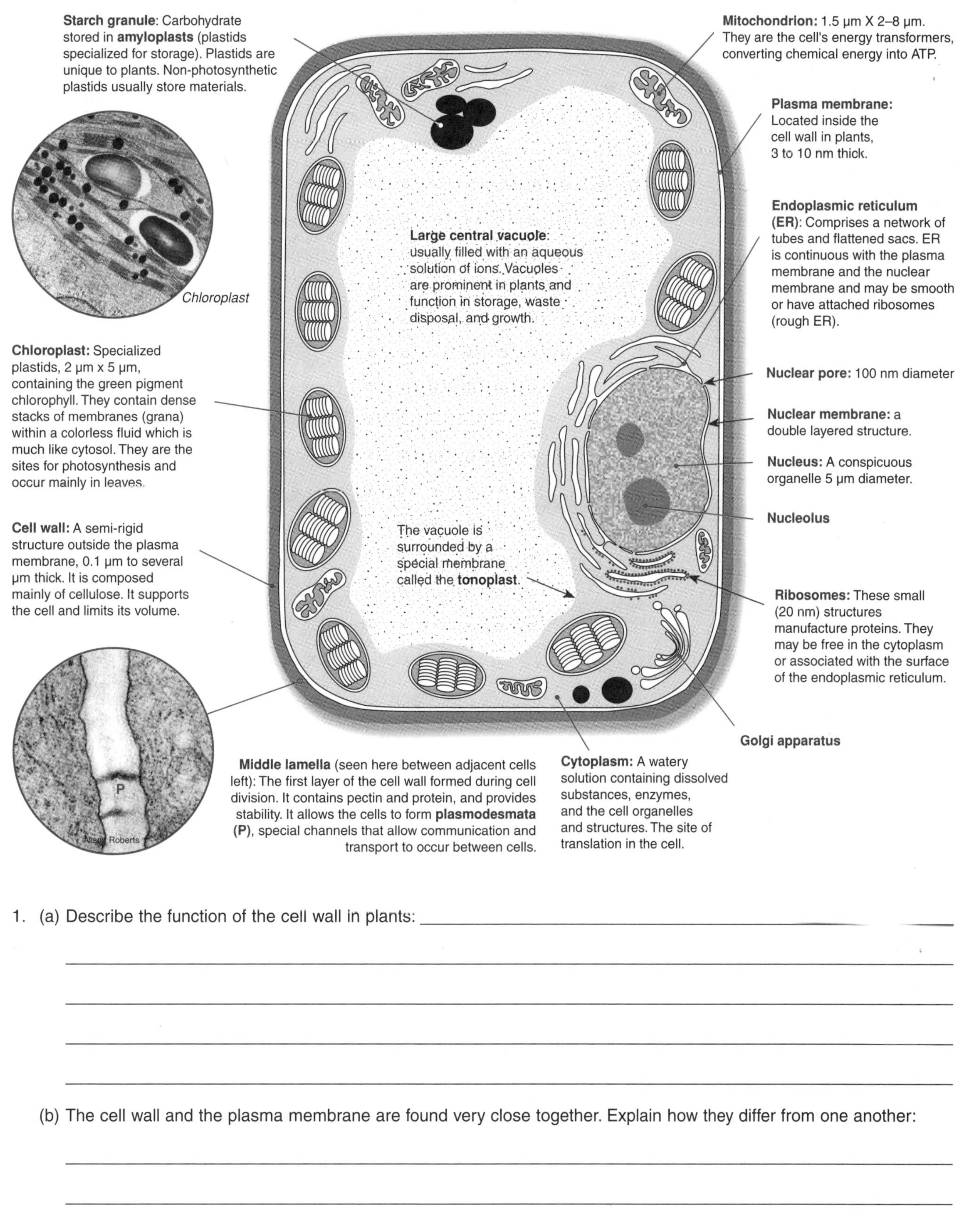
Starch granule: Carbohydrate stored in **amyloplasts** (plastids specialized for storage). Plastids are unique to plants. Non-photosynthetic plastids usually store materials.
Chloroplast
Chloroplast: Specialized plastids, 2 µm x 5 µm, containing the green pigment chlorophyll. They contain dense stacks of membranes (grana) within a colorless fluid which is much like cytosol. They are the sites for photosynthesis and occur mainly in leaves.
Cell wall: A semi-rigid structure outside the plasma membrane, 0.1 µm to several µm thick. It is composed mainly of cellulose. It supports the cell and limits its volume.
P
Alison Roberts
Large central vacuole: usually filled with an aqueous solution of ions. Vacuoles are prominent in plants and function in storage, waste disposal, and growth.
The vacuole is surrounded by a special membrane called the **tonoplast**.
Mitochondrion: 1.5 µm X 2–8 µm. They are the cell's energy transformers, converting chemical energy into ATP.
Plasma membrane: Located inside the cell wall in plants, 3 to 10 nm thick.
Endoplasmic reticulum (ER): Comprises a network of tubes and flattened sacs. ER is continuous with the plasma membrane and the nuclear membrane and may be smooth or have attached ribosomes (rough ER).
Nuclear pore: 100 nm diameter
Nuclear membrane: a double layered structure.
Nucleus: A conspicuous organelle 5 µm diameter.
Nucleolus
Ribosomes: These small (20 nm) structures manufacture proteins. They may be free in the cytoplasm or associated with the surface of the endoplasmic reticulum.
Golgi apparatus
Middle lamella (seen here between adjacent cells left): The first layer of the cell wall formed during cell division. It contains pectin and protein, and provides stability. It allows the cells to form **plasmodesmata** **(P)**, special channels that allow communication and transport to occur between cells.
Cytoplasm: A watery solution containing dissolved substances, enzymes, and the cell organelles and structures. The site of translation in the cell.

1. (a) Describe the function of the cell wall in plants: ______

(b) The cell wall and the plasma membrane are found very close together. Explain how they differ from one another:

______

ISBN: 978-1-927173-16-9

Plant Cells are Organized into Tissues

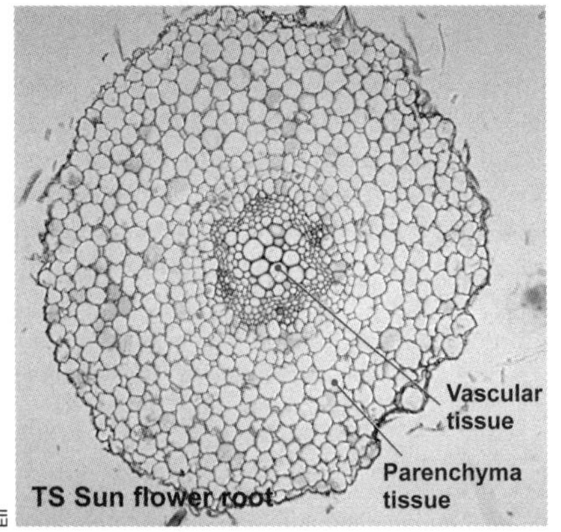
Vascular tissue
Parenchyma tissue
TS Sun flower root

EII

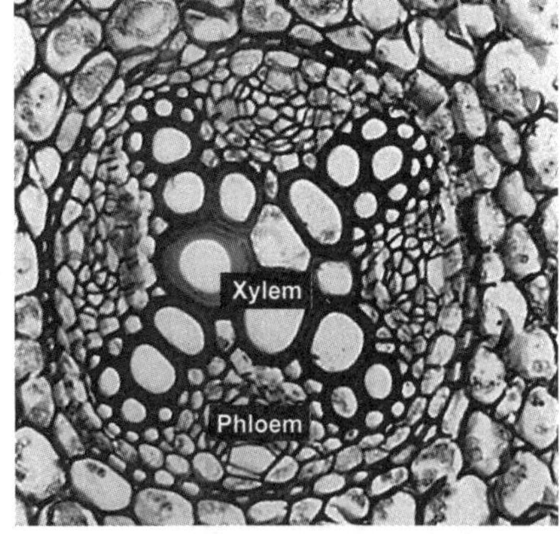
Xylem
Phloem

RCN

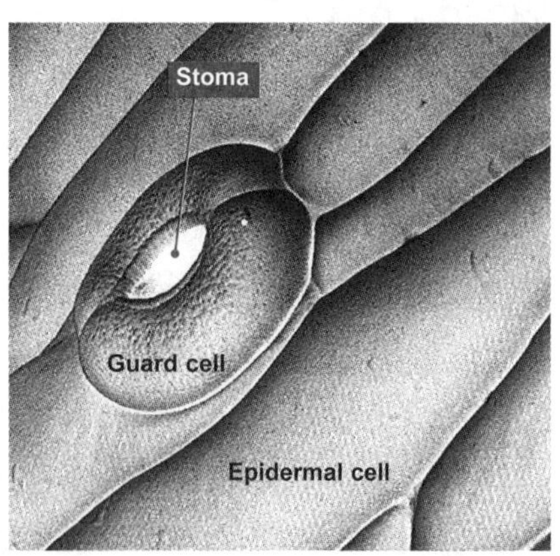
Stoma
Guard cell
Epidermal cell

EII

Simple Tissues

Simple tissues consists of only one or two cell types. **Parenchyma tissue** is the most common and involved in storage, photosynthesis, and secretion. **Collenchyma tissue** comprises thick-walled collenchyma cells alternating with layers of intracellular substances (pectin and cellulose) to provide flexible support. The cells of **sclerenchyma** tissue (fibers and sclereids) have rigid cell walls which provide support.

Complex Tissues

Xylem and phloem tissue (above left), which together make up the plant **vascular tissue** system, are complex tissues. Each comprises several tissue types including tracheids, vessel members, parenchyma and fibers in xylem, and sieve tube members, companion cells, parenchyma and sclerenchyma in phloem. **Dermal tissue** is also complex tissue and covers the outside of the plant. The composition of dermal tissue varies depending upon its location on the plant. Root epidermal tissue consist of epidermal cells which extend to root hairs (**trichomes**) for increasing surface area. In contrast, the epidermal tissue of leaves (above right) is covered by a waxy cuticle to reduce water loss, and specialized guard cells regulate water intake via the stomata (pores in the leaf through which gases enter and leave the leaf tissue).

2. Explain how organelles increase the efficiency of the cell: ______

3. Describe how simple tissues differ from complex tissues: ______

4. Plant cells range from 10 to 100 μm. O_2 diffuses through a cell at a rate of about 1 mm an hour. Calculate the following:

   (a) The time required for an oxygen molecule to reach the centre of a cell 10 μm in diameter: ______

   (b) The time required for an oxygen molecule to reach the centre of a cell 100 μm in diameter: ______

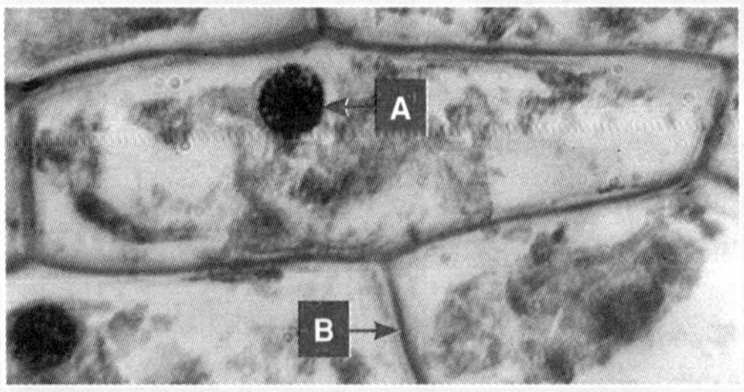

A
B

Onion epidermal cells

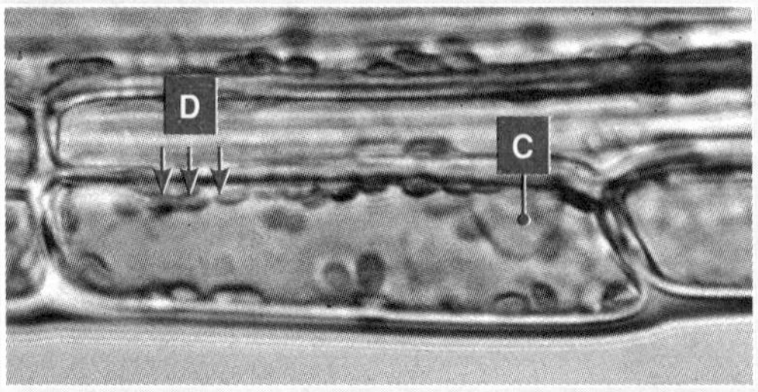

D
C

Photos: RCN

Elodea cells

5. The two photographs (left) show plant cells as seen by a light microscope. Identify the basic features labelled **A-D**:

   A: ______

   B: ______

   C: ______

   D: ______

6. Cytoplasmic streaming is a feature of eukaryotic cells, often clearly visible with a light microscope in plant (and algal) cells.

   (a) Explain what is meant by cytoplasmic streaming: ______

   (b) For the *Elodea* cell (lower, left), draw arrows to indicate cytoplasmic streaming movements.

7. Describe three structures/organelles present in generalized plant cells but absent from animal cells:

   (a) ______

   (b) ______

   (c) ______

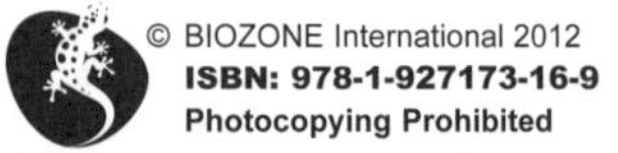

ISBN: 978-1-927173-16-9

Animal Cells

Although plant and animal cells have many features in common, animal cells do not have a regular shape, and some (such as phagocytic white blood cells) are quite mobile. The diagram below shows the ultrastructure of a **liver cell** or hepatocyte. It contains organelles common to most relatively unspecialized human cells. Hepatocytes make up 70-80% of the liver's mass. They are metabolically active, with a large central nucleus, many mitochondria, and large amounts of rough endoplasmic reticulum. Thin, cellular extensions called microvilli increase surface area of the cell, increasing its capacity for absorption.

Structures and Organelles in a Liver Cell

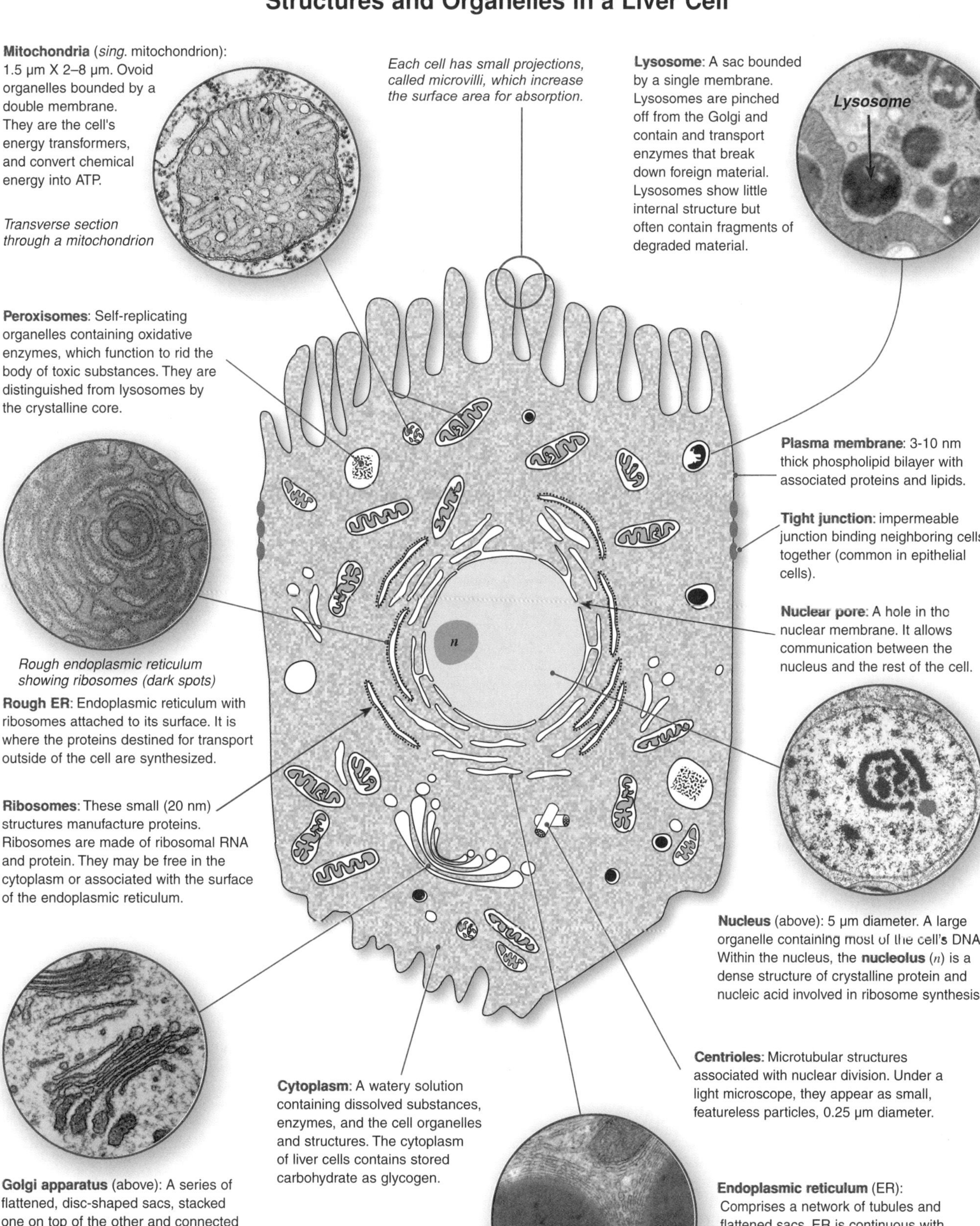
Mitochondria (*sing.* mitochondrion): 1.5 µm X 2–8 µm. Ovoid organelles bounded by a double membrane. They are the cell's energy transformers, and convert chemical energy into ATP.
Transverse section through a mitochondrion
Each cell has small projections, called microvilli, which increase the surface area for absorption.
Lysosome: A sac bounded by a single membrane. Lysosomes are pinched off from the Golgi and contain and transport enzymes that break down foreign material. Lysosomes show little internal structure but often contain fragments of degraded material.
Lysosome
Peroxisomes: Self-replicating organelles containing oxidative enzymes, which function to rid the body of toxic substances. They are distinguished from lysosomes by the crystalline core.
Plasma membrane: 3-10 nm thick phospholipid bilayer with associated proteins and lipids.
Tight junction: impermeable junction binding neighboring cells together (common in epithelial cells).
Nuclear pore: A hole in the nuclear membrane. It allows communication between the nucleus and the rest of the cell.
n
Rough endoplasmic reticulum showing ribosomes (dark spots)
Rough ER: Endoplasmic reticulum with ribosomes attached to its surface. It is where the proteins destined for transport outside of the cell are synthesized.
Ribosomes: These small (20 nm) structures manufacture proteins. Ribosomes are made of ribosomal RNA and protein. They may be free in the cytoplasm or associated with the surface of the endoplasmic reticulum.
Nucleus (above): 5 µm diameter. A large organelle containing most of the cell's DNA. Within the nucleus, the **nucleolus** (*n*) is a dense structure of crystalline protein and nucleic acid involved in ribosome synthesis.
Centrioles: Microtubular structures associated with nuclear division. Under a light microscope, they appear as small, featureless particles, 0.25 µm diameter.
Cytoplasm: A watery solution containing dissolved substances, enzymes, and the cell organelles and structures. The cytoplasm of liver cells contains stored carbohydrate as glycogen.
Golgi apparatus (above): A series of flattened, disc-shaped sacs, stacked one on top of the other and connected with the ER. The Golgi stores, modifies, and packages proteins. It 'tags' proteins so that they go to their correct destination.
Endoplasmic reticulum (ER): Comprises a network of tubules and flattened sacs. ER is continuous with the plasma membrane and the nuclear membrane. **Smooth ER**, as shown here, is a site for lipid and carbohydrate metabolism, including hormone synthesis.

Cell Biology

ISBN: 978-1-927173-16-9

Related activities: *Identifying Structures in an Animal Cell*
Weblinks: *Eukaryotic Cells Interactive Animation*

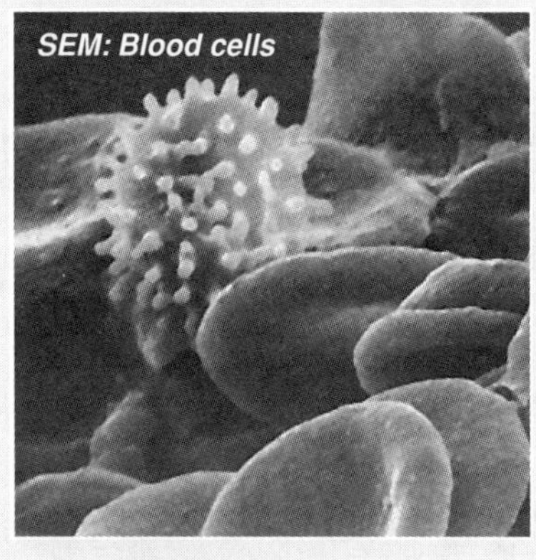
SEM: Blood cells

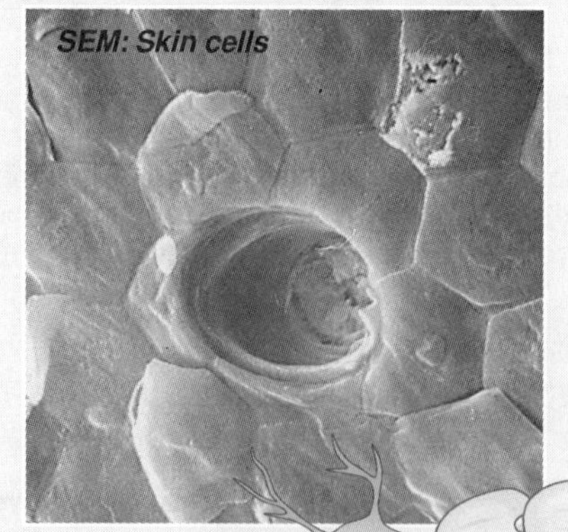
SEM: Skin cells

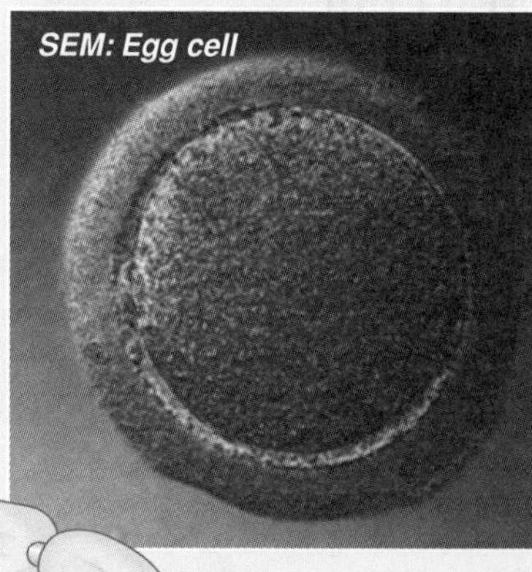
SEM: Egg cell

Many animal cells are specialized to carry out specific functions within the body. As a result, the morphology and physiology of animal cells are highly varied. Some examples are presented here.

Nerve cell

1. Explain what is meant by a generalized cell: ______________________

____

____

2. (a) Describe the features of the liver cell that make it relatively unspecialized compared to some other cells:

____

____

(b) What features of a liver cell are associated with it being metabolically very active: ______________________

____

____

____

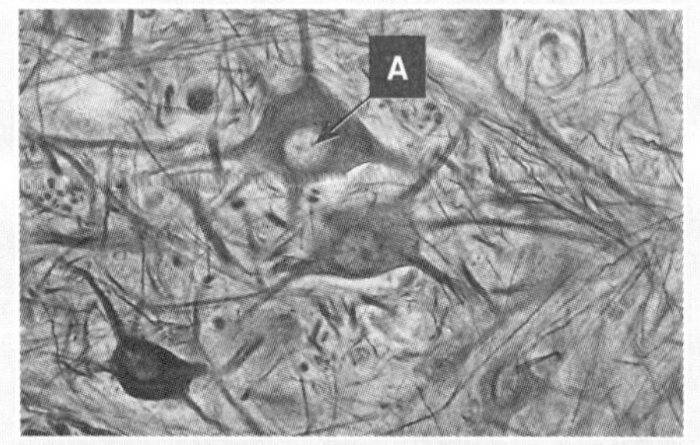
A

Neurons (nerve cells) in the spinal cord

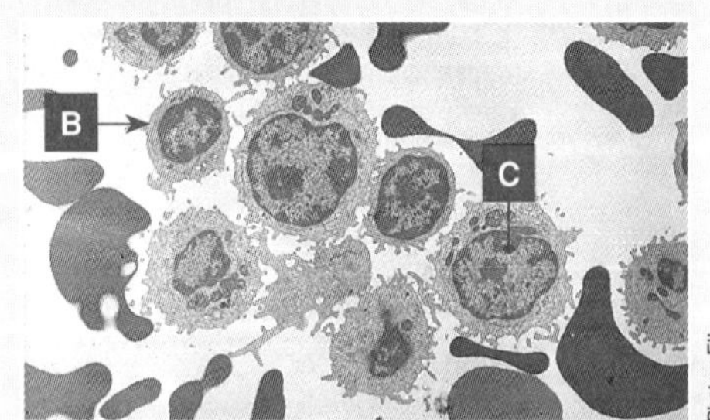
B
C
Photos: EII

White blood cells and red blood cells (blood smear)

3. The two photomicrographs (left) show several types of animal cells. Identify the features indicated by the letters **A**-**C**:

A: ______________________

B: ______________________

C: ______________________

4. White blood cells are mobile, phagocytic cells, whereas red blood cells are smaller than white blood cells and, in humans, lack a nucleus.

(a) In the photomicrograph (below, left), circle a white blood cell and a red blood cell:

(b) With respect to the features that you can see, explain how you made your decision.

____

____

____

____

5. Name and describe one structure or organelle present in generalized animal cells but absent from plant cells:

____

____

ISBN: 978-1-927173-16-9

Cell Structures and Organelles

The table below provides a format to summarize information about structures and organelles of typical eukaryotic cells. Complete the table using the list provided and by referring to a textbook and to other pages in this topic. Fill in the final three columns by writing either 'YES' or 'NO'. The first cell component has been completed for you as a guide and the log scale of measurements (top of next page) illustrates the relative sizes of some cellular structures. **List of structures and organelles**: *cell wall, mitochondrion, chloroplast, cell junctions, centrioles, ribosome, flagella, endoplasmic reticulum, Golgi apparatus, nucleus, flagella, cytoskeleton and vacuoles.*

Cell Component	Details	Present in: Plant cells	Present in: Animal cells	Visible under light microscope
(a) Double layer of phospholipids (called the lipid bilayer); Proteins	Name: Plasma (cell surface) membrane Location: Surrounding the cell Function: Gives the cell shape and protection. It also regulates the movement of substances into and out of the cell.	YES	YES	YES *(but not at the level of detail shown in diagram)*
(b) Large subunit; Small subunit	Name: Location: Function:			
(c) Outer membrane; Inner membrane; Matrix; Cristae	Name: Location: Function:			
(d) Secretory vesicles budding off; Cisternae; Transfer vesicles from the smooth endoplasmic reticulum	Name: Location: Function:			
(e) Ribosomes; transport pathway; Rough; Smooth; Vesicles budding off; Flattened membrane sacs	Name: Location: Function:			
(f) Grana comprise stacks of thylakoids; Stroma; Lamellae	Name: Location: Function:			

ISBN: 978-1-927173-16-9

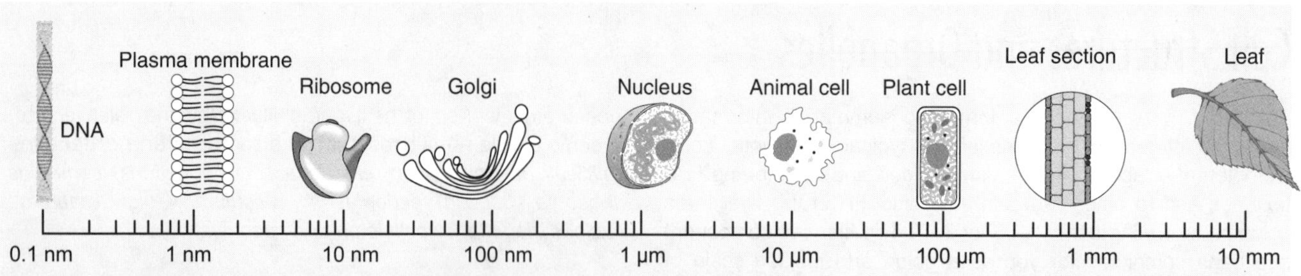

DNA
Plasma membrane
Ribosome
Golgi
Nucleus
Animal cell
Plant cell
Leaf section
Leaf
0.1 nm
1 nm
10 nm
100 nm
1 µm
10 µm
100 µm
1 mm
10 mm

Cell Component	Details	Present in: Plant cells	Present in: Animal cells	Visible under light microscope
(g) Food Vacuole; Digestion; Lysosome; Phagocytosis of food particle	Name: Lysosome and food vacuole Location: Function:			
(h) Nuclear membrane; Nuclear pores; Nucleolus; Genetic material	Name: Location: Function:			
(i) Microtubules	Name: Location: Function:			
(j) Two central, single microtubules; 9 doublets of microtubules in an outer ring; Extension of plasma membrane surrounding a core of microtubules in a 9+2 pattern; Basal body anchors the flagellum	Name: Location: Function:			

ISBN: 978-1-927173-16-9

Cell Component	Details	Present in		Visible under light microscope
		Plant cells	Animal cells	
(k) Plasma membrane Organelle Microtubule Intermediate filament Microfilament	Name: Location: Function:			
(l) Middle lamella Pectins Hemicelluloses Cellulose fibres	Name: Cellulose cell wall Location: Function:			
(m) Tight junction Desmosome Gap junction Extracellular matrix	Name: Cell junctions Location: Function:			

ISBN: 978-1-927173-16-9

Identifying Structures in an Animal Cell

Our current knowledge of cell ultrastructure has been made possible by the advent of electron microscopy. Transmission electron microscopy is the most frequently used technique for viewing cellular organelles. When viewing TEMs, the cellular organelles may appear to be quite different depending on whether they are in transverse or longitudinal section.

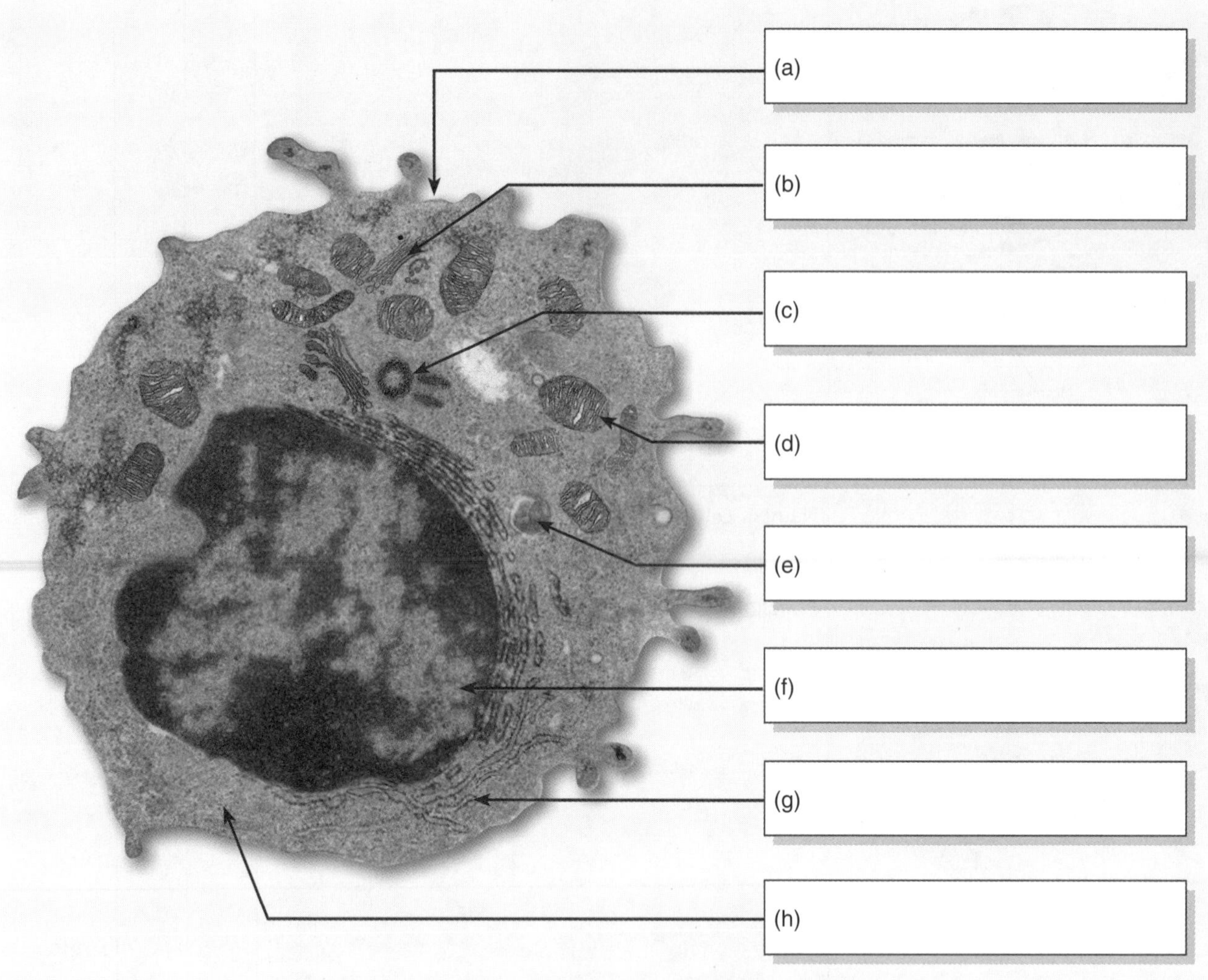

(a)
(b)
(c)
(d)
(e)
(f)
(g)
(h)

1. Identify and label the structures in the cell above using the following list of terms: *cytoplasm, plasma membrane, rough endoplasmic reticulum, mitochondrion, nucleus, centriole, Golgi apparatus, lysosome*

2. Which of the organelles in the EM above are shown in both transverse and longitudinal section?

3. Why do plants lack any of the mobile phagocytic cells typical of animals?

4. The animal pictured above is a lymphocyte. Describe the features that suggest to you that:

   (a) It has a role in producing and secreting proteins:

   (b) It is metabolically very active:

5. What features of the lymphocyte cell above identify it as eukaryotic?

Related activities: *Animal Cells*

ISBN: 978-1-927173-16-9

Identifying Structures in a Plant Cell

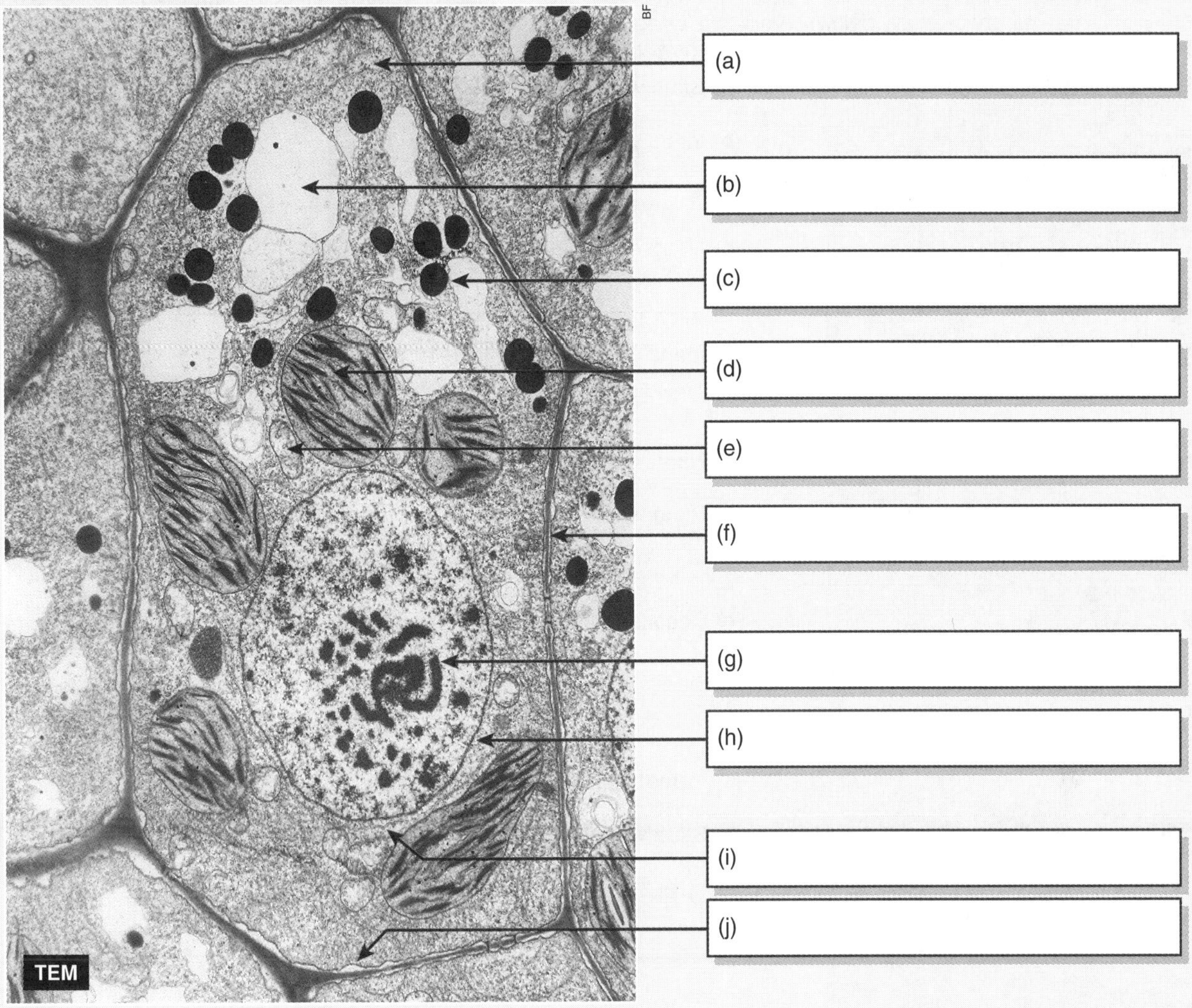

BF

TEM

(a)

(b)

(c)

(d)

(e)

(f)

(g)

(h)

(i)

(j)

1. Study the diagrams on the other pages in this chapter to familiarize yourself with the structures found in eukaryotic cells. Identify and label the ten structures in the cell above using the following list of terms: *nuclear membrane, cytoplasm, endoplasmic reticulum, mitochondrion, starch granules, chromosome, vacuole, plasma membrane, cell wall, chloroplast*

2. State how many cells, or parts of cells, are visible in the electron micrograph above: ____________________

3. Describe the features that identify this cell as a plant cell: ____________________

____

____

4. (a) Explain where cytoplasm is found in the cell: ____________________

____

(b) Describe what cytoplasm is made up of: ____________________

____

5. Describe two structures, pictured in the cell above, that are associated with storage:

(a) ____________________

____

(b) ____________________

____

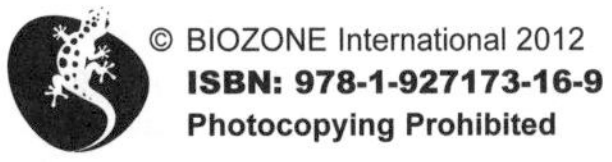

ISBN: 978-1-927173-16-9

Interpreting Electron Micrographs

Electron microscopes produce a magnified image at high **resolution** (they can distinguish between close together but separate objects). The photographs below were taken using a **transmission electron microscope** (TEM). They show the ultrastructure of some organelles. Remember that these photos are showing only **parts of cells**, **not whole cells**. Some of the photographs show more than one type of organelle. The questions refer to the main organelle indicated in the photo.

EII

1. (a) State which kind of cell this is: ______

   (b) Identify the structure labelled **A**: ______

   ______

   (c) Describe the function of this structure: ______

   ______

   (d) Identify the structure labelled **B**: ______

   ______

WMU

2. (a) Name this organelle (arrowed): ______

   (b) State which kind of cell(s) this organelle would be found in:

   ______

   (c) Describe the function of this organelle: ______

   ______

   ______

WMU

3. (a) Name the large, circular organelle: ______

   (b) State which kind of cell(s) this organelle would be found in:

   ______

   (c) Describe the function of this organelle: ______

   ______

   ______

   (d) Label **two** regions that can be seen **inside** this organelle.

WMU

4. (a) Name and label the ribbon-like organelle in this photograph (arrowed):

   ______

   (b) State which kind of cell(s) this organelle is found in:

   ______

   (c) Describe the function of these organelles: ______

   ______

   (d) Name the dark 'blobs' attached to the organelle you have labelled:

   ______

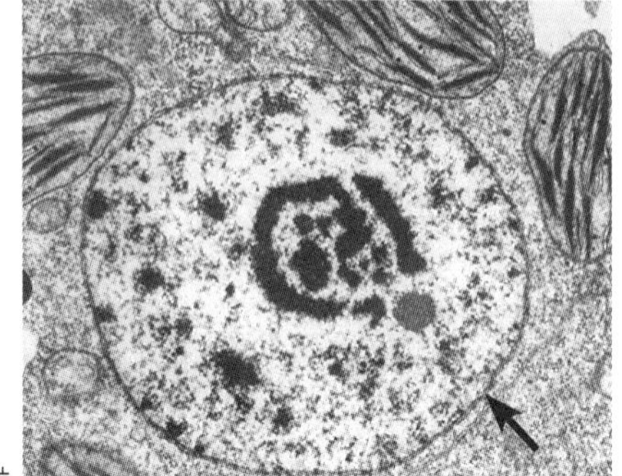

BF

5. (a) Name this large circular structure (arrowed): ______

   (b) State which kind of cell(s) this structure would be found in: ______

   ______

   (c) Describe the function of this structure: ______

   ______

   (d) Label three features relating to this structure in the photograph.

ISBN: 978-1-927173-16-9

Core Topic **2**

Cell **Processes**

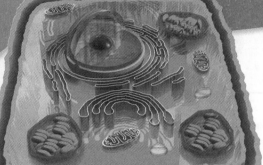

Key concepts

- Cellular metabolism depends on the transport of substances across cellular membranes.
- The plasma membrane forms a partially permeable barrier to entry and exit of substances into the cell.
- The fluid mosaic model satisfies the observed properties of cellular membranes.
- Substances in cells move by passive or active transport.
- New cells arise through cell division.

Key terms

Core

active transport
amphipathic
anaphase
aquaporin
carrier protein
cell wall
channel protein
concentration gradient
cytokinesis
cytosis
diffusion
endocytosis
exocytosis
facilitated diffusion
fluid mosaic model
glycolipid
glycoprotein
hypertonic
hypotonic
interphase
ion pump
isotonic
metaphase
mitosis
osmosis
partially permeable
passive transport
phospholipid
plasma membrane
prophase
surface area: volume ratio
telophase
transmembrane protein
tumor (tumour)

Periodicals:
Listings for this chapter are on page 398

Weblinks:
www.thebiozone.com/weblink/IB-3169.html

BIOZONE APP:
Student Review Series
Cell Membranes & Transport

Learning Objectives

- ☐ 1. Use the **KEY TERMS** to compile a glossary for this topic.

Membrane Structure and Function *(2.4)* pages 58-66

- ☐ 2. Outline the roles of **membranes** in cells. Describe the **fluid mosaic model** of the plasma membrane, including the significance of the **amphipathic** character of the phospholipids that make up the structural framework of the membrane and the role of **transmembrane proteins**, **glycoproteins**, and **glycolipids**.
- ☐ 3. Explain how the properties of the embedded proteins contribute to the partially permeable nature of the membrane. Include reference to **aquaporins**, and embedded **channel proteins** and **carrier proteins**.
- ☐ 4. Describe and explain **diffusion**, **facilitated diffusion**, and **osmosis**, identifying them as **passive transport** processes. Describe and explain factors affecting diffusion rates across membranes.
- ☐ 5. Explain the role of passive transport in the import of resources and export of waste products. Using an example, explain how membrane proteins are involved in the facilitated diffusion of charged and polar molecules across a membrane.
- ☐ 6. Recall the significance of **surface area to volume ratio** to cells. Explain why cell size is limited by the rate of diffusion.
- ☐ 7. Distinguish between passive transport and **active transport**, identifying the involvement of membrane proteins and **ATP** in active transport processes.
- ☐ 8. Using examples, describe and explain the role of **ion pumps** in the active transport of materials in and out of cells.
- ☐ 9. Describe **cytosis** in cells, recognizing it as a type of active transport. Explain how vesicles are used to transport materials within the cell between the rough endoplasmic reticulum, Golgi, and plasma membrane.
- ☐ 10. Describe how the fluid nature of the plasma membrane enables it to change shape, break, and reform during **endocytosis** and **exocytosis**.

Cell Division *(2.5)* pages 67-71

- ☐ 11. Describe the **cell cycle** in eukaryotes, recognizing **interphase**, **mitosis**, and **cytokinesis**. Describe the events in the stages of interphase: G_1, S, and G_2.
- ☐ 12. Explain how loss of regulation of the cell cycle can result in uncontrolled cell division and formation of **tumors**.
- ☐ 13. Describe mitosis as a continuous process, with distinct structural stages. Recognize and summarize the events occurring in each of the following stages in mitosis: **prophase**, **metaphase**, **anaphase**, and **telophase**.
- ☐ 14. Describe the outcome of mitotic division and outline the role of mitosis in growth and repair, and asexual reproduction.

The Role of Membranes in Cells

Many of the important structures and organelles in cells are composed of, or are enclosed by, membranes. These include: the endoplasmic reticulum, mitochondria, nucleus, Golgi body, chloroplasts, lysosomes, vesicles and the cell plasma membrane itself. All membranes within eukaryotic cells share the same basic structure as the plasma membrane that encloses the cell. Membranes have critical roles in the cell, creating compartments with different functions, controlling the entry and exit of substances, and enabling recognition and communication between cells. Some of these roles are described below.

Isolation of enzymes
Membrane-bound lysosomes contain enzymes for the destruction of wastes and foreign material. Peroxisomes are the site for destruction of the toxic and reactive molecule, hydrogen peroxide (formed as a result of some cellular reactions).

Role in lipid synthesis
The smooth ER is the site of lipid and steroid synthesis.

Containment of DNA
The nucleus is surrounded by a nuclear envelope of two membranes, forming a separate compartment for the cell's genetic material.

Role in protein and membrane synthesis
Some protein synthesis occurs on free ribosomes, but much occurs on membrane-bound ribosomes on the rough endoplasmic reticulum. Here, the protein is synthesized directly into the space within the ER membranes. The rough ER is also involved in membrane synthesis, growing in place by adding proteins and phospholipids.

Cell communication and recognition
The proteins embedded in the membrane act as receptor molecules for hormones and neurotransmitters. Glycoproteins and glycolipids stabilize the plasma membrane and act as cell identity markers, helping cells to organize themselves into tissues, and enabling foreign cells to be recognized.

Packaging and secretion
The Golgi apparatus is a specialized membrane-bound organelle which produces lysosomes and compartmentalizes the modification, packaging and secretion of substances such as proteins and hormones.

Transport processes
Channel and carrier proteins are involved in selective transport across the plasma membrane. The level of cholesterol in the membrane influences permeability and transport functions.

Entry and export of substances
The plasma membrane may take up fluid or solid material and form membrane-bound vesicles (or larger vacuoles) within the cell. Membrane-bound transport vesicles move substances to the inner surface of the cell where they can be exported from the cell by exocytosis.

Energy transfer
The reactions of cellular respiration (and photosynthesis in plants) take place in the membrane-bound energy transfer systems occurring in mitochondria and chloroplasts respectively.

1. Explain the crucial role of membrane systems and organelles in the following:

   (a) Providing compartments within the cell: ______________________

   ____

   ____

   (b) Increasing the total membrane surface area within the cell: ______________________

   ____

   ____

2. Explain the importance of the following components of cellular membranes:

   (a) Glycoproteins and glycolipids: ______________________

   ____

   (b) Channel proteins and carrier proteins: ______________________

   ____

3. Explain how cholesterol can play a role in membrane transport: ______________________

   ____

   ____

Related activities: *The Structure of Membranes, Active and Passive Transport Summary*
Weblinks: *Cell Membranes*

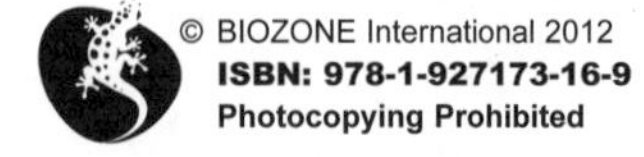

ISBN: 978-1-927173-16-9

The Structure of Membranes

All cells have a plasma membrane that forms the outer limit of the cell. Bacteria, fungi, and plant cells have a cell wall outside this, but it is quite distinct and outside the plasma membrane. Membranes are also found inside eukaryotic cells as part of membranous **organelles**. Current knowledge of membrane structure has been built up from many observations and experiments. The original model of membrane structure, proposed by Davson and Danielli, was the unit membrane (a lipid bilayer coated with protein). This model was later modified after the discovery that the protein molecules were embedded *within* the bilayer rather than coating the outside. The now-accepted model of membrane structure is the **fluid mosaic model** described below.

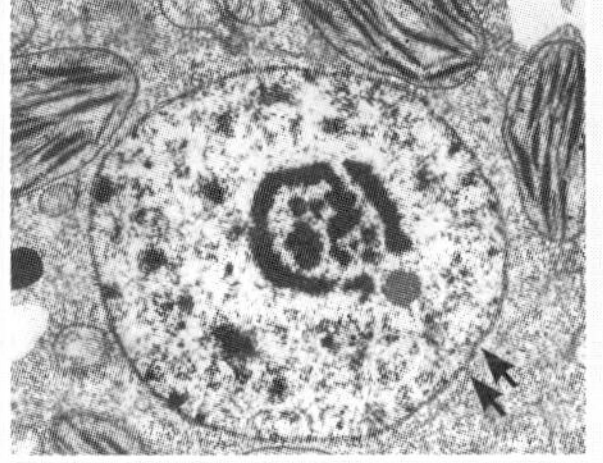
BF

The **nuclear membrane** that surrounds the nucleus helps to control the passage of genetic information to the cytoplasm. It may also serve to protect the DNA.

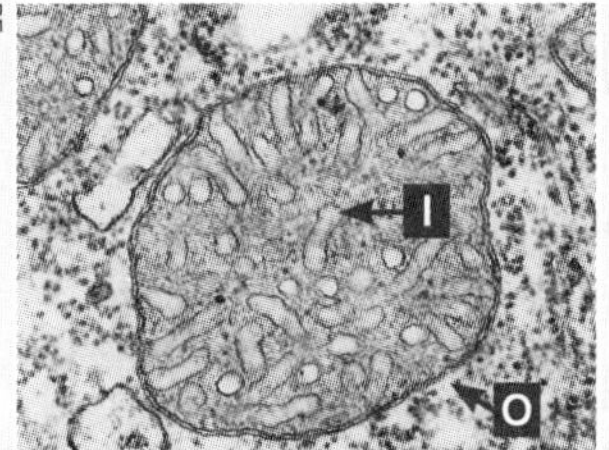
I
O
WMU

Mitochondria have an outer membrane (**O**) which controls the entry and exit of materials involved in aerobic respiration. Inner membranes (**I**) provide attachment sites for enzyme activity.

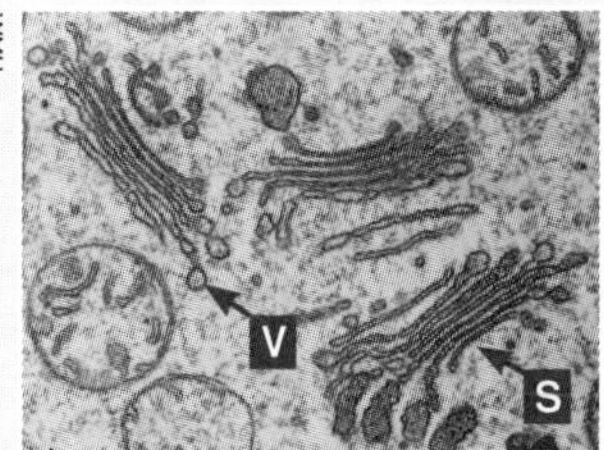
V
S
WMU

The **Golgi apparatus** comprises stacks of membrane-bound sacs (**S**). It is involved in packaging materials for transport or export from the cell as secretory vesicles (**V**).

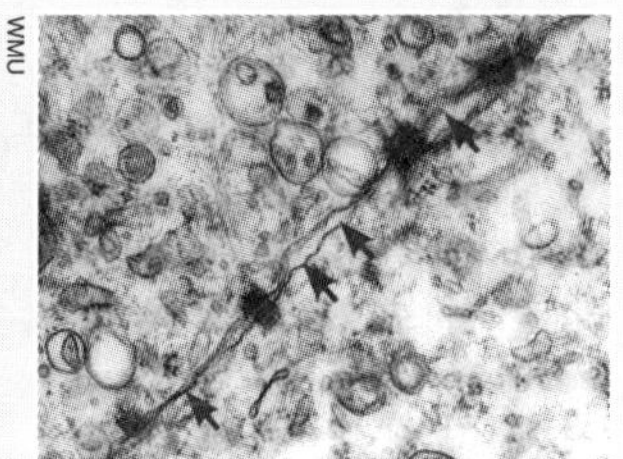
WMU

The cell is surrounded by a **plasma membrane** which controls the movement of most substances into and out of the cell. This photo shows two neighboring cells (arrows).

The Fluid Mosaic Model of Membrane Structure

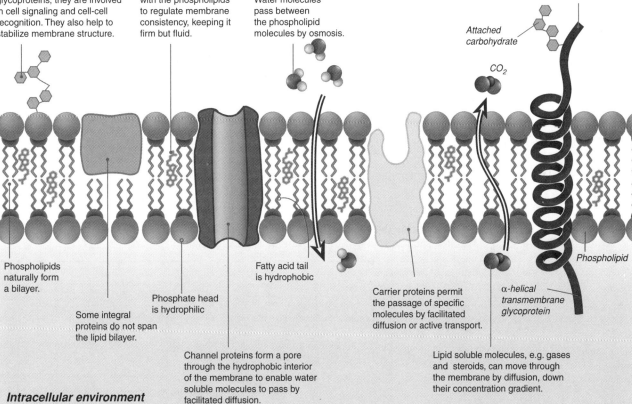

Glycolipids in membranes are phospholipids with attached carbohydrate. Like glycoproteins, they are involved in cell signaling and cell-cell recognition. They also help to stabilize membrane structure.

Cholesterol is a packing molecule and interacts with the phospholipids to regulate membrane consistency, keeping it firm but fluid.

Water molecules pass between the phospholipid molecules by osmosis.

Glycoproteins are proteins with attached carbohydrate. They are important in membrane stability, in cell-cell recognition, and in cell signaling, acting as receptors for hormones and neurotransmitters.

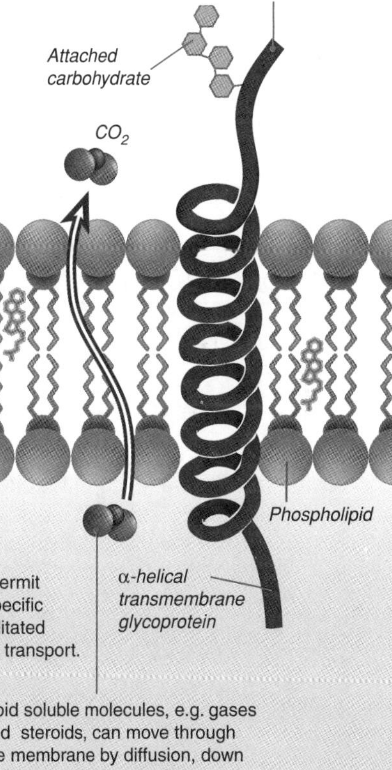
Attached carbohydrate
CO_2
Phospholipid
α-helical transmembrane glycoprotein

Phospholipids naturally form a bilayer.

Some integral proteins do not span the lipid bilayer.

Phosphate head is hydrophilic

Fatty acid tail is hydrophobic

Carrier proteins permit the passage of specific molecules by facilitated diffusion or active transport.

Channel proteins form a pore through the hydrophobic interior of the membrane to enable water soluble molecules to pass by facilitated diffusion.

Lipid soluble molecules, e.g. gases and steroids, can move through the membrane by diffusion, down their concentration gradient.

Intracellular environment

Based on a diagram in Biol. Sci. Review, Nov. 2009, pp. 20-21

1. Identify the component(s) of the plasma membrane involved in:

   (a) Facilitated diffusion: ______________________

   (b) Active transport: ______________________

   (c) Cell signaling: ______________________

   (d) Regulating membrane fluidity: ______________________

2. How do the properties of phospholipids contribute to their role in forming the structural framework of membranes?

____

____

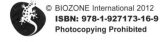

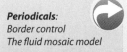

Cell Processes

3. (a) Describe the modern fluid mosaic model of membrane structure: ______

(b) Explain how the fluid mosaic model accounts for the observed properties of cellular membranes: ______

4. Discuss the various functional roles of membranes in cells: ______

5. (a) Name a cellular organelle that possesses a membrane: ______

(b) Describe the membrane's purpose in this organelle: ______

6. Describe the purpose of cholesterol in plasma membranes: ______

7. List three substances that need to be transported **into** all kinds of animal cells, in order for them to survive:

(a) ______ (b) ______ (c) ______

8. List two substances that need to be transported **out** of all kinds of animal cells, in order for them to survive:

(a) ______ (b) ______

9. Use the symbol for a phospholipid molecule (below) to draw a **simple labeled diagram** to show the structure of a plasma membrane (include features such as lipid bilayer and various kinds of proteins):

ISBN: 978-1-927173-16-9

Passive Transport Processes

The molecules that make up substances are constantly moving about in a random way. This random motion causes molecules to disperse from areas of high to low concentration. This movement is called **diffusion**. The molecules move down a **concentration gradient**. Diffusion and **osmosis** (diffusion of water molecules across a partially permeable membrane) are **passive** processes, and use no energy. Diffusion occurs freely across membranes, as long as the membrane is permeable to that molecule (partially permeable membranes allow the passage of some molecules but not others). Each type of molecule diffuses down its own concentration gradient. Diffusion of molecules in one direction does not hinder the movement of other molecules. Diffusion is important in allowing exchanges with the environment and in the regulation of cell water content.

Diffusion is the movement of particles from regions of high to low concentration (down a **concentration gradient**), with the end result being that the molecules become evenly distributed. In biological systems, diffusion often occurs across selectively permeable membranes.

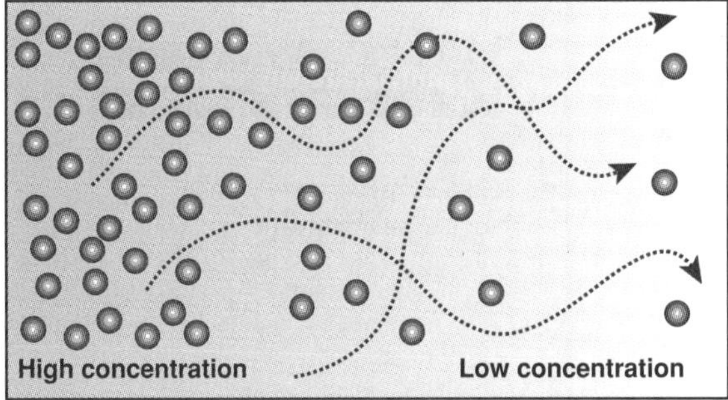
High concentration
Low concentration

Concentration gradient

If molecules are free to move, they move from high to low concentration until they are evenly dispersed.

Factors affecting rates of diffusion

Concentration gradient:	Diffusion rates will be higher when there is a greater difference in concentration between two regions.
The distance involved:	Diffusion over shorter distances occurs at a greater rate than diffusion over larger distances.
The area involved:	The larger the area across which diffusion occurs, the greater the rate of diffusion.
Barriers to diffusion:	Thicker barriers slow diffusion rate. Pores in a barrier enhance diffusion.

Fick's law

$$\frac{\text{Surface area of membrane} \times \text{Difference in concentration across the membrane}}{\text{Length of the diffusion path (thickness of the membrane)}}$$

These factors are expressed in Fick's law, which governs the rate of diffusion of substances within a system. Temperature also affects diffusion rates; at higher temperatures molecules have more energy and move more rapidly.

Diffusion Through Membranes

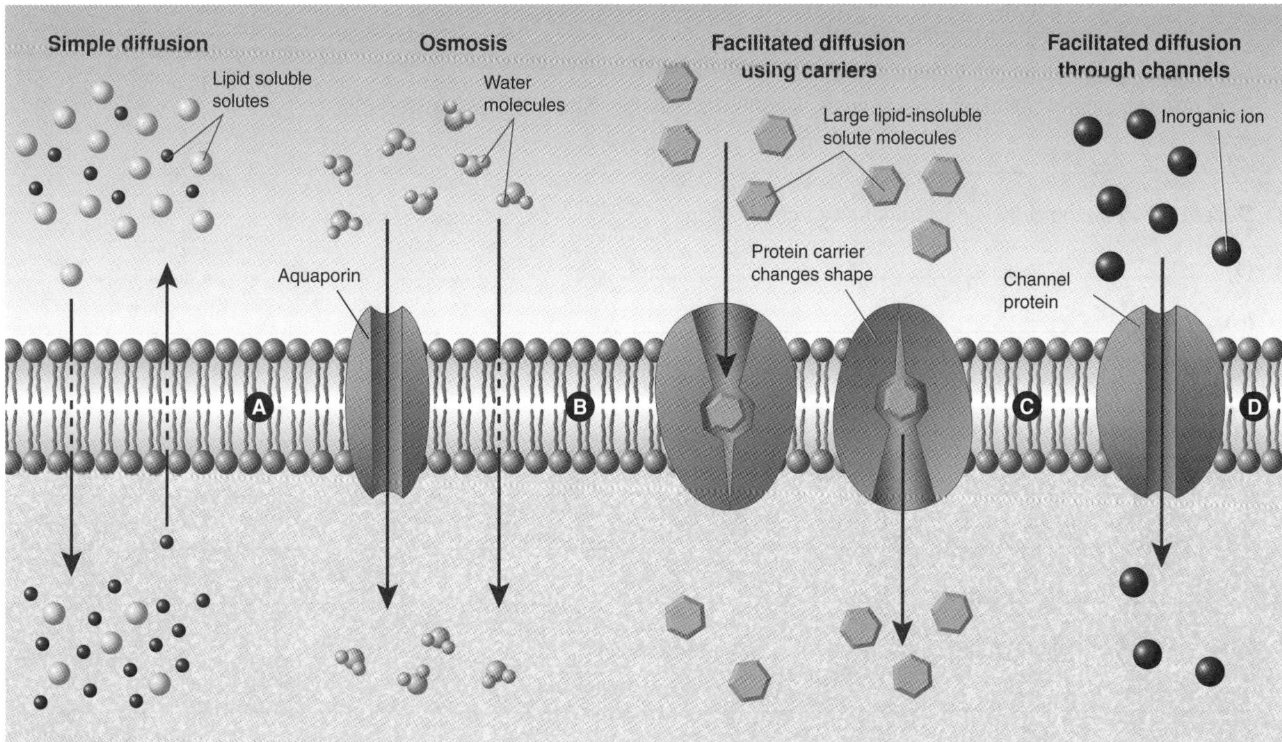
Simple diffusion
Lipid soluble solutes
Osmosis
Water molecules
Aquaporin
Facilitated diffusion using carriers
Large lipid-insoluble solute molecules
Protein carrier changes shape
Facilitated diffusion through channels
Inorganic ion
Channel protein
A
B
C
D

A: Some molecules (e.g. gases and lipid soluble molecules) diffuse directly across the plasma membrane. Two-way diffusion is common in biological systems, EXAMPLE: At the alveolar surface of the lung, oxygen diffuses into the blood and CO_2 diffuses out.

B: Osmosis is the diffusion of water across a selectively permeable membrane (in this case, the plasma membrane). Some water can diffuse directly through the lipid bilayer, but diffusion rate is increased by protein channels in the membrane called **aquaporins**.

C: A lipid-insoluble molecule is aided across the membrane by **carrier mediated facilitated diffusion**. This involves a trans-membrane carrier protein specific to the molecule being transported EXAMPLE: Glucose transport into red blood cells.

D: Small polar molecules and ions diffuse rapidly across the membrane by **channel-mediated facilitated diffusion**. Special channel proteins create hydrophilic pores that allow some solutes, usually inorganic ions, to pass through. EXAMPLE: Na^+ entering nerve cells.

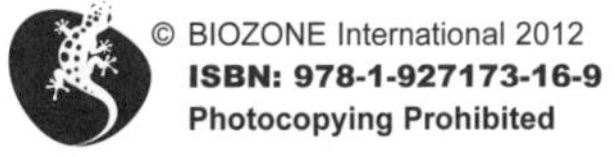

ISBN: 978-1-927173-16-9

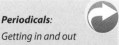

Periodicals: Getting in and out

Related activities: *Active and Passive Transport Summary*
Weblinks: *Cellular Transport*

Osmotic Gradients and Water Movement

Osmosis is the diffusion of water molecules, across a selectively permeable membrane, from higher to lower concentration of water molecules (sometimes described as from lower to higher solute concentration).Water always diffuses in this direction. The cytoplasm contains dissolved substances (**solutes**). When cells are placed in a solution of different concentration, there is an **osmotic gradient** between the external environment and the inside of the cell. In plant cells, the rigid cell wall is also important. When a plant cell takes up water, it swells until the cell contents exert a pressure on the cell wall. The cell wall is rigid and the pressure from the cytoplasm is called the wall pressure or **turgor pressure**. Turgor is important in plant support.

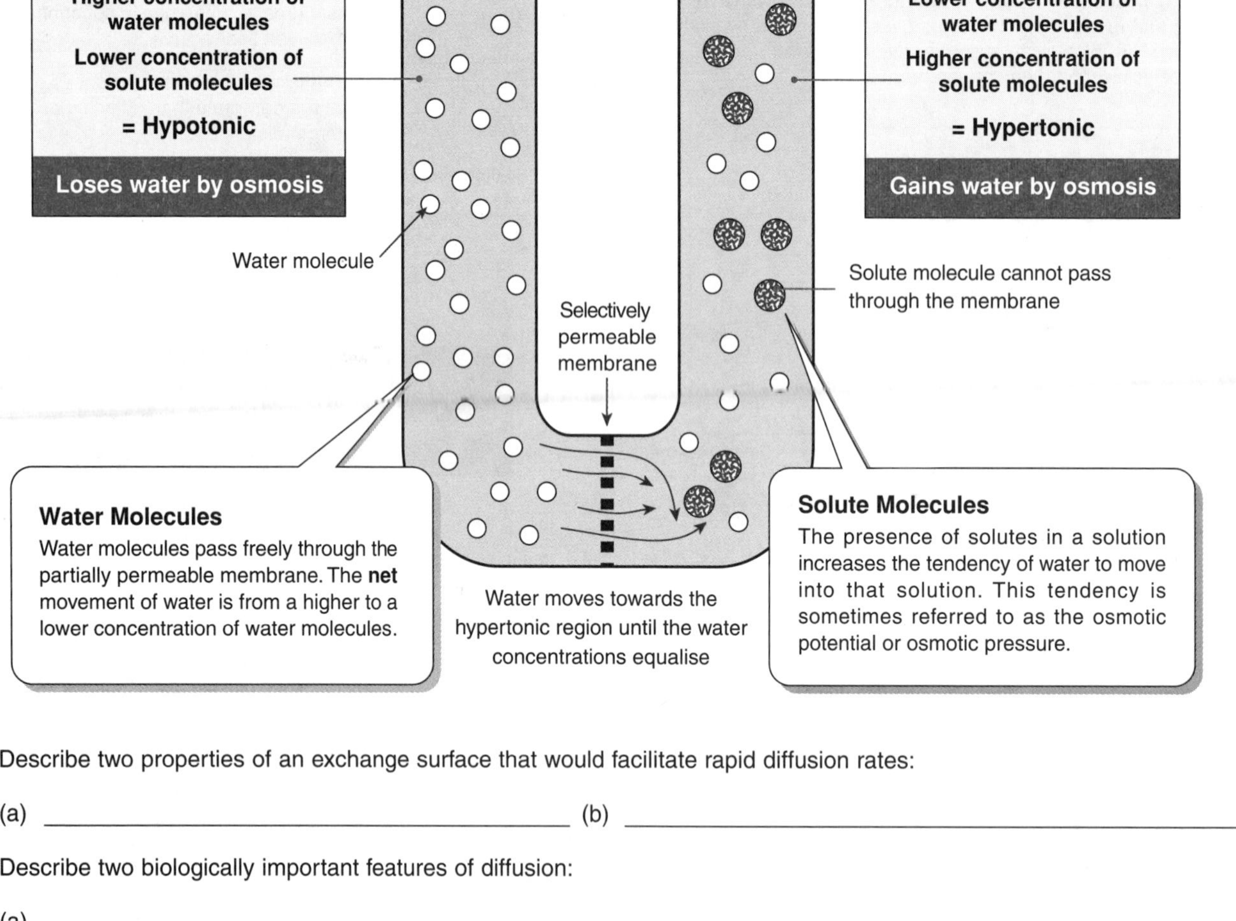

Higher concentration of water molecules
Lower concentration of solute molecules
= Hypotonic
Loses water by osmosis
Lower concentration of water molecules
Higher concentration of solute molecules
= Hypertonic
Gains water by osmosis
Water molecule
Solute molecule cannot pass through the membrane
Selectively permeable membrane
Water Molecules
Water molecules pass freely through the partially permeable membrane. The net movement of water is from a higher to a lower concentration of water molecules.
Water moves towards the hypertonic region until the water concentrations equalise
Solute Molecules
The presence of solutes in a solution increases the tendency of water to move into that solution. This tendency is sometimes referred to as the osmotic potential or osmotic pressure.

1. Describe two properties of an exchange surface that would facilitate rapid diffusion rates:

   (a) ______________________ (b) ______________________

2. Describe two biologically important features of diffusion:

   (a) ______________________

   (b) ______________________

3. Describe how facilitated diffusion is achieved for:

   (a) Small polar molecules and ions: ______________________

   (b) Glucose: ______________________

4. How are concentration gradients maintained across membranes? ______________________

5. Describe the role of aquaporins in the rapid movement of water through some cells: ______________________

6. (a) What happens if a cell takes up sucrose by active transport? ______________________

   (b) Describe a situation where this occurs in plants: ______________________

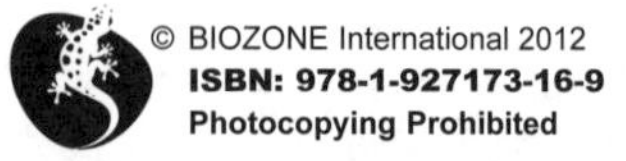

ISBN: 978-1-927173-16-9

Water Relations in Plant Cells

The plasma membrane of cells is a selectively permeable membrane and osmosis is the main way by which water enters and leaves the cell. When the external water concentration is the same as that of the cell there is no net movement of water. Two systems (cell and environment) with the same water concentration are termed isotonic. The diagram below illustrates two different situations: when the external water concentration is higher than the cell (**hypotonic**) and when it is lower than the cell (**hypertonic**).

Plasmolysis in a plant cell

Hypertonic salt solution

Water

Cell wall is freely permeable to water molecules.

Water concentration in the cell is higher than outside.

Cell contents more dilute than the external environment

Cytoplasm

Plasma membrane

In a **hypertonic** solution, the external water concentration is lower than the water concentration of the cell. Water leaves the cell and, because the cell wall is rigid, the cell membrane shrinks away from the cell wall. This process is termed **plasmolysis** and the cell becomes **flaccid** (turgor pressure = 0). Complete plasmolysis is irreversible; the cell cannot recover by taking up water.

Turgor in a plant cell

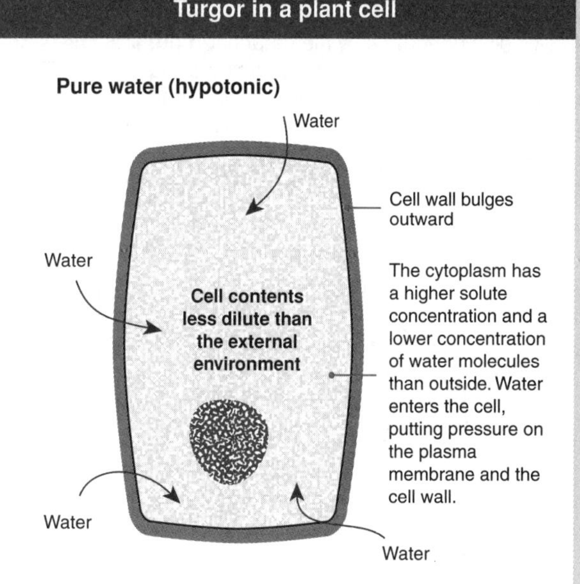
Pure water (hypotonic)
Water
Cell wall bulges outward
Water
Cell contents less dilute than the external environment
The cytoplasm has a higher solute concentration and a lower concentration of water molecules than outside. Water enters the cell, putting pressure on the plasma membrane and the cell wall.
Water
Water

In a **hypotonic** solution, the external water concentration is higher than the cell cytoplasm. Water enters the cell, causing it to swell tight. A wall (turgor) pressure is generated when enough water has been taken up to cause the cell contents to press against the cell wall. Turgor pressure rises until it offsets further net influx of water into the cell (the cell is turgid). The rigid cell wall prevents cell rupture.

7. Describe what would happen to an animal cell (e.g. a red blood cell) if it was placed into:

   (a) Pure water: __________

   (b) A hypertonic solution: __________

   (c) A hypotonic solution: __________

8. *Paramecium* is a freshwater protozoan. Describe the problem it has in controlling the amount of water inside the cell:

9. Fluid replacements are usually provided for heavily perspiring athletes after endurance events.

   (a) Identify the preferable tonicity of these replacement drinks (isotonic, hypertonic, or hypotonic): __________

   (b) Give a reason for your answer: __________

10. The malarial parasite lives in human blood. Relative to the tonicity of the blood, the parasite's cell contents would be hypertonic / isotonic / hypotonic (circle the correct answer).

11. (a) Explain the role of cell wall pressure in generating cell turgor in plants: __________

    (b) Discuss the role of cell turgor in plants: __________

ISBN: 978-1-927173-16-9

Ion Pumps

Diffusion alone cannot supply the cell's entire requirements for molecules (and ions). Some molecules (e.g. sucrose) are required by the cell in higher concentrations than occur outside the cell. Others (e.g. sodium) must be removed from the cell in order to maintain fluid balance. These molecules must be moved across the plasma membrane by **active transport** mechanisms, and this work is performed by membrane **ion pumps**. These are specific transport proteins in the membrane that utilize ATP to pump molecules from a low to a high concentration. The sodium-potassium pump is almost universal, and is often coupled to the cotransport of specific molecules such as glucose. Proton pumps can be utilized in the same way. The example below illustrates how plant cells use the gradient in hydrogen ion concentration to drive the active transport of sucrose. The transport protein couples the return of H^+ to the transport of sucrose into the phloem transfer cells (modified companion cells adjacent to the sieve tube cells). The sucrose rides with the H^+ as it diffuses down the concentration gradient maintained by the proton pump. From here, water follows by osmosis and generates a pressure-flow that moves plant sap by bulk flow in the phloem sieve tubes).

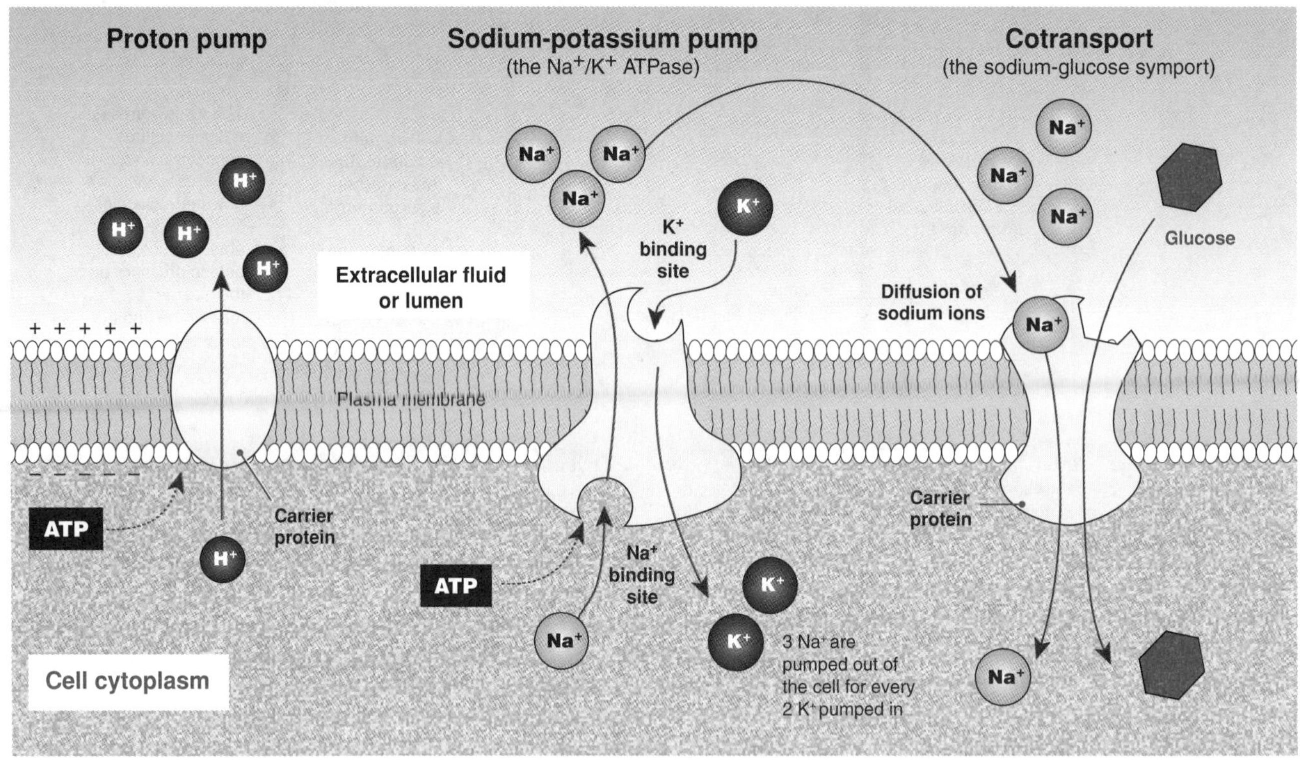
Proton pump
Sodium-potassium pump
(the Na+/K+ ATPase)
Cotransport
(the sodium-glucose symport)
H+
Na+
K+
K+ binding site
Extracellular fluid or lumen
Diffusion of sodium ions
Glucose
+ + + + +
Plasma membrane
– – – – –
ATP
Carrier protein
Na+ binding site
3 Na+ are pumped out of the cell for every 2 K+ pumped in
Cell cytoplasm

Proton pumps

ATP driven proton pumps use energy to remove hydrogen ions (H^+) from inside the cell to the outside. This creates a large difference in the proton concentration either side of the membrane, with the inside of the plasma membrane being negatively charged. This potential difference can be coupled to the transport of other molecules.

Sodium-potassium pump

The sodium-potassium pump is a specific protein in the membrane that uses energy in the form of ATP to exchange sodium ions (Na^+) for potassium ions (K^+) across the membrane. The unequal balance of Na^+ and K^+ across the membrane creates large concentration gradients that can be used to drive transport of other substances (e.g. cotransport of glucose).

Cotransport (coupled transport)

A gradient in sodium ions drives the active transport of **glucose** in intestinal epithelial cells. The specific transport protein couples the return of Na^+ down its concentration gradient to the transport of glucose into the intestinal epithelial cell. A low intracellular concentration of Na^+ (and therefore the concentration gradient) is maintained by a sodium-potassium pump.

1. Explain how the transport of molecules such as sucrose can be coupled to the activity of an ion pump: ______________

2. Why is ATP required for membrane pump systems to operate? ______________

3. Explain what is meant by cotransport: ______________

4. Describe two consequences of the extracellular accumulation of sodium ions: ______________

Related activities: *Active and Passive Transport Summary, Translocation* ***Weblinks:*** *Cellular Transport, Symport*

Periodicals: *How biological membranes achieve selective transport*

ISBN: 978-1-927173-16-9

Exocytosis and Endocytosis

Most cells carry out **cytosis**: a form of **active transport** involving the infolding or outfolding of the plasma membrane. The ability of cells to do this is a function of the flexibility of the plasma membrane. Cytosis results in bulk transport into or out of the cell and is achieved through the localized activity of microfilaments and microtubules in the cell cytoskeleton. **Endocytosis** involves material being engulfed. It typically occurs in protozoans and certain white blood cells of the mammalian defence system (phagocytes). **Exocytosis** is the reverse of endocytosis and involves the release of material from vesicles or vacuoles that have fused with the plasma membrane. Exocytosis is typical of cells that export material (secretory cells).

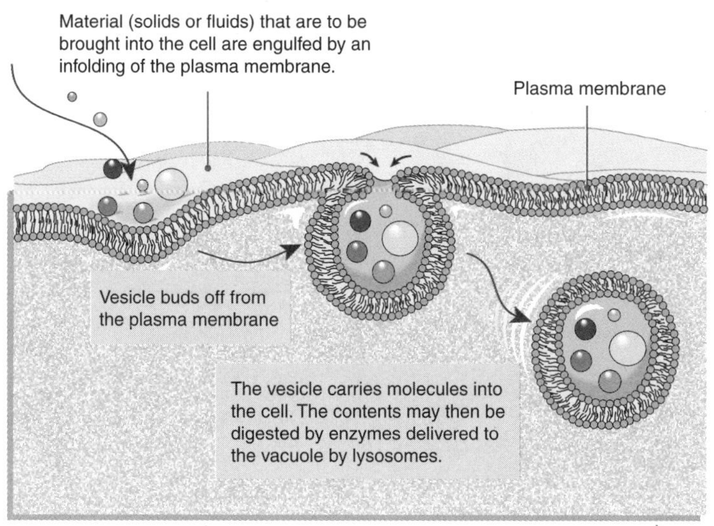
Material (solids or fluids) that are to be brought into the cell are engulfed by an infolding of the plasma membrane.
Plasma membrane
Vesicle buds off from the plasma membrane
The vesicle carries molecules into the cell. The contents may then be digested by enzymes delivered to the vacuole by lysosomes.

Endocytosis

Endocytosis (left) occurs by invagination (infolding) of the plasma membrane, which then forms vesicles or vacuoles that become detached and enter the cytoplasm. There are two main types of endocytosis:

Phagocytosis: 'cell-eating'
Phagocytosis involves the cell engulfing **solid material** to form large vesicles or vacuoles (e.g. food vacuoles). Examples: Feeding in *Amoeba*, phagocytosis of foreign material and cell debris by neutrophils and macrophages. Some endocytosis is **receptor mediated** and is triggered when receptor proteins on the extracellular surface of the plasma membrane bind to specific substances. Examples include the uptake of lipoproteins by mammalian cells.

Pinocytosis: 'cell-drinking'
Pinocytosis involves the non-specific uptake of **liquids** or fine suspensions into the cell to form small pinocytic vesicles. Pinocytosis is used primarily for absorbing extracellular fluid. Examples: Uptake in many protozoa, some cells of the liver, and some plant cells.

Both endocytosis and exocytosis require energy in the form of ATP

Areas of enlargement

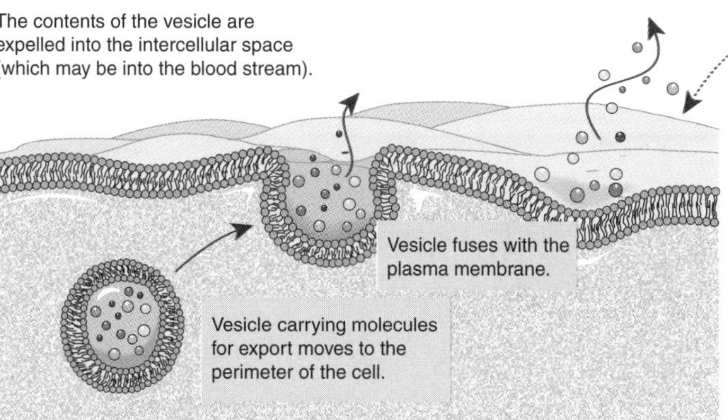
The contents of the vesicle are expelled into the intercellular space (which may be into the blood stream).
Vesicle fuses with the plasma membrane.
Vesicle carrying molecules for export moves to the perimeter of the cell.

Exocytosis

In multicellular organisms, several types of cells (e.g. lymphocytes) are specialized to manufacture and export products, such as proteins, from the cell to elsewhere in the body or outside it. Exocytosis (left) occurs by fusion of the vesicle membrane and the plasma membrane, followed by release of the vesicle's contents to the outside of the cell.

1. Distinguish between **phagocytosis** and **pinocytosis**: ______

2. Describe an example of phagocytosis and identify the cell type involved: ______

3. Describe an example of exocytosis and identify the cell type involved: ______

4. Why is cytosis affected by changes in oxygen level, whereas diffusion is not? ______

5. How does each of the following substances enter a living macrophage (for help, see *Passive Transport Processes*):

(a) Oxygen: ______ (c) Water: ______

(b) Cellular debris: ______ (d) Glucose: ______

Cell Processes

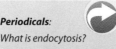

ISBN: 978-1-927173-16-9

Periodicals:
What is endocytosis?

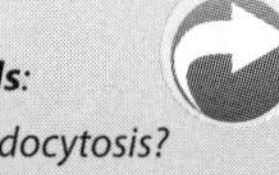

Related activities: *Passive Transport Processes, Active and Passive Transport Summary* ***Weblinks:*** *Cellular Transport*

Active and Passive Transport Summary

Cells have a need to move materials both into and out of the cell. Raw materials and other molecules necessary for metabolism must be accumulated from outside the cell. Some of these substances are scarce outside of the cell and some effort is required to accumulate them. Waste products and molecules for use in other parts of the body must be 'exported' out of the cell.

Some materials (e.g. gases and water) move into and out of the cell by **passive transport** processes, without the expenditure of energy on the part of the cell. Other molecules (e.g. sucrose) are moved into and out of the cell using **active transport**. Active transport processes involve the expenditure of energy in the form of ATP, and therefore use oxygen.

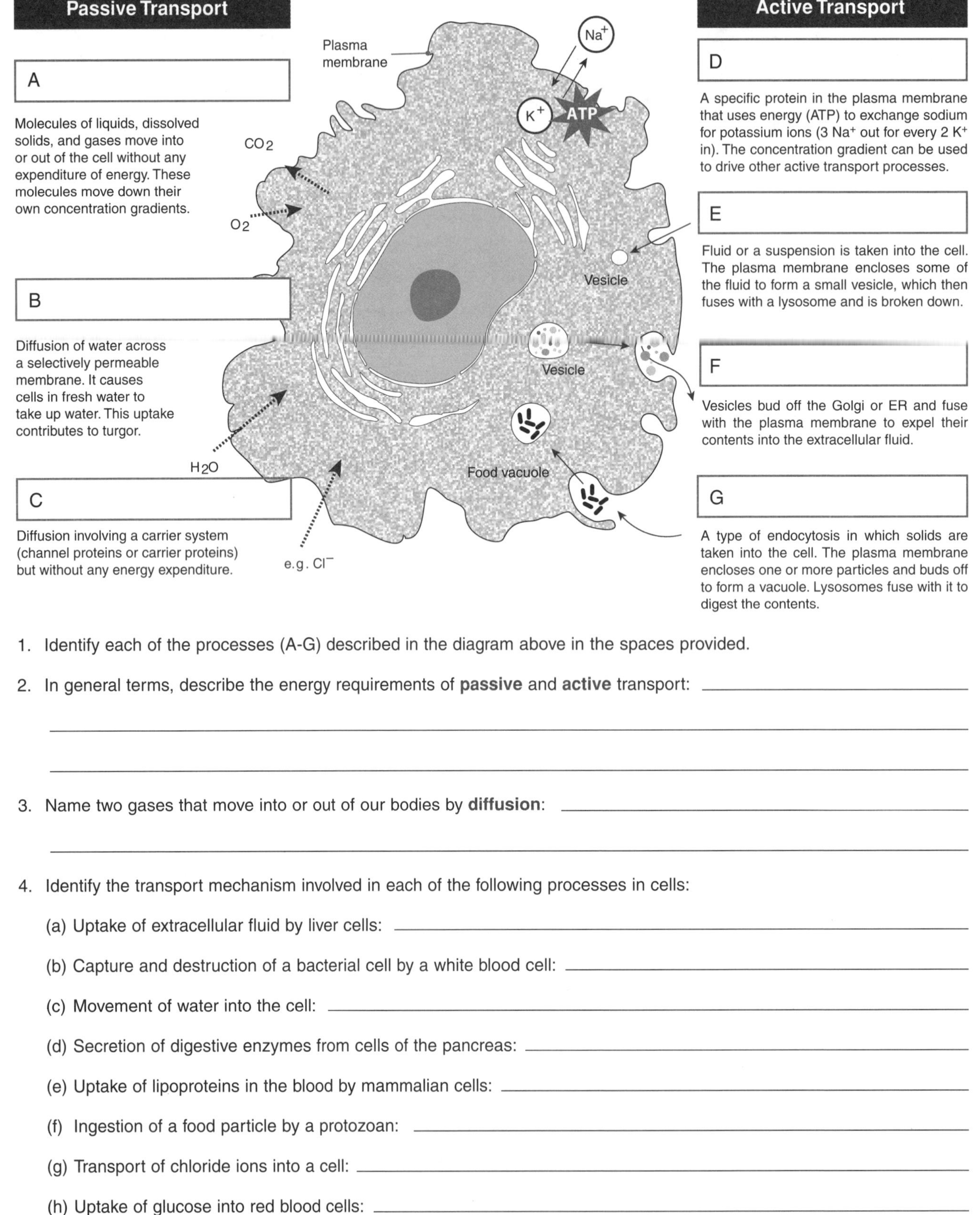

Passive Transport
A
Molecules of liquids, dissolved solids, and gases move into or out of the cell without any expenditure of energy. These molecules move down their own concentration gradients.
B
Diffusion of water across a selectively permeable membrane. It causes cells in fresh water to take up water. This uptake contributes to turgor.
C
Diffusion involving a carrier system (channel proteins or carrier proteins) but without any energy expenditure.
Plasma membrane
CO2
O2
H2O
e.g. Cl⁻
Na+
K+
ATP
Vesicle
Vesicle
Food vacuole
Active Transport
D
A specific protein in the plasma membrane that uses energy (ATP) to exchange sodium for potassium ions (3 Na+ out for every 2 K+ in). The concentration gradient can be used to drive other active transport processes.
E
Fluid or a suspension is taken into the cell. The plasma membrane encloses some of the fluid to form a small vesicle, which then fuses with a lysosome and is broken down.
F
Vesicles bud off the Golgi or ER and fuse with the plasma membrane to expel their contents into the extracellular fluid.
G
A type of endocytosis in which solids are taken into the cell. The plasma membrane encloses one or more particles and buds off to form a vacuole. Lysosomes fuse with it to digest the contents.

1. Identify each of the processes (A-G) described in the diagram above in the spaces provided.

2. In general terms, describe the energy requirements of **passive** and **active** transport: ____________________

3. Name two gases that move into or out of our bodies by **diffusion**: ____________________

4. Identify the transport mechanism involved in each of the following processes in cells:

   (a) Uptake of extracellular fluid by liver cells: ____________________

   (b) Capture and destruction of a bacterial cell by a white blood cell: ____________________

   (c) Movement of water into the cell: ____________________

   (d) Secretion of digestive enzymes from cells of the pancreas: ____________________

   (e) Uptake of lipoproteins in the blood by mammalian cells: ____________________

   (f) Ingestion of a food particle by a protozoan: ____________________

   (g) Transport of chloride ions into a cell: ____________________

   (h) Uptake of glucose into red blood cells: ____________________

   (i) Establishment of a potential difference across the membrane of a nerve cell:

Related activities: *Passive Transport Processes, Ion Pumps, Exocytosis and Endocytosis*
Weblinks: *Cellular Transport*

ISBN: 978-1-927173-16-9

Cell Division

The life cycle of a diploid sexually reproducing organism, such as a human, with **gametic meiosis** is illustrated below. In this life cycle, **gametogenesis** involves meiotic division to produce male and female gametes for the purpose of sexual reproduction. The life cycle in flowering plants is different in that the gametes are produced through mitosis in haploid gametophytes. The male gametes are produced inside the pollen grain and the female gametes are produced inside the embryo sac of the ovule. The gametophytes develop and grow from haploid spores, which are produced from meiosis.

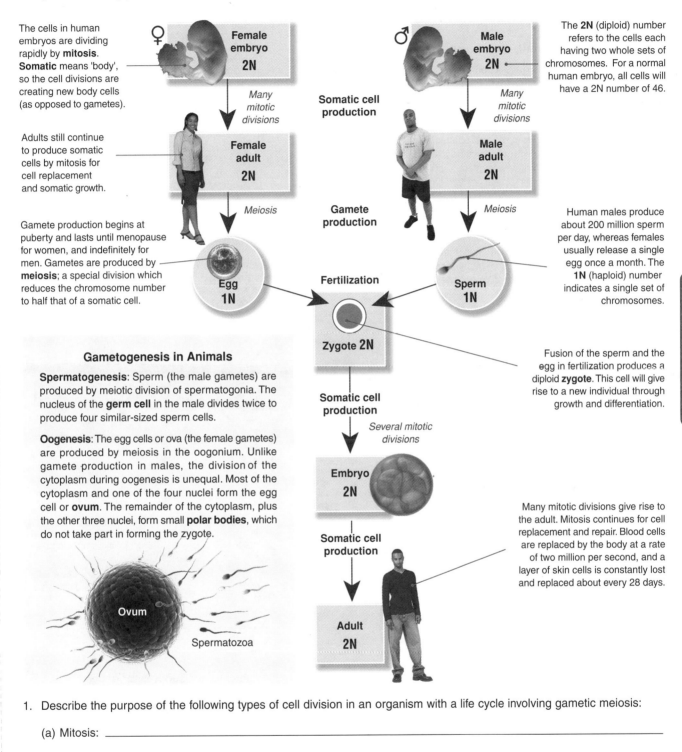

The cells in human embryos are dividing rapidly by **mitosis**. **Somatic** means 'body', so the cell divisions are creating new body cells (as opposed to gametes).
♀ Female embryo 2N
♂ Male embryo 2N
The **2N** (diploid) number refers to the cells each having two whole sets of chromosomes. For a normal human embryo, all cells will have a 2N number of 46.
Many mitotic divisions
Somatic cell production
Many mitotic divisions
Adults still continue to produce somatic cells by mitosis for cell replacement and somatic growth.
Female adult 2N
Male adult 2N
Meiosis
Gamete production
Meiosis
Gamete production begins at puberty and lasts until menopause for women, and indefinitely for men. Gametes are produced by **meiosis**; a special division which reduces the chromosome number to half that of a somatic cell.
Human males produce about 200 million sperm per day, whereas females usually release a single egg once a month. The **1N** (haploid) number indicates a single set of chromosomes.
Egg 1N
Fertilization
Sperm 1N
Zygote 2N
Fusion of the sperm and the egg in fertilization produces a diploid **zygote**. This cell will give rise to a new individual through growth and differentiation.
Somatic cell production
Several mitotic divisions
Embryo 2N
Many mitotic divisions give rise to the adult. Mitosis continues for cell replacement and repair. Blood cells are replaced by the body at a rate of two million per second, and a layer of skin cells is constantly lost and replaced about every 28 days.
Somatic cell production
Adult 2N

Gametogenesis in Animals

Spermatogenesis: Sperm (the male gametes) are produced by meiotic division of spermatogonia. The nucleus of the **germ cell** in the male divides twice to produce four similar-sized sperm cells.

Oogenesis: The egg cells or ova (the female gametes) are produced by meiosis in the oogonium. Unlike gamete production in males, the division of the cytoplasm during oogenesis is unequal. Most of the cytoplasm and one of the four nuclei form the egg cell or **ovum**. The remainder of the cytoplasm, plus the other three nuclei, form small **polar bodies**, which do not take part in forming the zygote.

Ovum
Spermatozoa

1. Describe the purpose of the following types of cell division in an organism with a life cycle involving gametic meiosis:

   (a) Mitosis: ________________

   (b) Meiosis: ________________

2. Describe the basic difference between the cell divisions involved in spermatogenesis and oogenesis:

   ____

   ____

3. How does gametogenesis differ between humans and flowering plants? ________________

   ____

ISBN: 978-1-927173-16-9

Related activities: *Mitosis and the Cell Cycle, Meiosis*

Cancer: Cells Out of Control

Normal cells do not live forever; they are programmed to die under certain circumstances, particularly during development. Cells that become damaged beyond repair will normally undergo this programmed cell death (called **apoptosis** or cell suicide). Cancer cells evade this control and become immortal, continuing to divide regardless of any damage incurred. **Carcinogens** are agents capable of causing cancer. Roughly 90% of carcinogens are also mutagens, i.e. they damage DNA. Chronic exposure to carcinogens accelerates the rate at which dividing cells make errors. Susceptibility to cancer is also influenced by genetic make-up. Any one or a number of cancer-causing factors (including defective genes) may interact to induce cancer.

Cancer: Cells out of Control

Cancerous transformation results from changes in the genes controlling normal cell growth and division. The resulting cells become immortal and no longer carry out their functional role. Two types of gene are normally involved in controlling the cell cycle: **proto-oncogenes**, which start the cell division process and are essential for normal cell development, and **tumor-suppressor genes**, which switch off cell division. In their normal form, both kinds of genes work as a team, enabling the body to perform vital tasks such as repairing defective cells and replacing dead ones. But mutations in these genes can disrupt these finely tuned checks and balances. Proto-oncogenes, through mutation, can give rise to **oncogenes**; genes that lead to uncontrollable cell division. Mutations to tumor-suppressor genes initiate most human cancers. The best studied tumor-suppressor gene is **p53**, which encodes a protein that halts the cell cycle so that DNA can be repaired before division.

The panel, right, shows the mutagenic action of some selected carcinogens on four of five codons of the **p53 gene**.

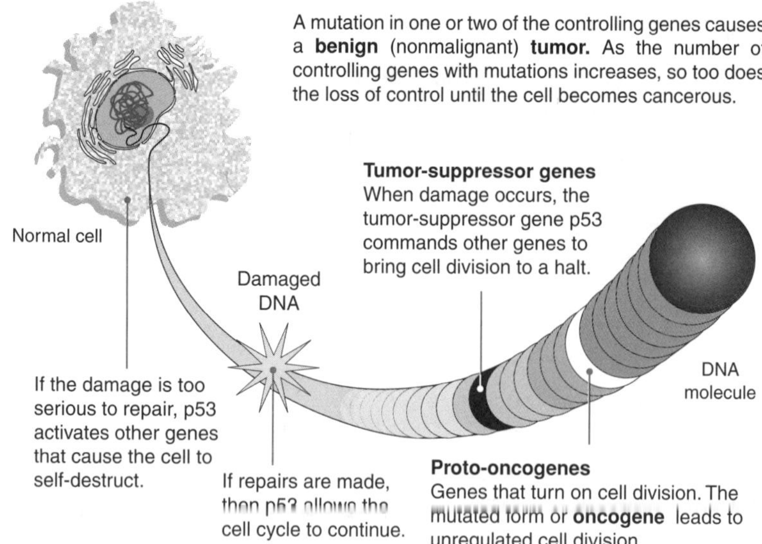

A mutation in one or two of the controlling genes causes a **benign** (nonmalignant) **tumor.** As the number of controlling genes with mutations increases, so too does the loss of control until the cell becomes cancerous.
Normal cell
Damaged DNA
Tumor-suppressor genes
When damage occurs, the tumor-suppressor gene p53 commands other genes to bring cell division to a halt.
DNA molecule
If the damage is too serious to repair, p53 activates other genes that cause the cell to self-destruct.
If repairs are made, then p53 allows the cell cycle to continue.
Proto-oncogenes
Genes that turn on cell division. The mutated form or **oncogene** leads to unregulated cell division.

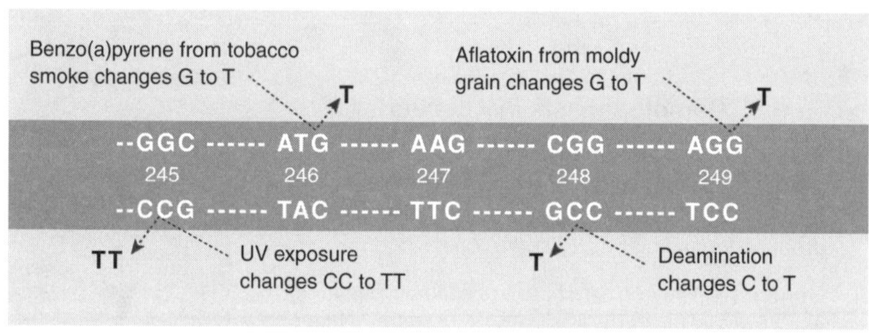

Benzo(a)pyrene from tobacco smoke changes G to T
T
Aflatoxin from moldy grain changes G to T
T
--GGC ------ ATG ------ AAG ------ CGG ------ AGG
245 246 247 248 249
--CCG ------ TAC ------ TTC ------ GCC ------ TCC
TT
UV exposure changes CC to TT
T
Deamination changes C to T

Features of Cancer Cells

The diagram right shows a single **lung cell** that has become cancerous. It no longer carries out the role of a lung cell, and instead takes on a parasitic lifestyle, taking from the body what it needs in the way of nutrients and contributing nothing in return. The rate of cell division is greater than in normal cells in the same tissue because there is no *resting phase* between divisions.

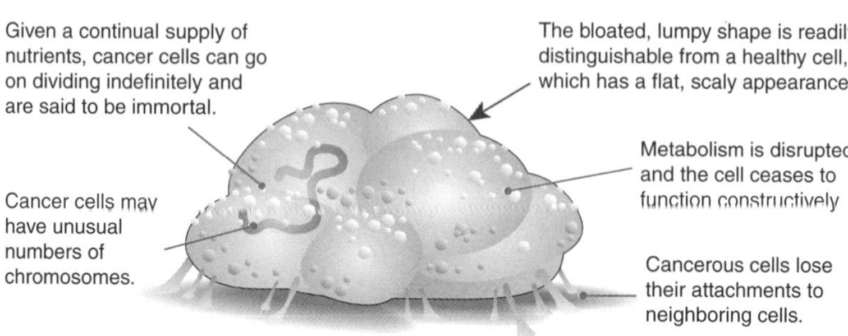

Given a continual supply of nutrients, cancer cells can go on dividing indefinitely and are said to be immortal.
The bloated, lumpy shape is readily distinguishable from a healthy cell, which has a flat, scaly appearance.
Metabolism is disrupted and the cell ceases to function constructively
Cancer cells may have unusual numbers of chromosomes.
Cancerous cells lose their attachments to neighboring cells.

1. How do cancerous cells differ from normal cells? ______________________________

____

2. Explain how the cell cycle is normally controlled, including reference to the role of **tumor-suppressor genes**:

____

____

____

____

3. With reference to the role of **oncogenes**, explain how the normal controls over the cell cycle can be lost:

____

____

____

ISBN: 978-1-927173-16-9

Mitosis and the Cell Cycle

Mitosis is part of the **cell cycle** in which an existing cell (the parent cell) divides into two (the daughter cells). Unlike meiosis, mitosis does not result in a change of chromosome numbers and the daughter cells are identical to the parent cell. Although mitosis is part of a continuous cell cycle, it is divided into stages (below). The example below illustrates the cell cycle in a plant cell. Note that **cytokinesis** in plant cells involves construction of a **cell plate** in the middle of the cell where Golgi vesicles release components for the construction of a new cell wall. In animal cells, cytokinesis involves the formation of a constriction that divides the cell in two. It is usually well underway by the end of telophase and does not involve the formation of a cell plate.

The Cell Cycle and Stages of Mitosis in Plants

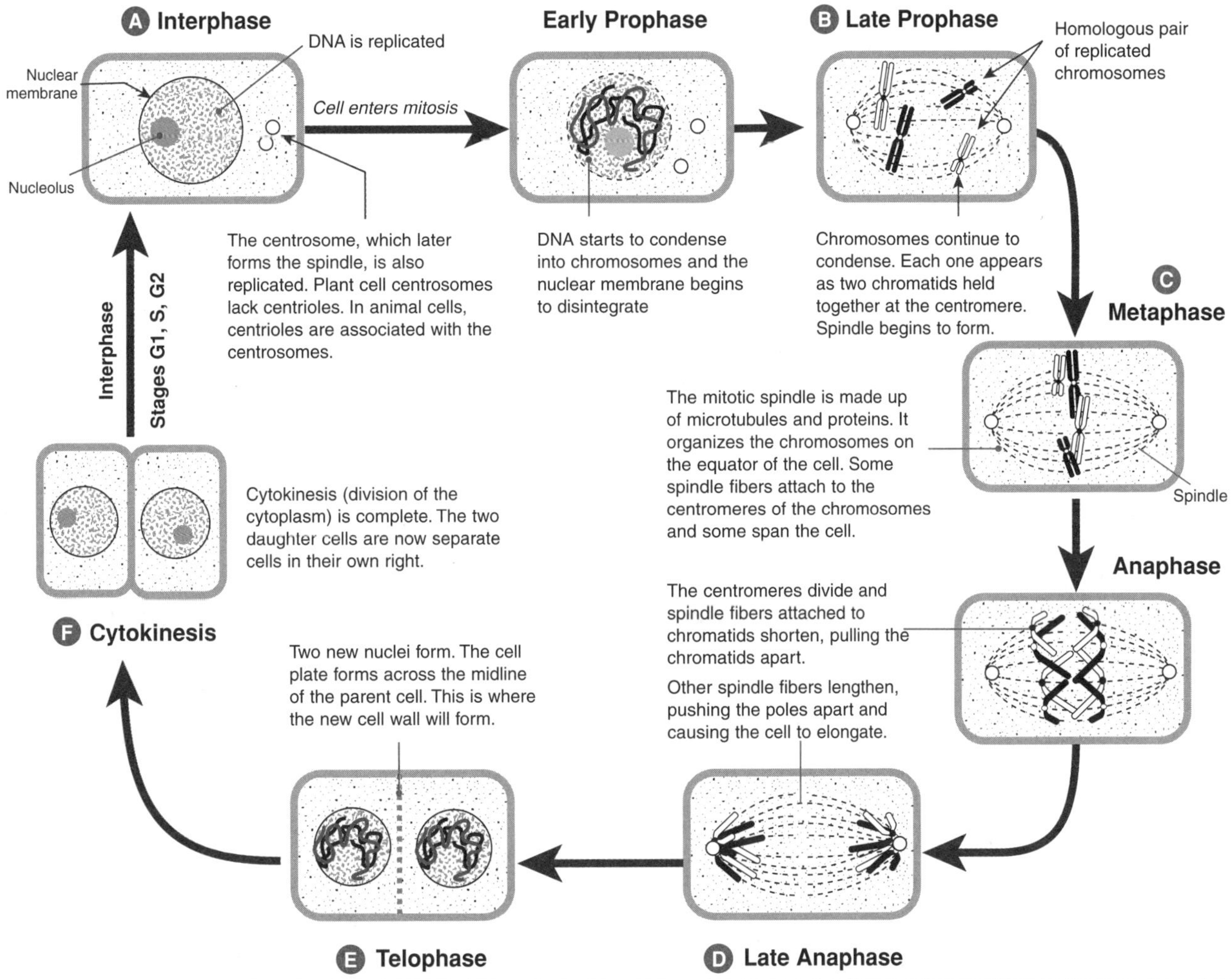

A Interphase

Nuclear membrane

Nucleolus

DNA is replicated

Cell enters mitosis

The centrosome, which later forms the spindle, is also replicated. Plant cell centrosomes lack centrioles. In animal cells, centrioles are associated with the centrosomes.

Early Prophase

DNA starts to condense into chromosomes and the nuclear membrane begins to disintegrate

B Late Prophase

Homologous pair of replicated chromosomes

Chromosomes continue to condense. Each one appears as two chromatids held together at the centromere. Spindle begins to form.

C Metaphase

The mitotic spindle is made up of microtubules and proteins. It organizes the chromosomes on the equator of the cell. Some spindle fibers attach to the centromeres of the chromosomes and some span the cell.

Spindle

Anaphase

The centromeres divide and spindle fibers attached to chromatids shorten, pulling the chromatids apart.

Other spindle fibers lengthen, pushing the poles apart and causing the cell to elongate.

D Late Anaphase

E Telophase

Two new nuclei form. The cell plate forms across the midline of the parent cell. This is where the new cell wall will form.

F Cytokinesis

Cytokinesis (division of the cytoplasm) is complete. The two daughter cells are now separate cells in their own right.

Interphase

Stages G1, S, G2

The Cell Cycle Overview

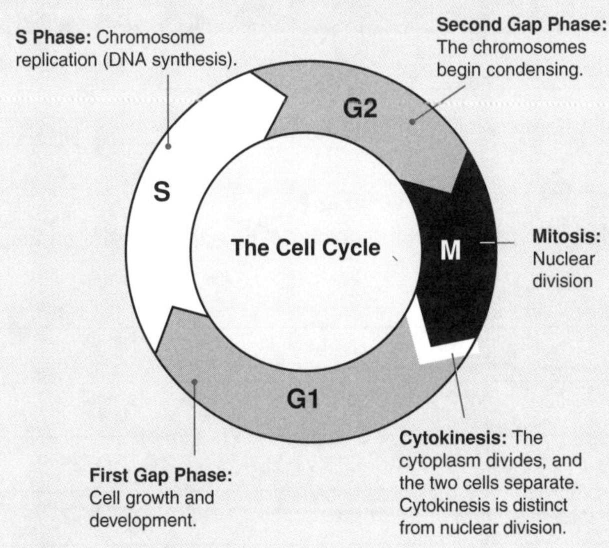

S Phase: Chromosome replication (DNA synthesis).
Second Gap Phase: The chromosomes begin condensing.
G2
S
The Cell Cycle
M
Mitosis: Nuclear division
G1
First Gap Phase: Cell growth and development.
Cytokinesis: The cytoplasm divides, and the two cells separate. Cytokinesis is distinct from nuclear division.

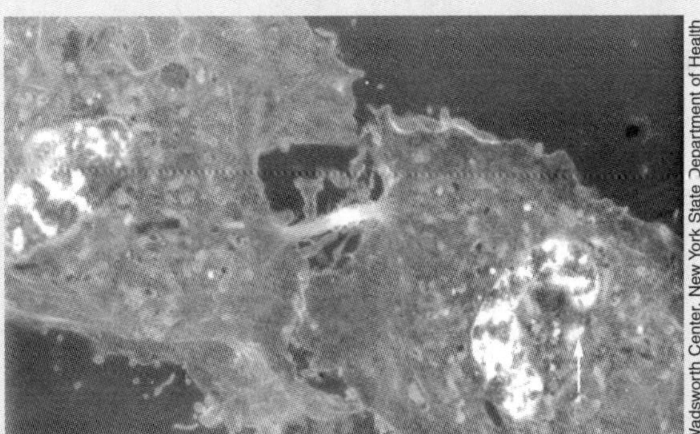

Wadsworth Center, New York State Department of Health

Animal cell cytokinesis (above) begins shortly after the sister chromatids have separated in anaphase of mitosis. A contractile ring of microtubular elements assembles in the middle of the cell, next to the plasma membrane, constricting it to form a **cleavage furrow**. In an energy-using process, the cleavage furrow moves inwards, forming a region of abscission where the two cells will separate. In the photograph above, an arrow points to a centrosome, which is still visible near the nucleus.

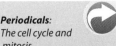

ISBN: 978-1-927173-16-9

Periodicals: *The cell cycle and mitosis*

Related activities: *Meiosis vs Mitosis*
Weblinks: *Mitosis in an Animal Cell*

Mitotic cell division has several purposes (below left). In multicellular organisms, mitosis repairs damaged cells and tissues, and produces the growth in an organism that allows it to reach its adult size. In unicellular organisms, and some small multicellular organisms, cell division allows organisms to reproduce asexually (as in the budding yeast cell cycle below).

The Functions of Mitosis

❶ Growth

In plants, cell division occurs in regions of **meristematic tissue**. In the plant root tip (right), the cells in the root apical meristem are dividing by mitosis to produce new cells. This elongates the root, resulting in **plant growth**.

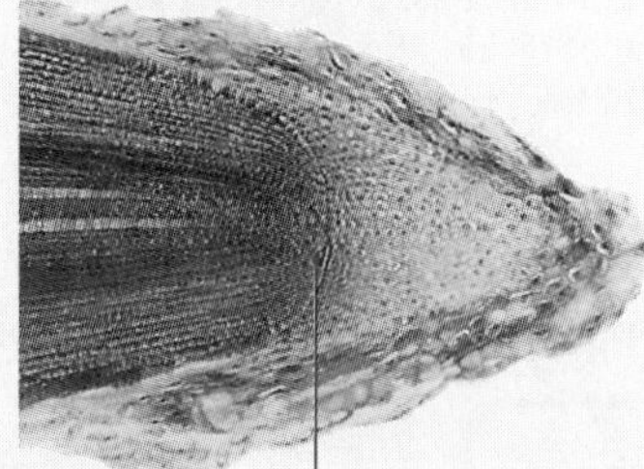

Root apical meristem

❷ Repair

Some animals, such as this skink (left), detach their limbs as a defense mechanism in a process called autotomy. The limbs can be **regenerated** via the mitotic process, although the tissue composition of the new limb differs slightly from that of the original.

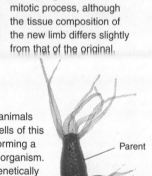

Photo: AB Sheldon

❸ Reproduction

Mitotic division enables some animals to reproduce **asexually**. The cells of this *Hydra* (left) undergo mitosis, forming a 'bud' on the side of the parent organism. Eventually the bud, which is genetically identical to its parent, detaches to continue the life cycle.

Parent

Bud

The Budding Yeast Cell Cycle

Yeasts can reproduce asexually through **budding**. In *Saccharomyces cerevisiae* (baker's yeast), budding involves mitotic division in the parent cell, with the formation of a daughter cell (or bud). As budding begins, a ring of chitin stabilizes the area where the bud will appear and enzymatic activity and turgor pressure act to weaken and extrude the cell wall. New cell wall material is incorporated during this phase. The nucleus of the parent cell also divides in two, to form a daughter nucleus, which migrates into the bud. The daughter cell is genetically identical to its parent cell and continues to grow, eventually separating from the parent cell.

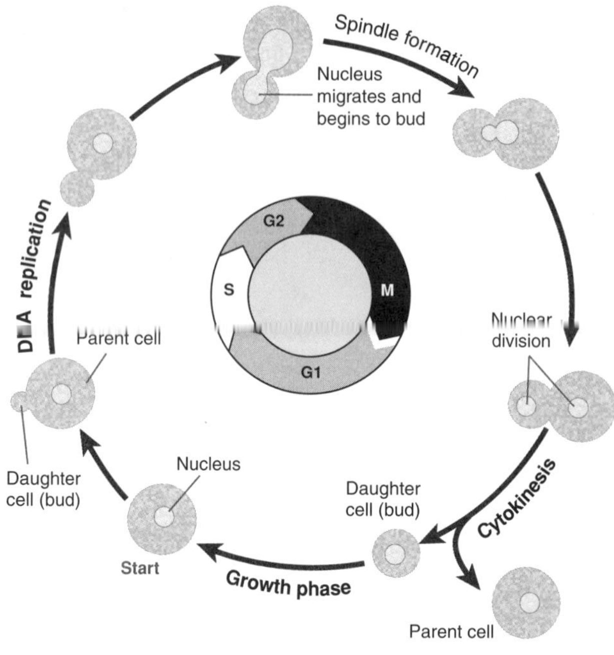
Spindle formation
Nucleus migrates and begins to bud
Nuclear division
Cytokinesis
Daughter cell (bud)
Parent cell
Growth phase
Start
Nucleus
Daughter cell (bud)
Parent cell
DNA replication
G2
S
M
G1

1. The photographs below were taken at various stages through mitosis in a plant cell. They are not in any particular order. Study the diagram on the previous page and determine the stage represented in each photograph (e.g. anaphase).

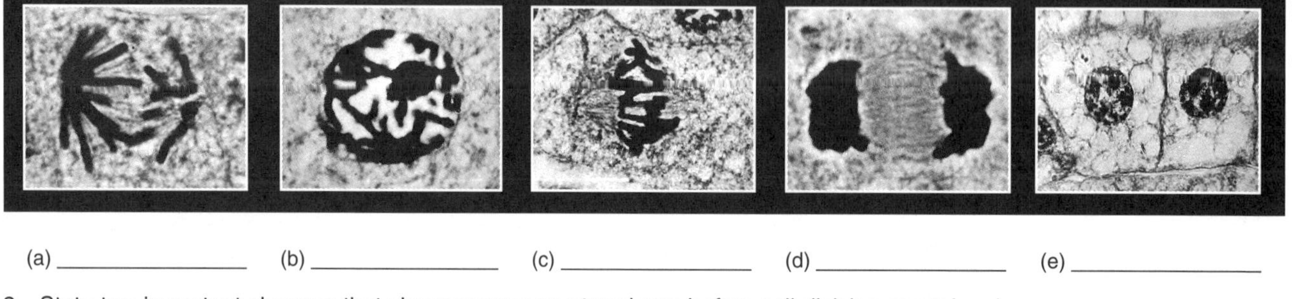

Photos: RCN

(a) ______ (b) ______ (c) ______ (d) ______ (e) ______

2. State two important changes that chromosomes must undergo before cell division can take place: ______

______

3. Briefly summarize the stages of the cell cycle by describing what is happening at the points (**A**-**F**) in the diagram on the previous page:

A. ______

B. ______

C. ______

D. ______

E. ______

F. ______

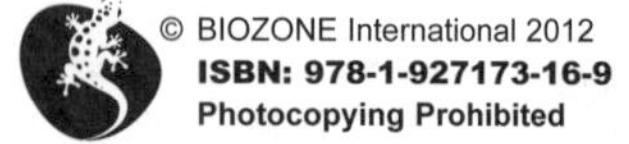

ISBN: 978-1-927173-16-9

Apoptosis: Programmed Cell Death

Apoptosis or programmed cell death (PCD) is a normal and necessary mechanism in multicellular organisms to trigger the death of a cell. Apoptosis has a number of crucial roles in the body, including the maintenance of adult cell numbers, and defence against damaged or dangerous cells, such as virus-infected cells and cells with DNA damage. Apoptosis also has a role in "sculpting" embryonic tissue during its development, e.g. in the formation of fingers and toes in a developing human embryo. Programmed cell death involves an orderly series of biochemical events that result in set changes in cell morphology and end in cell death. The process is carried out in such a way as to safely dispose of cell remains and fragments. This is in contrast to another type of cell death, called **necrosis**, in which traumatic damage to the cell results in spillage of cell contents. Apoptosis is tightly regulated by a balance between the factors that promote cell survival and those that trigger cell death. An imbalance between these regulating factors leads to defective apoptotic processes and is implicated in an extensive variety of diseases. For example, low rates of apoptosis result in uncontrolled proliferation of cells and cancers.

Stages in Apoptosis

Apoptosis is a normal cell suicide process in response to particular cell signals. It characterized by an overall compaction (shrinking) of the cell and its nucleus, and the orderly dissection of chromatin by endonucleases. Death is finalized by a rapid engulfment of the dying cell by phagocytosis. The cell contents remain membrane-bound and there is no inflammation.

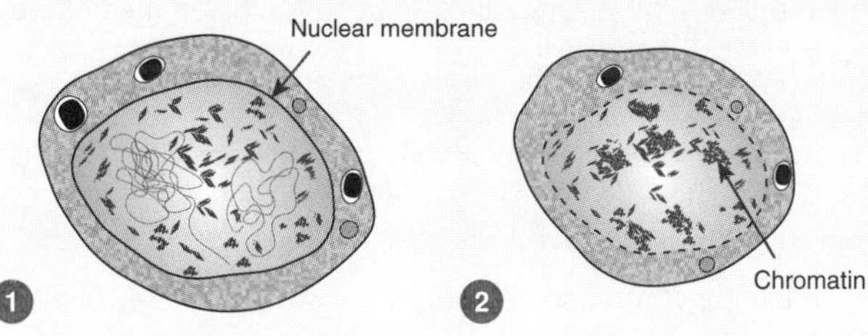
Nuclear membrane
Chromatin
1
2

1. The cell shrinks and loses contact with neighboring cells. The chromatin condenses and begins to degrade.

2. The nuclear membrane degrades. The cell loses volume. The chromatin clumps into **chromatin bodies**.

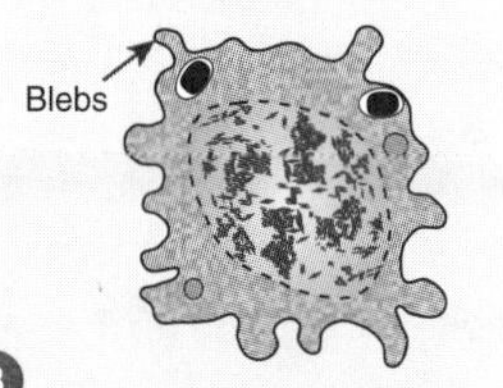
Blebs
3

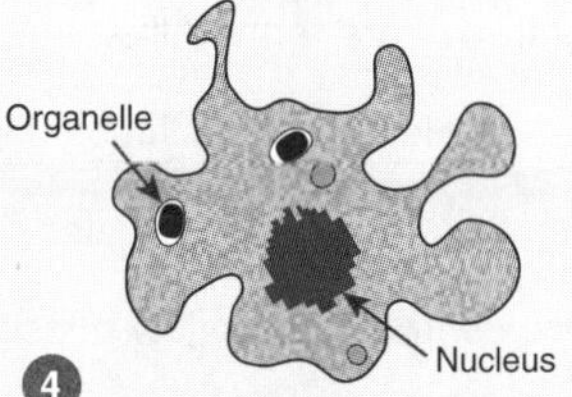
Organelle
Nucleus
4

3. **Zeiosis**: The plasma membrane forms bubble like **blebs** on its surface.

4. The nucleus collapses, but many membrane-bound organelles are not affected.

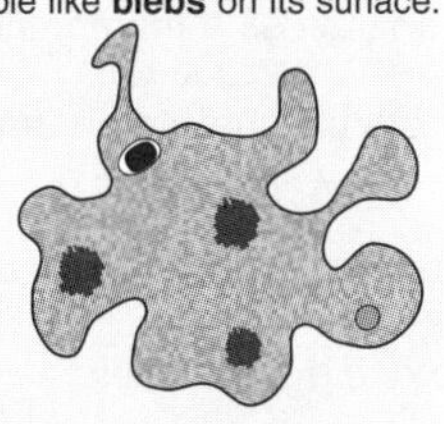
5

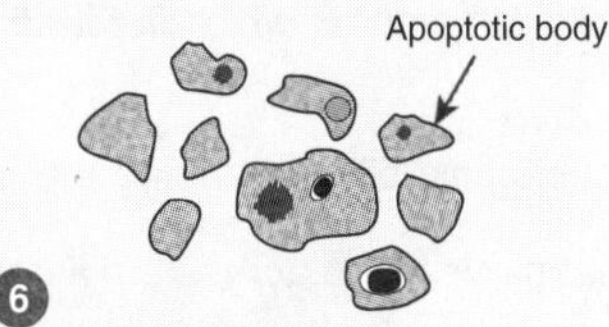
Apoptotic body
6

5. The nucleus breaks up into spheres and the DNA breaks up into small fragments.

6. The cell breaks into numerous **apoptotic bodies**, which are quickly resorbed by phagocytosis.

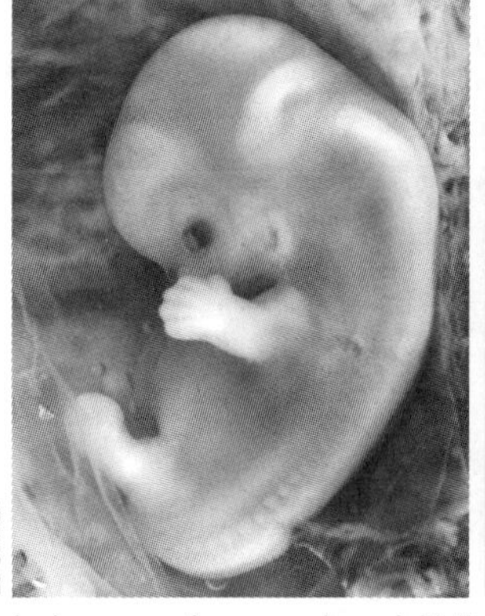

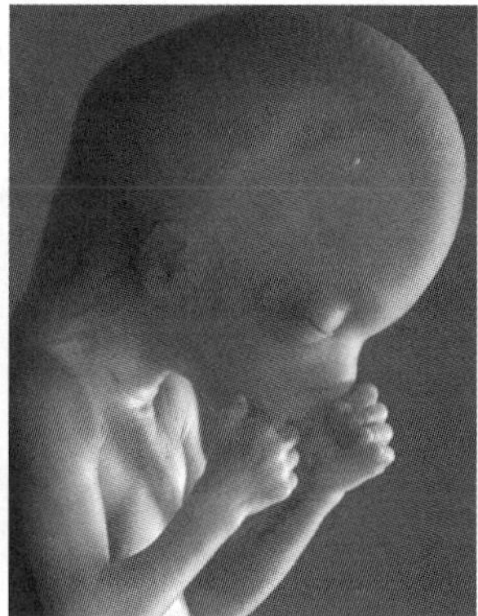

Ed Uthman

In humans, the mesoderm initially formed between the fingers and toes is removed by apoptosis. Forty one days after fertilization (top left), the digits of the hands and feet are webbed, making them look like small paddles. Apoptosis selectively destroys this superfluous webbing, sculpting them into digits when can be seen later in development (top right).

Cell Processes

Regulating Apoptosis

Apoptosis is a complicated and tightly controlled process, distinct from cell necrosis (uncontrolled cell death), when the cell contents are spilled. Apoptosis is regulated through both:

Positive signals, which prevent apoptosis and allow a cell to function normally. They include:

- interleukin-2
- bcl-2 protein and growth factors

Interleukin-2 is a positive signal for cell survival. Like other signaling molecules, it binds to cell surface receptors to regulate metabolism.

Negative signals (death activators), which trigger the changes leading to cell death. They include:

- inducer signals generated from within the cell itself in response to stress, e.g. DNA damage or cell starvation.
- signaling proteins and peptides such as lymphotoxin.

1. The photograph (right) shows a condition called syndactyly. Explain what might have happened during development to result in this condition:

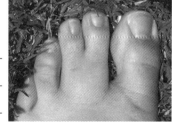

2. Describe one difference between apoptosis and necrosis:

3. Describe two situations, other than digit formation in development, in which apoptosis plays a crucial role:

(a)

(b)

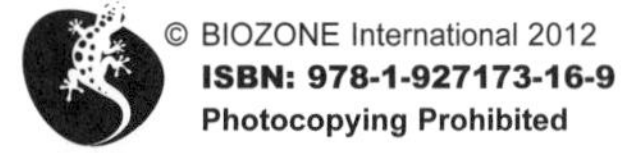

ISBN: 978-1-927173-16-9

Periodicals:
What is cell suicide?

Related activities: *Cancer: Cells Out of Control*
Weblinks: *Apoptosis: Dance of Death*

KEY TERMS: Mix and Match

INSTRUCTIONS: Test your vocabulary by matching each term to its definition, as identified by its preceding letter code.

active transport

amphipathic

anaphase

carrier protein

cell wall

channel protein

concentration gradient

diffusion

endocytosis

exocytosis

facilitated diffusion

fluid mosaic model

glycolipids

hypertonic

hypotonic

interphase

ion pump

isotonic

mitosis

metaphase

osmosis

partially permeable

passive transport

phagocytosis

pinocytosis

plasma membrane

plasmolysis

surface area: volume ratio

telophase

transmembrane protein

A Passive movement of water molecules across a selectively permeable membrane down a concentration gradient.

B The model for membrane structure which proposes a double phospholipid bilayer in which proteins and cholesterol are embedded.

C A type of passive transport, facilitated by transport proteins.

D Protein that spans the plasma membrane.

E The process in plant cells where the plasma membrane pulls away from the cell wall as a result of the loss of water through osmosis.

F The energy-requiring movement of substances across a biological membrane against a concentration gradient.

G A solution with lower solute concentration relative to another solution (across a membrane).

H Active transport in which molecules are engulfed by the plasma membrane, forming a phagosome or food vacuole within the cell.

I The gradual difference in the concentration of solutes as a function of distance through the solution.

J The last stage of mitosis, during which the chromosomes are at the poles and they begin to uncoil and become less condensed.

K Lipids with attached carbohydrates which serve as markers for cellular recognition.

L Solutions of equal solute concentration are termed this.

M This relationship determines capacity for effective diffusion in a cell.

N The uptake of liquids or fine suspensions by endocytosis.

O The passive movement of molecules from high to low concentration.

P The movement of substances across a biological membrane without energy expenditure.

Q A solution with higher solute concentration relative to another solution (across a membrane).

R A structure, present in plants and bacteria, which is found outside the plasma membrane and gives rigidity to the cell.

S A selectively-permeable phospholipid bilayer forming the boundary of all cells.

T A transmembrane protein that moves ions across a plasma membrane against their concentration gradient.

U A specific form of endocytosis involving the engulfment of solid particles by the plasma membrane.

V Protein that provides a channel through the plasma membrane for small polar molecules and ions.

W The stage of mitosis when chromosomes move to opposite poles of the cell

X Protein in the plasma membrane that facilitates the diffusion of a specific lipid insoluble molecule.

Y Active transport process by which membrane-bound secretory vesicles fuse with the plasma membrane and release the vesicle contents into the external environment.

Z A membrane that acts selectively to allow some substances, but not others, to pass.

AA Stage of mitosis in which chromosomes align in the middle of the cell.

BB Possessing both hydrophilic and hydrophobic (lipophilic) properties.

CC Phase of the cell cycle in which the cell spends most of its time and prepares for division.

DD The process by which a eukaryotic cell separates the chromosomes in its cell nucleus into two identical sets, in two separate nuclei.

ISBN: 978-1-927173-16-9

Core Topic
3
AHL Topic
7

The Chemistry of **Cells**

Key concepts

- Carbon, hydrogen, oxygen, and nitrogen are the most common elements in living systems.
- Water's properties make it essential to life.
- Proteins, carbohydrates, and lipids are three key groups of biological macromolecules.
- Condensation and hydrolysis reaction are important in building and breaking apart biological molecules.
- Proteins have a complex tertiary, and sometimes quaternary, structure, that determines function.

Key terms

Core

amino acid
carbohydrate
cellulose
condensation
disaccharide
fatty acids
fructose
galactose
glucose
glycogen
hydrolysis
inorganic compound
lactose
lipid
maltose
monosaccharide
organic compound
polypeptides
polysaccharide
protein
ribose
starch
sucrose
triglycerides
water

AHL only

non-polar amino acids
polar amino acids
primary structure
secondary structure
tertiary structure
prosthetic group
quaternary structure
globular protein
fibrous protein

Periodicals:
Listings for this chapter are on page 398

Weblinks:
www.thebiozone.com/weblink/IB-3169.html

BIOZONE APP:
Student Review Series
Molecules of Life

Learning Objectives

- ☐ 1. Use the **KEY TERMS** to compile a glossary for this topic.

Chemical Elements and Water *(3.1)* pages 74-76

- ☐ 2. Identify the most frequently occurring chemical elements in living systems and explain how they are used in the manufacture of **organic compounds**. Understand that other elements, including sulfur, calcium, phosphorus, iron, and sodium, are also required by living organisms and state at least one role for each.
- ☐ 3. Using an annotated diagram, describe the structure of **water**. Describe the physical and chemical properties of water that are important in biological systems, including its thermal, cohesive, and solvent properties.
- ☐ 4. Explain the relationship between water's properties and its functional roles in living things, e.g. as a coolant, and as a medium for metabolic reactions and transport.

Carbohydrates, Lipids, and Proteins *(3.2)* pages 77-82, 84

- ☐ 5. Distinguish between **organic** and **inorganic compounds**.
- ☐ 6. Identify **amino acids**, **glucose**, **ribose**, and **fatty acids** from diagrams, and describe the key features in their identification.
- ☐ 7. Identify three examples of **monosaccharides** and comment on their differences. Describe how **disaccharides** and **polysaccharides** are formed and broken apart by **condensation** and **hydrolysis**.
- ☐ 8. Describe examples of disaccharides, e.g. **maltose**, **lactose**, and **sucrose**.
- ☐ 9. Compare the structure of glucose polymers, **starch**, **cellulose**, and **glycogen**.
- ☐ 10. Describe one function of **glucose**, lactose, and glycogen in animals. Describe one function of **fructose**, sucrose, and cellulose in plants.
- ☐ 11. Describe three functions of **lipids** in biological systems. Describe how **triglycerides** are formed and broken apart by condensation and hydrolysis.
- ☐ 12. Explain how **polypeptides** are formed by condensation reactions between amino acids and how they are broken apart by hydrolysis.
- ☐ 13. Compare the use of carbohydrates and lipids in energy storage.

Protein Structure and Relationship to Function *(AHL: 7.5)* pages 83-86

- ☐ 14. Distinguish between a protein's **primary structure** and **secondary structure**. Explain the role of **hydrogen bonding** in the secondary structure.
- ☐ 15. Explain how the **tertiary structure** of a protein arises. Outline the differences between **fibrous** and **globular proteins**, with reference to two examples of each.
- ☐ 16. Using an example, e.g. hemoglobin, explain the **quaternary structure** of a protein.
- ☐ 17. Explain the significance of **polar** and **non-polar amino acids** in biological systems, e.g. in membranes and in enzymes.
- ☐ 18. Using examples, describe four functions of proteins in biological systems.

The Biochemical Nature of the Cell

Water is the main component of organisms, and provides an equable environment in which metabolic reactions can occur. Apart from water, most other substances in cells are compounds of **carbon**, **hydrogen**, **oxygen**, and **nitrogen**. The combination of carbon atoms with the atoms of other elements provides a huge variety of molecular structures, collectively called **organic molecules**. The organic molecules that make up living things can be grouped into four broad classes: carbohydrates, lipids, proteins, and nucleic acids. These are discussed in more detail in subsequent activities. In addition, a small number of **elements** and **inorganic ions** are also essential for life as components of larger molecules or extracellular fluids.

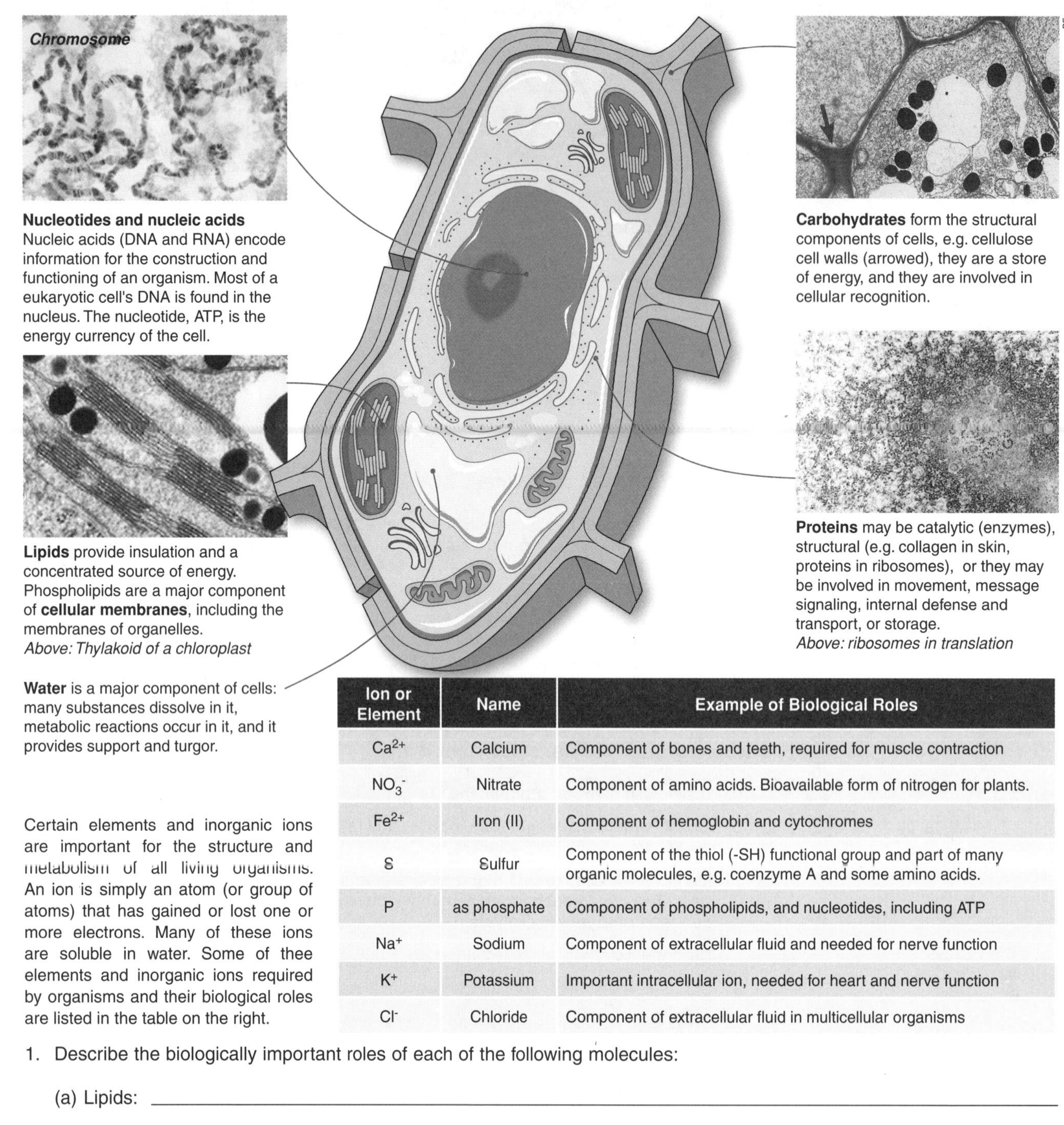
Chromosome
BF

Nucleotides and nucleic acids Nucleic acids (DNA and RNA) encode information for the construction and functioning of an organism. Most of a eukaryotic cell's DNA is found in the nucleus. The nucleotide, ATP, is the energy currency of the cell.

Carbohydrates form the structural components of cells, e.g. cellulose cell walls (arrowed), they are a store of energy, and they are involved in cellular recognition.

Lipids provide insulation and a concentrated source of energy. Phospholipids are a major component of **cellular membranes**, including the membranes of organelles.
Above: Thylakoid of a chloroplast

Proteins may be catalytic (enzymes), structural (e.g. collagen in skin, proteins in ribosomes), or they may be involved in movement, message signaling, internal defense and transport, or storage.
Above: ribosomes in translation

Water is a major component of cells: many substances dissolve in it, metabolic reactions occur in it, and it provides support and turgor.

Certain elements and inorganic ions are important for the structure and metabolism of all living organisms. An ion is simply an atom (or group of atoms) that has gained or lost one or more electrons. Many of these ions are soluble in water. Some of thee elements and inorganic ions required by organisms and their biological roles are listed in the table on the right.

Ion or Element	Name	Example of Biological Roles
Ca^{2+}	Calcium	Component of bones and teeth, required for muscle contraction
NO_3^-	Nitrate	Component of amino acids. Bioavailable form of nitrogen for plants.
Fe^{2+}	Iron (II)	Component of hemoglobin and cytochromes
S	Sulfur	Component of the thiol (-SH) functional group and part of many organic molecules, e.g. coenzyme A and some amino acids.
P	as phosphate	Component of phospholipids, and nucleotides, including ATP
Na^+	Sodium	Component of extracellular fluid and needed for nerve function
K^+	Potassium	Important intracellular ion, needed for heart and nerve function
Cl^-	Chloride	Component of extracellular fluid in multicellular organisms

1. Describe the biologically important roles of each of the following molecules:

   (a) Lipids: ___

   (b) Carbohydrates: ___

   (c) Proteins: ___

   (d) Nucleic acids: ___

2. Identify the most four most common chemical elements in living organisms: ___

3. Giving examples, describe the roles of some of the less common elements in living organisms: ___

Related activities: *Organic Molecules, The Role of Water*
Weblinks: *A Closer look at Water*

ISBN: 978-1-927173-16-9

Organic Molecules

Organic molecules are those chemical compounds containing carbon that are found in living things. Specific groups of atoms, called **functional groups**, attach to a carbon-hydrogen core and confer specific chemical properties on the molecule. Some organic molecules in organisms are small and simple, containing only one or a few functional groups, while others are large complex assemblies called **macromolecules**. The macromolecules that make up living things can be grouped into four classes: carbohydrates, lipids, proteins, and nucleic acids. An understanding of the structure and function of these molecules is necessary to many branches of biology, especially biochemistry, physiology, and molecular genetics. The diagram below illustrates some of the common ways in which biological molecules are portrayed. Note that the **molecular formula** expresses the number of atoms in a molecule, but does not convey its structure; this is indicated by the **structural formula**. Molecules can also be represented as **models**. A ball and stick model shows the arrangement and type of bonds while a space filling model gives a more realistic appearance of a molecule, showing how close the atoms really are.

Portraying Biological Molecules

The numbers next to the carbon atoms are used for identification when the molecule changes shape

$C_6H_{12}O_6$

Glucose

Molecular formula

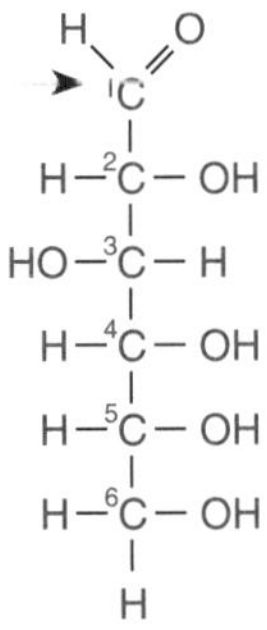

H–C=O (C1)
H–C2–OH
HO–C3–H
H–C4–OH
H–C5–OH
H–C6–OH
H

Structural formula
Glucose (straight form)

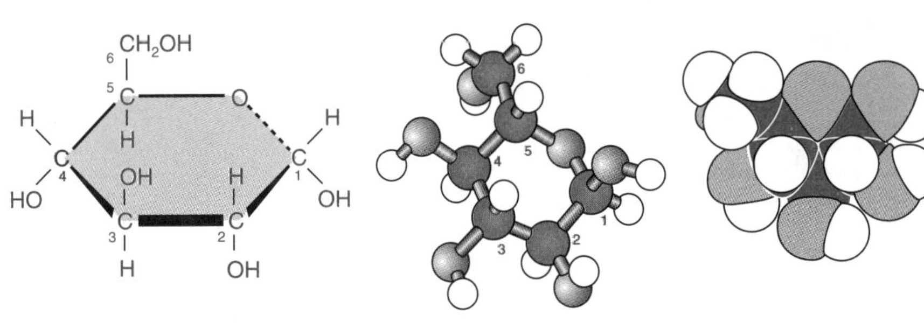

CH_2OH
6
5
4
3
2
1
H
O
HO
OH

Structural formula
α-glucose (ring form)

Ball and stick model
Glucose

Space filling model
β-D-glucose

Example of Biological Molecules

Biological molecules may also include atoms other than carbon, oxygen, and hydrogen atoms. Nitrogen and sulfur are components of molecules such as amino acids and nucleotides. Some molecules contain the C=O (carbonyl) group. If this group is joined to at least one hydrogen atom it forms an aldehyde. If it is located between two carbon atoms, it forms a ketone.

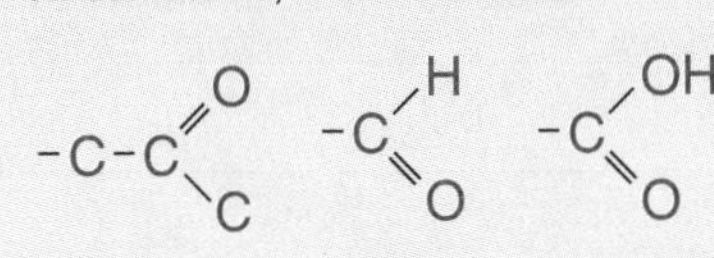

-C-C(=O)-C
-C(-H)=O
-C(-OH)=O

Ketone **Aldehyde** **Carboxyl**

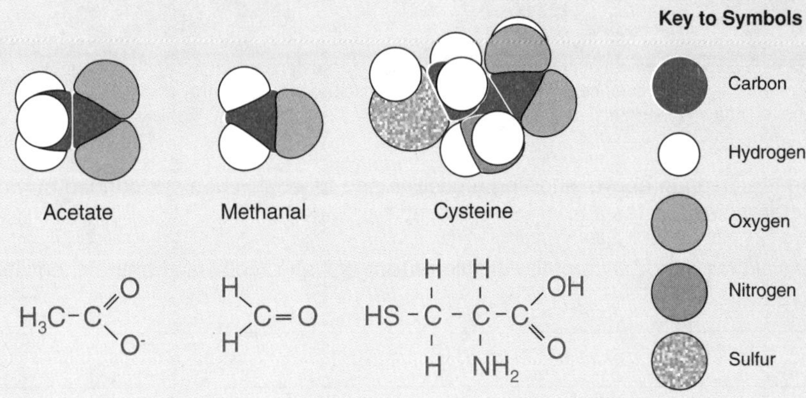

Acetate
Methanal
Cysteine
$H_3C-C(=O)-O^-$
$H_2C=O$
$HS-CH_2-CH(NH_2)-C(=O)OH$
Key to Symbols
Carbon
Hydrogen
Oxygen
Nitrogen
Sulfur

1. Identify the three main elements comprising the structure of organic molecules: ______

2. Name two other elements that are also frequently part of organic molecules: ______

3. State how many covalent bonds a carbon atom can form with neighboring atoms: ______

4. Distinguish between molecular and structural formulae for a given molecule: ______

______

______

5. Describe what is meant by a functional group: ______

______

6. Classify methanal according to the position of the C=O group: ______

7. Identify a functional group always present in amino acids: ______

8. Identify the significance of cysteine in its formation of disulfide bonds: ______

______

ISBN: 978-1-927173-16-9

Related activities: *The Biochemical Nature of the Cell, Amino Acids, Proteins*

The Role of Water

Water is the most abundant of the smaller molecules making up living things, and typically makes up about two-thirds of any organism. Water is a liquid at room temperature and many substances dissolve in it. It is a medium inside cells and for aquatic life. Water takes part in, and is a common product of, many reactions. Water molecules are **polar** and have a weak attraction for each other and inorganic ions, forming large numbers weak hydrogen bonds. It is this feature that gives water many of its unique properties, including its low viscosity and its chemical behavior as a **universal solvent**.

Important Properties of Water

A lot of energy is required before water will change state so aquatic environments are thermally stable and sweating and transpiration cause rapid cooling.

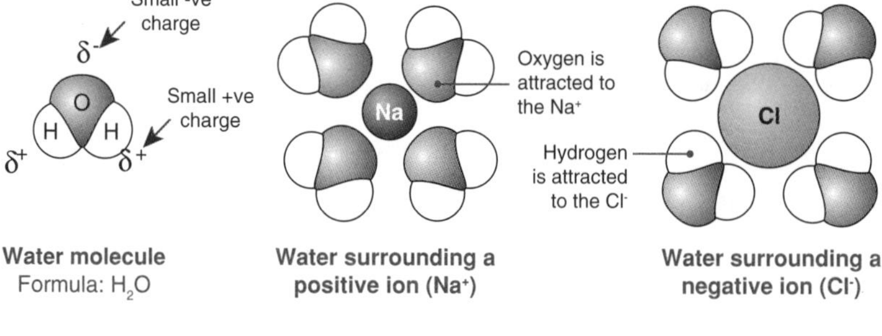

Small -ve charge
Small +ve charge
δ-
O
H
H
δ+
δ+
Na
Oxygen is attracted to the Na+
Hydrogen is attracted to the Cl-
Cl
Water molecule
Formula: H2O
Water surrounding a positive ion (Na+)
Water surrounding a negative ion (Cl-)

The most important feature of the chemical behavior of water is its dipole nature. It has a small positive charge on each of the two hydrogens and a small negative charge on the oxygen.

Water is colorless, with a high transmission of visible light, so light penetrates tissue and aquatic environments.

Ice is less dense than water. Consequently ice floats, insulating the underlying water and providing valuable habitat.

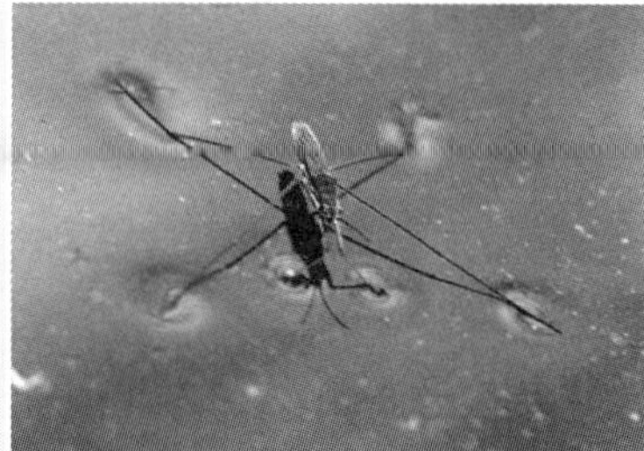

Water has low viscosity, strong cohesive properties, and high surface tension. It can flow freely through small spaces.

1. On the diagram above, showing a positive and a negative ion surrounded by water molecules, indicate the polarity of the water molecules (as shown in the example provided).

2. Explain the importance of the **dipole nature** of water molecules to the chemistry of life: ____________________

3. For (a)-(f), identify the important property of water, and describe an example of that property's biological significance:

(a) Property important in the clarity of seawater: __________

Biological significance: __________

(b) Property important in the transport of water in xylem: __________

Biological significance: __________

(c) Property important in the relatively stable temperature of water bodies: __________

Biological significance: __________

(d) Property important in the transport of glucose around the body: __________

Biological significance: __________

(e) Property important in the cooling effect of evaporation: __________

Biological significance: __________

(f) Property important in ice floating: __________

Biological significance: __________

ISBN: 978-1-927173-16-9

Monosaccharides and Disaccharides

Sugars (monosaccharides and disaccharides) are carbohydrates, which are a family of organic molecules with the general formula $C_m(H_2O)_n$. The most common arrangements found in sugars are hexose (6 sided) or pentose (5 sided) rings. Sugars play a central role in cells, providing energy and joining together to form carbohydrate macromolecules, such as starch and glycogen. Monosaccharide polymers form the major component of most plants (as cellulose). Monosaccharides are important as a primary energy source for cellular metabolism. Disaccharides (double-sugars) are important in human nutrition and are found in milk (lactose) table sugar (sucrose), and malt (maltose). They are formed from different combinations of monosaccharides.

Monosaccharides

Monosaccharides are used as a primary energy source for fuelling cell metabolism. They are single-sugar molecules and include glucose (grape sugar and blood sugar) and fructose (honey and fruit juices). The commonly occurring monosaccharides contain between three and seven carbon atoms in their carbon chains and, of these, the 6C hexose sugars occur most frequently. All monosaccharides are reducing sugars (i.e. they can participate in reduction reactions).

Single sugars (monosaccharides)

Triose

C
C
C

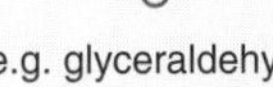

e.g. glyceraldehyde

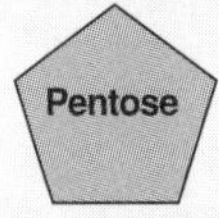
Pentose

e.g. ribose, deoxyribose

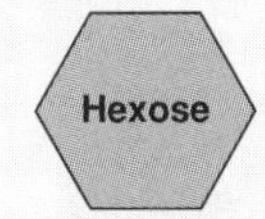
Hexose

e.g. glucose, fructose, galactose

Disaccharides

Disaccharides are double-sugar molecules and are used as energy sources and as building blocks for larger molecules. The type of disaccharide formed depends on the monomers involved and whether they are in their α- or β- form. Only a few disaccharides (e.g. lactose) are classified as reducing sugars.

Sucrose = α-glucose + β-fructose (simple sugar in plant sap)
Maltose = α-glucose + α-glucose (a product of starch hydrolysis)
Lactose = β-glucose + β-galactose (milk sugar)
Cellobiose = β-glucose + β-glucose (from cellulose hydrolysis)

Double sugars (disaccharides)

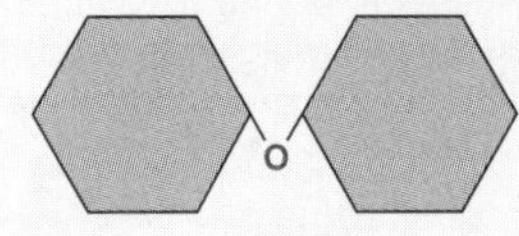
O

Examples sucrose, lactose, maltose, cellobiose

Lactose, a milk sugar, is made up of β-glucose + β-galactose. Milk contains 2-8% lactose by weight. It is the primary carbohydrate source for suckling mammalian infants.

Maltose is composed of two α-glucose molecules. These germinating wheat seeds contain maltose because the plant breaks down their starch stores to use it for food.

Sucrose (table sugar) is a simple sugar derived from plants such as sugar cane (above), sugar beet, or maple sap. It is composed of an α-glucose molecule and a β-fructose molecule.

1. Describe the two major functions of monosaccharides:

   (a) ______

   (b) ______

2. The breakdown of a disaccharide into its constituent monosaccharide units is an enzyme catalyzed hydrolysis (see next page). For each of the following common dissaccharides, identify the enzyme responsible for the catalysis and the products of the hydrolysis, and describe an example of where this enzyme might naturally occur:

   (a) Lactose: Enzyme: ______ Products of hydrolysis: ______

   Found: ______

   (b) Maltose: Enzyme: ______ Products of hydrolysis: ______

   Found: ______

   (c) Sucrose: Enzyme: ______ Products of hydrolysis: ______

   Found: ______

3. Use your understanding of disaccharide chemistry to suggest how the digestive disorder lactose intolerance arises:

______

______

______

ISBN: 978-1-927173-16-9

Periodicals: *Glucose & glucose containing carbohydrates*

Related activities: *Condensation and Hydrolysis of Sugars*
Weblinks: *Biomolecules: Carbohydrates*

Condensation and Hydrolysis of Sugars

Monomers are linked together by **condensation reactions**, so called because linking two units together results in the production of a water molecule. The reverse reaction, in which compound sugars are broken down into their constituent monosaccharides, is called **hydrolysis**. It splits polymers into smaller units by breaking the bond between two monomers. Hydrolysis literally means breaking with water, and so requires the addition of a water molecule to occur. Carbohydrates also exist as **isomers**. Isomers are compounds with the same molecular formula, but they have a different structural formula. Because of this they have different properties. For example, when α–glucose polymers are linked together they form starch, but β–glucose polymers form cellulose. In all carbohydrates, the structure is closely related to their functional properties.

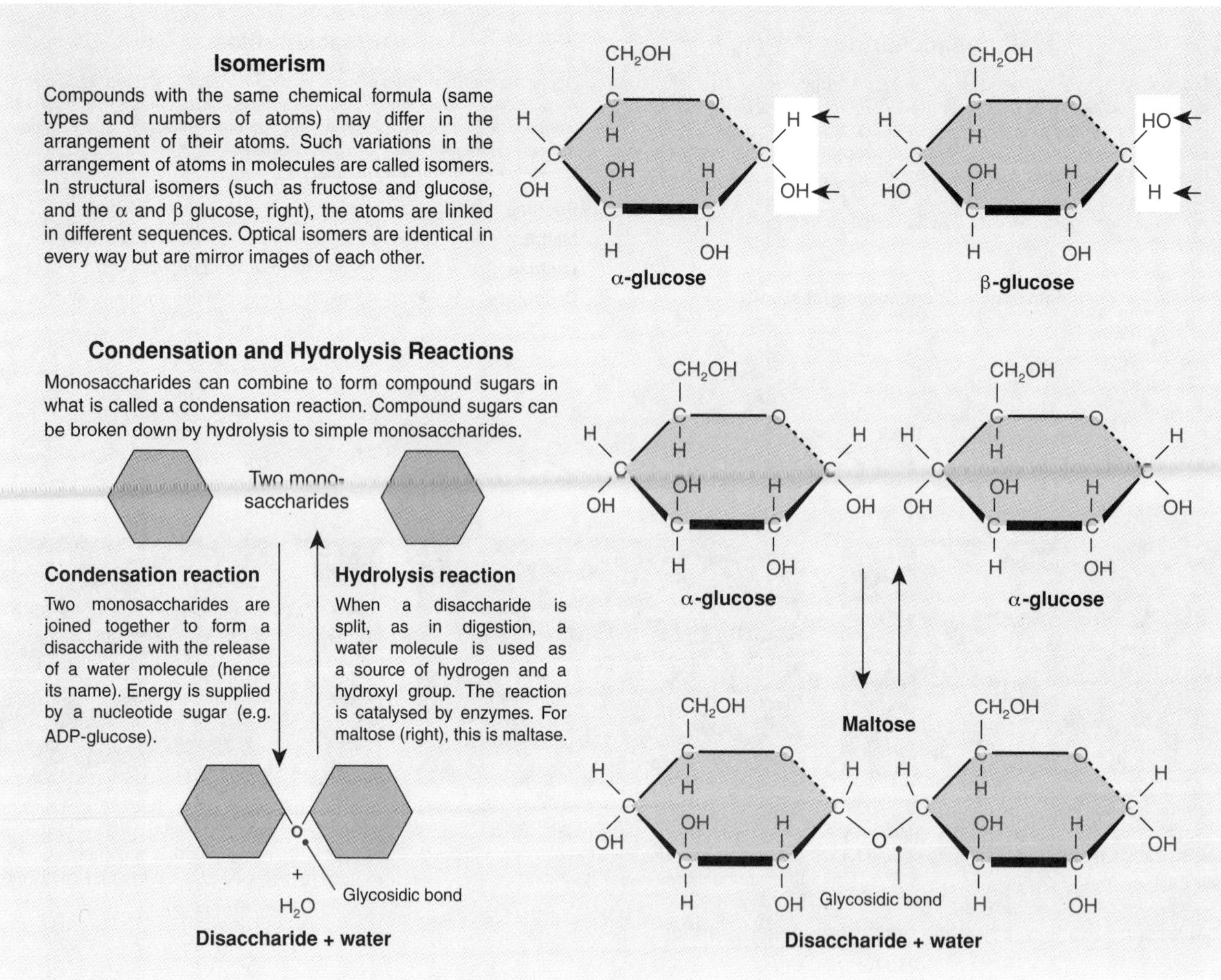
Isomerism
Compounds with the same chemical formula (same types and numbers of atoms) may differ in the arrangement of their atoms. Such variations in the arrangement of atoms in molecules are called isomers. In structural isomers (such as fructose and glucose, and the α and β glucose, right), the atoms are linked in different sequences. Optical isomers are identical in every way but are mirror images of each other.
α-glucose
β-glucose
Condensation and Hydrolysis Reactions
Monosaccharides can combine to form compound sugars in what is called a condensation reaction. Compound sugars can be broken down by hydrolysis to simple monosaccharides.
Two monosaccharides
Condensation reaction
Two monosaccharides are joined together to form a disaccharide with the release of a water molecule (hence its name). Energy is supplied by a nucleotide sugar (e.g. ADP-glucose).
Hydrolysis reaction
When a disaccharide is split, as in digestion, a water molecule is used as a source of hydrogen and a hydroxyl group. The reaction is catalysed by enzymes. For maltose (right), this is maltase.
α-glucose
α-glucose
Maltose
Glycosidic bond
Disaccharide + water
Glycosidic bond
Disaccharide + water

1. Distinguish between structural and optical isomers in carbohydrates, describing examples of each:

2. Explain briefly how compound sugars are formed and broken down:

3. Using examples, explain how the isomeric structure of a carbohydrate may affect its chemical behavior:

Related activities: Monosaccharides and Disaccharides, Polysaccharides
Weblinks: Condensation and Hydrolysis

ISBN: 978-1-927173-16-9

Polysaccharides

Polysaccharides or complex carbohydrates are straight or branched chains of many monosaccharides (sometimes many thousands) of the same or different types. The most common polysaccharides, cellulose, starch, and glycogen, contain only glucose, but their properties are very different. These differences are a function of the glucose isomer involved and the types of glycosidic linkages joining the glucose monomers. Different polysaccharides, based on the same sugar monomer, can thus be highly soluble and a source of readily available energy or a strong structural material that resists being digested.

Cellulose

Cellulose is a structural material in plants and is made up of unbranched chains of β-glucose molecules held together by 1,4 glycosidic links. As many as 10,000 glucose molecules may be linked together to form a straight chain. Parallel chains become cross-linked with hydrogen bonds and form bundles of 60-70 molecules called **microfibrils**. Cellulose microfibrils are very strong and are a major component of the structural components of plants, such as the cell wall (photo, right).

Starch

Starch is also a polymer of glucose, but it is made up of long chains of α-glucose molecules linked together. It contains a mixture of 25-30% amylose (unbranched chains linked by α-1,4 glycosidic bonds) and 70-75% amylopectin (branched chains with α-1, 6 glycosidic bonds every 24-30 glucose units). Starch is an energy storage molecule in plants and is found concentrated in insoluble starch granules within plant cells (see photo, right). Starch can be easily hydrolyzed by enzymes to soluble sugars when required.

Glycogen

Glycogen, like starch, is a branched polysaccharide. It is chemically similar to amylopectin, being composed of α-glucose molecules, but there are more α-1,6 glycosidic links mixed with α-1,4 links. This makes it more highly branched and water-soluble than starch. Glycogen is a storage compound in animal tissues and is found mainly in liver and muscle cells (photo, right). It is readily hydrolysed by enzymes to form glucose.

Chitin

Chitin is a tough modified polysaccharide made up of chains of β-glucose molecules. It is chemically similar to cellulose but each glucose has an amine group ($-NH_2$) attached. After cellulose, chitin is the second most abundant carbohydrate. It is found in the cell walls of fungi and is the main component of the exoskeleton of insects (right) and other arthropods.

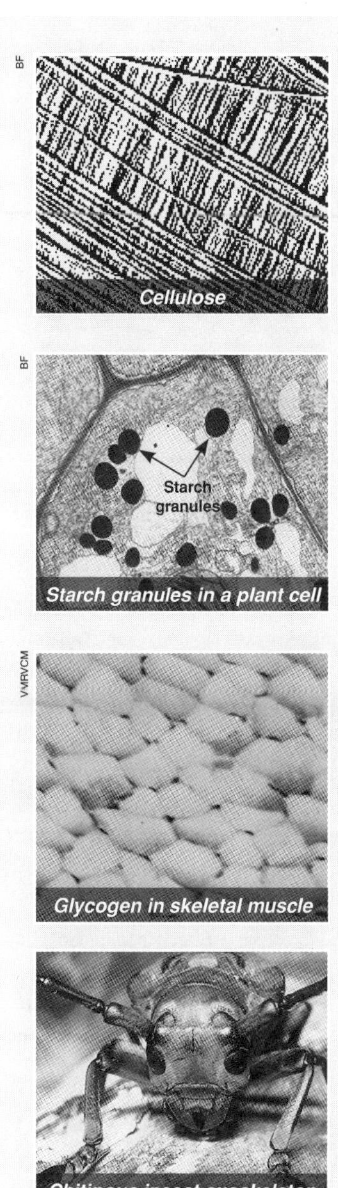
Starch granules

Cellulose

Starch granules in a plant cell

Glycogen in skeletal muscle

Chitinous insect exoskeleton

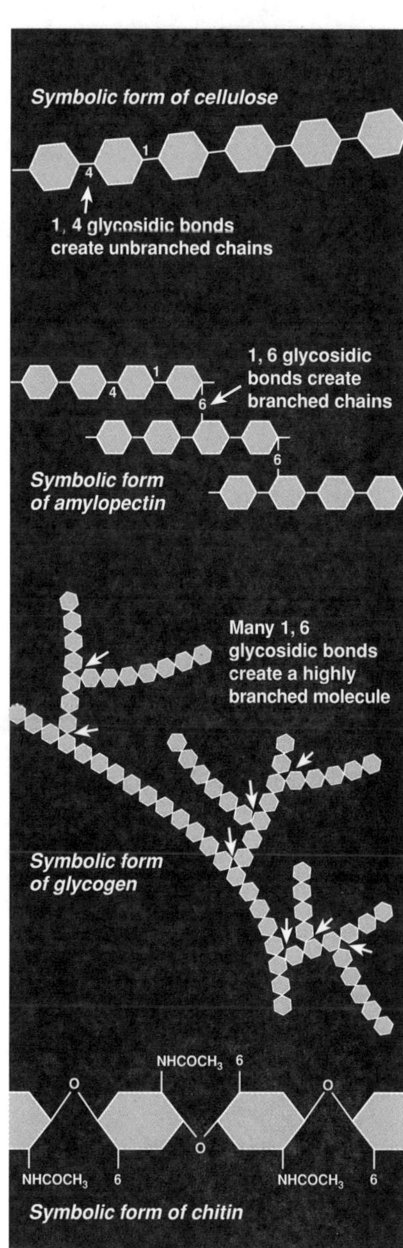
Symbolic form of cellulose
1
4
1, 4 glycosidic bonds create unbranched chains
1, 6 glycosidic bonds create branched chains
1
4
6
6
Symbolic form of amylopectin
Many 1, 6 glycosidic bonds create a highly branched molecule
Symbolic form of glycogen
NHCOCH3 6
O
O
O
NHCOCH3 6
NHCOCH3 6
Symbolic form of chitin

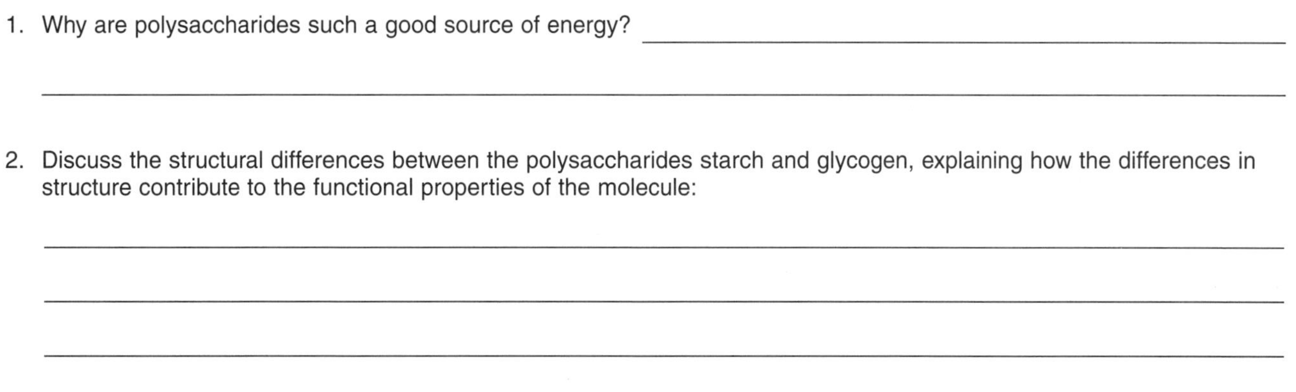

1. Why are polysaccharides such a good source of energy? ____

____

2. Discuss the structural differences between the polysaccharides starch and glycogen, explaining how the differences in structure contribute to the functional properties of the molecule:

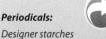

____

____

____

____

____

ISBN: 978-1-927173-16-9

Cellulose Structure and Function

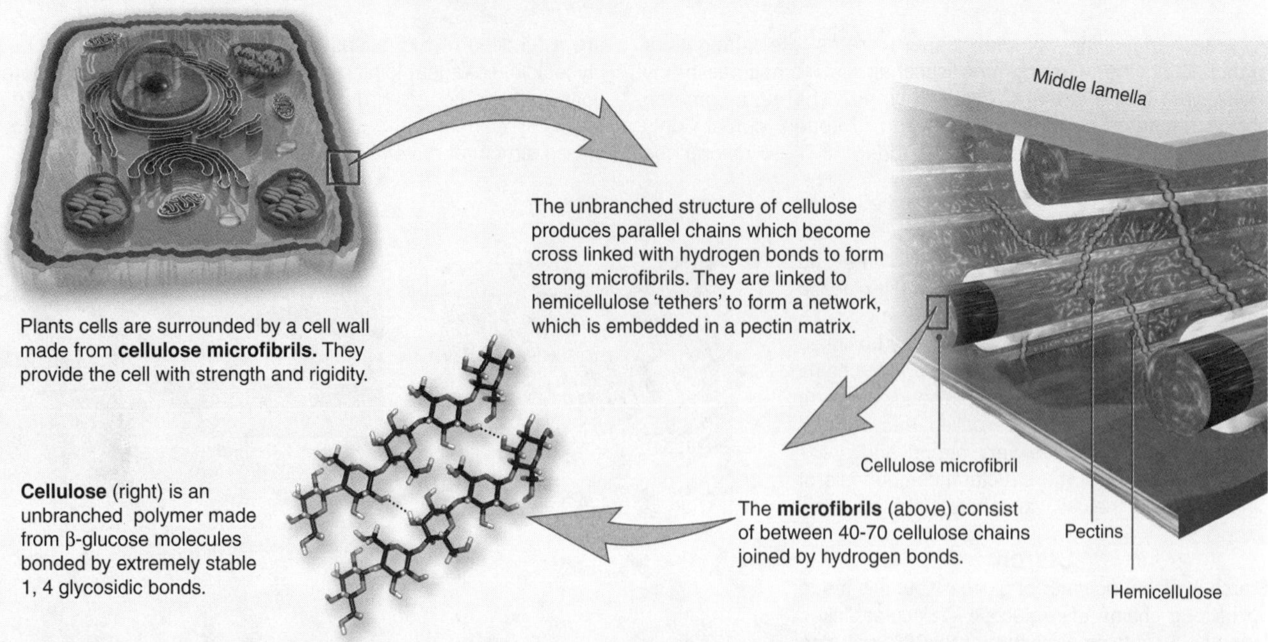
Middle lamella
Cellulose microfibril
Pectins
Hemicellulose

Plants cells are surrounded by a cell wall made from **cellulose microfibrils.** They provide the cell with strength and rigidity.

The unbranched structure of cellulose produces parallel chains which become cross linked with hydrogen bonds to form strong microfibrils. They are linked to hemicellulose 'tethers' to form a network, which is embedded in a pectin matrix.

Cellulose (right) is an unbranched polymer made from β-glucose molecules bonded by extremely stable 1, 4 glycosidic bonds.

The **microfibrils** (above) consist of between 40-70 cellulose chains joined by hydrogen bonds.

Starch Structure and Function

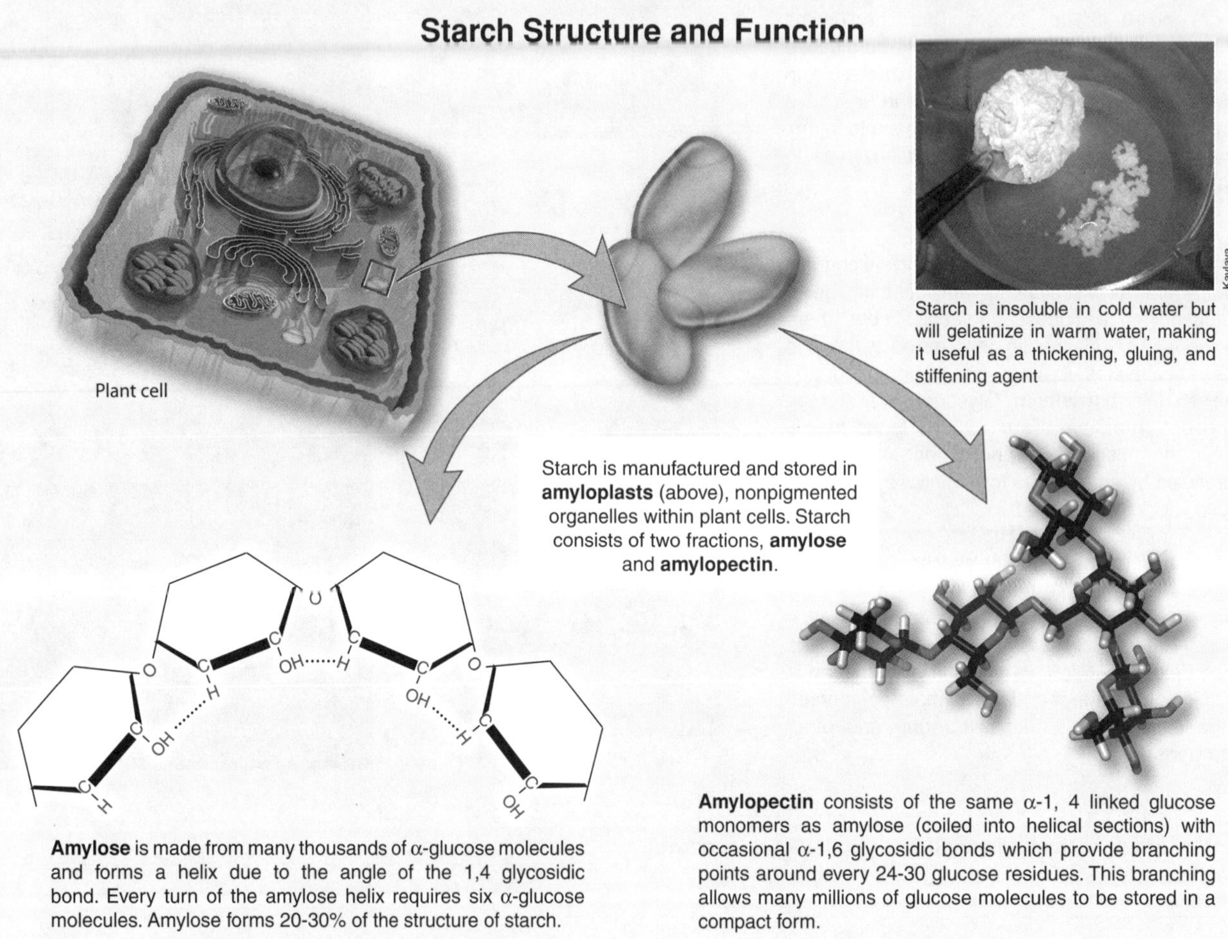
Plant cell
Kaylaya

Starch is insoluble in cold water but will gelatinize in warm water, making it useful as a thickening, gluing, and stiffening agent

Starch is manufactured and stored in **amyloplasts** (above), nonpigmented organelles within plant cells. Starch consists of two fractions, **amylose** and **amylopectin**.

Amylose is made from many thousands of α-glucose molecules and forms a helix due to the angle of the 1,4 glycosidic bond. Every turn of the amylose helix requires six α-glucose molecules. Amylose forms 20-30% of the structure of starch.

Amylopectin consists of the same α-1, 4 linked glucose monomers as amylose (coiled into helical sections) with occasional α-1,6 glycosidic bonds which provide branching points around every 24-30 glucose residues. This branching allows many millions of glucose molecules to be stored in a compact form.

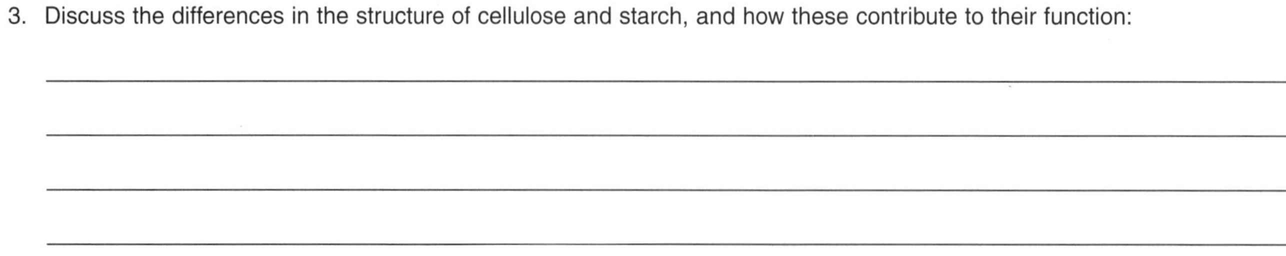

3. Discuss the differences in the structure of cellulose and starch, and how these contribute to their function:

ISBN: 978-1-927173-16-9

Lipids

Lipids are a group of organic compounds with an oily, greasy, or waxy consistency. They are relatively insoluble in water and tend to be water-repelling (e.g. cuticle on leaf surfaces). Lipids are important biological fuels, some are hormones, and some serve as structural components in plasma membranes. Proteins and carbohydrates may be converted into fats by enzymes and stored within cells of adipose tissue. During times of plenty, this store is increased, to be used during times of food shortage.

Neutral Fats and Oils

The most abundant lipids in living things are neutral fats. They make up the fats and oils found in plants and animals. Fats are an economical way to store fuel reserves because they yield more than twice as much energy as the same quantity of carbohydrate. **Neutral fats** are composed of a glycerol molecule attached to one (monoglyceride), two (diglyceride) or three (triglyceride) fatty acids. The fatty acid chains may be saturated or unsaturated (see below). Waxes are similar in structure to fats and oils, but they are formed with a complex alcohol instead of glycerol.

H_2C-O / $HC-O$ / H_2C-O 9 12 15

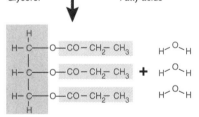

Glycerol Fatty acids

Triglyceride: an example of a neutral fat

Condensation

Glycerol + Fatty acids → Triglyceride + Water

Triglycerides form when glycerol bonds with three fatty acids. Glycerol is an alcohol containing three carbons. Each of these carbons is bonded to a hydroxyl (-OH) group.

When glycerol bonds with the fatty acid, an ester bond is formed and water is released. Three separate condensation reactions are involved in producing a triglyceride.

Saturated and Unsaturated Fatty Acids

Fatty acids are a major component of **neutral fats** and **phospholipids**. About 30 different kinds are found in animal lipids. **Saturated fatty acids** contain the maximum number of hydrogen atoms. **Unsaturated fatty acids** contain some carbon atoms that are double-bonded with each other and are not fully saturated with hydrogens. Lipids containing a high proportion of saturated fatty acids tend to be solids at room temperature (e.g. butter). Lipids with a high proportion of unsaturated fatty acids are oils and tend to be liquid at room temperature. This is because the unsaturation causes kinks in the straight chains so that the fatty acids do not pack closely together. Regardless of their degree of saturation, fatty acids yield a large amount of energy when oxidized.

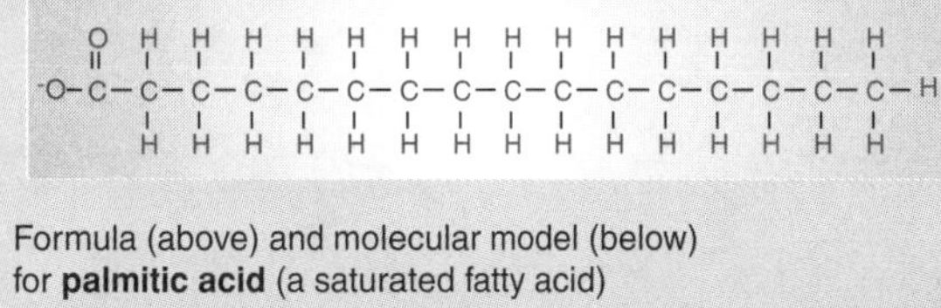

Formula (above) and molecular model (below) for **palmitic acid** (a saturated fatty acid)

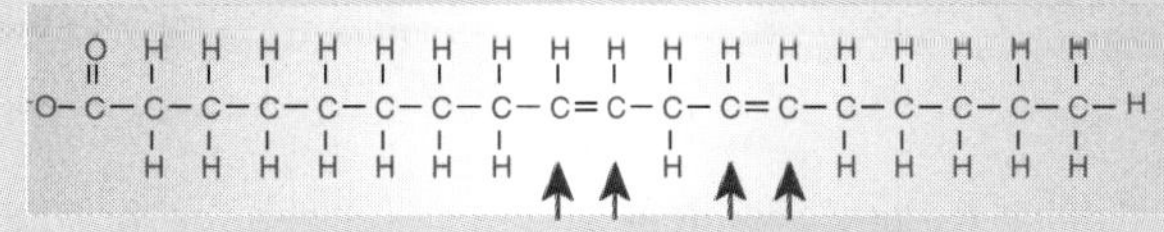

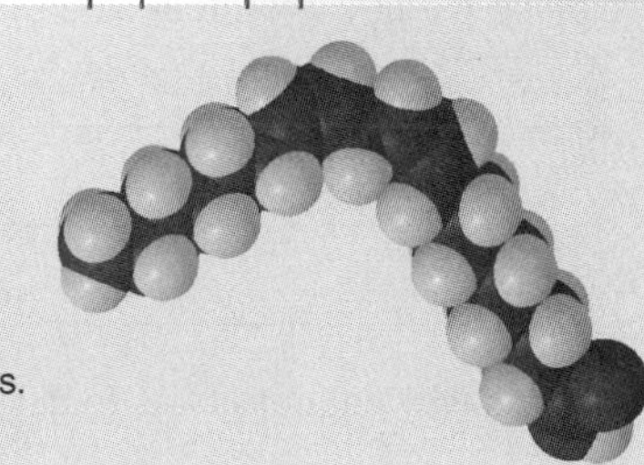

Formula (above) and molecular model (right) for **linoleic acid** (an unsaturated fatty acid). The arrows indicate double bonded carbon atoms that are not fully saturated with hydrogens.

1. (a) Distinguish between saturated and unsaturated fatty acids: ______________________

   (b) Explain how the type of fatty acid present in a neutral fat or phospholipid is related to that molecule's properties:

   ____

2. Explain how neutral fats can provide an animal with:

   (a) Energy: ______________________

   (b) Water: ______________________

   (c) Insulation: ______________________

ISBN: 978-1-927173-16-9

Phospholipids

Phospholipids are the main component of cellular membranes. They consist of a glycerol attached to two fatty acid chains and a phosphate (PO_4^{3-}) group. The phosphate end of the molecule is attracted to water (it is hydrophilic) while the fatty acid end is repelled (hydrophobic). The hydrophobic ends turn inwards in the membrane to form a **phospholipid bilayer.**

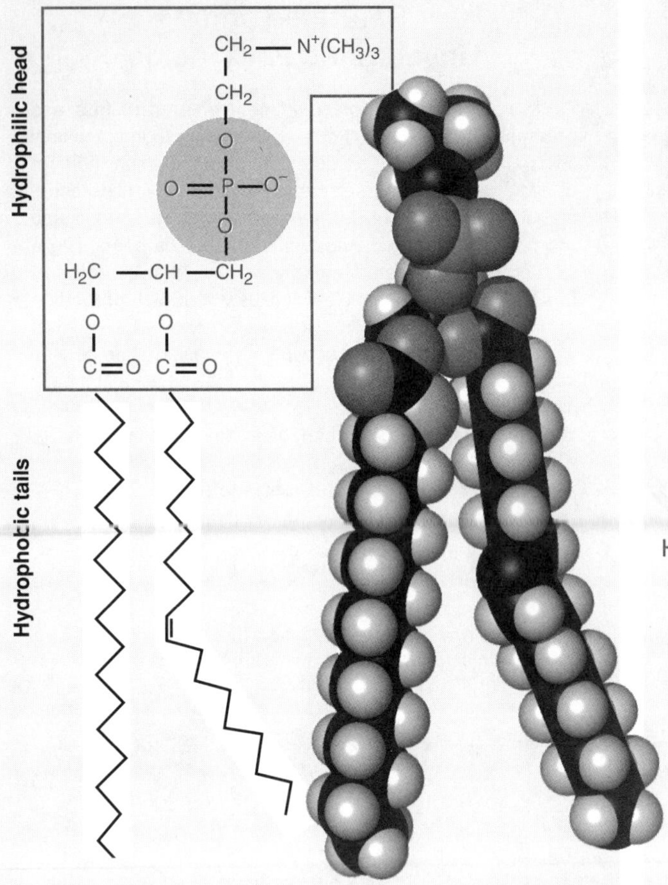

Steroids and Cholesterol

Although steroids are classified as lipids, their structure is quite different to that of other lipids. Steroids have a basic structure of three rings made of 6 carbon atoms each and a fourth ring containing 5 carbon atoms. Examples of steroids include the male and female sex hormones (testosterone and estrogen), and the hormones cortisol and aldosterone.

Cholesterol, while not a steroid itself, is a sterol lipid and is a precursor to several steroid hormones. It is present in the plasma membrane, where it regulates membrane fluidity by preventing the phospholipids from packing too closely together.

Like phospholipids, cholesterol is **amphipathic**. The hydroxyl (-OH) group on cholesterol interacts with the polar head groups of the membrane phospholipids, while the steroid ring and hydrocarbon chain tuck into the hydrophobic portion of the membrane. This helps to stabilise the outer surface of the membrane and reduce its permeability to small water-soluble molecules.

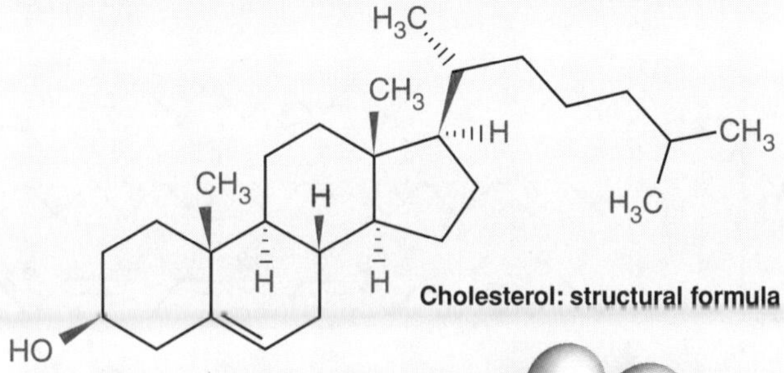

Cholesterol: structural formula

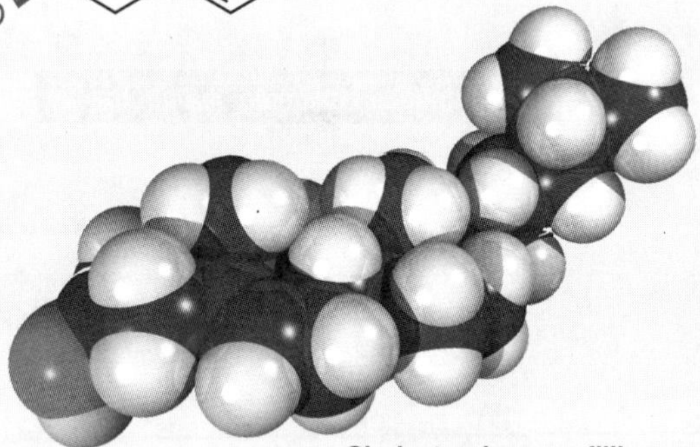

Cholesterol: space filling molecule

3. Outline the key **chemical** difference between a phospholipid and a triglyceride: ______________________

4. Explain why saturated fats (e.g. lard) are solid at room temperature: ______________________

5. (a) Relate the structure of phospholipids to their chemical properties and their functional role in cellular membranes:

(b) Suggest how the cell membrane structure of an Arctic fish might differ from that of tropical fish species:

6. Explain how the structure of cholesterol enables it to perform structural and functional roles within membranes:

ISBN: 978-1-927173-16-9

Amino Acids

Amino acids are the basic units from which proteins are made. Plants can manufacture all the amino acids they require from simpler molecules, but animals must obtain a certain number of ready-made amino acids (called **essential amino acids**) from their diet. Which amino acids are essential varies from species to species, as different metabolisms are able to synthesize different substances. The distinction between essential and non-essential amino acids is somewhat unclear though, as some amino acids can be produced from others and some are interconvertible by the urea cycle.

Structure of Amino Acids

There are over 150 amino acids found in cells, but only 20 occur commonly in proteins. The remaining, non-protein amino acids have roles as intermediates in metabolic reactions, or as neurotransmitters and hormones. All amino acids have a common structure (see right). The only difference between the different types lies with the 'R' group in the general formula. This group is variable, which means that it is different in each kind of amino acid.

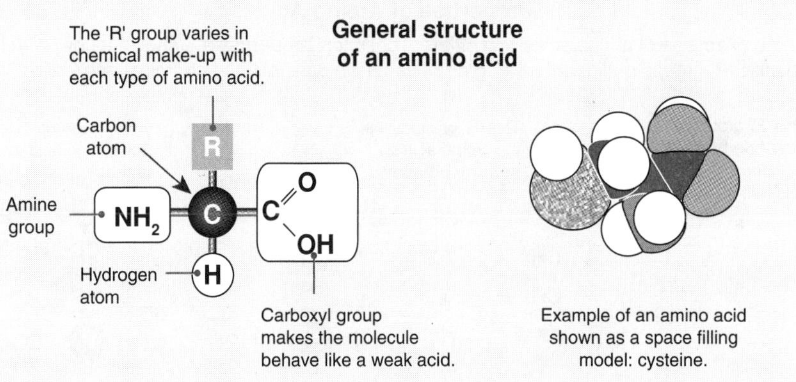

General structure of an amino acid
The 'R' group varies in chemical make-up with each type of amino acid.
Carbon atom
R
Amine group
NH_2
C
C
O
OH
Hydrogen atom
H
Carboxyl group makes the molecule behave like a weak acid.
Example of an amino acid shown as a space filling model: cysteine.

Condensation and Hydrolysis Reactions

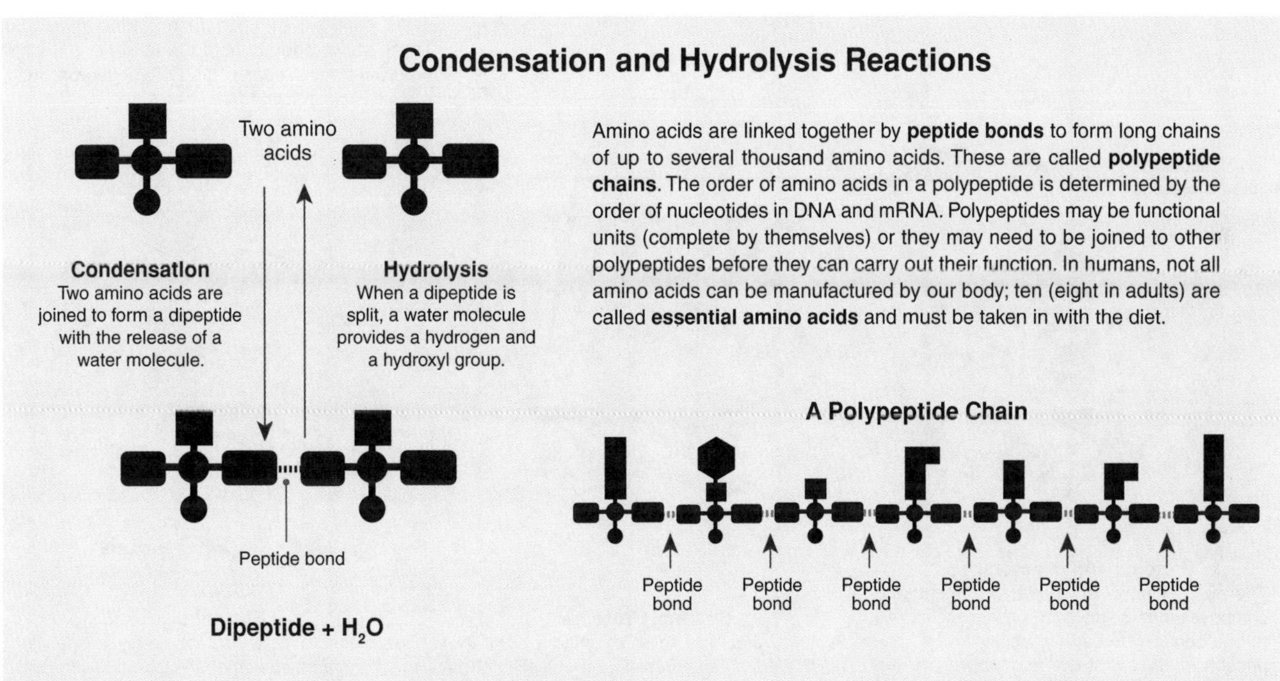

Two amino acids
Condensation
Two amino acids are joined to form a dipeptide with the release of a water molecule.
Hydrolysis
When a dipeptide is split, a water molecule provides a hydrogen and a hydroxyl group.
Peptide bond
Dipeptide + H_2O
A Polypeptide Chain
Peptide bond
Peptide bond
Peptide bond
Peptide bond
Peptide bond
Peptide bond

Amino acids are linked together by **peptide bonds** to form long chains of up to several thousand amino acids. These are called **polypeptide chains**. The order of amino acids in a polypeptide is determined by the order of nucleotides in DNA and mRNA. Polypeptides may be functional units (complete by themselves) or they may need to be joined to other polypeptides before they can carry out their function. In humans, not all amino acids can be manufactured by our body; ten (eight in adults) are called **essential amino acids** and must be taken in with the diet.

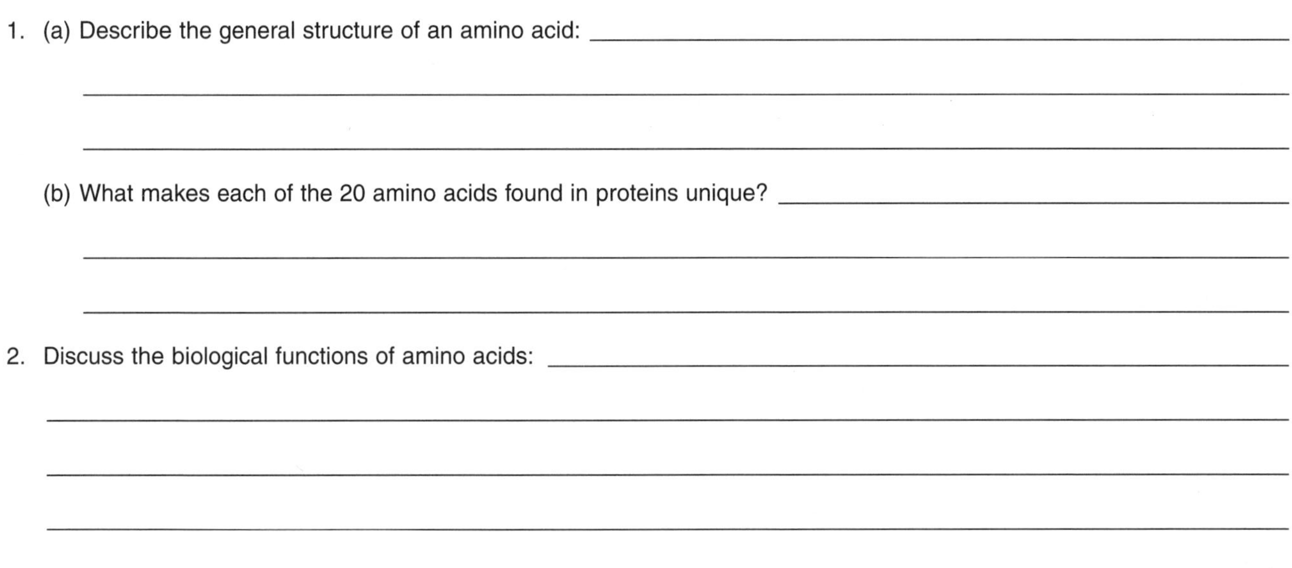

1. (a) Describe the general structure of an amino acid: ______________________

   (b) What makes each of the 20 amino acids found in proteins unique? ______________________

2. Discuss the biological functions of amino acids: ______________________

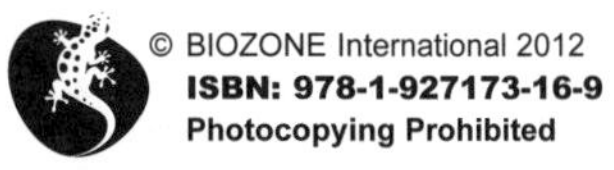

ISBN: 978-1-927173-16-9

Related activities: *Proteins, Genes to Proteins*
Weblinks: *Amino Acids and Proteins*

The Properties of Amino Acids

The twenty amino acids encoded by the genetic code to make proteins can be divided into **polar** or **non-polar** groups based on the properties of their R groups. These groups include polar side chains, electrically charged side chains, and hydrophobic side chains. The properties of the R groups are important in determining how the amino acid chain folds up into a functional protein, and how the protein interacts with other molecules, including other proteins. For example, the **hydrophobic** R groups of soluble protein are folded into the protein's interior, while the hydrophilic groups are presented to the outside.

Properties of Amino Acids

Three examples of amino acids with different chemical properties are shown below, with their specific 'R' groups outlined. The 'R' groups can have quite diverse chemical properties.

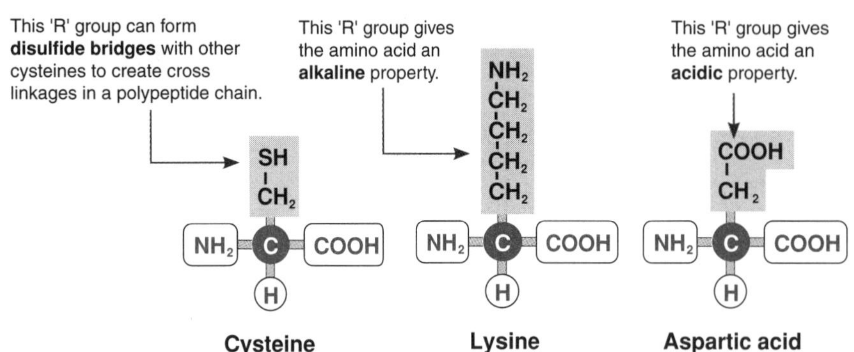
This 'R' group can form **disulfide bridges** with other cysteines to create cross linkages in a polypeptide chain.
This 'R' group gives the amino acid an **alkaline** property.
This 'R' group gives the amino acid an **acidic** property.
SH
CH_2
NH_2
C
COOH
H
NH_2
CH_2
CH_2
CH_2
CH_2
COOH
CH_2
Cysteine
Lysine
Aspartic acid

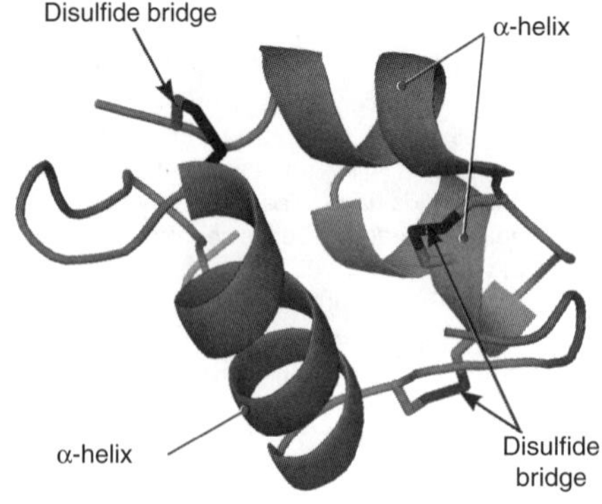
Disulfide bridge
α-helix
α-helix
Disulfide bridge

Insulin begins as a single amino acid chain. Interactions between the R groups of the amino acids cause it to form three coiled α-helices. Three sulfur bridges hold the chain in a 9-bend before it is cleaved into the final active product.

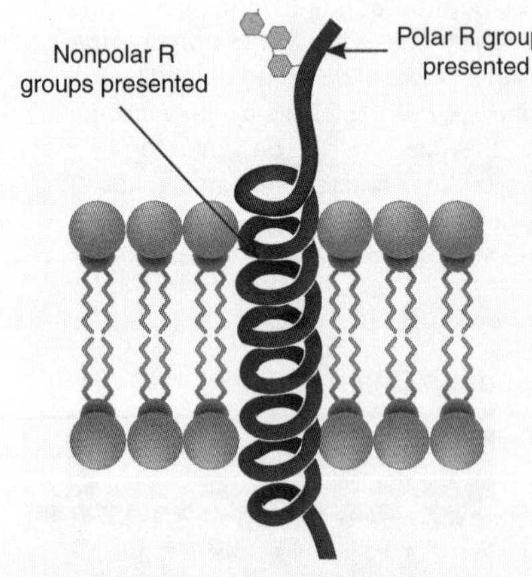
Nonpolar R groups presented
Polar R groups presented

Proteins and Membranes

The lipid interior of a plasma membrane comprises non-polar, hydrophobic molecules (the long hydrocarbon chains of the phospholipids). Proteins embedded in the membrane fold in such a way that they present hydrophobic, nonpolar R groups to the membrane. Above and below the membrane, the polar R groups are presented.

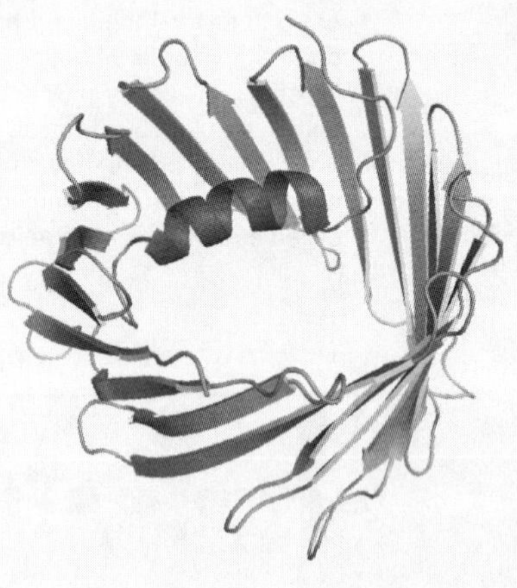

Channel Proteins

Proteins that fold to form channels in the plasma membrane present non-polar R groups to the membrane and polar R groups to the inside of the channel. Hydrophilic molecules and ions are then able to pass through these channels into the interior of the cell.

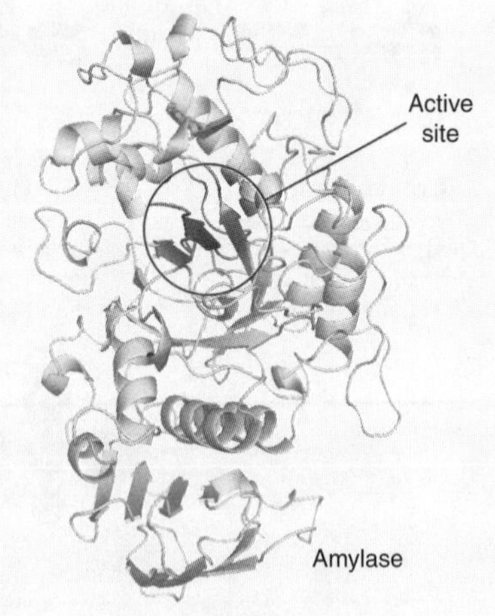
Active site
Amylase

Enzymes

Enzymes are proteins that catalyze specific reactions. Enzymes that are folded to present polar R groups at the active site will be specific for polar substances. Non-polar active sites will be specific for nonpolar substances.

1. Explain the importance of the amino acid sequence in protein folding: ____________________

2. Explain why channel proteins often fold with non-polar R groups to the channel's exterior and polar R groups to its interior:

3. What is the relationship between the R group polarity of an enzyme's active site and its substrate?

Related activities: Amino Acids
Weblinks: Amino Acids and Peptide Bond Formation

ISBN: 978-1-927173-16-9

Proteins

Proteins are large, complex **macromolecules**, built up from a linear sequence of repeating units called **amino acids**. Proteins are molecules of central importance in the chemistry of life. They account for more than 50% of the dry weight of most cells, and they are important in virtually every cellular process. The folding of a protein into its functional form creates a three dimensional arrangement of the active 'R' groups. It is this **tertiary structure** that gives a protein its unique chemical properties. If a protein loses this precise structure (through **denaturation**), it is usually unable to carry out its biological function.

Primary Structure - 1°
(amino acid sequence)

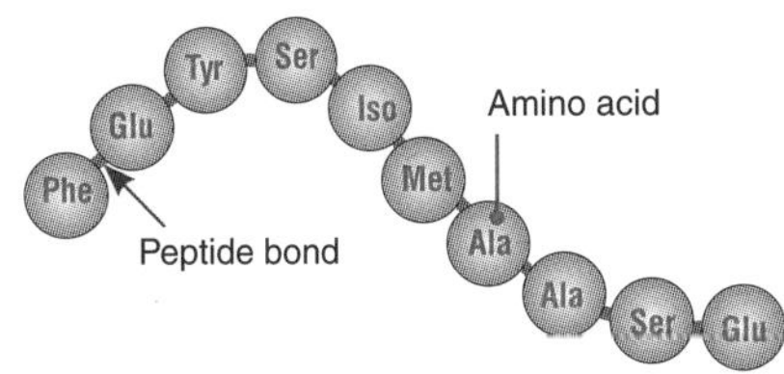
Phe
Glu
Tyr
Ser
Iso
Met
Ala
Ala
Ser
Glu
Amino acid
Peptide bond

Hundreds of amino acids are linked together by peptide bonds to form polypeptide chains. The attractive and repulsive charges on the amino acids determines how the protein is organized, and its biological function.

Secondary Structure - 2°
(α-helix or β-pleated sheet)

The helical shape is maintained with hydrogen bonds

α-helix

Two peptide chains

β-pleated sheet

Hydrogen bonds

Polypeptides fold into a secondary (2°) structure, usually either a coiled **α-helix** or a **β-pleated sheet**. Secondary structures are maintained with hydrogen bonds between neighboring CO and NH groups.

Tertiary Structure - 3°
(folding of the **2°** structure)

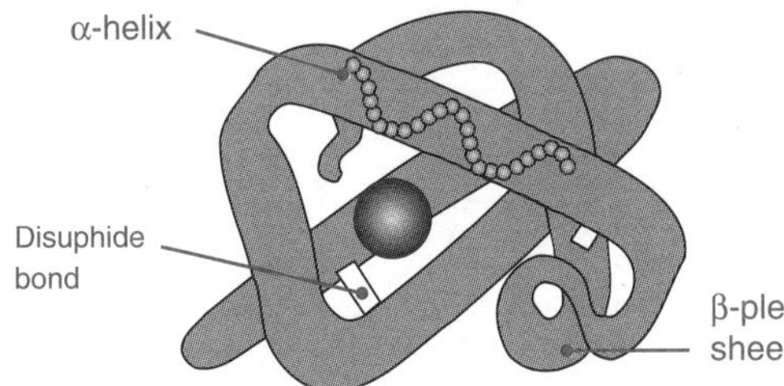
α-helix
Disuphide bond
β-pleated sheet

The tertiary structure of a protein is its three-dimensional structure, formed when the secondary structure folds up. Chemical bonds such as disulfide bridges between cysteine amino acids, ionic bonds, hydrogen bonds, and hydrophobic interactions result in folding of the protein. These bonds can be destroyed by the presence of heavy metals or some solvents, and unfavourable conditions such as pH and temperature.

Quaternary Structure - 4°

A hemoglobin molecule consists of four poplypeptide **subunits**: two identical alpha chains and two identical beta chains.

In hemoglobin, each polypeptide encloses an iron-containing prosthetic group.

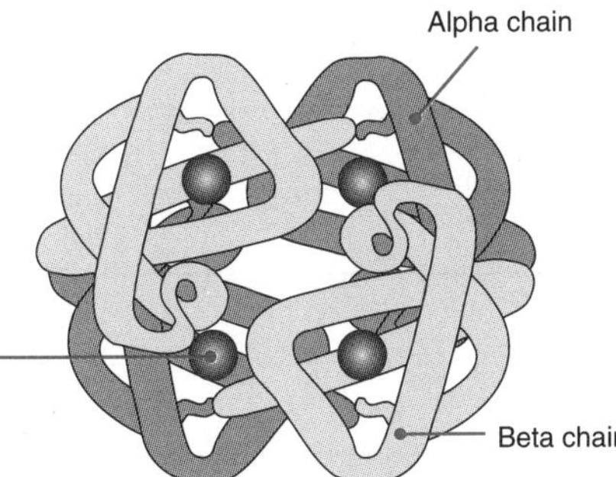
Alpha chain
Beta chain

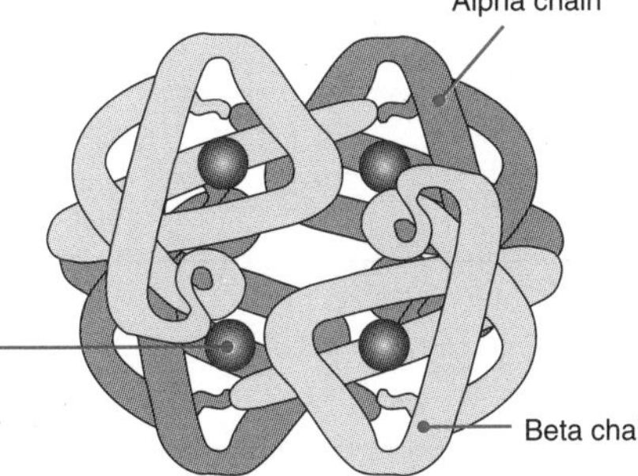

Many complex proteins exist as groups of polypeptide chains. The arrangement of the polypeptide chains into a functional protein is termed the quaternary structure. The example (above) shows hemoglobin.

1. Describe briefly the four main structures of a protein, and name any key factors contributing to the formation of each structure:

   (a) Primary structure: ______________________

   (b) Secondary structure: ______________________

   (c) Tertiary structure: ______________________

   (d) Quaternary structure: ______________________

2. How are proteins built up into a functional structure?

ISBN: 978-1-927173-16-9

Protein Structure and Function

Proteins can be classified according to their structure or their function. **Globular proteins** are roughly spherical and soluble in water. Many globular proteins are **enzymes** and have a **catalytic** role in regulating **metabolic pathways**. **Fibrous proteins** have an elongated structure and are not water soluble. They are often made up of repeating units and provide stiffness and rigidity to the more fluid components of cells and tissues. They have important **structural** and contractile roles.

Globular Proteins

Properties
- Easily water soluble
- Tertiary structure critical to function
- Polypeptide chains folded into a spherical shape

Function
- Catalytic *e.g. enzymes*
- Regulatory *e.g. hormones (insulin)*
- Transport *e.g. hemoglobin*
- Protective *e.g. antibodies*

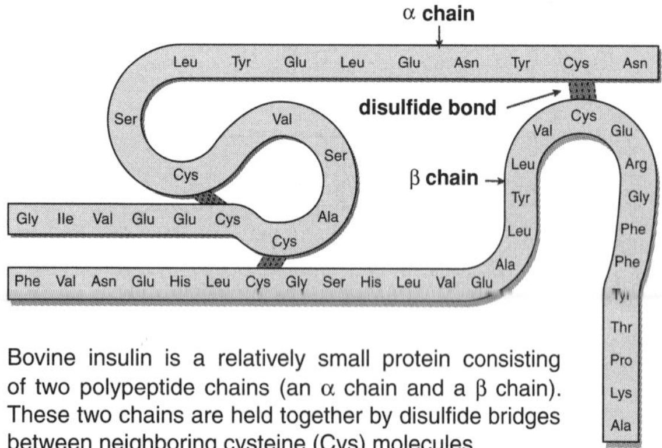

α chain
disulfide bond
β chain
Leu Tyr Glu Leu Glu Asn Tyr Cys Asn
Ser Val Ser Cys Ala Cys
Gly Ile Val Glu Glu Cys
Phe Val Asn Glu His Leu Cys Gly Ser His Leu Val Glu
Cys Val Glu Leu Arg Tyr Gly Leu Phe Ala Phe Tyr Thr Pro Lys Ala

Bovine insulin is a relatively small protein consisting of two polypeptide chains (an α chain and a β chain). These two chains are held together by disulfide bridges between neighboring cysteine (Cys) molecules.

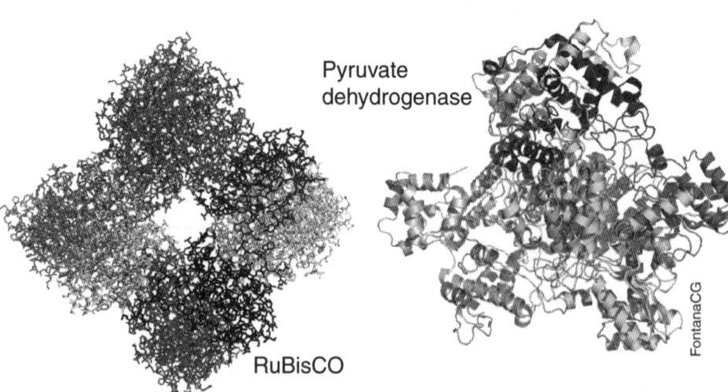

Pyruvate dehydrogenase
RuBisCO
FontanaCG

Pyruvate dehydrogenase is an enzyme involved in the conversion of pyruvate to acetyl-CoA, and so links the glycolysis pathway to the Krebs cycle. The enzyme **RuBisCo** is found in green plants and catalyzes the first step of carbon fixation in the Calvin cycle. RuBisCO is the most abundant protein in the world.

Fibrous Proteins

Properties
- Water insoluble
- Very tough physically; may be supple or stretchy
- Parallel polypeptide chains in long fibers or sheets

Function
- Structural role in cells and organisms e.g. *collagen found in connective tissue, cartilage, bones, tendons, and blood vessel walls.*
- Contractile e.g. *myosin, actin*

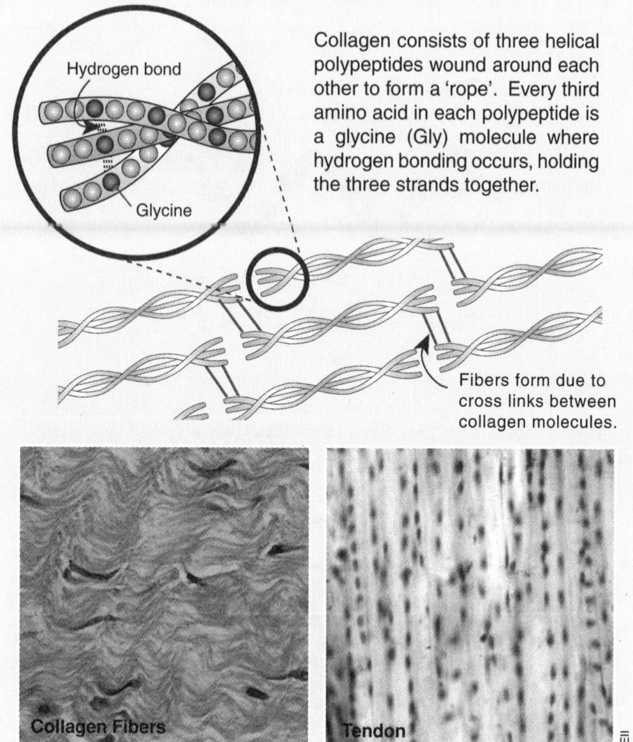

Hydrogen bond
Glycine
Fibers form due to cross links between collagen molecules.
Collagen Fibers
Tendon
EII

Collagen consists of three helical polypeptides wound around each other to form a 'rope'. Every third amino acid in each polypeptide is a glycine (Gly) molecule where hydrogen bonding occurs, holding the three strands together.

Collagen is the main component of connective tissue, and is mostly found in fibrous tissues (e.g. tendons, ligaments, and skin). **Tendons** connect muscle to bone and tendons are composed of closely packed collagen fibers lined parallel to each other.

1. How are proteins involved in the following roles? Give examples to help illustrate your answer:

   (a) Structural tissues of the body: ____

   (b) Catalyzing metabolic reactions in cells: ____

2. How does the shape of a fibrous protein relate to its functional role? ________________________________

3. How does the shape of a catalytic protein (enzyme) relate to its functional role? ________________________

ISBN: 978-1-927173-16-9

KEY TERMS: Mix and Match

INSTRUCTIONS: Test your vocabulary by matching each term to its definition, as identified by its preceding letter code.

amino acid ....

carbohydrate ....

cellulose ....

condensation ....

disaccharide ....

fatty acids ....

fructose ....

galactose ....

glucose ....

glycogen ....

hydrolysis ....

inorganic compound ....

lipld ....

maltose ....

monosaccharide ....

organic compound ....

polypeptides ....

polysaccharide ....

protein ....

ribose ....

starch ....

sucrose ....

triglycerides ....

water ....

A Unbranched molecule that makes up the majority of a cell wall. It comprises several thousand repeating β-glucose monomers.

B An organic compound, usually a linear polymer, made of amino acids linked together by peptide bonds.

C Organic molecule consisting only of carbon, hydrogen and oxygen that serves as a structural component in cells and as an energy source.

D A carbohydrate monomer. Examples include fructose and glucose.

E A building block of proteins.

F Chemical reaction that combines two molecules by the elimination of a smaller molecule, often water.

G A disaccharide consisting of two glucose monomers joined by an α 1-4 bond. Produced when amylase breaks down starch and commonly found in germinating seeds such as barely.

H Complex carbohydrates with structural and energy storage roles in cells. Examples include cellulose, starch, and glycogen.

I A class of organic compounds with an oily, greasy, or waxy consistency. Important as energy storage molecules and as components of cellular membranes.

J A disaccharide formed from glucose and fructose. Commonly known as table sugar.

K A long chain organic acid and a major component of the most abundant fats on Earth

L A compact branching molecule made up of glucose monomers used as a storage molecule in plants.

M A double sugar molecule used as an energy source and a building block of larger molecules. Examples are sucrose and lactose.

N A compound that does not contain carbon, e.g. NaCl

O A branched chain polysaccharide made up of α-glucose units, which acts as a storage compound in animal tissues.

P Macromolecules that form from the joining of multiple peptide subunits.

Q Chemical reaction in which a molecule is split by water (as H^+ and OH^-).

R Sometimes know as fruit sugar. An isomer of glucose but much sweeter.

S A compound that contains carbon, e.g. glucose.

T Monosaccharide sugar with the formula $C_6H_{12}O_6$. The α-form polymerizes to form starch, while the β-form polymerizes to form cellulose.

U Monosaccharide that does not usually occur freely in nature but combines with glucose to form lactose.

V Also called fat or oil. A biological compound made up of glycerol and fatty acid components.

W A monosaccharide with the formula $C_5H_{10}O_5$. It is a key structural component of RNA.

X Polar molecule consisting of two hydrogen atoms and one oxygen atom.

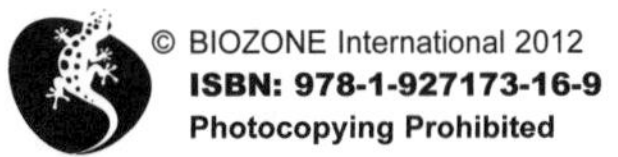

ISBN: 978-1-927173-16-9

Core Topic
3

AHL Topic
7

The Structure and Function of DNA

Key concepts

- Nucleic acids have a central role in storing genetic information and controlling the behaviour of cells.
- DNA is a self-replicating polynucleotide constructed according to strict base-pairing rules.
- DNA replication is a semi-conservative process controlled by enzymes.
- The universal genetic code, through gene expression, contains the information to construct proteins.

Key terms

Core

amino acid
anticodon
base-pairing rule
codon
DNA
DNA polymerase
DNA replication
double-helix
gene
gene expression
genetic code
helicase
messenger RNA (mRNA)
nucleic acids
nucleotide
polypeptide
protein
ribosome
RNA
RNA polymerase
semi-conservative
transcription
transfer RNA (tRNA)
translation

AHL only

antisense (=template) strand
DNA ligase
DNA polymerase (I and II)
exons
histone
hydrogen bonding
introns
nucleosome
Okazaki fragments
peptide bond
polysome
purine
pyrimidine
ribosomal RNA (rRNA)
sense (=coding) strand

Periodicals:

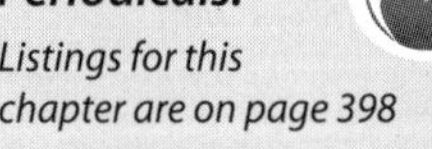

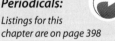

Listings for this chapter are on page 398

Weblinks:

www.thebiozone.com/weblink/IB-3169.html

BIOZONE APP:
Student Review Series
The Genetic Code

Learning Objectives

☐ 1. Use the **KEY TERMS** to help you understand and complete these objectives.

DNA Structure *(3.3 & 7.1)*

pages 89-98

☐ 2. Describe the structure of a DNA **nucleotide**, identifying the sugar, phosphate and base, and their relative positions. Identify the four bases found in DNA.

☐ 3. Describe how the nucleotides in DNA are linked together into a single strand and how the strands form a **double helix**. State the **base pairing rule**. Draw a simple annotated diagram of the molecular structure of **DNA**.

☐ 4. **AHL**: In more detail than above, describe the structure of the DNA double-helix, including its anti-parallel nature, the 3'-5' linkages, and the role of **hydrogen bonding** between **purine** and **pyrimidine bases**.

☐ 5. **AHL**: Describe the structure and role of **nucleosomes** and outline their role in packaging DNA, organizing chromosomes, and regulating transcription.

☐ 6. **AHL**: Distinguish between **introns** and **exons** in eukaryotic genes. Distinguish between unique sequences and highly repetitive sequences in nuclear DNA. How has our understanding of the role of these repetitive sequences changed?

DNA Replication *(3.4 & 7.2)*

pages 99-102

☐ 7. Explain the **semi-conservative replication** of DNA during interphase, including the role of **DNA polymerase** and **helicase**.

☐ 8. **AHL**: Describe DNA replication in prokaryotes. Explain the significance and consequences of DNA replication in the **5' → 3'** direction.

☐ 9. **AHL**: Recognize that DNA replication is initiated at many points in eukaryotic chromosomes and relate this to the larger size of chromosomes in eukaryotes.

Transcription and Translation *(3.5 & 7.3-7.4)*

pages 103-106

☐ 10. Compare and contrast the structure of RNA and DNA. Outline DNA transcription identifying the roles of **RNA polymerase** in synthesizing the mRNA molecule.

☐ 11. Describe the genetic code in terms of codons comprising triplets of bases.

☐ 12. **AHL**: Describe transcription in prokaryotes including the **5' → 3'** direction and the significance of the **sense** (coding) **strand** and **antisense** (template) **strands**.

☐ 13. **AHL**: Recognize that, in eukaryotes, introns are removed to form the mature mRNA.

☐ 14. Explain **translation** and the formation of a polypeptide. Include reference to the role of the **mRNA**, **tRNA**, **anticodons**, **ribosomes**, and **amino acids**.

☐ 15. **AHL**: Draw and label a diagram showing a **peptide bond** between two amino acids.

☐ 16. **AHL**: Explain the structure and function of tRNA molecules and ribosomes. Recognize the stages in translation and describe the role of start and stop codons. Explain the role of **polysomes** in simultaneously translating a mRNA molecule. Explain the significance of protein synthesis on free and on bound ribosomes.

Nucleotides and Nucleic Acids

Nucleic acids are a special group of chemicals in cells concerned with the transmission of inherited information. They have the capacity to store the information that controls cellular activity. The central nucleic acid is called **deoxyribonucleic acid** (DNA). DNA is a major component of chromosomes and is found primarily in the nucleus, although a small amount is found in mitochondria and chloroplasts. Other **ribonucleic acids** (RNA) are involved in the 'reading' of the DNA information. All nucleic acids are made up of simple repeating units called **nucleotides**, linked together to form chains or strands, often of great length. The strands vary in the sequence of the bases found on each nucleotide. It is this sequence which provides the 'genetic code' for the cell. In addition to nucleic acids, certain nucleotides and their derivatives are also important as suppliers of energy (**ATP**) or as hydrogen ion and electron carriers in respiration and photosynthesis (NAD, NADP, and FAD).

Chemical Structure of a Nucleotide

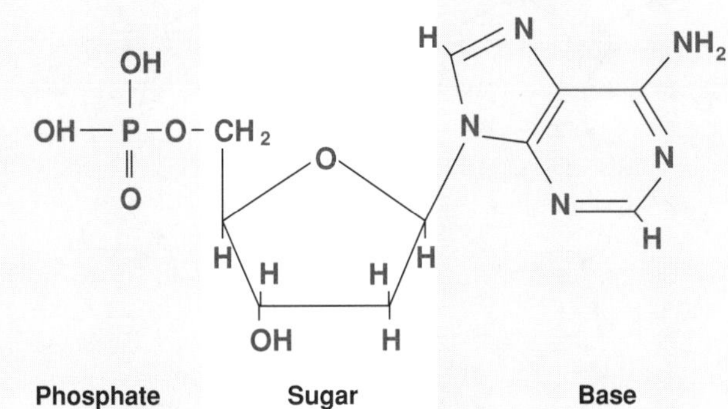
OH
OH—P-O-CH2
O
O
H
N
NH2
N
N
N
H
H
H
H
H
OH
H
Phosphate
Sugar
Base

Symbolic Form of a Nucleotide

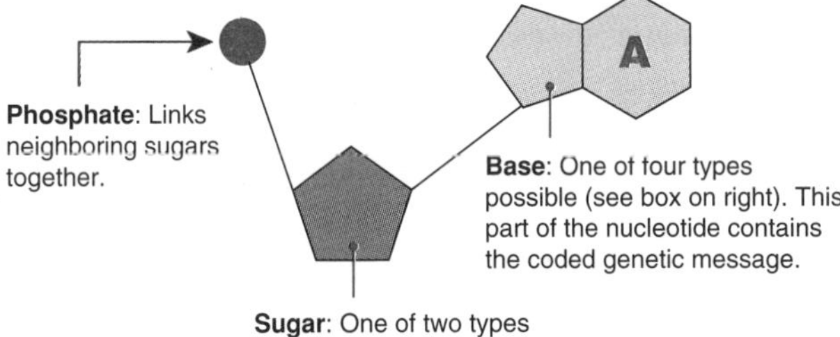
A
Phosphate: Links neighboring sugars together.
Base: One of four types possible (see box on right). This part of the nucleotide contains the coded genetic message.
Sugar: One of two types possible: ribose in RNA and deoxyribose in DNA.

Nucleotides are the building blocks of DNA. Their precise sequence in a DNA molecule provides the genetic instructions for the organism to which it governs. Accidental changes in nucleotide sequences are a cause of mutations, usually harming the organism, but occasionally providing benefits.

Bases

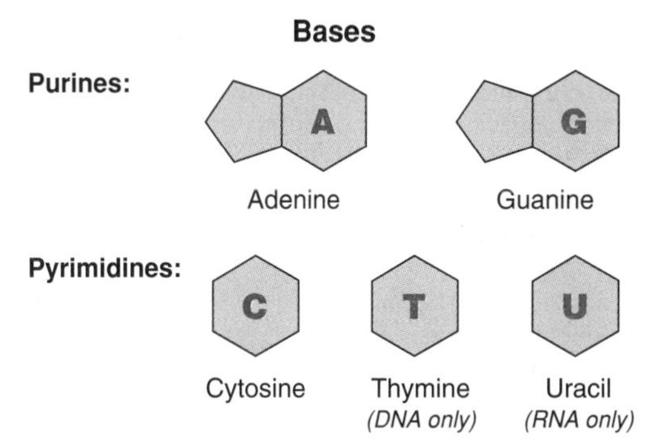
Purines:
A
G
Adenine
Guanine
Pyrimidines:
C
T
U
Cytosine
Thymine
(DNA only)
Uracil
(RNA only)

The two-ringed bases above are **purines**. The single-ringed bases are **pyrimidines**. Although only one of four kinds of base can be used in a nucleotide, **uracil** is found only in RNA, replacing **thymine**. DNA contains A, T, G, and C, while RNA contains A, U, G, and C.

Sugars

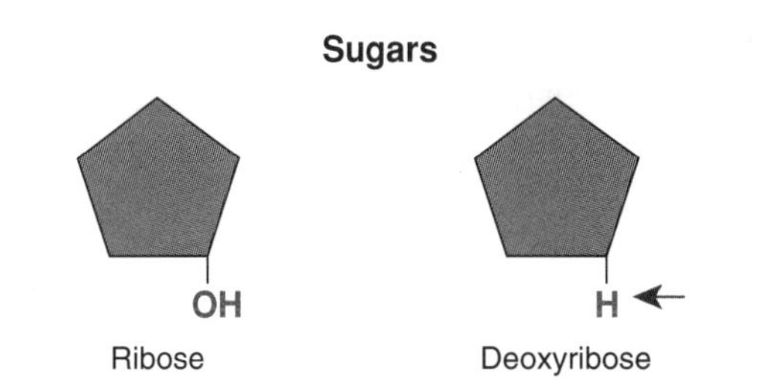
OH
H
Ribose
Deoxyribose

Deoxyribose sugar is found only in DNA. It differs from **ribose** sugar, found in RNA, by the lack of a single oxygen atom (arrowed).

RNA Molecule

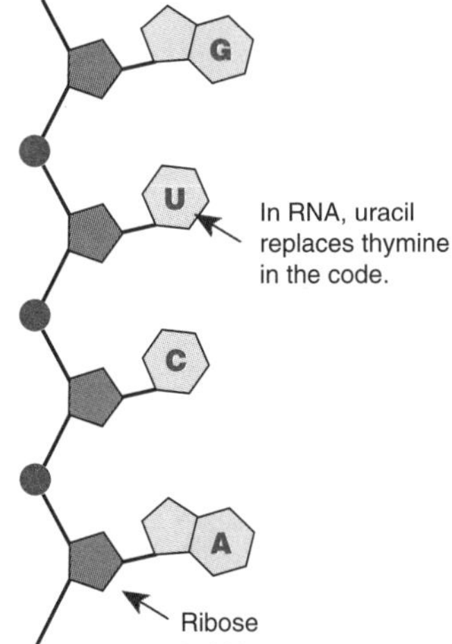
G
U
In RNA, uracil replaces thymine in the code.
C
A
Ribose sugar

DNA Molecule

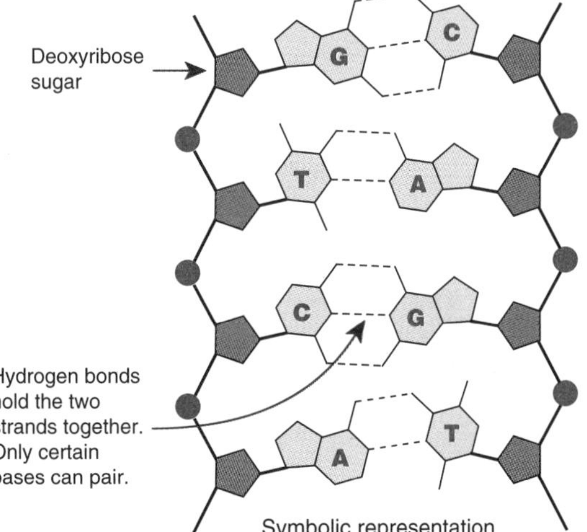
Deoxyribose sugar
G
C
T
A
C
G
A
T
Hydrogen bonds hold the two strands together. Only certain bases can pair.

Symbolic representation

DNA Molecule

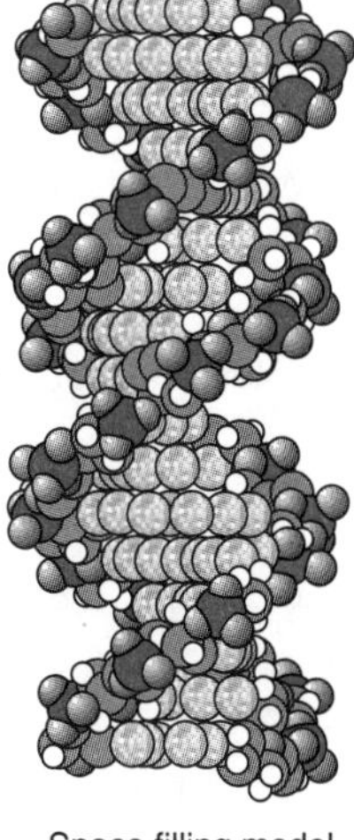

Space filling model

Ribonucleic acid (RNA) comprises a *single strand* of nucleotides linked together.

Deoxyribonucleic acid (DNA) comprises a *double strand* of nucleotides linked together. It is shown unwound in the symbolic representation (left). The DNA molecule takes on a twisted, double-helix shape as shown in the space filling model on the right.

ISBN: 978-1-927173-16-9

Formation of a nucleotide

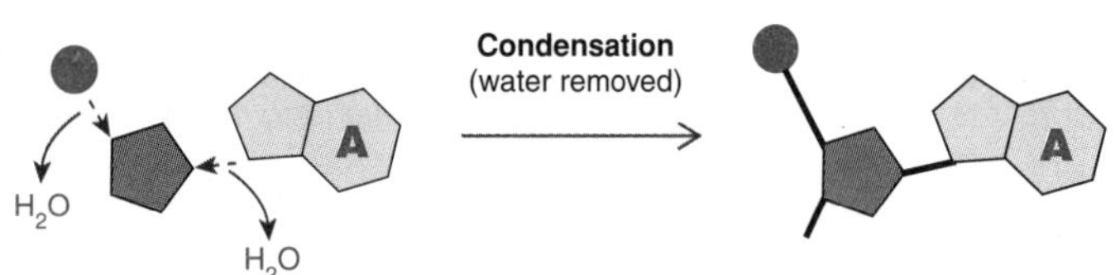
Condensation
(water removed)
H_2O
H_2O
A
A

A nucleotide is formed when phosphoric acid and a base are chemically bonded to a sugar molecule. In both cases, water is given off, and they are therefore condensation reactions. In the reverse reaction, a nucleotide is broken apart by the addition of water **(hydrolysis)**.

Formation of a dinucleotide

H_2O

T

C

Two nucleotides are linked together by a condensation reaction between the phosphate of one nucleotide and the sugar of another.

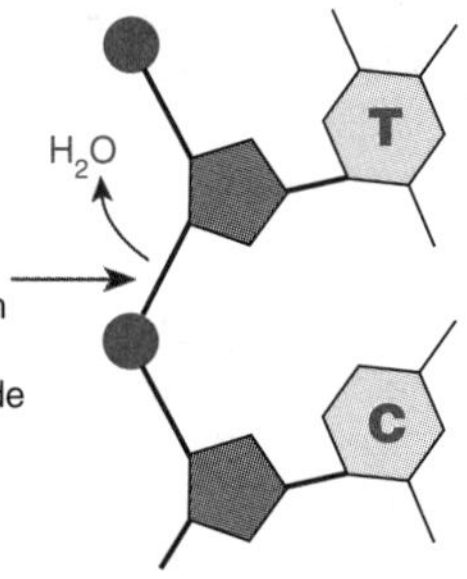

Double-Stranded DNA

The double-helix structure of DNA is like a ladder twisted into a corkscrew shape around its longitudinal axis. It is 'unwound' here to show the relationships between the bases.

- The DNA backbone is made up from alternating phosphate and sugar molecules, giving the DNA molecule an asymmetrical structure.
- The asymmetrical structure gives a DNA strand a **direction**. Each strand runs in the opposite direction to the other.
- The ends of a DNA strand are labeled the 5' (five prime) and 3' (three prime) ends. The **5'** end has a terminal phosphate group (off carbon 5), the **3'** end has a terminal hydroxyl group (off carbon 3).
- The way the pairs of bases come together to form hydrogen bonds is determined by the number of bonds they can form and the configuration of the bases.

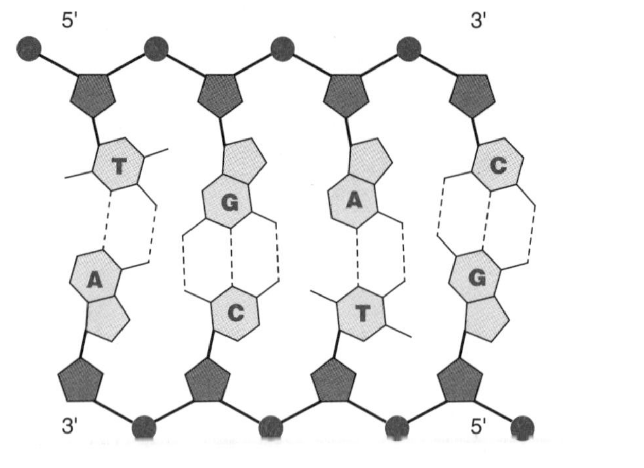
5'
3'
T
G
A
C
A
C
T
G
3'
5'

1. The diagram above depicts a double-stranded DNA molecule. Label the following parts on the diagram:
   (a) **Sugar** (deoxyribose)
   (b) **Phosphate**
   (c) **Hydrogen bonds** (between bases)
   (d) **Purine** bases
   (e) **Pyrimidine** bases

2. (a) Explain the **base-pairing rule** that applies in double-stranded DNA: ______________________

   (b) How is the base-pairing rule for mRNA different? ______________________

   (c) What is the purpose of the hydrogen bonds in double-stranded DNA? ______________________

3. Describe the functional role of nucleotides: ______________________

4. (a) Why do the DNA strands have an asymmetrical structure? ______________________

   (b) What are the differences between the 5' and 3' ends of a DNA strand? ______________________

5. Complete the following table summarizing the differences between DNA and RNA molecules:

	DNA	RNA
Sugar present		
Bases present		
Number of strands		
Relative length		

ISBN: 978-1-927173-16-9

Packaging DNA in the Nucleus

The chromosomes of eukaryote cells (such as those from plants and animals) are complex in their structure compared to those of prokaryotes. The illustration below shows a chromosome during the early stage of meiosis. Here it exists as a chromosome consisting of two chromatids. A non-dividing cell would have chromosomes with the 'equivalent' of a single chromatid only. The chromosome consists of a protein coated strand which coils in three ways during the time when the cell prepares to divide.

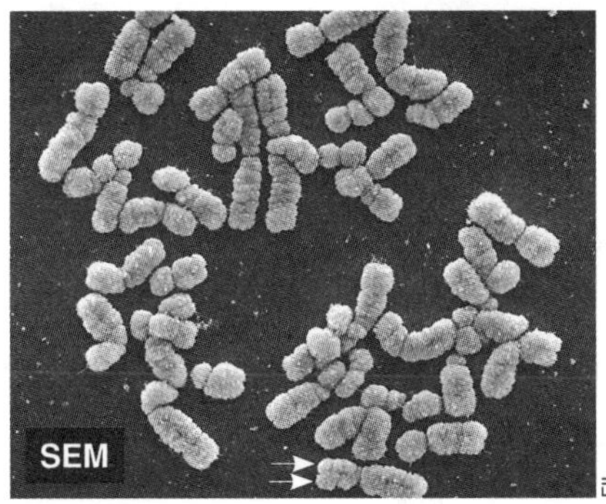
SEM
EII

A cluster of human chromosomes seen during metaphase of cell division. Individual chromatids (arrowed) are difficult to discern on these double chromatid chromosomes.

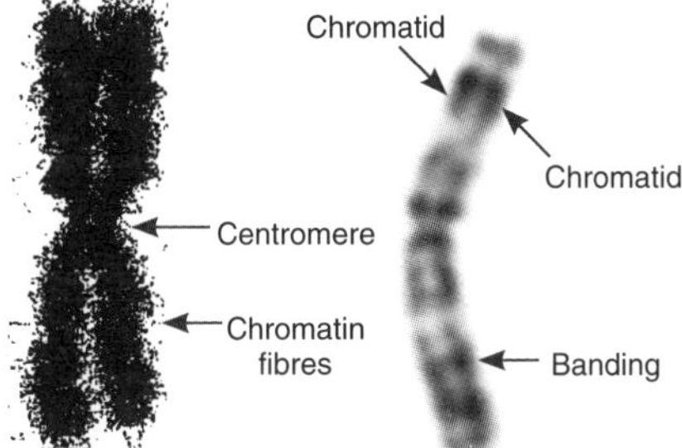
Chromatid
Chromatid
Centromere
Chromatin fibres
Banding
Chromosome TEM
Human chromosome 3

A human chromosome from a dividing white blood cell (above left). Note the compact organisation of the chromatin in the two chromatids. The LM photograph (above right) shows the banding visible on human chromosome 3.

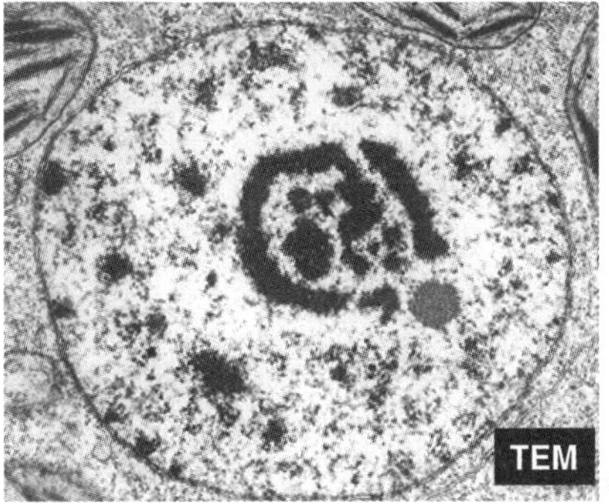
TEM
BF

In non-dividing cells, chromosomes exist as single-armed structures. They are not visible as coiled structures, but are 'unwound' to make the genes accessible for transcription (above).

Looped domains
EII

The evidence for the existence of looped domains comes from the study of giant lampbrush chromosomes in amphibian oocytes (above). Under electron microscopy, the lateral loops of the DNA-protein complex appear brushlike.

The Packaging of Chromatin

Chromatin structure is based on successive levels of DNA packing. **Histone proteins** are responsible for packing the DNA into a compact form. Without them, the DNA could not fit into the nucleus. Five types of histone proteins form a complex with DNA, in a way that resembles "beads on a string". These beads, or **nucleosomes**, form the basic unit of DNA packing.

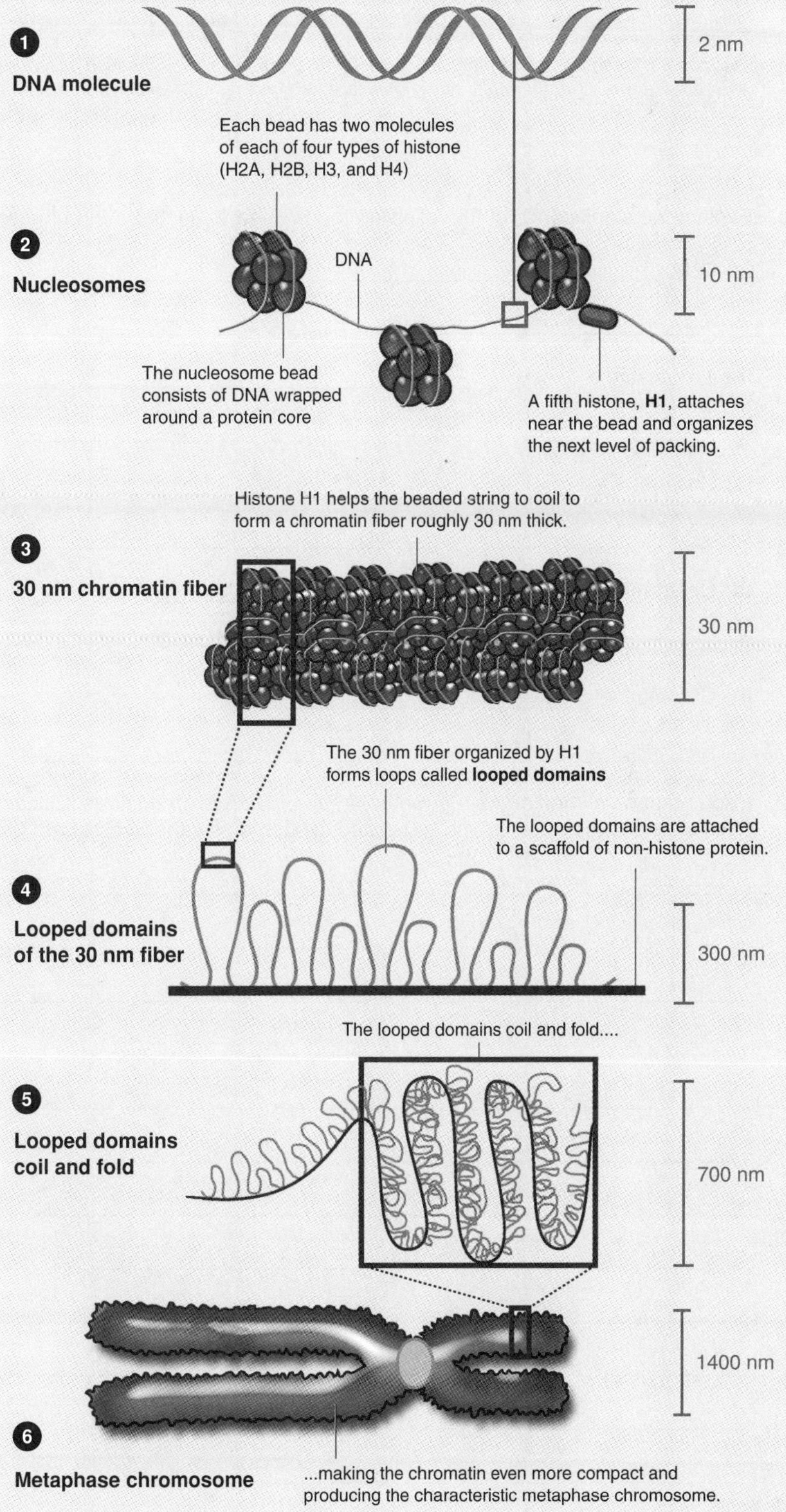

1 **DNA molecule** — 2 nm

Each bead has two molecules of each of four types of histone (H2A, H2B, H3, and H4)

2 **Nucleosomes** — 10 nm

DNA

The nucleosome bead consists of DNA wrapped around a protein core

A fifth histone, **H1**, attaches near the bead and organizes the next level of packing.

Histone H1 helps the beaded string to coil to form a chromatin fiber roughly 30 nm thick.

3 **30 nm chromatin fiber** — 30 nm

The 30 nm fiber organized by H1 forms loops called **looped domains**

The looped domains are attached to a scaffold of non-histone protein.

4 **Looped domains of the 30 nm fiber** — 300 nm

The looped domains coil and fold....

5 **Looped domains coil and fold** — 700 nm

1400 nm

6 **Metaphase chromosome**

...making the chromatin even more compact and producing the characteristic metaphase chromosome.

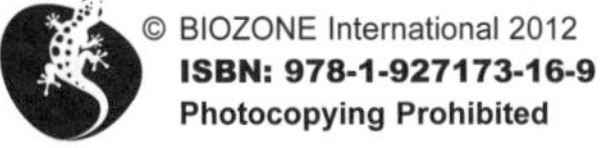

ISBN: 978-1-927173-16-9

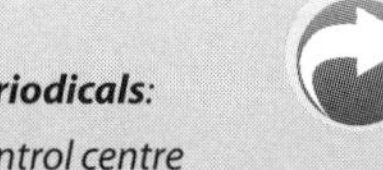

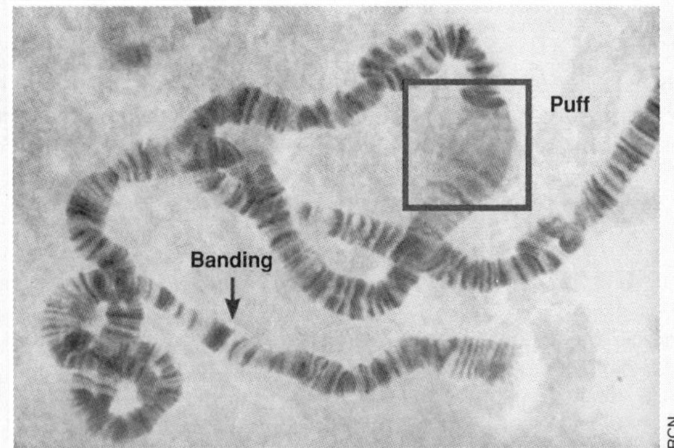
Puff
Banding
RCN

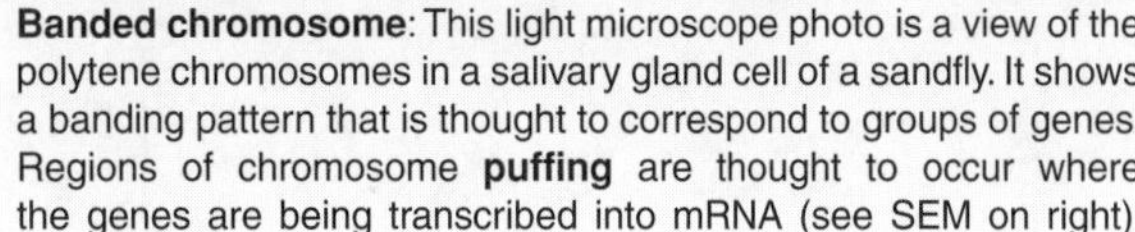
Banded chromosome: This light microscope photo is a view of the polytene chromosomes in a salivary gland cell of a sandfly. It shows a banding pattern that is thought to correspond to groups of genes. Regions of chromosome **puffing** are thought to occur where the genes are being transcribed into mRNA (see SEM on right).

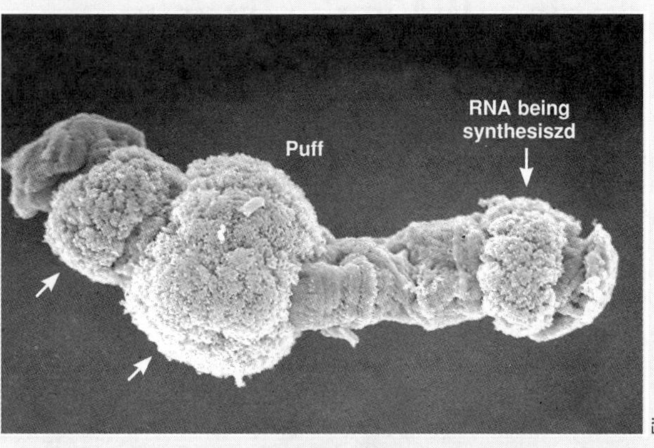
Puff
RNA being synthesiszd
EII

A **polytene chromosome** viewed with a scanning electron microscope (SEM). The arrows indicate localized regions of the chromosome that are uncoiling to expose their genes (puffing) to allow transcription of those regions. Polytene chromosomes are a type of chromosome consisting of a large bundle of chromatids bound tightly together.

1. Explain the significance of the following terms used to describe the structure of chromosomes:

   (a) DNA: ______________________________

   (b) Chromatin: ______________________________

   (c) Histone: ______________________________

   (d) Centromere: ______________________________

   (e) Chromatid: ______________________________

2. Each human cell has about a 1 meter length of DNA in its nucleus. Discuss the mechanisms by which this DNA is packaged into the nucleus and organized in such a way that it does not get ripped apart during cell division:

____

ISBN: 978-1-927173-16-9

DNA Molecules

Even the smallest DNA molecules are extremely long. The DNA from the small *Polyoma* virus, for example, is 1.7 µm long; about three times longer than the longest proteins. The DNA in a bacterial chromosome is 1000 times longer than the cell it is packed in. The amount of DNA in the nucleus of a eukaryotic cell varies widely. In vertebrate sex cells, the quantity of DNA ranges from 40,000 **kb** to 80,000,000 **kb**, with humans about in the middle of the range. The traditional focus of DNA research has been on protein-coding DNA sequences, yet protein-coding DNA accounts for less than 2% of the DNA in human chromosomes. The rest of the DNA was once called 'junk', meaning it did not code for anything. We now know that much of it codes for regulatory RNA molecules and it is not junk at all. The genomes of more complex organisms contain more of these RNA-only 'hidden' genes than the genomes of simple organisms. The sequences are short and difficult to identify, but they are highly conserved (they do not change much through evolution). This tells us that they must have an important role.

Total length of DNA in viruses, bacteria, and eukayotes

Taxon	Organism	Base pairs (in 1000s, or kb)	Length
Viruses	Polyoma or SV40	5.1	1.7 µm
	Lambda phage	48.6	17 µm
	T2 phage	166	56 µm
	Vaccinia	190	65 µm
Bacteria	Mycoplasma	760	260 µm
	E.coli (from human gut)	4600	1.56 mm
Eukaryotes	Yeast	13 500	4.6 mm
	Drosophila (fruit fly)	165 000	5.6 cm
	Human	2 900 000	99 cm

Kilobase (kb)

A kilobase (kb) is 1000 base pairs of a double-stranded nucleic acid molecule (or 1000 bases of a single-stranded molecule). One kb of double stranded DNA has a length of approximately 0.34 µm (1 µm = 1/1000 mm).

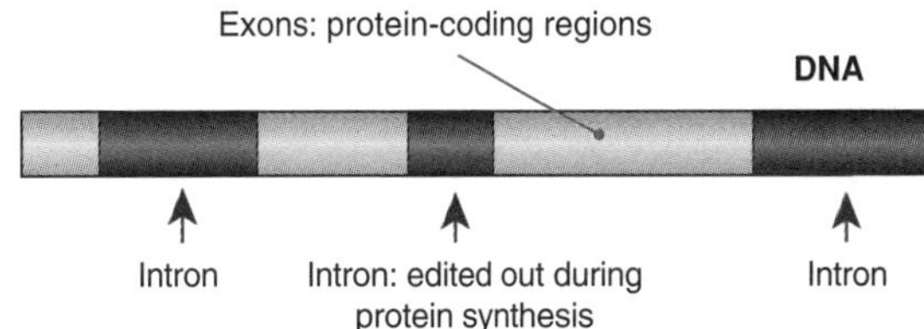
Exons: protein-coding regions
DNA
Intron
Intron: edited out during protein synthesis
Intron

Most protein-coding genes in eukaryotic DNA are not continuous. The protein-coding regions (**exons**) are interrupted by non-protein-coding regions called **introns**. Introns range in frequency from 1 to over 30 in a single 'gene' and also in size (100 to more than 10,000 bases). Introns are edited out of the protein-coding sequence during protein synthesis. After processing, the introns may go on to serve a regulatory function.

Giant Lampbrush Chromosomes

Lampbrush chromosomes are large chromosomes found in amphibian eggs, with lateral loops of DNA that produce a brushlike appearance under the microscope. The two scanning electron micrographs (below and right) show minute strands of DNA giving a fuzzy appearance in the high power view.

Loops of DNA
EII

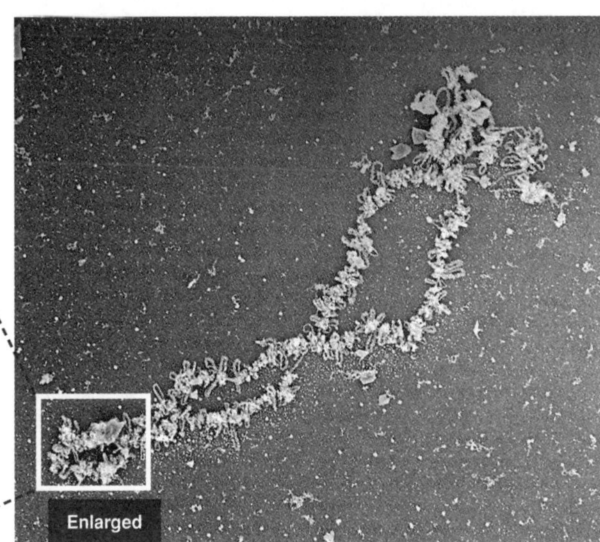
Enlarged
EII

1. Consult the table above and make the following comparisons. Determine how much more DNA is present in:

   (a) The bacterium *E. coli* compared to the Lambda Phage virus: ______

   (b) Human cells compared to the bacteria *E. coli:* ______

2. What proportion of DNA in a eukaryotic cell is used to code for proteins? ______

3. (a) Describe the organization of protein-coding regions in eukaryotic DNA: ______

   ______

   ______

   (b) What might be the purpose of the introns? ______

   ______

   ______

The Structure and Function of DNA

ISBN: 978-1-927173-16-9

Periodicals: *DNA: 50 years of the double helix*

***Related activities**: Genes to Proteins*

The Genetic Code

The genetic code consists of the sequence of bases arranged along the DNA molecule. It consists of a four-letter alphabet (from the four kinds of bases) and is read as three-letter words (called the triplet code on the DNA, or the codon on the mRNA). Each of the different kinds of triplet codes for a specific amino acid. This code is represented at the bottom of the page in the **mRNA-amino acid table**. There are 64 possible combinations of the four kinds of bases making up three-letter words. As a result, there is some **degeneracy** in the code: a specific amino acid may have several triplet codes (which leads to redundancy).

1. (a) Use the base-pairing rule for DNA replication to create the complementary strand for the template strand below.

   (b) For the same DNA strand, determine the mRNA sequence and then use the mRNA–amino acid table to determine the corresponding amino acid sequence. Note that in mRNA, uracil (U) replaces thymine (T) and pairs with adenine.

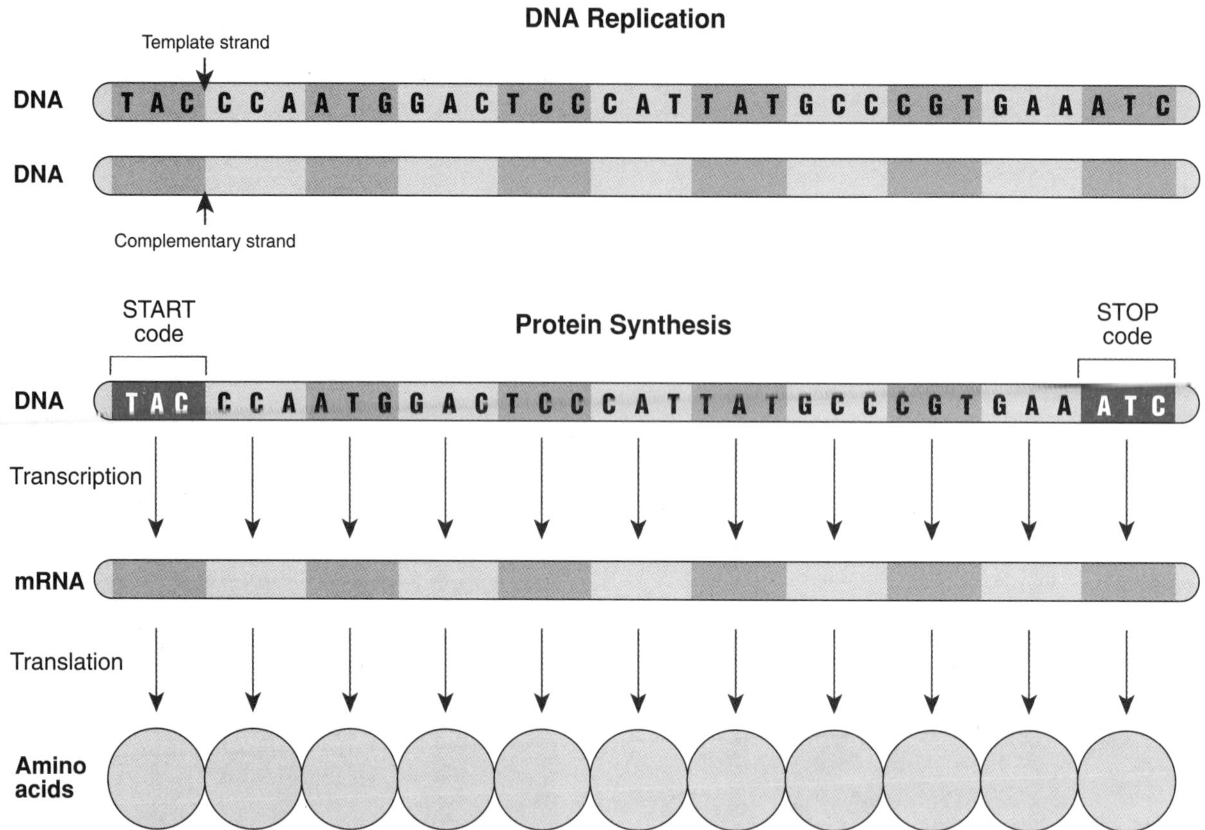
DNA Replication
Template strand
DNA
TAC CCA ATG GAC TCC CAT TAT GCC CGT GAA ATC
DNA
Complementary strand
START code
Protein Synthesis
STOP code
DNA
TAC CCA ATG GAC TCC CAT TAT GCC CGT GAA ATC
Transcription
mRNA
Translation
Amino acids

mRNA - Amino Acid Table

How to read the table

The table on the right is used to 'decode' the genetic code as a sequence of amino acids in a polypeptide chain, from a given mRNA sequence. The amino acid names are shown as three letter abbreviations (e.g. Ser = serine). To work out which amino acid is coded for by a codon (3 bases in the mRNA), carry out the following steps:

i Look for the first letter of the codon in the row on the left hand side of the table.

ii Look for the column that intersects the same row from above that matches the second base.

iii Locate the third base in the codon by looking along the row on the right hand side that matches your codon.

Example: **GAU** codes for Asp (asparagine)

Read second letter here. Read first letter here. Read third letter here.

First Letter	Second Letter: U	C	A	G	Third Letter
U	**UUU** Phe	**UCU** Ser	**UAU** Tyr	**UGU** Cys	U
	UUC Phe	**UCC** Ser	**UAC** Tyr	**UGC** Cys	C
	UUA Leu	**UCA** Ser	**UAA** STOP	**UGA** STOP	A
	UUG Leu	**UCG** Ser	**UAG** STOP	**UGG** Trp	G
C	**CUU** Leu	**CCU** Pro	**CAU** His	**CGU** Arg	U
	CUC Leu	**CCC** Pro	**CAC** His	**CGC** Arg	C
	CUA Leu	**CCA** Pro	**CAA** Gin	**CGA** Arg	A
	CUG Leu	**CCG** Pro	**CAG** Gin	**CGG** Arg	G
A	**AUU** Ile	**ACU** Thr	**AAU** Asn	**AGU** Ser	U
	AUC Ile	**ACC** Thr	**AAC** Asn	**AGC** Ser	C
	AUA Ile	**ACA** Thr	**AAA** Lys	**AGA** Arg	A
	AUG Met	**ACG** Thr	**AAG** Lys	**AGG** Arg	G
G	**GUU** Val	**GCU** Ala	**GAU** Asp	**GGU** Gly	U
	GUC Val	**GCC** Ala	**GAC** Asp	**GGC** Gly	C
	GUA Val	**GCA** Ala	**GAA** Glu	**GGA** Gly	A
	GUG Val	**GCG** Ala	**GAG** Glu	**GGG** Gly	G

2. (a) State the mRNA START and STOP codons: ______________________________

   (b) Describe the function of the START and STOP codons in a mRNA sequence: ______________________________

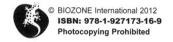

   ____

***Related activities**: Genes to Proteins*

ISBN: 978-1-927173-16-9

Creating a DNA Model

Although DNA molecules can be enormous in terms of their molecular size, they are made up of simple repeating units called **nucleotides**. A number of factors control the way in which these nucleotide building blocks are linked together. These factors cause the nucleotides to join together in a predictable way. This is referred to as the **base pairing rule** and can be used to construct a complementary DNA strand from a template strand, as illustrated in the exercise below:

DNA Base Pairing Rule			
Adenine	is always attracted to	**Thymine**	A ⟷ T
Thymine	is always attracted to	**Adenine**	T ⟷ A
Cytosine	is always attracted to	**Guanine**	C ⟷ G
Guanine	is always attracted to	**Cytosine**	G ⟷ C

1. Cut out page 97 and separate each of the 24 nucleotides by cutting along the columns and rows (see arrows indicating cutting points). Although drawn as geometric shapes, these symbols represent chemical structures.
2. Place one of each of the four kinds of nucleotide on their correct spaces below:

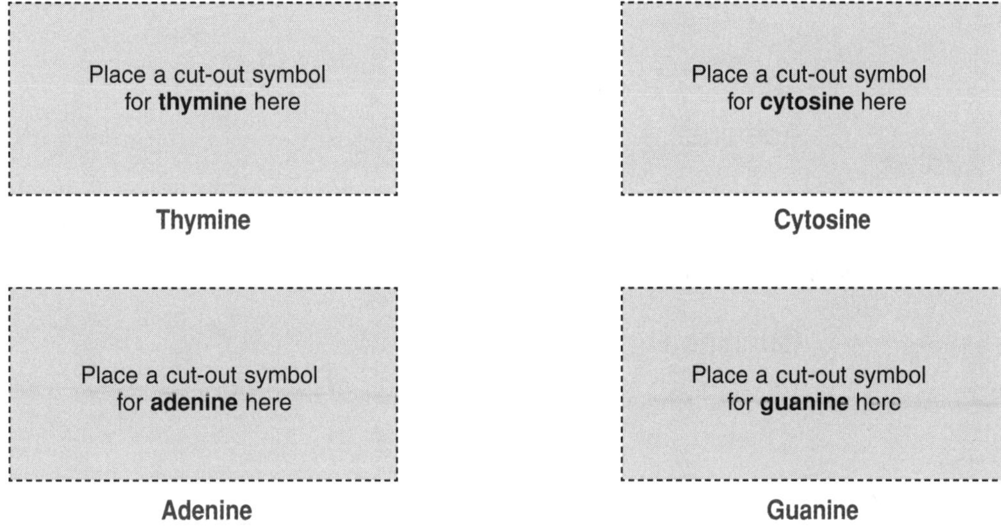
Place a cut-out symbol for **thymine** here
Thymine
Place a cut-out symbol for **cytosine** here
Cytosine
Place a cut-out symbol for **adenine** here
Adenine
Place a cut-out symbol for **guanine** here
Guanine

3. Identify and **label** each of the following features on the *adenine* nucleotide immediately above: **phosphate**, **sugar**, **base**, **hydrogen bonds**
4. Create one strand of the DNA molecule by placing the 9 correct 'cut out' nucleotides in the labelled spaces on the following page (DNA molecule). Make sure these are the right way up (with the **P** on the left) and are aligned with the left hand edge of each box. Begin with thymine and end with guanine.
5. Create the complementary strand of DNA by using the base pairing rule above. Note that the nucleotides have to be arranged upside down.
6. Under normal circumstances, it is not possible for adenine to pair up with guanine or cytosine, nor for any other mismatches to occur. Describe the two factors that prevent a mismatch from occurring:

   (a) Factor 1: ______________________________

   ____

   (b) Factor 2: ______________________________

   ____

7. Once you have checked that the arrangement is correct, you may glue, paste or tape these nucleotides in place.

NOTE:	There may be some value in keeping these pieces loose in order to practice the base pairing rule. For this purpose, *removable tape* would be best.

ISBN: 978-1-927173-16-9

DNA Molecule

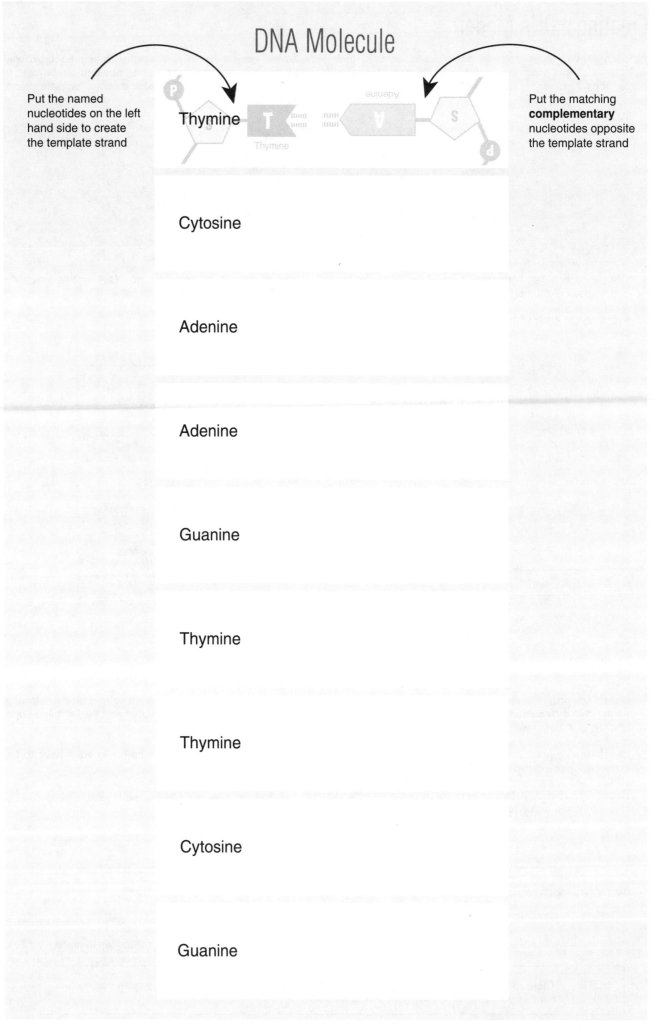

Put the named nucleotides on the left hand side to create the template strand

Put the matching **complementary** nucleotides opposite the template strand

Thymine
T
Thymine
A
Adenine
S
P

Cytosine

Adenine

Adenine

Guanine

Thymine

Thymine

Cytosine

Guanine

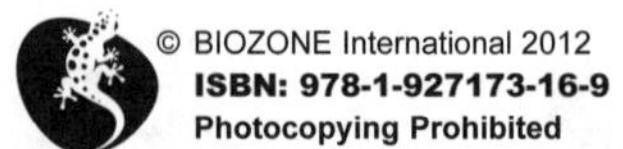

ISBN: 978-1-927173-16-9

Nucleotides

Tear out this page along the perforation and separate each of the 24 nucleotides by cutting along the columns and rows (see arrows indicating the cutting points).

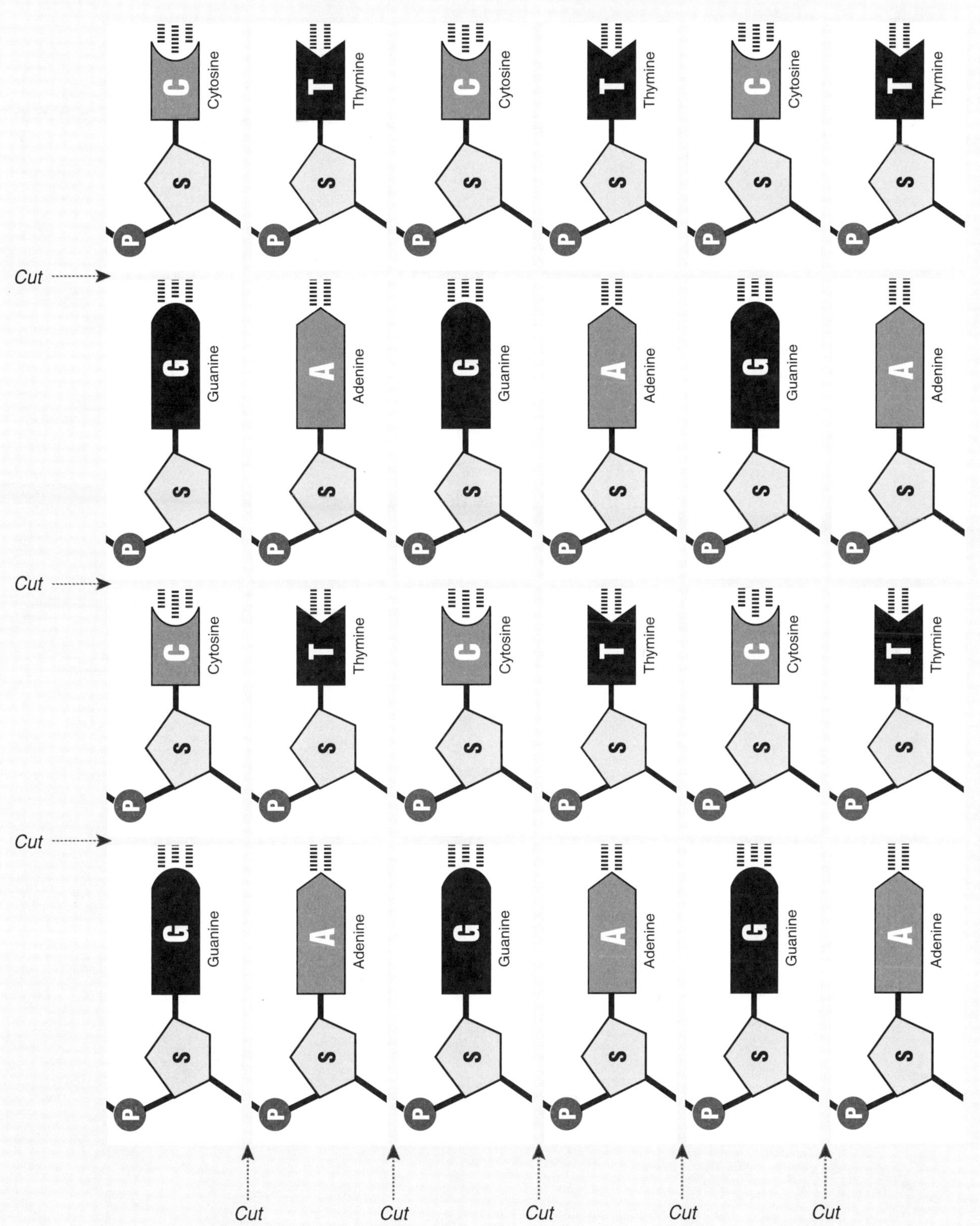

C
Cytosine
T
Thymine
G
Guanine
A
Adenine
S
P
Cut

ISBN: 978-1-927173-16-9

This page is left blank deliberately

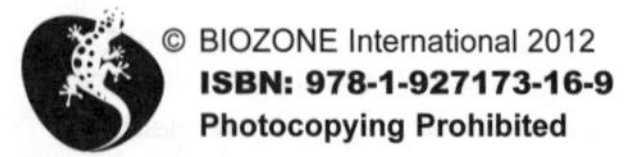

ISBN: 978-1-927173-16-9

DNA Replication

Cells carry out the process of **DNA replication** (DNA duplication) prior to cell division (mitosis and meiosis). This process ensures that each resulting cell is able to receive a complete set of genes from the original cell. After the DNA has replicated, each chromosome is made up of two chromatids, which are joined at the centromere. DNA replication is **semi-conservative**; each chromatid contains half original (parent) DNA and half new (daughter) DNA. The two chromatids will become separated during cell division to form two separate chromosomes. During DNA replication, new nucleotides become added at a region called the **replication fork**. The position of the replication fork moves along the chromosome as the replication progresses. This whole process occurs simultaneously for each chromosome of a cell and the entire process is tightly controlled by enzymes.

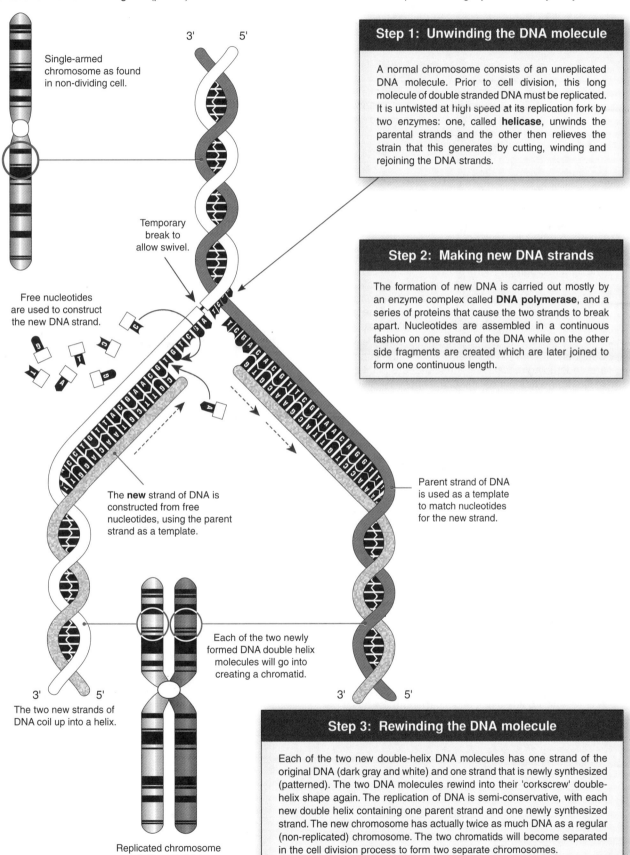

Single-armed chromosome as found in non-dividing cell.

3' 5'

Step 1: Unwinding the DNA molecule

A normal chromosome consists of an unreplicated DNA molecule. Prior to cell division, this long molecule of double stranded DNA must be replicated. It is untwisted at high speed at its replication fork by two enzymes: one, called **helicase**, unwinds the parental strands and the other then relieves the strain that this generates by cutting, winding and rejoining the DNA strands.

Temporary break to allow swivel.

Step 2: Making new DNA strands

The formation of new DNA is carried out mostly by an enzyme complex called **DNA polymerase**, and a series of proteins that cause the two strands to break apart. Nucleotides are assembled in a continuous fashion on one strand of the DNA while on the other side fragments are created which are later joined to form one continuous length.

Free nucleotides are used to construct the new DNA strand.

The **new** strand of DNA is constructed from free nucleotides, using the parent strand as a template.

Parent strand of DNA is used as a template to match nucleotides for the new strand.

Each of the two newly formed DNA double helix molecules will go into creating a chromatid.

3' 5'

The two new strands of DNA coil up into a helix.

3' 5'

Replicated chromosome ready for cell division.

Step 3: Rewinding the DNA molecule

Each of the two new double-helix DNA molecules has one strand of the original DNA (dark gray and white) and one strand that is newly synthesized (patterned). The two DNA molecules rewind into their 'corkscrew' double-helix shape again. The replication of DNA is semi-conservative, with each new double helix containing one parent strand and one newly synthesized strand. The new chromosome has actually twice as much DNA as a regular (non-replicated) chromosome. The two chromatids will become separated in the cell division process to form two separate chromosomes.

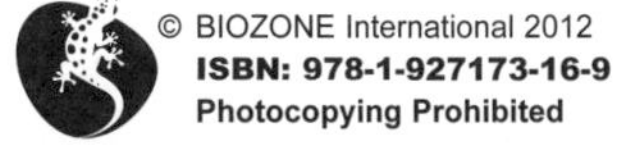

ISBN: 978-1-927173-16-9

Periodicals:
DNA polymerase

Related activities: *Mitosis and the Cell Cycle, The Genetic Code*
Weblinks: *DNA Replication*

1. State the purpose of DNA replication: ________________________

2. Summarize the three main steps involved in DNA replication:

   (a) ________________________

   (b) ________________________

   (c) ________________________

3. For a cell with 22 chromosomes, state how many chromatids would exist following DNA replication: ____________

4. Discuss the importance of enzymes in DNA replication: ________________________

5. DNA replication occurs during the S (synthesis) phase of the **cell cycle**. This is part of a larger phase called interphase. It is the phase in which the cell is not dividing (in mitosis).

   The light micrograph (right) shows a section of cells in an onion root tip. These cells have a cell cycle of approximately 24 hours. The cells can be seen to be in various stages of the cell cycle. By counting the number of cells in the various stages it is possible to calculate how long the cell spends in each stage of the cycle.

   Count and record the number of cells in the image which are undergoing mitosis and those that are in interphase. Estimate the amount of time a cell spends in each phase.

Stage	No. of cells	% of total cells	Estimated time in stage
Interphase			
Mitosis			
Total		100	

Onion Root Tip Cells

6. Match the statements in the table below to form complete sentences, then put the sentences in order to make a coherent paragraph about DNA replication and its role:

The enzymes also proofread the DNA during replication...	...is required before mitosis or meiosis can occur.
DNA replication is the process by which the DNA molecule...	...by enzymes.
Replication is tightly controlled...	...to correct any mistakes.
After replication, the chromosome...	...and half new (daughter) DNA.
DNA replication...	...during mitosis
The chromatids separate...	...is copied to produce to identical DNA strands.
A chromatid contains half original (parent) ...	...is made up of two chromatids.

Write the complete paragraph here: ________________________

ISBN: 978-1-927173-16-9

Enzyme Control of DNA Replication

The sequence of enzyme controlled events in DNA replication is shown below (1-5). Although shown as separate, many of the enzymes are found clustered together as enzyme complexes. These enzymes are also able to 'proof-read' the new DNA strand as it is made and correct mistakes. The polymerase enzyme can only work in one direction, so that one new strand is constructed as a continuous length (the **leading strand**) while the other new strand is made in short segments to be later joined together (the **lagging strand**). Note that the nucleotides are present as deoxynucleoside triphosphates. When they are hydrolyzed, energy released is used to incorporate the nucleotide into the growing strand.

DNA replication occurs during interphase of the cell cycle at an astounding rate. As many as 4000 nucleotides per second are replicated. This explains how under ideal conditions, bacterial cells with as many as 4 million nucleotides, can complete a cell cycle in about 20 minutes.

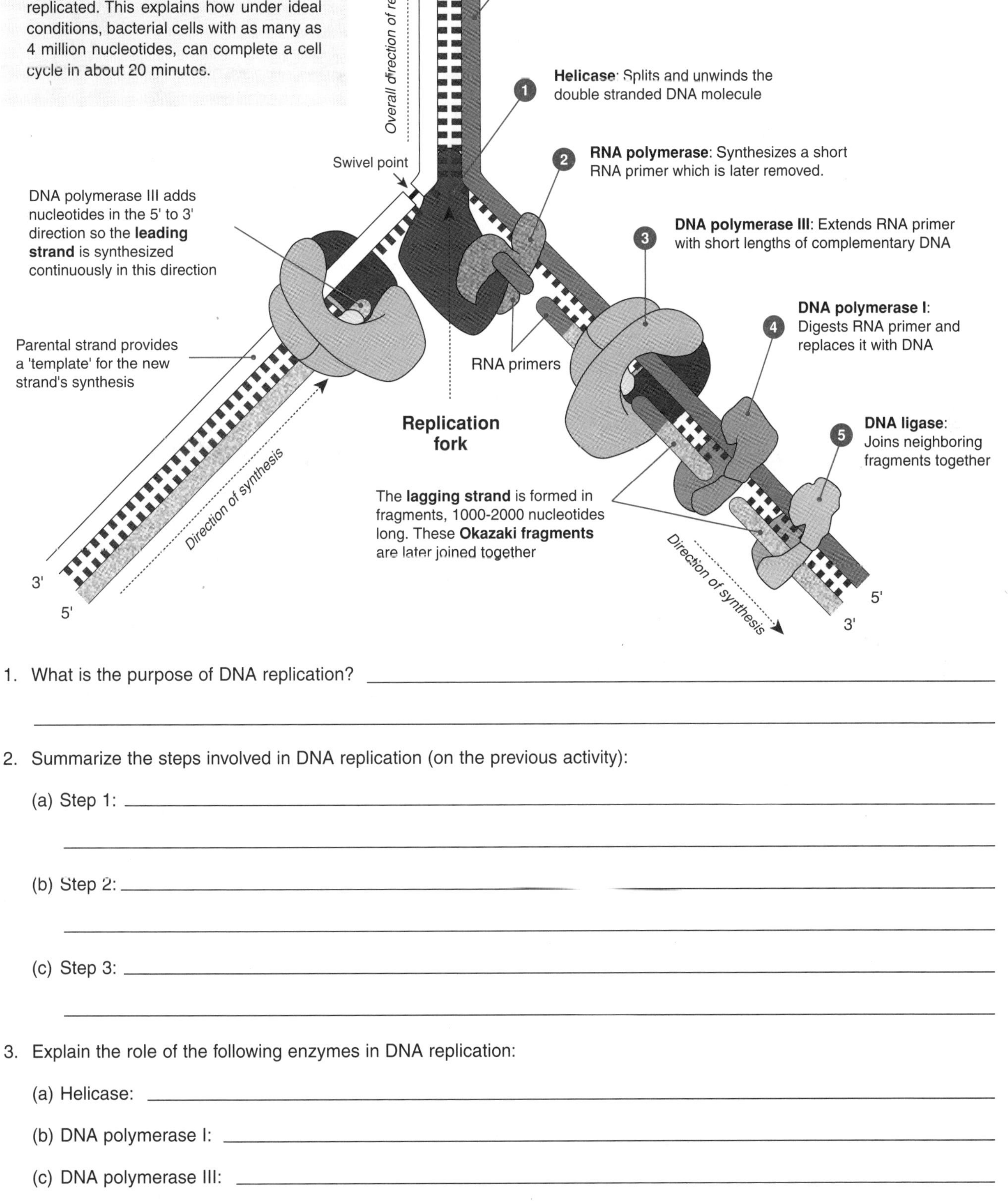

5'
3'
Overall direction of replication
Double strand of original (parental) DNA
1 Helicase: Splits and unwinds the double stranded DNA molecule
Swivel point
2 RNA polymerase: Synthesizes a short RNA primer which is later removed.
DNA polymerase III adds nucleotides in the 5' to 3' direction so the leading strand is synthesized continuously in this direction
3 DNA polymerase III: Extends RNA primer with short lengths of complementary DNA
4 DNA polymerase I: Digests RNA primer and replaces it with DNA
Parental strand provides a 'template' for the new strand's synthesis
RNA primers
Replication fork
5 DNA ligase: Joins neighboring fragments together
Direction of synthesis
The lagging strand is formed in fragments, 1000-2000 nucleotides long. These Okazaki fragments are later joined together
Direction of synthesis
3'
5'
5'
3'

1. What is the purpose of DNA replication? __________

2. Summarize the steps involved in DNA replication (on the previous activity):

   (a) Step 1: __________

   (b) Step 2: __________

   (c) Step 3: __________

3. Explain the role of the following enzymes in DNA replication:

   (a) Helicase: __________

   (b) DNA polymerase I: __________

   (c) DNA polymerase III: __________

   (d) Ligase: __________

4. Determine the time it would take for a bacteria to replicate its DNA (see note in diagram above): __________

ISBN: 978-1-927173-16-9

Related activities: DNA Replication
Weblinks: DNA Replication (Advanced)

Review of DNA Replication

The diagram below summarizes the main steps in DNA replication. You should use this activity to test your understanding of the main features of DNA replication, using the knowledge gained in the previous activity to fill in the missing information. You should attempt this from what you have learned, but refer to the previous activity if you require help.

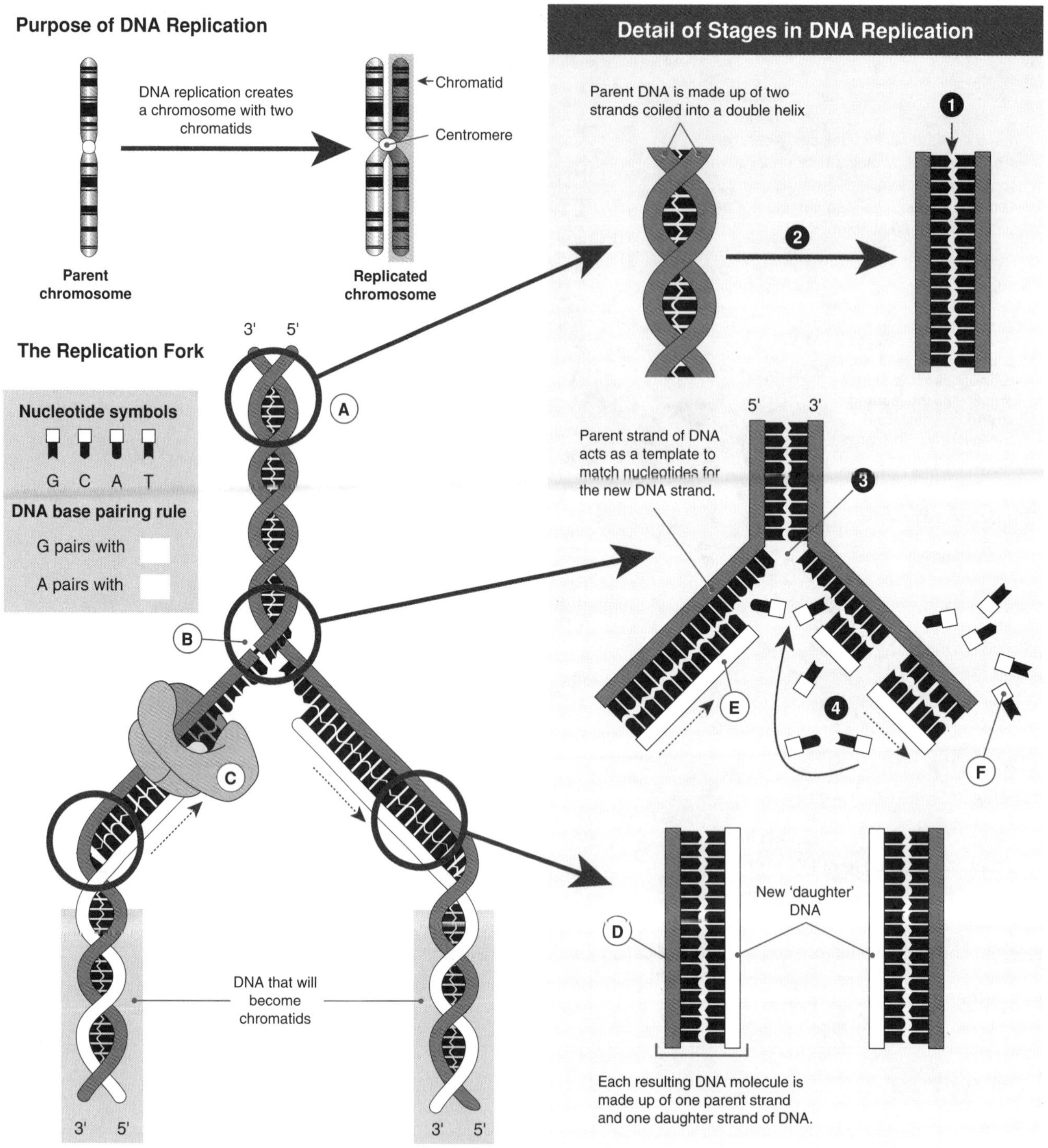

Purpose of DNA Replication

DNA replication creates a chromosome with two chromatids

Chromatid

Centromere

Parent chromosome

Replicated chromosome

The Replication Fork

3' 5'

Nucleotide symbols

G C A T

DNA base pairing rule

G pairs with

A pairs with

A

B

C

DNA that will become chromatids

3' 5'

3' 5'

Detail of Stages in DNA Replication

Parent DNA is made up of two strands coiled into a double helix

1

2

5' 3'

Parent strand of DNA acts as a template to match nucleotides for the new DNA strand.

3

4

E

F

D

New 'daughter' DNA

Each resulting DNA molecule is made up of one parent strand and one daughter strand of DNA.

1. In the white boxes provided in the diagram, state the base pairing rule for making a strand of DNA:

2. Identify each of the structures marked with a letter. (A-F):

A: ____________ C: ____________ E: ____________

B: ____________ D: ____________ F: ____________

3. Match each of the processes (1-4) to the correct summary of the process provided below:

☐ Unwinding of parent DNA double helix

☐ Unzipping of parent DNA

☐ Free nucleotides occupy spaces alongside exposed bases

☐ DNA strands are joined by base pairing

Related activities: *DNA Replication*
Weblinks: *DNA Replication*

ISBN: 978-1-927173-16-9

Genes to Proteins

The traditionally held view of genes was as sections of DNA coding only for protein. This view has been revised in recent years with the discovery that much of the nonprotein-coding DNA encodes functional RNAs; it is not all non-coding "junk" DNA as was previously assumed. In fact, our concept of what constitutes a gene is changing rapidly and now encompasses all those segments of DNA that are transcribed (to RNA). This activity considers only the simplest scenario: one in which the gene codes for a functional protein. **Nucleotides**, the basic unit of genetic information, are read in groups of three (**triplets**). Some triplets have a special controlling function in the making of a polypeptide chain. The equivalent of the triplet on the mRNA molecule is the **codon**. Three codons can signify termination of the amino acid chain (UAG, UAA and UGA in the mRNA code). The codon AUG is found at the beginning of every gene (on mRNA) and marks the starting point for reading the gene. The genes required to form a functional end-product (in this case, a functional protein) are collectively called a **transcription unit**.

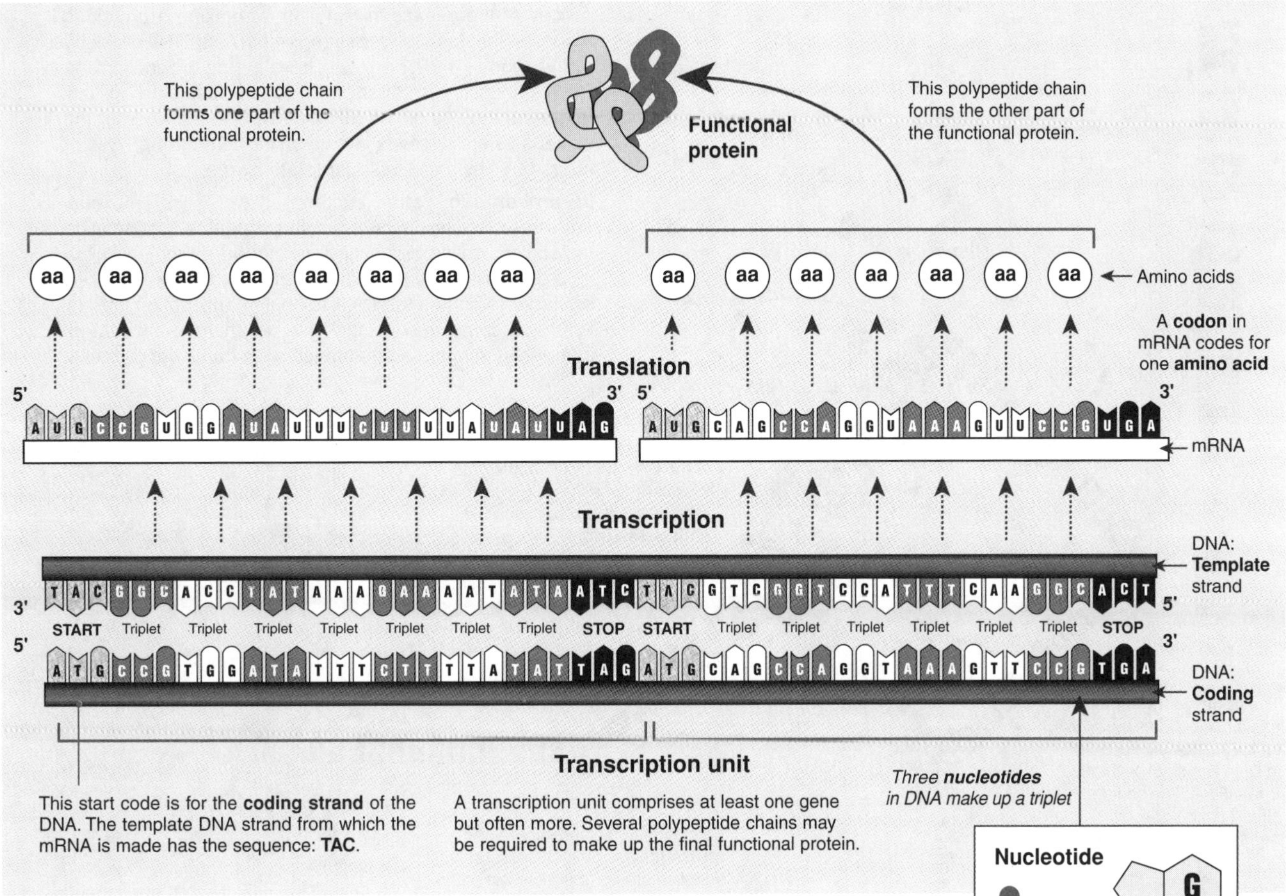

This polypeptide chain forms one part of the functional protein.

Functional protein

This polypeptide chain forms the other part of the functional protein.

aa aa aa aa aa aa aa aa | aa aa aa aa aa aa aa ← Amino acids

A **codon** in mRNA codes for one **amino acid**

Translation

5' AUGCCGUGGAUAUUUCUUUUAUAUUAG 3' 5' AUGCAGCCAGGUAAAGUUCCGUGA 3' ← mRNA

Transcription

3' TACGGCACCTATAAAGAAAATATAATCTACGTCGGTCCATTTCAAGGCACT 5' ← DNA: **Template** strand

START Triplet Triplet Triplet Triplet Triplet Triplet Triplet **STOP** **START** Triplet Triplet Triplet Triplet Triplet Triplet **STOP**

5' ATGCCGTGGATATTTCTTTTATATTAGATGCAGCCAGGTAAAGTTCCGTGA 3' ← DNA: **Coding** strand

Transcription unit

This start code is for the **coding strand** of the DNA. The template DNA strand from which the mRNA is made has the sequence: **TAC**.

A transcription unit comprises at least one gene but often more. Several polypeptide chains may be required to make up the final functional protein.

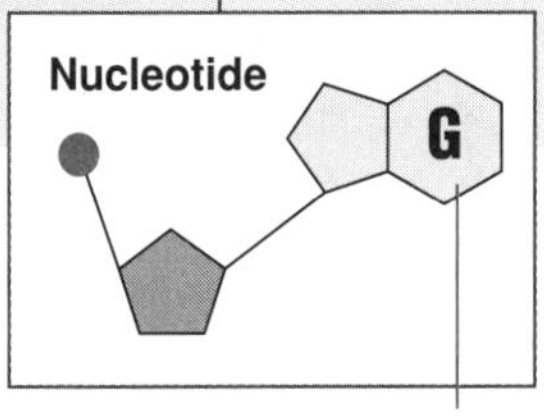

Three ***nucleotides*** *in DNA make up a triplet*

Nucleotide

G

In models of nucleic acids, nucleotides are denoted by their base letter. (In this case: **G** is for guanine)

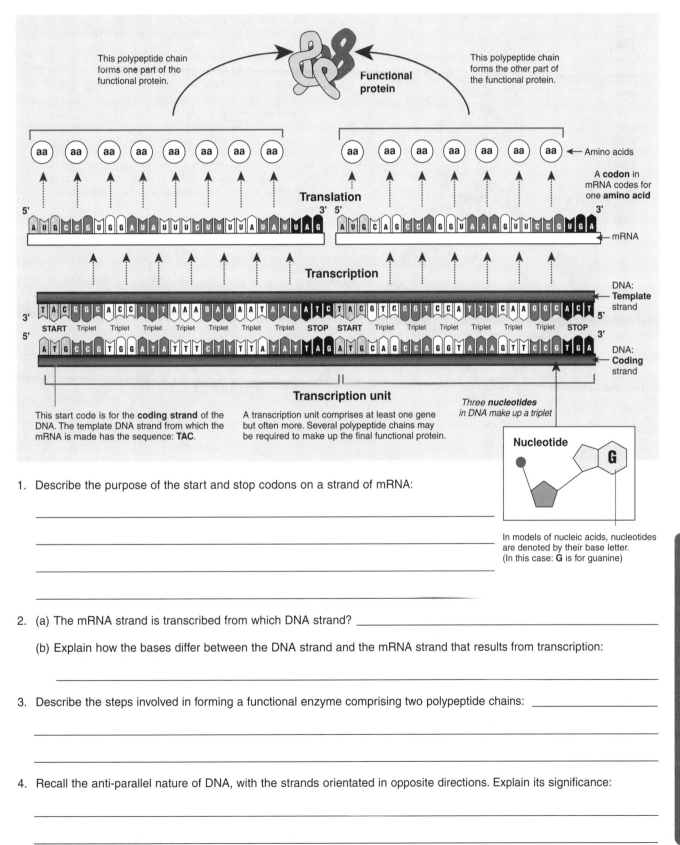

1. Describe the purpose of the start and stop codons on a strand of mRNA:

2. (a) The mRNA strand is transcribed from which DNA strand?

   (b) Explain how the bases differ between the DNA strand and the mRNA strand that results from transcription:

3. Describe the steps involved in forming a functional enzyme comprising two polypeptide chains:

4. Recall the anti-parallel nature of DNA, with the strands orientated in opposite directions. Explain its significance:

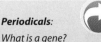

ISBN: 978-1-927173-16-9

Transcription

Transcription is the process by which the code contained in the DNA molecule is transcribed (rewritten) into a **mRNA** molecule. Transcription is under the control of the cell's metabolic processes, which must activate a gene before this process can begin. The enzyme that directly controls the process is **RNA polymerase**. It makes a strand of mRNA using the single strand of DNA (the **template strand**) as a template (hence the term). The enzyme transcribes only a gene length of DNA at a time, recognizing start and stop signals (codes) at the beginning and end of the gene. Only RNA polymerase is involved in mRNA synthesis and it also causes the unwinding of the DNA. It is common to find several RNA polymerase enzyme molecules on the same gene at any one time, allowing a high rate of mRNA synthesis to occur. Before the mRNA can be translated into a protein, non-coding sections called **introns** must first be removed and the remaining **exons** spliced together to form mature mRNA.

A copy of the genetic information for making a protein is made in the form of **messenger RNA** (mRNA). Many mRNA copies may be made from a single gene on the DNA molecule. In **eukaryotic cells** (depicted in this diagram), the completed mRNA travels to the edge of the nucleus where it moves into the cytoplasm through one of numerous tiny holes called **nuclear pores**.

In **prokaryotic cells** (bacteria) there is no nucleus, and transcription occurs in the cytoplasm alongside translation, so protein synthesis can be extremely rapid. In prokaryotes, the RNA polymerase binds directly to a promoter region on the DNA to begin transcription. mRNA synthesis continues until the RNA polymerase encounters a termination sequence, at which point transcription stops.

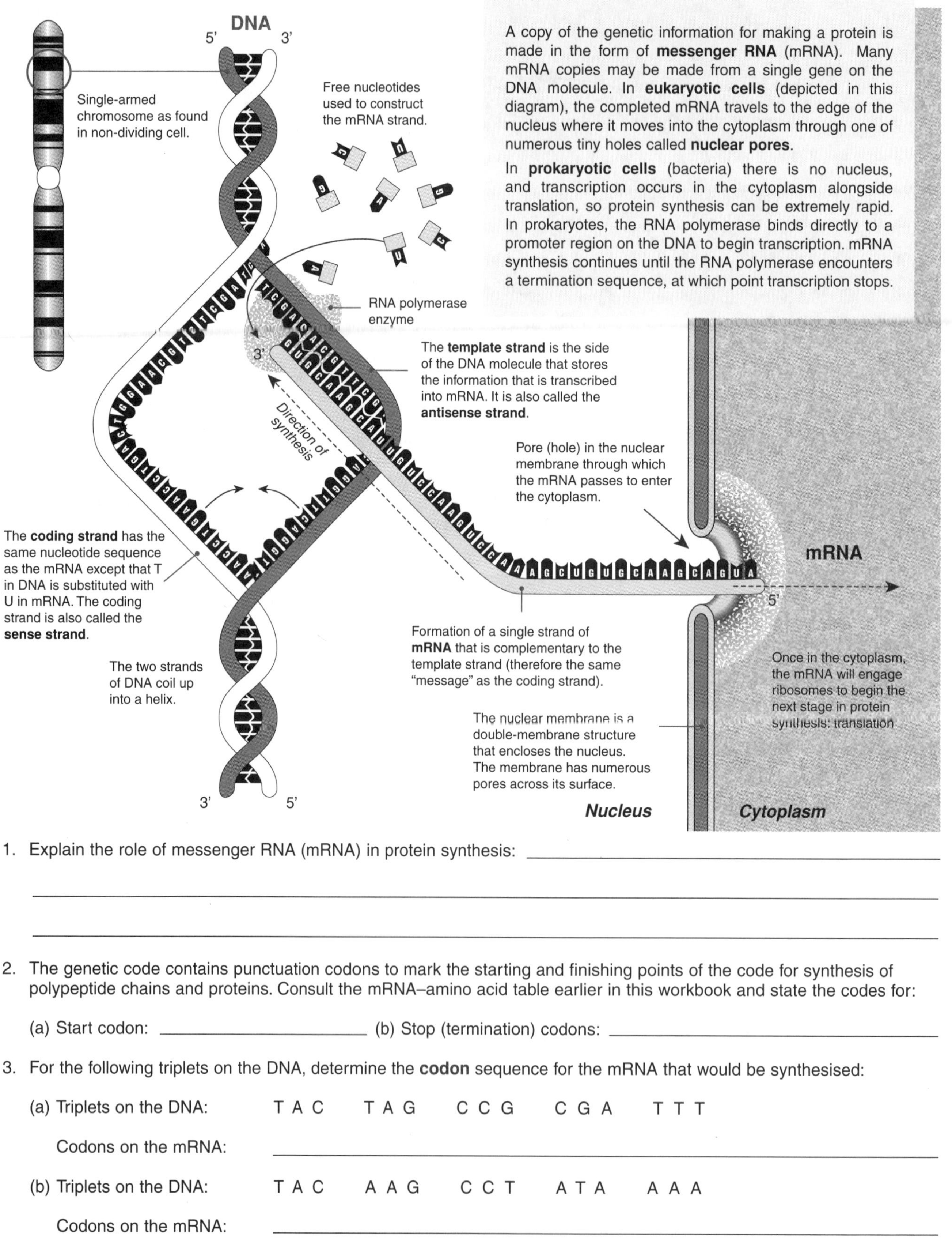
DNA
5'
3'
Single-armed chromosome as found in non-dividing cell.
Free nucleotides used to construct the mRNA strand.
RNA polymerase enzyme
The **template strand** is the side of the DNA molecule that stores the information that is transcribed into mRNA. It is also called the **antisense strand**.
Direction of synthesis
Pore (hole) in the nuclear membrane through which the mRNA passes to enter the cytoplasm.
mRNA
The **coding strand** has the same nucleotide sequence as the mRNA except that T in DNA is substituted with U in mRNA. The coding strand is also called the **sense strand**.
The two strands of DNA coil up into a helix.
Formation of a single strand of **mRNA** that is complementary to the template strand (therefore the same "message" as the coding strand).
The nuclear membrane is a double-membrane structure that encloses the nucleus. The membrane has numerous pores across its surface.
Once in the cytoplasm, the mRNA will engage ribosomes to begin the next stage in protein synthesis: translation
Nucleus
Cytoplasm
3'
5'

1. Explain the role of messenger RNA (mRNA) in protein synthesis: ____________________

2. The genetic code contains punctuation codons to mark the starting and finishing points of the code for synthesis of polypeptide chains and proteins. Consult the mRNA–amino acid table earlier in this workbook and state the codes for:

   (a) Start codon: ____________ (b) Stop (termination) codons: ____________

3. For the following triplets on the DNA, determine the **codon** sequence for the mRNA that would be synthesised:

   (a) Triplets on the DNA: T A C T A G C C G C G A T T T

   Codons on the mRNA: ____________________

   (b) Triplets on the DNA: T A C A A G C C T A T A A A A

   Codons on the mRNA: ____________________

ISBN: 978-1-927173-16-9

Translation

The diagram below shows the translation phase of protein synthesis. The scene shows how a single mRNA molecule can be 'serviced' by many ribosomes at the same time. The ribosome on the right is in a more advanced stage of constructing a polypeptide chain because it has 'translated' more of the mRNA than the ribosome on the left. The anticodon at the base of each tRNA must make a perfect complementary match with the codon on the mRNA before the amino acid is released. Once released, the amino acid is added to the growing polypeptide chain by enzymes.

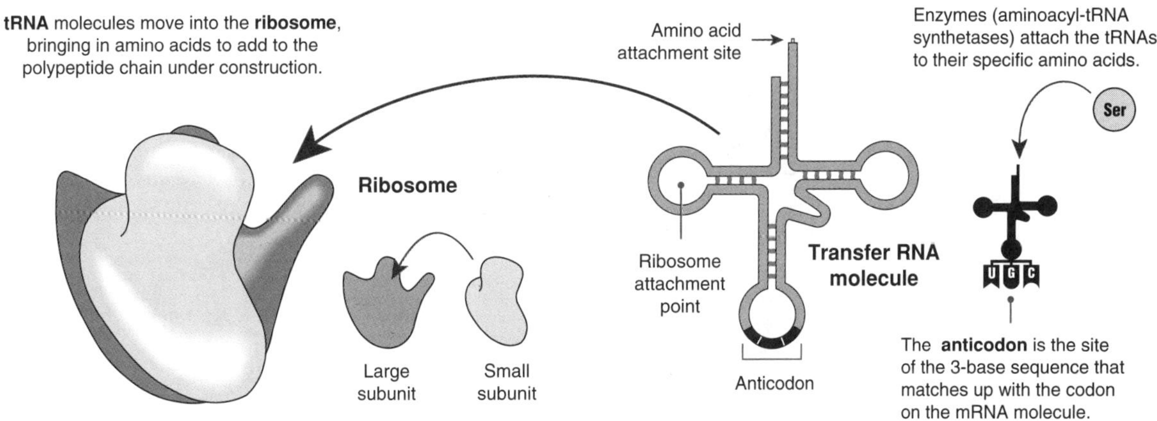

tRNA molecules move into the **ribosome**, bringing in amino acids to add to the polypeptide chain under construction.
Ribosome
Large subunit
Small subunit
Amino acid attachment site
Ribosome attachment point
Transfer RNA molecule
Anticodon
Enzymes (aminoacyl-tRNA synthetases) attach the tRNAs to their specific amino acids.
Ser
U G C
The **anticodon** is the site of the 3-base sequence that matches up with the codon on the mRNA molecule.

Ribosomes are made up of a complex of ribosomal RNA (rRNA) and proteins. They exist as two separate sub-units (above) until they are attracted to a binding site on the mRNA molecule, when they join together. Ribosomes have binding sites that attract transfer RNA (**tRNA**) molecules loaded with amino acids. The tRNA molecules are about 80 nucleotides in length and are made under the direction of genes in the chromosomes. There is a different tRNA molecule for each of the different possible anticodons (see the diagram below) and, because of the degeneracy of the genetic code, there may be up to six different tRNAs carrying the same amino acid.

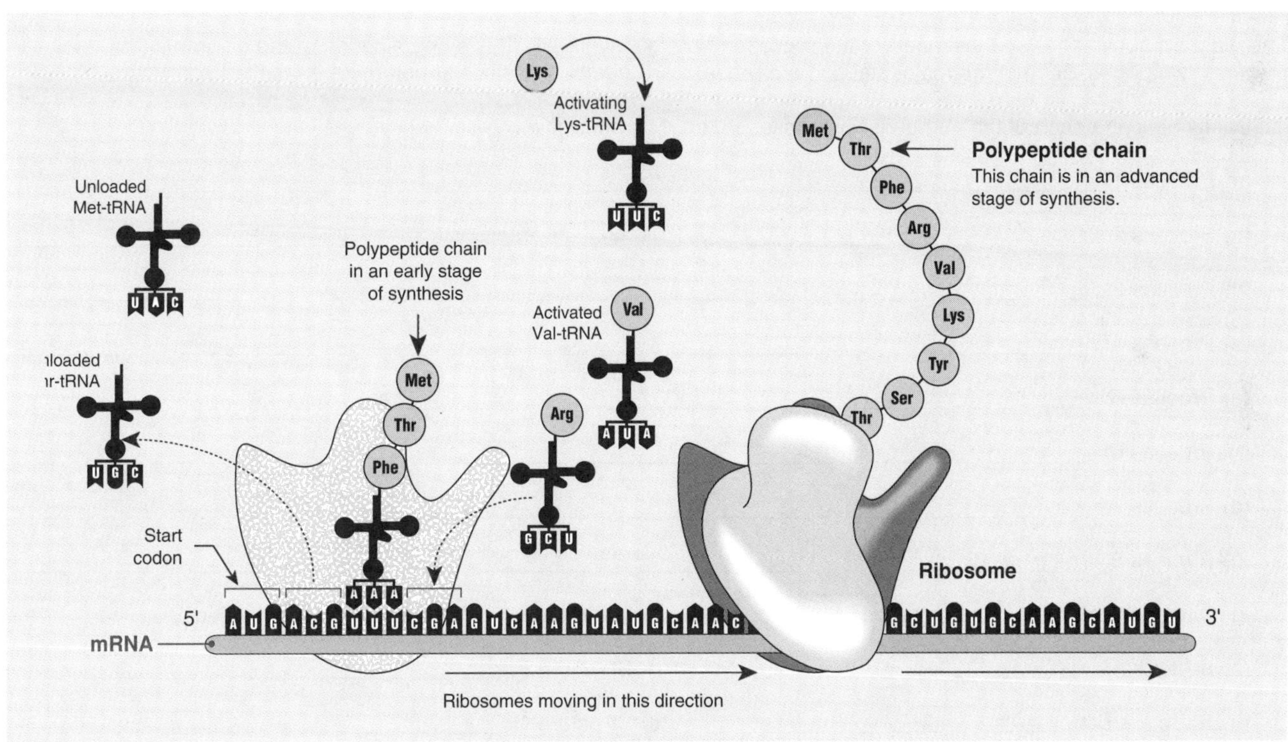

Lys
Activating Lys-tRNA
U U C
Unloaded Met-tRNA
U A C
Polypeptide chain in an early stage of synthesis
Activated Val-tRNA
Val
A U A
Met
Thr
Phe
Arg
G C U
U G C
Start codon
A A A
Polypeptide chain
This chain is in an advanced stage of synthesis.
Met
Thr
Phe
Arg
Val
Lys
Tyr
Ser
Thr
Ribosome
5'
A U G A C G U U U C G A G U C A A G U A U G C A A C
G C U G U G C A A G C A U G U
3'
mRNA
Ribosomes moving in this direction

1. For the following codons on the mRNA, determine the **anticodons** for each tRNA that would deliver the amino acids:

   Codons on the mRNA: U A C U A G C C G C G A U U U

   Anticodons on the tRNAs: ______________________________

2. There are many different types of tRNA molecules, each with a different anticodon (HINT: see the mRNA table).

   (a) How many different tRNA types are there, each with a unique anticodon? ______________________________

   (b) Explain your answer: ______________________________

   ____

   ____

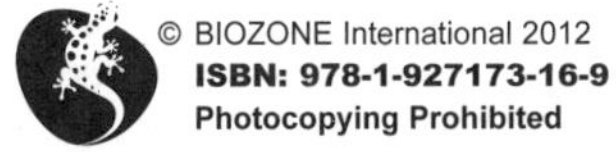

ISBN: 978-1-927173-16-9

Periodicals: *Transfer RNA*

Related activities: *The Genetic Code, Genes to Proteins*
Weblinks: *Polyribosomes, Review of Protein Synthesis*

Protein Synthesis Summary

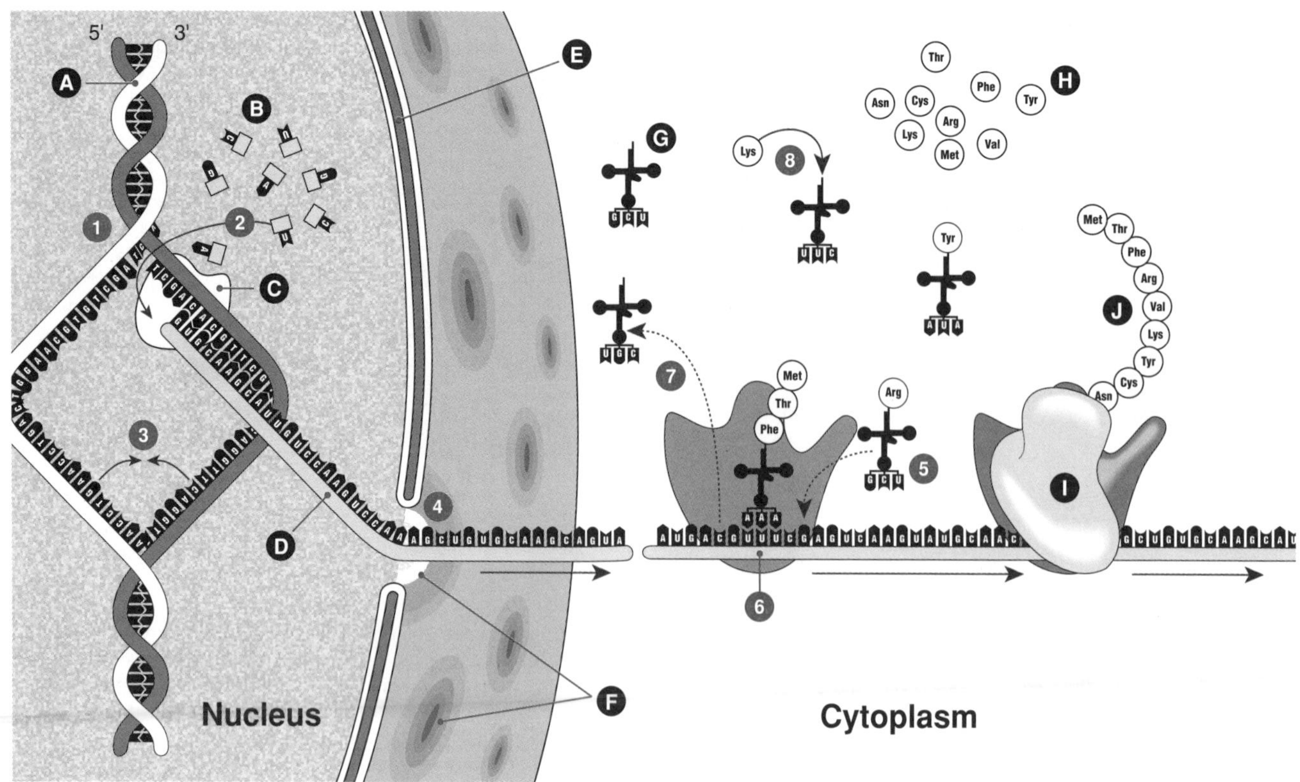

5'
3'
A
B
C
D
E
F
G
H
I
J
1
2
3
4
5
6
7
8
Nucleus
Cytoplasm
Thr
Asn
Cys
Phe
Tyr
Arg
Lys
Met
Val
Lys
Tyr
Met
Thr
Phe
Arg
Val
Lys
Tyr
Cys
Asn
Met
Thr
Phe
Arg

The diagram above shows an overview of the process of protein synthesis. It is a combination of the diagrams from the previous two pages. Each of the major steps in the process are numbered, while structures are labelled with letters.

1. Briefly describe each of the numbered processes in the diagram above:

(a) Process 1: ______

(b) Process 2: ______

(c) Process 3: ______

(d) Process 4: ______

(e) Process 5: ______

(f) Process 6: ______

(g) Process 7: ______

(h) Process 8: ______

2. Identify each of the structures marked with a letter and write their names below in the spaces provided:

(a) Structure A: ______ (f) Structure F: ______

(b) Structure B: ______ (g) Structure G: ______

(c) Structure C: ______ (h) Structure H: ______

(d) Structure D: ______ (i) Structure I: ______

(e) Structure E: ______ (j) Structure J: ______

3. Describe two factors that would determine whether or not a particular protein is produced in the cell:

(a) ______

(b) ______

Related activities: *Transcription, Translation*
Weblinks: *Review of Protein Synthesis*

Periodicals: *Gene structure and expression*

ISBN: 978-1-927173-16-9

KEY TERMS: Word find

Use the clues below to find the relevant key terms in the WORD FIND grid

A	N	F	J	D	N	A	R	E	P	L	I	C	A	T	I	O	N	J	I	Y	Y	R	F	R
T	S	I	H	G	G	S	E	N	S	E	S	T	R	A	N	D	B	K	W	F	U	S	R	N
N	U	C	L	E	O	T	I	D	E	S	Z	J	D	F	F	I	X	L	G	L	L	V	G	A
J	V	T	S	E	P	H	Y	D	R	O	G	E	N	B	O	N	D	I	N	G	W	V	P	K
W	T	R	A	N	S	L	A	T	I	O	N	G	E	X	E	X	O	N	S	Q	S	I	R	D
E	R	J	G	E	N	E	E	X	P	R	E	S	S	I	O	N	Q	W	M	A	J	U	H	U
S	F	U	E	O	V	N	A	M	I	N	O	A	C	I	D	Y	N	J	Z	I	T	C	P	F
G	E	N	E	T	I	C	C	O	D	E	D	N	A	L	I	G	A	S	E	F	X	N	I	V
A	H	G	D	K	X	O	K	A	Z	A	K	I	F	R	A	G	M	E	N	T	S	Y	I	X
N	F	W	C	P	C	T	H	S	M	M	U	S	X	K	P	I	N	T	R	O	N	S	M	M
T	D	E	C	X	Y	M	D	L	O	B	A	S	E	P	A	I	R	I	N	G	R	U	L	E
I	A	M	G	B	C	S	O	P	D	X	D	N	A	P	O	L	Y	M	E	R	A	S	E	F
C	S	B	O	N	O	Y	A	N	T	I	S	E	N	S	E	S	T	R	A	N	D	W	L	R
O	W	R	D	F	D	K	T	Q	Q	X	Z	N	U	C	L	E	I	C	A	C	I	D	S	V
D	Z	A	S	M	O	P	R	O	T	E	I	N	S	M	A	K	P	N	H	A	K	M	D	L
O	F	X	J	Z	N	J	S	T	H	E	L	I	C	A	S	E	O	R	I	R	Q	O	X	B
N	H	T	R	A	N	S	C	R	I	P	T	I	O	N	O	Y	D	N	A	S	F	F	H	X

CORE CLUES

Single stranded nucleic acid that consists of nucleotides that contain ribose sugar.

Organic macromolecules composed of linear chains of amino acids joined together by peptide bonds and then organized, e.g. through folding, into a functional structure.

The process by which genetic information is used to produce a functional gene product.

The rule governing the pairing of complementary bases in DNA. .

The region of a transfer RNA with a sequence of three bases that are complementary to a codon in the messenger RNA.

The semi-conservative process by which two identical DNA molecules are produced from a single double-stranded DNA molecule.

The process of creating an equivalent RNA copy of a sequence of DNA.

An enzyme that separates two annealed DNA strands using energy from ATP hydrolysis.

Form of intermolecular bonding between hydrogen and an electro-negative atom such as oxygen.

Universally found macromolecules composed of chains of nucleotides. These molecules carry genetic information within cells.

An enzyme that catalyzes the incorporation of deoxyribonucleotides into a DNA strand.

The structural units of nucleic acids, DNA and RNA.

A sequence of three adjacent nucleotides constituting the code for an amino acid.

Organic compound consisting of a carboxyl, an amine and an R group (where R may be one of 20 different atomic groupings). Polymerized by peptide bonds to form proteins.

Macromolecule consisting of many millions of units containing a phosphate group, sugar and a base (A,T, C or G). Stores the genetic information of the cell.

The stage of gene expression in which mRNA is decoded to produce a polypeptide.

A set of rules by which information encoded in DNA or mRNA is translated into proteins.

AHL ONLY CLUES

The sequence of DNA that is read during the synthesis of mRNA.

DNA regions within a gene that are not translated into protein.

The DNA strand with the same base sequence as the RNA transcript produced (although with thymine replaced by uracil in mRNA).

Nucleic acid sequences that are represented in the mature form of an RNA molecule.

Relatively short pieces of DNA created on the lagging strand during DNA replication.

An enzyme that links together two DNA strands that have double-strand break.

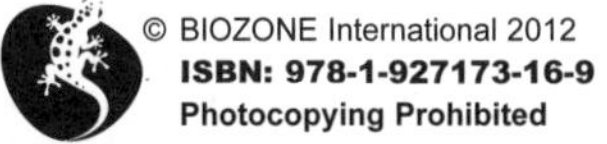

ISBN: 978-1-927173-16-9

Core Topic **3**

AHL Topic **7 & 8**

Enzymes and **Metabolism**

Key concepts

- Enzymes catalyze reactions by lowering the energy required for the reaction to take place.
- ATP is the universal energy currency in cells.
- Both cell respiration and photosynthesis involve the use of ATP and electron carriers.
- Cell respiration involves the stepwise oxidation of glucose in the mitochondria.
- Photosynthesis uses light energy to fix carbon in organic compounds. It occurs in the chloroplasts.

Key terms

Core

alcoholic fermentation
ATP
biological catalyst
cell respiration
chlorophyll
denaturation
enzyme
enzyme-substrate specificity
fermentation
glycolysis
lactase
lactic acid fermentation
lock and key model
mitochondrion
photosynthesis
photosynthetic rate
pyruvate
substrate

AHL only

absorption spectrum
action spectrum
allostery
Calvin cycle
chemiosmosis
competitive inhibition
electron transport chain
end-product inhibition
induced fit model
Krebs cycle
light dependent phase
light independent phase
link reaction
metabolic pathway
NAD^+
non-competitive inhibition
oxidation
oxidative phosphorylation
photophosphorylation
reduction

Periodicals:
Listings for this chapter are on page 398

Weblinks:
www.thebiozone.com/weblink/IB-3169.html

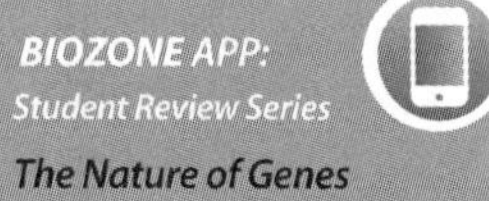
BIOZONE APP:
Student Review Series
The Nature of Genes

Learning Objectives

☐ 1. Use the **KEY TERMS** to help you understand and complete these objectives.

Enzymes *(3.6 & 7.6)* pages 109-115

☐ 2. Define the terms **enzyme** and **active site**. Use the **lock and key model** as a basis for explaining **enzyme-substrate specificity**.

☐ 3. Explain the effect of temperature, pH, and substrate concentration on enzyme activity. Explain what is meant by **denaturation** and describe how extremes of temperature and pH cause loss of a protein's biological function.

☐ 4. Describe the use of **lactase** in the production of lactose-free milk.

☐ 5. **AHL**: Explain how enzymes lower the activation energy required for a reaction to proceed. Describe the **induced-fit model** of enzyme function and explain why it is a more suitable model for enzyme function than the lock-and-key model.

☐ 6. **AHL**: Using examples, distinguish between **competitive** and **non-competitive inhibition**. How would you distinguish them in graphs of enzyme activity?

☐ 7. **AHL**: Explain what is mean by a **metabolic pathway**. Explain the role of **end-product inhibition** and **allostery** in the regulation of metabolic pathways.

Cell Respiration *(3.7 & 8.1)* pages 116-124

☐ 8. Define **cell respiration** as a catabolic process which releases energy as **ATP**. Recognize **glycolysis** as a universal anaerobic process occurring in the cytoplasm, which produces pyruvate with a small yield of ATP. Describe the complete breakdown of **pyruvate** to carbon dioxide and water in the **mitochondrion**. Contrast the ATP yield from glycolysis and aerobic respiration. Describe **fermentation** in muscle and yeast.

☐ 9. **AHL**: Define **oxidation** and **reduction**. Outline the process of **glycolysis**.

☐ 10. **AHL**: Describe the structure of a **mitochondrion**, and relate its structure to its function. Describe aerobic respiration, including the **link reaction**, **Krebs cycle**, the role of **NAD^+** + H^+ and the **electron transport chain**, and the role of oxygen.

☐ 11. **AHL**: Explain how ATP is generated by **chemiosmosis** in **oxidative phosphorylation**.

Photosynthesis *(3.8 & 8.2)* pages 125-130

☐ 12. Understand that **photosynthesis** is an anabolic process that converts light energy to chemical energy. Describe the nature of light energy from the Sun in terms of wavelengths. Explain the role of **chlorophyll** in light capture by green plants and outline differences in absorption of red, blue, and green light by chlorophyll.

☐ 13. Describe the role of ATP and hydrogen in the fixation of carbon in organic molecules.

☐ 14. Explain direct and indirect methods to measure the rate of photosynthesis. Describe and explain factors affecting **photosynthetic rate**.

☐ 15. **AHL**: Use a labeled diagram to describe the structure of a chloroplast as seen in electron micrographs. Relate chloroplast structure to function. Explain what is meant by the **absorption spectrum** and **action spectrum** of photosynthetic pigments.

☐ 16. **AHL**: Describe and explain **photosynthesis** in a C_3 plant, including reference to:
- The generation of ATP and $NADPH_2$ in the **light dependent phase**.
- The **Calvin cycle** and the fixation of CO_2 in the **light independent phase**.

☐ 17. **AHL**: Using specific examples, explain the concept of limiting factors in photosynthesis.

Enzymes

Most enzymes are proteins. They are capable of catalyzing (speeding up) biochemical reactions and are therefore called biological **catalysts**. Enzymes act on one or more compounds (called the **substrate**). They may break down a single substrate molecule into simpler substances, or join two or more substrate molecules together. The enzyme itself is unchanged in the reaction; its presence merely allows the reaction to take place more rapidly. The part of the enzyme into which the substrate binds and undergoes reaction is the **active site**. It is a function of the polypeptide's complex tertiary structure.

Enzyme Structure

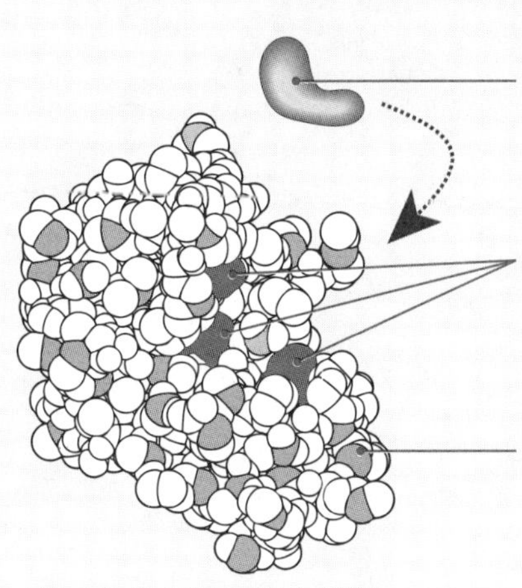

Substrate molecule: Substrate molecules are the chemicals that an enzyme acts on. They are drawn into the cleft of the enzyme.

Active site. These attraction points draw the substrate to the enzyme's surface. Substrate molecule(s) are positioned in a way to promote a reaction: either joining two molecules together or splitting up a larger one (as in this case).

Enzyme molecule: The complexity of the active site is what makes each enzyme so specific for the substrate it acts on.

Source: After *Biochemistry*, (1981) by Lubert Stryer

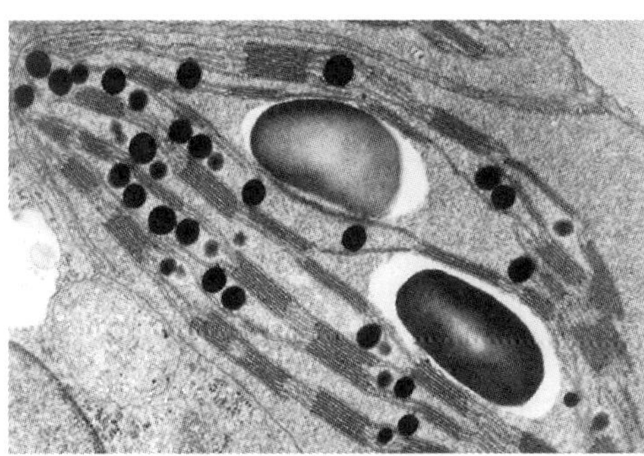

In the first stage of photosynthesis, enzymes catalyze the production of ATP and NADPH. These provide the energy and hydrogen molecules for the second stage of photosynthesis. Enzymes also catalyze the steps that fix carbon from CO_2 to produce carbohydrates.

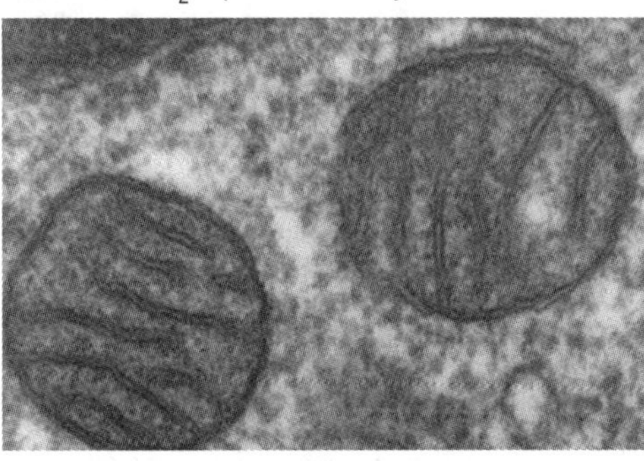

The breakdown of glucose is catalyzed by numerous enzymes. Glycolysis uses ten different enzymes, one for each step of the process. The Krebs cycle involves another eight enzymes. The electron transport chain moves H^+ across a membrane to create the proton gradient required to produce ATP using the enzyme ATP synthase.

The Lock and Key Model of Enzyme Function

The **lock and key** model proposed earlier last century suggested that the (perfectly fitting) substrate was simply drawn into a matching cleft on the enzyme molecule (below). This model was supported by early X-ray crystallography but has since been modified to recognize the flexibility of enzymes.

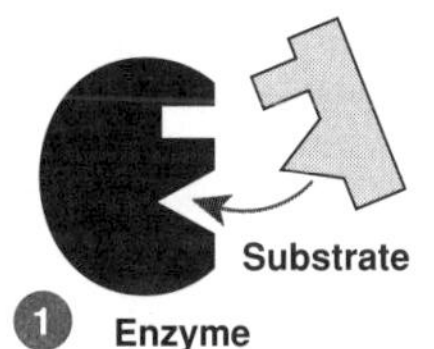
Substrate
1
Enzyme

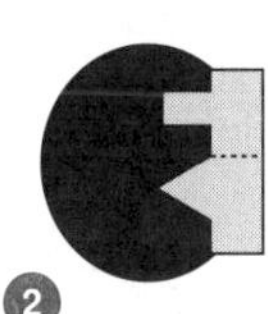
2

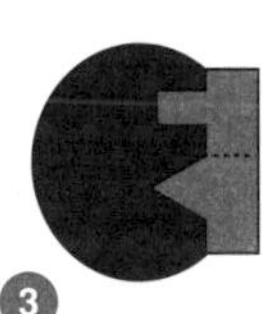
3

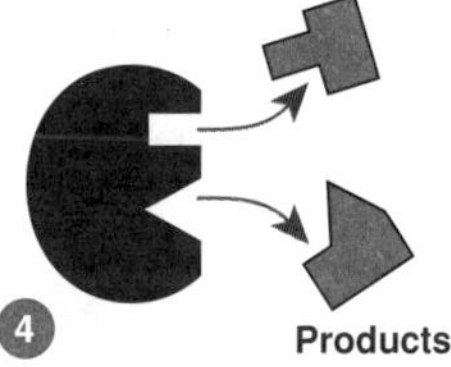
4
Products

1. Explain what is meant by the **active site** of an enzyme and relate it to the enzyme's tertiary structure:

2. What might happen to an enzyme's activity if a mutation caused a change to the shape of the active site?

3. Describe the key features of the '**lock and key**' model of enzyme action:

4. Using examples, explain the role of enzymes in metabolic processes:

ISBN: 978-1-927173-16-9

Periodicals: *Enzymes: nature's catalytic machines*

Related activities: *The Role of the Digestive System, Enzyme Reaction Rates*

How Enzymes Work

Chemical reactions in cells are accompanied by energy changes. Any reaction, even an **exothermic reaction**, needs to raise the energy of the substrate to an unstable **transition state** before the reaction will proceed (below). The amount of energy required to do this is the **activation energy** (E_a). Enzymes work by lowering the activation energy for any given chemical reaction. They do this by orienting the substrate in a particular way, or by adding charges or otherwise inducing strain in the substrate so that bonds are destabilized and the substrate is more reactive (the induced fit model).

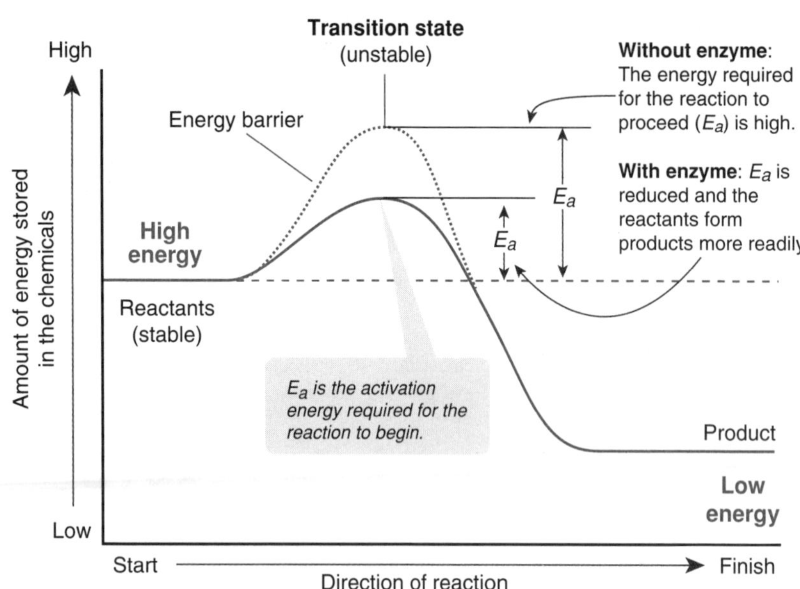
Enzymes Lower the Activation Energy
Transition state
(unstable)
High
Energy barrier
Without enzyme: The energy required for the reaction to proceed (E_a) is high.
With enzyme: E_a is reduced and the reactants form products more readily.
E_a
E_a
Amount of energy stored in the chemicals
High energy
Reactants (stable)
E_a is the activation energy required for the reaction to begin.
Product
Low energy
Low
Start
Direction of reaction
Finish

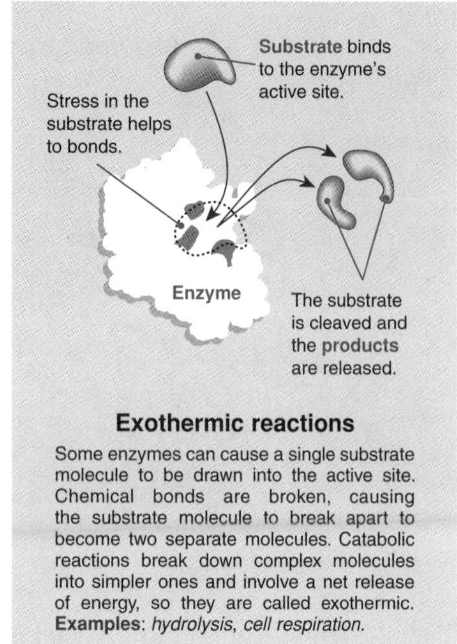
Substrate binds to the enzyme's active site.
Stress in the substrate helps to bonds.
Enzyme
The substrate is cleaved and the products are released.

Exothermic reactions

Some enzymes can cause a single substrate molecule to be drawn into the active site. Chemical bonds are broken, causing the substrate molecule to break apart to become two separate molecules. Catabolic reactions break down complex molecules into simpler ones and involve a net release of energy, so they are called exothermic. **Examples**: *hydrolysis, cell respiration.*

The Current Model: Induced Fit

An enzyme's interaction with its substrate is best regarded as an induced fit (below). The shape of the enzyme changes when the substrate fits into the cleft. The reactants become bound to the enzyme by weak chemical bonds. This binding can weaken bonds within the reactants themselves, allowing the reaction to proceed more readily. The current induced-fit model of enzyme function is supported by studies of enzyme inhibitors, which show that enzymes are flexible and change shape when interacting with the substrate.

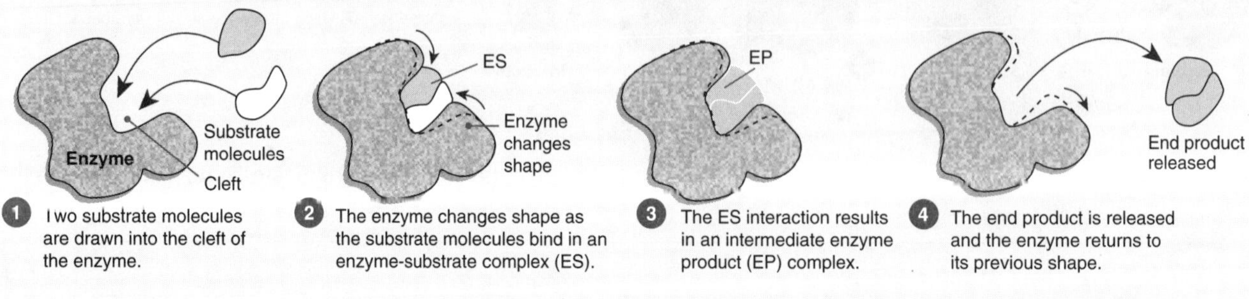
Enzyme
Substrate molecules
Cleft
ES
Enzyme changes shape
EP
End product released

1. Two substrate molecules are drawn into the cleft of the enzyme.
2. The enzyme changes shape as the substrate molecules bind in an enzyme-substrate complex (ES).
3. The ES interaction results in an intermediate enzyme product (EP) complex.
4. The end product is released and the enzyme returns to its previous shape.

1. Explain how enzymes act as **biological catalysts**: ______________________________

____

____

____

2. Describe the current '**induced fit**' model of enzyme action, explaining how it differs from the lock and key model:

____

____

____

____

3. Explain why it was necessary to modify the lock and key model of enzyme function: ______________________________

____

ISBN: 978-1-927173-16-9

Enzyme Reaction Rates

Enzymes are sensitive molecules. They often have a narrow range of conditions under which they operate properly. For most of the enzymes associated with plant and animal metabolism, there is little activity at low temperatures. As the temperature increases, so too does the enzyme activity, until the point is reached where the temperature is high enough to damage the enzyme's structure. At this point, the enzyme ceases to function. This phenomenon called enzyme or protein **denaturation**.

Extremes in pH can also cause the protein structure of enzymes to denature. Poisons often work by denaturing enzymes or occupying the enzyme's active site so that it does not function. In some cases, enzymes will not function without cofactors, such as vitamins or trace elements. In the four graphs below, the *rate of reaction* or *degree of enzyme activity* is plotted against each of four factors that affect enzyme performance. Answer the questions relating to each graph:

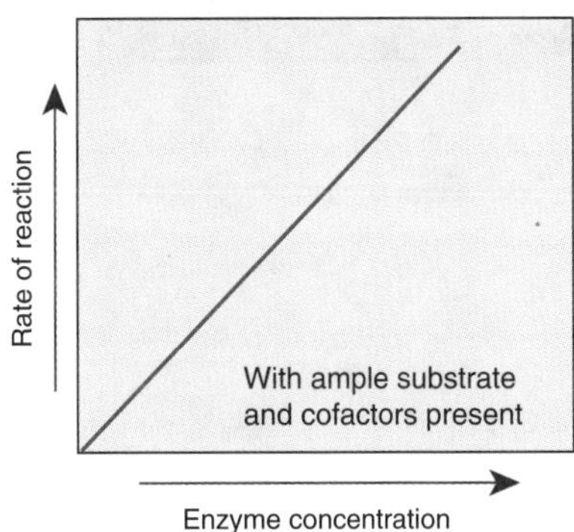
Rate of reaction
With ample substrate and cofactors present
Enzyme concentration

1. **Enzyme concentration**

(a) Describe the change in the rate of reaction when the enzyme concentration is increased (assuming there is plenty of the substrate present):

____

____

(b) Suggest how a cell may vary the amount of enzyme present in a cell:

____

____

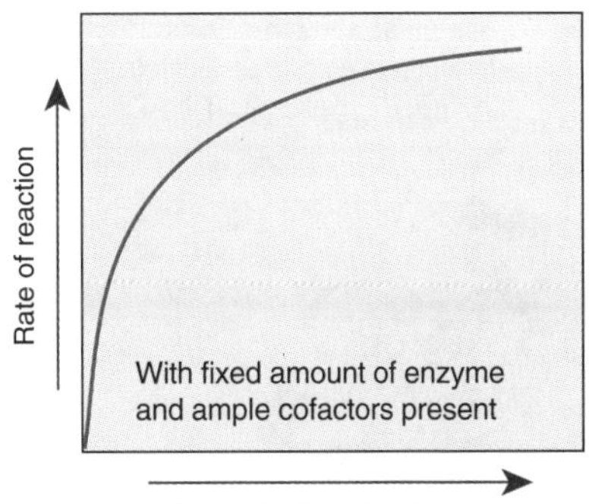
Rate of reaction
With fixed amount of enzyme and ample cofactors present
Concentration of substrate

2. **Substrate concentration**

(a) Describe the change in the rate of reaction when the substrate concentration is **increased** (assuming a fixed amount of enzyme and ample cofactors):

____

____

(b) Explain why the rate changes the way it does: ______________________________

____

____

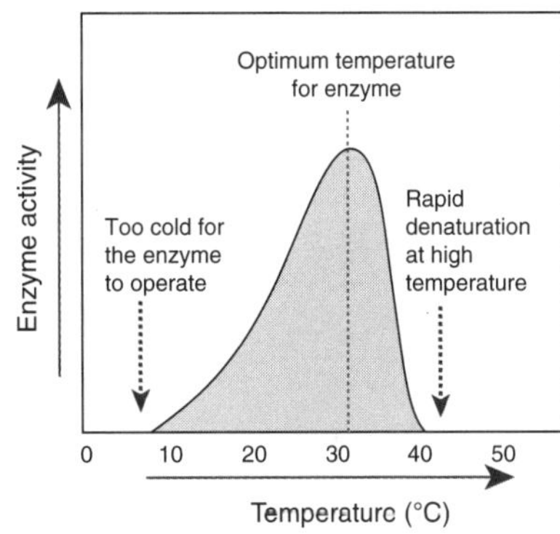
Optimum temperature for enzyme
Enzyme activity
Too cold for the enzyme to operate
Rapid denaturation at high temperature
0
10
20
30
40
50
Temperature (°C)

3. **Temperature**

Higher temperatures speed up all reactions, but few enzymes can tolerate temperatures higher than 50–60°C. The rate at which enzymes are **denatured** (change their shape and become inactive) increases with higher temperatures.

(a) Describe what is meant by an *optimum temperature* for enzyme activity:

____

____

(b) Explain why most enzymes perform poorly at low temperatures:

____

____

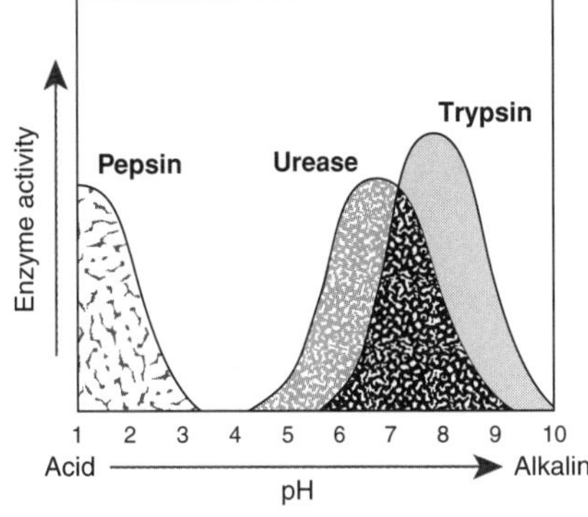
Enzyme activity
Pepsin
Urease
Trypsin
1
2
3
4
5
6
7
8
9
10
Acid
Alkaline
pH

4. **Acidity or Alkalinity (pH)**

Like all proteins, enzymes are **denatured** by *extremes* of **pH** (very acid or alkaline). Within these extremes, most enzymes are still influenced by pH. Each enzyme has a preferred pH range for optimum activity.

(a) State the optimum pH for each of the enzymes:

Pepsin: ____________ Trypsin: ____________ Urease: ____________

(b) Pepsin acts on proteins in the stomach. Explain how its optimum pH is suited to its working environment:

____

____

ISBN: 978-1-927173-16-9

Periodicals: *Enzymes*

Related activities: *Enzyme Cofactors, Enzyme Inhibitors*

Enzyme Cofactors and Inhibitors

Nearly all enzymes are made of protein, although RNA has been demonstrated to have enzymatic properties. Some enzymes (e.g. pepsin) consist of only protein. Other enzymes require the addition of extra non-protein components to complete their catalytic properties. In these cases, the protein portion is called the **apoenzyme**, and the additional chemical component is called a **cofactor**. Neither the apoenzyme nor the cofactor has catalytic activity on its own. Cofactors may be organic molecules (e.g. vitamin C and the coenzymes in the respiratory chain) or inorganic ions (e.g. Ca^{2+}, Zn^{2+}). They also may be tightly or loosely bound to the enzyme. Permanently bound cofactors are called **prosthetic groups**, whereas temporarily attached molecules, which detach after a reaction are called **coenzymes**. Some cofactors include both an organic and a non-organic component. Examples include the heme prosthetic groups, which consist of an iron atom in the centre of a porphyrin ring.

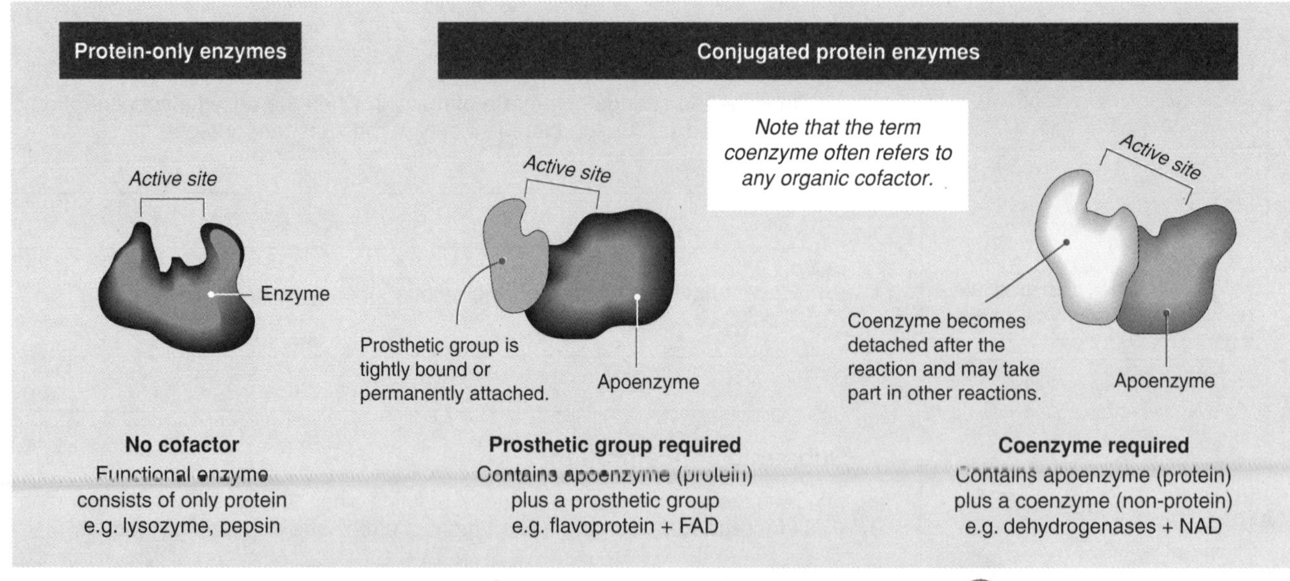

Protein-only enzymes
Conjugated protein enzymes
Note that the term coenzyme often refers to any organic cofactor.
Active site
Enzyme
Active site
Prosthetic group is tightly bound or permanently attached.
Apoenzyme
Active site
Coenzyme becomes detached after the reaction and may take part in other reactions.
Apoenzyme
No cofactor
Functional enzyme consists of only protein e.g. lysozyme, pepsin
Prosthetic group required
Contains apoenzyme (protein) plus a prosthetic group e.g. flavoprotein + FAD
Coenzyme required
Contains apoenzyme (protein) plus a coenzyme (non-protein) e.g. dehydrogenases + NAD

Enzyme cofactor	Enzyme
Cupric (copper ion)	Cytochrome oxidase
Ferrous or ferric (iron ion)	Catalase and cytochrome (via heme)
Selenium	Glutathione peroxidase
Magnesium	Glucose 6-phosphatase
Flavin	NADH dehydrogenase
Heme L (derived from heme B)	Peroxidases, e.g. thyroid peroxidase

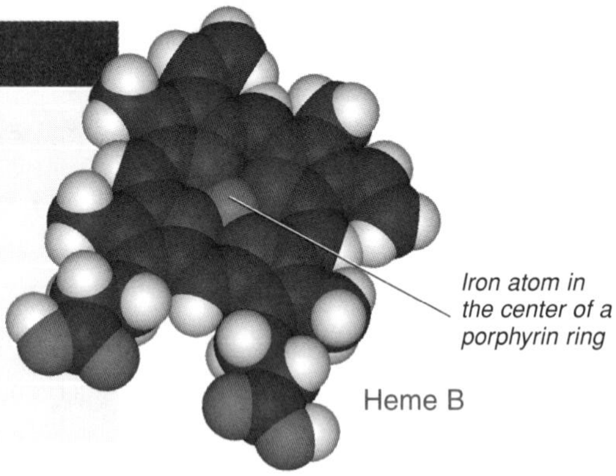

Iron atom in the center of a porphyrin ring

Heme B

1. Describe the general role of **cofactors** in enzyme activity: ____________________

2. Explain exactly how cofactors enable an enzyme's catalytic activity: ____________________

3. Distinguish between the apoenzyme and the cofactor: ____________________

4. Identify the two broad categories of cofactors and describe an example of each: ____________________

5. Describe the importance of adequate vitamin and mineral intake in the diet: ____________________

Related activities: *Enzymes, Enzyme Reaction Rates*
Weblinks: *Science in the Box: Enzymes*

Periodicals:
Enzymes

ISBN: 978-1-927173-16-9

Competitive Inhibition

Competitive inhibitors compete with the normal substrate for the enzyme's active site.

A competitive inhibitor occupies the active site only temporarily and so the inhibition is reversible.

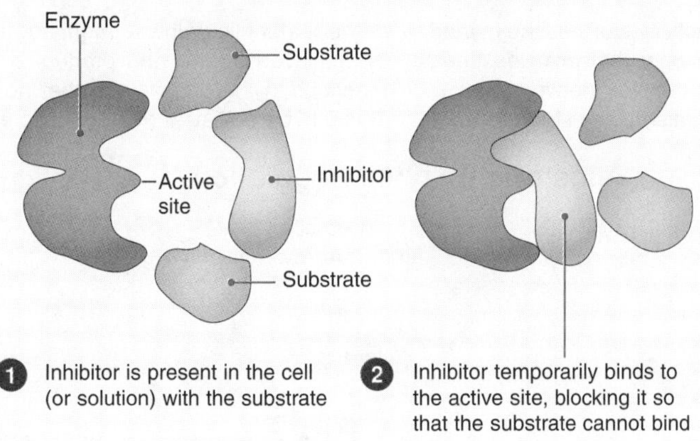
Enzyme
Substrate
Active site
Inhibitor
Substrate

1 Inhibitor is present in the cell (or solution) with the substrate

2 Inhibitor temporarily binds to the active site, blocking it so that the substrate cannot bind

Fig.1 Effect of competitive inhibition on enzyme reaction rate at different substrate concentration

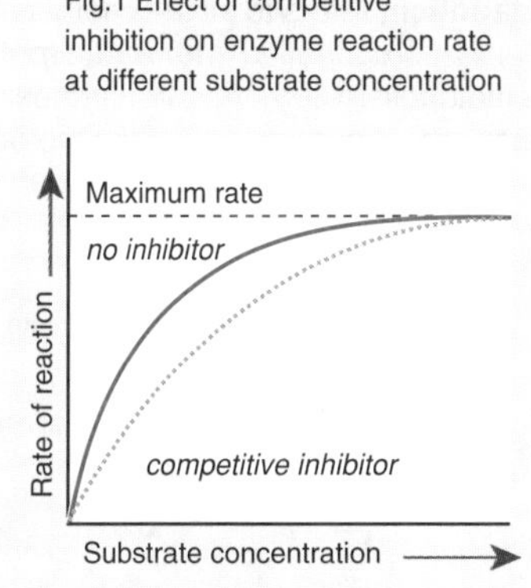
Maximum rate
no inhibitor
competitive inhibitor
Rate of reaction
Substrate concentration

Non-competitive Inhibition

Non-competitive inhibitors bind with the enzyme at a site other than the active site. They inactivate the enzyme by altering its shape.

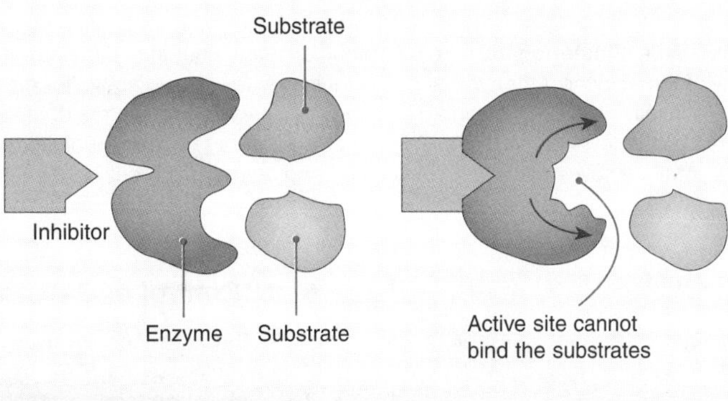
Substrate
Inhibitor
Enzyme
Substrate
Active site cannot bind the substrates

1 Without the inhibitor bound, the enzyme can bind the substrate

2 When the inhibitor binds, the enzyme changes shape.

Fig.2 Effect of non-competitive inhibition on enzyme reaction rate at different substrate concentration

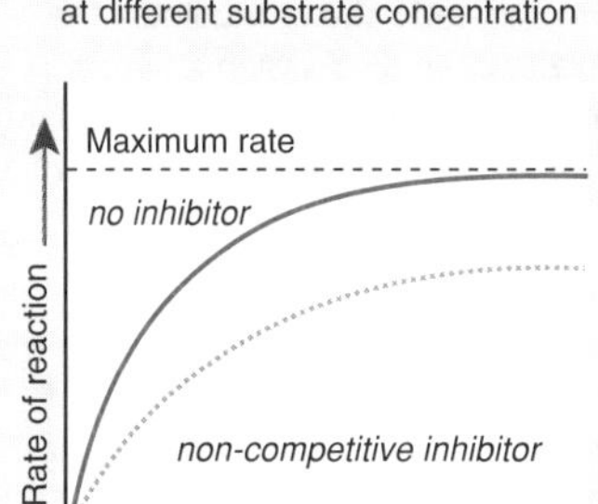
Maximum rate
no inhibitor
non-competitive inhibitor
Rate of reaction
Substrate concentration

Enzymes may be deactivated, temporarily or permanently, by chemicals called enzyme inhibitors. **Competitive inhibitors** compete directly with the substrate for the active site, and their effect can be overcome by increasing the concentration of available substrate. A **non-competitive inhibitor** does not occupy the active site, but distorts it so that the substrate and enzyme can no longer interact.

6. Distinguish between **competitive** and **non-competitive** inhibition: ______________________

____

____

____

____

7. (a) Compare and contrast the effect of competitive and non-competitive inhibition on the relationship between the substrate concentration and the rate of an enzyme controlled reaction (figs 1 and 2 above):

____

____

____

____

____

(b) How could you distinguish between competitive and non-competitive inhibition in an isolated system?

____

____

____

ISBN: 978-1-927173-16-9

Control of Metabolic Pathways

Metabolism refers to all the chemical activities of life. The myriad of enzyme-controlled **metabolic pathways** that are described as metabolism form a tremendously complex network that is necessary in order to 'maintain' the organism. Often the products of the metabolic pathway regulate the pathway itself. This might be achieved by the end product of the pathway inhibiting the reactions in the pathway so that no more product is produced. This can be achieved by **allosteric enzyme regulation** (below).

Allosteric Enzyme Regulation

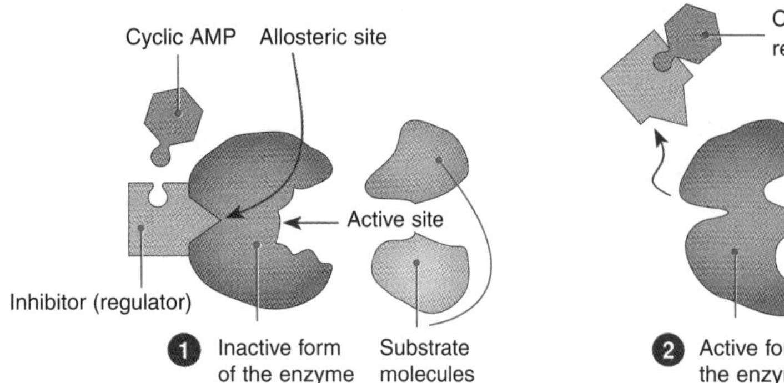
Cyclic AMP
Allosteric site
Active site
Inhibitor (regulator)
1 Inactive form of the enzyme
Substrate molecules
Cyclic AMP removes inhibitor
2 Active form of the enzyme
Protein kinase A has many roles in the cell, including regulating glycogen, sugar, and lipid metabolism.
3 Enzyme-substrate complex

Allosteric regulators have a receptor site, called the **allosteric site**, on a part of the enzyme other than the active site. When a substance binds to the allosteric site, it regulates the activity of the enzyme. Often the action is inhibitory (as shown above for protein kinase A), but allosteric regulators can also switch an enzyme from its inactive to its active form. Thus, they can serve as regulators of metabolic pathways. The activity of the enzyme **protein kinase A** is regulated by the level of **cyclic AMP** in the cell. When a regulatory inhibitor protein binds reversibly to its allosteric site, the enzyme is inactive. Cyclic AMP removes the allosteric inhibitor and activates the enzyme.

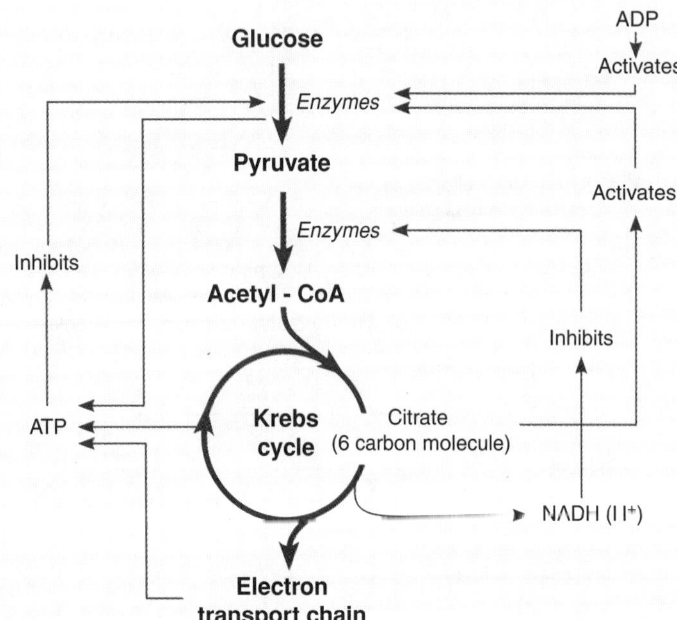
ADP
Glucose
Activates
Enzymes
Pyruvate
Activates
Enzymes
Inhibits
Acetyl - CoA
Inhibits
ATP
Krebs cycle
Citrate (6 carbon molecule)
NADH (H+)
Electron transport chain

Control of Glucose Catabolism

The reactions involved in the catabolism of glucose form an example of **feedback inhibition** (negative feedback loop).

The two key control points involve the enzymes involved in glycolysis and the enzyme involved in the link reaction.

- High levels of ATP from glycolysis, the Krebs cycle, and the electron transport chain inhibit the enzymes involved in glycolysis. As ATP is used up, the levels of ADP rise and cause the reactivation of glycolysis.
- Low levels of citrate (the first molecule produced in the Krebs cycle) also activate glycolysis. $NADH^+$ production (an electron carrier) inhibits the production of acetyl CoA, which lowers the levels of citrate.

1. Explain how an allosteric regulator controls an enzyme's activity: ____________________

2. Explain why **end product inhibition** can also be called feedback inhibition: ____________________

3. Describe how the catabolism of glucose is regulated: ____________________

Related activities: *Enzymes*

ISBN: 978-1-927173-16-9

Applications of Enzymes

Milk is a high quality food containing protein, fat, carbohydrates, minerals and vitamins. Milk contains the disaccharide **lactose**. A large proportion of the world's population show some degree of **lactose intolerance** (an inability to digest lactose) as they grow older. These people often avoid milk products, and so lose out on the nutritional benefits of consuming milk. The enzyme **lactase** can be used to remove lactose from milk. This allows people with lactose intolerance to continue to include milk in their diet as adults, and gain the nutritional benefits (e.g. good source of calcium) from its consumption.

Lactose Removal

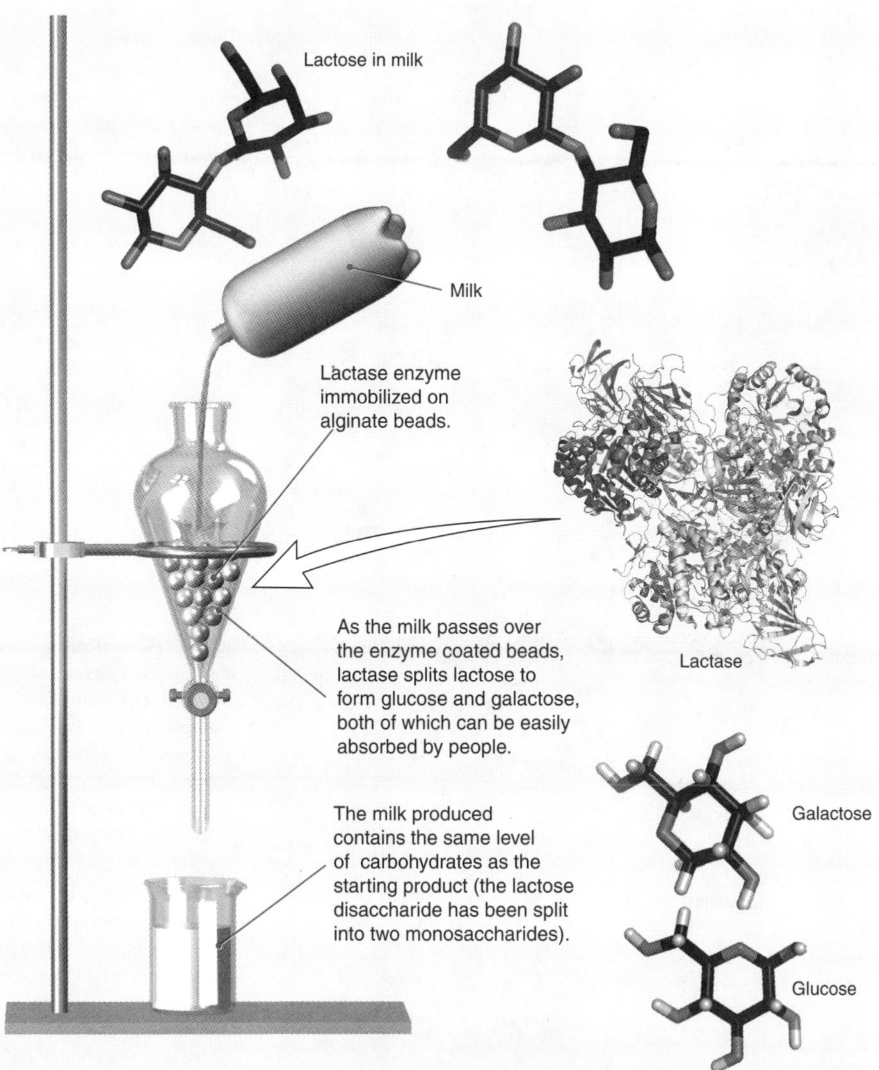
Lactose in milk
Milk
Lactase enzyme immobilized on alginate beads.
As the milk passes over the enzyme coated beads, lactase splits lactose to form glucose and galactose, both of which can be easily absorbed by people.
Lactase
The milk produced contains the same level of carbohydrates as the starting product (the lactose disaccharide has been split into two monosaccharides).
Galactose
Glucose

Lactase and Humans

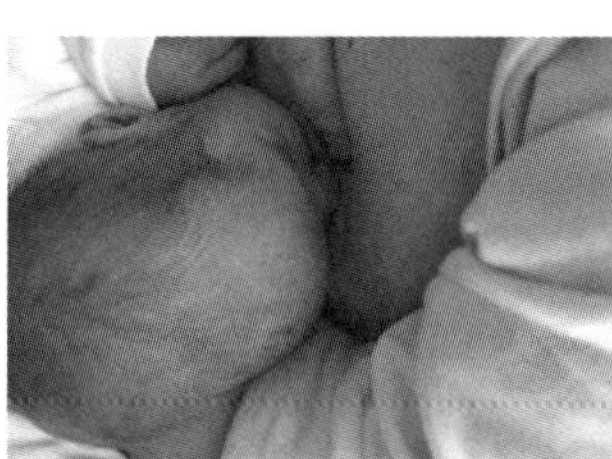

Lactose is a disaccharide found in milk. It is less sweet than glucose. All infant humans produce the enzyme **lactase**, which hydrolyzes lactose into glucose and galactose.

As humans become older, their production of lactase gradually declines and they lose their ability to hydrolyze lactose. As adults, they are **lactose intolerant**, and feel bloated after drinking milk.

In humans of mainly European, East African, or Indian descent, lactase production continues into adulthood. But people of mainly Asian descent cease production early in life and become lactose intolerant.

1. Explain why being able to continue to drink milk throughout life is of benefit to humans: ______

2. Describe how lactase is used in the production of lactose-free milk: ______

3. Explain why lactose-free milk often has a slightly sweeter taste that ordinary milk: ______

ISBN: 978-1-927173-16-9

ATP and Metabolism

All organisms require energy to be able to perform the metabolic processes required for them to function and reproduce. This energy is obtained by **cell respiration**, a set of metabolic reactions which ultimately convert biochemical energy from 'food' into the nucleotide **adenosine triphosphate** (**ATP**). The details of this process are provided in the exercise "*Cell Respiration*". ATP is considered to be a universal energy carrier, transporting chemical energy within the cell for use in metabolic processes such as biosynthesis, cell division, cell signaling, thermoregulation, cell mobility, and active transport of substances across membranes.

Adenosine Triphosphate (ATP)

The ATP molecule consists of three components; a purine base (**adenine**), a pentose sugar (**ribose**), and **three phosphate groups** which attach to the 5' carbon of the pentose sugar. The three dimensional structure of ATP is described below.

ATP acts as a store of energy within the cell. The bonds between the phosphate groups are **high-energy bonds**, meaning that a large amount of free energy is released when they are hydrolyzed. Typically, this hydrolysis involves the removal of one phosphate group from the ATP molecule resulting in the formation of adenosine diphosphate (ADP).

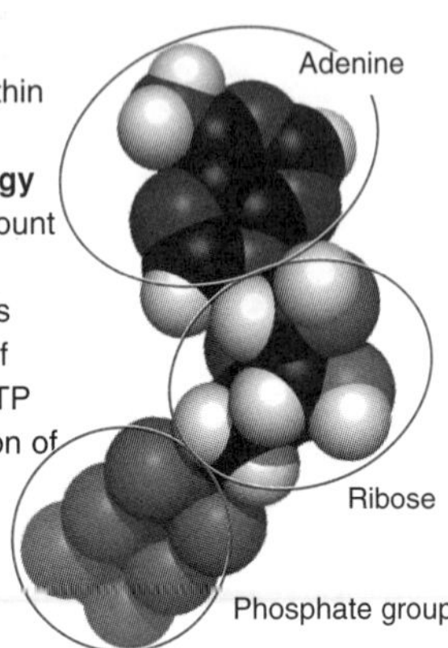
Adenine
Ribose
Phosphate groups

The Mitochondrion

Cell respiration and ATP production occur in the mitochondria. A mitochondrion is bound by a double membrane. The inner and outer membranes are separated by an intermembrane space, compartmentalizing the regions of the mitochondrion in which the different reactions of cell respiration take place.

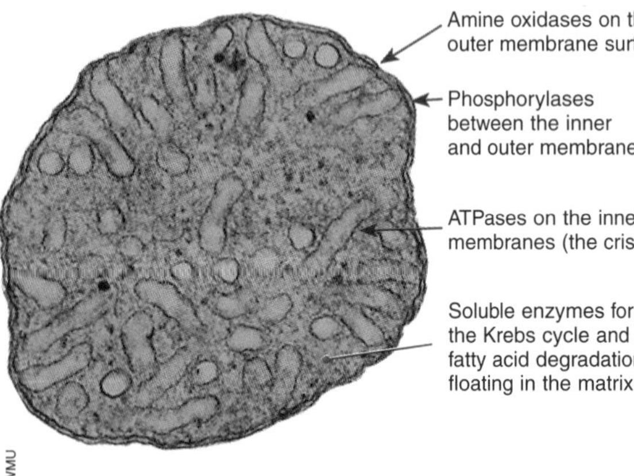
Amine oxidases on the outer membrane surface
Phosphorylases between the inner and outer membranes
ATPases on the inner membranes (the cristae)
Soluble enzymes for the Krebs cycle and fatty acid degradation floating in the matrix
WMU

ATP Powers Metabolism

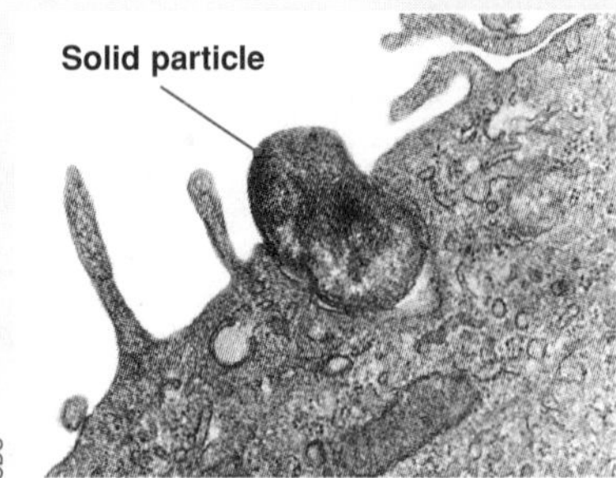
Solid particle
CDC

The energy released from the removal of a phosphate group of ATP is used to actively transport molecules and substances across the cellular membrane. **Phagocytosis** (left), which involves the engulfment of solid particles, is one such example.

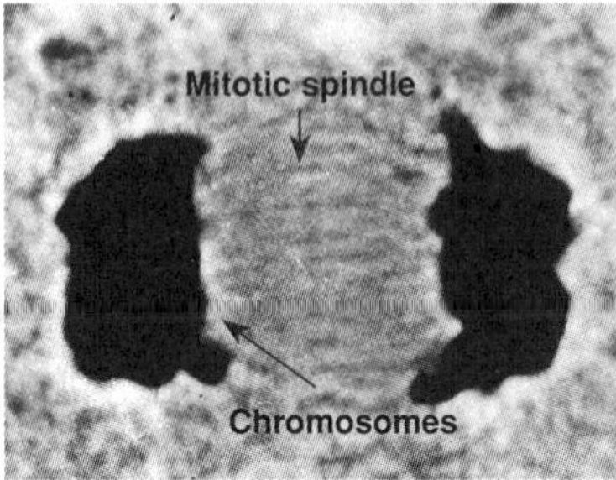
Mitotic spindle
Chromosomes

Cell division (mitosis), as observed in this onion cell, requires ATP to proceed. Formation of the mitotic spindle and chromosome separation are two aspects of cell division which require energy from ATP hydrolysis to occur.

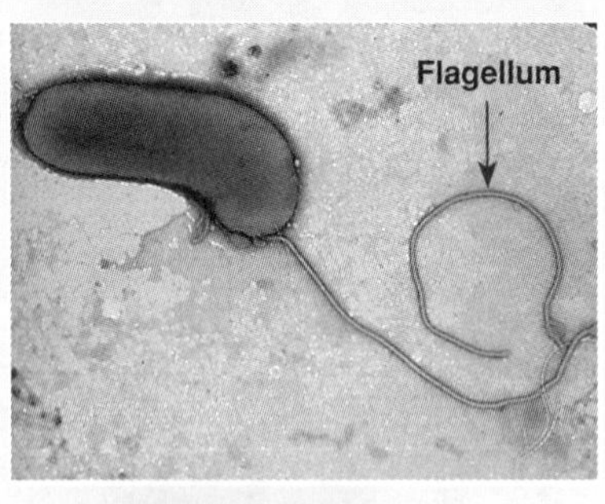
Flagellum
EII

The hydrolysis of ATP provides the energy for motile cells to achieve movement via a tail-like structure called a flagellum. For example, the bacterium, *Helicobacter pylori* (left), is motile. Likewise, mammalian sperm must be able to move to the ovum to fertilize it.

The maintenance of body temperature requires energy. To maintain body heat, muscular activity increases (e.g. shivering, erection of body hairs). Cooling requires expenditure of energy too. For example, sweating is an energy requiring process involving secretion from glands in the skin.

1. Explain why organisms need to respire: ______

2. (a) Describe the general role of mitochondria in cell respiration: ______

   (b) Explain the importance of compartmentalization in the mitochondrion: ______

3. Explain why thermoregulation is associated with energy expenditure: ______

Related activities: *The Role of ATP in Cells, Cell Respiration*
Weblinks: *ATP in Metabolism*

ISBN: 978-1-927173-16-9

The Role of ATP in Cells

The molecule **ATP** (adenosine triphosphate) is the universal energy carrier for the cell. ATP can release its energy quickly; only one chemical reaction (hydrolysis of the terminal phosphate) is required. This reaction is catalyzed by the enzyme ATPase. Once ATP has released its energy, it becomes ADP (adenosine diphosphate), a low energy molecule that can be recharged by adding a phosphate. This requires energy, which is supplied by the controlled breakdown of respiratory substrates in **cell respiration**. The most common respiratory substrate is glucose, but other molecules (e.g. fats or proteins) may also be used.

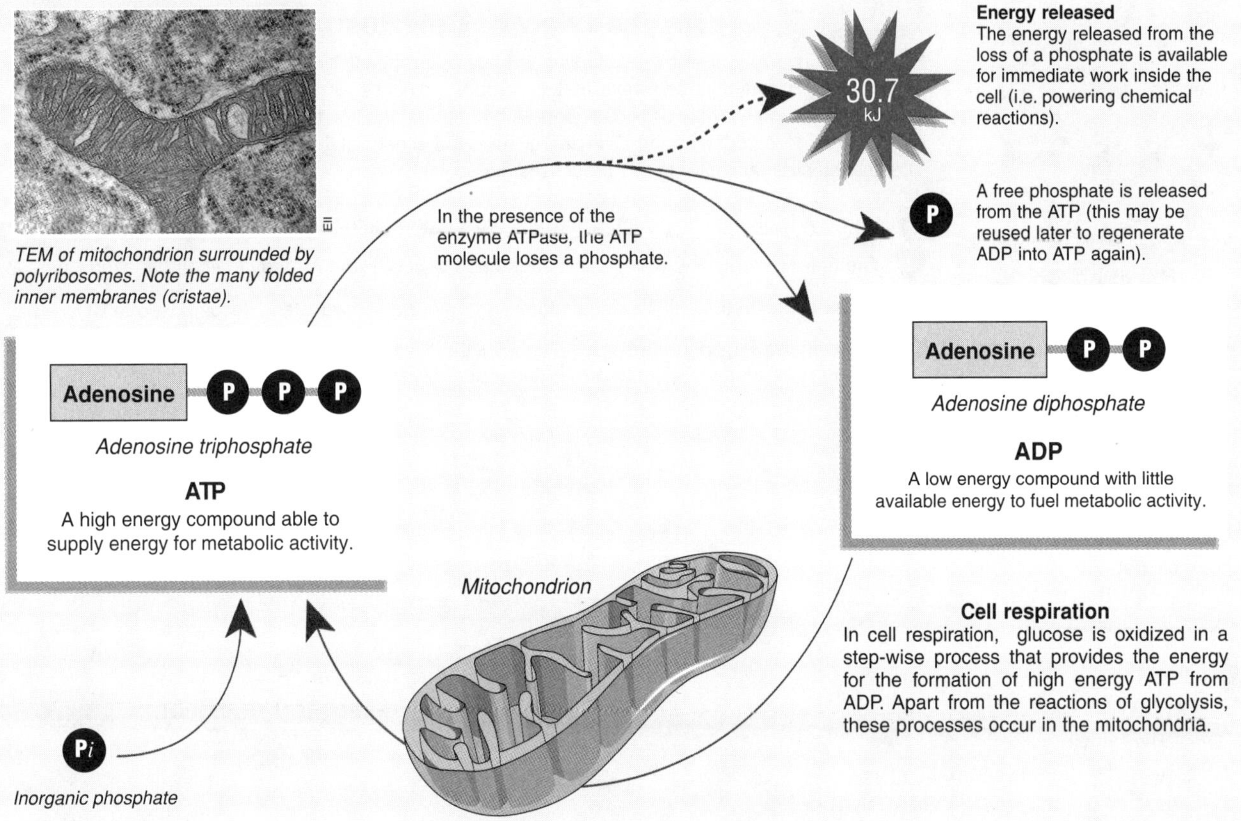

TEM of mitochondrion surrounded by polyribosomes. Note the many folded inner membranes (cristae).

In the presence of the enzyme ATPase, the ATP molecule loses a phosphate.

Energy released
The energy released from the loss of a phosphate is available for immediate work inside the cell (i.e. powering chemical reactions).

30.7 kJ

P

A free phosphate is released from the ATP (this may be reused later to regenerate ADP into ATP again).

Adenosine P P P

Adenosine triphosphate

ATP

A high energy compound able to supply energy for metabolic activity.

Adenosine P P

Adenosine diphosphate

ADP

A low energy compound with little available energy to fuel metabolic activity.

Mitochondrion

Cell respiration
In cell respiration, glucose is oxidized in a step-wise process that provides the energy for the formation of high energy ATP from ADP. Apart from the reactions of glycolysis, these processes occur in the mitochondria.

Pi

Inorganic phosphate

1. Describe how ATP acts as a supplier of energy to power metabolic reactions: ______________________________

2. Name the immediate source of energy used to reform ATP from ADP molecules: ______________________________

3. Name the process of re-energizing ADP into ATP molecules: ______________________________

4. Name the ultimate source of energy for plants: ______________________________

5. Name the ultimate source of energy for animals: ______________________________

6. Explain in what way the ADP/ATP system can be likened to a rechargeable battery: ______________________________

7. In the following table, use brief statements to contrast photosynthesis and cell respiration in terms of the following:

Feature	Photosynthesis	Cell respiration
Starting materials		
Waste products		
Role of hydrogen carriers: NAD, NADP		
Role of ATP		
Overall biological role		

Cell Respiration

Cell respiration is the process by which organisms break down energy rich molecules (e.g. glucose) to release the energy in a useable form (ATP). All living cells respire in order to exist, although the substrates they use may vary. **Aerobic respiration** requires oxygen. Forms of cell respiration that do not require oxygen are said to be **anaerobic**. Some plants and animals can generate ATP anaerobically for short periods of time. Other organisms use only anaerobic respiration and live in oxygen-free environments. For these organisms, there is some other final electron acceptor other than oxygen (e.g. nitrate or Fe^{2+}).

An Overview of Cell Respiration

Respiration involves three metabolic stages (plus a link reaction) summarized below. The first two stages are the catabolic pathways that decompose glucose and other organic fuels. In the third stage, the electron transport chain accepts electrons from the first two stages and passes these from one electron acceptor to another. The energy released at each stepwise transfer is used to make ATP. The final electron acceptor in this process is molecular oxygen.

1. **Glycolysis**. In the cytoplasm, glucose is broken down into two molecules of pyruvate.
2. **The link reaction**. Pyruvate is split and added to coenzyme A ready to enter the Krebs cycle.
3. **Krebs cycle**. In the mitochondrial matrix, a derivative of pyruvate is decomposed to CO_2.
4. **Electron transport chain**. This occurs in the inner membranes of the mitochondrion and accounts for almost 90% of the ATP generated by respiration.

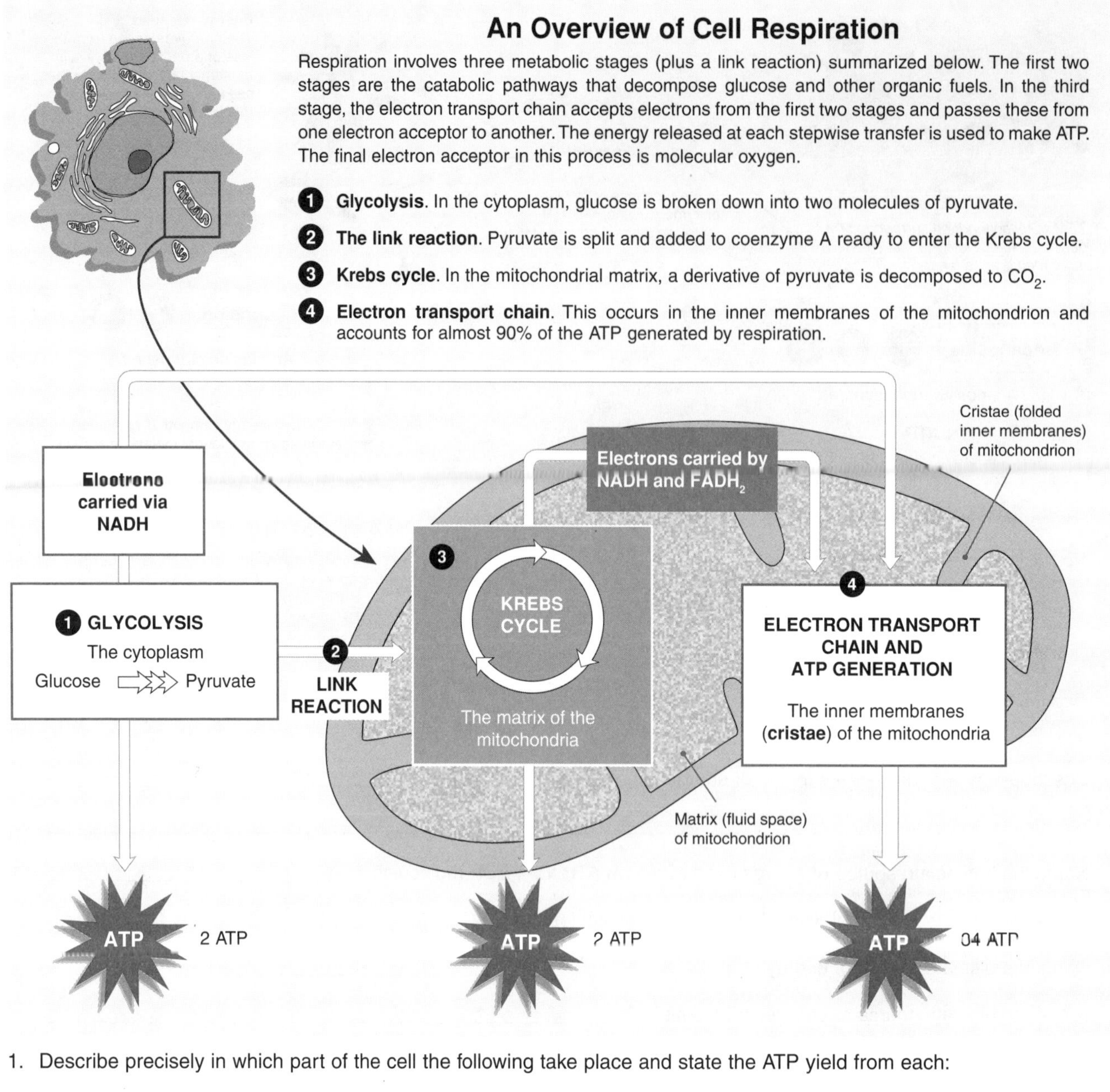
Electrons carried via NADH
Electrons carried by NADH and $FADH_2$
Cristae (folded inner membranes) of mitochondrion
1 GLYCOLYSIS
The cytoplasm
Glucose
Pyruvate
2 LINK REACTION
3 KREBS CYCLE
The matrix of the mitochondria
4 ELECTRON TRANSPORT CHAIN AND ATP GENERATION
The inner membranes (cristae) of the mitochondria
Matrix (fluid space) of mitochondrion
ATP
2 ATP
ATP
2 ATP
ATP
34 ATP

1. Describe precisely in which part of the cell the following take place and state the ATP yield from each:

   (a) Glycolysis: ______________________________

   (b) Krebs cycle reactions: ______________________________

   (c) Electron transport chain: ______________________________

2. Summarize the events occurring in each of the following:

   (a) Glycolysis: ______________________________

   ____

   (b) Krebs cycle: ______________________________

   ____

   (c) Electron transport: ______________________________

   ____

ISBN: 978-1-927173-16-9

Related activities: The Biochemistry of Respiration

The Biochemistry of Respiration

Cell respiration is a catabolic, energy yielding pathway. The breakdown of glucose and other organic fuels (such as fats and proteins) to simpler molecules is **exothermic** and releases energy for the synthesis of ATP. As summarized in the previous activity, respiration involves glycolysis, the Krebs cycle, and electron transport. The diagram below provides a more detailed overview of the events in each of these stages. Glycolysis and the Krebs cycle supply electrons (via NADH) to the electron transport chain, which drives **oxidative phosphorylation**. Glycolysis nets two ATP, produced by **substrate-level phosphorylation**.

The conversion of pyruvate (the end product of glycolysis) to **acetyl CoA** links glycolysis to the Krebs cycle. One "turn" of the cycle releases carbon dioxide, forms one ATP by substrate level phosphorylation, and passes electrons to three NAD^+ and one FAD. Most of the ATP generated in cell respiration is produced by oxidative phosphorylation when NADH and $FADH_2$ donate electrons to the series of electron carriers in the electron transport chain. At the end of the chain, electrons are passed to molecular oxygen, reducing it to water. Electron transport is coupled to ATP synthesis by **chemiosmosis** (see next activity).

Glycogen → Glucose (6C) (a)

Fats → Glycerol → phosphorylated 6C sugar (hexose bisphosphate)

2 ADP

4 ATP are produced but 2 are used in the process

2 ATP

2 x 3C triose phosphate

Proteins → Amino acids → Pyruvate* (b)

NADH + H^+

1C lost as carbon dioxide

Fats → Fatty acids → Acetyl Coenzyme A (c)

CoA

(d)

NADH + H^+

NADH + H^+

oxaloacetate (f)

Krebs cycle

2C lost as carbon dioxide

(e)

α-ketoglutarate

ATP

2NADH + $2H^+$

*$FAD.H_2$

Electron transport chain

e^- … e^-

17 ADP → 17 ATP

$\frac{1}{2}O_2$

Oxygen is used as a terminal electron acceptor

$2H^+$ → Water

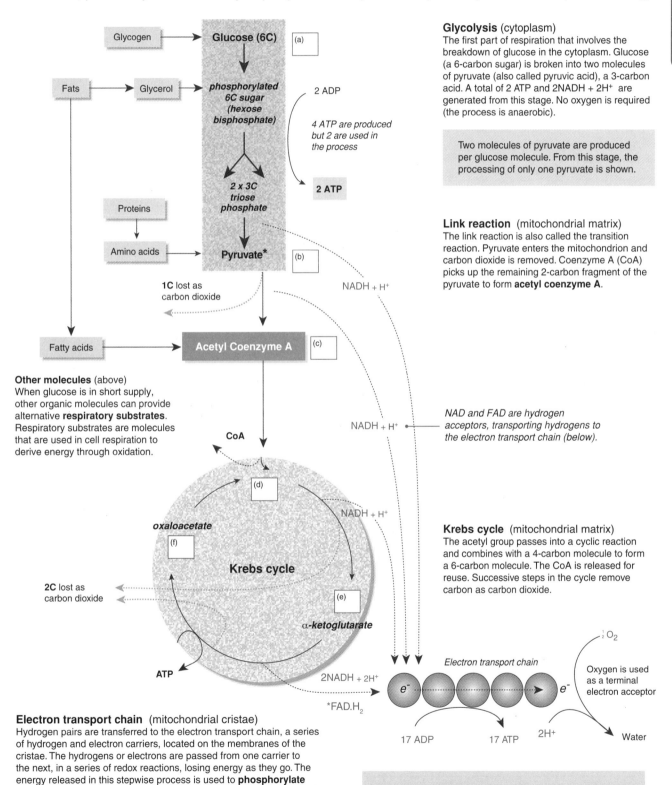

Glycolysis (cytoplasm)

The first part of respiration that involves the breakdown of glucose in the cytoplasm. Glucose (a 6-carbon sugar) is broken into two molecules of pyruvate (also called pyruvic acid), a 3-carbon acid. A total of 2 ATP and 2NADH + $2H^+$ are generated from this stage. No oxygen is required (the process is anaerobic).

Two molecules of pyruvate are produced per glucose molecule. From this stage, the processing of only one pyruvate is shown.

Link reaction (mitochondrial matrix)

The link reaction is also called the transition reaction. Pyruvate enters the mitochondrion and carbon dioxide is removed. Coenzyme A (CoA) picks up the remaining 2-carbon fragment of the pyruvate to form **acetyl coenzyme A**.

Other molecules (above)

When glucose is in short supply, other organic molecules can provide alternative **respiratory substrates**. Respiratory substrates are molecules that are used in cell respiration to derive energy through oxidation.

NAD and FAD are hydrogen acceptors, transporting hydrogens to the electron transport chain (below).

Krebs cycle (mitochondrial matrix)

The acetyl group passes into a cyclic reaction and combines with a 4-carbon molecule to form a 6-carbon molecule. The CoA is released for reuse. Successive steps in the cycle remove carbon as carbon dioxide.

Electron transport chain (mitochondrial cristae)

Hydrogen pairs are transferred to the electron transport chain, a series of hydrogen and electron carriers, located on the membranes of the cristae. The hydrogens or electrons are passed from one carrier to the next, in a series of redox reactions, losing energy as they go. The energy released in this stepwise process is used to **phosphorylate** ADP to form ATP. Oxygen is the final electron acceptor and is reduced to water (hence the term **oxidative phosphorylation**).
Note FAD enters the electron transport chain at a lower energy level than NAD, and only 2ATP are generated per $FAD.H_2$.

Total ATP yield per glucose
Glycolysis: 2 ATP, *Krebs cycle*: 2 ATP, *Electron transport*: 34 ATP

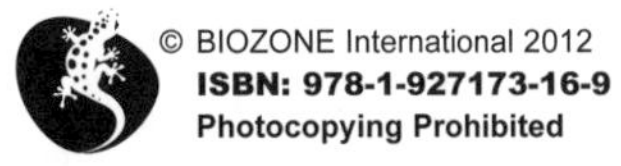

ISBN: 978-1-927173-16-9

Glycolysis and Fermentation	Aerobic Respiration
Occurs in the cytoplasm of the cell.	Occurs in the mitochondria of eukaryotic cells.
Oxygen is not required.	Oxygen is required.
Glycolysis: glucose is converted to pyruvic acid with a net production of 2ATP molecules (a very low energy yield).	Pyruvic acid is converted to carbon dioxide, water, and a further 36 ATP molecules from the Krebs cycle and electron transport chain (a high energy yield per glucose molecule).
Very inefficient production of energy.	An energy efficient process.
In the absence of oxygen, pyruvic acid cannot enter the mitochondrion. It is converted to ethanol and carbon dioxide in plants, and lactic acid in animals.	When oxygen is present, pyruvic acid can be oxidized further via the Krebs cycle and electron transport chain.

1. Explain the purpose of the link reaction: ______________________________

2. On the diagram of cell respiration (previous page), state the number of carbon atoms in each of the molecules (a)-(f):

3. Determine how many ATP molecules **per molecule of glucose** are generated during the following stages of respiration:

   (a) Glycolysis: ________ (b) Krebs cycle: ________ (c) Electron transport chain: ________ (d) Total: ________

4. Explain what happens to the carbon atoms lost during respiration: ______________________________

5. Describe the role of each of the following in cell respiration:

   (a) Hydrogen atoms: ______________________________

   (b) NAD and FAD: ______________________________

   (c) Oxygen: ______________________________

   (d) Acetyl coenzyme A: ______________________________

6. Explain what happens when the supply of glucose for cell respiration is limited: ______________________________

7. Distinguish between reduction and oxidation: ______________________________

8. Explain what happens during oxidative phosphorylation: ______________________________

ISBN: 978-1-927173-16-9

Chemiosmosis

Chemiosmosis is the process in which electron transport is coupled to ATP synthesis. It takes place in the membranes of the mitochondria of all eukaryotic cells, the chloroplasts of plants, and across the plasma membrane of bacteria. Chemiosmosis involves the establishment of a proton (hydrogen) gradient across biological membranes (shown below for cell respiration). The concentration gradient is a form of potential energy, which in chemiosmosis is used to generate ATP. Chemiosmosis has two key components: an **electron transport chain** (ETC) sets up a proton gradient as electrons pass along it to a final electron acceptor, and an enzyme called **ATP synthase** which uses the proton gradient to catalyze ATP synthesis In cell respiration, electron transport carriers on the inner membrane of the mitochondrion oxidize NADH + H^+ and $FADH_2$. Energy from this process forces protons to move, against their concentration gradient, from the mitochondrial matrix into the space between the two membranes. The protons then flow back into the matrix via ATP synthase molecules in the membrane. As the protons flow down their concentration gradient, energy is released and ATP is synthesized. In the chloroplasts of green plants, the process is similar as ATP is produced when protons pass from the thylakoid lumen to the chloroplast stroma via ATP synthase.

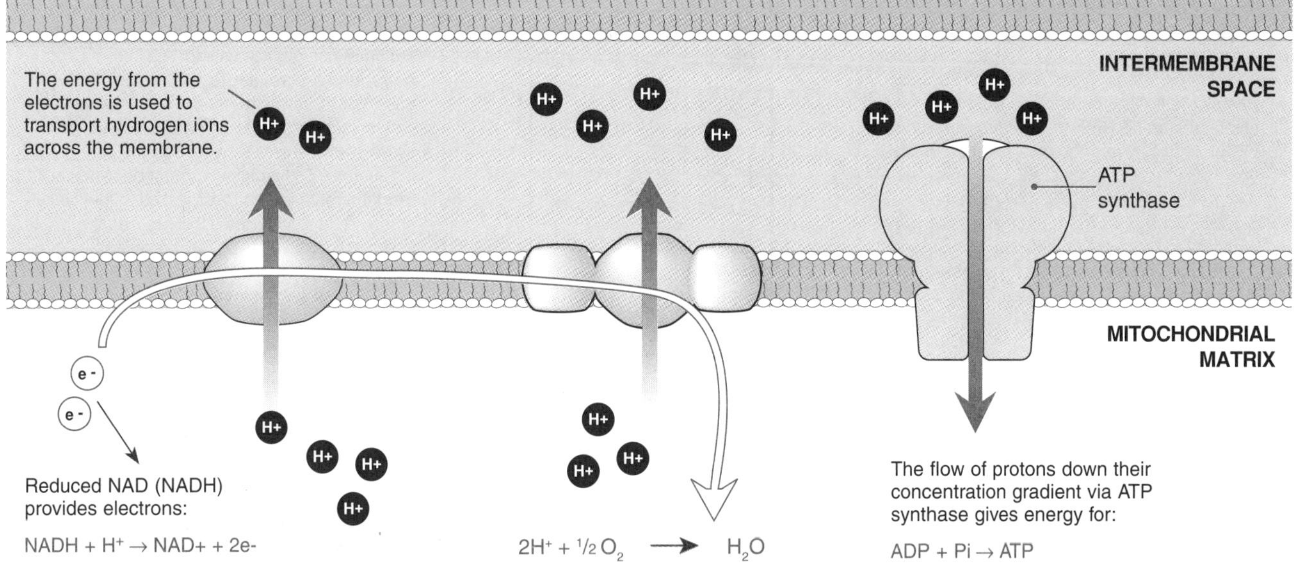
INTERMEMBRANE SPACE
The energy from the electrons is used to transport hydrogen ions across the membrane.
H+
ATP synthase
MITOCHONDRIAL MATRIX
e-
Reduced NAD (NADH) provides electrons:
NADH + H⁺ → NAD+ + 2e-
2H⁺ + ½ O₂ → H₂O
The flow of protons down their concentration gradient via ATP synthase gives energy for:
ADP + Pi → ATP

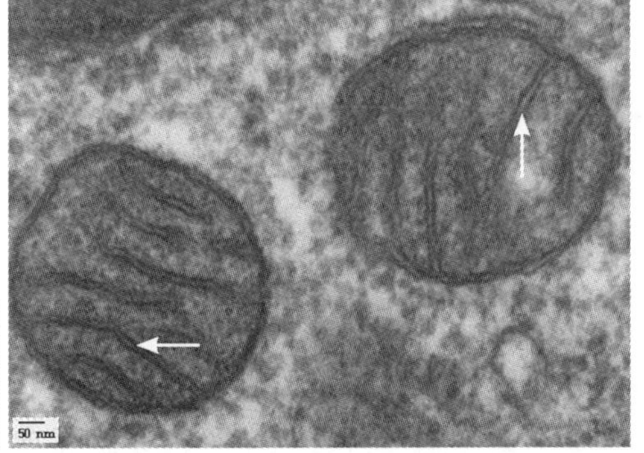
Louisa Howard, DEMF
50 nm
The intermembrane spaces can be seen (arrows) in this transverse section of miotchondria.

The Evidence for Chemiosmosis

The British biochemist Peter Mitchell proposed the chemiosmotic hypothesis in 1961. He proposed that, because living cells have membrane potential, electrochemical gradients could be used to do work, i.e. provide the energy for ATP synthesis. Scientists at the time were skeptical, but the evidence for chemiosmosis was extensive and came from studies of isolated mitochondria and chloroplasts. Evidence included:

- The outer membranes of mitochondria were removed leaving the inner membranes intact. Adding protons to the treated mitochondria increased ATP synthesis.
- When isolated chloroplasts were illuminated, the medium in which they were suspended became alkaline.
- Isolated chloroplasts were kept in the dark and transferred first to a low pH medium (to acidify the thylakoid interior) and then to an alkaline medium (low protons). They then spontaneously synthesized ATP (no light was needed).

1. Summarize the process of chemiosmosis: ______________________________

2. Why did the addition of protons to the treated mitochondria increase ATP synthesis? ______________________________

3. Why did the suspension of isolated chloroplasts become alkaline when illuminated? ______________________________

4. (a) What was the purpose of transferring the chloroplasts first to an acid then to an alkaline medium? ______________________________

   (b) Why did ATP synthesis occur spontaneously in these treated chloroplasts? ______________________________

ISBN: 978-1-927173-16-9

Anaerobic Pathways

All organisms can metabolize glucose anaerobically (without oxygen) using **glycolysis** in the cytoplasm, but the energy yield from this process is low and few organisms can obtain sufficient energy for their needs this way. In the absence of oxygen, glycolysis soon stops unless there is an alternative acceptor for the electrons produced from the glycolytic pathway. In yeasts and the root cells of higher plants this acceptor is ethanal, and the pathway is called **alcoholic fermentation**. In the skeletal muscle of mammals, the acceptor is pyruvate itself and the end product is **lactic acid**. In both cases, the duration of the fermentation is limited by the toxic effects of the organic compound produced. Although fermentation is often used synonymously with anaerobic respiration, they are not the same. Respiration always involves hydrogen ions passing down a chain of carriers to a terminal acceptor, and this does not occur in fermentation. In anaerobic respiration, the terminal H^+ acceptor is a molecule other than oxygen, e.g. Fe^{2+} or nitrate.

Alcoholic Fermentation

In alcoholic fermentation, the H^+ acceptor is ethanal which is reduced to ethanol with the release of CO_2. Yeasts respire aerobically when oxygen is available but can use alcoholic fermentation when it is not. At levels above 12-15%, the ethanol produced by alcoholic fermentation is toxic to the yeast cells and this limits their ability to use this pathway indefinitely. The root cells of plants also use fermentation as a pathway when oxygen is unavailable but the ethanol must be converted back to respiratory intermediates and respired aerobically.

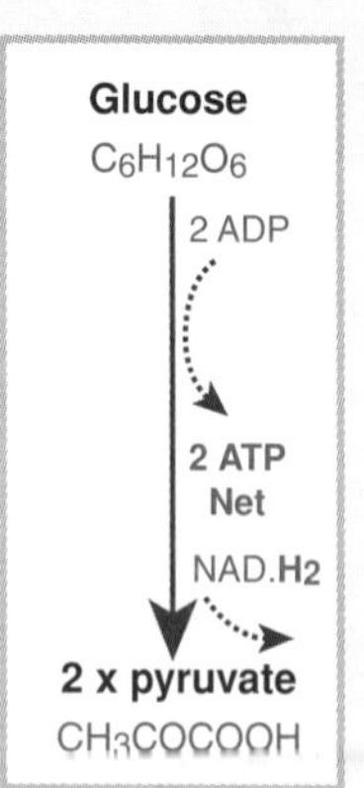
Glucose
$C_6H_{12}O_6$
2 ADP
2 ATP Net
NAD.H2
2 x pyruvate
$CH_3COCOOH$

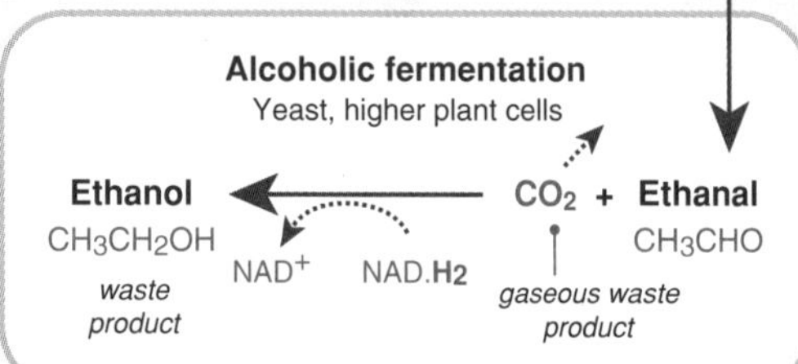
Alcoholic fermentation
Yeast, higher plant cells
Ethanol
CH_3CH_2OH
waste product
NAD^+
NAD.H2
CO_2 + Ethanal
CH_3CHO
gaseous waste product

Lactic Acid Fermentation

In the absence of oxygen, the skeletal muscle cells of mammals are able to continue using glycolysis for ATP production by reducing pyruvate to lactic acid (the H^+ acceptor is pyruvate itself). This process is called lactic acid fermentation. Lactic acid is toxic and this pathway cannot continue indefinitely. The lactic acid must be removed from the muscle and transported to the liver, where it is converted back to respiratory intermediates and respired aerobically.

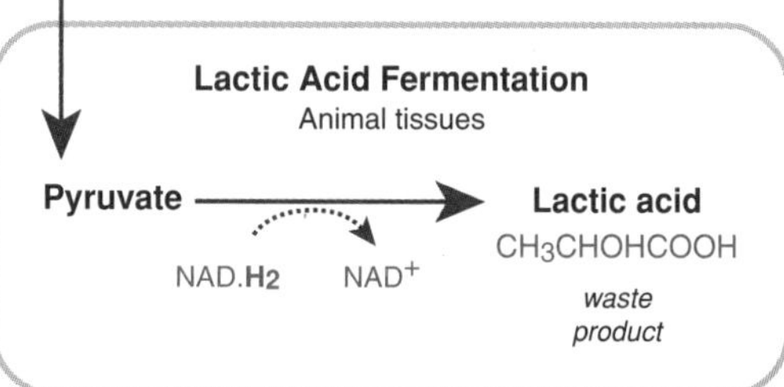
Glucose
$C_6H_{12}O_6$
2 ADP
2 ATP Net
NAD.H2
2 x pyruvate
$CH_3COCOOH$
Lactic Acid Fermentation
Animal tissues
Pyruvate
Lactic acid
$CH_3CHOHCOOH$
NAD.H2
NAD^+
waste product

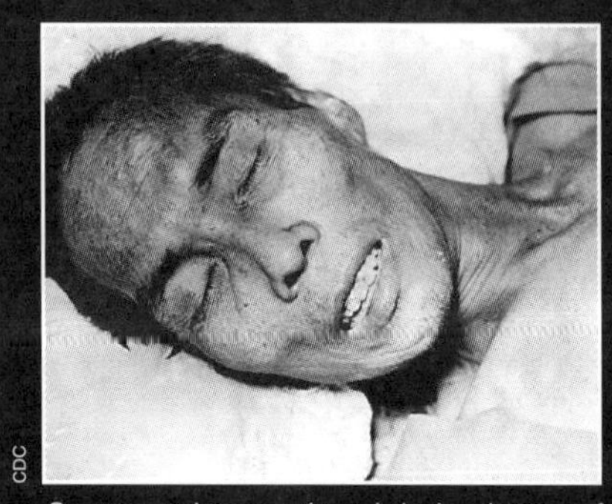
CDC

Some organisms respire only in the absence of oxygen and are known as obligate anaerobes. Many of these organisms are bacterial pathogens and cause diseases such as tetanus (above), gangrene, and botulism.

Vertebrate skeletal muscle is facultatively anaerobic because it has the ability to generate ATP for a short time in the absence of oxygen. The energy from this pathway comes from glycolysis and the yield is low.

The products of alcoholic fermentation have been utilized by humans for centuries. The alcohol and carbon dioxide produced from this process form the basis of the brewing and baking industries.

1. Describe the key difference between aerobic respiration and fermentation: ______________________

____

2. (a) Refer to page 119 and determine the efficiency of fermentation compared to aerobic respiration: __________ %

(b) Why is the efficiency of these anaerobic pathways so low? ______________________

____

3. Why can't fermentation go on indefinitely? ______________________

____

____

Related activities: *The Biochemistry of Respiration*
Weblinks: *Lactate and Alcoholic Fermentation*

Periodicals: *Lactic acid: who needs it?*

ISBN: 978-1-927173-16-9

Investigating Yeast Fermentation

Any practical investigation requires you to critically evaluate your results in the light of your own hypothesis and your biological knowledge. A critical evaluation of any study involves analyzing, presenting, and discussing the results, as well as accounting for any deficiencies in your procedures and erroneous results. This activity describes an experiment comparing different carbohydrates for their effectiveness as substrates for fermentation. Brewer's yeast is a **facultative anaerobe** (meaning it can respire aerobically or use fermentation). It will preferentially use alcoholic fermentation when sugars are in excess. One would expect glucose to be the preferred substrate, as it is the starting molecule in cell respiration, but yeast are capable of utilizing a variety of sugars, including disaccharides that can be broken down into single units. Completing this activity, which involves a critical evaluation of the second-hand data provided, will help to prepare you for your own evaluative task.

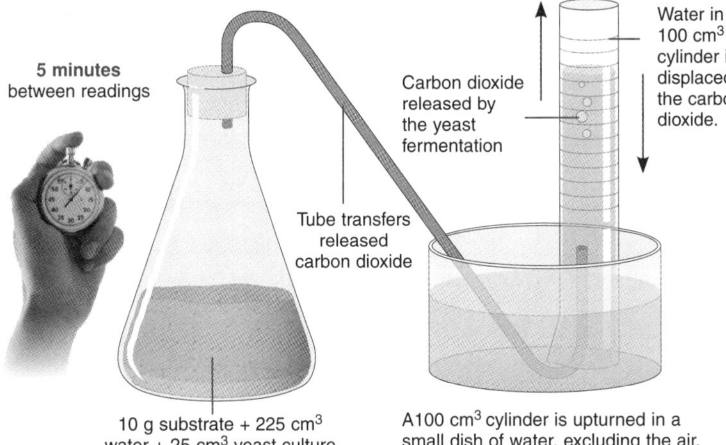
5 minutes between readings
Carbon dioxide released by the yeast fermentation
Water in the 100 cm^3 cylinder is displaced by the carbon dioxide.
Tube transfers released carbon dioxide
10 g substrate + 225 cm^3 water + 25 cm^3 yeast culture
A100 cm^3 cylinder is upturned in a small dish of water, excluding the air.

The Apparatus

In this experiment, all substrates tested used the same source culture of 30 g active yeast dissolved in 150 cm^3 of room temperature (24°C) tap water. For each substrate, 25 g of the substrate to be tested was added to 225 cm^3 room temperature (24°C) tap water buffered to pH 4.5. Then 25 cm^3 of source culture was added to the test solution. The control contained yeast solution but no substrate:

Substrate / Time (min)	Group 1: Volume of carbon dioxide collected (cm^3)				
	None	**Glucose**	**Maltose**	**Sucrose**	**Lactose**
0	0	0	0	0	0
5	0	0	0.8	0	0
10	0	0	0.8	0	0
15	0	0	0.8	0.1	0
20	0	0.5	2.0	0.8	0
25	0	1.2	3.0	1.8	0
30	0	2.8	3.6	3.0	0.5
35	0	4.2	5.4	4.8	0.5
40	0	4.6	5.6	4.8	0.5
45	0	7.4	8.0	7.2	1.0
50	0	10.8	8.9	7.6	1.3
55	0	13.6	9.6	7.7	1.3
60	0	16.1	10.4	9.6	1.3
65	0	22.0	12.1	10.2	1.8
70	0	23.8	14.4	12.0	1.8
75	0	26.7	15.2	12.6	2.0
80	0	32.5	17.3	14.3	2.1
85	0	37.0	18.7	14.9	2.4
90	0	39.9	21.6	17.2	2.6

Substrate / Time (min)	Group 2: Volume of carbon dioxide collected (cm^3)				
	None	**Glucose**	**Maltose**	**Sucrose**	**Lactose**
90	0	24.4	19.0	17.5	0

The Aim

To investigate the suitability of different mono- and disaccharide sugars as substrates for alcoholic fermentation in yeast.

Background

The rate at which brewer's or baker's yeast (*Saccharomyces cerevisiae*) metabolizes carbohydrate substrates is influenced by factors such as temperature, solution pH, and type of carbohydrate available. The literature describes yeast metabolism as optimal in warm, slightly acid environments and this experiment was set up to investigate the pH optimum for the active yeast which is readily available at supermarkets.

Note: High levels of sugars suppress aerobic respiration in yeast, so yeast will preferentially use the fermentation pathway in the presence of excess substrate.

1. Write the equation for the fermentation of glucose by yeast:

____

2. Calculate the rate of carbon dioxide production per minute for each substrate in group 1's results:

(a) None: _______________

(b) Glucose: _______________

(c) Maltose: _______________

(d) Sucrose: _______________

(e) Lactose: _______________

3. A second group of students performed the same experiment. Their results are summarized, below left. Calculate the rate of carbon dioxide production per minute for each substrate in group 2's results:

(a) None: _______________

(b) Glucose: _______________

(c) Maltose: _______________

(d) Sucrose: _______________

(e) Lactose: _______________

Experimental design and results adapted from Tom Schuster, Rosalie Van Zyl, & Harold Coller , California State University Northridge 2005

ISBN: 978-1-927173-16-9

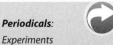

4. What assumptions are being made in this experimental design and do you think they were reasonable?

5. Use the tabulated data to plot an appropriate graph of group 1's results on the grid provided:

6. (a) Summarize the results of group 1's fermentation experiment:

(b) Explain the findings based on your understanding of cell respiration and carbohydrate chemistry:

7. (a) Plot a column chart to compare the results of the two groups in the volume of CO_2 collected after 90 minutes for each substrate (axes have been completed):

(b) Compare the results of the two groups:

Vol. CO_2 produced in 90 minutes (cm^3)
0
5
10
15
20
25
30
35
40
45
50
None
Glucose
Maltose
Sucrose
Lactose
Substrate type

(c) Provide a probable explanation for any differences in the results:

(d) Describe one improvement you could make to the experiment in order to generate more reliable data:

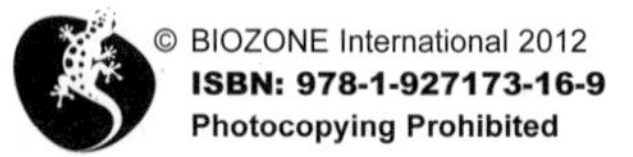

ISBN: 978-1-927173-16-9

Photosynthesis

Photosynthesis is of fundamental importance to living things because it transforms sunlight energy into chemical energy stored in molecules, releases free oxygen gas, and absorbs carbon dioxide (a waste product of cellular metabolism). Photosynthetic organisms use special pigments, called **chlorophylls**, to absorb light of specific wavelengths and thereby capture the light energy.

Visible light is a small fraction of the total **electromagnetic radiation** reaching Earth from the sun. Of the visible spectrum, only certain wavelengths (red and blue) are absorbed by chlorophyll. Other wavelengths, particularly green, are reflected or transmitted. The diagram below summarizes the process of photosynthesis.

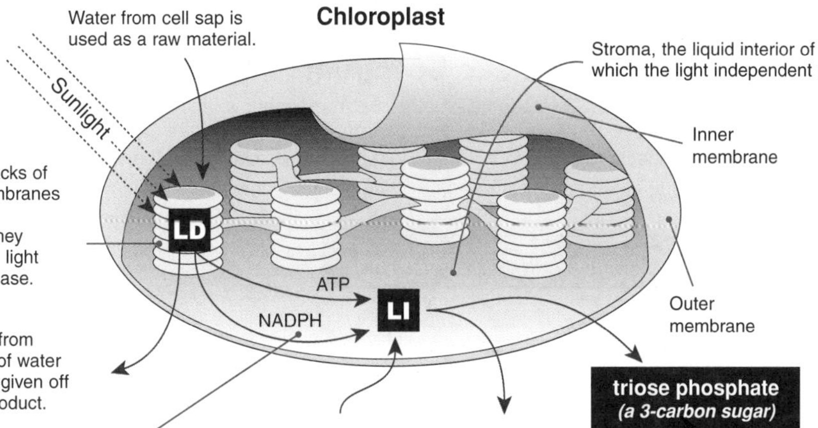

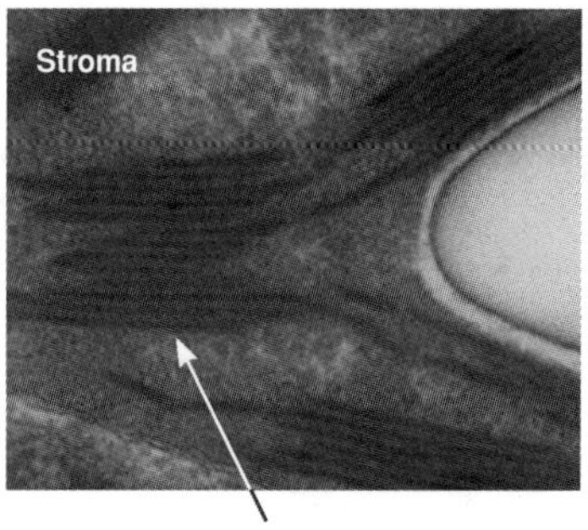

Chloroplast

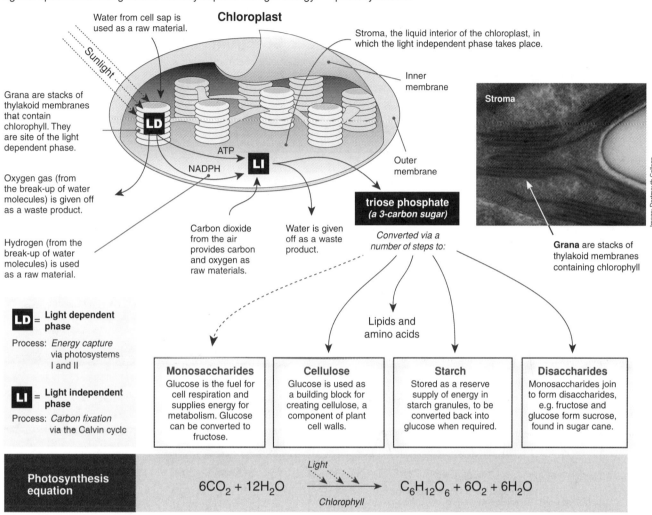

Sunlight

Water from cell sap is used as a raw material.

Stroma, the liquid interior of the chloroplast, in which the light independent phase takes place.

Inner membrane

Outer membrane

Grana are stacks of thylakoid membranes that contain chlorophyll. They are site of the light dependent phase.

LD

ATP

NADPH

LI

Oxygen gas (from the break-up of water molecules) is given off as a waste product.

Hydrogen (from the break-up of water molecules) is used as a raw material.

Carbon dioxide from the air provides carbon and oxygen as raw materials.

Water is given off as a waste product.

triose phosphate *(a 3-carbon sugar)*

Converted via a number of steps to:

Stroma

Grana are stacks of thylakoid membranes containing chlorophyll

Image: Dartmouth College

Lipids and amino acids

LD = **Light dependent phase**
Process: *Energy capture* via photosystems I and II

LI = **Light independent phase**
Process: *Carbon fixation* via the Calvin cycle

Monosaccharides	**Cellulose**	**Starch**	**Disaccharides**
Glucose is the fuel for cell respiration and supplies energy for metabolism. Glucose can be converted to fructose.	Glucose is used as a building block for creating cellulose, a component of plant cell walls.	Stored as a reserve supply of energy in starch granules, to be converted back into glucose when required.	Monosaccharides join to form disaccharides, e.g. fructose and glucose form sucrose, found in sugar cane.

Photosynthesis equation

$$6CO_2 + 12H_2O \xrightarrow[\text{Chlorophyll}]{\text{Light}} C_6H_{12}O_6 + 6O_2 + 6H_2O$$

1. Distinguish between the two different regions of a chloroplast and describe the biochemical processes that occur in each:

   (a) ______________________________

   ____

   (b) ______________________________

   ____

2. State the origin and fate of the following molecules involved in photosynthesis:

   (a) Carbon dioxide: ______________________________

   (b) Oxygen: ______________________________

   (c) Hydrogen: ______________________________

3. How might scientists have determined the fate of these molecules? ______________________________

   ____

4. Explain why the leaves of most plants look green: ______________________________

   ____

Pigments and Light Absorption

As light meets matter, it may be reflected, transmitted, or absorbed. Substances that absorb visible light are called **pigments**, and different pigments absorb light of different wavelengths. The ability of a pigment to absorb particular wavelengths of light can be measured with a spectrophotometer. The light absorption vs the wavelength is called the **absorption spectrum** of that pigment. The absorption spectrum of different photosynthetic pigments provides clues to their role in photosynthesis, since light can only perform work if it is absorbed. An **action spectrum** profiles the effectiveness of different wavelength light in fuelling photosynthesis. It is obtained by plotting wavelength against some measure of photosynthetic rate (e.g. CO_2 production). Some features of photosynthetic pigments and their light absorbing properties are outlined below.

The Electromagnetic Spectrum

Light is a form of energy known as electromagnetic radiation. The segment of the electromagnetic spectrum most important to life is the narrow band between about 380 nm and 750 nm. This radiation is known as visible light because it is detected as colors by the human eye (although some other animals, such as insects, can see in the UV range). It is the visible light that drives photosynthesis.

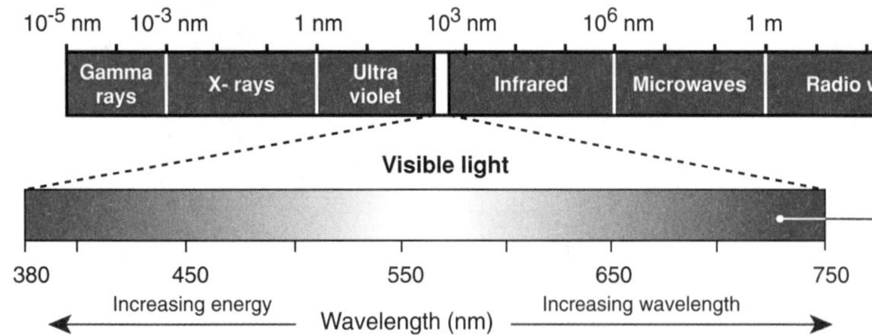

10^{-5} nm
10^{-3} nm
1 nm
10^3 nm
10^6 nm
1 m
10^3 m
Gamma rays
X- rays
Ultra violet
Infrared
Microwaves
Radio waves
Visible light
380
450
550
650
750
Increasing energy
Wavelength (nm)
Increasing wavelength

Electromagnetic radiation (EMR) travels in waves, where wavelength provides a guide to the energy of the photons; the greater the wavelength of EMR, the lower the energy of the photons in that radiation.

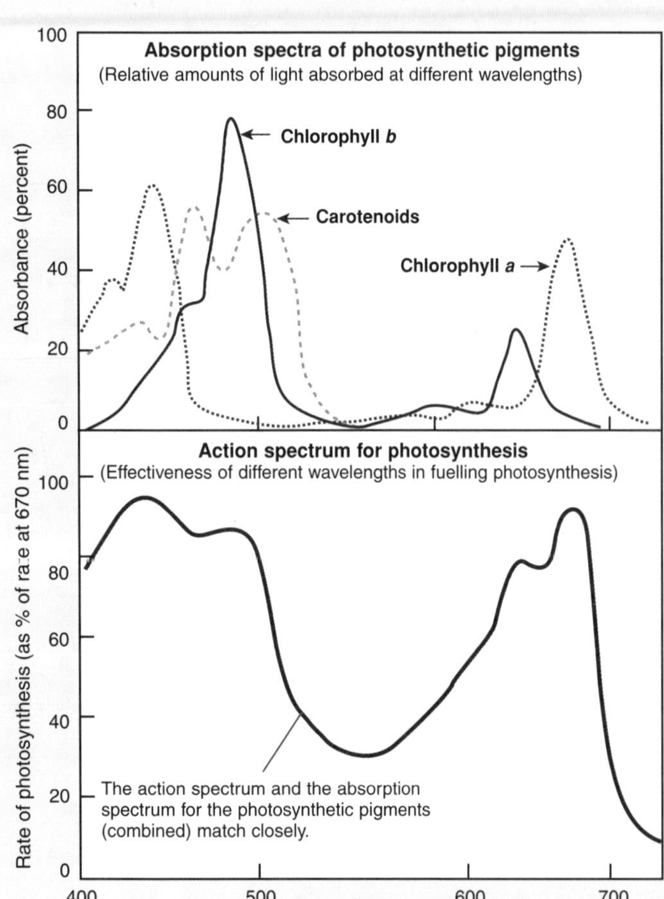

Absorption spectra of photosynthetic pigments
(Relative amounts of light absorbed at different wavelengths)
Absorbance (percent)
Chlorophyll b
Carotenoids
Chlorophyll a
Action spectrum for photosynthesis
(Effectiveness of different wavelengths in fuelling photosynthesis)
Rate of photosynthesis (as % of rate at 670 nm)
The action spectrum and the absorption spectrum for the photosynthetic pigments (combined) match closely.
Wavelength (nm)
400
500
600
700

The photosynthetic pigments of plants

The photosynthetic pigments of plants fall into two categories: chlorophylls (which absorb red and blue-violet light) and carotenoids (which absorb strongly in the blue-violet and appear orange, yellow, or red). The pigments are located on the chloroplast membranes (the thylakoids) and are associated with membrane transport systems.

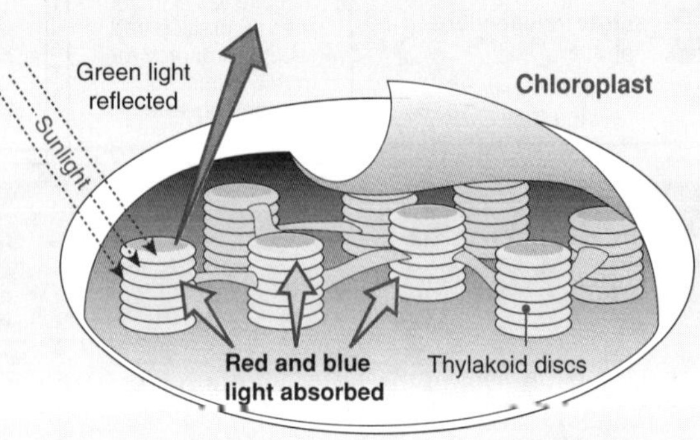

Green light reflected
Sunlight
Chloroplast
Red and blue light absorbed
Thylakoid discs

The pigments of chloroplasts in higher plants (above) absorb blue and red light, and the leaves therefore appear green (which is reflected). Each photosynthetic pigment has its own characteristic absorption spectrum (left, top graph). Although only chlorophyll a can participate directly in the light reactions of photosynthesis, the accessory pigments (chlorophyll *b* and carotenoids) can absorb wavelengths of light that chlorophyll *a* cannot. The accessory pigments pass the energy (photons) to chlorophyll a, thus broadening the spectrum that can effectively drive photosynthesis.

Left: Graphs comparing absorption spectra of photosynthetic pigments compared with the action spectrum for photosynthesis.

1. What is meant by the absorption spectrum of a pigment? ______________________

2. Why doesn't the **action spectrum** for photosynthesis exactly match the absorption spectrum of chlorophyll *a*?

ISBN: 978-1-927173-16-9

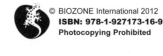

Factors Affecting Photosynthetic Rate

The rate at which plants make carbohydrate (the photosynthetic rate) is dependent on environmental factors, the most important of which are the availability of **light** and **carbon dioxide** (CO_2). and **temperature**. The effect of these factors can be tested experimentally by altering one of the factors while holding the others constant (a controlled experiment). In reality, a plant in its natural environment is subjected to variations in many different environmental factors, all of which will influence, directly or indirectly, the rate at which photosynthesis can occur.

A: Light intensity vs photosynthetic rate
Rate of photosynthesis (mm3 CO2 cm−2 h−1)
90
80
70
60
50
40
1
2
3
4
5
6
7
Units of light intensity (arbitrary scale)

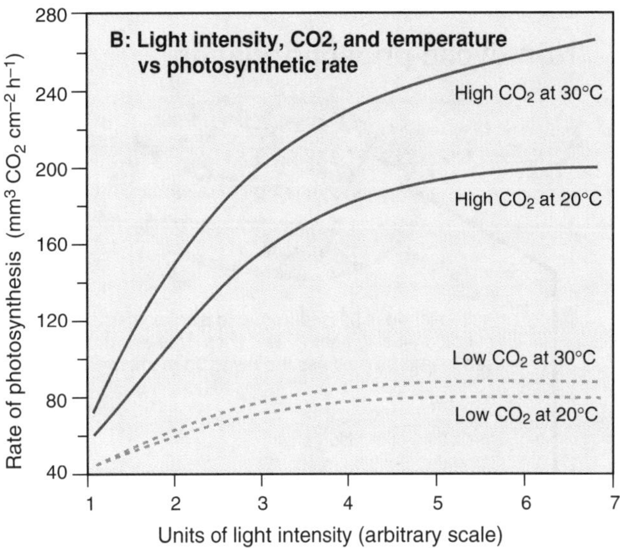
B: Light intensity, CO2, and temperature vs photosynthetic rate
High CO2 at 30°C
High CO2 at 20°C
Low CO2 at 30°C
Low CO2 at 20°C
Rate of photosynthesis (mm3 CO2 cm−2 h−1)
280
240
200
160
120
80
40
1
2
3
4
5
6
7
Units of light intensity (arbitrary scale)

These figures illustrate the effect of different limiting factors on the rate of photosynthesis in cucumber plants. Figure A shows the effect of different light intensities when the temperature and carbon dioxide (CO_2) level are kept constant. Figure B shows the effect of different light intensities at two temperatures and two CO_2 concentrations. In each of these experiments, either CO_2 level or temperature was changed at each light intensity in turn.

1. Based on the figures above, summarize and explain the effect of each of the following factors on photosynthetic rate:

   (a) CO_2 concentration: ______________________________

   ____

   (b) Light intensity: ______________________________

   ____

   (c) Temperature: ______________________________

   ____

2. Explain why photosynthetic rate declines when the CO_2 level is reduced: ______________________________

   ____

   ____

3. (a) In figure B, explain how the effects of CO_2 concentration were distinguished from the effects of temperature:

   ____

   (b) Identify which factor (CO_2 or temperature) had the greatest effect on photosynthetic rate: ______________

   (c) Explain how you can tell this from the graph: ______________________________

   ____

4. Explain how glasshouses can be used to create an environment in which photosynthetic rates are maximised:

   ____

   ____

   ____

5. Design an experiment to demonstrate the effect of temperature on photosynthetic rate. You should include a hypothesis, list of equipment, and methods. Staple your experiment to this page.

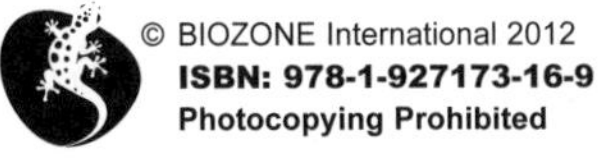

ISBN: 978-1-927173-16-9

Light Dependent Reactions

Like cellular respiration, photosynthesis is a redox process, but in photosynthesis, water is split, and electrons and hydrogen ions, are transferred from water to CO_2, reducing it to sugar. The electrons increase in potential energy as they move from water to sugar. The energy to do this is provided by light. Photosynthesis has two phases. In the **light dependent reactions**, light energy is converted to chemical energy (ATP and NADPH). In the **light independent reactions**, the chemical energy is used to synthesize carbohydrate. The light dependent reactions most commonly involve **non-cyclic phosphorylation**, which produces ATP and NADPH in roughly equal quantities. The electrons lost are replaced from water. In **cyclic phosphorylation**, the electrons lost from photosystem II are replaced by those from photosystem I. ATP is generated, but not NADPH.

Non-cyclic phosphorylation

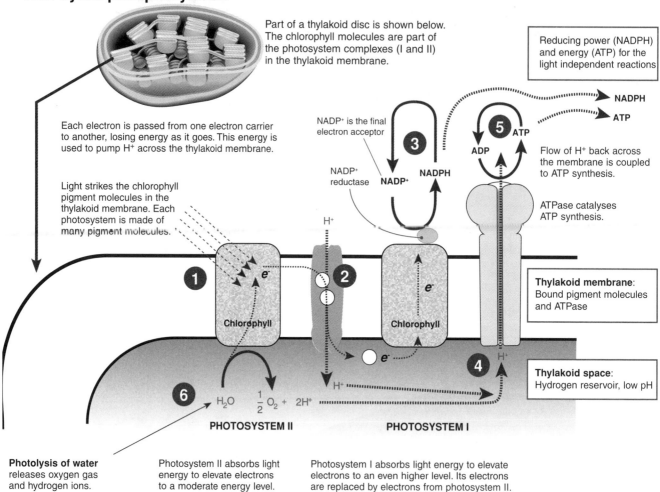

Part of a thylakoid disc is shown below. The chlorophyll molecules are part of the photosystem complexes (I and II) in the thylakoid membrane.

Reducing power (NADPH) and energy (ATP) for the light independent reactions

NADPH

ATP

Each electron is passed from one electron carrier to another, losing energy as it goes. This energy is used to pump H^+ across the thylakoid membrane.

$NADP^+$ is the final electron acceptor

3

$NADP^+$ reductase

$NADP^+$

NADPH

5

ADP

ATP

Flow of H^+ back across the membrane is coupled to ATP synthesis.

Light strikes the chlorophyll pigment molecules in the thylakoid membrane. Each photosystem is made of many pigment molecules.

ATPase catalyses ATP synthesis.

H^+

1

e^-

2

e^-

Chlorophyll

Chlorophyll

Thylakoid membrane: Bound pigment molecules and ATPase

e^-

H^+

4

Thylakoid space: Hydrogen reservoir, low pH

6

H_2O $\frac{1}{2}O_2$ + $2H^+$

H^+

PHOTOSYSTEM II

PHOTOSYSTEM I

Photolysis of water releases oxygen gas and hydrogen ions.

Photosystem II absorbs light energy to elevate electrons to a moderate energy level.

Photosystem I absorbs light energy to elevate electrons to an even higher level. Its electrons are replaced by electrons from photosystem II.

Cyclic phosphorylation

Cyclic phosphorylation involves only photosystem I and NADPH is not generated. Electrons from photosystem I are shunted back to the electron carriers in the membrane. This pathway produces ATP only. The Calvin cycle uses more ATP than NADPH, so cyclic phosphorylation makes up the difference. It is activated when NADPH levels build up, and remains active until enough ATP is made to meet demand.

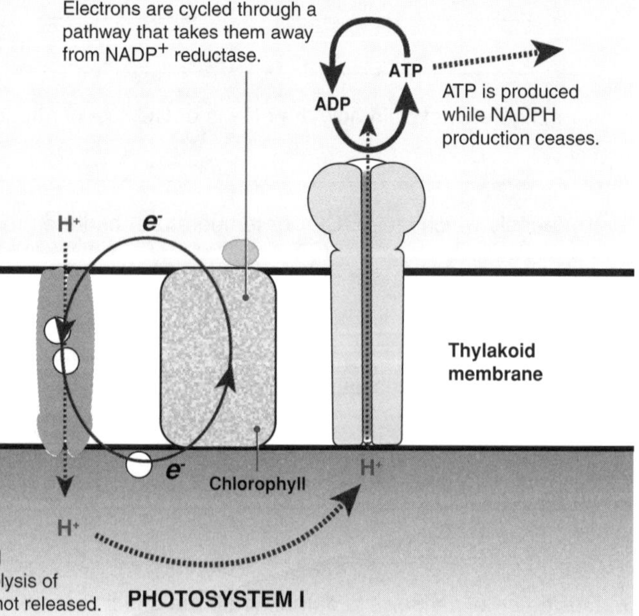
Electrons are cycled through a pathway that takes them away from NADP+ reductase.
ADP
ATP
ATP is produced while NADPH production ceases.
H+
e-
Thylakoid membrane
e-
Chlorophyll
H+
H+
PHOTOSYSTEM I

Chlorophyll

PHOTOSYSTEM II is not active. Photolysis of water stops. O_2 is not released.

Related activities: *Chloroplasts, Light Independent Reactions*
Web links: *Photosynthesis Light Reactions, Photosystem II*

Periodicals: *Photosynthesis..most hated topic?*

ISBN: 978-1-927173-16-9

1. Describe the role of the carrier molecule **NADP** in photosynthesis:

2. Explain the role of chlorophyll molecules in photosynthesis:

3. Summarize the events of the light dependent reactions and identify where they occur:

4. Describe how ATP is produced as a result of light striking chlorophyll molecules during the light dependent phase:

5. (a) Explain what you understand by the term **non-cyclic phosphorylation**:

   (b) Suggest why this process is also known as non-cyclic **photo**phosphorylation:

6. (a) Describe how **cyclic photophosphorylation** differs from non-cyclic photophosphorylation:

   (b) Both cyclic and noncyclic pathways operate to varying degrees during photosynthesis. Since the non-cyclic pathway produces both ATP and NAPH, explain the purpose of the cyclic pathway of electron flow:

7. Explain how the independence of photosystem I gives a mechanism for evolution of the photosynthetic pathway:

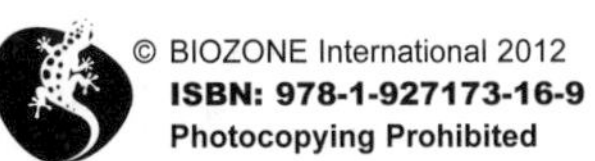

ISBN: 978-1-927173-16-9

Light Independent Reactions

The **light independent reactions** of photosynthesis (the **Calvin cycle**) take place in the stroma of the chloroplast, and do not require light to proceed. Here, hydrogen (H^+) is added to CO_2 and a 5C intermediate to make carbohydrate. The H^+ and ATP are supplied by the light dependent reactions. The Calvin cycle uses more ATP than NADPH, but the cell uses cyclic phosphorylation (which does not produce NADPH) when it runs low on ATP to make up the difference.

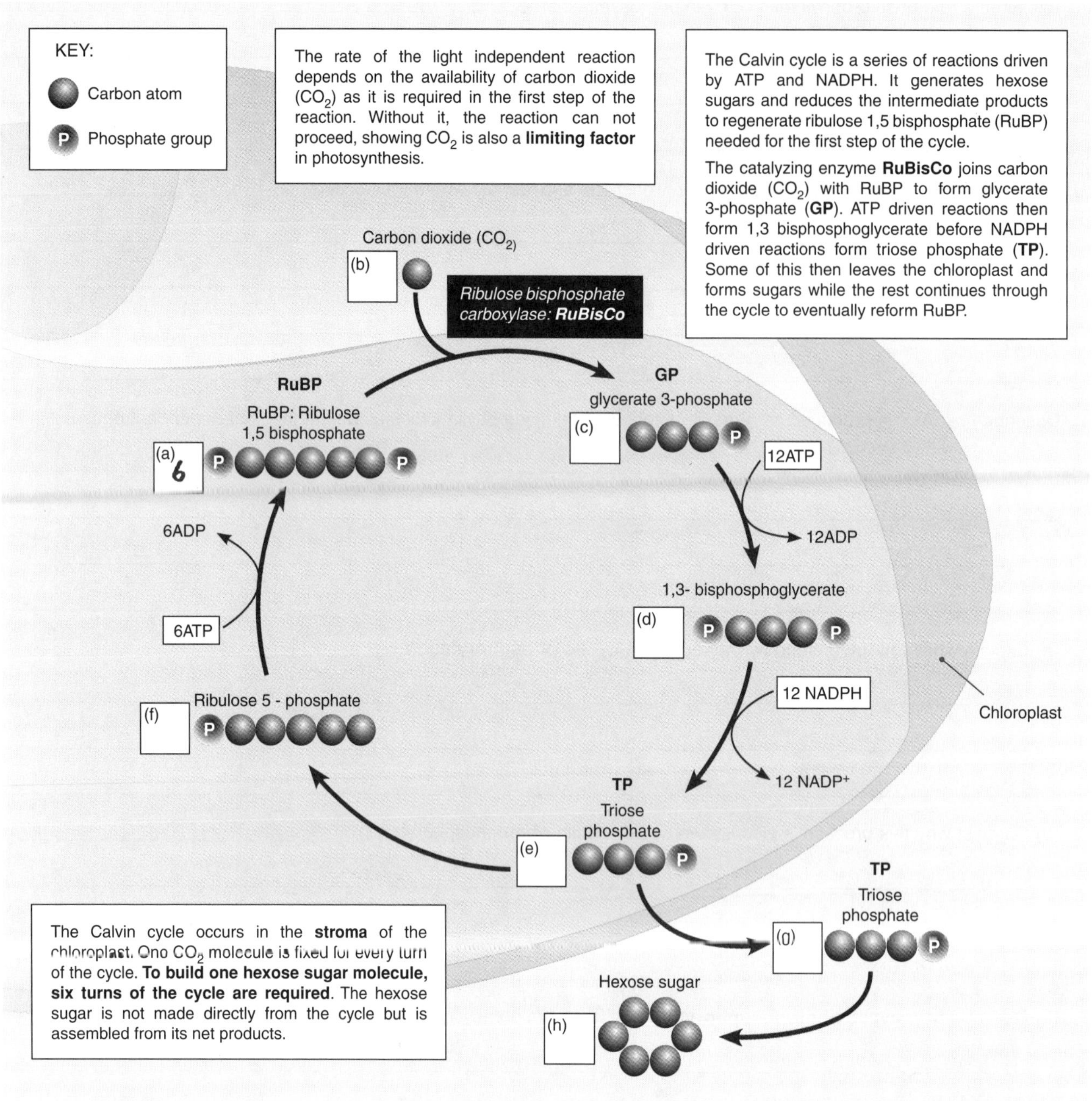

KEY:
Carbon atom
P Phosphate group

The rate of the light independent reaction depends on the availability of carbon dioxide (CO_2) as it is required in the first step of the reaction. Without it, the reaction can not proceed, showing CO_2 is also a **limiting factor** in photosynthesis.

The Calvin cycle is a series of reactions driven by ATP and NADPH. It generates hexose sugars and reduces the intermediate products to regenerate ribulose 1,5 bisphosphate (RuBP) needed for the first step of the cycle.

The catalyzing enzyme **RuBisCo** joins carbon dioxide (CO_2) with RuBP to form glycerate 3-phosphate (**GP**). ATP driven reactions then form 1,3 bisphosphoglycerate before NADPH driven reactions form triose phosphate (**TP**). Some of this then leaves the chloroplast and forms sugars while the rest continues through the cycle to eventually reform RuBP.

Carbon dioxide (CO_2)
(b)
Ribulose bisphosphate carboxylase: ***RuBisCo***
RuBP
RuBP: Ribulose 1,5 bisphosphate
(a) 6
GP
glycerate 3-phosphate
(c)
12ATP
12ADP
1,3- bisphosphoglycerate
(d)
12 NADPH
12 NADP+
6ADP
6ATP
Ribulose 5 - phosphate
(f)
Chloroplast
TP
Triose phosphate
(e)
TP
Triose phosphate
(g)
Hexose sugar
(h)

The Calvin cycle occurs in the **stroma** of the chloroplast. One CO_2 molecule is fixed for every turn of the cycle. **To build one hexose sugar molecule, six turns of the cycle are required**. The hexose sugar is not made directly from the cycle but is assembled from its net products.

1. In the boxes on the diagram above, write the number of molecules formed at each step during the formation of **one hexose sugar molecule.** The first one has been done for you:

2. Explain the importance of RuBisCo in the Calvin cycle: ____________________

____

3. Identify the actual end product on the Calvin cycle: ____________________

4. Write the equation for the production of one hexose sugar molecule from carbon dioxide: ____________________

____

5. Explain why the Calvin cycle is likely to cease in the dark for most plants, even though it is independent of light:

____

Related activities: *Light Dependent Reactions*
Web links: *Photosynthesis Biochemistry*

Periodicals:
Rubisco

ISBN: 978-1-927173-16-9

KEY TERMS: Mix and Match

INSTRUCTIONS: Test your vocabulary by matching each term to its definition, as identified by its preceding letter code.

Core only

ATP

biological catalyst

cell respiration

chlorophyll

denaturation

enzyme

fermentation

glycolysis

lock and key model

mitochondrion

photosynthesis

pyruvate

substrate

A The loss of a protein's three-dimensional functional structure is called this.

B A model of enzyme activity where the substrate slots into a rigid active site.

C Process of deriving energy from oxidation of organic compounds using an endogenous acceptor rather than an exogenous one (such as oxygen).

D The organelle in the cytoplasm of eukaryotic cells, where cellular respiration and ATP production occur.

E A nucleotide comprising a purine base, a pentose sugar, and three phosphate groups, which acts as the cell's energy carrier.

F The catabolic process in which the chemical energy in complex organic molecules is coupled to ATP production

G A photosynthetic pigment which strongly absorbs red and blue-violet light and appears green in color.

H First stage of glucose catabolism occurring in the cytoplasm.

I The compound on which an enzyme acts.

J A globular protein which acts as a catalyst to speed up a specific biological reaction.

K The biochemical process that uses light energy to convert carbon dioxide and water into glucose molecules and oxygen.

L A three carbon compound produced as the end product of glycolysis.

M A substance or molecule that lowers the activation energy of a reaction but is itself not used up during the reaction. In biological systems this function is carried out by enzymes.

AHL only

absorption spectrum

action spectrum

Calvin cycle

chemiosmosis

competitive inhibition

electron transport chain

induced fit model

Krebs cycle

light dependent phase

non-competitive inhibition

oxidative phosphorylation

A The process where the synthesis of ATP is coupled to electron transport and the movement of protons.

B A type of enzyme inhibition where the substance and inhibitor compete to bind in the active site.

C The currently accepted model for enzyme function.

D A series of biochemical reactions, occurring in the stroma of chloroplasts, in which CO_2 is incorporated into carbohydrates. Also called the light independent phase.

E Also known as the citric acid cycle. A metabolic pathway in which acetate (as acetyl-CoA) is consumed, NAD+ is reduced to NADH, and carbon dioxide is produced.

F The term to describe the light absorption vs the wavelength of a pigment.

G The chain of enzyme based redox reactions which pass electrons from high to low redox potentials. The energy released is used to pump protons across a membrane and produce ATP.

H The process in cell respiration involving the oxidation of glucose by a series of redox reactions that provide the energy for the formation of ATP.

I A type of enzyme inhibition where the inhibitor does not occupy the active site but binds to some other part of the enzyme.

J The phase in photosynthesis when light energy is absorbed by the photosystems in the thylakoid membranes of the chloroplast, generating NADPH and ATP.

K The relative effectiveness of different wavelengths of light generating electrons from a pigment.

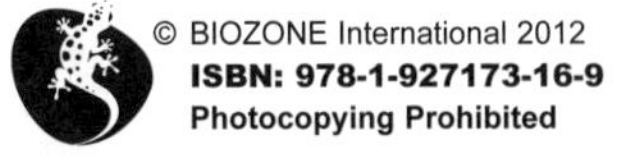

ISBN: 978-1-927173-16-9

Core Topic
4

AHL Topic
10

Chromosomes and **Meiosis**

Key concepts

- ▶ Meiosis is a reduction division that produces haploid gametes.
- ▶ The events in meiosis give rise to variation in the gametes and therefore the offspring.
- ▶ Karyotyping is a diagnostic tool for detecting chromosomal abnormalities such as trisomy.
- ▶ Mutations alter the base sequence of DNA and can result in detrimental phenotypic effects.

Key terms

Core

amniocentesis
base substitution
chorionic villus sampling
chromosome
crossing over
diploid
haploid
homologous chromosome (homolog)
interphase
karyotype
meiosis
mutagen
mutation
non-disjunction
pre-natal diagnosis
reduction division
sickle cell anemia
spontaneous mutation
trisomy

AHL only

anaphase
bivalent
chiasma
chromatid
fertilization
independent assortment
interphase
maternal chromosome
metaphase
paternal chromosome
prophase
synapsis
telophase

Periodicals:
Listings for this chapter are on page 399

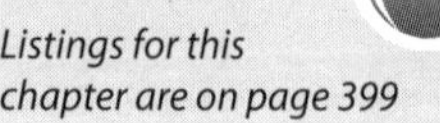

Weblinks:
www.thebiozone.com/weblink/IB-3169.html

BIOZONE APP:
Student Review Series
Processes in the Nucleus

Learning Objectives

- ☐ 1. Use the **KEY TERMS** to help you understand and complete these objectives.

Chromosomes, Genes, Alleles, & Mutation *(4.1)* pages 93, 133-34, 141-42

- ☐ 2. Understand that eukaryotic chromosomes consist of DNA and proteins and that the proteins have a role in packing the genetic material into the nucleus.
- ☐ 3. Define the terms **gene**, **allele**, and **genome** and use them appropriately as required.
- ☐ 4. Define the term **mutation** and appreciate that mutations can arise spontaneously (through errors in DNA replication) or through the action of **mutagens**.
- ☐ 5. Using the example of the mutation causing **sickle cell anemia**, explain the consequences of **base substitution** mutations in relation to transcription and translation of the genetic code.

Chromosomes and Meiosis *(4.2 & 10.1)* pages 135-140, 143-147

- ☐ 6. Know that **meiosis**, like mitosis, involves DNA replication during **interphase** in the parent cell, but that this is followed by two cycles of nuclear division. Know that meiosis is a **reduction division** and explain what this means.
- ☐ 7. Define the term **homologous chromosome** (or **homolog**) and explain the significance of chromosomes existing as homologs in **diploid** organisms.
- ☐ 8. Summarize the events in meiosis, including the pairing of homologs and **crossing over**, followed by two divisions and the production of four **haploid** cells.
- ☐ 9. Explain what is meant by **non-disjunction** and outline its consequences with reference to specific examples, e.g. Down syndrome (**trisomy** 21).
- ☐ 10. Describe karyotyping, including the procedures involved in preparation of the karyotype and the diagnostic significance of **karyotype** analysis. Include reference to the use of cells obtained by **chorionic villus sampling** or **amniocentesis**, for **pre-natal diagnosis** of chromosome abnormalities. Analyze a human karyotype to determine gender and the occurrence (or otherwise) of non-disjunction.
- ☐ 11. **AHL**: Summarize the principal events in meiosis and their significance, including:
  - **Synapsis** and the formation of **bivalents**.
  - **Chiasma** formation and exchange of genetic material between **chromatids** in the first, (reduction) division.
  - Separation of chromatids (second division) and production of haploid cells.
- ☐ 12. **AHL**: Describe the behavior of **homologous chromosomes** (and their associated **alleles**) during meiosis and **fertilization**, with reference to:
  - The **recombination** of segments of maternal and paternal homologous chromosomes in **crossing over** during prophase I.
  - Random orientation of homologous pairs during metaphase I and the **independent assortment** of **maternal** and **paternal chromosomes**.
  - The random fusion of gametes during **fertilization**.
- ☐ 13. **AHL**: Explain the genetic consequences of the events in meiosis.

Genomes

Genome research has become an important field of genetics. A **genome** is the entire haploid complement of genetic material of a cell or organism. Each species has a unique genome, although there is a small amount of genetic variation between individuals within a species. For example, in humans the average genetic difference is one in every 500-1000 bases. Every cell in an individual has a complete copy of the genome. The base sequence shown below is the total DNA sequence for the genome of a virus. There are nine genes in the sequence, coding for nine different proteins. At least 2000 times this amount of DNA would be found in a single bacterial cell. Half a million times the quantity of DNA would be found in the genome of a single human cell. The first gene has been highlighted blue, while the start and stop codes are in dark blue rectangles.

Genome for the ϕX174 bacterial virus

Start

The blue area represents the nucleotide sequence for a single gene

CCGTCAGGATTGACACCCTCCCAATTGTATGTTTTCATGCCTCCAAATCTTGGAGGCTTTT**ATGTGGTTCGTTCTTATTACCCTTCTGAATGTCACGCTG**
ACGAATACCTTCGGTTCGTAACCCCTAACTCTTTCTCATCTTTACGGTGTTCGGAGTTATCGTCCAAATTCTGGGAGCTATGGAGTTTCAGTTTTATTA
GATGGATAACCGCATCAAGCTCTTGGAAGAGATTCTGTCTTTTCGTATGCAGGCCCTTGAGTTCGATAATGGTGATATGTATGTTGACGGCCATAAGGCT
ACAATAATTATAGTTCAACCCCCTCGTGTAACATCGTAACACGGTTAAGTAGGTAATTGAAGAGTCATTGTCTATGTTTGAGTAGTGCTTGGAGTCTTGG
CTATAGACCACCGCCCCGAAGGGGACGAAAAATGGTTTTTACAGAACGAGAAGACGGTTACGCAGTTTTGCCGCAAGCTGGCTGGTGAACGCCCTCTTAA
TTTCGGACATGCGCTATAGAATCAGGTCCGGACCTCGTTAGAACTTGTGAGTAGGAATTATGGAAAGAAAAACCCCATTAATATGAGTAGCGCTTATAGG
GCTATTCAGCGTTTGATGAATGCAATGCGACAGGCTCATGCTGATGGTTGGTTTATCGTTTTTGACAGTCTCACGTTGGCTGACGACCGATTAGAGGCGT
GTGAGGCGCACAGTTAGTAATCGGAACGCTCGGAGCCGTGGTTCTTGGTATGCTGGTTATAGTGCTTTTATCAGTGCGTTTCGTAACCCTAATAGTATTT
GTATCAGTATTTTTGTGTGGCTGAGTATCGTAGAGCTAATGGCCGTCTTCATTTCCATGCGGTGCACTTTATGGGGACACTTCGTAGAGGTAGCGTTGAG
CGCACATGGGTTGAGGCTACCCGTATGAGATTGGTATTCCGGTGCATAAAAGGTTCGATAAATTGAGCCGCGGTAACGCATAGGCTGCTGGTTTTAATCG
AGGACGCTTTTTCACGTTGTCGTTGGTTGTGGCCTGTTGATGCTAAAGGTGAGGCGCTTAAAGGTACCAGTTATATGGCTGTTGGTTTCTATGTGGCTAA
GAAGGCTTCATGGGTGTCGAACCAAAAATCAGTCAACAAGGTAAGAAATGGAACAACTCGAAATGGTCGTTCCAGGTATAGAGTGAAAAAGAATTGCATA
AAGGTGTTCAGAATCAGAATGAGCCGCAAGTTCGGGATGAAAATGGTCACAATGAGAAATCTCTCCACGGAGTGGTTAATCGAAGTTAGCAAGGTGGGTT
ACGCGGCGGTTTTGCAGCCGATGTCATTGAAAAGGGTCGGAGTTAGAGTACAGAGAAAAAGGCAAGACGAAGTTATAGACCAACTTGCCGCAGCGCAGCA
AGCTGTGACGAGAAATCTGGTCAAATTTATGCGCGCTTGGATAAAAATGATTGGCGTATCCAACCTGGAGAGTTTTATCGGTTCCATGAGGCAGAAGTTA
AAGGGGGTCGTCAGGTGAAGGTAAATTAAGCATTTGTTCGTCATCATTAAGGACGAAATAGTTCTATTAAAAAGGTGAGTAGTCTTTATAGGCTTTCACA
AATGAGAAAATTCGACCTATCCTTGCGCAGCTCGAGAAGCTCTTAGTTTGCGACGTTTCGCCATCAACTAACGATTGTGTCAAAAAGTGACGCGTTGGAT
AATTTTAGAGTTGTTGTGTTAGAGATGGTACTTGTTTTACAGTGACTATAGATTTGGTCAGGAAGTGCTTGCAGGGTTCGTATAATTCGGTGAAGAGGAG
AAGAGCGTGGATTACTATCTGAGTCCGATGCTGTTCAAGGAGTAATAGGTAAGAAATCATGAGTCAAGTTAGTGAACAATCCGTACGTTTCCAGACCGCT
GCAGTCATCGTTAGGTTTGAAACAATGAGGAGTCTTTTAGCTTTAGTAGAAGCCAATTTAGGTTTTGCCGTCTTCGGACTTACTCGAATTATCTCCGGTT
GCTCTCGTGCTCGTCGCTGCGTTGAGGCTTGCGTTTATGGTACGCT
GCCTGCGAGGTGCGGTAATTATTACAAAAGGATTTAAGTCGCGGA
GTTAAAGCCGGTGAATTGTTCGGGTTTACCTTGCGTGTAGGCGGA
TCGCGGAAATGGGAACGGAAATCATCGAGGGTTGCCGACGGCTGCT
CGTGTTTGGTATGTAGGTGGTCAACAATTTTAATTGCAGGGGCTT
TAGAGCTTCGTCAGCGGTCGCTATTGGCCTCATCAACTTTAGCAT
GGACGCCGTTGGCGCTGTCCGTCTTTGTGGATTGCGTCGTGGCGT
GTTCTTTTCGCCGTAGGAGTTATATTGGTCATCACAATTCTCAGC
GCACGATTAAGCCTGATACCAATAAAATCCCTAAGCATTTGTTTCA
GAGTCCTCGTTCGCGTGGTCAGGTTTACAAAAAGTGTAGGGTGGT
ACTGAGCTTTCTGGCCAAATGAGGAGTTCTACCAGATCTATTGACA
GTCTCTAATGTCGCGTACTGTTCATTTCGTGCGAACAGTCGCAGTA
GGCATCTGGGTATGATGTTGATGGAACTGAGCAAACGTCGTTAGG
TTCAGTTTCGTGGAAATCGCAATTCCATGACTTAGAGAAATCAGC
ATACCGATATTGGTGGCGACCCTGTTTTGTATGGCAACTTGGCGC
TCTTCGGCCAAGGAGTTACTTAGCCTTCGGAAGTTCTTCCACTAT
GGTGATTTGCAAGAACGCGTACTTATTCGCAACCATGATTATGAC
GGCCGTTTTAATTTTAAAAATGGCGAAGCGAATATTGGAGTGTGAGTTAGAAAATAGTGGTTCAGTAGTAACTTAGCGGTCACCAGCCGTGTAACGCTA
TGAGGGGTTGACCAAGCGAAGCGGGGTAGGTTTTGTGCTTAGGAGTTTAATCATGTTTCAGACTTTTATTTGTCGCCACAATTGAAAGTTTTTTTCTGAT
ATTGGGAGTTTGATAGTTTTATATTGGAACTGGTACATCGAAATCCACAGACATTTTGTCCACGGCTTCTTCGAGGTGATTGTCTTCAGTCTTGGTGGAA
ATGCTGGTAATGGTGGTTTTCTTCATTGCATTCAGATGGATACATCTGTCAAGGCCGCTAATCAGGTTGTTTCAGTTGGTGCTGATATTGCTTTTGATGG
TACGGCTGGTAGGTTTCCTATTTGTAGTATCCGTCAGCCCTCCCATCAGCCTTGGTTCTTCTGAGTTTCGGTTGGTTTGTCGGTTTTTTAAATCCCAGC
GATGGTGGTTATTATACGGTCAAGGACTGTGTGACTATTGACGTCCTTCCCC
TCGTGGTTTGTATTTAGTGGAGTGAATTCACCGACCTCTGTTTATTAGAGAAATTATTGGACTAAGTCGCTTTGGTTAGGGGCCGTAAATCATCGCCATT
ATTGGTGGCGGTATTGCTTGTGCTCTTGGTGGTGGCGCCATGTCTAAATTGTTTGGAGCCGGTCAAAAAGCGGCCTCCGGTGGCATTCAAGGTGATGTGG
TTTGATCCCCGCCGGACTAGTCCCAATCCTTGTAATGTCGGAAGTTACCGTCTAAATTATGGTCGTAGTGGGTACGGATGTCATAAGAATAGCCATCGTT
TGTTTCTGGTGGTATGGGTAAAGGTCGTAAAGGAGTTCTTGAAGGTACGTTGGAGGGTGGGAGTTCTGCCGTTTGTGATAAGTTGGTTGATTTGGTTGGA
GTACTCGTGGTCGTGGGAGGGTTCGTAATTCGAGTCCCTTAGGTCGTCGTTCTATTAGTGCTCATAGGAAAGGAAATAGTCGGCGTCTGAACGGTGGTTC
GTTCGTCTGCTGGTATCGTTGACGGCGGATTTGAGAATCAAAAAGAGCTTAGTAAAATGGAACTGGACAATCAGAAAGAGATTGCCGAGATGCAAAATGA
GAAGACAACTATTCGTTCGTAGAGTAAAAGACGTATATGGACCAGAAAGCATAAGACCGCACTTGAGCGGGTGAGTTACGGTCGTTAGAGAAAAAGTCAG
GAGTCTAGTGCTCGGCTTGCGTCTATTATGGAAAAGACCAATCTTTCCAAGCAACAGCAGGTTTCCGAGATTATGGGCGAAATGCTTACTCAAGCTCAAA
ATGGGGAGTAAGAGGCAAAGGACTACTTGATTCAGTTGGAGTCCTGATTGGAACGGTCAGTAAAGAAAGTAAACCAGTAACCATTTTATGACTGGTCGGC
TGGCTGTTGTCATATTGGCGCTAGTGCAAAGGATATTTCTAATGTCGTCACTGATGCTGCTTCTGGTGTGGTTGATATTTTTCATCGTATTGATAAAGCT
AATAAAGGATGTGTTTAATCTCGGTTATGGTAGTGGAAATGGGAGAAAGGTCTTTAACAAGGTTCATAGCCGTTG

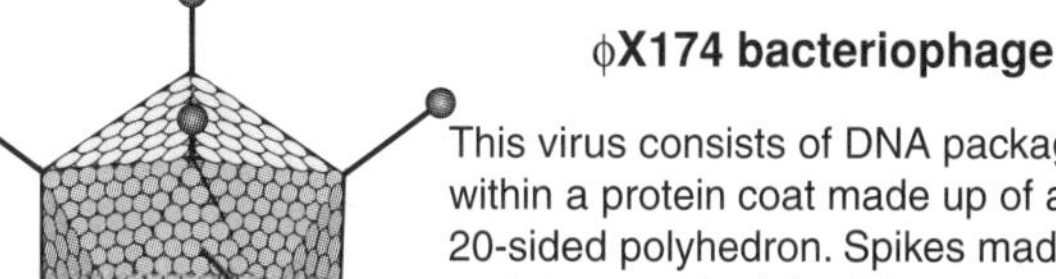

ϕX174 bacteriophage

This virus consists of DNA packaged within a protein coat made up of a 20-sided polyhedron. Spikes made of protein at each of the 12 corners are used to attach itself to a bacterial cell.

The entire DNA sequence for the virus is made up of just 9 genes

1. Explain what is meant by the **genome** of an organism: ____________________

____

2. Determine the number of bases, kilobases, and megabases in this genome (100 bases in each row, except the last):

1 kb = 1 kilobase = 1000 bases **1 Mb** = 1 megabase = 1,000,000 bases

(a) Bases: __________ (b) Kilobases: __________ (c) Megabases: __________

3. Determine how many bases are present in the gene shown above (in the blue area): __________

4. State whether the genome of the virus above is **small, average** or **large** in size compared to those viruses listed in the table on the earlier page *DNA Molecules*:

____

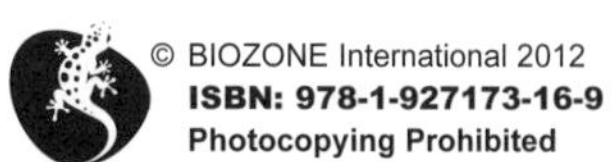

ISBN: 978-1-927173-16-9

Related activities: *DNA Molecules, Genome Projects, The Human Genome Project*

Alleles

Sexually reproducing organisms in nearly all cases have paired sets of chromosomes. One set comes from the mother and one from the father. The equivalent chromosomes that form a pair are called **homologues**. They contain equivalent sets of genes on them. But there is the potential for different versions of a gene to exist in a population and these are termed **alleles**.

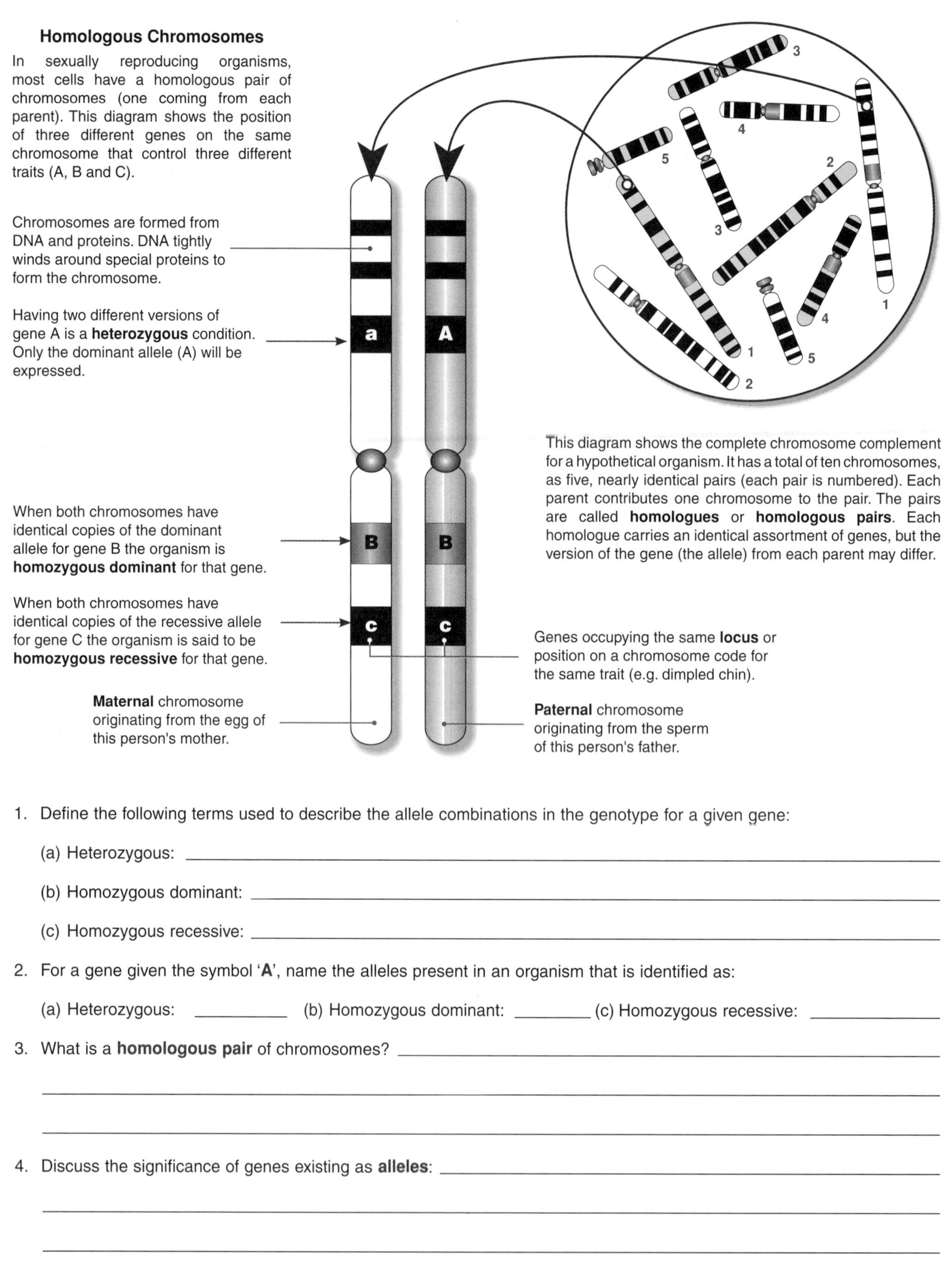

Homologous Chromosomes
In sexually reproducing organisms, most cells have a homologous pair of chromosomes (one coming from each parent). This diagram shows the position of three different genes on the same chromosome that control three different traits (A, B and C).
Chromosomes are formed from DNA and proteins. DNA tightly winds around special proteins to form the chromosome.
Having two different versions of gene A is a heterozygous condition. Only the dominant allele (A) will be expressed.
When both chromosomes have identical copies of the dominant allele for gene B the organism is homozygous dominant for that gene.
When both chromosomes have identical copies of the recessive allele for gene C the organism is said to be homozygous recessive for that gene.
Maternal chromosome originating from the egg of this person's mother.
a
A
B
B
c
c
Genes occupying the same locus or position on a chromosome code for the same trait (e.g. dimpled chin).
Paternal chromosome originating from the sperm of this person's father.
3
4
5
2
3
1
4
1
5
2
This diagram shows the complete chromosome complement for a hypothetical organism. It has a total of ten chromosomes, as five, nearly identical pairs (each pair is numbered). Each parent contributes one chromosome to the pair. The pairs are called homologues or homologous pairs. Each homologue carries an identical assortment of genes, but the version of the gene (the allele) from each parent may differ.

1. Define the following terms used to describe the allele combinations in the genotype for a given gene:

   (a) Heterozygous: ____________________

   (b) Homozygous dominant: ____________________

   (c) Homozygous recessive: ____________________

2. For a gene given the symbol '**A**', name the alleles present in an organism that is identified as:

   (a) Heterozygous: ________ (b) Homozygous dominant: ________ (c) Homozygous recessive: ________

3. What is a **homologous pair** of chromosomes? ____________________

   ____

   ____

4. Discuss the significance of genes existing as **alleles**: ____________________

   ____

   ____

   ____

   ____

ISBN: 978-1-927173-16-9

Mitosis vs Meiosis

Cell division is fundamental to all life, as cells arise only by the division of existing cells. All types of cell division begin with replication of the cell's DNA. In eukaryotes, this is followed by division of the nucleus. There are two forms of nuclear division: **mitosis** and **meiosis**, and they have quite different purposes and outcomes. Mitosis is the simpler of the two and produces two identical daughter cells from each parent cell. Mitosis is responsible for growth and repair processes in multicellular organisms and reproduction in single-celled and asexual eukaryotes. Meiosis involves a **reduction division** in which haploid gametes are produced for the purposes of sexual reproduction. Fusion of haploid gametes in fertilization restores the diploid cell number in the **zygote**. These two fundamentally different types of cell division are compared below.

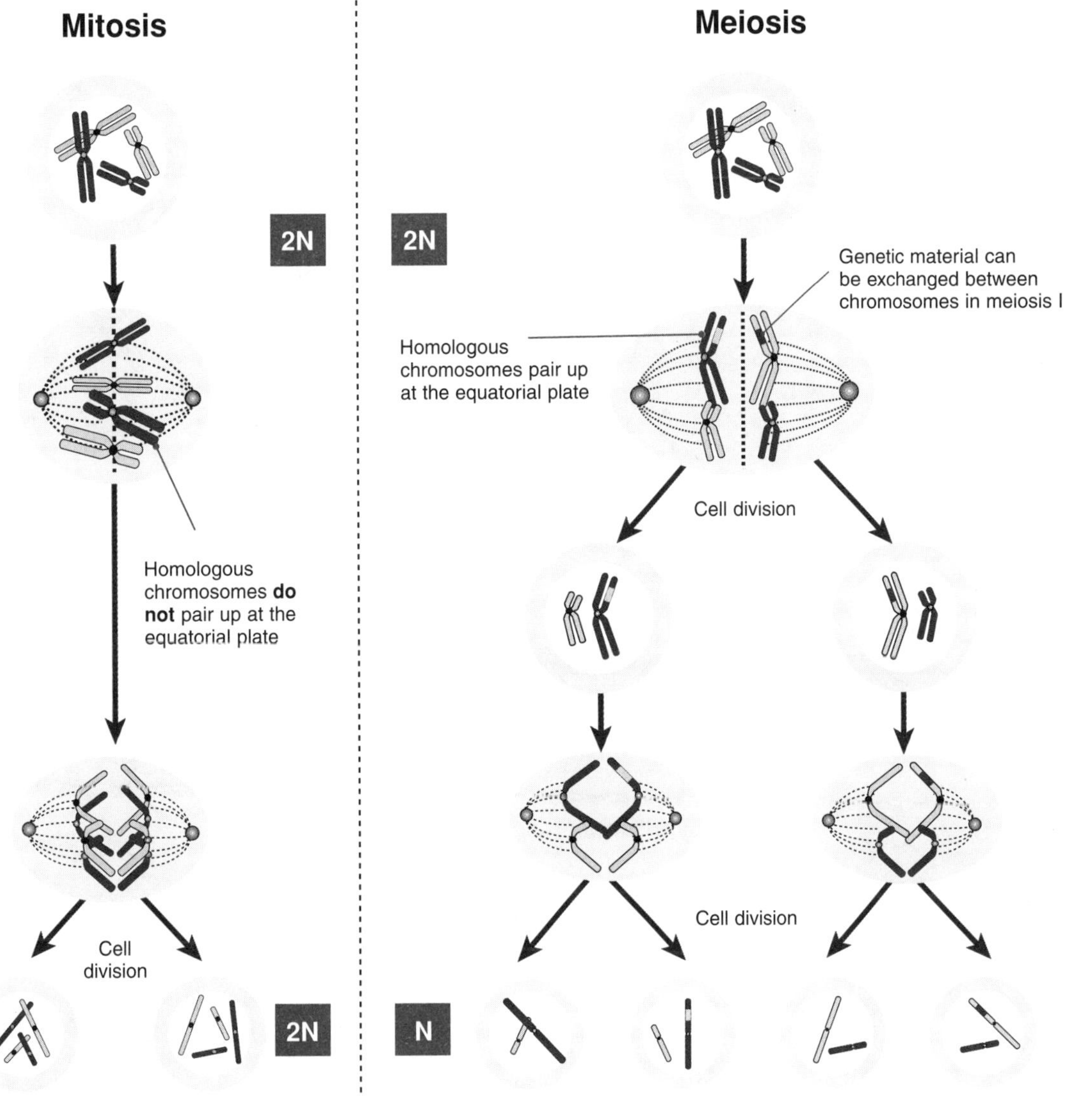
Mitosis
Meiosis
2N
2N
Genetic material can be exchanged between chromosomes in meiosis I
Homologous chromosomes pair up at the equatorial plate
Homologous chromosomes **do not** pair up at the equatorial plate
Cell division
Cell division
Cell division
2N
N
Meiosis I: Reduction division
Meiosis II: 'Mitotic' division

1. Explain how mitosis conserves chromosome number while meiosis reduces the number from diploid to haploid:

2. Describe a fundamental difference between the first and second divisions of meiosis:

3. Explain how meiosis introduces genetic variability into gametes and offspring (following gamete fusion in fertilization):

ISBN: 978-1-927173-16-9

Non-Disjunction in Meiosis

The meiotic spindle normally distributes chromosomes to daughter cells without error. However, mistakes can occur in which the homologous chromosomes fail to separate properly at anaphase during meiosis I, or sister chromatids fail to separate during meiosis II. In these cases, one gamete receives two of the same type of chromosome and the other gamete receives no copy. This mishap is called **non-disjunction** and it results in abnormal numbers of chromosomes passing to the gametes. If either of the aberrant gametes unites with a normal one at fertilization, the offspring will have an abnormal chromosome number, known as an **aneuploidy**. One example of nondisjunction is **trisomy 21** (three copies of chromosome 21) which results in Down syndrome.

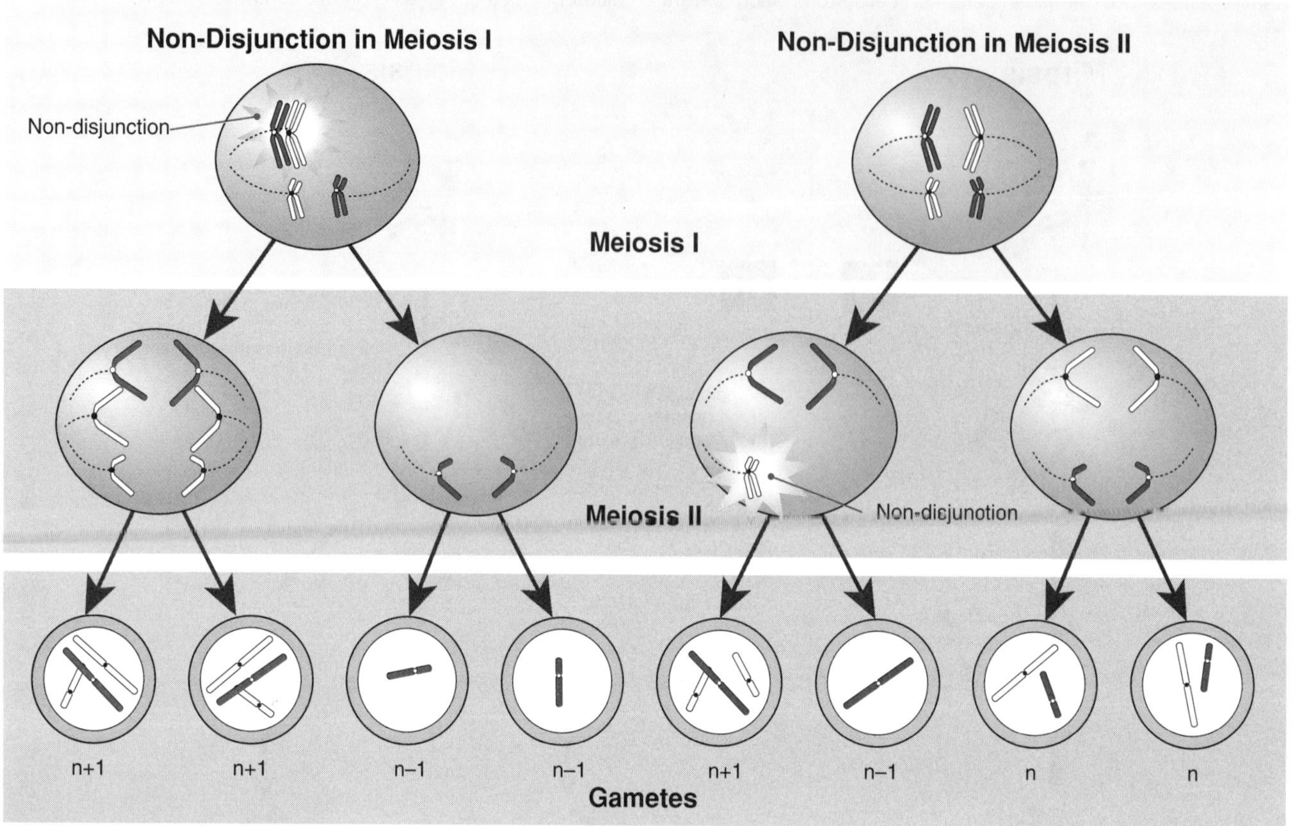
Non-Disjunction in Meiosis I
Non-Disjunction in Meiosis II
Non-disjunction
Meiosis I
Meiosis II
Non-disjunction
n+1
n+1
n–1
n–1
n+1
n–1
n
n
Gametes

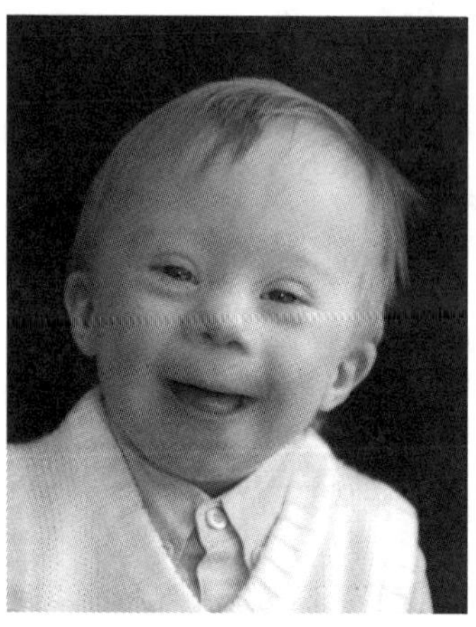

Down Syndrome (Trisomy 21)

Down syndrome is the most common of the human aneuploidies. The incidence rate in humans is about 1 in 800 births for women aged 30 to 31 years, with a **maternal age effect** (the rate increases rapidly with maternal age).

Nearly all cases (approximately 95%) result from **non-disjunction** of chromosome 21 during **meiosis**. When this happens, a gamete (most commonly the oocyte) ends up with 24 rather than 23 chromosomes, and fertilization produces a trisomic offspring.

Left: Down syndrome phenotype.

Right: The karyotype of an individual with trisomy 21. The chromosomes are circled.

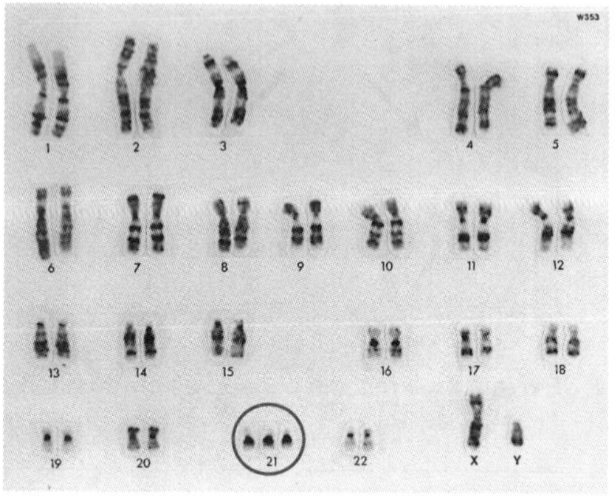
1
2
3
4
5
6
7
8
9
10
11
12
13
14
15
16
17
18
19
20
21
22
X
Y

Photo: Waikato Hospital

1. Describe the consequences of non-disjunction during meiosis: ______________________

____

____

2. Explain why non-disjunction in meiosis I results in a higher proportion of faulty gametes than non-disjunction in meiosis II:

____

____

3. What is the maternal age effect and what are its consequences? ______________________

____

ISBN: 978-1-927173-16-9

Stages in Meiosis

Meiosis is a special type of cell division concerned with producing sex cells (gametes) for the purpose of sexual reproduction. It involves a single chromosomal duplication followed by two successive nuclear divisions, and results in a halving of the diploid chromosome number. Meiosis occurs in the sex organs of plants and animals. If genetic mistakes (**gene** and **chromosome mutations**) occur here, they will be passed on to the offspring (they will be inherited).

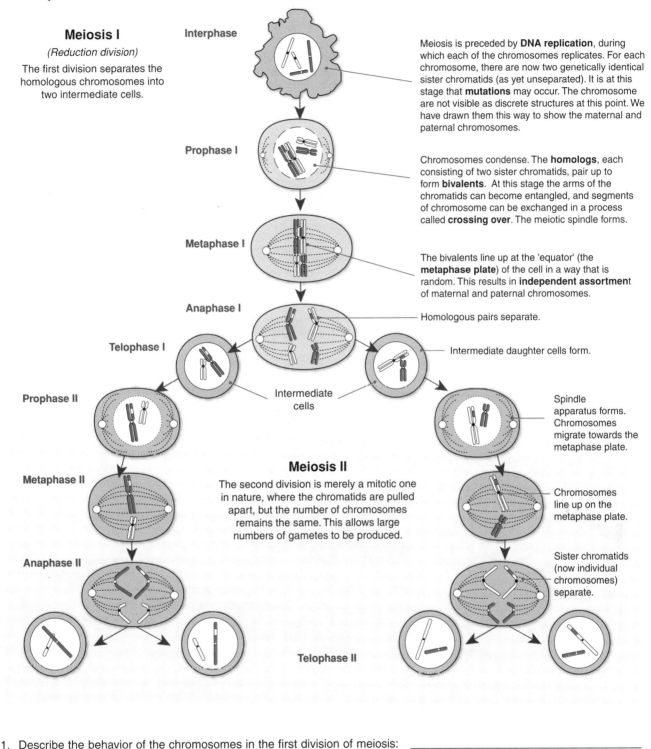

Meiosis I

(Reduction division)

The first division separates the homologous chromosomes into two intermediate cells.

Interphase

Meiosis is preceded by **DNA replication**, during which each of the chromosomes replicates. For each chromosome, there are now two genetically identical sister chromatids (as yet unseparated). It is at this stage that **mutations** may occur. The chromosome are not visible as discrete structures at this point. We have drawn them this way to show the maternal and paternal chromosomes.

Prophase I

Chromosomes condense. The **homologs**, each consisting of two sister chromatids, pair up to form **bivalents**. At this stage the arms of the chromatids can become entangled, and segments of chromosome can be exchanged in a process called **crossing over**. The meiotic spindle forms.

Metaphase I

The bivalents line up at the 'equator' (the **metaphase plate**) of the cell in a way that is random. This results in **independent assortment** of maternal and paternal chromosomes.

Anaphase I

Homologous pairs separate.

Telophase I

Intermediate daughter cells form.

Intermediate cells

Meiosis II

The second division is merely a mitotic one in nature, where the chromatids are pulled apart, but the number of chromosomes remains the same. This allows large numbers of gametes to be produced.

Prophase II

Spindle apparatus forms. Chromosomes migrate towards the metaphase plate.

Metaphase II

Chromosomes line up on the metaphase plate.

Anaphase II

Sister chromatids (now individual chromosomes) separate.

Telophase II

1. Describe the behavior of the chromosomes in the first division of meiosis: ______________________________

____

____

____

2. Describe the behavior of the chromosomes in the second division of meiosis: ______________________________

____

____

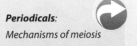

ISBN: 978-1-927173-16-9

Crossing Over

Crossing over refers to the mutual exchange of pieces of chromosome and involves the swapping of whole groups of genes between the **homologous** chromosomes. This process can occur only during **prophase I** in the first division of **meiosis**. Errors in crossing over can result in detrimental **chromosome mutations**. Recombination as a result of crossing over is an important mechanism to increase genetic variability in the offspring and has the general effect of allowing genes to move independently of each other through the generations in a way that allows concentration of beneficial alleles.

Pairing of Homologous Chromosomes

Every somatic cell contains a pair of each type of chromosome, one from each parent. These are called **homologous pairs** or **homologs**. In prophase of meiosis I, the homologs pair up to form **bivalents**. This process is called **synapsis** and it brings the chromatids of the homologs into close contact.

Chiasma Formation and Crossing Over

Synapsis allows the homologous, non-sister chromatids to become entangled and the chromosomes exchange segments. This exchange occurs at regions called **chiasmata** (*sing.* chiasma). In the diagram (centre), a chiasma is forming and the exchange of pieces of chromosome has not yet taken place. Numerous chiasmata may develop between homologs.

Separation

Crossing over produces new allele combinations, a phenomenon known as **recombination**. When the homologs separate in anaphase of meiosis I, each of the chromosomes pictured will have a new mix of alleles that will be passed into the gametes soon to be formed. Recombination is an important source of variation in population gene pools.

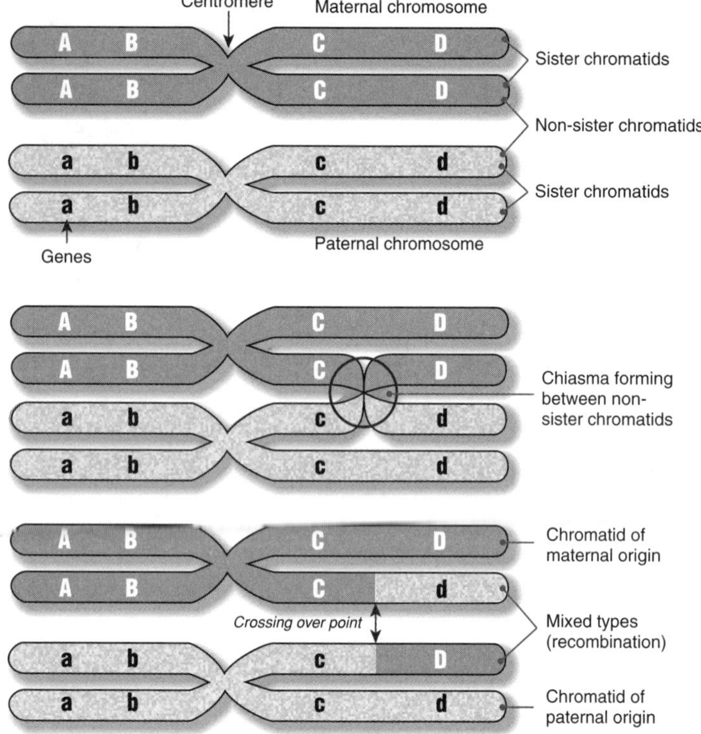
Centromere
Maternal chromosome
Sister chromatids
Non-sister chromatids
Sister chromatids
Paternal chromosome
Genes
A B C D
a b c d
Chiasma forming between non-sister chromatids
Chromatid of maternal origin
Crossing over point
Mixed types (recombination)
Chromatid of paternal origin

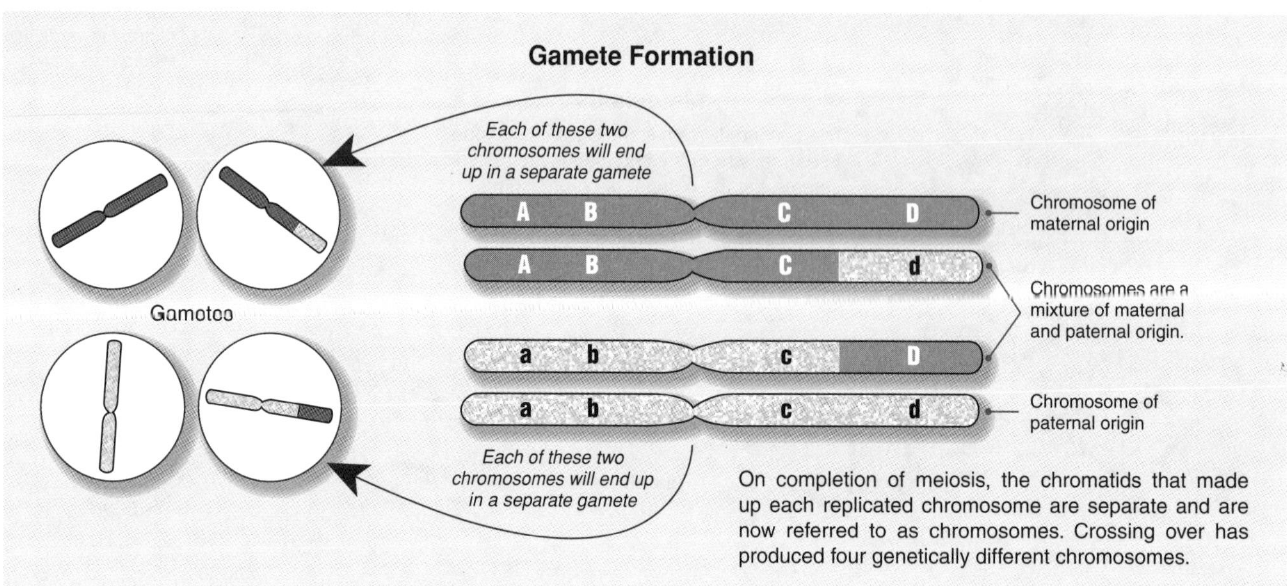
Gamete Formation
Each of these two chromosomes will end up in a separate gamete
Gametes
Chromosome of maternal origin
Chromosomes are a mixture of maternal and paternal origin.
Chromosome of paternal origin
Each of these two chromosomes will end up in a separate gamete
A B C D
a b c d

On completion of meiosis, the chromatids that made up each replicated chromosome are separate and are now referred to as chromosomes. Crossing over has produced four genetically different chromosomes.

1. (a) In a general way, describe how crossing over alters the genotype of gametes: ____________________

   (b) What is the consequence of this? ____________________

2. What is the significance of crossing over in the evolution of sexually producing populations? ____________________

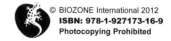

Related activities: *Meiosis*

ISBN: 978-1-927173-16-9

Modelling Meiosis

This practical activity uses ice-block sticks to simulate the production of gametes (sperm and eggs) by **meiosis** and shows you how **crossing over** increases genetic variability. This is demonstrated by studying how two of your own alleles are inherited by the child produced at the completion of the activity. Completing this activity will help you to visualize and understand meiosis, and demonstrate how crossing over contributes to genetic variation in the offspring. It will take 25-45 minutes.

Background

Each of your somatic cells contain 46 chromosomes. You received 23 chromosomes from your mother (**maternal chromosomes**), and 23 chromosomes from your father (**paternal chromosomes**). Therefore, you have 23 homologous (same) pairs. For simplicity, the number of chromosomes studied in this exercise has been reduced to four (two homologous pairs). To study the effect of crossing over on genetic variability, you will look at the inheritance of two of your own traits: the ability to **tongue roll** and **handedness**.

Chromosome #	Phenotype	Genotype
10	Tongue roller	TT, Tt
10	Non-tongue roller	tt
2	Right handed	RR, Rr
2	Left handed	rr

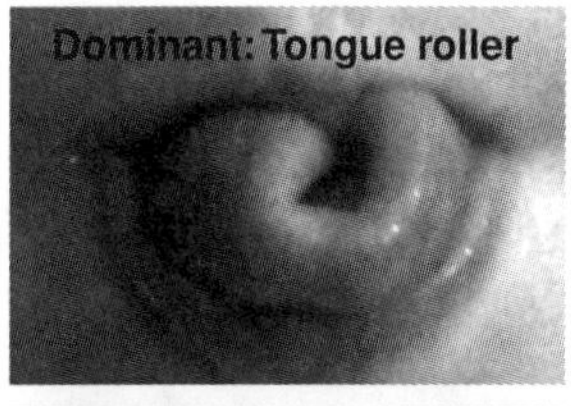
Dominant: Tongue roller

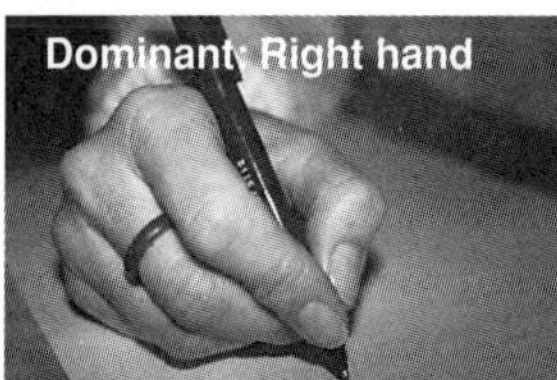
Dominant: Right hand

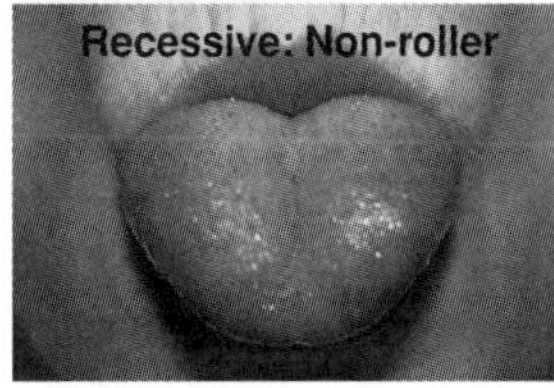
Recessive: Non-roller

Recessive: Left hand

Trait	Phenotype	Genotype
Handedness		
Tongue rolling		

Record your phenotype and genotype for each trait in the table (right). **NOTE:** If you have a dominant trait, you will not know if you are heterozygous or homozygous for that trait, so you can choose either genotype for this activity.

BEFORE YOU START THE SIMULATION: Partner up with a classmate. Your gametes will combine with theirs (fertilization) at the end of the activity to produce a child. Decide who will be the female, and who will be the male. You will need to work with this person again at step 6.

1. Collect four ice-blocks sticks. These represent four chromosomes. Color two sticks blue or mark them with a P. These are the paternal chromosomes. The plain sticks are the maternal chromosomes. Write your initial on each of the four sticks. Label each chromosome with their chromosome number (right).

   Label four sticky dots with the alleles for each of your phenotypic traits, and stick it onto the appropriate chromosome. For example, if you are heterozygous for tongue rolling, the sticky dots with have the alleles **T** and **t**, and they will be placed on chromosome 10. If you are left handed, the alleles will be **r** and **r** and be placed on chromosome 2 (right).

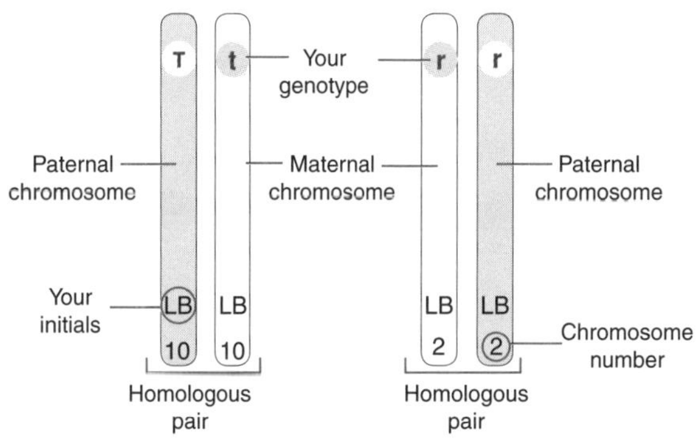
T
t
Your genotype
r
r
Paternal chromosome
Maternal chromosome
Paternal chromosome
Your initials
LB
LB
LB
LB
10
10
2
2
Chromosome number
Homologous pair
Homologous pair

2. Randomly drop the chromosomes onto a table. This represents a cell in either the testes or ovaries. **Duplicate** your chromosomes (to simulate DNA replication) by adding four more identical ice-block sticks to the table (below). This represents **interphase**.

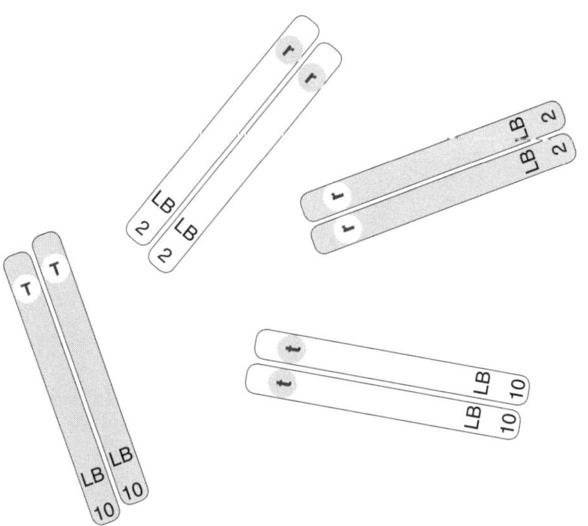
T
T
t
t
r
r
r
r
LB
LB
10
10
2
2

3. Simulate **prophase I** by lining the duplicated chromosome pair with their homologous pair (below). For each chromosome number, you will have four sticks touching side-by-side (A). At this stage **crossing over** occurs. Simulate this by swapping sticky dots from adjoining homologs (B).

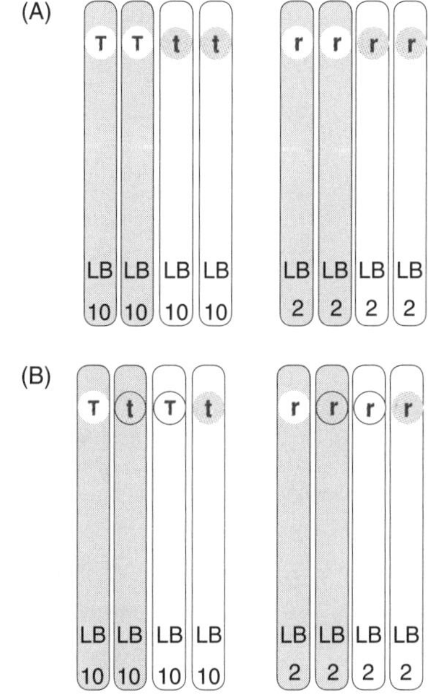
(A)
T T t t
r r r r
LB LB LB LB
10 10 10 10
LB LB LB LB
2 2 2 2
(B)
T t T t
r r r r
LB LB LB LB
10 10 10 10
LB LB LB LB
2 2 2 2

ISBN: 978-1-927173-16-9

Periodicals: *Mechanisms of meiosis*

Related activities: *Stages in Meiosis, Crossing Over*
Weblinks: *Meiosis Tutorial*

4. Randomly align the homologous chromosome pairs to simulate alignment on the metaphase plate (as occurs in **metaphase I**). Simulate **anaphase I** by separating chromosome pairs. For each group of four sticks, two are pulled to each pole.

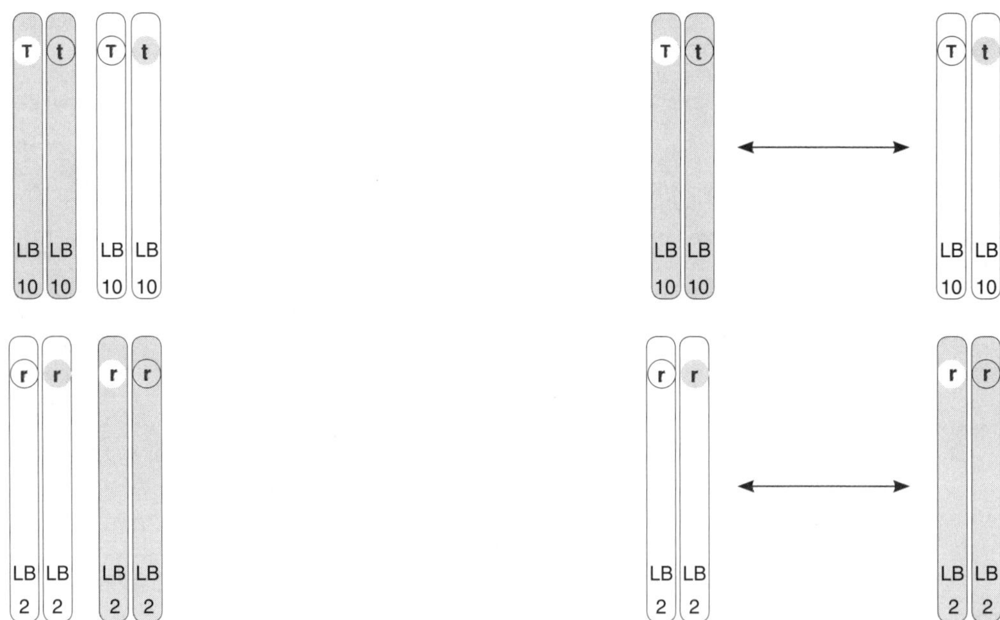

T
t
T
t
LB 10
r
r
r
r
LB 2

5. **Telophase I:** Two intermediate cells are formed. If you have been random in the previous step, each intermediate cell will contain a mixture of maternal and paternal chromosomes. This is the end of **meiosis 1**.

   Now that meiosis 1 is completed, your cells need to undergo **meiosis 2.** Carry out prophase II, metaphase II, anaphase II, and telophase II. Remember, there is no crossing over in meiosis II. At the end of the process each intermediate cell will have produced two haploid gametes (below).

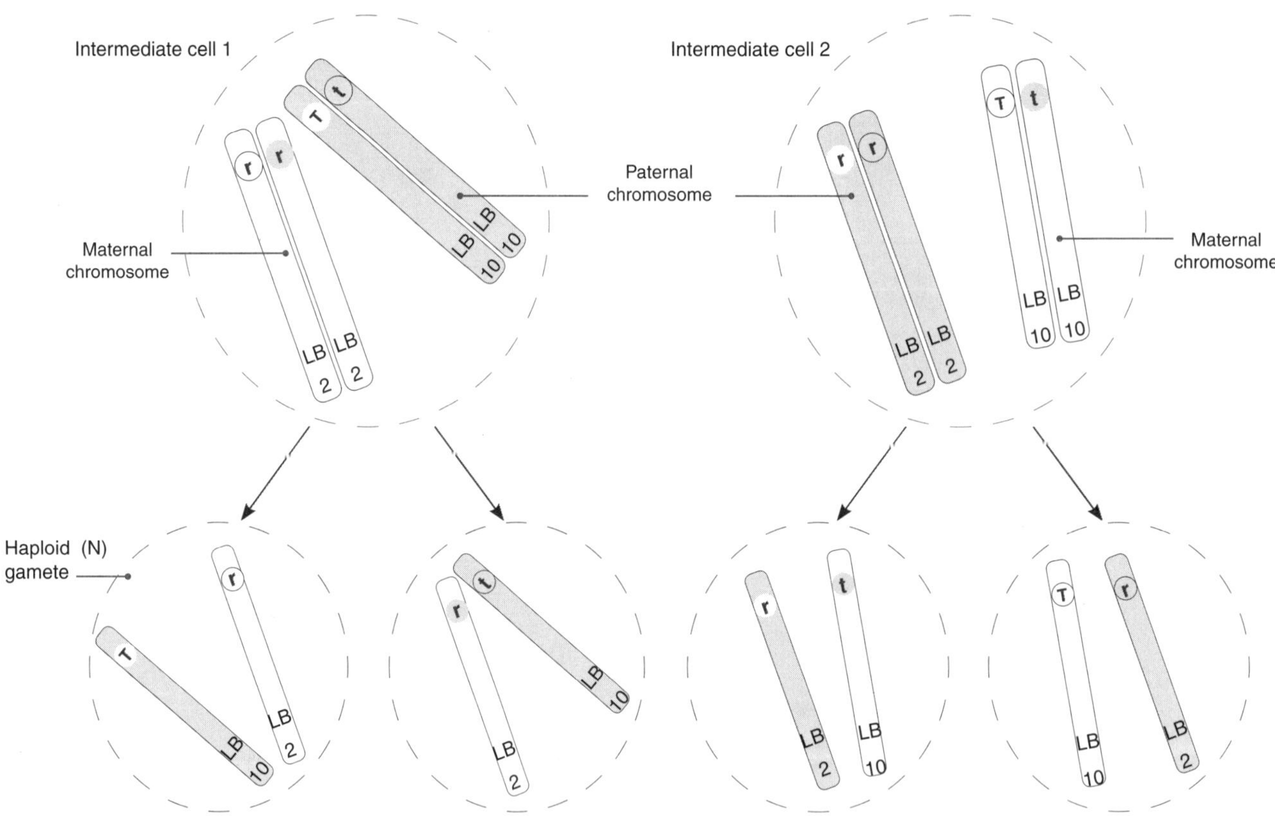

Intermediate cell 1
Intermediate cell 2
Paternal chromosome
Maternal chromosome
Maternal chromosome
Haploid (N) gamete
LB 10
LB 2

6. Pair up with the partner you chose at the beginning of the exercise to carry out **fertilization**. Randomly select one sperm and one egg cell. The unsuccessful gametes can be removed from the table. Combine the chromosomes of the successful gametes. You have created a child! Fill in the following chart to describe your child's genotype and phenotype for tongue rolling and handedness.

Trait	Phenotype	Genotype
Handedness		
Tongue rolling		

ISBN: 978-1-927173-16-9

Changes to the DNA Sequence

The DNA sequence is not unchangeable. **Mutagens** and errors in copying can cause DNA changes. Every time a DNA molecule is copied (DNA replication), there is a chance that a base or series of bases will be copied incorrectly. DNA replication has a low **error rate**, with only one mistake for every billion base pairs copied. Errors that have no effect on the organism or its offspring are called **neutral mutations**. Other errors may create new **alleles**, many of which will be detrimental. One example is the most common form of genetic hearing loss (called NSRD), and accounts for up to 50% of childhood deafness.

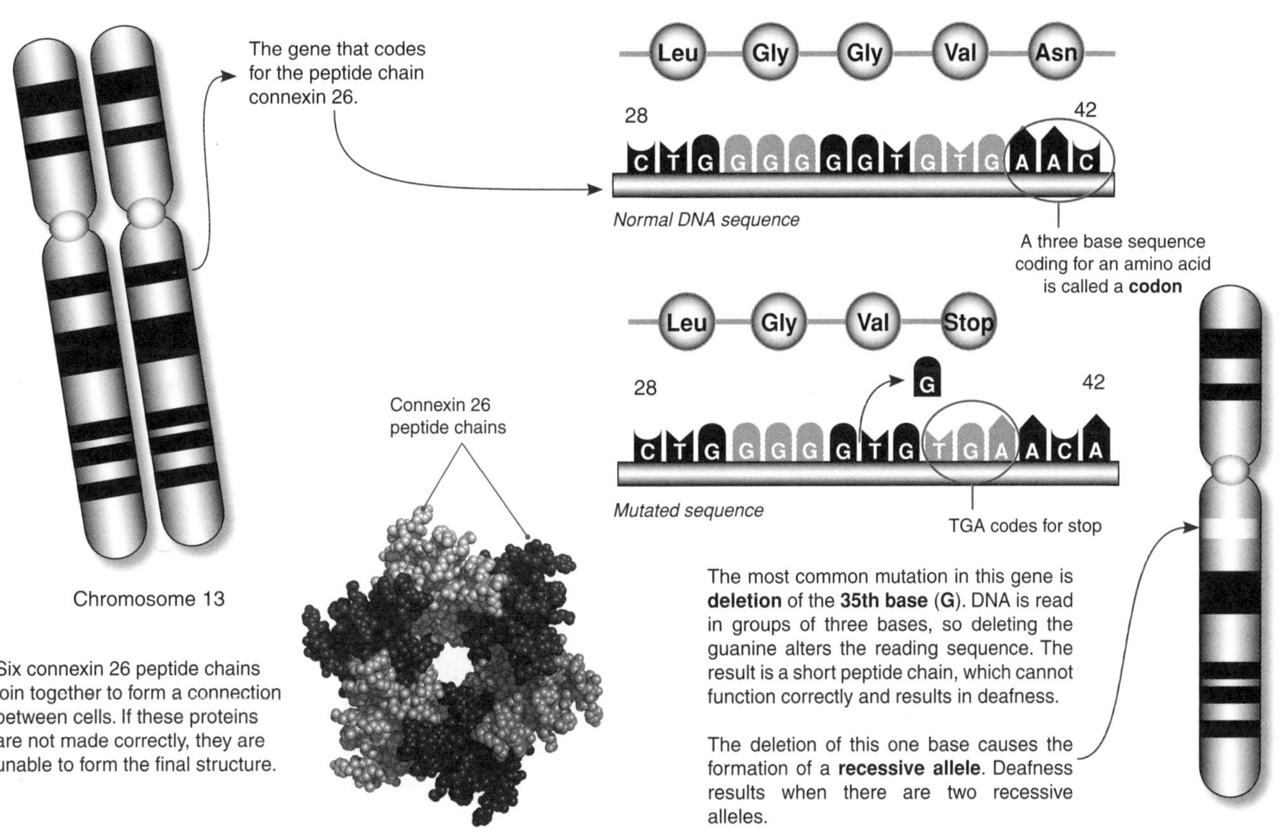

The gene that codes for the peptide chain connexin 26.
Leu Gly Gly Val Asn
28
42
C T G G G G G G T G T G A A C
Normal DNA sequence
A three base sequence coding for an amino acid is called a **codon**
Leu Gly Val Stop
G
28
42
C T G G G G G T G T G A A C A
Mutated sequence
TGA codes for stop
Connexin 26 peptide chains
Chromosome 13

The most common mutation in this gene is **deletion** of the **35th base** (**G**). DNA is read in groups of three bases, so deleting the guanine alters the reading sequence. The result is a short peptide chain, which cannot function correctly and results in deafness.

The deletion of this one base causes the formation of a **recessive allele**. Deafness results when there are two recessive alleles.

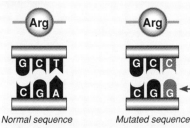

Six connexin 26 peptide chains join together to form a connection between cells. If these proteins are not made correctly, they are unable to form the final structure.

Arg
G C T
C G A
Normal sequence
Arg
G C C
C G G
Mutated sequence

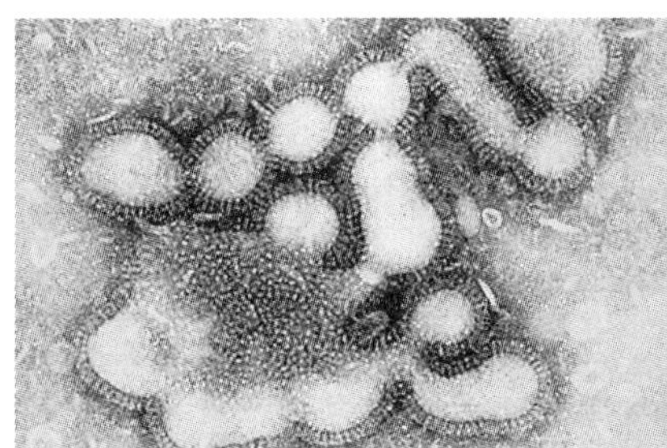

Harmful Mutations

Most mutations cause harmful effects, usually because they stop or alter the production of a protein (often an enzyme). Albinism (above) is one of the more common mutations in nature, and leaves an animal with no pigmentation.

Silent Mutations

Silent mutations do not change the amino acid sequence nor the final protein. In the genetic code, several codons may code for the same amino acid. Silent mutations are also **neutral** if they do not alter the fitness of the organism.

Beneficial Mutations

Sometimes mutations help the survival of an organism. In viruses (such as the *Influenzavirus* above) genes coding for the glycoprotein coat are constantly mutating, producing new strains that avoid detection by the host's immune system.

1. How can changes in a DNA sequence occur? ______

2. How can a mutation in a single base be as damaging as a mutation in a sequence of bases? ______

3. Explain how mutation can be harmful or beneficial: ______

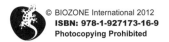

ISBN: 978-1-927173-16-9

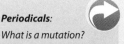

Periodicals: What is a mutation?

Related activities: The Genetic Code, Mutagens

Weblinks: Mutation by Base Substitution

Sickle Cell Mutation

Sickle cell anemia (now called sickle cell disease) is an inherited disorder caused by a gene mutation which codes for a faulty beta (β) chain hemoglobin (Hb) protein. This in turn causes the red blood cells to deform causing a whole range of medical problems. The DNA sequence below is the beginning of the transcribing sequence for the **normal** β-chain Hb molecule.

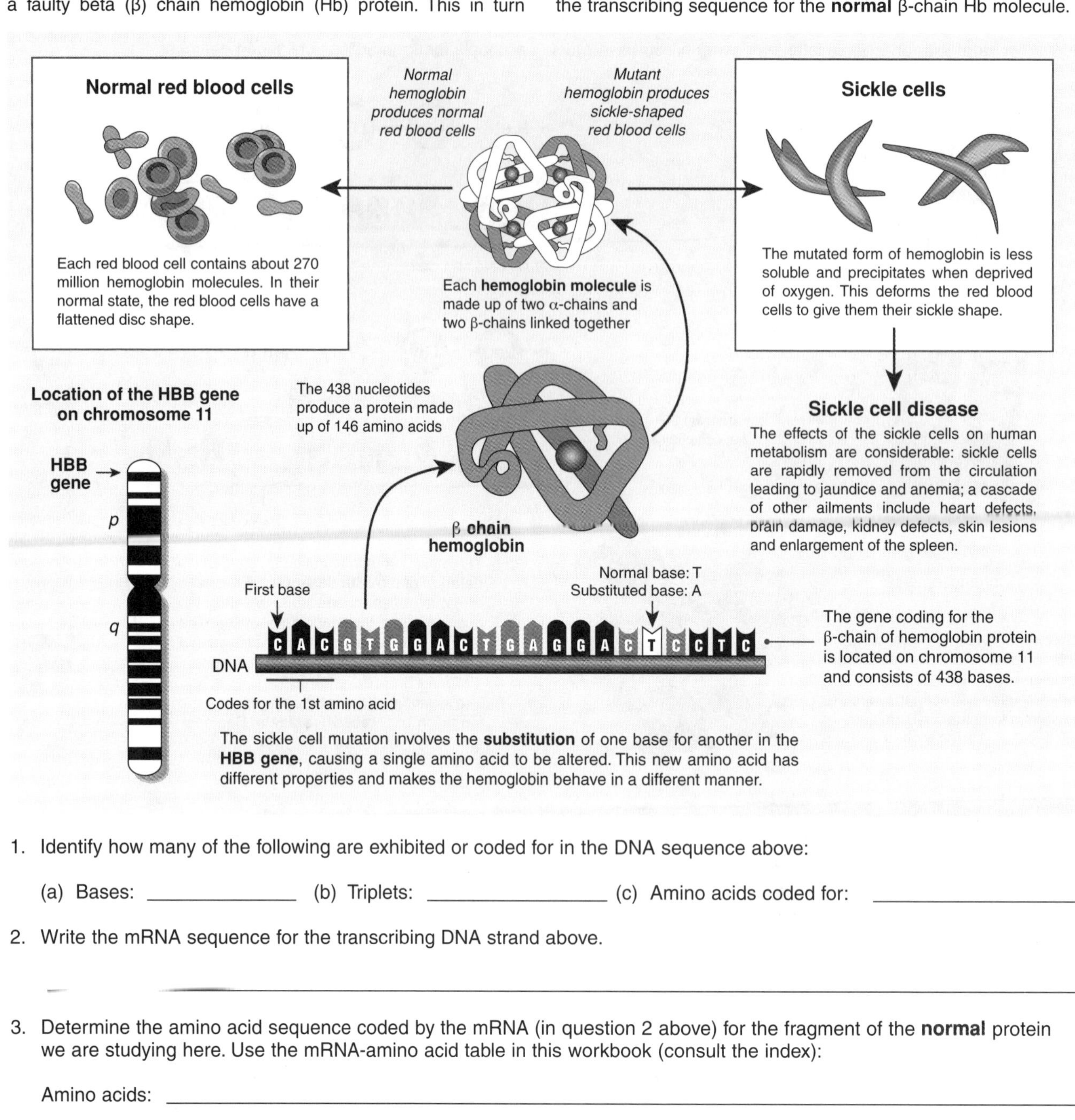

Normal red blood cells
Normal hemoglobin produces normal red blood cells
Mutant hemoglobin produces sickle-shaped red blood cells
Sickle cells
Each red blood cell contains about 270 million hemoglobin molecules. In their normal state, the red blood cells have a flattened disc shape.
Each hemoglobin molecule is made up of two α-chains and two β-chains linked together
The mutated form of hemoglobin is less soluble and precipitates when deprived of oxygen. This deforms the red blood cells to give them their sickle shape.
Location of the HBB gene on chromosome 11
HBB gene
p
q
The 438 nucleotides produce a protein made up of 146 amino acids
β chain hemoglobin
Sickle cell disease
The effects of the sickle cells on human metabolism are considerable: sickle cells are rapidly removed from the circulation leading to jaundice and anemia; a cascade of other ailments include heart defects, brain damage, kidney defects, skin lesions and enlargement of the spleen.
Normal base: T
Substituted base: A
First base
C A C G T G G A C T G A G G A C T C C T C
DNA
Codes for the 1st amino acid
The gene coding for the β-chain of hemoglobin protein is located on chromosome 11 and consists of 438 bases.
The sickle cell mutation involves the substitution of one base for another in the HBB gene, causing a single amino acid to be altered. This new amino acid has different properties and makes the hemoglobin behave in a different manner.

1. Identify how many of the following are exhibited or coded for in the DNA sequence above:

   (a) Bases: ______________ (b) Triplets: ________________ (c) Amino acids coded for: ___________________

2. Write the mRNA sequence for the transcribing DNA strand above.

   ____

3. Determine the amino acid sequence coded by the mRNA (in question 2 above) for the fragment of the **normal** protein we are studying here. Use the mRNA-amino acid table in this workbook (consult the index):

   Amino acids: ____

4. Rewrite the transcribing DNA sequence above with the 17th nucleotide (base) changed from a **T** to **A.** This is the mutation that causes sickle cell disease.

   Mutant DNA: ____ Type of mutation: ________________

5. Write the mRNA sequence for the **mutant** DNA strand above.

   ____

6. Determine the amino acid sequence coded by the mRNA (in question 5 above) for the fragment of the **mutant** protein we are studying here. Use the mRNA-amino acid table in this workbook (consult the index):

   ____

7. Explain how the sickle cell mutation results in the symptoms of the disease: ______________________

   ____

   ____

Related activities: *The Genetic Code*
Weblinks: *Sickle Cell Disease*

Periodicals:
Genetics of sickle cell anaemia

ISBN: 978-1-927173-16-9

Karyotypes

The diagram below shows the **karyotype** of a normal human. Karyotypes are prepared from the nuclei of cultured white blood cells that are 'frozen' at the metaphase stage of mitosis (see the photo circled on the next page). A photograph of the chromosomes is then cut up and the chromosomes are rearranged on a grid so that the homologous pairs are placed together. Homologous pairs are identified by their general shape, length, and the pattern of banding produced by a special staining technique. Karyotypes for a human male and female are shown below. The **male karyotype** has 44 autosomes, a single X chromosome, and a Y chromosome (written as 44 + XY), whereas the **female karyotype** shows two X chromosomes (written as 44 + XX).

Typical Layout of a Human Karyotype

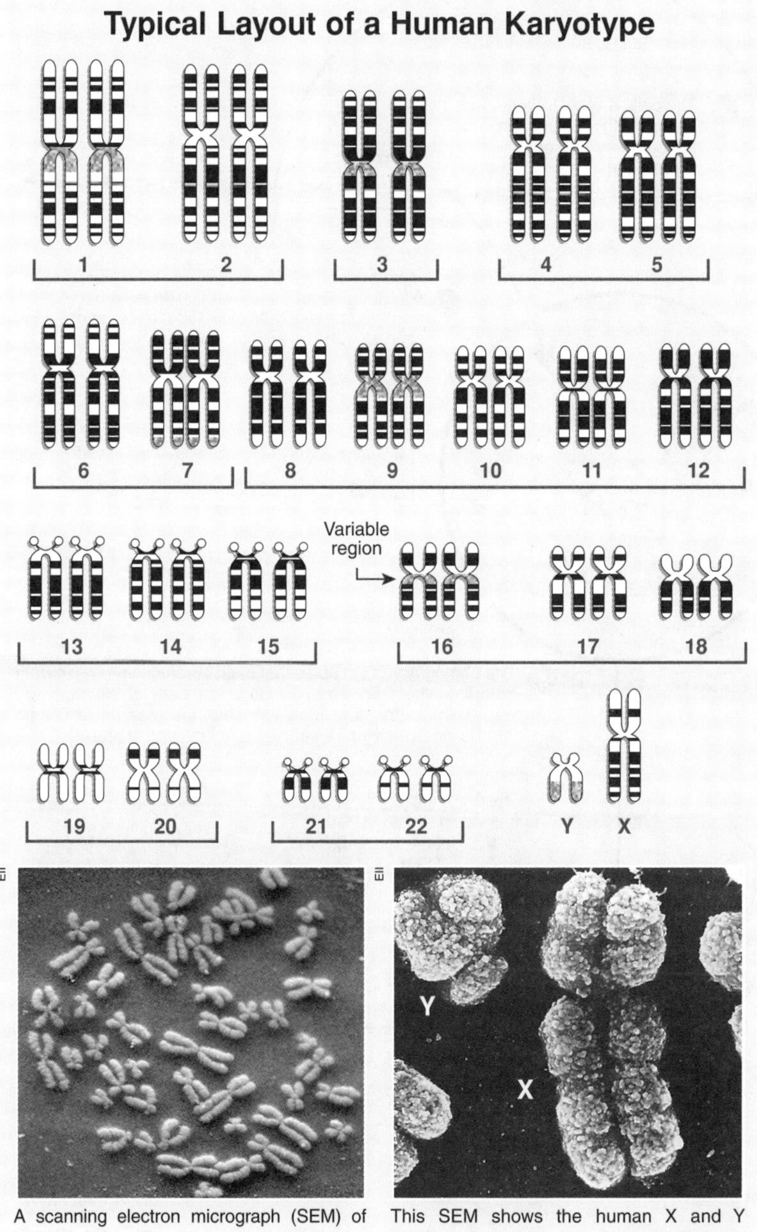
1
2
3
4
5
6
7
8
9
10
11
12
Variable region
13
14
15
16
17
18
19
20
21
22
Y
X
Y
X

A scanning electron micrograph (SEM) of human chromosomes clearly showing their double chromatids.

This SEM shows the human X and Y chromosomes. Although these two are the sex chromosomes, they are not homologous.

Karyotypes for different species

The term **karyotype** refers to the chromosome complement of a cell or a whole organism. In particular, it shows the number, size, and shape of the chromosomes as seen during metaphase of mitosis. The diagram on the left depicts the human karyotype. Chromosome numbers vary considerably among organisms and may differ markedly between closely related species:

Organism	Chromosome number (2N)
Vertebrates	
human	46
chimpanzee	48
gorilla	48
horse	64
cattle	60
dog	78
cat	38
rabbit	44
rat	42
turkey	82
goldfish	94
Invertebrates	
fruit fly, *Drosophila*	8
housefly	12
honey bee	32 or 16
Hydra	32
Plants	
cabbage	18
broad bean	12
potato	48
orange	18, 27 or 36
barley	14
garden pea	14
Ponderosa pine	24

NOTE: The number of chromosomes is not a measure of the quantity of genetic information.

1. (a) What is a **karyotype**? ______________________________

   (b) What information can it provide? ______________________________

2. Distinguish between **autosomes** and **sex chromosomes**: ______________________________

ISBN: 978-1-927173-16-9

Preparing a Karyotype

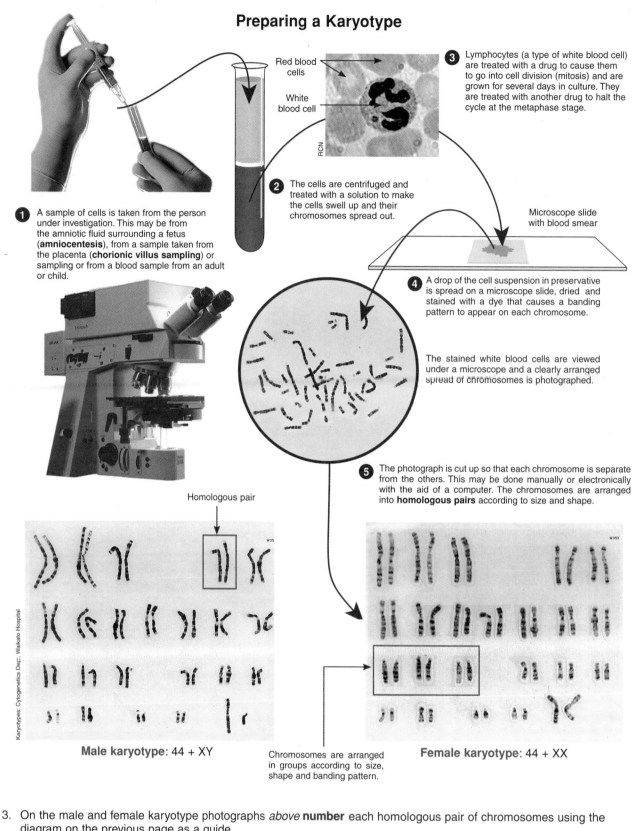

1 A sample of cells is taken from the person under investigation. This may be from the amniotic fluid surrounding a fetus (**amniocentesis**), from a sample taken from the placenta (**chorionic villus sampling**) or sampling or from a blood sample from an adult or child.

2 The cells are centrifuged and treated with a solution to make the cells swell up and their chromosomes spread out.

Red blood cells

White blood cell

RCN

3 Lymphocytes (a type of white blood cell) are treated with a drug to cause them to go into cell division (mitosis) and are grown for several days in culture. They are treated with another drug to halt the cycle at the metaphase stage.

Microscope slide with blood smear

4 A drop of the cell suspension in preservative is spread on a microscope slide, dried and stained with a dye that causes a banding pattern to appear on each chromosome.

The stained white blood cells are viewed under a microscope and a clearly arranged spread of chromosomes is photographed.

5 The photograph is cut up so that each chromosome is separate from the others. This may be done manually or electronically with the aid of a computer. The chromosomes are arranged into **homologous pairs** according to size and shape.

Homologous pair

Karyotypes: Cytogenetics Dept., Waikato Hospital

Male karyotype: 44 + XY

Chromosomes are arranged in groups according to size, shape and banding pattern.

Female karyotype: 44 + XX

3. On the male and female karyotype photographs *above* **number** each homologous pair of chromosomes using the diagram on the previous page as a guide.

4. **Circle** the sex chromosomes (**X** and **Y**) in the female karyotype and male karyotype.

5. Write down the number of *autosomes* and the arrangement of *sex chromosomes* for each sex:

   (a) **Female**: No. of autosomes: ________________ Sex chromosomes: ________________

   (b) **Male**: No. of autosomes: ________________ Sex chromosomes: ________________

6. State how many chromosomes are found in a:

   (a) Normal human (somatic) body cell: ______________ (b) Normal human sperm or egg cell: ______________

ISBN: 978-1-927173-16-9

Human Karyotype Exercise

Each chromosome has distinctive features that make it distinguishable from others. Chromosomes are stained in a special technique that gives them a banded appearance in which the banding pattern represents regions of the chromosome containing up to many hundreds of genes. Cut out the chromosomes below and arrange them on the *Record Sheet* in order to determine the sex and chromosome condition of the individual whose karyotype is shown. The karyotypes presented on the previous pages and the hints on how to recognise chromosome pairs can be used to help you complete this activity.

Distinguishing Characteristics of Chromosomes

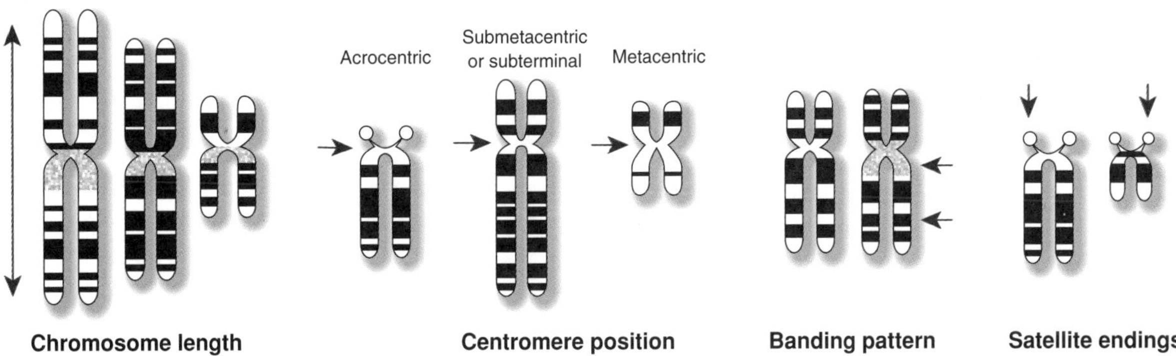
Acrocentric
Submetacentric or subterminal
Metacentric
Chromosome length
Centromere position
Banding pattern
Satellite endings

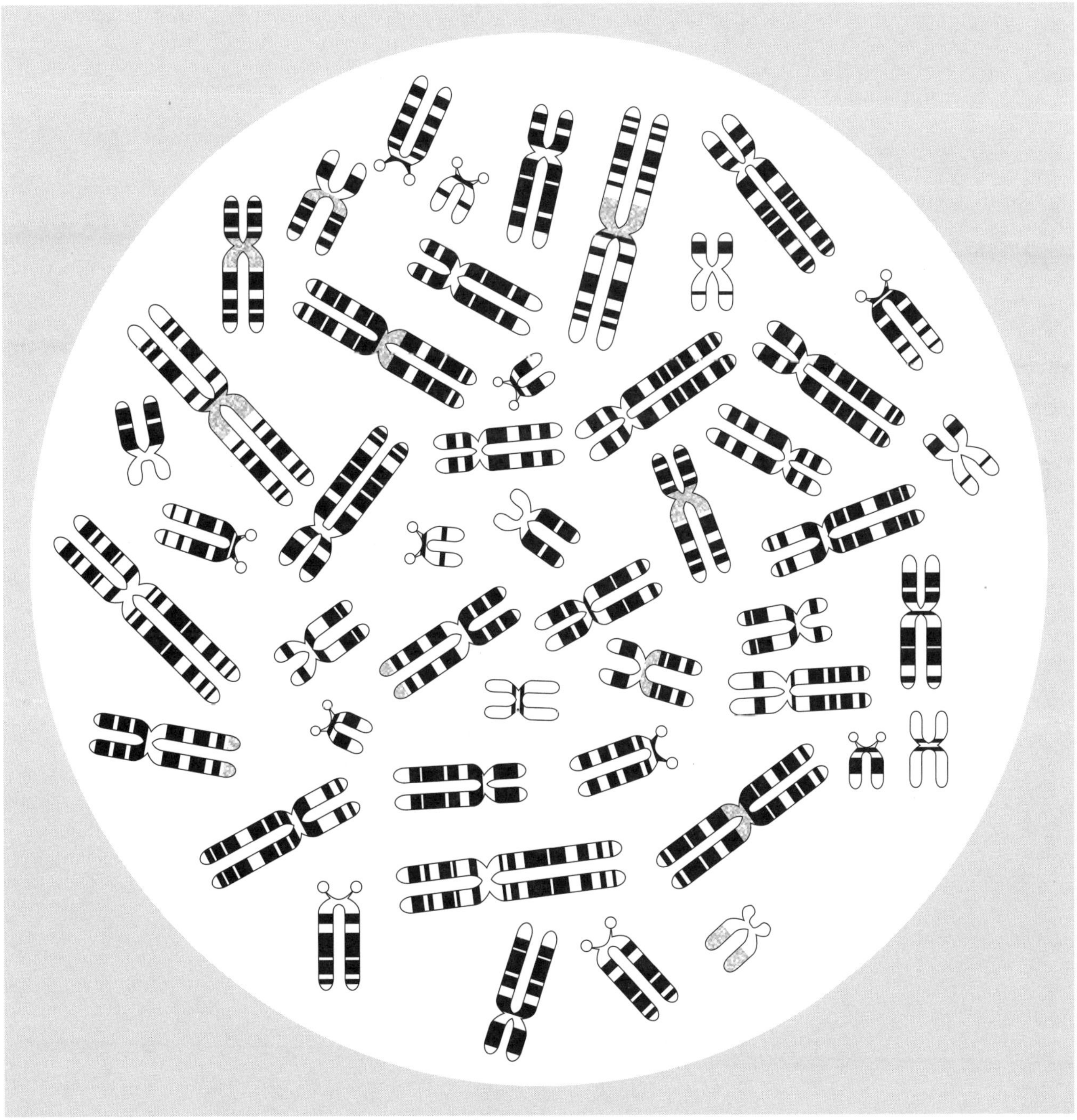

ISBN: 978-1-927173-16-9

This page is left blank deliberately

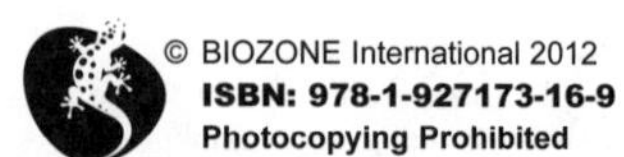

ISBN: 978-1-927173-16-9

1. Cut out the chromosomes on page 145 and arrange them on the record sheet below in their homologous pairs.
2. (a) Determine the sex of this individual: **male** or **female** (circle one)

   (b) State whether the individual's *chromosome arrangement* is: **normal** or **abnormal** (circle one)

   (c) If the arrangement is *abnormal*, state in what way and name the syndrome displayed: ______________________

   ____

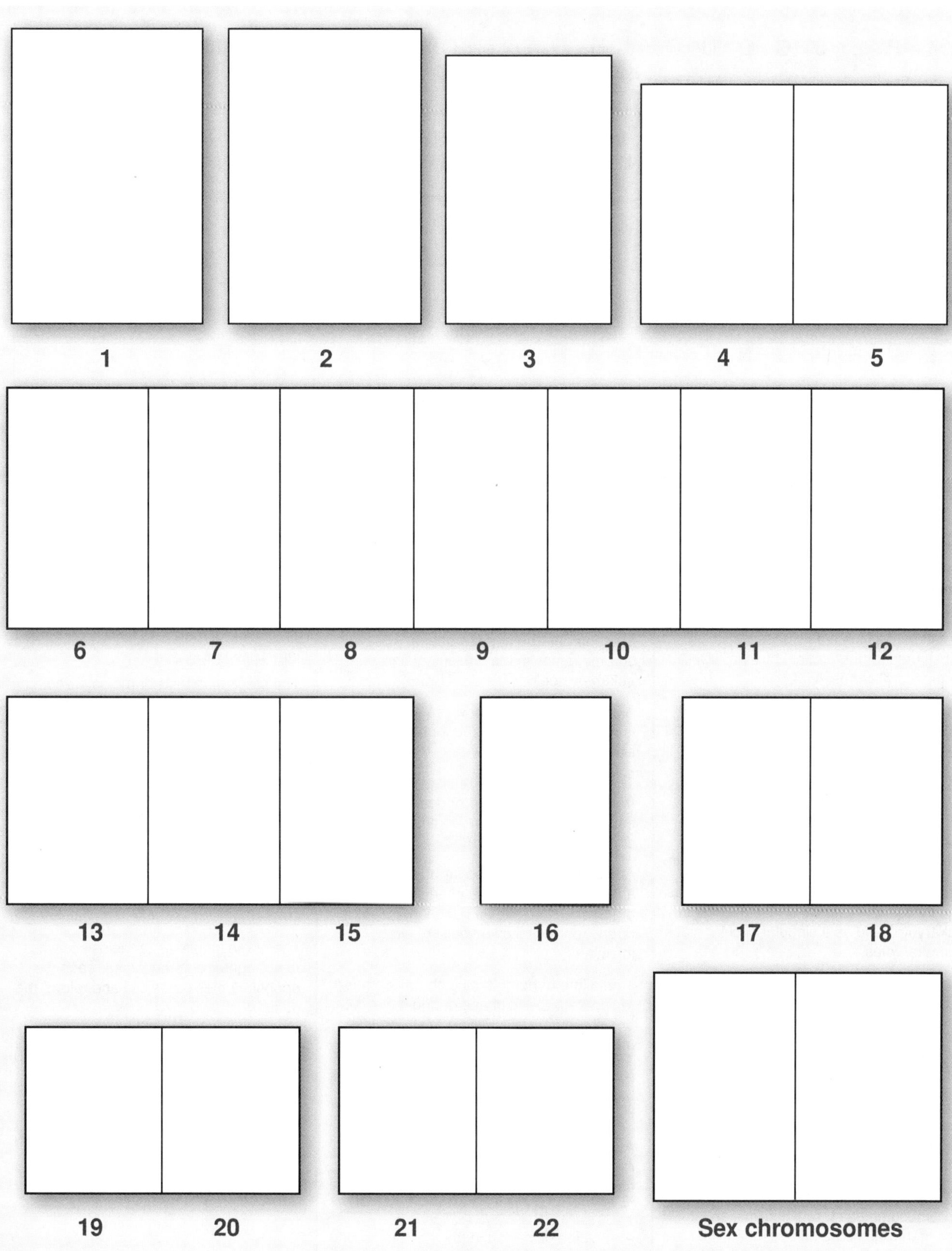
1
2
3
4
5
6
7
8
9
10
11
12
13
14
15
16
17
18
19
20
21
22
Sex chromosomes

ISBN: 978-1-927173-16-9

KEY TERMS: Word Find

Use the clues below to find the relevant key terms in the WORD FIND grid

K	Q	B	I	N	D	E	P	E	N	D	E	N	T	A	S	S	O	R	T	M	E	N	T	N
A	T	I	N	B	H	O	M	O	L	O	G	O	U	S	C	H	R	O	M	O	S	O	M	E
R	E	S	P	O	N	T	A	N	E	O	U	S	M	U	T	A	T	I	O	N	J	M	S	U
Y	L	G	N	U	H	H	F	U	I	B	I	V	G	H	V	M	M	U	T	A	G	E	N	Z
O	O	M	S	U	L	C	N	C	H	R	O	M	O	S	O	M	E	L	K	E	X	T	I	I
T	P	E	J	G	Q	U	Q	W	H	B	Q	M	G	T	R	I	S	O	M	Y	D	A	V	E
Y	H	I	A	M	N	I	O	C	E	N	T	E	S	I	S	F	M	L	Q	H	L	P	M	F
P	A	O	H	C	H	R	O	M	A	T	I	D	V	A	N	A	P	H	A	S	E	H	L	A
E	S	S	N	D	M	N	O	N	D	I	S	J	U	N	C	T	I	O	N	V	R	A	Y	H
E	E	I	I	H	I	B	A	S	E	S	U	B	S	T	I	T	U	T	I	O	N	S	D	U
S	J	S	D	C	R	E	D	U	C	T	I	O	N	D	I	V	I	S	I	O	N	E	D	M
A	L	Q	F	Z	D	A	I	D	C	R	O	S	S	I	N	G	O	V	E	R	L	W	G	Q
C	O	C	H	O	R	I	O	N	I	C	V	I	L	L	U	S	S	A	M	P	L	I	N	G
E	V	R	N	I	N	T	E	R	P	H	A	S	E	X	I	C	H	A	P	L	O	I	D	P
F	W	G	X	S	I	C	K	L	E	C	E	L	L	A	N	E	M	I	A	S	V	V	S	C
D	I	P	L	O	I	D	C	E	L	M	U	T	A	T	I	O	N	R	X	L	T	X	A	J
P	R	O	P	H	A	S	E	M	Y	Y	F	E	R	T	I	L	I	Z	A	T	I	O	N	I

CORE CLUES

A change to the DNA sequence of an organism, e.g. a deletion or substitution of a base in the DNA sequence.

Mutation in which a new base is substituted for the original base in a DNA sequence.(2 words: 4, 12)

A prenatal diagnostic test where a small amount of placental tissue is extracted and analyzed for chromosomal abnormalities. (3 words: 9, 6, 8)

A type of aneuploidy where the chromosome number for the affected chromosome is 2n+1.

Single piece of DNA that contains many genes and associated regulatory elements and proteins. Found within the nucleus in eukaryotes and as a singular circular piece in prokaryotes.

Event during meiosis where two homologous chromosomes exchange genetic material. (2 words: 8, 4)

Having two homologous copies of each chromosome (2N), usually one from the mother and one from the father.

Having only N chromosomes in the nucleus of the cell. The number of chromosomes in the gamete of an organism.

Chromosome pairs, one paternal and one maternal, of the same length, centromere position, and staining pattern with genes for the same characteristics at corresponding loci. (2 words: 10, 10)

The stage in the cell cycle between divisions.

The number and appearance of chromosomes in the nucleus of a eukaryotic cell.

A prenatal genetic test where amniotic fluid is extracted and analyzed for chromosomal abnormalities.

The process of double nuclear division (reduction division) to produce four nuclei, each containing half the original number of chromosomes (haploid).

Any chemical, influence, or object that is able to increase the mutation rate of an organism. This may be radiation, industrial or environmental chemicals, or viral infection.

The failure of homologous chromosomes to separate at anaphase during meiosis I or sister chromatids failure to separate in meiosis II.

Nuclear division in which the daughter cells have half the number of chromosomes as the parent. (2 words: 9, 8)

An inherited disorder caused by a gene mutation which codes for a faulty beta (β) chain hemoglobin (Hb) protein. (3 words: 6, 4, 6)

A mutation occurring in the absence of any mutagenic influence. (2 words: 11, 8)

AHL ONLY CLUES

One of two identical DNA strands forming the chromosome and held together by the centromere after DNA replication.

The union of male and female gametes to form a zygote.

The stage in mitosis or meiosis when the chromosomes have become aligned on the equator of the cell with all the centromeres lying along the spindle equator.

A stage in mitosis or meiosis, that involves the separation of chromosomal material to give two groups of chromosomes which will eventually become new cell nuclei.

The first stage in meiosis or mitosis in which the replicated chromosomes condense and become visible as double structures.

The stage in mitosis or meiosis in which the daughter chromosomes reach the opposite poles of the cell and reform the nuclei.

The random allocation of chromosomes to newly forming gametes. (2 words: 10, 10)

ISBN: 978-1-927173-16-9

Core Topic **4.3**

AHL Topic **10**

Heredity

Key concepts

- Meiosis and sexual reproduction introduce variation in the offspring: the raw material for natural selection.
- The dominance of alleles can be inferred from the genetic outcomes of crosses.
- The inheritance of some traits is dependent on gender.
- Polygenic inheritance results in continuous phenotypic variation for a characteristic.

Key terms

Core

autosome
carrier
codominant alleles
dominant allele
genetic counseling
genotype
heterozygous
homozygous
locus
monohybrid cross
multiple alleles
pedigree chart
phenotype
Punnett square
recessive allele
sex chromosome
sex linkage
test cross

AHL only

autosomes
continuous variation
dihybrid cross
linkage group
linked genes
polygenes (= multiple genes)
polygenic inheritance
recombinant offspring
sex chromosomes

Learning Objectives

☐ 1. Use the **KEY TERMS** to compile a glossary for this topic.

Theoretical Genetics *(4.3 & 10.1)* pages 150-161, 168-171, 174-176, 179

☐ 2. **AHL**: State Mendel's **law of independent assortment**. Explain the relationship between Mendel's law of independent assortment and meiosis.

☐ 3. Define the following terms and use them appropriately: **dominant allele**, **recessive allele**, **codominant alleles**, **locus**, **heterozygous**, **homozygous**, **genotype**, **phenotype**, **test cross**, **carrier**.

☐ 4. Use a **Punnett square** to solve problems involving monohybrid inheritance with a simple **dominant-recessive** inheritance pattern.

☐ 5. Recognize that **multiple alleles** may exist for some genes. Describe and explain inheritance involving **codominance**, including codominance in a multiple allele system (e.g. ABO blood groups).

☐ 6. Describe gender determination in humans with reference to the X and Y chromosomes.

☐ 7. Define the term **sex linkage** and explain its effects on the inheritance of alleles, as illustrated by the examples of red-green color blindness and hemophilia.

☐ 8. Explain how females can be carriers for sex linked characteristics.

☐ 9. Predict the genotypic and phenotypic outcomes of monohybrid crosses involving codominance or sex linkage.

☐ 10. Use **pedigree charts** to determine the genotypes and phenotypes of individuals in a family tree. Demonstrate use of appropriate terminology throughout.

☐ 11. Discuss the social and ethical issues surrounding the use of **genetic counseling** and genetic screening for known genetic disorders or predispositions (*TOK*).

Dihybrid Crosses and Gene Linkage *(10.2)* pages 162-167, 172-173

☐ 12. **AHL**: Calculate and predict the genotype and phenotype ratios of offspring of **dihybrid crosses** involving unlinked, autosomal genes for two independent characteristics.

☐ 13. **AHL**: Distinguish **sex chromosomes** from **autosomes**.

☐ 14. **AHL**: Recall how crossing over between homologous chromosomes in prophase I of meiosis can result in an exchange of alleles. Define the term **linkage group**.

☐ 15. **AHL**: Explain examples involving dihybrid inheritance of **linked genes**. Know that the probability of linked genes being inherited together as a unit is a function of the distance between them. Identify **recombinant offspring** in a dihybrid cross involved linked genes.

Polygenic Inheritance *(10.3)* pages 177-178

☐ 16. **AHL**: Define the term **polygenic inheritance**.

☐ 17. **AHL**: Explain examples of inheritance involving **polygenes** (**multiple genes**). Show how polygeny can contribute to **continuous variation** in phenotypes. Compare with the qualitative traits which are determined by only two alleles in any one individual.

Periodicals:
Listings for this chapter are on page 399

Weblinks:
www.thebiozone.com/weblink/IB-3169.html

BIOZONE APP:
Student Review Series
Inheritance

A Gene That Can Tell Your Future?

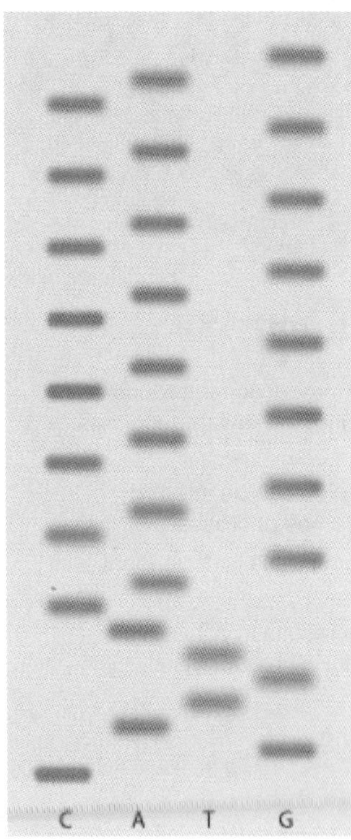
C
A
T
G

Huntington's disease (HD) is a genetic neuro-degenerative disease that normally does not affect people until about the age of 40. Its symptoms usually appear first as a shaking of the hands and an awkward gait. Later manifestations of the disease include serious loss of muscle control and mental function, often ending in dementia and premature death.

All humans have the huntingtin (**HTT**) gene, which in its normal state produces a protein with roles in gene transcription, synaptic transmission, and brain cell survival. The mutant gene (**mHTT**) causes changes to and death of the cells of the cerebrum, the hippocampus, and cerebellum, resulting in the atrophy (reduction) of brain matter. The gene was discovered by Nancy Wexler in 1983 after ten years of research working with cell samples and family histories of more than 10,000 people from the town of San Luis in Venezuela, where around 1% of the population have the disease (compared to about 0.01% in the rest of the world). Ten years later the exact location of the gene on the chromosome 4 was discovered.

The identification of the HD gene began by looking for a gene probe that would bind to the DNA of people who had HD, and not to those who didn't. Eventually a marker for HD, called **G8**, was found. The next step was to find which chromosome carried the marker and where on the chromosome it was. The researchers hybridized human cells with those of mice so that each cell contained only one human chromosome, a different chromosome in each cell. The hybrid cell with chromosome 4 was the one with the G8 marker. They then found a marker that overlapped G8 and then another marker that overlapped that marker. By repeating this many times, they produced a map of the genes on chromosome 4. The researchers then sequenced the genes and found people who had HD had one gene that was considerably longer than people who did not have HD. Moreover the increase in length was caused by the repetition of the base sequence CAG.

The HD mutation (mHTT) is called a trinucleotide repeat expansion. In the case of mHTT, the base sequence CAG is repeated multiple times on the short arm of chromosome 4. The normal number of CAG repeats is between 6 and 30. The mHTT gene causes the repeat number to be 35 or more and the size of the repeat often increases from generation to generation, with the severity of the disease increasing with the number of repeats. Individuals who have 27 to 35 CAG repeats in the HTT gene do not develop Huntington disease, but they are at risk of having children who will develop the disorder. The mutant allele, mHTT, is also dominant, so those who are homozygous or heterozygous for the allele are both at risk of developing HD.

New research has shown that the mHTT gene activates an enzyme called JNK3, which is expressed only in the neurons and causes a drop in nerve cell activity. While a person is young and still growing, the neurons can compensate for the accumulation of JNK3. However, when people get older and neuron growth stops, the effects of JNK3 become greater and the physical signs of HD become apparent. Because of mHTT's dominance, an affected person has a 50% chance of having offspring who are also affected. Genetic testing for the disease is relatively easy now that the genetic cause of the disease is known. While locating and counting the CAG repeats does not give a date for the occurrence of HD, it does provide some understanding of the chances of passing on the disease.

1. Describe the physical effects of Huntington's disease: ______________________

2. Describe how the mHTT gene was discovered: ______________________

3. Discuss the cause of Huntington's disease and its pattern of increasing severity with each generation: ______________________

ISBN: 978-1-927173-16-9

Variation

Variation refers to the diversity within and between species. The genetic variation in species is largely due to meiosis and sexual reproduction, which shuffles existing genetic material into new combinations as it is passed from generation to generation. **Mutation** is also a source of variation as it may create new alleles. Variation gives species more opportunity to adapt to a changing environment because, at any one time, some individuals will have higher fitness (leave more offspring) than others. Variation in a population can be continuous or discontinuous. Traits determined by a single gene (e.g. ABO blood groups) show **discontinuous variation**, with a very limited number of variants present in the population. In contrast, traits determined by a large number of genes (e.g. skin color) show **continuous variation**, and the number of phenotypes is very large. Environmental influences (differences in diet for example) also contribute to the observable variation in a population.

Mutations

gene mutations, chromosome mutations

Mutations are the source of all **new** alleles. Existing genes are modified by base substitutions and deletions, causing the formation of new alleles. Silent mutations may escape selection pressure until conditions change.

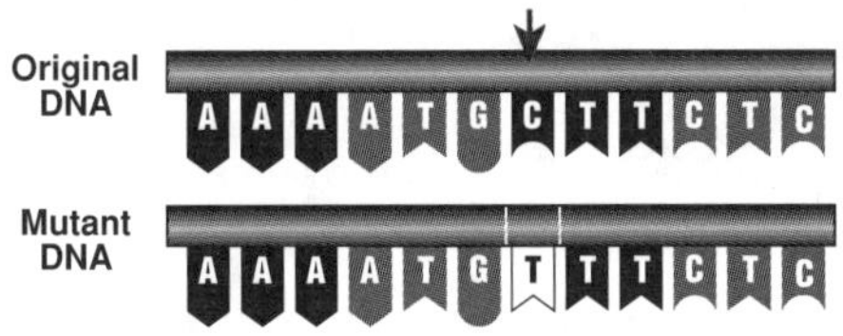

Mutation: Substitute **T** instead of **C**

Original DNA: A A A A T G C T T C T C

Mutant DNA: A A A A T G T T T C T C

Sexual Reproduction

independent assortment, crossing over and recombination, mate selection

Sexual reproduction rearranges and reshuffles the genetic material into new combinations.

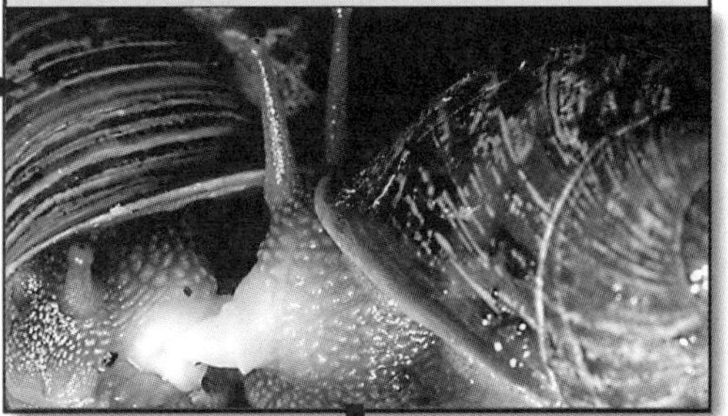

Genotype

Determines the **genetic potential** of an individual

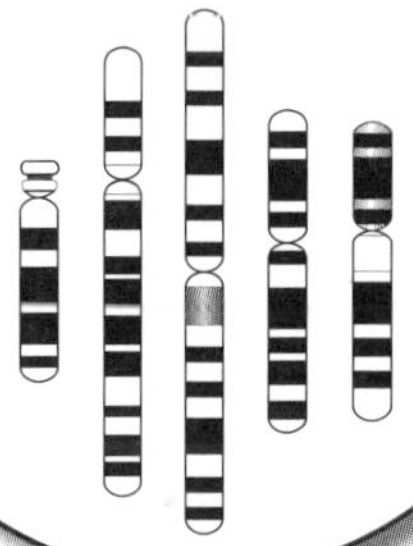

Dominant, recessive, codominant and multiple allele systems, as well as interactions between genes, combine in their effects.

Phenotype

An individual's phenotype is the result of the interaction of genetic and environmental factors during its lifetime (including during development). The genetic instructions for creating the individual may be modified by the internal and external environment both during and after development. Phenotypic plasticity in response to environment can be adaptive if it increases fitness.

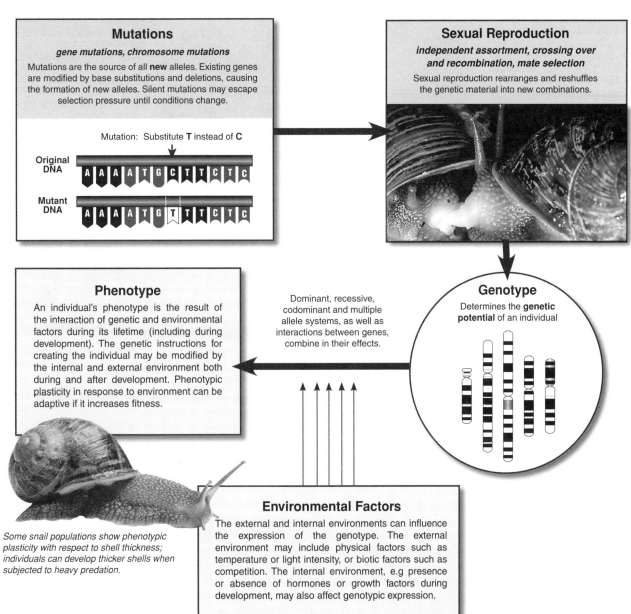

Some snail populations show phenotypic plasticity with respect to shell thickness; individuals can develop thicker shells when subjected to heavy predation.

Environmental Factors

The external and internal environments can influence the expression of the genotype. The external environment may include physical factors such as temperature or light intensity, or biotic factors such as competition. The internal environment, e.g presence or absence of hormones or growth factors during development, may also affect genotypic expression.

1. Using examples, explain how the environment of a particular genotype can affect the phenotype: ______________________

2. Discuss the significance of variation in selection: ______________________

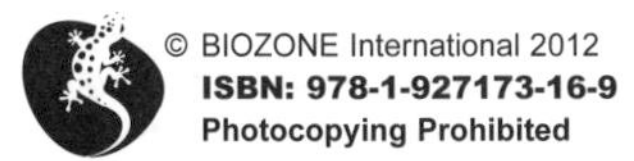

ISBN: 978-1-927173-16-9

Periodicals: *What is variation?*

Related activities: *Changes to the DNA Sequence, Stages in Meiosis*

Albinism (above) is the result of the inheritance of recessive alleles for melanin production. Those with the albino phenotype lack melanin pigment in the eyes, skin, and hair.

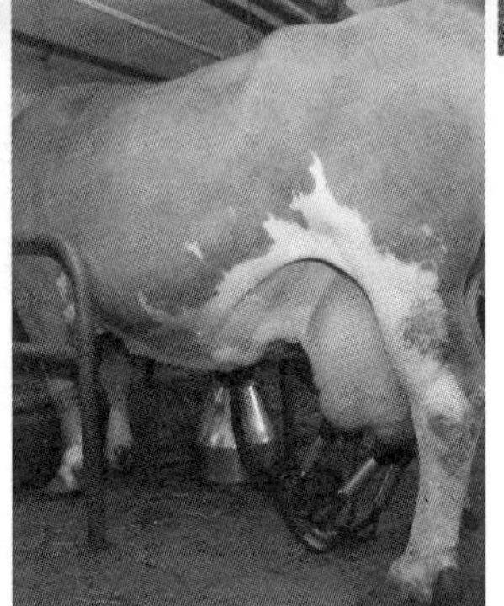

Quantitative traits are characterized by **continuous variation**, with individuals falling somewhere on a normal distribution curve of the phenotypic range. Typical examples include skin color and height in humans (left), grain yield in corn (above), growth in pigs (above, left), and milk production in cattle (far left). Quantitative traits are determined by genes at many loci (polygenic) but most are also influenced by environmental factors.

Comb shape in poultry is a **qualitative trait** and birds have one of four phenotypes depending on which combination of four alleles they inherit. The dash (missing allele) indicates that the allele may be recessive or dominant.

Single comb	Walnut comb	Pea comb	Rose comb
rrpp	**R_P_**	**rrP_**	**R_pp**

Flower color in snapdragons (right) is also a **qualitative trait** determined by two alleles (red and white). The alleles show incomplete dominance and the heterozygote (C^RC^W) exhibits an intermediate phenotype between the two homozygotes.

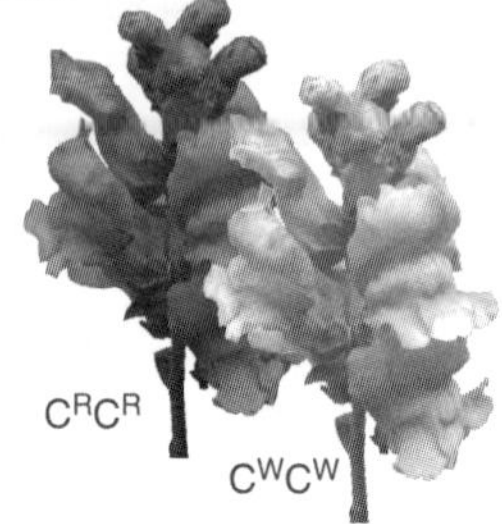
C^RC^R
C^WC^W

3. Describe three ways in which sexual reproduction can provide genetic variation in individuals:

   (a) ____

   (b) ____

   (c) ____

4. (a) What is a neutral mutation? ____

   (b) What is the significance of neutral mutations? ____

5. Describe the differences between **continuous** and **discontinuous** variation, giving examples to illustrate your answer:

____

6. Identify each of the following phenotypic traits as continuous (quantitative) or discontinuous (qualitative):

   (a) Wool production in sheep: ______________
   (b) Hand span in humans: ______________
   (c) Blood groups in humans: ______________
   (d) Albinism in mammals: ______________
   (e) Body weight in mice: ______________
   (f) Flower color in snapdragons: ______________

ISBN: 978-1-927173-16-9

Mendel's Pea Plant Experiments

Gregor Mendel (1822-84), pictured right, was an Austrian monk who carried out the pioneering studies of inheritance. Mendel bred pea plants so he could study the inheritance patterns of a number of **traits** (specific characteristics). He showed that characters could be masked in one generation but could reappear in later generations and proposed that inheritance involved the transmission of discrete units of inheritance from one generation to the next. We now call these units of inheritance **genes**.
Mendel examined six phenotypic traits and found that they were inherited in predictable ratios, depending on the phenotypes of the parents. Some of his results from crossing heterozygous plants are tabulated below. The numbers in the results column represent how many offspring had those phenotypic features. Mendel's results have sometimes been reported as being "too good", with some statisticians suggesting he may in some way (consciously or unconsciously) have biased his results.

1. Study the **results** for each of the six experiments below. Determine which of the two phenotypes is dominant, and which is the recessive. Place your answers in the spaces in the **dominance** column in the table below.
2. Calculate the ratio of dominant phenotypes to recessive phenotypes (to two decimal places). The first one has been done for you (5474 ÷ 1850 = 2.96). Place your answers in the spaces provided in the table below:

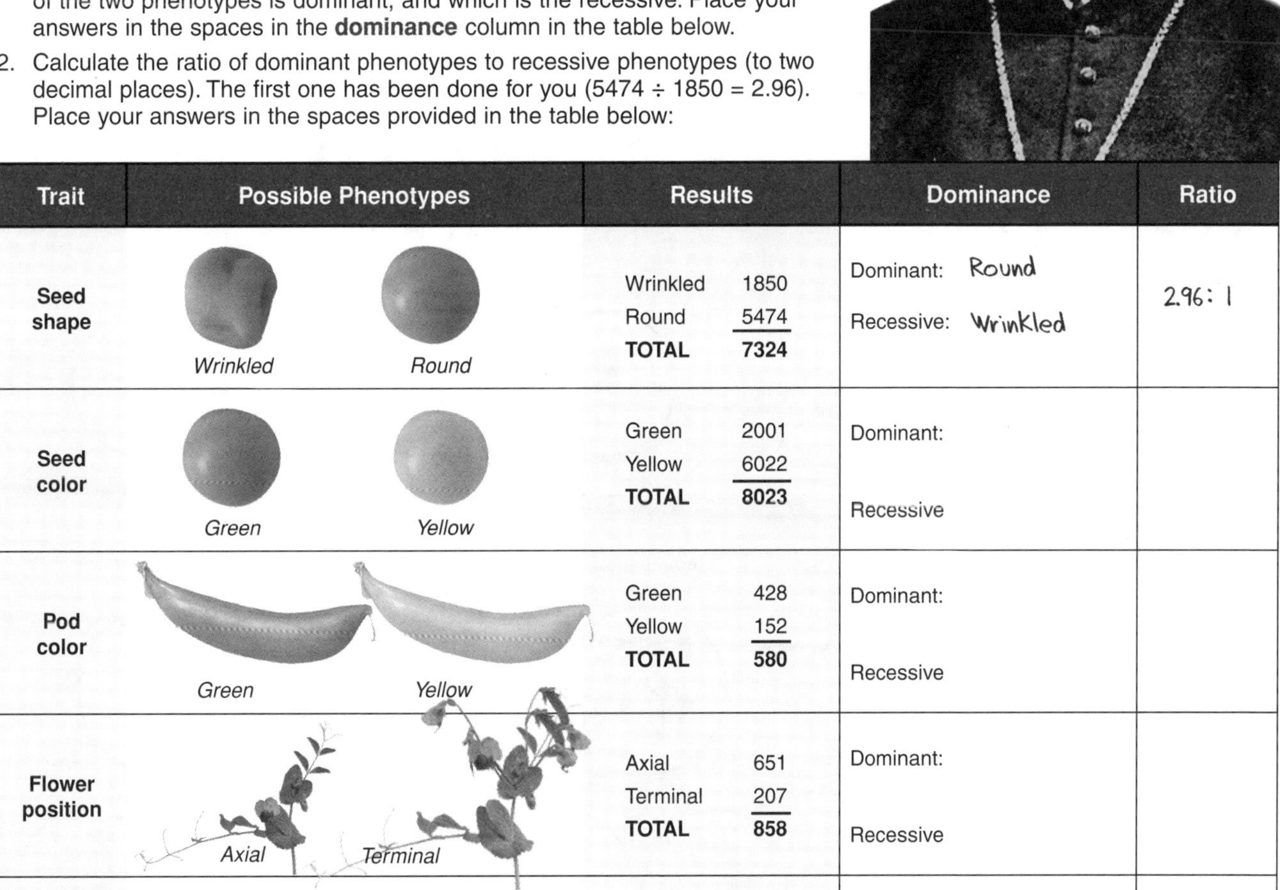

Trait	Possible Phenotypes	Results	Dominance	Ratio
Seed shape	*Wrinkled* / *Round*	Wrinkled 1850 Round 5474 **TOTAL 7324**	Dominant: Round Recessive: Wrinkled	2.96 : 1
Seed color	*Green* / *Yellow*	Green 2001 Yellow 6022 **TOTAL 8023**	Dominant: Recessive	
Pod color	*Green* / *Yellow*	Green 428 Yellow 152 **TOTAL 580**	Dominant: Recessive	
Flower position	*Axial* / *Terminal*	Axial 651 Terminal 207 **TOTAL 858**	Dominant: Recessive	
Pod shape	*Constricted* / *Inflated*	Constricted 299 Inflated 882 **TOTAL 1181**	Dominant: Recessive	
Stem length	*Tall* / *Dwarf*	Tall 787 Dwarf 277 **TOTAL 1064**	Dominant: Recessive	

3. Mendel's experiments identified that two heterozygous parents should produce offspring in the ratio of three times as many dominant offspring to those showing the recessive phenotype.

   (a) Which three of Mendel's experiments provided ratios closest to the theoretical 3:1 ratio?

   ____

   (b) Suggest why these results deviated less from the theoretical ratio than the others: ______________________________

   ____

   ____

Heredity

ISBN: 978-1-927173-16-9

Periodicals: *Mendel's legacy*

Related activities: *Alleles, Mendel's Laws of Inheritance*
Weblinks: *Basic Principles of Genetics*

Mendel's Laws of Inheritance

From his work on the inheritance of phenotypic traits in peas, Mendel formulated a number of ideas about the inheritance of characters. These were later given formal recognition as Mendel's laws of inheritance. These are outlined below.

The Theory of Particulate Inheritance

Characteristics of both parents are passed on to the next generation as discrete entities (genes).

This model explained many observations that could not be explained by the idea of blending inheritance, which was universally accepted prior to this theory. The trait for flower color (right) appears to take on the appearance of only one parent plant in the first generation, but reappears in later generations.

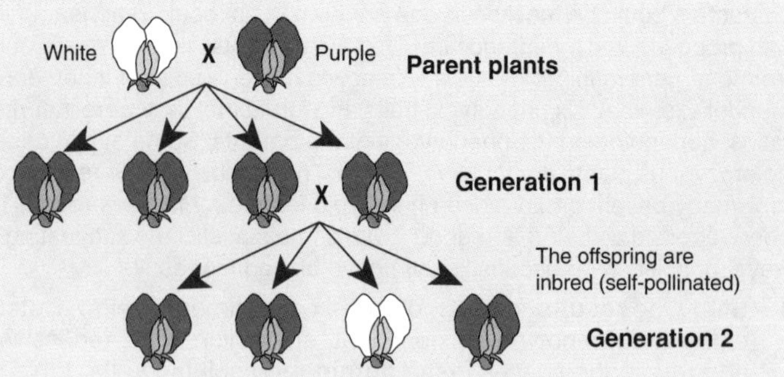
White
X
Purple
Parent plants
Generation 1
X
The offspring are inbred (self-pollinated)
Generation 2

Law of Segregation

During gametic meiosis, the two members of any pair of alleles segregate unchanged and are passed into different gametes.

These gametes are eggs (ova) and sperm cells. The allele in the gamete will be passed on to the offspring.

NOTE: This diagram has been simplified, omitting the stage where the second chromatid is produced for each chromosome.

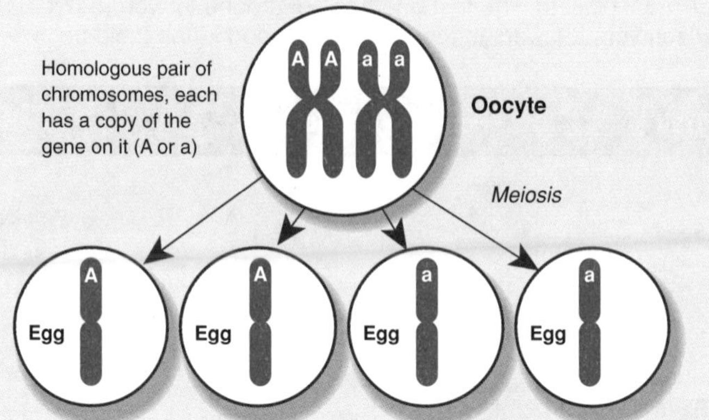
Homologous pair of chromosomes, each has a copy of the gene on it (A or a)
A A a a
Oocyte
Meiosis
Egg
A
Egg
A
Egg
a
Egg
a

Law of Independent Assortment

Allele pairs separate independently during gamete formation, and traits are passed on to offspring independently of one another (this is only true for unlinked genes).

This diagram shows two genes (A and B) that code for different traits. Each of these genes is represented twice, one copy (allele) on each of two homologous chromosomes. The genes A and B are located on different chromosomes and, because of this, they will be inherited independently of each other i.e. the gametes may contain any combination of the parental alleles.

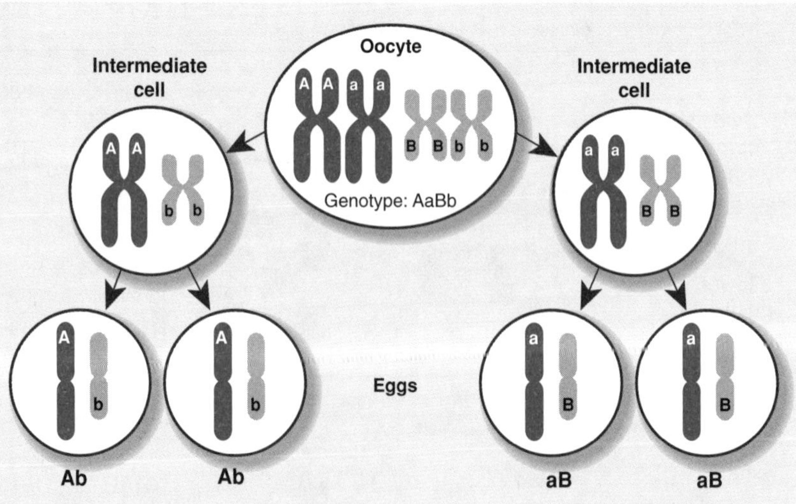
Oocyte
A A a a
B B b b
Genotype: AaBb
Intermediate cell
A A
b b
Intermediate cell
a a
B B
Eggs
Ab
Ab
aB
aB

1. State the **property of genetic inheritance** that allows parent pea plants of different flower color to give rise to flowers of a single color in the first generation, with both parental flower colors reappearing in the following generation:

   ____

2. The oocyte is the egg producing cell in the ovary of an animal. In the diagram illustrating the **law of segregation** above:

   (a) State the genotype for the oocyte (adult organism): ______________________________

   (b) State the genotype of each of the **four** gametes: ______________________________

   (c) State how many different kinds of gamete can be produced by this oocyte: ______________________________

3. The diagram illustrating the **law of independent assortment** (above) shows only one possible result of the random sorting of the chromosomes to produce: Ab and aB in the gametes.

   (a) List another possible combination of genes (on the chromosomes) ending up in gametes from the same oocyte:

   ____

   (b) How many different gene combinations are possible for the oocyte? ______________________________

Related activities: *Alleles, Mendel's Pea Plant Experiments*

ISBN: 978-1-927173-16-9

Basic Genetic Crosses

As a simple introduction to Mendelian inheritance, examine the diagrams below on monohybrid crosses. HL students should also complete the exercise for dihybrid (two gene) inheritance. The F_1 generation by definition describes the offspring of a cross between distinctly different, true-breeding (homozygous) parents. A **back cross** refers to any cross between an offspring and one of its parents. If the cross is to a homozygous recessive and is diagnostic, it is a test cross.

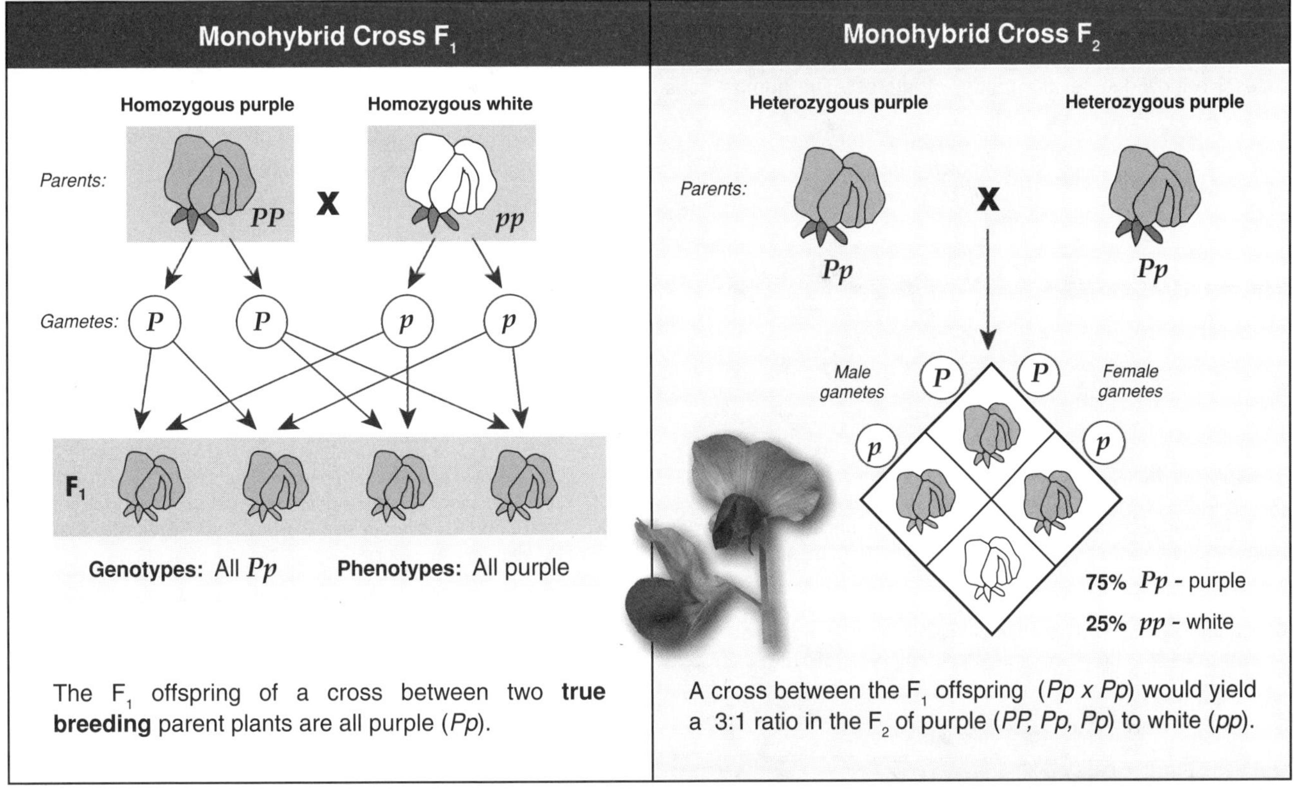

Monohybrid Cross F1
Homozygous purple
Homozygous white
Parents:
PP
X
pp
Gametes:
P
P
p
p
F1
Genotypes: All Pp
Phenotypes: All purple
Monohybrid Cross F2
Heterozygous purple
Heterozygous purple
Parents:
Pp
X
Pp
Male gametes
Female gametes
P
P
p
p
75% Pp - purple
25% pp - white

The F_1 offspring of a cross between two **true breeding** parent plants are all purple (*Pp*).

A cross between the F_1 offspring (*Pp x Pp*) would yield a 3:1 ratio in the F_2 of purple (*PP, Pp, Pp*) to white (*pp*).

Dihybrid Cross

A dihbrid cross studies the inheritance patterns of two genes. In pea seeds, yellow color (*Y*) is dominant to green (*y*) and round shape (*R*) is dominant to wrinkled (r). Each **true breeding** parental plant has matching alleles for each of these characters (*YYRR* or *yyrr*). F_1 offspring will all have the same genotype and phenotype (yellow-round: *YyRr*).

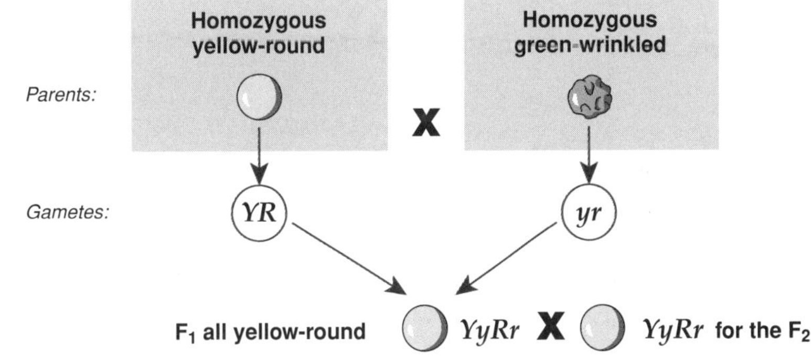

Homozygous yellow-round
Homozygous green-wrinkled
Parents:
X
Gametes:
YR
yr
F1 all yellow-round
YyRr
X
YyRr
for the F2

1. Fill in the Punnett square (below right) to show the genotypes of the F_2 generation.

2. In the boxes below, use fractions to indicate the numbers of each phenotype produced from this cross.

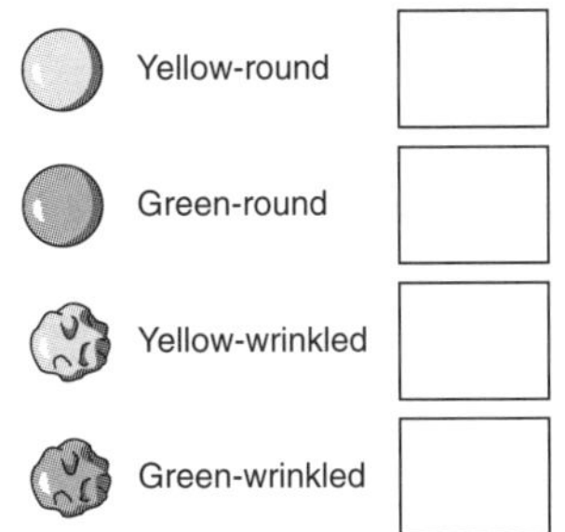

Yellow-round
Green-round
Yellow-wrinkled
Green-wrinkled

3. Express these numbers as a ratio:

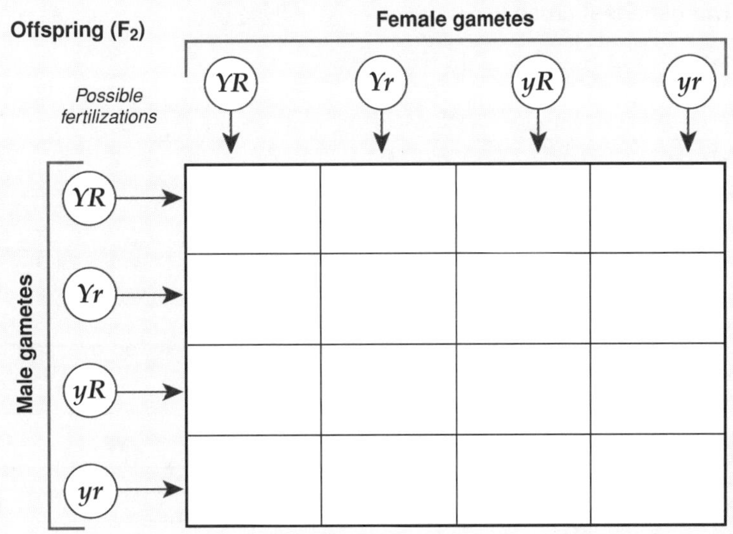

Offspring (F2)
Female gametes
Possible fertilizations
YR
Yr
yR
yr
Male gametes
YR
Yr
yR
yr

ISBN: 978-1-927173-16-9

Related activities: *The Test Cross, Monohybrid Cross, Dihybrid Cross*

The Test Cross

It is not always possible to determine an organism's genotype by its appearance because the expression of genes is complicated by patterns of dominance and by gene interactions. The **test cross** was developed by Gregor Mendel as a way to establish the genotype of an organism with the dominant phenotype for a particular trait. The principle of the test cross is simple. The individual with the unknown genotype is bred with a homozygous recessive individual for the trait(s) of interest. The homozygous recessive can produce only one type of allele (recessive), so the phenotypes of the resulting offspring will reveal the genotype of the unknown parent. For example, if the unknown individual is homozygous for the trait, all of the offspring will display the dominant phenotype. However, if the offspring display both dominant and recessive phenotypes, then the unknown must be heterozygous for that trait. The test cross can be used to determine the genotype of single genes or multiple genes.

Parent 1
Unknown genotype
(but with dominant traits)

?
?
?
?

X

Parent 2
Homozygous recessive genotype
(no dominant traits)

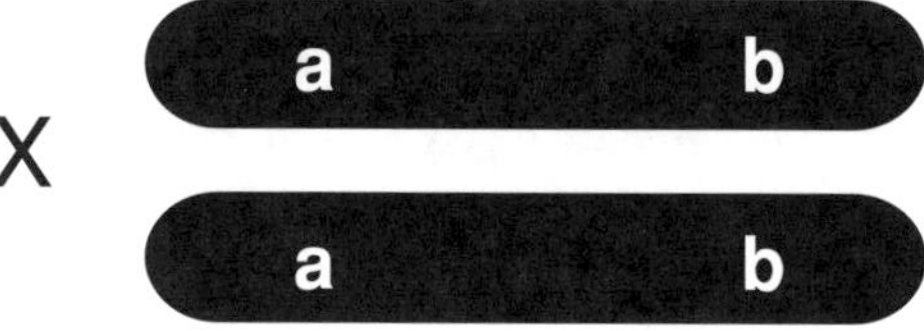
a
b
a
b

The common fruit fly (*Drosophila melanogaster*) is often used to illustrate basic principles of inheritance because it has several genetic markers whose phenotypes are easily identified. Once such phenotype is body color. Wild type (normal) *Drosophila* have yellow-brown bodies. The allele for yellow-brown body color (E) is dominant. The allele for an ebony colored body (e) is recessive. The test crosses below show the possible outcomes for an individual with homozygous and heterozygous alleles for ebony body color.

A. A homozygous recessive female (ee) with an ebony body is crossed with a homozyogous dominant male (EE).

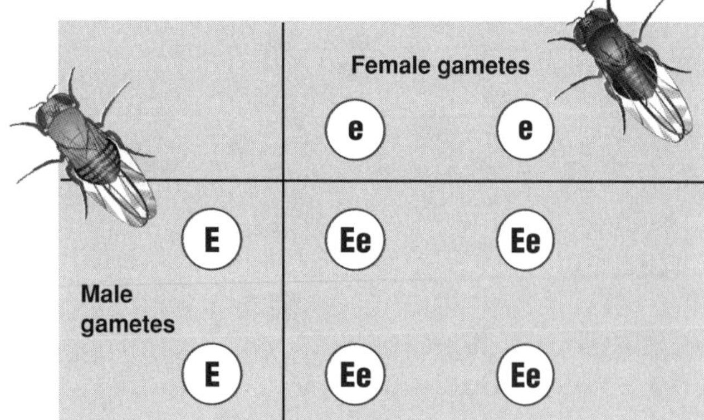
Female gametes
e
e
Male gametes
E
E
Ee
Ee
Ee
Ee

Cross A:

(a) Genotype frequency: 100% Ee

(b) Phenotype frequency: 100% yellow-brown

B. A homozygous recessive female (ee) with an ebony body is crossed with a heterozygous male (Ee).

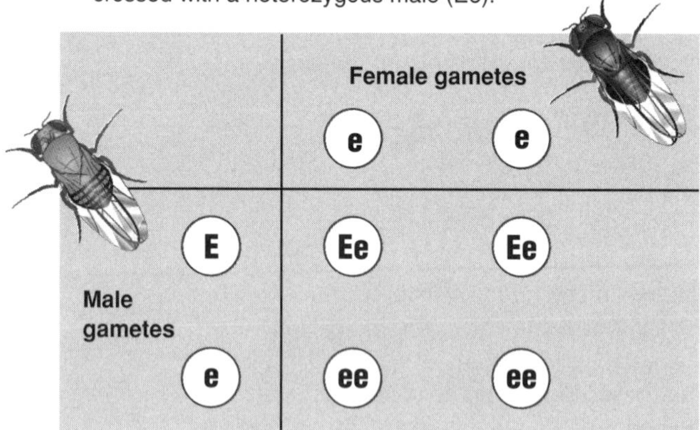
Female gametes
e
e
Male gametes
E
e
Ee
Ee
ee
ee

Cross B:

(a) Genotype frequency: 50% Ee, 50% ee

(b) Phenotype frequency: 50% yellow-brown, 50% ebony

1. In *Drosophila*, the allele for brown eyes (b) is recessive, while the red eye allele (B) is dominant. How would you set up a **two gene test cross** to determine the genotype of a male who has a normal body color and red eyes?

2. List all of the **possible genotypes** for the male *Drosophila*:

3. 50% of the resulting progeny are yellow-brown bodies with red eyes, and 50% have ebony bodies with red eyes.

   (a) What is the genotype of the male *Drosophila*?

   (b) Explain your answer:

Related activities: Monohybrid Cross

ISBN: 978-1-927173-16-9

Monohybrid Cross

The study of **single-gene inheritance** is achieved by performing **monohybrid crosses**. The six basic types of matings possible among the three genotypes can be observed by studying a pair of alleles that govern coat color in the guinea pig. A dominant allele: given the symbol **B** produces **black** hair, and its recessive allele: **b**, produces white. Each of the parents can produce two types of gamete by the process of **meiosis** (in reality there are four, but you get identical pairs). Determine the **genotype** and **phenotype frequencies** for the crosses below (enter the frequencies in the spaces provided). For crosses 3 to 6, you must also determine gametes produced by each parent (write these in the circles), and offspring (F_1) genotypes and phenotypes (write in the genotype inside the offspring and state if black or white).

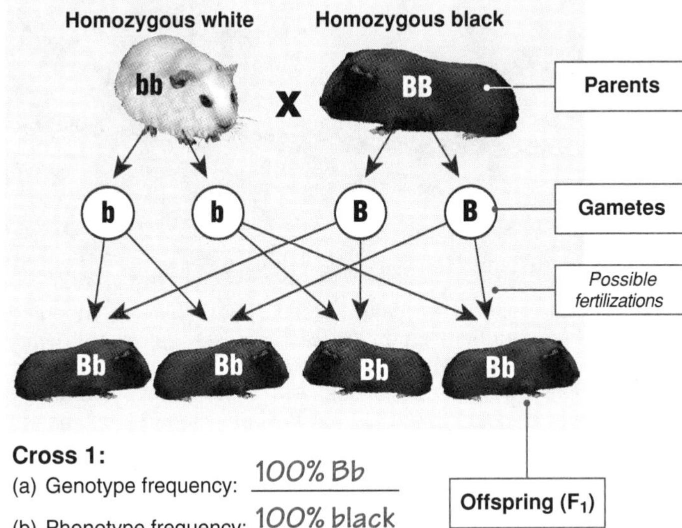

Homozygous white
Homozygous black
bb
X
BB
Parents
b
b
B
B
Gametes
Possible fertilizations
Bb
Bb
Bb
Bb
Offspring (F_1)

Cross 1:

(a) Genotype frequency: 100% Bb

(b) Phenotype frequency: 100% black

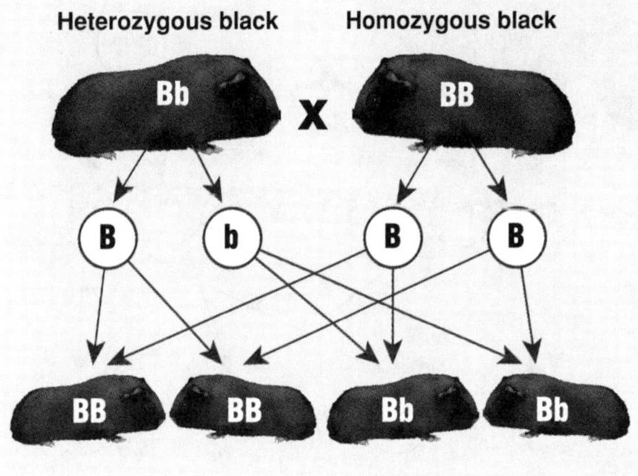

Heterozygous black
Homozygous black
Bb
X
BB
B
b
B
B
BB
BB
Bb
Bb

Cross 2:

(a) Genotype frequency: ____________________

(b) Phenotype frequency: ____________________

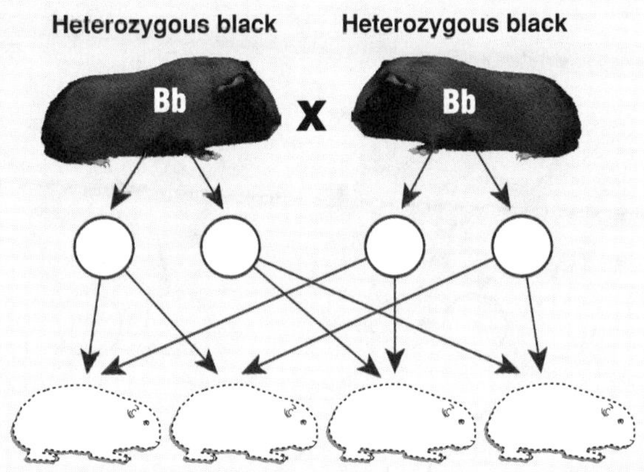

Heterozygous black
Heterozygous black
Bb
X
Bb

Cross 3:

(a) Genotype frequency: ____________________

(b) Phenotype frequency: ____________________

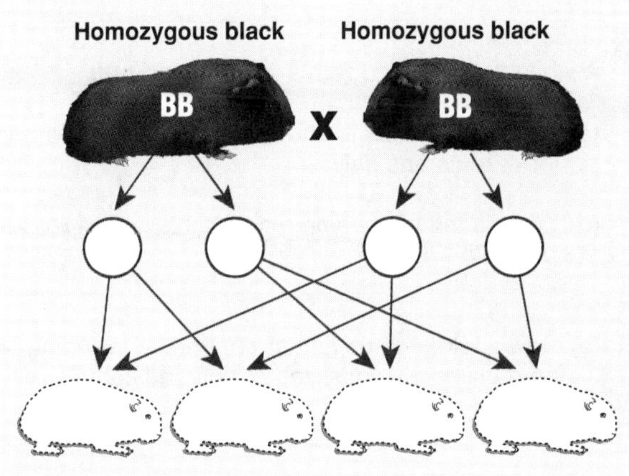

Homozygous black
Homozygous black
BB
X
BB

Cross 4:

(a) Genotype frequency: ____________________

(b) Phenotype frequency: ____________________

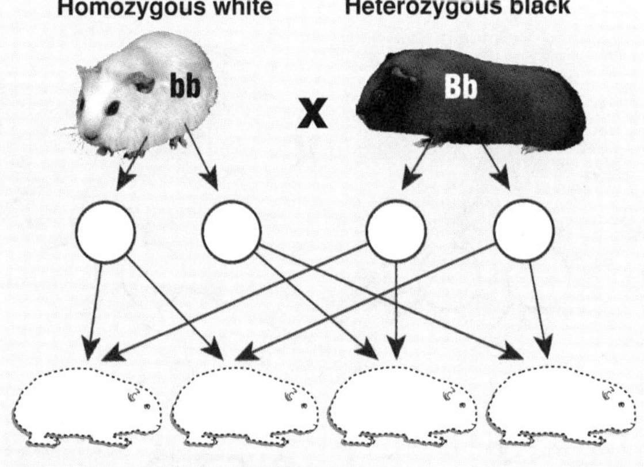

Homozygous white
Heterozygous black
bb
X
Bb

Cross 5:

(a) Genotype frequency: ____________________

(b) Phenotype frequency: ____________________

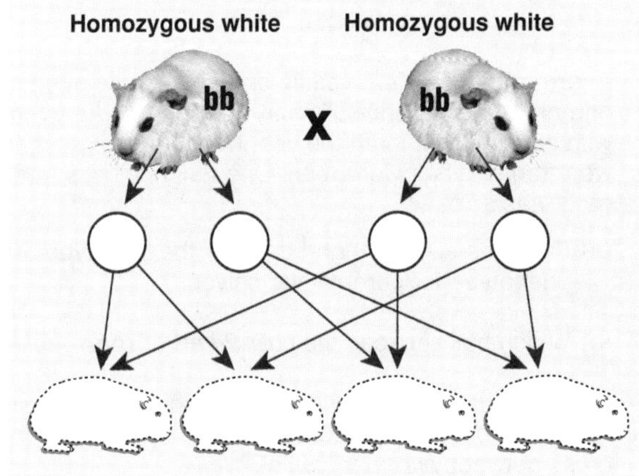

Homozygous white
Homozygous white
bb
X
bb

Cross 6:

(a) Genotype frequency: ____________________

(b) Phenotype frequency: ____________________

Heredity

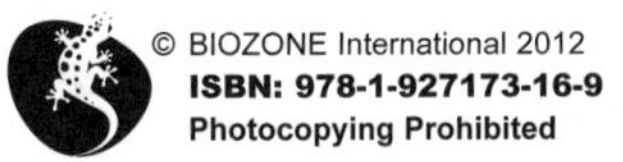

ISBN: 978-1-927173-16-9

Related activities: *Basic Genetic Crosses*
Weblinks: *Drag and Drop Genetics*

Codominance of Alleles

Codominance refers to an inheritance pattern in which both alleles in a heterozygote contribute to the phenotype. Both alleles are **independently** and **equally expressed**. One example includes the human blood group AB which is the result of two alleles: A and B, both being equally expressed. Other examples include certain coat colors in horses and cattle. Reddish coat color is equally dominant with white. Animals that have both alleles have coats that are roan-colored (coats with a mix of red and white hairs). The red hairs and white hairs are expressed equally and independently (not blended to produce pink).

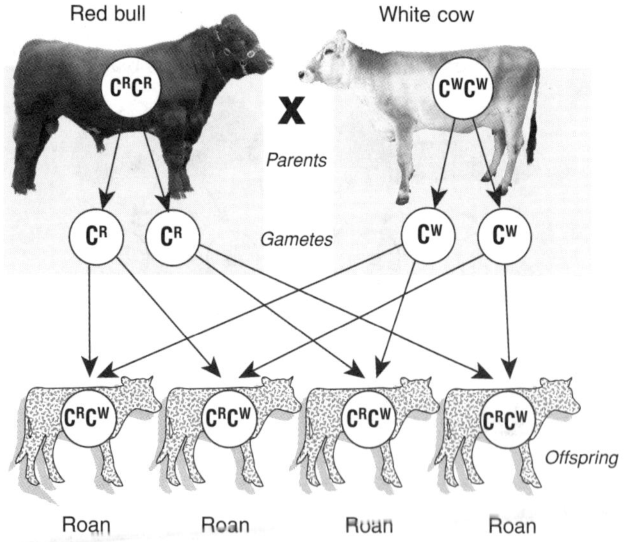

Red bull
White cow
C^RC^R
X
C^WC^W
Parents
C^R
C^R
Gametes
C^W
C^W
C^RC^W
C^RC^W
C^RC^W
C^RC^W
Offspring
Roan
Roan
Roan
Roan

A roan shorthorn heifer

In the shorthorn cattle breed, coat color is inherited. White shorthorn parents always produce calves with white coats. Red parents always produce red calves. However, when a red parent mates with a white one, the calves have a coat color that is different from either parent; a mixture of red and white hairs, called roan. Use the example (left) to help you to solve the problems below.

1. Explain how codominance of alleles can result in offspring with a phenotype that is different from either parent:

2. A white bull is mated with a roan cow (right):

   (a) Fill in the spaces to show the genotypes and phenotypes for parents and calves:

   (b) What is the phenotype ratio for this cross?

   (c) How could a cattle farmer control the breeding so that the herd ultimately consisted of only red cattle?

White bull

X

Roan cow

3. A farmer has only roan cattle on his farm. He suspects that one of the neighbors' bulls may have jumped the fence to mate with his cows earlier in the year because half the calves born were red and half were roan. One neighbor has a red bull, the other has a roan.

   (a) Fill in the spaces (right) to show the genotype and phenotype for parents and calves.

   (b) Which bull serviced the cows? **red** or **roan** (*delete one*)

4. Describe the classical phenotypic ratio for a codominant gene resulting from the cross of two heterozygous parents (e.g. a cross between two roan cattle):

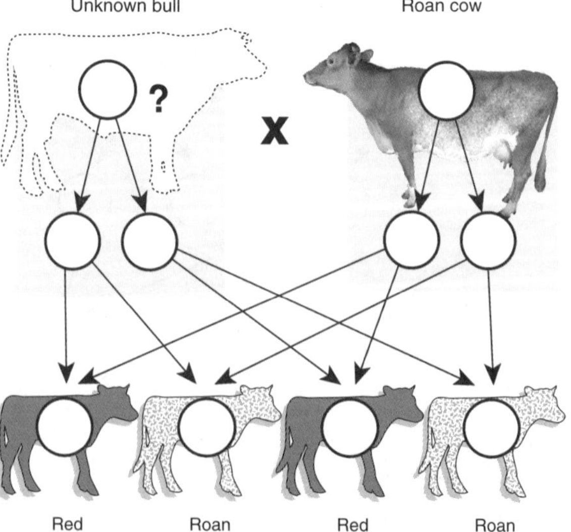

Unknown bull
Roan cow
?
X
Red
Roan
Red
Roan

ISBN: 978-1-927173-16-9

Codominance in Multiple Allele Systems

The four common blood groups of the human 'ABO blood group system' are determined by three alleles: ***A***, ***B***, and ***O*** (also represented in some textbooks as: I^A, I^B, and i^O or just **i**). This is an example of a **multiple allele** system for a gene. The ABO antigens consist of sugars attached to the surface of red blood cells. The alleles code for enzymes (proteins) that join these sugars together. The allele ***O*** produces a non-functioning enzyme that is unable to make any changes to the basic antigen (sugar) molecule. The other two alleles *(**A**, **B**)* are **codominant** and are expressed equally. They each produce a different functional enzyme that adds a different, specific sugar to the basic sugar molecule. The blood group A and B antigens are able to react with antibodies present in the blood from other people and must be matched for transfusion.

Recessive allele: ***O*** produces a non-functioning protein
Dominant allele: ***A*** produces an enzyme which forms **A antigen**
Dominant allele: ***B*** produces an enzyme which forms **B antigen**

If a person has the ***AO*** allele combination then their blood group will be group **A**. The presence of the recessive allele has no effect on the blood group in the presence of a dominant allele. Another possible allele combination that can create the same blood group is ***AA***.

Blood group (phenotype)	Possible genotypes	Frequency*		
		White	Black	Native American
O	OO	45%	49%	79%
A	AA AO	40%	27%	16%
B		11%	20%	4%
AB		4%	4%	1%

* Frequency is based on North American population
Source: www.kcom.edu/faculty/chamberlain/Website/MSTUART/Lect13.htm

1. Use the information above to complete the table for the possible genotypes for blood group B and group AB.

2. Below are six crosses possible between couples of various blood group types. The first example has been completed for you. Complete the genotype and phenotype for the other five crosses below:

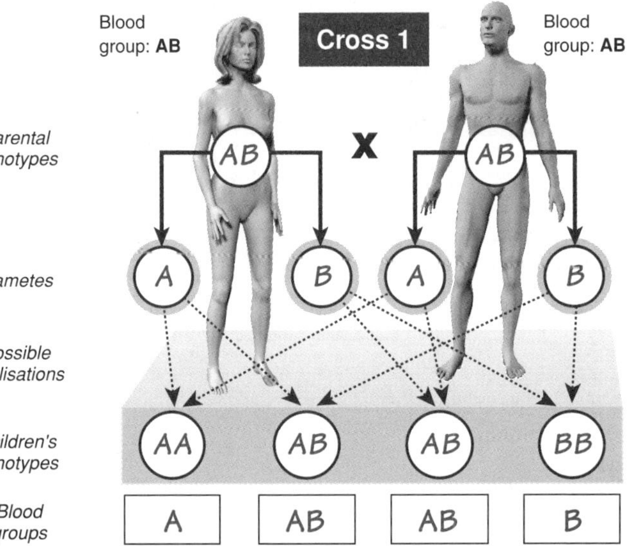
Blood group: AB
Cross 1
Blood group: AB
Parental genotypes
AB
X
AB
Gametes
A
B
A
B
Possible fertilisations
Children's genotypes
AA
AB
AB
BB
Blood groups
A
AB
AB
B

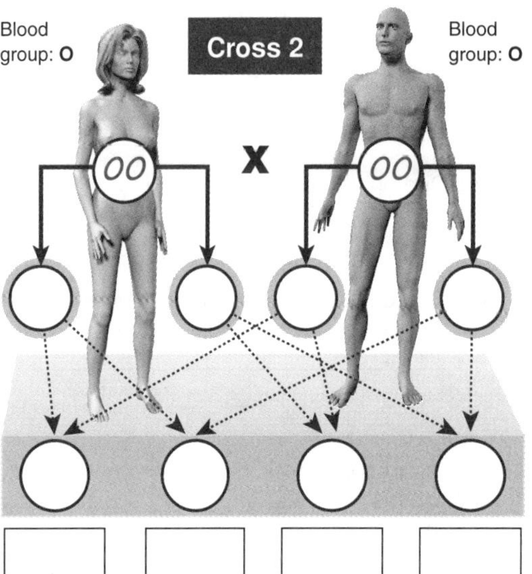
Blood group: O
Cross 2
Blood group: O
OO
X
OO

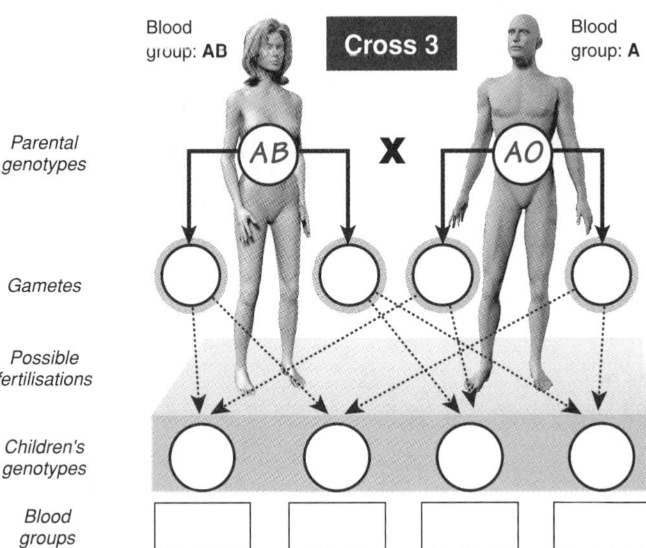
Blood group: AB
Cross 3
Blood group: A
Parental genotypes
AB
X
AO
Gametes
Possible fertilisations
Children's genotypes
Blood groups

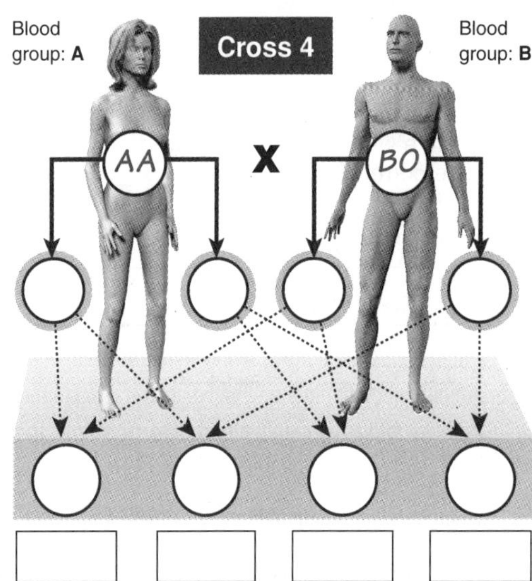
Blood group: A
Cross 4
Blood group: B
AA
X
BO

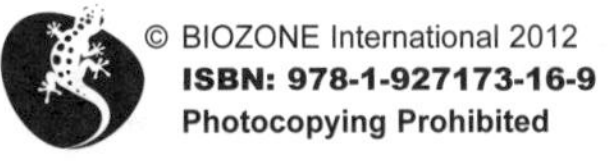

ISBN: 978-1-927173-16-9

Related activities: *Codominance of Alleles*

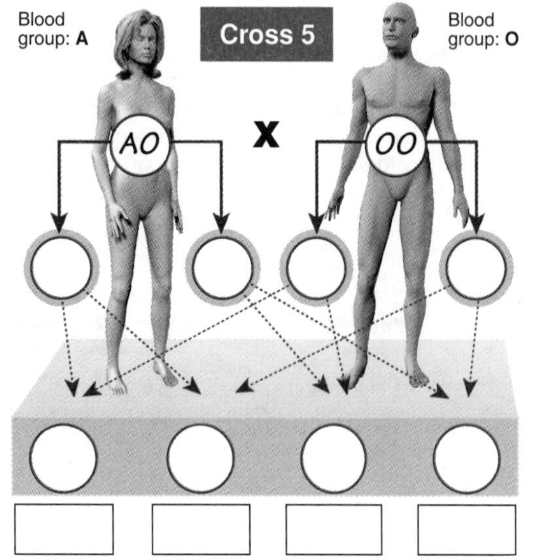
Blood group: A
Cross 5
Blood group: O
AO
X
OO

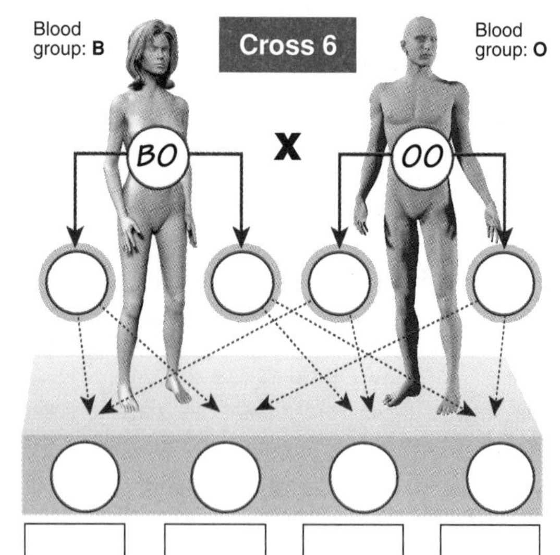
Parental genotypes
Gametes
Possible fertilisations
Children's genotypes
Blood groups
Blood group: B
Cross 6
Blood group: O
BO
X
OO

3. A wife is heterozygous for blood group **A** and the husband has blood group **O**.

   (a) Give the genotypes of each parent (fill in spaces on the diagram on the right).

   Determine the probability of:

   (b) One child having blood group **O**:

   ____

   (c) One child having blood group **A**:

   ____

   (d) One child having blood group **AB**:

   ____

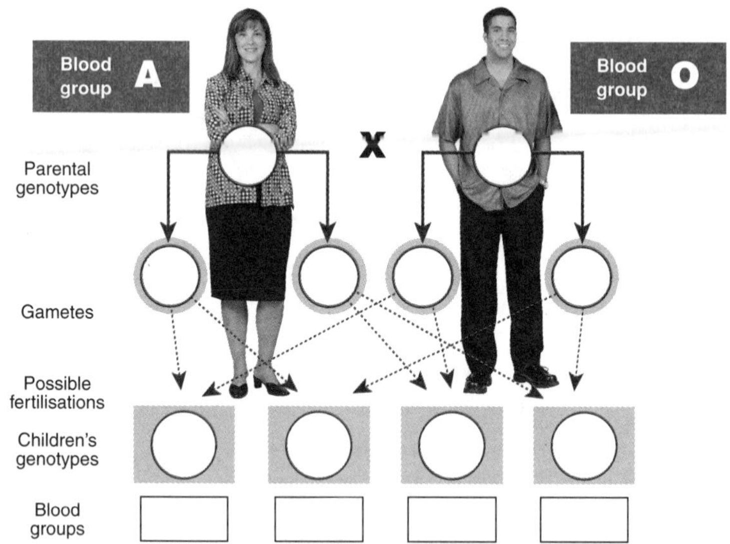
Blood group A
Blood group O
X
Parental genotypes
Gametes
Possible fertilisations
Children's genotypes
Blood groups

4. In a court case involving a paternity dispute (i.e. who is the father of a child) a man claims that a male child (blood group **B**) born to a woman is his son and wants custody. The woman claims that he is not the father.

   (a) If the man has a blood group **O** and the woman has a blood group **A**, could the child be his son? Use the diagram on the right to illustrate the genotypes of the three people involved.

   (b) State with reasons whether the man can be correct in his claim:

   ____

   ____

   ____

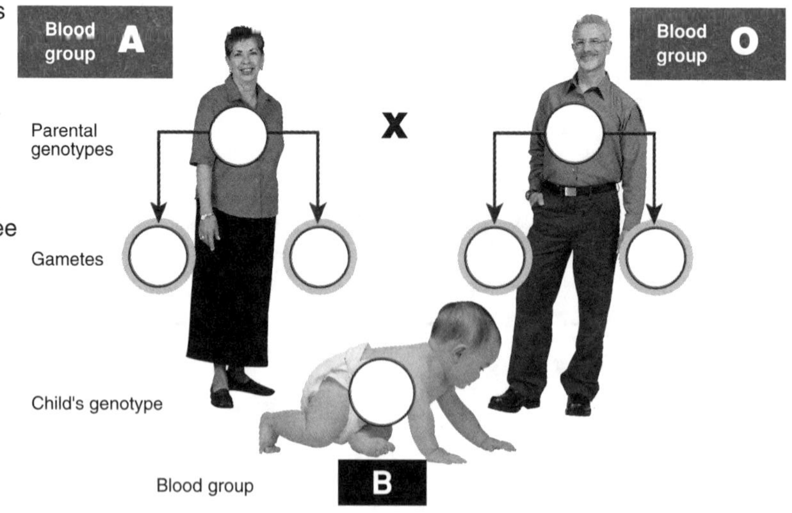
Blood group A
Blood group O
X
Parental genotypes
Gametes
Child's genotype
Blood group
B

5. Give the blood groups which are possible for children of the following parents (remember that in some cases you don't know if the parent is homozygous or heterozygous).

   (a) Mother is group **AB** and father is group **O**: ______________________________

   (b) Father is group **B** and mother is group **A**: ______________________________

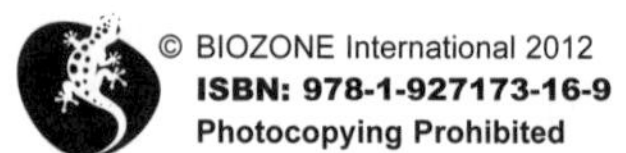

ISBN: 978-1-927173-16-9

Problems Involving Monohybrid Inheritance

The following problems involve Mendelian crosses. The alleles involved are associated with various phenotypic traits controlled by a single gene. The problems are to give you practice in problem solving using Mendelian genetics.

1. A dominant gene (**W**) produces wire-haired texture in dogs; its recessive allele (**w**) produces smooth hair. A group of heterozygous wire-haired individuals are crossed and their F_1 progeny are then test-crossed. Determine the expected genotypic and phenotypic ratios among the **test cross** progeny:

2. In sheep, black wool is due to a recessive allele (**b**) and white wool to its dominant allele (**B**). A white ram is crossed to a white ewe. Both animals carry the black allele (b). They produce a white ram lamb, which is then back crossed to the female parent. Determine the probability of the **back cross** offspring being black:

3. A recessive allele, a, is responsible for albinism, an inability to produce or deposit melanin in tissues. Humans and a variety of other animals can exhibit this phenotype. In each of the following cases, determine the possible genotypes of the mother and father, and of their children:

   (a) Both parents have normal phenotypes; some of their children are albino and others are unaffected:

   (b) Both parents are albino and have only albino children:

   (c) The woman is unaffected, the man is albino, and they have one albino child and three unaffected children:

Heredity

4. Chickens with shortened wings and legs are called creepers. When creepers are mated to normal birds, they produce creepers and normals with equal frequency. When creepers are mated to creepers they produce two creepers to one normal. Crosses between normal birds produce only normal progeny. Explain these results:

5. In a dispute over parentage, the mother of a child with blood group O identifies a male with blood group A as the father. The mother is blood group B. Draw Punnett squares to show possible genotype/phenotype outcomes to determine if the male is the father and the reasons (if any) for further dispute:

ISBN: 978-1-927173-16-9

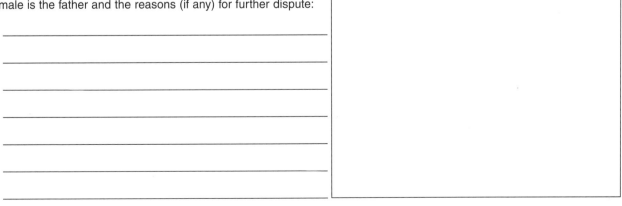

Dihybrid Cross

A cross (or mating) between two organisms where the inheritance patterns of **two genes** are studied is called a **dihybrid cross** (compared with the study of one gene in a monohybrid cross). There are a greater number of gamete types produced when two genes are considered (four types). Remember that the genes described are being carried by separate chromosomes and are sorted independently of each other during meiosis (that is why you get four kinds of gamete). The two genes below control two unrelated characteristics, **hair color** and **coat length**. Black and short are dominant.

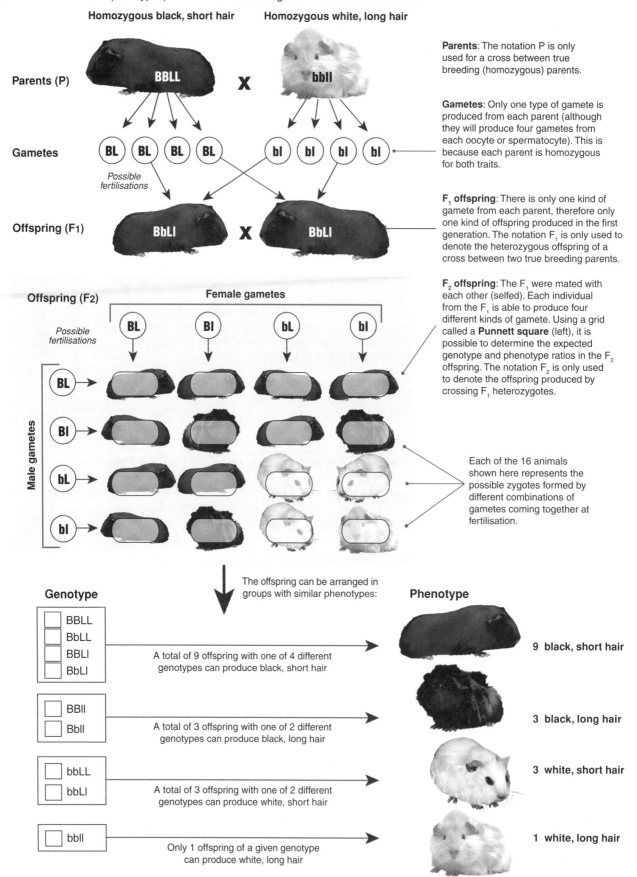

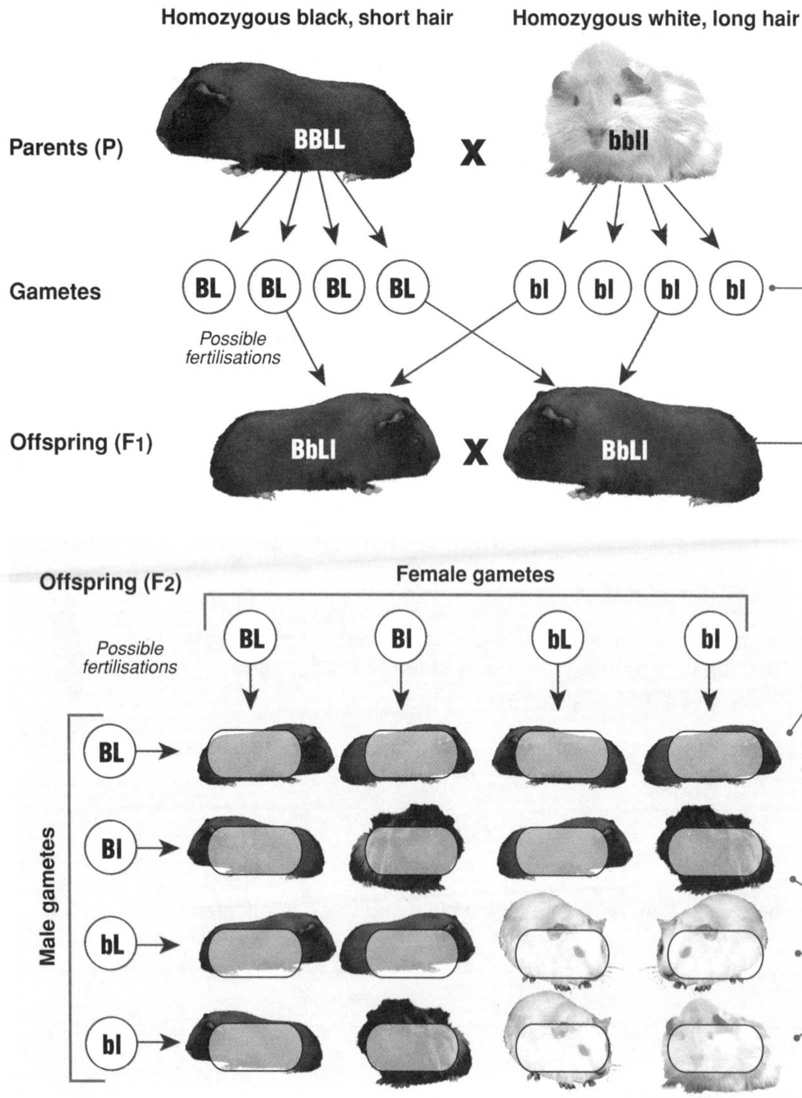

Homozygous black, short hair **Homozygous white, long hair**

Parents (P): BBLL X bbll

Gametes: BL, BL, BL, BL; bl, bl, bl, bl

Possible fertilisations

Offspring (F1): BbLl X BbLl

Parents: The notation P is only used for a cross between true breeding (homozygous) parents.

Gametes: Only one type of gamete is produced from each parent (although they will produce four gametes from each oocyte or spermatocyte). This is because each parent is homozygous for both traits.

F_1 offspring: There is only one kind of gamete from each parent, therefore only one kind of offspring produced in the first generation. The notation F_1 is only used to denote the heterozygous offspring of a cross between two true breeding parents.

Offspring (F2)

Female gametes: BL, Bl, bL, bl

Possible fertilisations

Male gametes: BL, Bl, bL, bl

F_2 offspring: The F_1 were mated with each other (selfed). Each individual from the F_1 is able to produce four different kinds of gamete. Using a grid called a **Punnett square** (left), it is possible to determine the expected genotype and phenotype ratios in the F_2 offspring. The notation F_2 is only used to denote the offspring produced by crossing F_1 heterozygotes.

Each of the 16 animals shown here represents the possible zygotes formed by different combinations of gametes coming together at fertilisation.

The offspring can be arranged in groups with similar phenotypes:

Genotype		Phenotype
☐ BBLL ☐ BbLL ☐ BBLl ☐ BbLl	A total of 9 offspring with one of 4 different genotypes can produce black, short hair	**9 black, short hair**
☐ BBll ☐ Bbll	A total of 3 offspring with one of 2 different genotypes can produce black, long hair	**3 black, long hair**
☐ bbLL ☐ bbLl	A total of 3 offspring with one of 2 different genotypes can produce white, short hair	**3 white, short hair**
☐ bbll	Only 1 offspring of a given genotype can produce white, long hair	**1 white, long hair**

1. Complete the Punnett square above and use it to fill in the number of each genotype in the boxes (above left).

Related activities: Basic Genetic Crosses

ISBN: 978-1-927173-16-9

Inheritance of Linked Genes

Linkage refers to genes that are located on the same chromosome. Linked genes tend to be inherited together, and the extent of crossing over depends on how close together they are on the chromosome. Linkage generally reduces the variety of offspring that can be produced. In genetic crosses, linkage is indicated when a greater proportion of the offspring resulting from a cross are of the parental type (than would be expected if the alleles were assorting independently). If the genes in question had been on separate chromosomes, there would have been more genetic variation in the gametes and therefore in the offspring. Note that in the hypothetical example below there are only two possible genotype outcomes, both the same as the parent type. If the alleles were assorting independently (on different chromosomes) there would be four outcomes.

Overview of Linkage

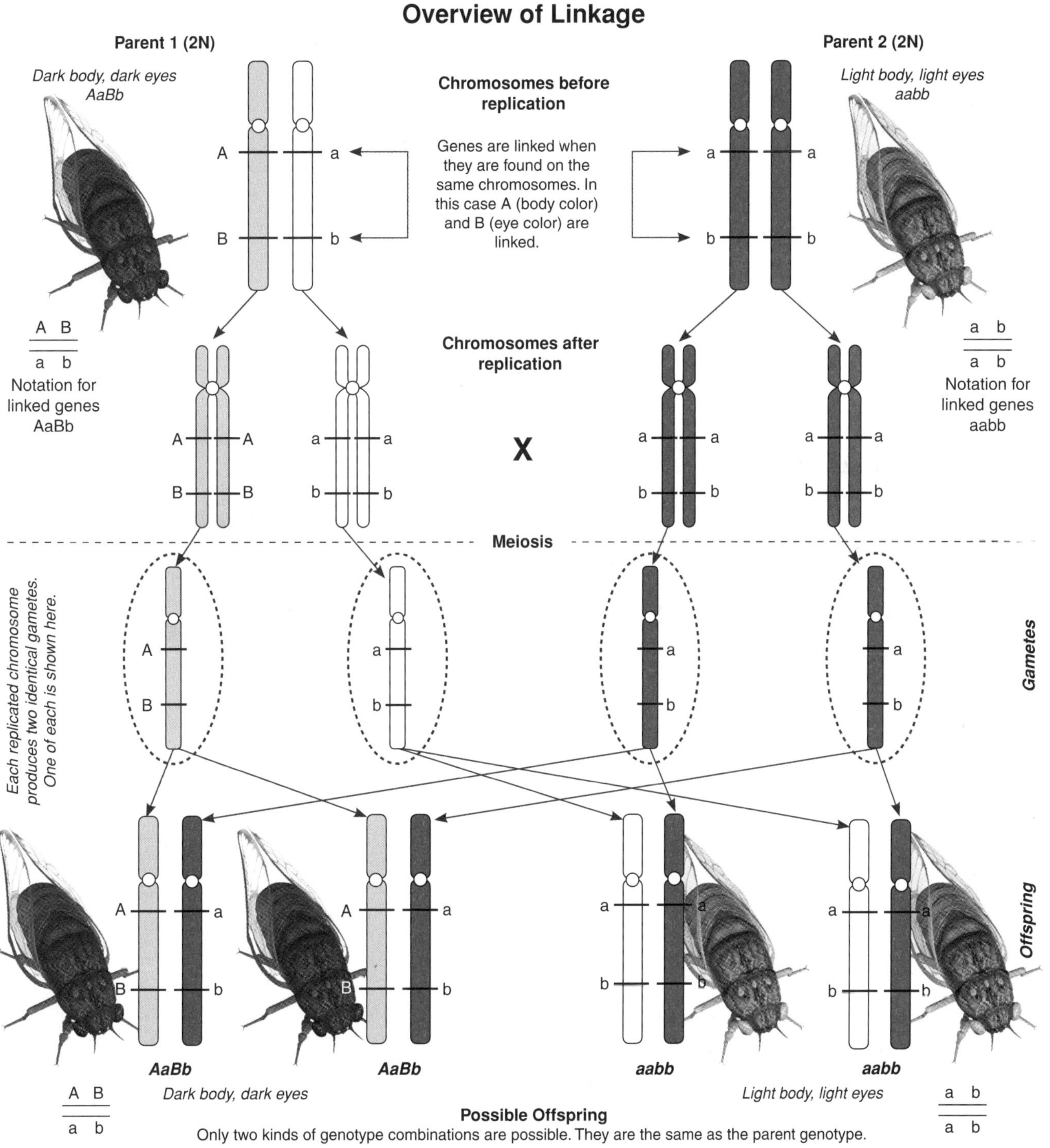

Parent 1 (2N)
Dark body, dark eyes
AaBb
Parent 2 (2N)
Light body, light eyes
aabb
Chromosomes before replication
Genes are linked when they are found on the same chromosomes. In this case A (body color) and B (eye color) are linked.
A B / a b
Notation for linked genes AaBb
a b / a b
Notation for linked genes aabb
Chromosomes after replication
X
Meiosis
Each replicated chromosome produces two identical gametes. One of each is shown here.
Gametes
Offspring
AaBb
AaBb
aabb
aabb
Dark body, dark eyes
Light body, light eyes

Possible Offspring
Only two kinds of genotype combinations are possible. They are the same as the parent genotype.

1. What is the effect of **linkage** on the inheritance of genes? ______________________________

2. Explain how linkage decreases the amount of genetic variation in the offspring: ______________________________

ISBN: 978-1-927173-16-9

An Example of Linked Genes in *Drosophila*

The genes for wing shape and body color are linked (they are on the same chromosome).

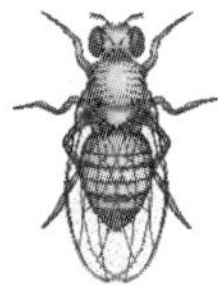

Parent	**Wild type female**	**Mutant male**
Phenotype	Straight wing Gray body	Curled wing Ebony body
Genotype	Cucu Ebeb	cucu ebeb

Linkage

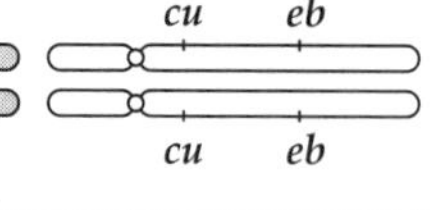

Cu Eb
cu eb
cu eb
cu eb

Meiosis

Gametes from female fly (N) ***Gametes from male fly (N)***

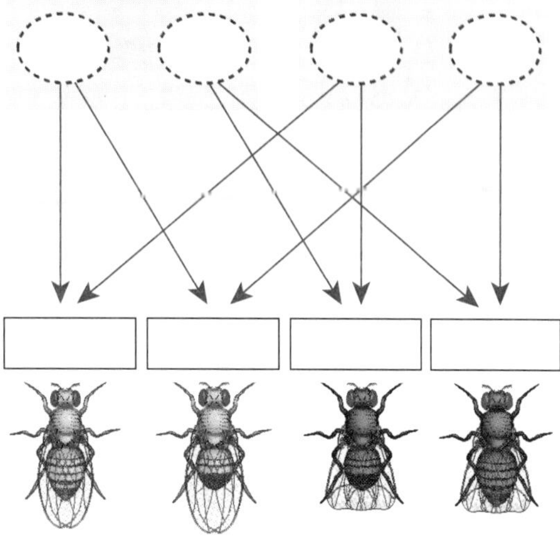

Sex of offspring is irrelevant in this case

Contact **Newbyte Educational Software** for details of their superb *Drosophila Genetics* software package which includes coverage of linkage and recombination. *Drosophila* images © Newbyte Educational Software.

Drosophila and linked genes

In the example shown left, wild type alleles are dominant and are given an upper case symbol of the mutant phenotype (Cu or Eb). This symbology used for *Drosophila* departs from the convention of using the dominant gene to provide the symbol. This is necessary because there are many mutant alternative phenotypes to the wild type (e.g. curled and vestigial wings). A lower case symbol of the wild type (e.g. ss for straight wing) would not indicate the mutant phenotype involved.

Drosophila melanogaster is known as a model organism. Model organisms are used to study particular biological phenomena, such as mutation. *Drosophila melanogaster* is particularly useful because it produces such a wide range of heritable mutations. Its short reproduction cycle, high offspring production, and low maintenance make it ideal for studying in the lab.

W
KTBN

Drosophila melanogaster examples showing variations in eye and body color. The wild type is marked with a w in the photo above.

3. Complete the linkage diagram above by adding the gametes in the ovals and offspring genotypes in the rectangles.

4. (a) List the possible genotypes in the offspring (above, left) if genes Cu and Eb had been on **separate chromosomes**:

   ____

   (b) If the female *Drosophila* had been homozygous for the dominant wild type alleles (CuCu EbEb), state:

   The genotype(s) of the F_1: ______________ The phenotype(s) of the F_1: ______________

5. A second pair of *Drosophila* are mated. The female genotype is Vgvg EbEb (straight wings, gray body), while the male genotype is vgvg ebeb (straight wings, ebony body). Assuming the genes are linked, carry out the cross and list the genotypes and phenotypes of the offspring. Note vg = vestigial (no) wings:

   ____

   ____

   ____

   The genotype(s) of the F_1: ______________ The phenotype(s) of the F_1: ______________

6. Explain why *Drosophila* are often used as model organisms in the study of genetics: ______________

   ____

   ____

ISBN: 978-1-927173-16-9

Recombination and Dihybrid Inheritance

Genetic recombination refers to the exchange of alleles between homologous chromosomes as a result of **crossing over**. The alleles of parental linkage groups separate and new associations of alleles are formed in the gametes. Offspring formed from these gametes show new combinations of characteristics and are known as **recombinants** (offspring with genotypes unlike either parent). The proportion of recombinants in the offspring can be used to calculate the frequency of recombination (crossover value). These values are fairly constant for any given pair of alleles and can be used to produce gene maps indicating the relative positions of genes on a chromosome. In contrast to linkage, recombination increases genetic variation. Recombination between the alleles of parental linkage groups is indicated by the appearance of non-parental types in the offspring, although not in the numbers that would be expected had the alleles been on separate chromosomes (independent assortment).

Overview of Recombination

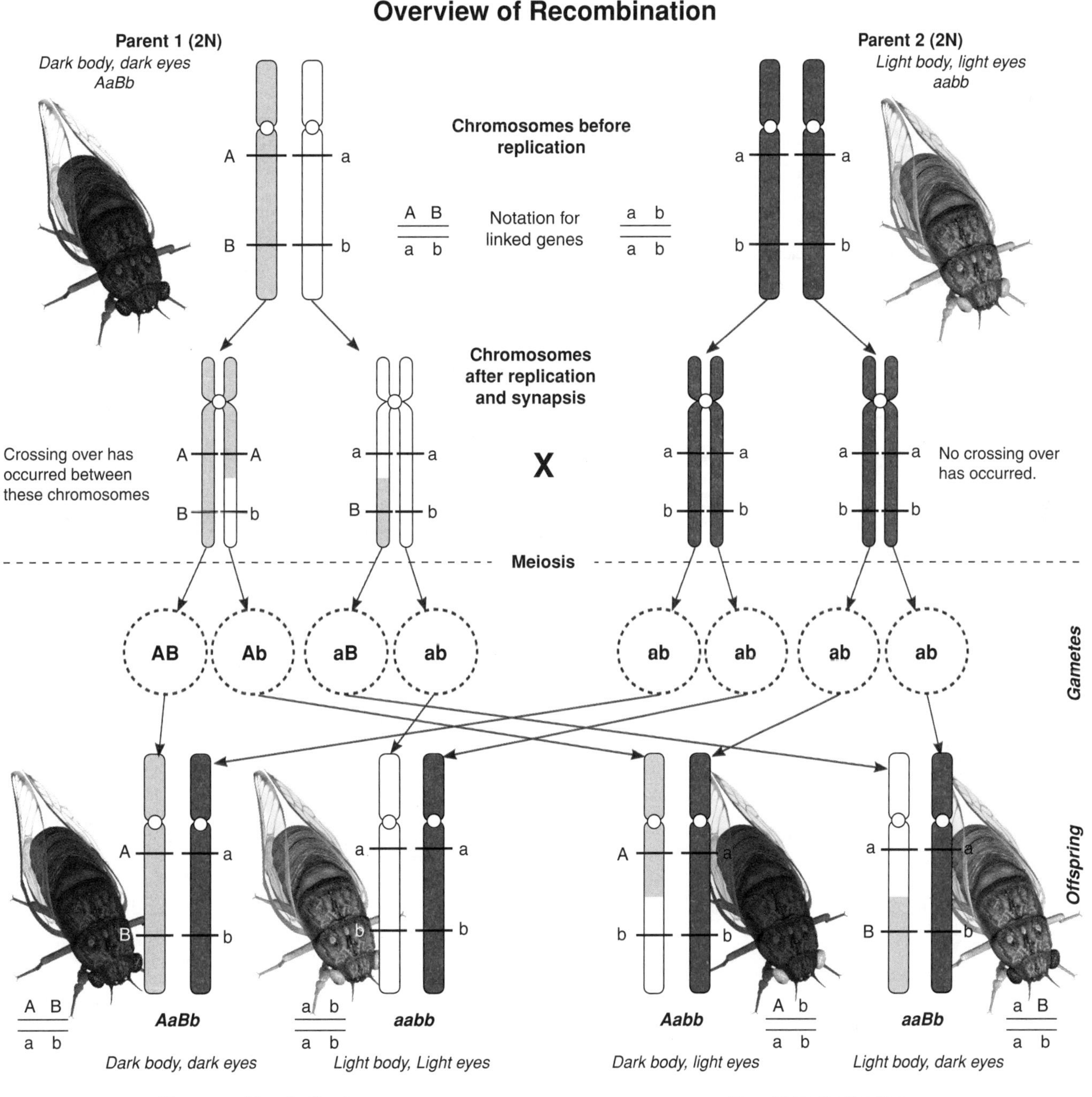
Parent 1 (2N)
Dark body, dark eyes
AaBb
Parent 2 (2N)
Light body, light eyes
aabb
Chromosomes before replication
Notation for linked genes
A B / a b
a b / a b
Chromosomes after replication and synapsis
Crossing over has occurred between these chromosomes
No crossing over has occurred.
X
Meiosis
Gametes
AB
Ab
aB
ab
ab
ab
ab
ab
Offspring
AaBb
aabb
Aabb
aaBb
Dark body, dark eyes
Light body, Light eyes
Dark body, light eyes
Light body, dark eyes
Heredity

Non-recombinant offspring

These two offspring exhibit allele combinations that are expected as a result of independent assortment during meiosis. Also called parental types.

Recombinant offspring

These two offspring exhibit unexpected allele combinations. They can only arise if one of the parent's chromosomes has undergone crossing over.

Possible Offspring

Offspring with four kinds of genotype combinations are produced instead of the two kinds expected (AaBb and aabb) if no crossing over had occurred.

1. Describe the effect of **recombination** on the inheritance of genes: ______

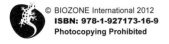

______

______

______

ISBN: 978-1-927173-16-9

An Example of Recombination

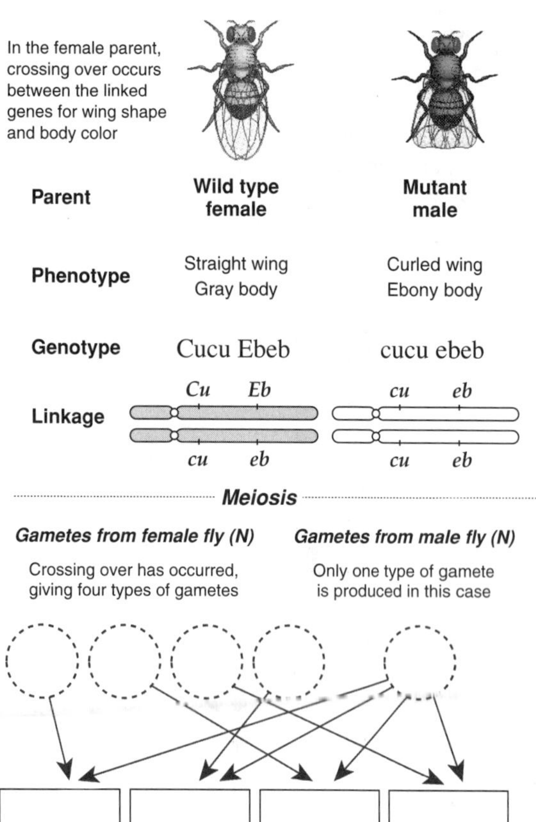
In the female parent, crossing over occurs between the linked genes for wing shape and body color
Parent
Wild type female
Mutant male
Phenotype
Straight wing
Gray body
Curled wing
Ebony body
Genotype
Cucu Ebeb
cucu ebeb
Linkage
Cu Eb
cu eb
cu eb
cu eb
Meiosis
Gametes from female fly (N)
Crossing over has occurred, giving four types of gametes
Gametes from male fly (N)
Only one type of gamete is produced in this case

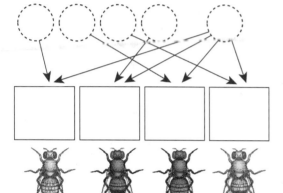

Non-recombinant offspring **Recombinant offspring**

The sex of the offspring is irrelevant in this case

Contact **Newbyte Educational Software** for details of their superb *Drosophila Genetics* software package which includes coverage of linkage and recombination. *Drosophila* images © Newbyte Educational Software.

The cross (left) uses the same genotypes as the previous activity but, in this case, crossing over occurs between the alleles in a linkage group in one parent. The symbology used is the same.

Recombination Produces Variation

If crossing over does not occur, the possible combinations in the gametes remains limited. Crossing over and recombination increase the variation in the offspring. In humans, even without crossing over, there are approximately $(2^{23})^2$ or 70 trillion genetically different zygotes that could form for every couple. Taking crossing over and recombination into account produces $(4^{23})^2$ or 5000 trillion trillion genetically different zygotes for every couple.

Family members may resemble each other, but they'll never be identical (except for identical twins).

Using Recombination

Analyzing recombination gave geneticists a way to map the genes on a chromosome. Crossing over is less likely to occur between genes that are close together on a chromosome than between genes that are far apart. By counting the number of offspring of each phenotype, you can calculate the **frequency of recombination**. The higher the frequency of recombination between two genes, the further apart they must be on the chromosome.

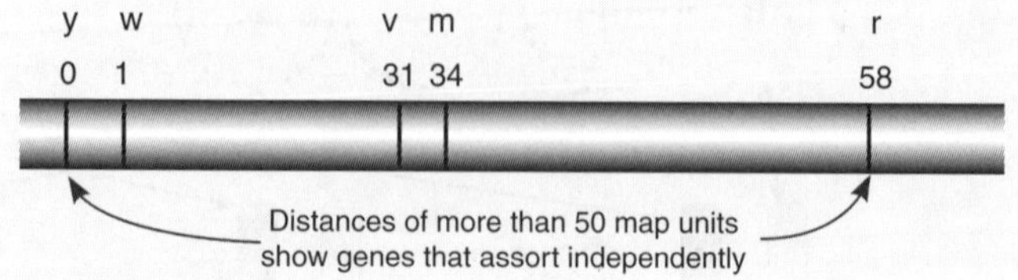
y w v m r
0 1 31 34 58
Distances of more than 50 map units show genes that assort independently

Map of the X chromosome of *Drosophila*, showing the relative distances between five different genes (in map units).

2. Complete the recombination diagram above, adding the gametes in the ovals and offspring genotypes and phenotypes in the rectangles:

3. Explain how recombination increases the amount of genetic variation in offspring: ______

4. Explain why it is not possible to have a recombination frequency of greater than 50% (half recombinant progeny):

5. A second pair of *Drosophila* are mated. The female is Cucu YY (straight wing, gray body), while the male is Cucu yy (straight wing, yellow body). Assuming recombination, perform the cross and list the offspring genotypes and phenotypes:

ISBN: 978-1-927173-16-9

Detecting Linkage in Dihybrid Inheritance

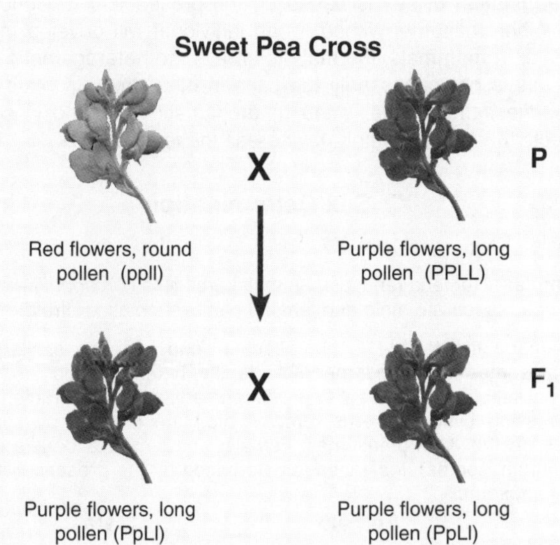
Sweet Pea Cross
X
P
Red flowers, round pollen (ppll)
Purple flowers, long pollen (PPLL)
X
F_1
Purple flowers, long pollen (PpLl)
Purple flowers, long pollen (PpLl)

In sweet peas, purple flowers (P) are dominant to red (p), and long pollen grains (L) are dominant to round (l). Bateson and Punnett crossed pure breeding red-flowered and round-grain sweet peas with pure breeding purple-flowered and long-grained sweet peas. If these genes were unlinked, the outcome of an F_1 cross should have been a 9:3:3:1 ratio.

Table 1: Sweet Pea Cross Results

	Observed	Expected
Purple long (P_L_)	284	
Purple round (P_ll)	21	
Red long (ppL_)	21	
Red round (ppll)	55	
Total	381	381

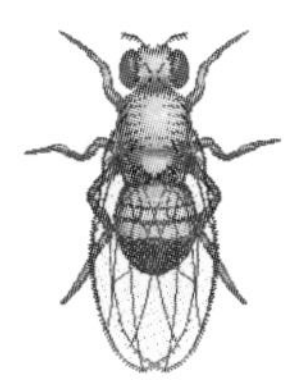

X

Red eyed normal winged (pr^+pr vg^+vg)

Purple eyed vestigial winged (prpr vgvg)

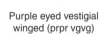

Morgan performed experiments to investigate linked genes in *Drosophila*. He crossed a heterozygous red-eyed normal-winged (pr^+pr vg^+vg) fly with a homozygous purple-eyed vestigial-winged (prpr vgvg) fly. The table (below) shows the outcome of the cross.

Table 2: *Drosophila* Cross Results

Genotype	Observed	Expected	Gamete type
pr^+pr vg^+vg	1339	710	Parental
prpr vg^+vg	152		
pr^+pr vgvg	154		
prpr vgvg	1195		
Total	2840	2840	

Shortly after the rediscovery of Mendel's work early in the 20th century, it became apparent that his ratios of 9:3:3:1 for heterozygous dihybrid crosses did not always hold true. Experiments carried out on sweet peas by William Bateson and Reginald Punnett, and on *Drosophila* by Thomas Hunt Morgan, found that there appeared to be some kind of coupling between genes, but that the coupling did not follow any genetic relationship known at the time.

1. Fill in the missing numbers in the **expected** column of **Table 1**, remembering that a 9:3:3:1 ratio is expected:

2. (a) Fill in the missing numbers in the **expected** column of **Table 2**, remembering that a 1:1:1:1 ratio is expected:

   (b) Add the gamete type (parental/recombinant) to the gamete type column in Table 2:

   (c) What type of cross did Morgan perform here?

3. (a) From the pedigree chart below, determine if nail-patella syndrome is dominant or recessive, giving reasons for your choice:

   (b) What evidence is there that nail-patella syndrome is linked to the ABO blood group locus?

   (c) Suggest a likely reason why individual III-3 is not affected despite carrying the B allele:

Pedigree for Nail-Patella Syndrome

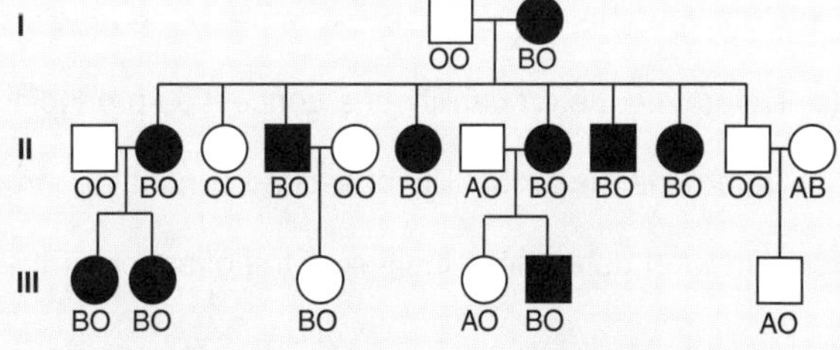
I
OO BO
II
OO BO OO BO OO BO AO BO BO BO OO AB
III
BO BO BO AO BO AO

Individual with nail-patella syndome 

Blood types OO, BO, AO, AB

Linked genes can be detected by pedigree analysis. The diagram above shows the pedigree for the inheritance of nail-patella syndrome, which results in small, poorly developed nails and kneecaps in affected people. Other body parts such as elbows, chest, and hips can also be affected. The nail-patella syndrome gene is linked to the ABO blood group locus.

Heredity

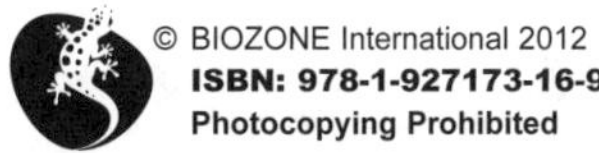

ISBN: 978-1-927173-16-9

Sex Determination

The determination of the sex of an organism is controlled in most cases by the sex chromosomes provided by each parent. These have evolved to regulate the ratios of males and females produced and preserve the genetic differences between the sexes. In humans, males are the **heterogametic sex** because each somatic cell has one X and one Y chromosome. The determination of sex is based on the presence or absence of the Y chromosome; without it, an individual will develop into a female. In mammals, the male is always the heterogametic sex, but this is not necessarily the case in other taxa. In birds and butterflies, the female is the heterogametic sex, and in some insects the male is simply X whereas the female is XX.

Sex Determination in Humans

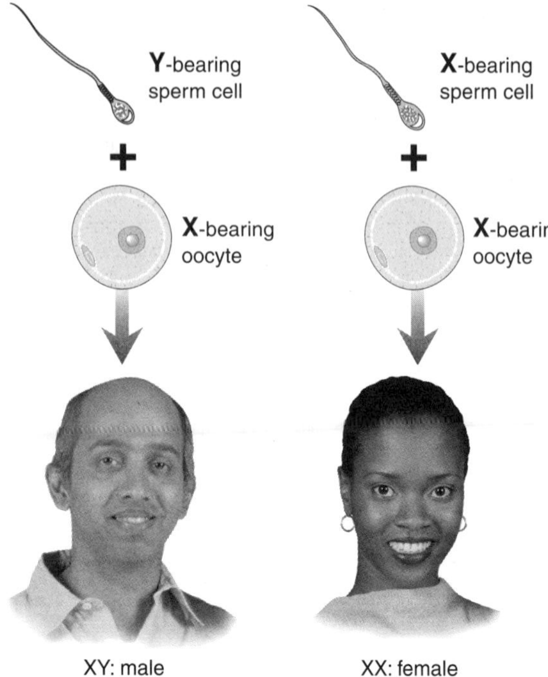
Y-bearing sperm cell
X-bearing sperm cell
+
+
X-bearing oocyte
X-bearin
oocyte
XY: male
XX: female

XY Sex Determination

Female: **XX** Male: **XY**

Examples: Humans (and all mammals), fruit flies (*Drosophila*), some dioecious (separate male and female) plants such as kiwifruit.

In humans the female is homogametic and has two similar sex chromosomes (XX) and the male is the heterogametic sex with two unlike chromosomes (XY). The primary sexual characteristics (possessing ovaries, uterus, breasts etc.) are initiated by special genes on the X chromosomes. Females must have a double dose (2X chromosomes). Maleness is determined by the presence of the Y chromosome.

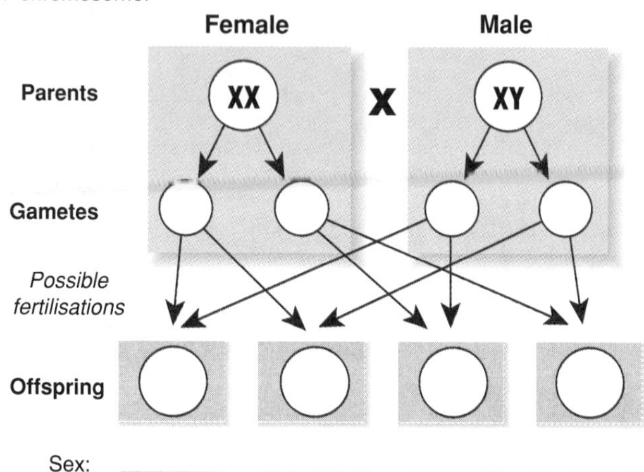
Female
Male
Parents
XX
X
XY
Gametes
Possible fertilisations
Offspring
Sex:

X and Y Chromosomes

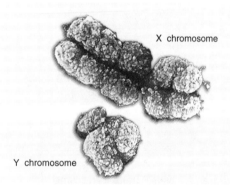

X chromosome

Y chromosome

The X chromosome is much larger than the Y and contains many more genes. As these are not present in the Y chromosome, they will be expressed in the male whether they are dominant or recessive.

The Y chromosome has lost approximately 96% of its original estimated genetic material, currently possessing 86 genes and producing just 23 proteins.

One of the more important genes located on the Y chromosome is the SRY gene (sex determining region Y) which produces the TDF protein (or SRY protein) that initiates determination of the male sex.

1. (a) Complete the diagrams above, to show the resulting gametes, genotype and sex of the offspring:

   (b) Determine the probability of a conception producing a male child: __________

   (c) Determine the probability of second conception producing a female child: __________

2. Explain what determines the sex of the offspring at the moment of conception in humans: __________

3. Explain why in males, many genes on the X chromosome will be expressed whether they are dominant or recessive:

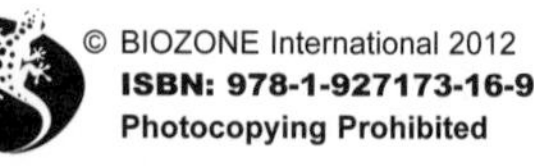

ISBN: 978-1-927173-16-9

Sex Linkage

Sex linkage is a special case of linkage occurring when a gene is located on a sex chromosome (usually the X). The result of this is that the character encoded by the gene is usually seen only in one sex (the heterogametic sex) and occurs rarely in the homogametic sex. In humans, recessive sex linked genes are responsible for a number of heritable disorders in males, e.g. hemophilia. Women who have the recessive alleles on their chromosomes are said to be **carriers**.

Hemophilia is an inherited genetic disorder linked to the X-chromosome that results in ineffective blood clotting when a blood vessel is damaged. The most common type, hemophilia A, occurs in 1 in 5000 male births. Any male who carries the gene will express the phenotype. Hemophilia is extremely rare in women.

1. A couple wish to have children. The woman knows she a carrier for hemophilia. The man is not a hemophiliac. Use the notation X_h for hemophilia and X_H for the dominant allele to complete the diagram on the right including the parent genotypes, gametes and possible fertilizations. Summarize the genotypes and phenotypes in the table below.

	Genotypes	Phenotypes
Male children		
Female children		

2. (a) A second couple also wish to have children. The woman knows her maternal grandfather was a hemophiliac but neither her mother or father were. Determine the probability she is a carrier (X_HX_h) Use the Punnett squares, right, to help you:

____

____

(b) The man is a normal non-hemophiliac male. Determine the probability that their first male child will have hemophilia. Use the Punnett squares, right, to help you:

____

____

3. The gene for color vision is carried on the X chromosome. If the gene is faulty color blindness (X_b) will occur in males. Color blindness occurs in about 8% of males but in less than 1% of females.

A color blind man has children with woman who is not color blind. The couple have four children. Their phenotypes are: 1 non color blind son, 1 color blind son, 2 non color blind daughters. Describe the mother's:

(a) Genotype: ______________________

(b) Phenotype: ______________________

(c) Identify the genotype not possessed by any of the children:

____

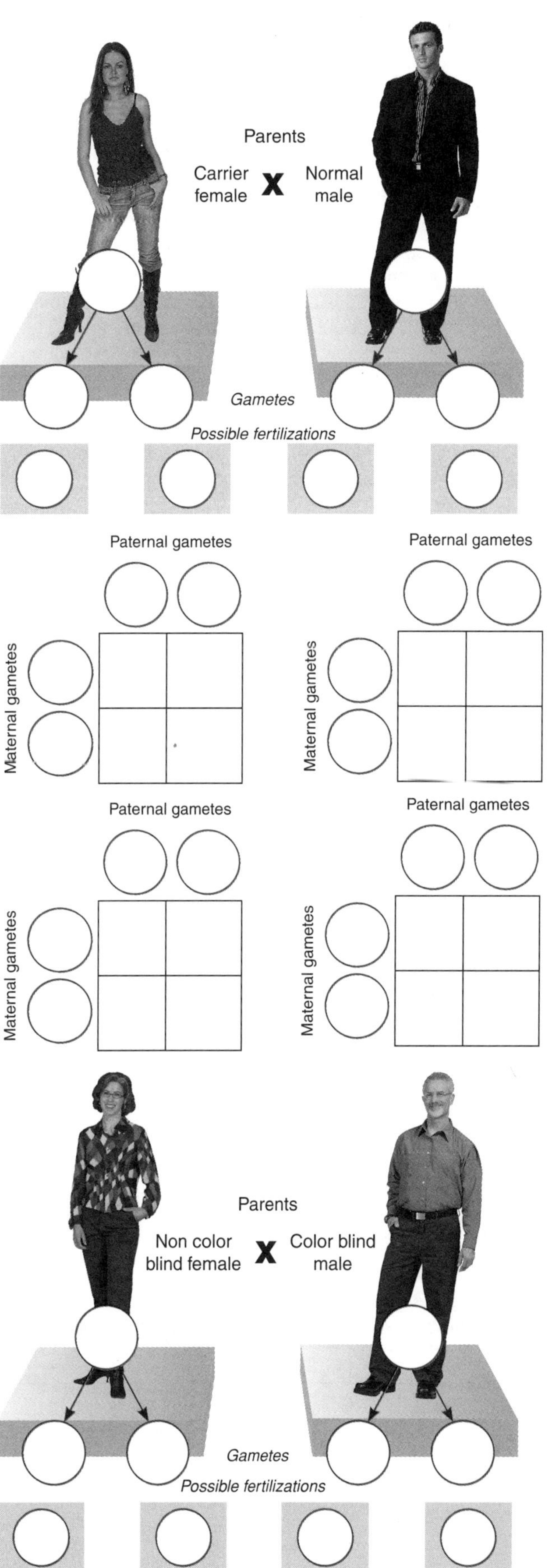
Parents
Carrier female
X
Normal male
Gametes
Possible fertilizations
Paternal gametes
Maternal gametes
Paternal gametes
Maternal gametes
Paternal gametes
Maternal gametes
Paternal gametes
Maternal gametes
Parents
Non color blind female
X
Color blind male
Gametes
Possible fertilizations

Heredity

ISBN: 978-1-927173-16-9

Related activities: Inheritance Patterns, Pedigree Analysis
Weblinks: X Linked Inheritance

Dominant allele in humans

A rare form of rickets in humans is determined by a **dominant** allele of a gene on the **X chromosome** (it is not found on the Y chromosome). This condition is not successfully treated with vitamin D therapy. The allele types, genotypes, and phenotypes are as follows:

Allele types		Genotypes		Phenotypes
X_R = affected by rickets		X_RX_R, X_RX	=	Affected female
X = normal		X_RY	=	Affected male
		XX, XY	=	Normal female, male

As a genetic counsellor you are presented with a married couple where one of them has a family history of this disease. The husband is affected by this disease and the wife is normal. The couple, who are thinking of starting a family, would like to know what their chances are of having a child born with this condition. They would also like to know what the probabilities are of having an affected boy or affected girl. Use the symbols above to complete the diagram right and determine the probabilities stated below (expressed as a proportion or percentage).

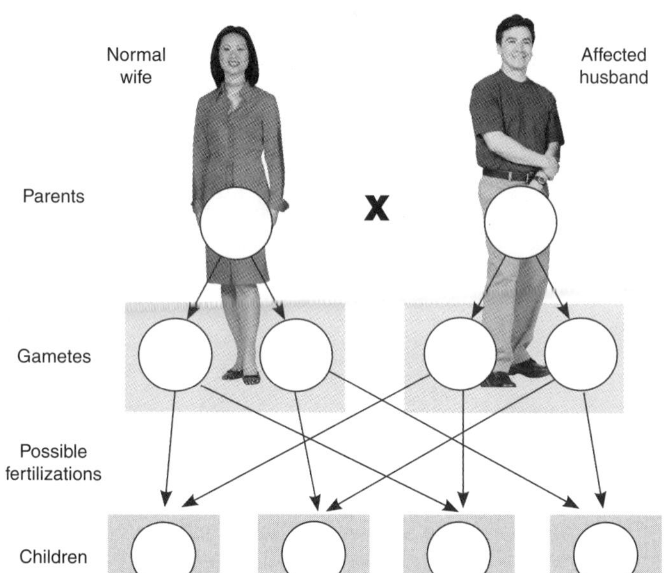
Normal wife
Affected husband
Parents
X
Gametes
Possible fertilizations
Children

4. Determine the probability of having:

   (a) Affected children: ____________

   (b) An affected girl: ____________

   (c) An affected boy: ____________

Another couple with a family history of the same disease also come in to see you to obtain genetic counseling. In this case the husband is normal and the wife is affected. The wife's father was not affected by this disease. Determine what their chances are of having a child born with this condition. They would also like to know what the probabilities are of having an affected boy or affected girl. Use the symbols above to complete the diagram right and determine the probabilities stated below (expressed as a proportion or percentage).

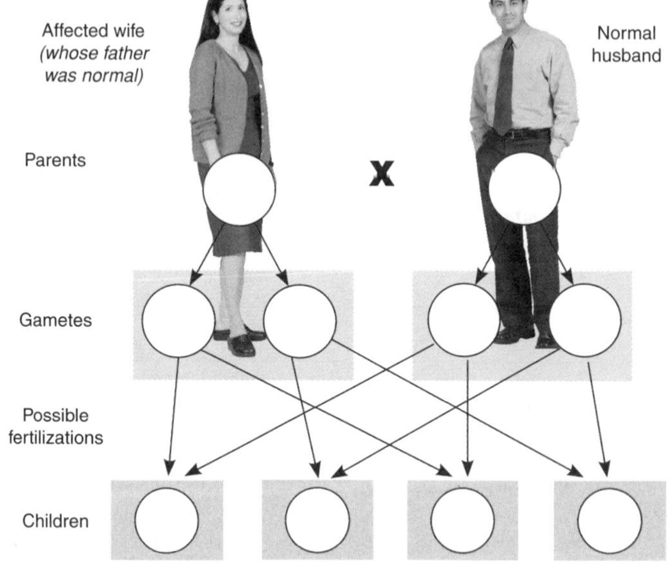
Affected wife
(whose father was normal)
Normal husband
Parents
X
Gametes
Possible fertilizations
Children

5. Determine the probability of having:

   (a) Affected children: ____________

   (b) An affected girl: ____________

   (c) An affected boy: ____________

6. Describing examples other than those above, discuss the role of **sex linkage** in the inheritance of genetic disorders:

____

____

____

____

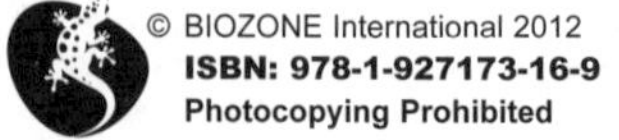

ISBN: 978-1-927173-16-9

Inheritance Patterns

Complete the following monohybrid crosses for different types of inheritance patterns in humans: autosomal recessive, autosomal dominant, sex linked recessive, and sex linked dominant inheritance.

1. **Inheritance of autosomal recessive traits**
   Example: *Red-green color blindness*

   Red-green color blindness is a sex-linked disorder. Males require just one recessive allele to be color blind.

   Using the codes: X_BX_B (normal) X_BY Nomal male
   X_BX_b (carrier) X_bY Color blind male
   X_bX_b (color blind female)

   (a) Enter the parent phenotypes and complete the Punnett square for a cross between two carrier genotypes.
   (b) Give the ratios for the phenotypes from this cross.

   Phenotype ratios: ______________________________

   ____

2. **Inheritance of autosomal dominant traits**
   Example: *Woolly hair*

   Woolly hair is inherited as an autosomal dominant allele. Each affected individual will have at least one affected parent.

   Using the codes: **WW** (woolly hair)
   Ww (woolly hair, heterozygous)
   ww (normal hair)

   (a) Enter the parent phenotypes and complete the Punnett square for a cross between two heterozygous individuals.
   (b) Give the ratios for the phenotypes from this cross.

   Phenotype ratios: ______________________________

   ____

3. **Inheritance of sex linked recessive traits**
   Example: *Hemophilia*

   Inheritance of hemophilia is sex linked. Males with the recessive (hemophilia) allele, are affected. Females can be carriers.

   Using the codes: **XX** (normal female)
   XX_h (carrier female)
   X_hX_h (hemophiliac female)
   XY (normal male)
   X_hY (hemophiliac male)

   (a) Enter the parent phenotypes and complete the Punnett square for a cross between a normal male and a carrier female.
   (b) Give the ratios for the phenotypes from this cross.

   Phenotype ratios: ______________________________

   ____

4. **Inheritance of sex linked dominant traits**
   Example: *Sex linked form of rickets*

   A rare form of rickets is inherited on the X chromosome.

   Using the codes: **XX** (normal female); **XY** (normal male)
   X_RX (affected heterozygote female)
   X_RX_R (affected female)
   X_RY (affected male)

   (a) Enter the parent phenotypes and complete the Punnett square for a cross between an affected male and heterozygous female.
   (b) Give the ratios for the phenotypes from this cross.

   Phenotype ratios: ______________________________

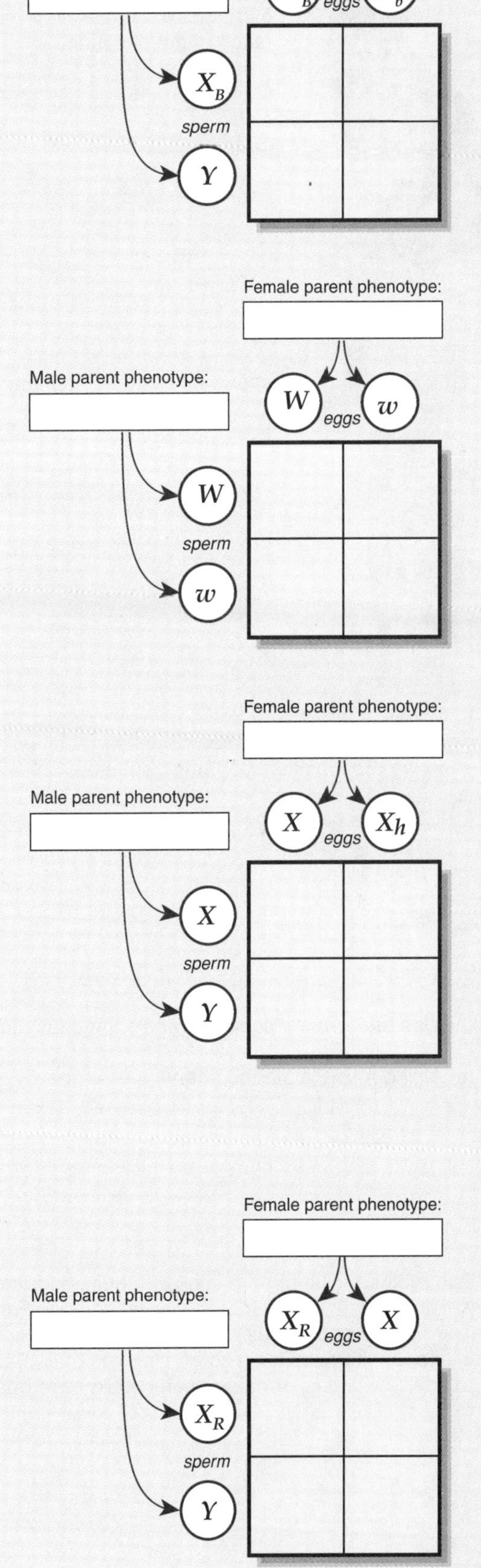

Female parent phenotype:
Male parent phenotype:
X_B eggs X_b
X_B sperm Y
Female parent phenotype:
Male parent phenotype:
W eggs w
W sperm w
Female parent phenotype:
Male parent phenotype:
X eggs X_h
X sperm Y
Female parent phenotype:
Male parent phenotype:
X_R eggs X
X_R sperm Y

Heredity

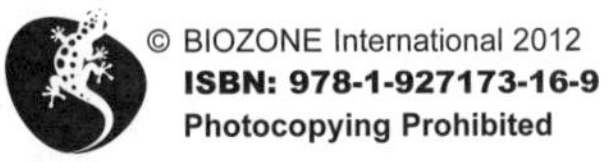

ISBN: 978-1-927173-16-9

Problems Involving Dihybrid Inheritance

Dihybrid inheritance can involve genes in which there is no interaction between them (such as genes for the wrinkliness and color of pea seeds). Other dihybrid crosses can involve genes that do interact with each other and the combination of dominant and recessive alleles can have an outcome on a single phenotype.

1. In cats, the following alleles are present for coat characteristics: black (B), brown (b), short (L), long (l), tabby (T), blotched tabby (tb). Use the information to complete the dihybrid crosses below:

   (a) A black short haired (BBLl) male is crossed with a black long haired (Bbll) female. Determine the genotypic and phenotypic ratios of the offspring:

   Genotype ratio: ______

   Phenotype ratio: ______

   (b) A tabby, short haired male (TtbLl) is crossed with a blotched tabby, short haired (tbtbLl) female. Determine ratios of the offspring:

   Genotype ratio: ______

   Phenotype ratio: ______

2. In rabbits, spotted coat **S** is dominant to solid color **s,** while for coat color, black **B** is dominant to brown **b**. A brown spotted rabbit is mated with a solid black one and all the offspring are black spotted (the genes are not linked).

   (a) State the genotypes:

   Parent 1: ______

   Parent 2: ______

   Offspring: ______

   (b) Use the Punnett square to show the outcome of a cross between the F_1 (the F_2):

   (c) Using ratios, state the phenotypes of the F_2 generation: ______

3. Two mothers give birth to sons at a busy hospital. The son of the first couple has hemophilia, which is a recessive, X-linked disease (see the activity *Inheritance Patterns* in the workbook). Neither parent from couple #1 has the disease. The second couple has a normal (unaffected) son, despite the fact that the father has hemophilia. The two couples challenge the hospital in court, claiming their babies must have been swapped at birth. Your job is to provide expertise to advise whether or not the sons could have been swapped. Explain what would you tell the jury:

ISBN: 978-1-927173-16-9

4. Male domestic cats may be black or orange. Females may be black, tortoise-shell pattern or orange.

 (a) If these colors are sex-linked, explain these results:

 (b) Using appropriate symbols, determine the phenotypes expected in the offspring from the mating of an orange female to a black male:

 (c) In breeding experiments, a reciprocal cross is a breeding experiment designed to test the role of parental sex on a given inheritance pattern. The parents must be true-breeding. In one cross, a female expressing the trait of interest will be crossed with a male not expressing the trait. In the other (reciprocal) cross, a male expressing the trait is crossed with a female that is not expressing the trait. Determine the phenotypes expected in a **reciprocal cross** of (b):

 (d) A certain kind of mating produces females, half of which are black and half of which are tortoiseshell; half the males are black and half are orange. Determine the phenotypes of the parents in such crosses:

 (e) Another kind of mating produces offspring, 1/4 of which are orange males, 1/4 orange females, 1/4 black males, and 1/4 tortoise-shell females. Determine the phenotypes of the parents in such crosses:

5. The sex of fishes is determined by the same XY system as in humans. An allele of one locus on the Y-chromosome for *Lebistes* (guppy) causes a pigmented spot to occur on the dorsal fin. A 'spotted' male fish mates with a female fish that has an unspotted dorsal fin. Describe the phenotypes of the F_1 and F_2 generations from this cross:

6. The Himalayan color-pointed, long-haired cat is a breed developed by crossing a pedigree (true-breeding), uniform-colored, long-haired Persian with a pedigree color-pointed (darker face, ears, paws, and tail) short-haired Siamese. The genes controlling hair coloring and length are on separate chromosomes: uniform color **U**, color pointed **u**, short hair **S**, long hair **s**.

 (a) Using the symbols above, indicate the genotype of each breed below its photograph (above, right).

 (b) State the genotype of the F_1 (Siamese X Persian):

 (c) State the phenotype of the F_1:

 (d) Use the Punnett square to show the outcome of a cross between the F_1 (the F_2):

 (e) State the ratio of the F_2 that would be Himalayan:

 (f) State whether the Himalayan would be true breeding:

 (g) State the ratio of the F_2 that would be color-point, short-haired cats:

 (h) Explain how two F_2 cats of the same phenotype could have different genotypes:

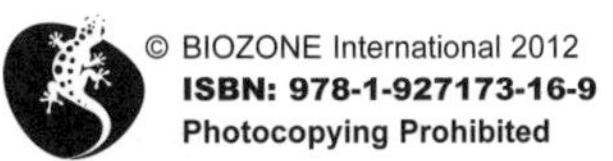

ISBN: 978-1-927173-16-9

Pedigree Analysis

Sample Pedigree Chart

Pedigree charts are a way of graphically illustrating inheritance patterns over a number of generations. They are used to study the inheritance of genetic disorders. The key (below the chart) should be consulted to make sense of the various symbols. Particular individuals are identified by their generation number and their order number in that generation. For example, **II-6** is the sixth person in the second row. The arrow indicates the **propositus**; the person through whom the pedigree was discovered (i.e. who reported the condition).

If the chart on the right were illustrating a human family tree, it would represent three generations: grandparents (I-1 and I-2) with three sons and one daughter. Two of the sons (II-3 and II-4) are identical twins, but did not marry or have any children. The other son (II-1) married and had a daughter and another child (sex unknown). The daughter (II-5) married and had two sons and two daughters (plus a child that died in infancy).

For the particular trait being studied, the grandfather was expressing the phenotype (showing the trait) and the grandmother was a carrier. One of their sons and one of their daughters also show the trait, together with one of their granddaughters.

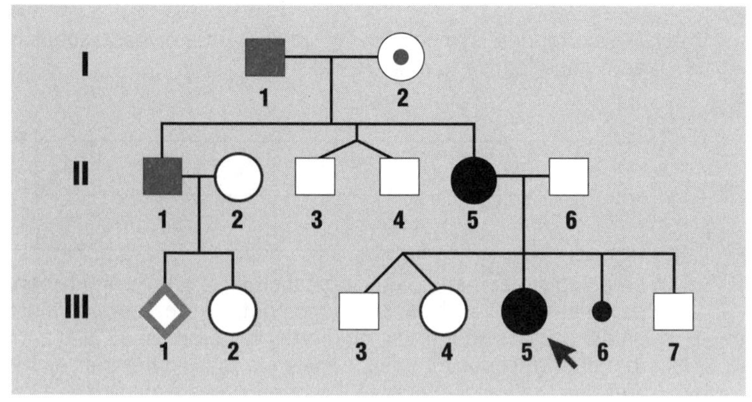
I
1
2
II
1
2
3
4
5
6
III
1
2
3
4
5
6
7

Key to Symbols

Symbol	Meaning	Symbol	Meaning
○	Normal female	◇	Sex unknown
□	Normal male	● (small)	Died in infancy
●	Affected female		Identical twins
■	Affected male		Non-identical twins
⊙	Carrier (heterozygote)		
1, 2, 3	Children (in order of birth)	**I, II, III**	Generations

1. **Pedigree chart of your family**
   Using the symbols in the key above and the example illustrated as a guide, construct a pedigree chart of your own family (or one that you know of) starting with the parents of your mother and/or father on the first line. Your parents will appear on the second line (II) and you will appear on the third line (III). There may be a fourth generation line (IV) if one of your brothers or sisters has had a child. Use a ruler to draw up the chart carefully.

Related activities: *Sex Linkage, Inheritance Patterns*
Web links: *Patterns of Inheritance*

ISBN: 978-1-927173-16-9

2. **Autosomal recessive traits**
Albinos lack pigment in the hair, skin and eyes. This trait is inherited as an autosomal recessive allele (i.e. it is not carried on the sex chromosome).

(a) Write the genotype for each of the individuals on the chart using the following letter codes: **PP** normal skin color; **P-** normal, but unknown if homozygous; **Pp** carrier; **pp** albino.

(b) Why must the parents (II-3) and (II-4) be **carriers** of a **recessive** allele:

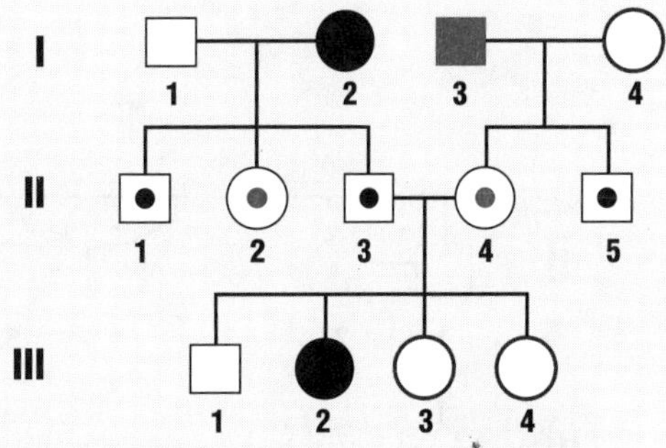
Albinism in humans
I
1
2
3
4
II
1
2
3
4
5
III
1
2
3
4

3. **Sex linked recessive traits**
Hemophilia is a disease where blood clotting is affected. A person can die from a simple bruise (which is internal bleeding). The clotting factor gene is carried on the X chromosome.

(a) Write the genotype for each of the individuals on the chart using the codes: **XY** normal male; X_hY affected male; **XX** normal female; X_hX female carrier; X_hX_h affected female:

(b) Why can males never be carriers?

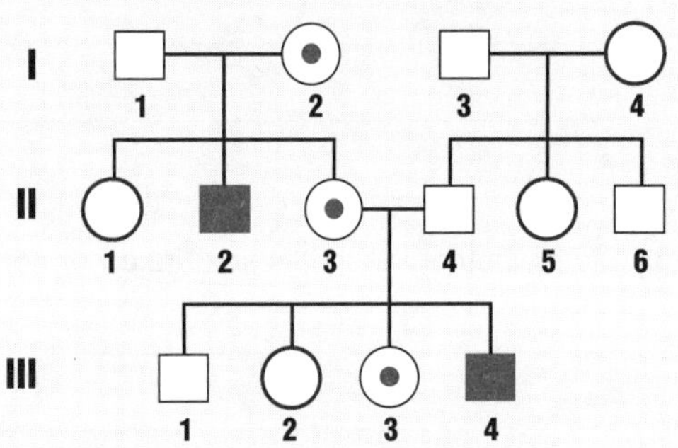
Hemophilia in humans
I
1
2
3
4
II
1
2
3
4
5
6
III
1
2
3
4

4. **Autosomal dominant traits**
An unusual trait found in some humans is woolly hair (not to be confused with curly hair). Each affected individual will have at least one affected parent.

(a) Write the genotype for each of the individuals on the chart using the following letter codes:
WW woolly hair; **Ww** woolly hair (heterozygous); **W-** woolly hair, but unknown if homozygous; **ww** normal hair

(b) Describe a feature of this inheritance pattern that suggests the trait is the result of a **dominant** allele:

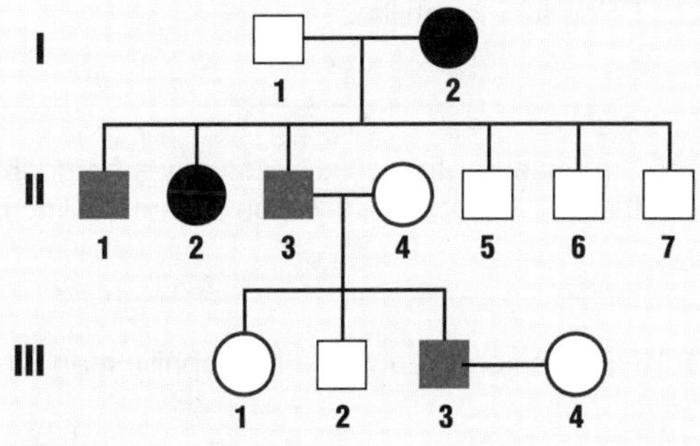
Woolly hair in humans
I
1
2
II
1
2
3
4
5
6
7
III
1
2
3
4

5. **Sex linked dominant traits**
A rare form of rickets is inherited on the X chromosome. All daughters of affected males will be affected. More females than males will show the trait.

(a) Write the genotype for each of the individuals on the chart using the following letter codes:
XY normal male; X_RY affected male; **XX** normal female; X_{R-} female (unknown if homozygous); X_RX_R affected female.

(b) Why will more females than males be affected?

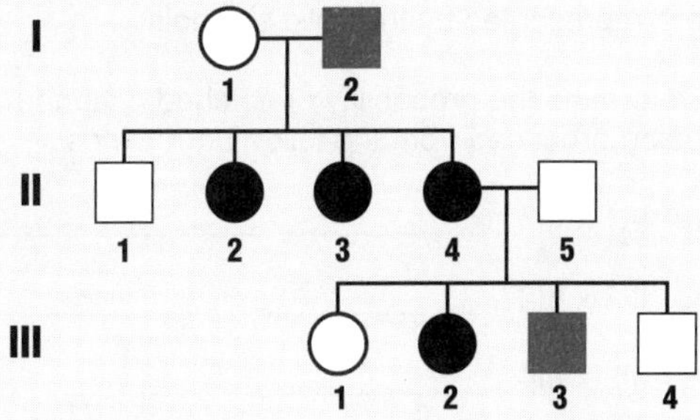
A rare form of rickets in humans
I
1
2
II
1
2
3
4
5
III
1
2
3
4

Heredity

ISBN: 978-1-927173-16-9

6. The pedigree chart below illustrates the inheritance of a trait (darker symbols) in two families joined in marriage.

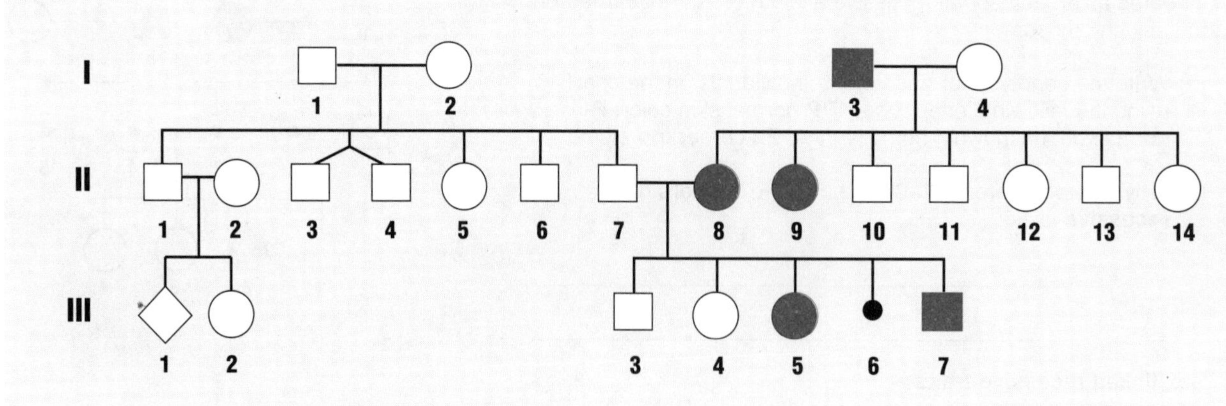
I
1
2
3
4
II
1
2
3
4
5
6
7
8
9
10
11
12
13
14
III
1
2
3
4
5
6
7

(a) State whether the trait is **dominant** or **recessive**, and explain your reasoning: ______________________

(b) State whether the trait is **sex linked** or not, and explain your reasoning: ______________________

7. The recessive sex linked gene (h) prolongs the blood-clotting time, resulting in the genetically inherited disease called hemophilia. From the information in the pedigree chart (right), answer the following questions:

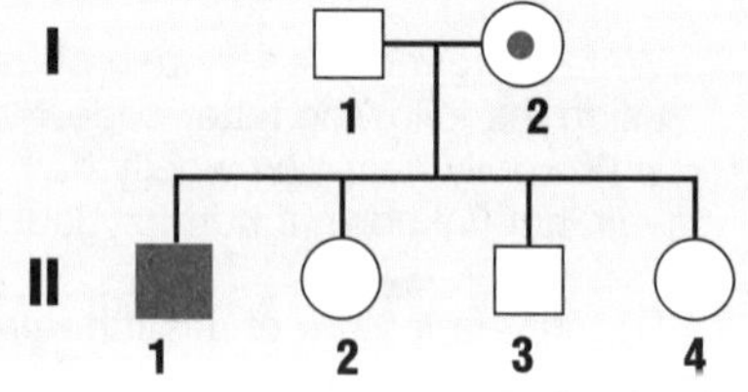
Hemophilia in humans
I
1
2
II
1
2
3
4

(a) If **II2** marries a normal man, determine the probability of her first child being a hemophiliac:

____

(b) Suppose her first child is actually a hemophiliac. Determine the chance that her second child will be a hemophiliac boy:

____

(c) If **II4** has children with a hemophilic man, determine is the probability of her first child being phenotypically normal:

____

(d) If the mother of **I2** was phenotypically normal, state the phenotype of her father: ______________________

8. The phenotypic expression of a dominant gene in Ayrshire cattle is a notch in the tips of the ears. In the pedigree chart on the right, notched animals are represented by the solid symbols.

Determine the probability of notched offspring being produced from the following matings:

(a) III1 x III3 ______________________

(b) III3 x III2 ______________________

(c) III3 x III4 ______________________

(d) III1 x III5 ______________________

(e) III2 x III5 ______________________

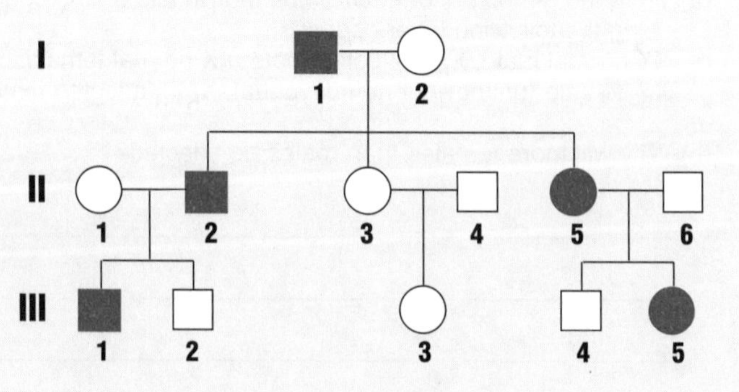
Ear notches in Ayrshire cattle
I
1
2
II
1
2
3
4
5
6
III
1
2
3
4
5

ISBN: 978-1-927173-16-9

Polygenes

Some phenotypes (e.g. kernel color in maize and skin color in humans) are determined by more than one gene and show **continuous variation** in a population. The production of the skin pigment melanin in humans is controlled by at least three genes. The amount of melanin produced is directly proportional to the number of dominant alleles for either gene (from 0 to 6).

Very pale	Light	Medium light	Medium	Medium dark	Dark	Black
0	1	2	3	4	5	6

A light-skinned person

A dark-skinned person

There are seven shades skin color ranging from very dark to very pale, with most individual being somewhat intermediate in skin color. No dominant allele results in a lack of dark pigment (aabbcc). Full pigmentation (black) requires six dominant alleles (AABBCC).

1. Complete the Punnett square for the F_2 generation (below) by entering the genotypes and the number of dark alleles resulting from a cross between two individuals of intermediate skin color. Color-code the offspring appropriately for easy reference.

   (a) How many of the 64 possible offspring of this cross will have darker skin than their parents:

   ____

   (b) How many genotypes are possible for this type of gene interaction:

   ____

2. Explain why we see many more than seven shades of skin color in reality:

   ____

   ____

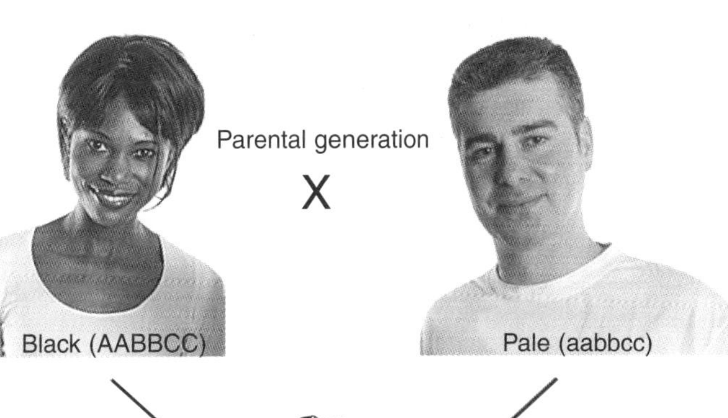
Parental generation
X
Black (AABBCC)
Pale (aabbcc)

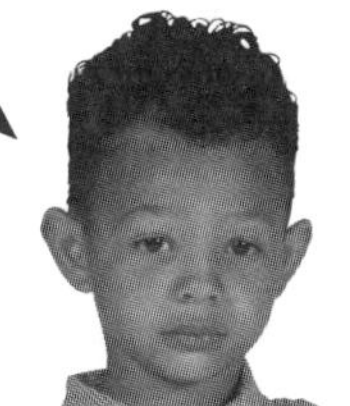
Medium (AaBbCc)

F_2 generation (AaBbCc X AaBbCc)

GAMETES	ABC	ABc	AbC	Abc	aBC	aBc	abC	abc
ABC								
ABc								
AbC								
Abc								
aBC								
aBc								
abC								
abc								

ISBN: 978-1-927173-16-9

3. Discuss the differences between **continuous** and **discontinuous** variation, giving examples to illustrate your answer:

4 From a sample of no less than 30 adults, collect data (by request or measurement) for one continuous variable (e.g. height, weight, shoe size, or hand span). Record and tabulate your results in the space below, and then plot a frequency histogram of the data on the grid below:

Raw data

Tally Chart (frequency table)

Variable: ....

Frequency

(a) Calculate each of the following for your data. See *Descriptive Statistics* if you need help and attach your working:

Mean: ______ **Mode**: ______ **Median**: ______

Standard deviation: ______

(b) Describe the pattern of distribution shown by the graph, giving a reason for your answer:

(c) What is the genetic basis of this distribution?

(d) What is the importance of a large sample size when gathering data relating to a continuous variable?

ISBN: 978-1-927173-16-9

Genetic Counseling

Genetic counseling is an analysis of the risk of producing offspring with known gene defects within a family. Counsellors identify families at risk, investigate the problem present in the family, interpret information about the disorder, analyze inheritance patterns and risks of recurrence, and review available options with the family. Increasingly, there are DNA tests for the identification of specific defective genes. People usually consider genetic counseling if they have a family history of a genetic disorder, or if a routine prenatal screening test yields an unexpected result. While screening for many genetic disorders is now recommended, the use of presymptomatic tests for adult-onset disorders, such as Alzheimer's, is still controversial.

Autosomal Recessive Conditions

Common inherited disorders caused by recessive alleles on autosomes. Recessive conditions are evident only in homozygous recessive genotypes.

Cystic fibrosis: Malfunction of the pancreas and other glands; thick mucus leads to pneumonia and emphysema. Death usually occurs in childhood. CF is the most frequent lethal genetic disorder in childhood (about 1 case in 3700 live births).

Maple syrup urine disease: Mental and physical retardation produced by a block in amino acid metabolism. Isoleucine in the urine produces the characteristic odor.

Tay-Sachs disease: A lipid storage disease which causes progressive developmental paralysis, mental deterioration, and blindness. Death usually occurs by three years of age.

Autosomal Dominant Conditions

Inherited disorders caused by dominant alleles on autosomes. Dominant conditions are evident both in heterozygotes and in homozygous dominant individuals.

Huntington disease: Involuntary movements of the face and limbs with later general mental deterioration. The beginning of symptoms is highly variable, but occurs usually between 30 to 40 years of age.

?

Genetic testing may involve biochemical tests for gene products such as enzymes and other proteins, microscopic examination of stained or fluorescent chromosomes, or examination of the DNA molecule itself. Various types of genetic tests are performed for various reasons, including:

Carrier Screening

Identifying unaffected individuals who carry one copy of a gene for a disease that requires two copies for the disease to be expressed.

Preimplantation Genetic Diagnosis

Screens for genetic flaws in embryos used for *in vitro* fertilization. The results of the analysis are used to select mutation-free embryos.

Prenatal Diagnostic Testing

Tests for chromosomal abnormalities such as Down syndrome.

Newborn Screening

Newborn babies are screened for a variety of enzyme-based disorders.

Presymptomatic Testing

Testing before symptoms are apparent is important for estimating the risk of developing adult-onset disorders, including Huntington's, cancers, and Alzheimer's disease.

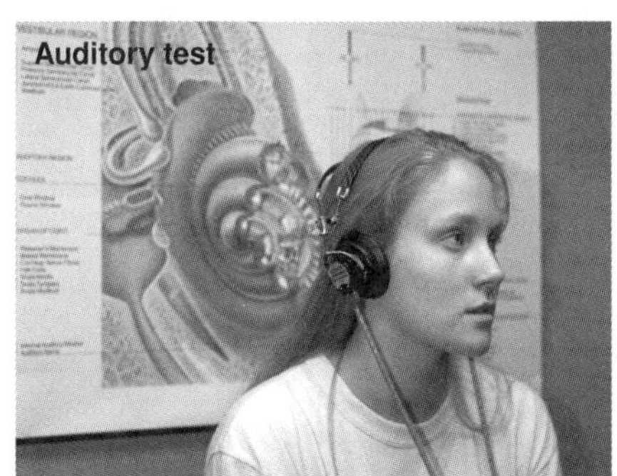
Auditory test

About half of the cases of childhood deafness are the result of an autosomal recessive disorder. Early identification of the problem prepares families and allows early appropriate treatment.

Genetic counseling provides information to families who have members with birth defects or genetic disorders, and to families who may be at risk for a variety of inherited conditions.

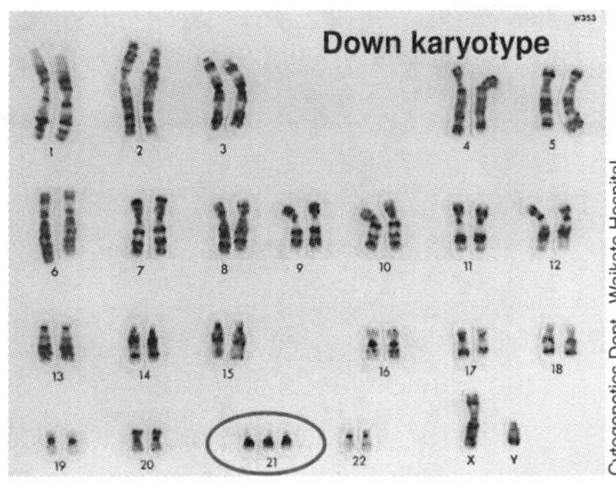
Down karyotype
1
2
3
4
5
6
7
8
9
10
11
12
13
14
15
16
17
18
19
20
21
22
X
Y
Cytogenetics Dept., Waikato Hospital

Most pregnant women in developed countries will have a prenatal test to detect chromosomal abnormalities such as Down syndrome and developmental anomalies such as neural defects.

1. Outline the benefits of **carrier screening** to a couple with a family history of a genetic disorder:

2. (a) Suggest why Huntington disease persists in the human population when it is caused by a lethal, dominant allele:

   (b) How could presymptomatic genetic testing change this?

ISBN: 978-1-927173-16-9

Related activities: *A Gene that can Tell Your Future?*
Weblinks: *Patterns of Inheritance*

KEY TERMS: Crossword

Complete the crossword below, which will test your understanding of key terms in this chapter and their meanings

Clues Across

2. Alleles which both contribute to the phenotype when in the heterozygous combination.
6. Analysis of the risk of producing offspring with known gene defects within a family. (2 words: 7, 10)
7. The position of a gene on a chromosome.
8. The specific allele combination of an organism.
11. A chromosome that is not a sex chromosome.
13. A way of graphically illustrating inheritance patterns over a number of generations. (2 words: 8, 5)
14. The chromosome that carries the genes for determination of sex in individual organisms. (2 words: 3, 10)
15. A person who has inherited and can pass on a genetic trait but does not display its symptoms.
16. Possessing two of the same alleles of a particular gene, one inherited from each parent.
17. The physical appearance of the genotype.
18. The locating of genes on sex chromosomes so that their inheritance is linked to a particular sex. (2 words: 3, 7)

Clues Down

1. A graphical way of illustrating the outcome of a cross.
3. A cross involving only one pair of alleles. (2 words: 10, 5,)
4. Allele that expresses its trait irrespective of the other allele.
5. Some genes have more than two types of alleles, such as the ABO alleles in humans. These are called __________ alleles.
9. A diagnostic cross involving breeding an unknown genotype with a homozygous recessive. (2 words: 4, 5)
10. Possessing two different alleles of a particular gene, one inherited from each parent.
12. Allele that expresses its characteristics only when in the homozygous combination.

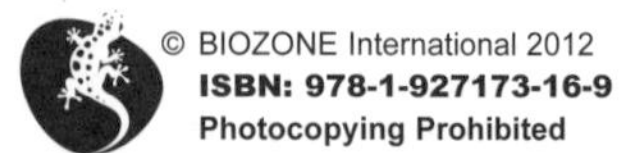

ISBN: 978-1-927173-16-9

Core Topic **4.4**

Genetic Engineering and **Biotechnology**

Key terms

Core

bioinformatics
clone
differentiated cell
DNA amplification
DNA ligase
DNA ligation
DNA polymerase
DNA profiling
DNA sequencing
embryonic stem cell
forensic analysis
gel electrophoresis
genetic modification
Human Genome Project
paternity
plasmid
polymerase chain reaction (= PCR)
primer
recombinant DNA
restriction enzyme (endonuclease)
somatic cell nuclear transfer
therapeutic cloning
transgenic organism

Key concepts

- Biotechnology makes use of a few basic techniques but has many diverse applications.
- DNA profiling is a diagnostic and forensic tool that utilizes identifiable diversity in the human genome.
- Genetic modification potentially offers huge benefits, but presents ethical concerns for many.
- Cloning technology has many applications, including in the rapid production of transgenics.
- Therapeutic cloning offers a way to provide immune-compatible tissues for medicine.

Learning Objectives

☐ 1. Use the **KEY TERMS** to compile a glossary for this topic.

PCR and Gel Electrophoresis *(4.4.1-4.4.3)* pages 182-185

☐ 2. Recognize that the same, relatively few, basic techniques are used in a range of different processes and applications in gene technology.

☐ 3. Describe and explain the role of **polymerase chain reaction** (**PCR**) in **DNA amplification**, including the role of **primers** and **DNA polymerase**.

☐ 4. Explain the principle of **gel electrophoresis** and explain how it used to separate DNA fragments on the basis of size. Outline the uses of gel electrophoresis in gene technology, including its use in **DNA profiling**.

DNA Profiling and Sequencing *(4.4.4-4.4.6)* pages 186-191

☐ 5. Describe the use of DNA profiling as a tool to determine **paternity** and in **forensic analysis**. Analyze DNA profiles to draw conclusions about paternity or forensic investigations. Understand why profiling provides a better diagnostic tool for determining paternity than blood group analysis (*TOK*).

☐ 6. Explain what is meant by **bioinformatics**. Describe the application of modern **DNA sequencing** techniques to our understanding of the human genome. Outline three outcomes of the **Human Genome Project**. Appreciate the wider implications of our increasing understanding of the human genome (TOK).

Genetic Engineering *(4.4.7-4.4.10)* pages 192-208

☐ 7. Explain why the transfer of genetic material between species is possible.

☐ 8. Explain the use of **restriction enzymes** (restriction endonucleases) and **DNA ligation** in creating **recombinant DNA**. Outline the basic technique for gene transfer using **plasmids**. Including reference to the host cell (e.g. yeast, bacterium, or other cell), restriction enzymes and **DNA ligase**.

☐ 9. Define the term **transgenic organism**. Describe two current applications of genetically modified crops or animals. Examples could include genetic engineering for salt tolerance in crops, synthesis of beta-carotene in rice, and synthesis of useful human proteins such as insulin and blood clotting factors.

☐ 10. Using one specific example, discuss the potential harms and benefits of genetic modification. Discuss the risks *vs* the benefits of the technology and evaluate the net outcome in terms of gain or loss (*TOK*).

Genetic Engineering *(4.4.11-4.4.13)* pages 209-211

☐ 11. Define the term **clone** with reference to both cells and whole organisms.

☐ 12. Outline a technique, e.g. **somatic cell nuclear transfer** (SCNT) for cloning animals using **differentiated cells**.

☐ 13. Define the term **embryonic stem cell**. Explain what is meant by **therapeutic cloning** and discuss the issue of this technology.

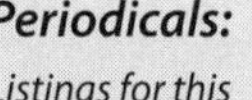

Periodicals: *Listings for this chapter are on page 399*

Weblinks: *www.thebiozone.com/weblink/IB-3169.html*

BIOZONE APP: *Student Review Series* ***Gene Technology***

Amazing Organisms, Amazing Enzymes

Before the 1980s scientists knew of only a few organisms that could survive in extreme conditions. Indeed, many scientists believed that life in highly saline or high temperature and pressure environments was impossible. That view changed with the discovery of bacteria inhabiting the deep sea hydrothermal vents. They tolerate temperatures over 110°C and pressures of over 200 atmospheres. Bacteria were also found in volcanic hot pools on land, some surviving at temperatures in excess of 80°C. Most enzymes are denatured at temperatures above 40°C, but these **thermophilic** bacteria have enzymes that are fully functional at high temperatures. This discovery led to the development of one of the most important techniques in biotechnology, the **polymerase chain reaction** (PCR).

PCR is a technique, first described in the 1970s, that allows scientists to copy and multiply a piece of DNA millions of times. The DNA is heated to 98°C so that it separates into single strands and polymerase enzyme is added to synthesize new DNA strands from supplied free nucleotides. This earlier technique was labor intensive and expensive because the polymerase denatured at the high temperatures and had to be replaced every cycle. In 1985, a thermophilic polymerase (***Taq* polymerase**) was isolated from the bacterium *Thermophilus aquaticus,* which inhabited the hot springs of Yellowstone National Park. Isolating this enzyme enabled automation of the PCR process, because the polymerase was stable throughout multiple cycles of synthesis. This led to a rapid growth in biotechnology, and gene technology in particular, because DNA samples could be easily copied for sequencing.

Searching for novel compounds in organisms from extreme environments is important in the development of new biotechnologies. Organisms must have compounds that can work in their specific environment, and the identification and extraction of these may allow them to be adapted for human use. For example, the Antarctic sea sponge *Kirkpatrickia variolosa* produces an alkaloid excreted as a toxic defence to prevent other organisms growing nearby. Tests indicate that this same chemical may have biological activity against cancer cells. Compounds from other sponge species are currently being assessed to treat a range of diseases including cancer, AIDS, tuberculosis and other bacterial infections, and cystic fibrosis.

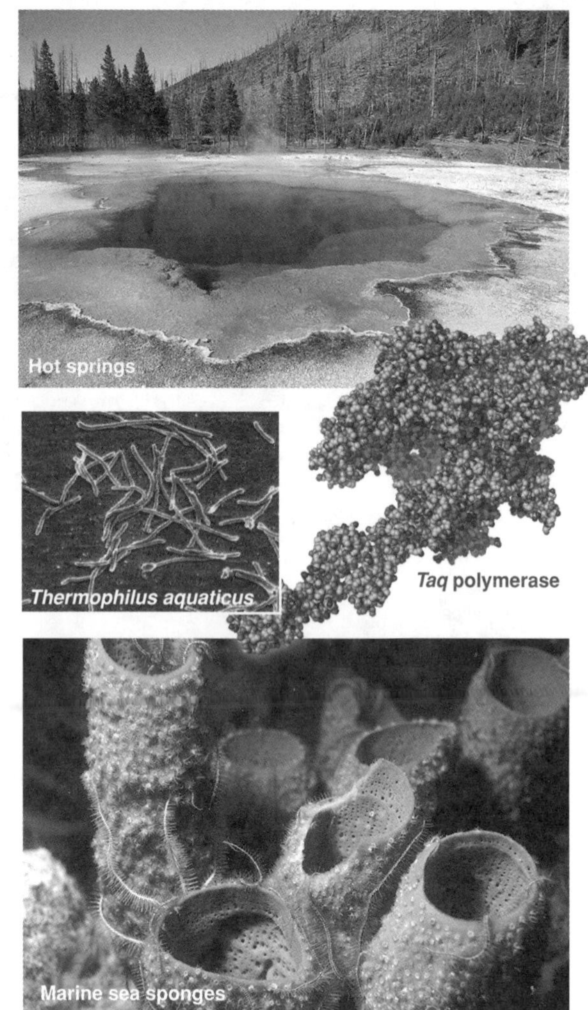
Hot springs
Thermophilus aquaticus
Taq polymerase
Marine sea sponges

1. Why was PCR not a viable technique until the mid 1980s? ______________________

2. Explain why *Taq* polymerase was so important in the development of PCR: ______________________

3. Explain how investigating the lifestyles of other organisms can lead to advances in unrelated areas of science: ______________________

Related activities: *Polymerase Chain Reaction*

ISBN: 978-1-927173-16-9

Polymerase Chain Reaction

Many procedures in DNA technology (such as DNA sequencing and DNA profiling) require substantial amounts of DNA to work with. Some samples, such as those from a crime scene or fragments of DNA from a long extinct organism, may be difficult to get in any quantity. The diagram below describes the laboratory process called **polymerase chain reaction** (**PCR**).

Using this technique, vast quantities of DNA identical to the original samples can be created. This process is often termed **DNA amplification**. Although only one cycle of replication is shown below, the following cycles replicate DNA at an exponential rate. PCR can be used to make billions of copies of DNA in only a few hours.

A Single Cycle of the Polymerase Chain Reaction

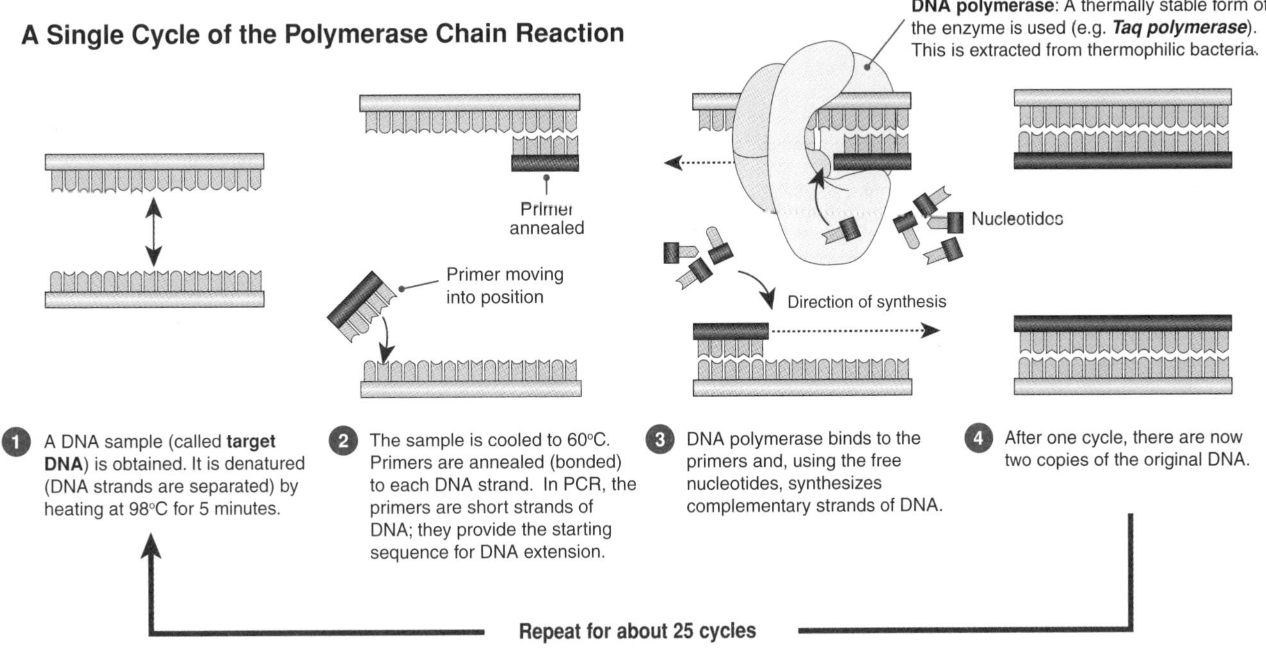
DNA polymerase: A thermally stable form of the enzyme is used (e.g. ***Taq polymerase***). This is extracted from thermophilic bacteria.
Primer annealed
Primer moving into position
Nucleotides
Direction of synthesis
1 A DNA sample (called **target DNA**) is obtained. It is denatured (DNA strands are separated) by heating at 98°C for 5 minutes.
2 The sample is cooled to 60°C. Primers are annealed (bonded) to each DNA strand. In PCR, the primers are short strands of DNA; they provide the starting sequence for DNA extension.
3 DNA polymerase binds to the primers and, using the free nucleotides, synthesizes complementary strands of DNA.
4 After one cycle, there are now two copies of the original DNA.
Repeat for about 25 cycles

Repeat cycle of heating and cooling until enough copies of the target DNA have been produced

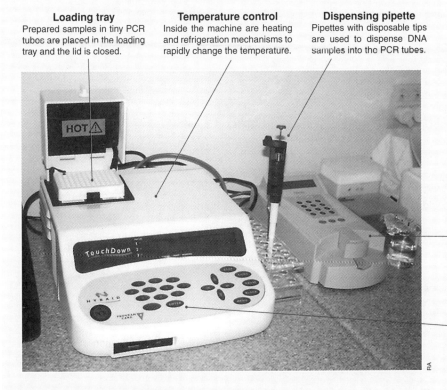

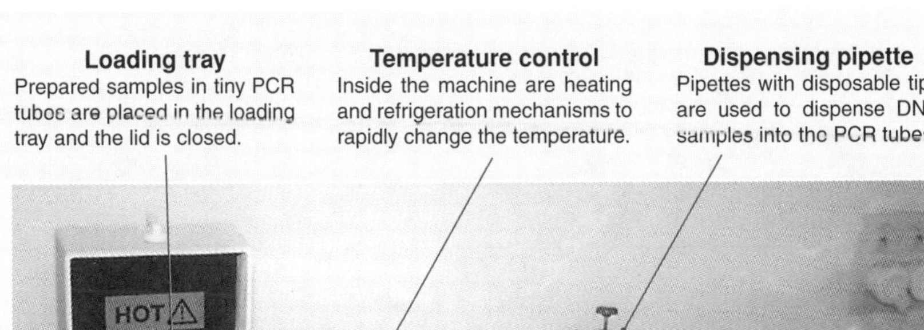
Loading tray
Prepared samples in tiny PCR tubes are placed in the loading tray and the lid is closed.
Temperature control
Inside the machine are heating and refrigeration mechanisms to rapidly change the temperature.
Dispensing pipette
Pipettes with disposable tips are used to dispense DNA samples into the PCR tubes.
HOT
TouchDown
HYBAID
DNA quantitation
The amount of DNA in a sample can be determined by placing a known volume in this quantitation machine. For many genetic engineering processes, a minimum amount of DNA is required.
Controls
The control panel allows a number of different PCR programmes to be stored in the machine's memory. Carrying out a PCR run usually just involves starting one of the stored programmes.
RA

Thermal Cycler

Amplification of DNA can be carried out with simple-to-use machines called thermal cyclers. Once a DNA sample has been prepared, in just a few hours the amount of DNA can be increased billions of times. Thermal cyclers are in common use in the biology departments of universities, as well as other kinds of research and analytical laboratories. The one pictured on the left is typical of this modern piece of equipment.

1. Explain the purpose of PCR: ____________________

____

____

____

ISBN: 978-1-927173-16-9

2. Describe how the **polymerase chain reaction** works: ___

3. Describe three situations where only very small DNA samples may be available for sampling and PCR could be used:

   (a) ___

   (b) ___

   (c) ___

4. After only two cycles of replication, four copies of the double-stranded DNA exist. Calculate how much a DNA sample will have increased after:

   (a) 10 cycles: ___ (b) 25 cycles: ___

5. The risk of contamination in the preparation for PCR is considerable.

   (a) Describe the effect of having a single molecule of unwanted DNA in the sample prior to PCR:

   ___

   (b) Describe two possible sources of DNA contamination in preparing a PCR sample:

   Source 1: ___

   Source 2: ___

   (c) Describe two precautions that could be taken to reduce the risk of DNA contamination:

   Precaution 1: ___

   Precaution 2: ___

6. Describe two other genetic engineering/genetic manipulation procedures that require PCR amplification of DNA:

   (a) ___

   (b) ___

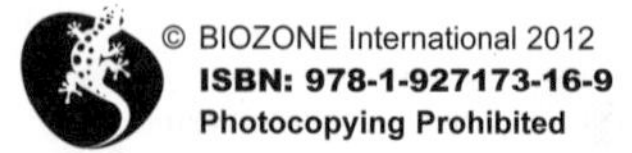

ISBN: 978-1-927173-16-9

Gel Electrophoresis

Gel electrophoresis is a method that separates large molecules (including nucleic acids or proteins) on the basis of size, electric charge, and other physical properties. Such molecules possess a slight electric charge (see DNA below). To prepare DNA for gel electrophoresis the DNA is often cut up into smaller pieces. This is done by mixing DNA with restriction enzymes in controlled conditions for about an hour. Called **restriction digestion**, it produces a range of DNA fragments of different lengths. During electrophoresis, molecules are forced to move through the pores of a **gel** (a jelly-like material), when the electrical current is applied. Active electrodes at each end of the gel provide the driving force. The electrical current from one electrode repels the molecules while the other electrode simultaneously attracts the molecules. The frictional force of the gel resists the flow of the molecules, separating them by size. Their rate of migration through the gel depends on the strength of the electric field, size and shape of the molecules, and on the ionic strength and temperature of the buffer in which the molecules are moving. After staining, the separated molecules in each lane can be seen as a series of bands spread from one end of the gel to the other.

Analyzing DNA using Gel Electrophoresis

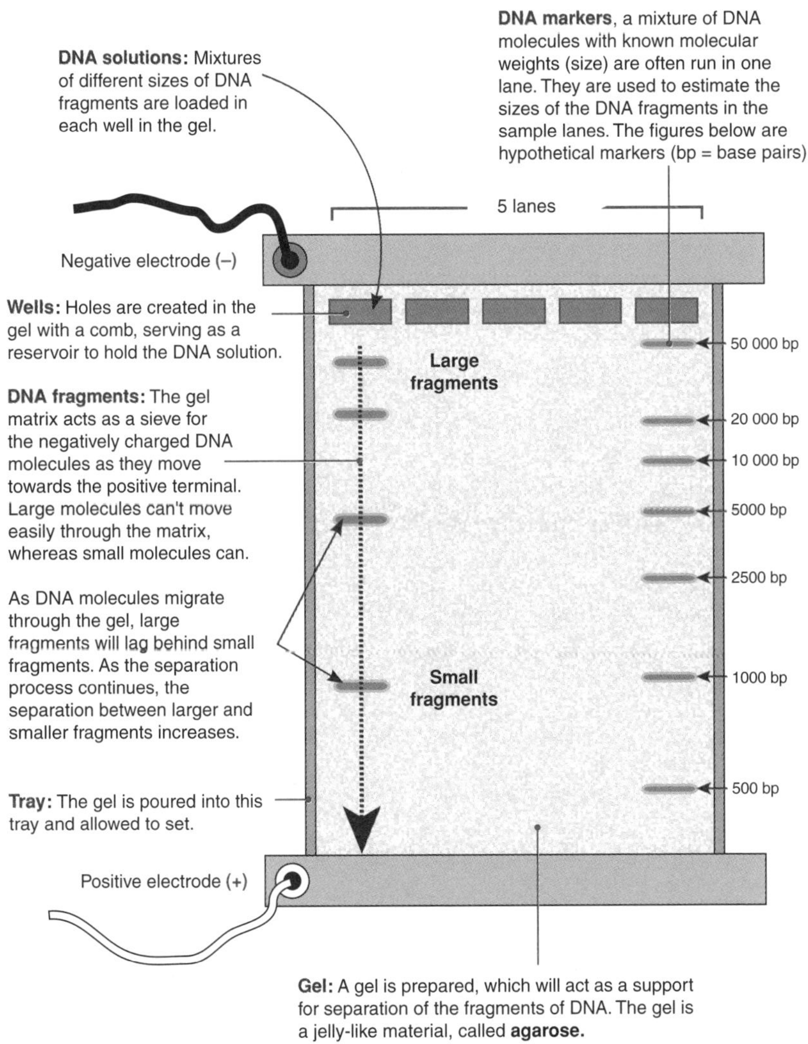

DNA solutions: Mixtures of different sizes of DNA fragments are loaded in each well in the gel.
DNA markers, a mixture of DNA molecules with known molecular weights (size) are often run in one lane. They are used to estimate the sizes of the DNA fragments in the sample lanes. The figures below are hypothetical markers (bp = base pairs).
5 lanes
Negative electrode (–)
Wells: Holes are created in the gel with a comb, serving as a reservoir to hold the DNA solution.
Large fragments
DNA fragments: The gel matrix acts as a sieve for the negatively charged DNA molecules as they move towards the positive terminal. Large molecules can't move easily through the matrix, whereas small molecules can.
50 000 bp
20 000 bp
10 000 bp
5000 bp
2500 bp
As DNA molecules migrate through the gel, large fragments will lag behind small fragments. As the separation process continues, the separation between larger and smaller fragments increases.
Small fragments
1000 bp
500 bp
Tray: The gel is poured into this tray and allowed to set.
Positive electrode (+)

Gel: A gel is prepared, which will act as a support for separation of the fragments of DNA. The gel is a jelly-like material, called **agarose.**

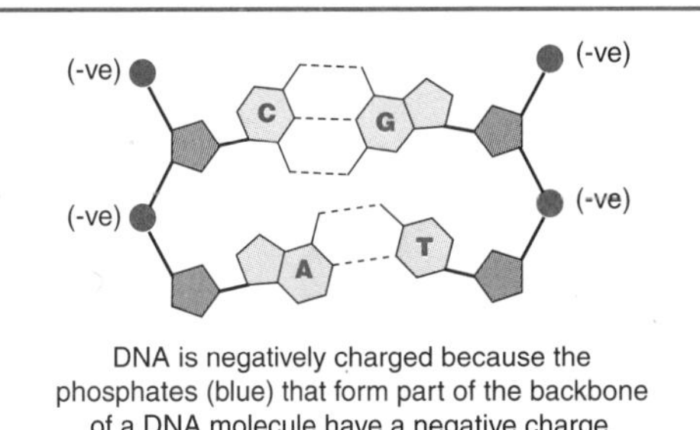

(-ve)
(-ve)
(-ve)
(-ve)
C
G
A
T

DNA is negatively charged because the phosphates (blue) that form part of the backbone of a DNA molecule have a negative charge.

Steps in the process of gel electrophoresis of DNA

1. A tray is prepared to hold the gel matrix.
2. A gel comb is used to create holes in the gel. The gel comb is placed in the tray.
3. Agarose gel powder is mixed with a buffer solution (this carries the DNA in a stable form). The solution is heated until dissolved and poured into the tray and allowed to cool.
4. The gel tray is placed in an electrophoresis chamber and the chamber is filled with buffer, covering the gel. This allows the electric current from electrodes at either end of the gel to flow through the gel.
5. DNA samples are mixed with a "loading dye" to make the DNA sample visible. The dye also contains glycerol or sucrose to make the DNA sample heavy so that it will sink to the bottom of the well.
6. A safety cover is placed over the gel, electrodes are attached to a power supply and turned on.
7. When the dye marker has moved through the gel, the current is turned off and the gel is removed from the tray.
8. DNA molecules are made visible by staining the gel with **methylene blue** or ethidium bromide which binds to DNA and will fluoresce in UV light.

1. Explain the purpose of gel electrophoresis: ______

2. Describe the two forces that control the speed at which fragments pass through the gel:

   (a) ______

   (b) ______

3. Explain why the smallest fragments travel through the gel the fastest: ______

ISBN: 978-1-927173-16-9

DNA Profiling Using PCR

In chromosomes, some of the DNA contains simple, repetitive sequences. These *noncoding* nucleotide sequences repeat themselves over and over again and are found scattered throughout the genome. Some repeating sequences are short (2-6 base pairs) called **microsatellites** or **short tandem repeats** (STRs) and can repeat up to 100 times. The human genome has numerous different microsatellites. Equivalent sequences in different people vary considerably in the numbers of the repeating unit. This phenomenon has been used to develop **DNA profiling**, which identifies the natural variations found in every person's DNA. Identifying such differences in the DNA of individuals is a useful tool for forensic investigations. In 1998, the FBI's Combined Offender DNA Index System (CODIS) was established, providing a national database of DNA samples from convicted criminals, suspects, and crime scenes. In the USA, there are many laboratories approved for forensic DNA testing. Increasingly, these are targeting the 13 core STR loci recommended by the FBI; enough to guarantee that the odds of someone else sharing the same result are extremely unlikely (less than one in a thousand million). The CODIS may be used to solve previously unsolved crimes and to assist in current or future investigations. DNA profiling can also be used to establish genetic relatedness (e.g. in paternity or pedigree disputes), or when searching for a specific gene (e.g. screening for disease).

Microsatellites (Short Tandem Repeats)

Microsatellites consist of a variable number of tandem repeats of a 2 to 6 base pair sequence. In the example below it is a two base sequence (CA) that is repeated.

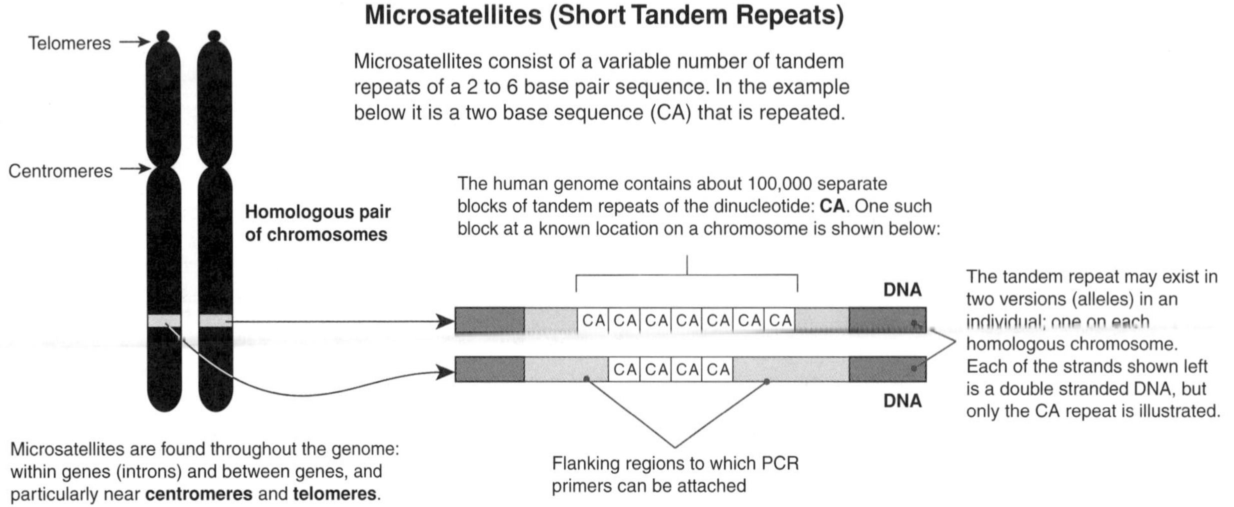

Telomeres
Centromeres
Homologous pair of chromosomes
The human genome contains about 100,000 separate blocks of tandem repeats of the dinucleotide: **CA**. One such block at a known location on a chromosome is shown below:
DNA
CA CA CA CA CA CA CA
CA CA CA CA
DNA
The tandem repeat may exist in two versions (alleles) in an individual; one on each homologous chromosome. Each of the strands shown left is a double stranded DNA, but only the CA repeat is illustrated.
Flanking regions to which PCR primers can be attached

Microsatellites are found throughout the genome: within genes (introns) and between genes, and particularly near **centromeres** and **telomeres**.

How short tandem repeats are used in DNA profiling

This diagram shows how three people can have quite different microsatellite arrangements at the same point (locus) in their DNA. Each will produce a different DNA profile using gel electrophoresis:

1 Extract DNA from sample

A sample collected from the tissue of a living or dead organism is treated with chemicals and enzymes to extract the DNA, which is separated and purified.

2 Amplify microsatellite using PCR

Specific primers (arrowed) that attach to the flanking regions (light gray) either side of the microsatellite are used to make large quantities of the micro-satellite and flanking regions sequence only (no other part of the DNA is amplified/replicated).

3 Visualize fragments on a gel

The fragments are separated by length, using **gel electrophoresis**. DNA, which is negatively charged, moves toward the positive terminal. The smaller fragments travel faster than larger ones.

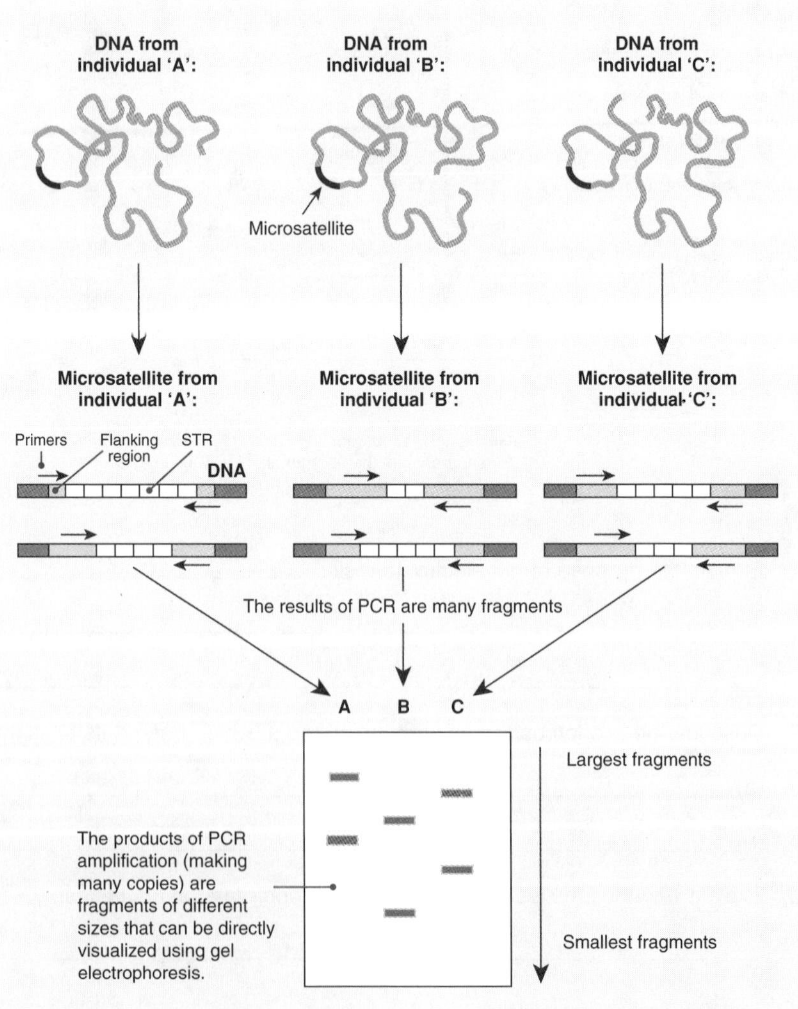

DNA from individual 'A':
DNA from individual 'B':
DNA from individual 'C':
Microsatellite
Microsatellite from individual 'A':
Microsatellite from individual 'B':
Microsatellite from individual 'C':
Primers
Flanking region
STR
DNA
The results of PCR are many fragments
A
B
C
Largest fragments
Smallest fragments
The products of PCR amplification (making many copies) are fragments of different sizes that can be directly visualized using gel electrophoresis.

Related activities: Gel Electrophoresis
Weblinks: DNA Applications: Profiling

Periodicals: Bioinformatics: what use is it?

ISBN: 978-1-927173-16-9

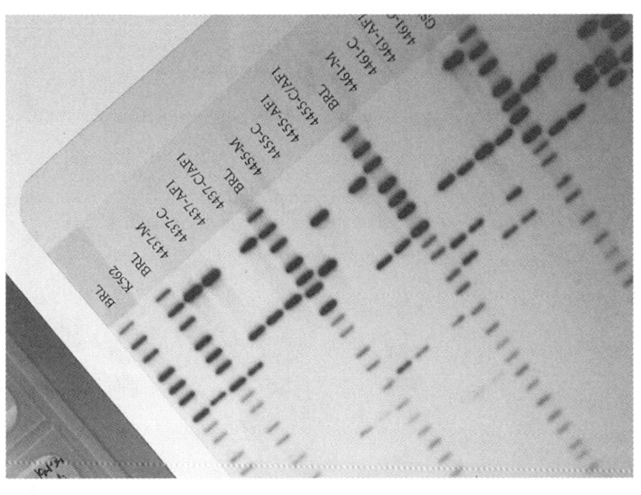

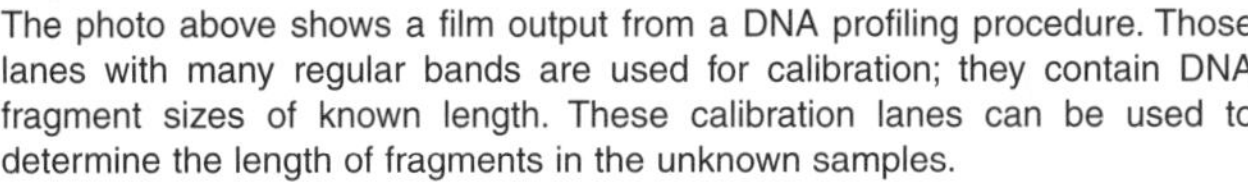

The photo above shows a film output from a DNA profiling procedure. Those lanes with many regular bands are used for calibration; they contain DNA fragment sizes of known length. These calibration lanes can be used to determine the length of fragments in the unknown samples.

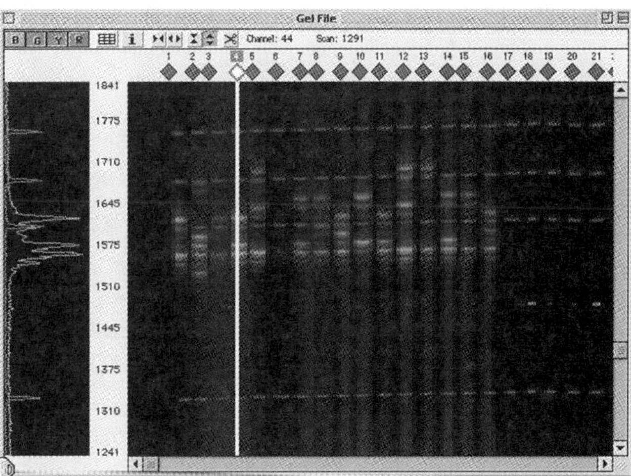
Gel File
Channel: 44
Scan: 1291
1841
1775
1710
1645
1575
1510
1445
1375
1310
1241

DNA profiling can be automated in the same way as DNA sequencing. Powerful computer software is able to display the results of many samples that are run at the same time. In the photo above, the sample in lane 4 has been selected and displays fragments of different length on the left of the screen.

1. Describe the properties of **short tandem repeats** that are important to the application of **DNA profiling** technology:

2. Explain the role of each of the following techniques in the process of DNA profiling:

   (a) Gel electrophoresis:

   (b) PCR:

3. Describe the three main steps in DNA profiling using PCR:

   (a)

   (b)

   (c)

4. Explain why as many as 10 STR sites are used to gain a DNA profile for forensic evidence:

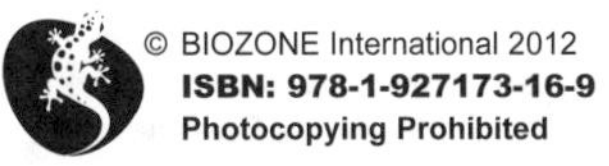

ISBN: 978-1-927173-16-9

Forensic Applications of DNA Profiling

The use of DNA as a tool for solving crimes such as homicide is well known, but it can also be used to as a solution to many other problems. DNA evidence has been used to identify body parts, solve cases of industrial sabotage and contamination, for paternity testing, and even in identifying animal products illegally made from endangered species.

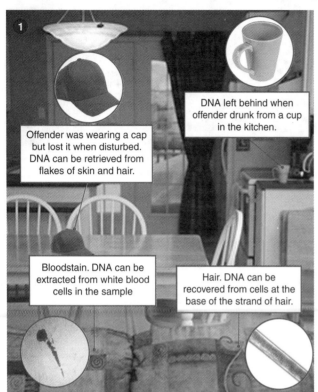

1

Offender was wearing a cap but lost it when disturbed. DNA can be retrieved from flakes of skin and hair.

DNA left behind when offender drunk from a cup in the kitchen.

Bloodstain. DNA can be extracted from white blood cells in the sample

Hair. DNA can be recovered from cells at the base of the strand of hair.

During the initial investigation, samples of material that may contain DNA are taken for analysis. At a crime scene, this may include blood and body fluids as well as samples of clothing or objects that the offender might have touched. Samples from the victim are also taken to eliminate them as a possible source of contamination.

2 DNA is isolated and profiles are made from all samples and compared to known DNA profiles such as that of the victim.

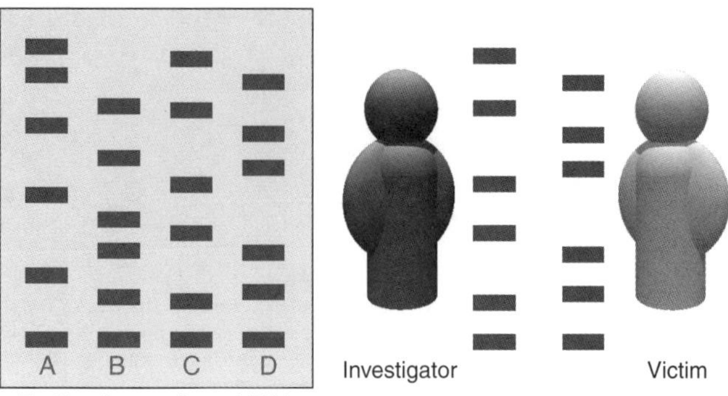
A
B
C
D
Investigator
Victim

Profiles from collected DNA

3 Unknown DNA samples are compared to DNA databases of convicted offenders and to the DNA of the alleged offender.

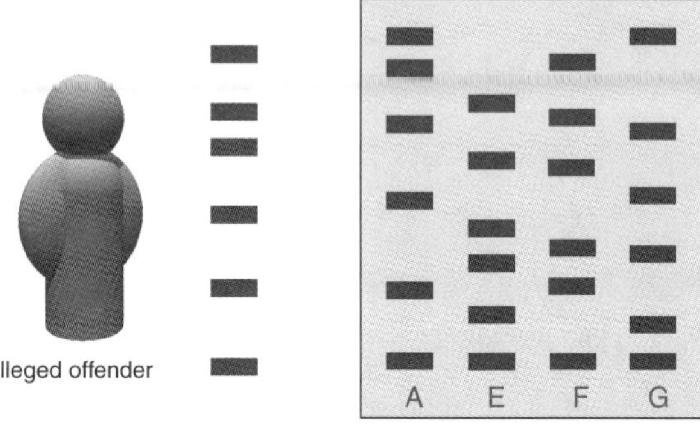
Alleged offender
A
E
F
G

Profiles from DNA database

4 Although it does not make a complete case, DNA profiling, in conjunction with other evidence, is one of the most powerful tools in identifying offenders or unknown tissues.

1. In the above case, two sets of profiles are shown. What is the purpose of lane A in each set of profiles?

2. Why are DNA profiles obtained for both the victim and investigator?

3. Use the evidence to decide if the alleged offender is innocent or guilty and explain your decision:

4. How could DNA profiling be used to refute official claims of the number of whales being captured and sold in fish markets?

Whale DNA: Tracking Illegal Slaughter

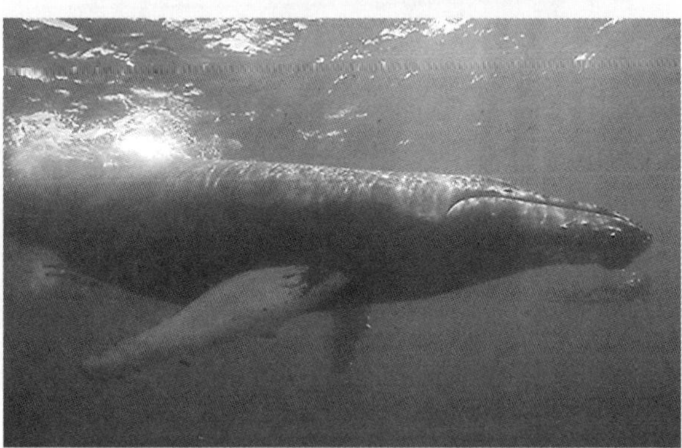

Under International Whaling Commission regulations, some species of whales can be captured for scientific research and their meat sold legally. Most, including humpback and blue whales, are fully protected and to capture or kill them for any purpose is illegal. Between 1999 and 2003 Scott Baker and associates from Oregon State University's Marine Mammal Institute investigated whale meat sold in markets in Japan and South Korea. Using DNA profiling techniques, they found around 10% of the samples tested were from fully protected whales including western gray whales and humpbacks. They also found that many more whales were being killed than were being officially reported.

ISBN: 978-1-927173-16-9

Finding the Connection

During the last decade, the study of **species relationships** has been revolutionized by the use of **DNA technology**. Through genomic analyses, many so-called species that were once thought to be separate have been shown to be one. Conversely, taxa that were thought to be monospecific have now been shown to include more than one genetically distinct group. There are a number of different sheep breeds, each of which has been bred for a specific purpose (e.g. meat of wool production). While some of these breeds are very old, others are recent developments. As new sheep breeds are developed by selective breeding, older breeds become less profitable to farm and are eventually replaced. These older breeds are becoming increasingly important because they carry traits that could be valuable but are now absent in more recent breeds. The relationships between the older and newer breeds are important as they show the development of breeds and help farmers and breeders to plan for the future development of their flocks.

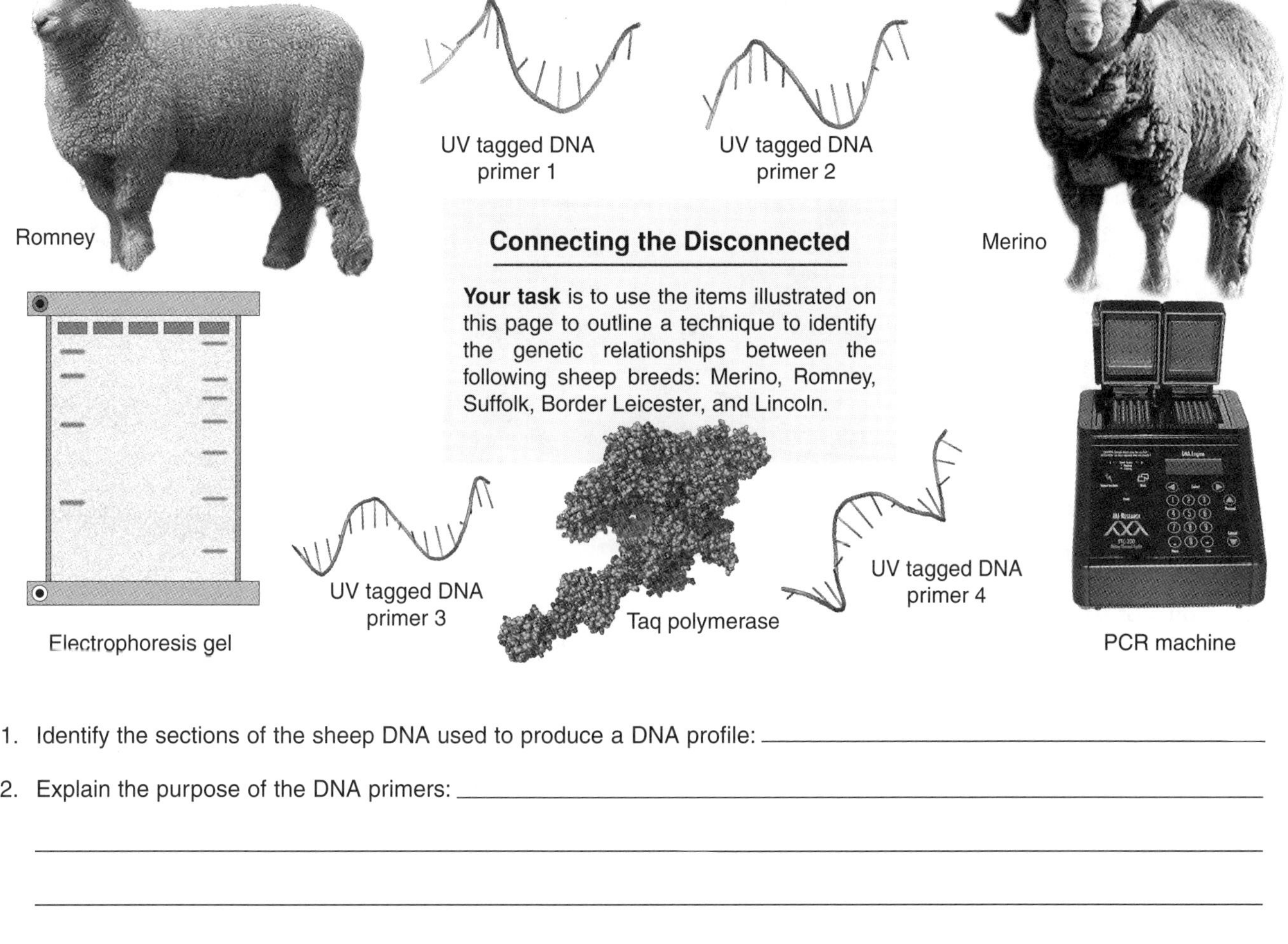
UV tagged DNA primer 1
UV tagged DNA primer 2
Romney
Merino
Electrophoresis gel
UV tagged DNA primer 3
Taq polymerase
UV tagged DNA primer 4
PCR machine

Connecting the Disconnected

Your task is to use the items illustrated on this page to outline a technique to identify the genetic relationships between the following sheep breeds: Merino, Romney, Suffolk, Border Leicester, and Lincoln.

1. Identify the sections of the sheep DNA used to produce a DNA profile: __________

2. Explain the purpose of the DNA primers: __________

3. Describe briefly how identifiable pieces of DNA are isolated. __________

4. Explain the purpose of the electrophoresis gel in relation to this investigation: __________

5. Explain how the DNA profile produced can be used to identify relationships between the breeds of sheep:

ISBN: 978-1-927173-16-9

***Related activities**: DNA Profiling using PCR, Forensic Applications of DNA Profiling*

The Human Genome Project

The **Human Genome Project** (HGP) is a publicly funded venture involving many different organizations throughout the world. In 1998, Celera Genomics in the USA began a competing project, as a commercial venture, in a race to be the first to determine the human genome sequence. In 2000, both organizations reached the first draft stage, and the entire genome is now available as a high quality (golden standard) sequence. In addition to determining the order of bases in the human genome, genes are being identified, sequenced, and mapped (their specific chromosomal location identified). The next challenge is to assign functions to the identified genes. By identifying and studying the protein products of genes (a field known as **proteomics**), scientists can develop a better understanding of genetic disorders. Long term benefits of the HGP are both medical and non-medical (see next page). Many biotechnology companies have taken out patents on gene sequences. This practice is controversial because it restricts the use of the sequence information to the patent holders. Other genome sequencing projects have arisen as a result of the initiative to sequence the human one. In 2002 the International HapMap Project was started with the aim of developing a haplotype map (HapMap) of the human genome. Initially data was gathered from four populations with African, Asian and European ancestry and additional populations may be included as analysis of human genetic variation continues.

Gene Mapping

This process involves determining the precise position of a gene on a chromosome. Once the position is known, it can be shown on a diagram.

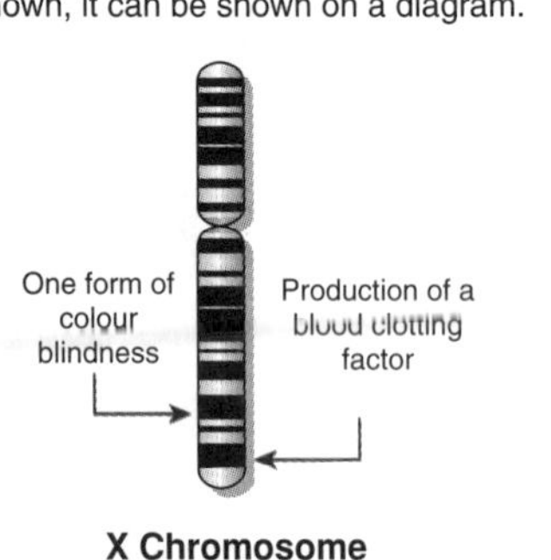

One form of colour blindness
Production of a blood clotting factor

X Chromosome

Equipment used for DNA Sequencing

Genesis Research and Development Corp, Auckland

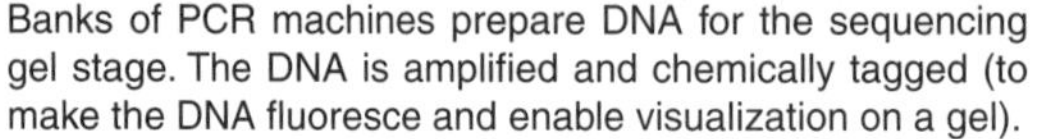

Banks of PCR machines prepare DNA for the sequencing gel stage. The DNA is amplified and chemically tagged (to make the DNA fluoresce and enable visualization on a gel).

HGSI

Banks of DNA sequencing gels and powerful computers are used to determine the base order in DNA.

Count of Mapped Genes

The aim of the HGP was to produce a continuous block of sequence information for each chromosome. Initially the sequence information was obtained to draft quality, with an error rate of 1 in 1000 bases. The **Gold Standard** sequence, with an error rate of <1 per 100,000 bases, was completed in October 2004. This table shows the length and number of mapped genes for each chromosome.

Chromosome	Length (Mb)	No. of Mapped Genes
1	263	1873
2	255	1113
3	214	965
4	203	614
5	194	782
6	183	1217
7	171	995
8	155	591
9	145	804
10	144	872
11	144	1162
12	143	894
13	114	290
14	109	1013
15	106	510
16	98	658
17	92	1034
18	85	302
19	67	1129
20	72	599
21	50	386
22	56	501
X	164	1021
Y	59	122
	Total:	**19 447**

Data to March 2008 from gdb.org (now offline)

Examples of Mapped Genes

The positions of an increasing number of genes have been mapped onto human chromosomes (see below). Sequence variations can cause or contribute to identifiable disorders. Note that chromosome 21 (the smallest human chromosome) has a relatively low gene density, while others are gene rich. This is possibly why trisomy 21 (Down syndrome) is one of the few viable human autosomal trisomies.

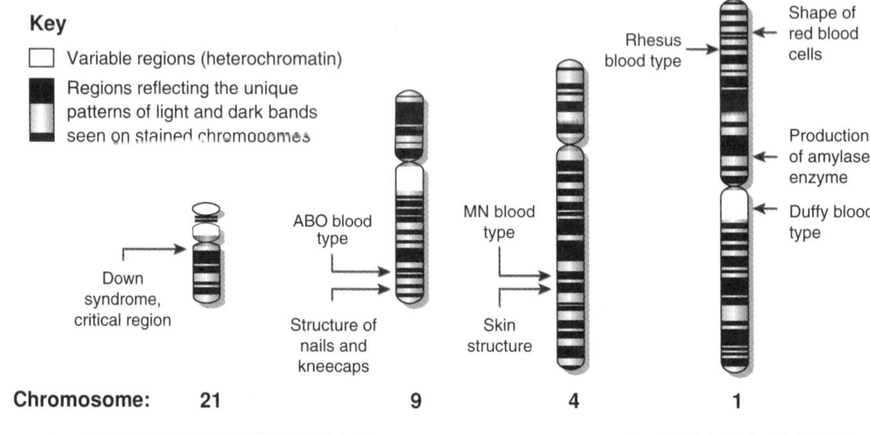

Key
Variable regions (heterochromatin)
Regions reflecting the unique patterns of light and dark bands seen on stained chromosomes
Down syndrome, critical region
ABO blood type
Structure of nails and kneecaps
MN blood type
Skin structure
Rhesus blood type
Shape of red blood cells
Production of amylase enzyme
Duffy blood type
Chromosome: 21 9 4 1

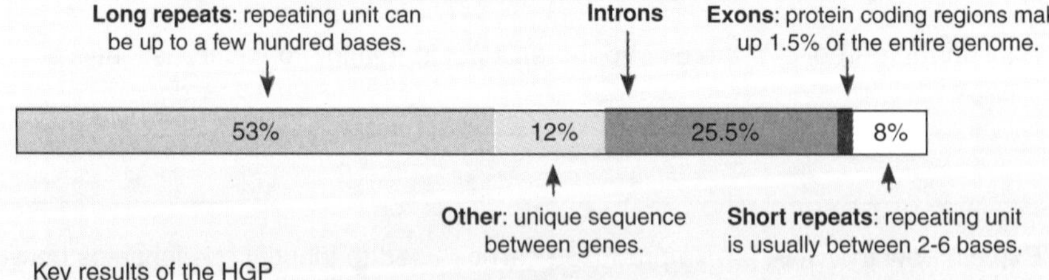

Long repeats: repeating unit can be up to a few hundred bases.
Introns
Exons: protein coding regions make up 1.5% of the entire genome.
53%
12%
25.5%
8%
Other: unique sequence between genes.
Short repeats: repeating unit is usually between 2-6 bases.

Key results of the HGP

- There are perhaps only 20,000-25,000 protein-coding genes in our human genome.
- It covers 99% of the gene containing parts of the genome and is 99.999% accurate.
- The new sequence correctly identifies almost all known genes (99.74%).
- Its accuracy and completeness allows systematic searches for causes of disease.

Related activities: The Genome

Periodicals: What is Genomics?, Revolution postponed

ISBN: 978-1-927173-16-9

Benefits and ethical issues arising from the Human Genome Project

Medical benefits

- Improved **diagnosis** of disease and predisposition to disease by genetic testing.
- Better identification of disease carriers, through genetic testing.
- Better **drugs** can be designed using knowledge of protein structure (from gene sequence information) rather than by trial and error.
- Greater possibility of successfully using **gene therapy** to correct genetic disorders.

Non-medical benefits

- Greater knowledge of **family relationships** through genetic testing, e.g. paternity testing in family courts.
- Advances **forensic science** through analysis of DNA at crime scenes.
- Improved knowledge of the evolutionary relationships between humans and other organisms, which will help to develop better, more accurate classification systems.

Possible ethical issues

- It is unclear whether third parties, e.g. health insurers, have rights to genetic test results.
- If treatment is unavailable for a disease, genetic knowledge about it may have no use.
- Genetic tests are costly, and there is no easy answer as to who should pay for them.
- Genetic information is hereditary so knowledge of an individual's own genome has implications for members of their family.

Couples can already have a limited range of genetic tests to determine the risk of having offspring with some disease-causing mutations.

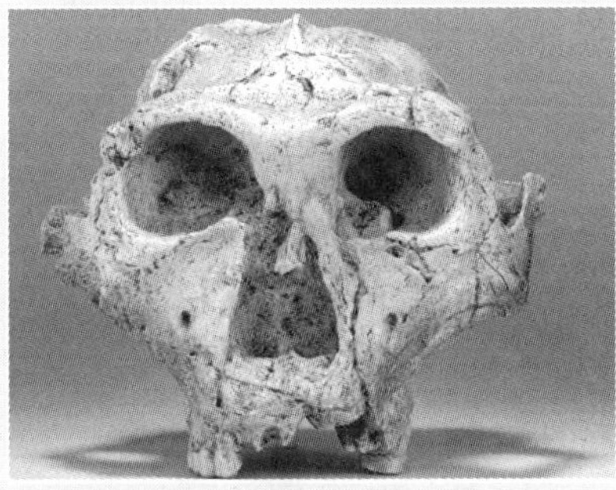

When DNA sequences are available for humans and their ancestors, comparative analysis may provide clues about human evolution.

Legislation is needed to ensure that there is no discrimination on the basis of genetic information, e.g. at work or for health insurance.

1. Briefly describe the objectives of the Human Genome Project (HGP) and the Human Genome Diversity Project (HGDP):

   HGP: ______________________________

   ____

   HGDP: ______________________________

2. Suggest a reason why indigenous peoples around the world are reluctant to provide DNA samples for the HGDP:

   ____

   ____

3. Describe two possible **benefits** of Human Genome Project (HGP):

   (a) Medical: ______________________________

   ____

   (b) Non-medical: ______________________________

   ____

4. Explain what is meant by **proteomics** and explain its significance to the HGP and the ongoing benefits arising from it:

   ____

   ____

   ____

   ____

5. Suggest two possible points of view for one of the **ethical issues** described in the list above:

   (a) ______________________________

   ____

   (b) ______________________________

   ____

Genetic Engineering and Biotechnology

ISBN: 978-1-927173-16-9

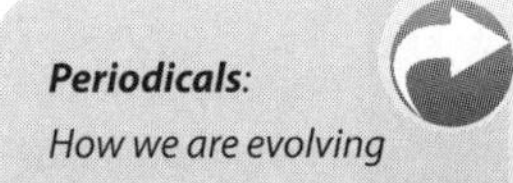

Periodicals:
How we are evolving

What is Genetic Modification?

The genetic modification of organisms is a vast industry, and the applications of the technology are exciting and far reaching. It brings new hope for medical cures, promises to increase yields in agriculture, and has the potential to help solve the world's pollution and resource crises. Organisms with artificially altered DNA are referred to as **genetically modified organisms** or **GMOs**. They may be modified in one of three ways (outlined below). Some of the current and proposed applications of gene technology raise complex ethical and safety issues. The benefits of their use must be carefully weighed against the risks to human health, as well as the health and well-being of other organisms and the environment as a whole.

Producing Genetically Modified Organisms (GMOs)

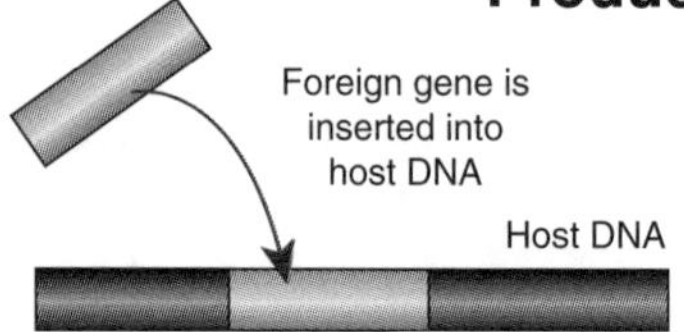
Foreign gene is inserted into host DNA
Host DNA

Add a foreign gene

A novel (foreign) gene is inserted from another species. This will enable the GMO to express the trait coded by the new gene. Organisms genetically altered in this way are referred to as **transgenic**.

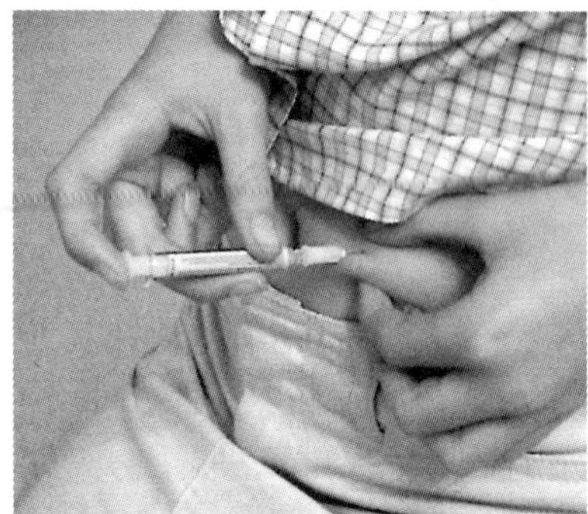

Human insulin, used to treat diabetic patients, is now produced using transgenic bacteria.

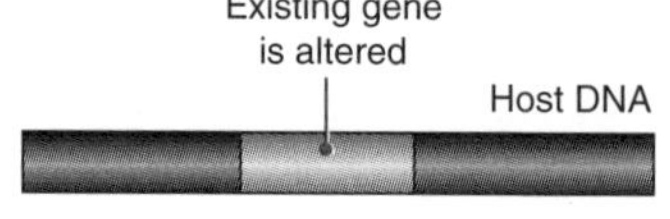
Existing gene is altered
Host DNA

Alter an existing gene

An existing gene may be altered to make it express at a higher level (e.g. produce more growth hormone) or in a different way (in tissue that would not normally express it). This method is also used for gene therapy.

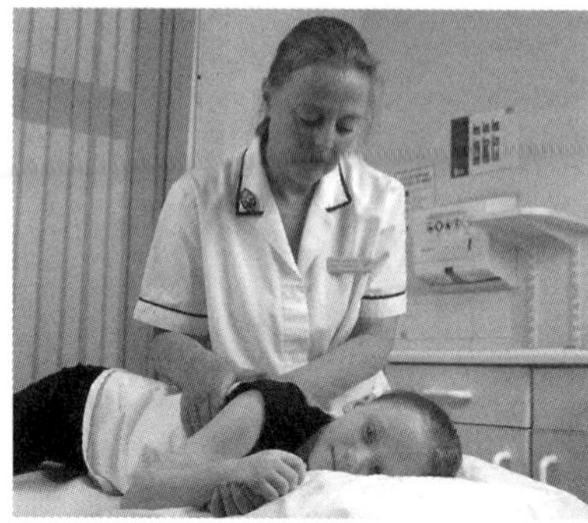

Gene therapy could be used treat genetic disorders, such as cystic fibrosis.

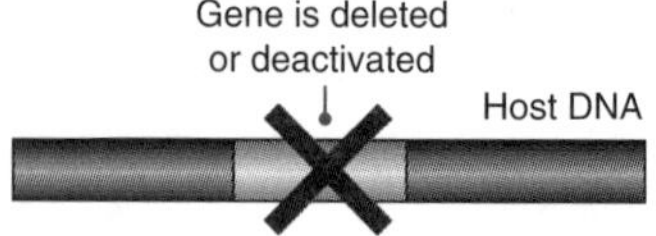
Gene is deleted or deactivated
Host DNA

Delete or 'turn off' a gene

An existing gene may be deleted or deactivated (switched off) to prevent the expression of a trait (e.g. the deactivation of the ripening gene in tomatoes produced the Flavr-Savr tomato).

Manipulating gene action is one way in which to control processes such as ripening in fruit.

1. Using examples, discuss the ways in which an organism may be genetically modified (to produce a GMO):

2. Explain how human needs or desires have provided a stimulus for the development of the following biotechnologies:

   (a) Gene therapy:

   (b) The production and use of transgenic organisms:

   (c) The extension of shelf life in stored produce:

Related activities: *Applications of GMOs*
Weblinks: *Biotechnology Timeline*

ISBN: 978-1-927173-16-9

Restriction Enzymes

One of the essential tools of genetic engineering is a group of special **restriction enzymes** (also known as restriction endonucleases). These have the ability to cut DNA molecules at very precise sequences of 4 to 8 base pairs called **recognition sites**. These enzymes are the "molecular scalpels" that allow genetic engineers to cut up DNA in a controlled way. Although first isolated in 1970, these enzymes were discovered earlier in many bacteria. The purified forms of these bacterial restriction enzymes are used today as tools to cut DNA (see table on the next page for examples). Enzymes are named according to the bacterial species from which they were first isolated. By using a 'tool kit' of over 400 restriction enzymes recognizing about 100 recognition sites, genetic engineers can isolate, sequence, and manipulate individual genes derived from any type of organism. The sites at which the fragments of DNA are cut may result in overhanging "sticky ends" or non-overhanging "blunt ends". Pieces may later be joined together using an enzyme called **DNA ligase** in a process called **ligation**.

Sticky End Restriction Enzymes

1. A **restriction enzyme** cuts the double-stranded DNA molecule at its specific **recognition** site (see the table on the following page for a representative list of restriction enzymes and their recognition sites).

2. The cuts produce a DNA fragment with two **sticky ends** (ends with exposed nucleotide bases at each end). The piece it is removed from is also left with sticky ends.

Restriction enzymes may cut DNA leaving an overhang or sticky end, without its complementary sequence opposite. DNA cut in such a way is able to be joined to other exposed end fragments of DNA with matching sticky ends. Such joins are specific to their recognition sites.

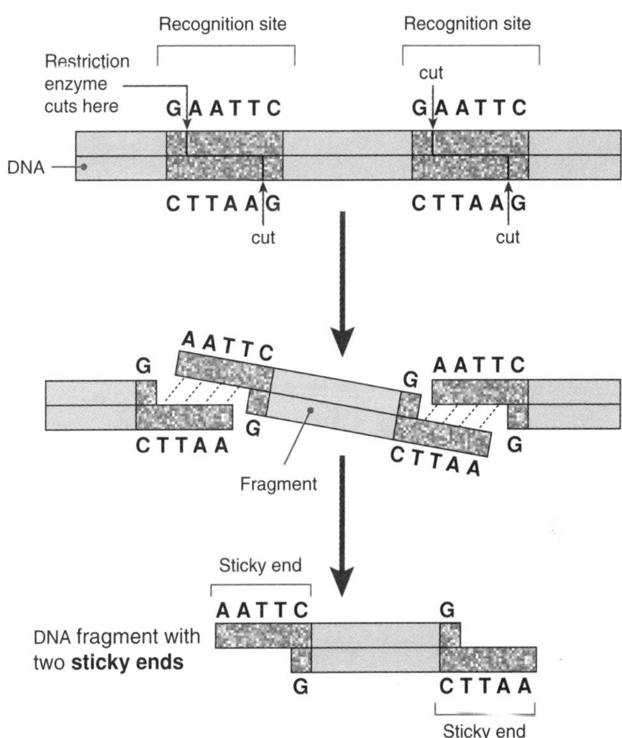

Recognition site
Recognition site
Restriction enzyme cuts here
cut
G A A T T C
G A A T T C
DNA
C T T A A G
C T T A A G
cut
cut
A A T T C
G
G
A A T T C
C T T A A
G
C T T A A
G
Fragment
Sticky end
A A T T C
G
DNA fragment with two **sticky ends**
G
C T T A A
Sticky end

Blunt End Restriction Enzymes

1. A **restriction enzyme** cuts the double-stranded DNA molecule at its specific recognition site (see the table opposite for a representative list of restriction enzymes and their recognition sites).

2. The cuts produce a DNA fragment with two **blunt ends** (ends with no exposed nucleotide bases at each end). The piece it is removed from is also left with blunt ends.

It is possible to use restriction enzymes that cut leaving no overhang. DNA cut in such a way is able to be joined to any other blunt end fragment, but tends to be nonspecific because there are no sticky ends as recognition sites.

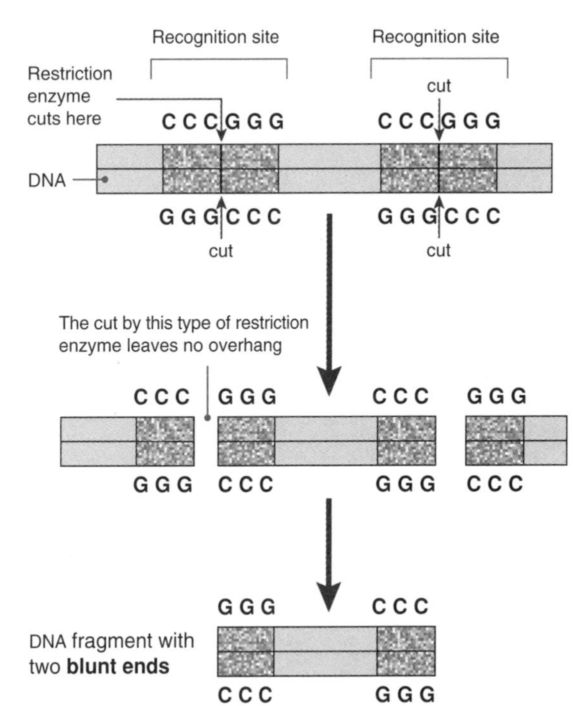

Recognition site
Recognition site
Restriction enzyme cuts here
cut
C C C G G G
C C C G G G
DNA
G G G C C C
G G G C C C
cut
cut
The cut by this type of restriction enzyme leaves no overhang
C C C
G G G
C C C
G G G
G G G
C C C
G G G
C C C
G G G
C C C
DNA fragment with two **blunt ends**
C C C
G G G

ISBN: 978-1-927173-16-9

Related activities: *Ligation*
Weblinks: *DNA Interactive: Cut and Paste*

Origin of Restriction Enzymes

Restriction enzymes have been isolated from many bacteria. It was observed that certain *bacteriophages* (viruses that infect bacteria) could not infect bacteria other than their usual hosts. The reason was found to be that other potential hosts could destroy almost all of the phage DNA using *restriction enzymes* present naturally in their cells; a defence mechanism against the entry of foreign DNA. Restriction enzymes are named according to the species they were first isolated from, followed by a number to distinguish different enzymes isolated from the same organism.

Recognition sites for selected restriction enzymes

Enzyme	Source	Recognition sites
***Eco*RI**	*Escherichia coli* RY13	G A A T T C
***Bam*HI**	*Bacillus amyloliquefaciens* H	G G A T C C
***Hae*III**	*Haemophilus aegyptius*	G G C C
***Hind*III**	*Haemophilus influenzae* Rd	A A G C T T
***Hpa*I**	*Haemophilus parainfluenzae*	G T T A A C
***Hpa*II**	*Haemophilus parainfluenzae*	C C G G
***Mbo*I**	*Moraxella bovis*	G A T C
***Not*I**	*Norcardia otitidis-caviarum*	G C G G C C G C
***Taq*I**	*Thermus aquaticus*	T C G A

1. Explain the following terms, identifying their role in recombinant DNA technology:

   (a) Restriction enzyme:

   (b) Recognition site:

   (c) Sticky end:

   (d) Blunt end:

2. The action of a specific sticky end restriction enzyme is illustrated on the previous page (top). Use the table above to:

   (a) Name the **restriction enzyme** used:

   (b) Name the organism from which it was first isolated:

   (c) State the **base sequence** for this restriction enzyme's recognition site:

3. A genetic engineer wants to use the restriction enzyme ***Bam*HI** to cut the DNA sequence below:

   (a) Consult the table above and state the recognition site for this enzyme:

   (b) Circle every **recognition site** on the DNA sequence below that could be cut by the enzyme ***Bam*HI**:

```
          10         20         30         40         50         60
|AATGGGTACG|CACAGTGGAT|CCACGTAGTA|TGCGATGCGT|AGTGTTTATG|GAGAGAAGAA|
          70         80         90        100        110        120
|AACGCGTCGC|CTTTTATCGA|TGCTGTACGG|ATGCGGAAGT|GGCGATGAGG|ATCCATGCAA|
         130        140        150        160        170        180
|TCGCGGCCGA|TCGXGTAATA|TATCGTGGCT|GCGTTTATTA|TCGTGACTAG|TAGCAGTATG|
         190        200        210        220        230        240
|CGATGTGACT|GATGCTATGC|TGACTATGCT|ATGTTTTTAT|GCTGGATCCA|GCGTAAGCAT|
         250        260        270        280        290        300
|TTCGCTGCGT|GGATCCCATA|TCCTTATATG|CATATATTCT|TATACGGATC|GCGCACGTTT|
```

   (c) State how many fragments of DNA were created by this action:

4. When restriction enzymes were first isolated in 1970, there were not many applications for them. Now, they are an important tool in genetic engineering. Describe the human needs and demands that have driven the development and use of restriction enzymes in genetic engineering:

ISBN: 978-1-927173-16-9

Ligation

DNA fragments produced using restriction enzymes may be reassembled by a process called **ligation**. Pieces are joined together using an enzyme called **DNA ligase**. DNA of different origins produced in this way is called **recombinant DNA** (because it is DNA that has been recombined from different sources). The combined techniques of using restriction enzymes and ligation are the basic tools of genetic engineering (also known as recombinant DNA technology).

Creating a Recombinant DNA Plasmid

1. If two pieces of DNA are cut by the same restriction enzyme, they will produce fragments with matching **sticky ends** (ends with exposed nucleotide bases at each end).

2. When two such matching sticky ends come together, they can join by base-pairing. This process is called **annealing.** This can allow DNA fragments from a different source, perhaps a **plasmid**, to be joined to the DNA fragment.

3. The joined fragments will usually form either a linear molecule or a circular one, as shown here for a **plasmid**. However, other combinations of fragments can occur.

4. The fragments of DNA are joined together by the enzyme **DNA ligase**, producing a molecule of **recombinant DNA.**

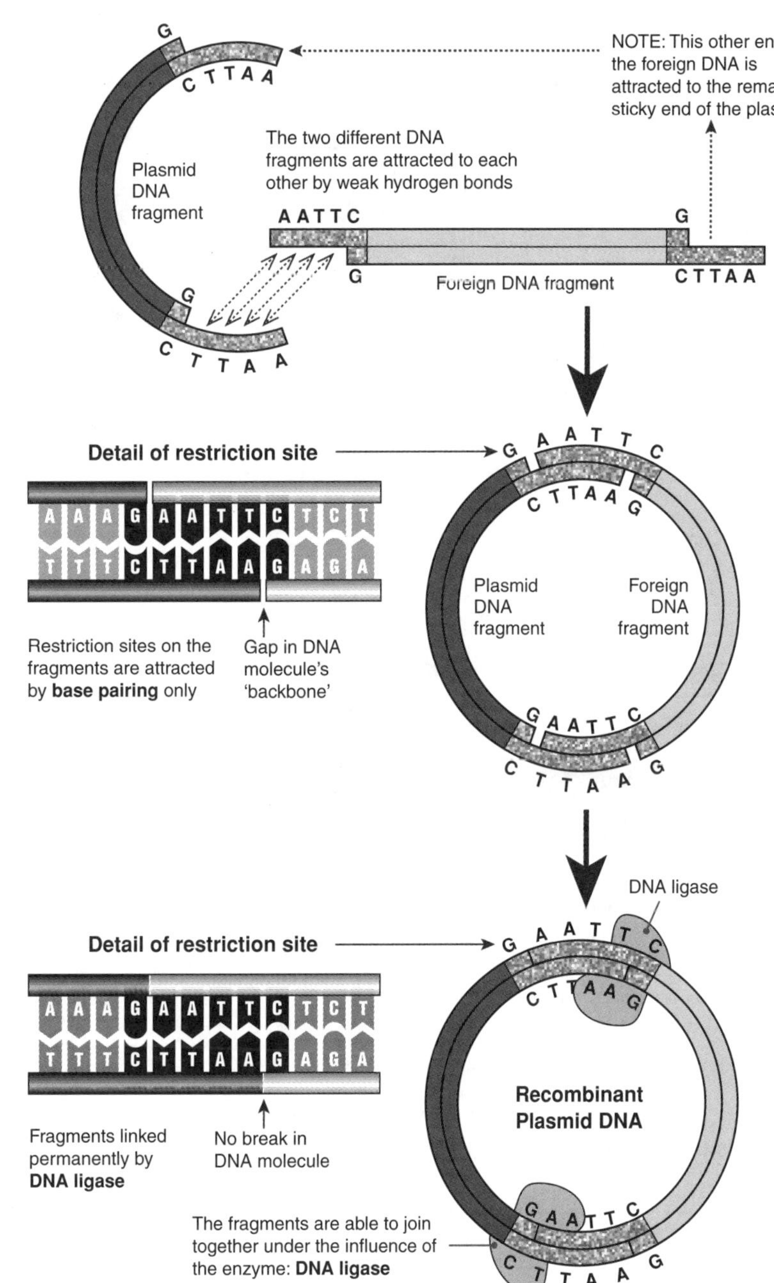

NOTE: This other end of the foreign DNA is attracted to the remaining sticky end of the plasmid
The two different DNA fragments are attracted to each other by weak hydrogen bonds
Plasmid DNA fragment
Foreign DNA fragment
AATTC
CTTAA
Detail of restriction site
Restriction sites on the fragments are attracted by base pairing only
Gap in DNA molecule's 'backbone'
Plasmid DNA fragment
Foreign DNA fragment
DNA ligase
Fragments linked permanently by DNA ligase
No break in DNA molecule
Recombinant Plasmid DNA
The fragments are able to join together under the influence of the enzyme: DNA ligase

1. Explain in your own words the two main steps in the process of joining two DNA fragments together:

(a) Annealing: ______________________________

(b) DNA ligase: ______________________________

2. Refer to the activity "DNA Replication", earlier in this book, and state the usual role of DNA ligase in a cell:

____

3. Explain why **ligation** can be considered the *reverse* of the **restriction enzyme** process: ______________________________

ISBN: 978-1-927173-16-9

Applications of GMOs

Techniques for genetic manipulation are now widely applied throughout modern biotechnology: in food and enzyme technology, in industry and medicine, and in agriculture and horticulture. Microorganisms are among the most widely used GMOs, with applications ranging from pharmaceutical production and vaccine development to environmental clean-up. Crop plants are also popular candidates for genetic modification although their use, as with much of genetic engineering of higher organisms, is controversial and sometimes problematic.

Application of GMOs

Extending shelf life

Some fresh produce (e.g. tomatoes) have been engineered to have an extended keeping quality. In the case of tomatoes, the gene for ripening has been switched off, delaying the process of softening in the fruit.

Pest or herbicide resistance

Plants can be engineered to produce their own insecticide and become pest resistant. Genetically engineered herbicide resistance is also common. In this case, chemical weed killers can be used freely without crop damage.

Crop improvement

Gene technology is now an integral part of the development of new crop varieties. Crops can be engineered to produce higher protein levels or to grow in inhospitable conditions (e.g. salty or arid conditions).

Environmental clean-up

Some bacteria have been engineered to thrive on waste products, such as liquefied newspaper pulp or oil. As well as degrading pollutants and wastes, the bacteria may be harvested as a commercial protein source.

Biofactories

Transgenic bacteria are widely used to produce desirable products: often hormones or proteins. Large quantities of a product can be produced using bioreactors (above). Examples: insulin production by recombinant yeast, production of bovine growth hormone.

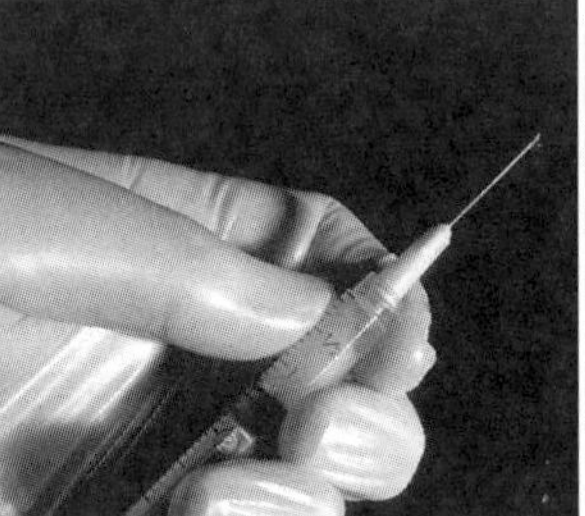

Vaccine development

The potential exists for multipurpose vaccines to be made using gene technology. Genes coding for vaccine components (e.g. viral protein coat) are inserted into an unrelated live vaccine (e.g. polio vaccine), and deliver proteins to stimulate an immune response.

Livestock improvement using transgenic animals

Transgenic sheep have been used to enhance wool production in flocks (above, left). The keratin protein of wool is largely made of a single amino acid, cysteine. Injecting developing sheep with the genes for the enzymes that generate cysteine produces woollier transgenic sheep. In some cases, transgenic animals have been used as biofactories. Transgenic sheep carrying the human gene for a protein, α-1-antitrypsin produce the protein in their milk. The antitrypsin is extracted from the milk and used to treat hereditary emphysema.

1. Discuss the potential benefits and disadvantages of using GMOs for one of the applications described above:

Related activities: *The Ethics of GMO Technology, Golden Rice, Production of Insulin*

Periodicals: *Birds, bees and super-weeds*

ISBN: 978-1-927173-16-9

In Vivo Gene Cloning

It is possible to use the internal replication machinery of a cell to clone a gene, or even many genes, at once. By using cells to copy desired genes, it is also possible to produce any protein product the genes may code for. Recombinant DNA techniques (restriction digestion and ligation) are used to insert a gene of interest into the DNA of a vector (e.g. plasmid or viral DNA). This produces a **recombinant DNA molecule** that can used to transmit the gene of interest to another organism. To be useful, all vectors must be able to replicate inside their host organism, they must have one or more sites at which a restriction enzyme can cut, and they must have some kind of **genetic marker** that allows them to be easily identified. Viruses, and organisms such as bacteria and yeasts have DNA that behaves in this way. Bacterial plasmids are commonly used because they are easy to manipulate, their restriction sites are well known, and they are readily taken up by cells in culture. Once the recombinant plasmid vector (containing the desired gene) has been taken up by bacterial cells, and those cells are identified, the gene can be replicated many times as the bacteria grow and divide.

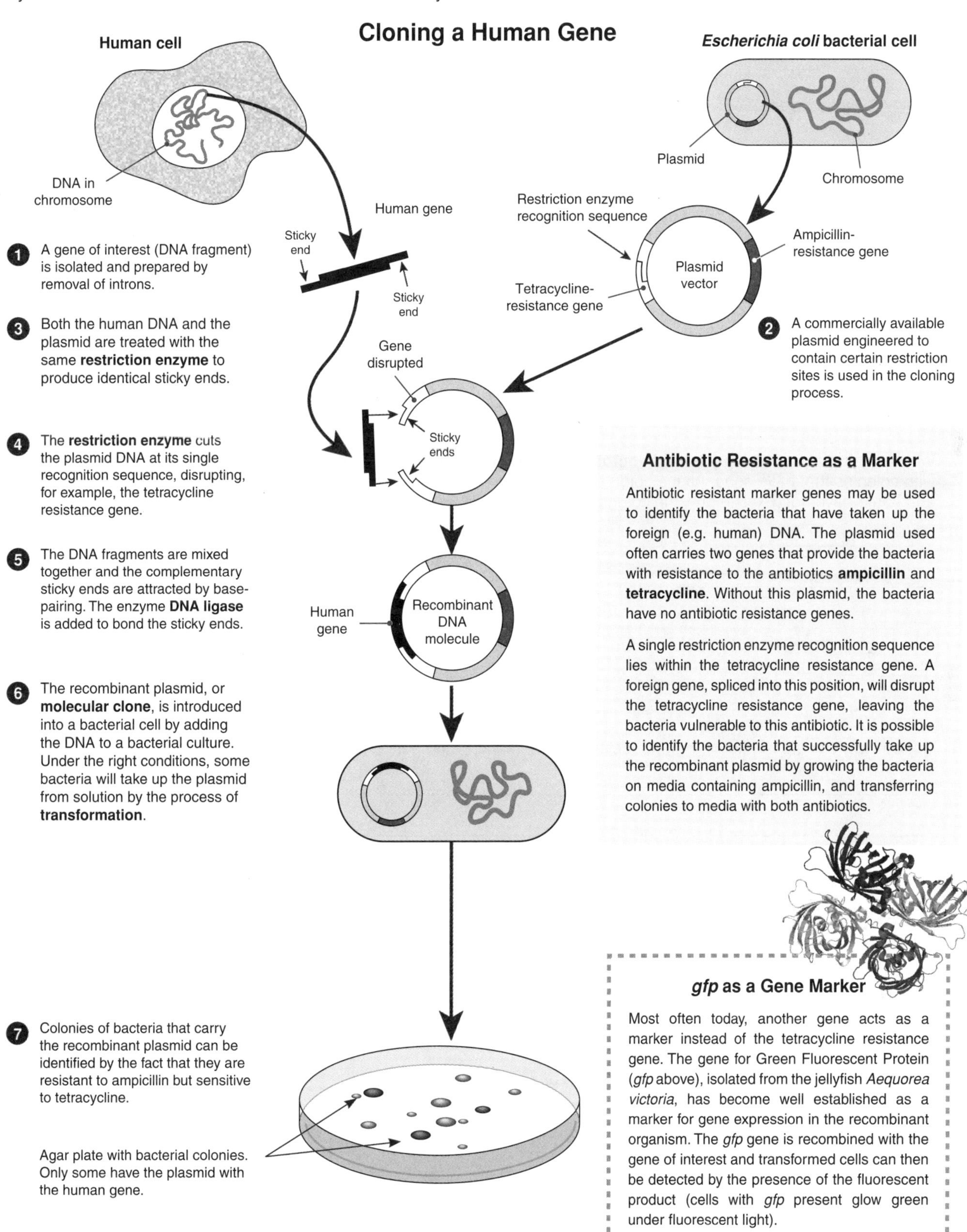
Cloning a Human Gene
Human cell
DNA in chromosome
Escherichia coli bacterial cell
Plasmid
Chromosome
Human gene
Sticky end
Sticky end
Restriction enzyme recognition sequence
Ampicillin-resistance gene
Plasmid vector
Tetracycline-resistance gene
1 A gene of interest (DNA fragment) is isolated and prepared by removal of introns.
2 A commercially available plasmid engineered to contain certain restriction sites is used in the cloning process.
3 Both the human DNA and the plasmid are treated with the same restriction enzyme to produce identical sticky ends.
Gene disrupted
Sticky ends
4 The restriction enzyme cuts the plasmid DNA at its single recognition sequence, disrupting, for example, the tetracycline resistance gene.
5 The DNA fragments are mixed together and the complementary sticky ends are attracted by base-pairing. The enzyme DNA ligase is added to bond the sticky ends.
Human gene
Recombinant DNA molecule
6 The recombinant plasmid, or molecular clone, is introduced into a bacterial cell by adding the DNA to a bacterial culture. Under the right conditions, some bacteria will take up the plasmid from solution by the process of transformation.
7 Colonies of bacteria that carry the recombinant plasmid can be identified by the fact that they are resistant to ampicillin but sensitive to tetracycline.
Agar plate with bacterial colonies. Only some have the plasmid with the human gene.

Antibiotic Resistance as a Marker

Antibiotic resistant marker genes may be used to identify the bacteria that have taken up the foreign (e.g. human) DNA. The plasmid used often carries two genes that provide the bacteria with resistance to the antibiotics **ampicillin** and **tetracycline**. Without this plasmid, the bacteria have no antibiotic resistance genes.

A single restriction enzyme recognition sequence lies within the tetracycline resistance gene. A foreign gene, spliced into this position, will disrupt the tetracycline resistance gene, leaving the bacteria vulnerable to this antibiotic. It is possible to identify the bacteria that successfully take up the recombinant plasmid by growing the bacteria on media containing ampicillin, and transferring colonies to media with both antibiotics.

gfp as a Gene Marker

Most often today, another gene acts as a marker instead of the tetracycline resistance gene. The gene for Green Fluorescent Protein (*gfp* above), isolated from the jellyfish *Aequorea victoria*, has become well established as a marker for gene expression in the recombinant organism. The *gfp* gene is recombined with the gene of interest and transformed cells can then be detected by the presence of the fluorescent product (cells with *gfp* present glow green under fluorescent light).

Genetic Engineering and Biotechnology

ISBN: 978-1-927173-16-9

Related activities: *Restriction Enzymes*
Weblinks: *Gene Cloning*

1. Explain why it might be desirable to use *in vivo* methods to clone genes rather than PCR:

2. Explain when it may not be desirable to use bacteria to clone genes:

3. Explain how a human gene is removed from a chromosome and placed into a plasmid.

4. A bacterial plasmid replicates at the same rate as the bacteria. If a bacteria containing a recombinant plasmid replicates and divides once every thirty minutes, calculate the number of plasmid copies there will be after twenty four hours:

5. When cloning a gene using **plasmid vectors**, the bacterial colonies containing the recombinant plasmids are mixed up with colonies that have none. All the colonies look identical, but some have taken up the plasmids with the human gene, and some have not. Explain how the colonies with the recombinant plasmids are identified:

6. Explain why the *gfp* marker is a more desirable gene marker than genes for antibiotic resistance:

7. Bacteriophages are viruses that infect bacteria:

   (a) What feature of bacteriophages make them useful for genetic engineering?

   (b) How could a bacteriophage be used to clone a gene?

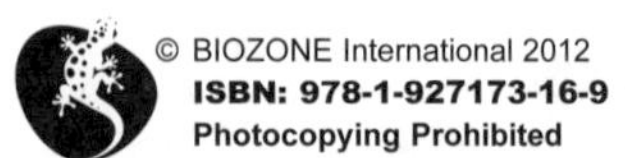

ISBN: 978-1-927173-16-9

Using Recombinant Bacteria

The Issue

- **Chymosin** (also known as **rennin**) is an enzyme that digests milk proteins. It is the active ingredient in rennet, a substance used by cheesemakers to clot milk into curds.
- Traditionally rennin is extracted from "chyme", i.e. the stomach secretions of suckling calves (hence its name of chymosin).
- By the 1960s, a shortage of chymosin was limiting the volume of cheese produced.
- Enzymes from fungi were used as an alternative but were unsuitable because they caused variations in the cheese flavor.

Concept 1	Concept 2	Concept 3	Concept 4	Concept 5
Enzymes are proteins made up of amino acids. The amino acid sequence of chymosin can be determined and the mRNA coding sequence for its translation identified.	**Reverse transcriptase** can be used to synthesize a DNA strand from the mRNA. This process produces DNA without the introns, which cannot be processed by bacteria.	DNA can be cut at specific sites using **restriction enzymes** and rejoined using **DNA ligase.** New genes can be inserted into self-replicating bacterial **plasmids**.	Under certain conditions, bacteria are able to lose or take up plasmids from their environment. Bacteria are readily grown in vat cultures at little expense.	The protein in made by the bacteria in large quantities.

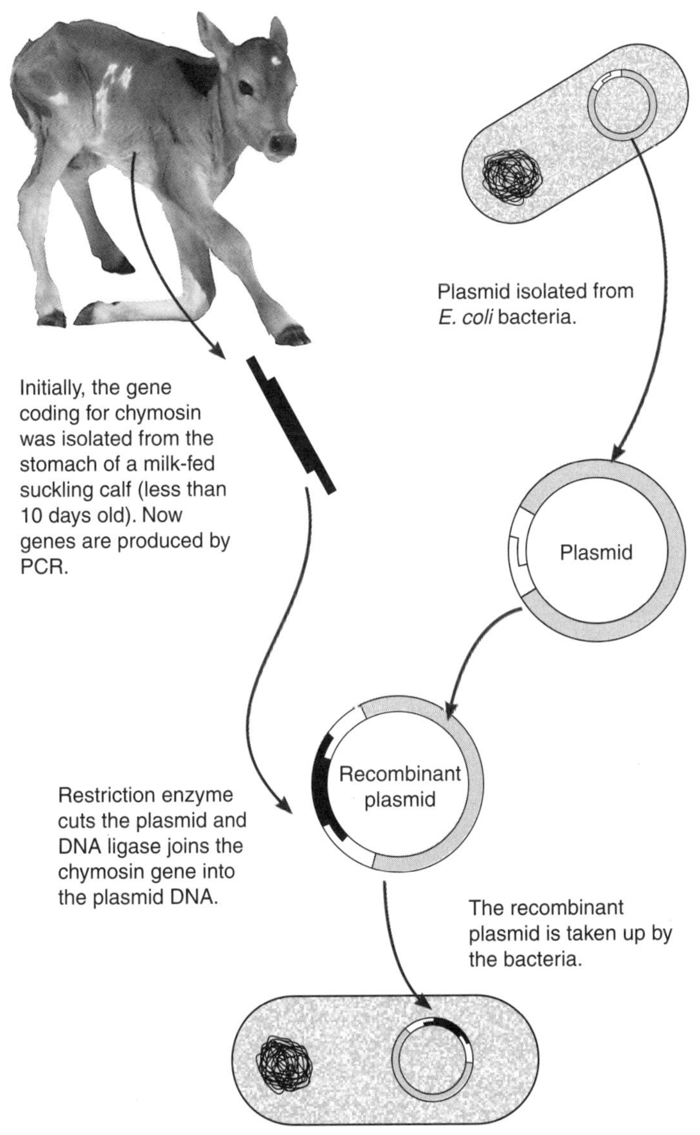

Plasmid isolated from *E. coli* bacteria.
Initially, the gene coding for chymosin was isolated from the stomach of a milk-fed suckling calf (less than 10 days old). Now genes are produced by PCR.
Plasmid
Recombinant plasmid
Restriction enzyme cuts the plasmid and DNA ligase joins the chymosin gene into the plasmid DNA.
The recombinant plasmid is taken up by the bacteria.
Transformed bacterial cells are grown in a vat culture

Techniques

The amino acid sequence of chymosin is first determined and the RNA codons for each amino acid identified.

mRNA matching the identified sequence is isolated from the stomach of young calves. **Reverse transcriptase** is used to transcribe mRNA into DNA. The DNA sequence can also be made synthetically once the sequence is determined.

The DNA is amplified using PCR.

Plasmids from *E. coli* bacteria are isolated and cut using **restriction enzymes.** The DNA sequence for chymosin is inserted using **DNA ligase**.

Plasmids are returned to *E. coli* by placing the bacteria under conditions that induce them to take up plasmids.

Outcomes

The transformed bacteria are grown in vat culture. Chymosin is produced by *E. coli* in packets within the cell that are separated during the processing and refining stage.

Recombinant chymosin entered the marketplace in 1990. It established a significant market share because cheesemakers found it to be cost effective, of high quality, and in consistent supply. Most cheese is now produced using recombinant chymosin such as CHY-MAX.

Further Applications

A large amount of processing is required to extract chymosin from *E.coli*. There are now a number of alternative bacteria and fungi that have been engineered to produce the enzyme. Most chymosin is now produced using the fungi ***Aspergillus niger*** and ***Kluyveromyces lactis***. Both are produced in a similar way as that described for *E. coli*.

ISBN: 978-1-927173-16-9

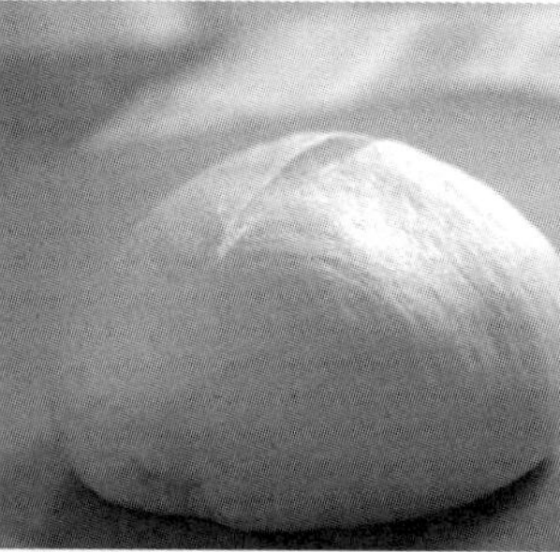

Enzymes from GMOs are widely used in the baking industry. Maltogenic alpha amylase from *Bacillus subtilis* bacteria is used as an anti-staling agent to prolong shelf life. Hemicellulases from *B. subtilis* and xylanase from the fungus *Aspergillus oryzae* are used for improvement of dough, crumb structure, and volume during the baking process.

Romain Behar

Lipase from *Aspergillus oryzae* is used in processing of palm oil to produce low cost cocoa butter substitutes (above), which have a similar 'mouth feel' to cocoa butter.

Dual Freq

Acetolactate decarboxylase from *B. subtilis* is one of several enzymes used in the brewing industry. It reduces maturation time of the beer by by-passing a rate-limiting step.

1. Describe the main use of chymosin: ______________________________

2. What was the traditional source of chymosin? ______________________________

3. Summarize the key concepts that led to the development of the technique for producing chymosin:

   (a) Concept 1: ______________________________

   (b) Concept 2: ______________________________

   (c) Concept 3: ______________________________

   (d) Concept 4: ______________________________

   (e) Concept 5: ______________________________

4. Discuss how the gene for chymosin was isolated and how the technique could be applied to isolating other genes:

5. Describe three advantages of using chymosin produced by GE bacteria over chymosin from traditional sources:

   (a) ______________________________

   (b) ______________________________

   (c) ______________________________

6. Explain why the fungus *Aspergillus niger* is now more commonly used to produce chymosin instead of *E. coli*:

ISBN: 978-1-927173-16-9

Golden Rice

The Issue

- **Beta-carotene** (β-carotene) is a precursor to **vitamin A** which is involved in many functions including vision, immunity, fetal development, and skin health.
- Vitamin A deficiency is common in developing countries where up to 500,000 children suffer from night blindness, and death rates due to infections are high due to a lowered immune response.
- Providing enough food containing useful quantities of β-carotene is difficult and expensive in many countries.

Concept 1

Rice is a staple food in many developing countries. It is grown in large quantities and is available to most of the population, but it lacks many of the essential nutrients required by the human body for healthy development. It is low in β-carotene.

Concept 2

Rice plants produce β-carotene but not in the edible rice **endosperm**. Engineering a new biosynthetic pathway would allow β-carotene to be produced in the endosperm. Genes expressing enzymes for carotene synthesis can be inserted into the rice genome.

Concept 3

The enzyme **carotene desaturase (CRT1)** in the soil bacterium *Erwinia uredovora,* catalyses multiple steps in carotenoid biosynthesis. **Phytoene synthase (PSY)** overexpresses a colorless carotene in the daffodil plant *Narcissus pseudonarcissus.*

Concept 4

DNA can be inserted into an organism's genome using a suitable **vector**. ***Agrobacterium tumefaciens*** is a tumor-forming bacterial plant pathogen that is commonly used to insert novel DNA into plants.

The Development of Golden Rice

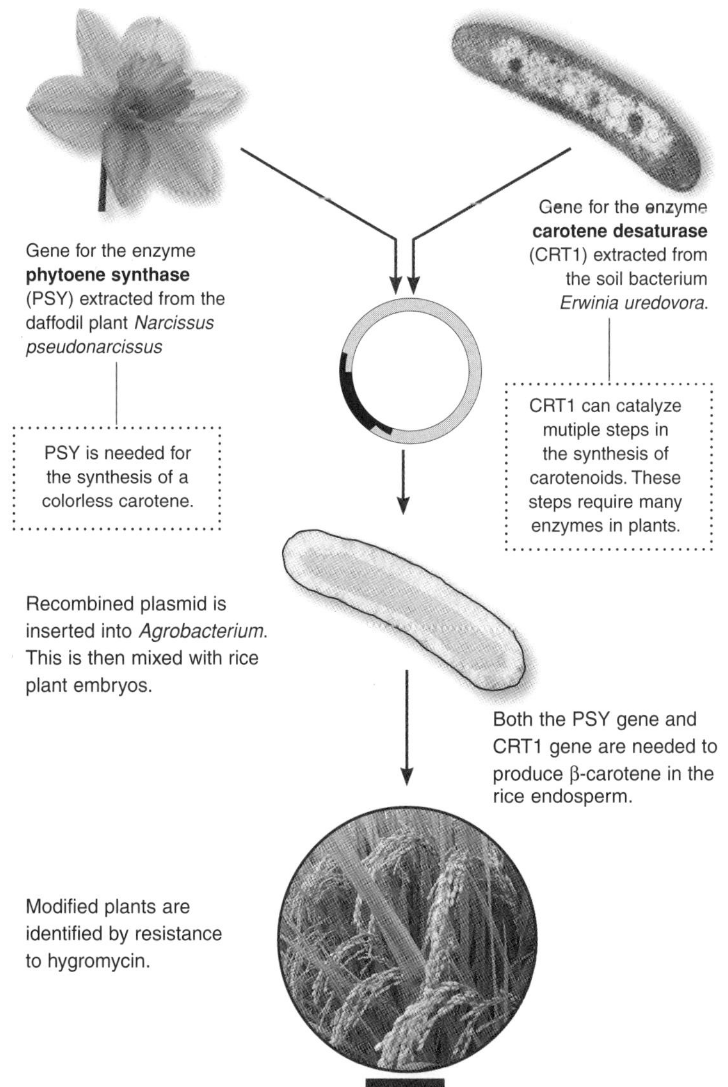
Gene for the enzyme **phytoene synthase** (PSY) extracted from the daffodil plant *Narcissus pseudonarcissus*
PSY is needed for the synthesis of a colorless carotene.
Gene for the enzyme **carotene desaturase** (CRT1) extracted from the soil bacterium *Erwinia uredovora.*
CRT1 can catalyze mutiple steps in the synthesis of carotenoids. These steps require many enzymes in plants.
Recombined plasmid is inserted into *Agrobacterium.* This is then mixed with rice plant embryos.
Both the PSY gene and CRT1 gene are needed to produce β-carotene in the rice endosperm.
Modified plants are identified by resistance to hygromycin.
SGR1

Techniques

The **PSY** gene from daffodils and the **CRT1** gene from *Erwinia uredovora* are sequenced.

DNA sequences are synthesized into packages containing the CRT1 or PSY gene, terminator sequences, and **endosperm specific promoters** (these ensure expression of the gene only in the edible portion of the rice).

The ***Ti* plasmid** from *Agrobacterium* is modified using restriction enzymes and DNA ligase to delete the tumor-forming gene and insert the synthesized DNA packages. A gene for resistance to the antibiotic **hygromycin** is also inserted so that transformed plants can be identified later. The parts of the *Ti* plasmid required for plant transformation are retained.

Modified *Ti* plasmid is inserted into the bacterium.

Agrobacterium is incubated with rice plant embryo. Transformed embryos are identified by their resistance to hygromycin.

Outcomes

The rice produced had endosperm with a distinctive yellow color. Under greenhouse conditions golden rice (**SGR1**) contained 1.6 µg per g of carotenoids. Levels up to five times higher were produced in the field, probably due to improved growing conditions.

Further Applications

Further research on the action of the PSY gene identified more efficient methods for the production of β-carotene. The second generation of golden rice now contains up to 37 µg per g of carotenoids. Golden rice was the first instance where a complete biosynthetic pathway was engineered. The procedures could be applied to other food plants to increase their nutrient levels.

ISBN: 978-1-927173-16-9

Periodicals:
Rice, risk, and regulations,
The engineering of crop plants

Related activities: *In Vivo Gene Cloning*
Weblinks: *Genetically Modified Plants*

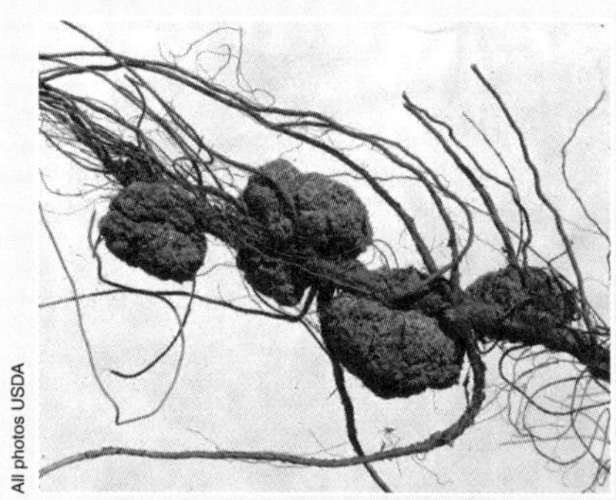
All photos USDA

The ability of *Agrobacterium* to transfer genes to plants is exploited for crop improvement. The tumor-inducing *Ti* plasmid is modified to delete the tumor-forming gene and insert a gene coding for a desirable trait. The parts of the *Ti* plasmid required for plant transformation are retained.

Soybeans are one of the many food crops that have been genetically modified for broad spectrum herbicide resistance. The first GM soybeans were planted in the US in 1996. By 2007, nearly 60% of the global soybean crop was genetically modified; the highest of any other crop plant.

GM cotton was produced by inserting the gene for the BT toxin into its genome. The bacterium *Bacillus thuringiensis* naturally produces BT toxin, which is harmful to a range of insects, including the larvae that eat cotton. The BT gene causes cotton to produce this insecticide in its tissues.

1. Describe the basic methodology used to create golden rice: ____________________

2. Explain how scientists ensured β-carotene was produced in the endosperm: ____________________

3. What property of *Agrobacterium tumefaciens* makes it an ideal vector for introducing new genes into plants?

4. (a) How could this new variety of rice reduce disease in developing countries? ____________________

   (b) Absorption of vitamin A requires sufficient dietary fat. Explain how this could be problematic for the targeted use of golden rice in developing countries:

5. As well as increasing nutrient content as in golden rice, other traits of crop plants are also desirable. For each of the following traits, suggest features that could be desirable in terms of increasing yield:

   (a) Grain size or number: ____________________

   (b) Maturation rate: ____________________

   (c) Pest resistance: ____________________

ISBN: 978-1-927173-16-9

Production of Insulin

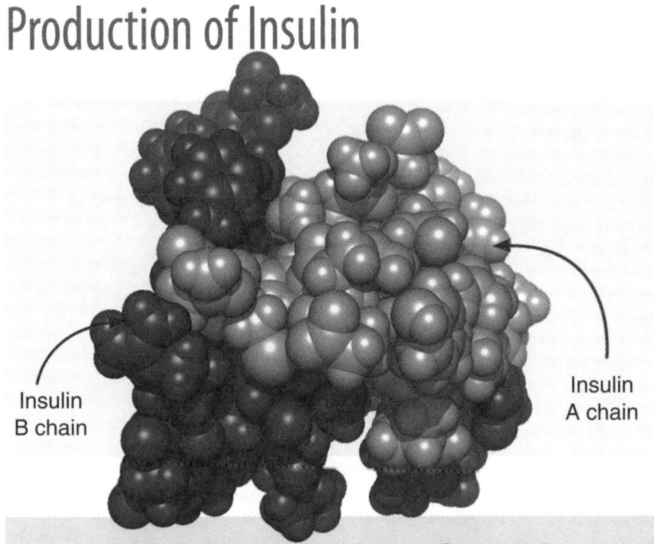
Insulin
B chain
Insulin
A chain

The Issue

- **Type I diabetes mellitus** is a metabolic disease caused by a lack of **insulin**. Around 25 people in every 100,000 suffer from type I diabetes.
- It is treatable only with injections of insulin.
- In the past, insulin was taken from the pancreases of cows and pigs and purified for human use. The method was expensive and some patients had severe allergic reactions to the foreign insulin or its contaminants.

Concept 1

DNA can be cut at specific sites using **restriction enzymes** and joined together using **DNA ligase**. New genes can be inserted into self-replicating bacterial **plasmids** at the point where the cuts are made.

Concept 2

Plasmids are small, circular pieces of DNA found in some bacteria. They usually carry genes useful to the bacterium. *E. coli* plasmids can carry promoters required for the transcription of genes.

Concept 3

Under certain conditions, Bacteria are able to lose or pick up plasmids from their environment. Bacteria can be readily grown in vat cultures at little expense.

Concept 4

The DNA sequences coding for the production of the two polypeptide chains (A and B) that form human insulin can be isolated from the human genome.

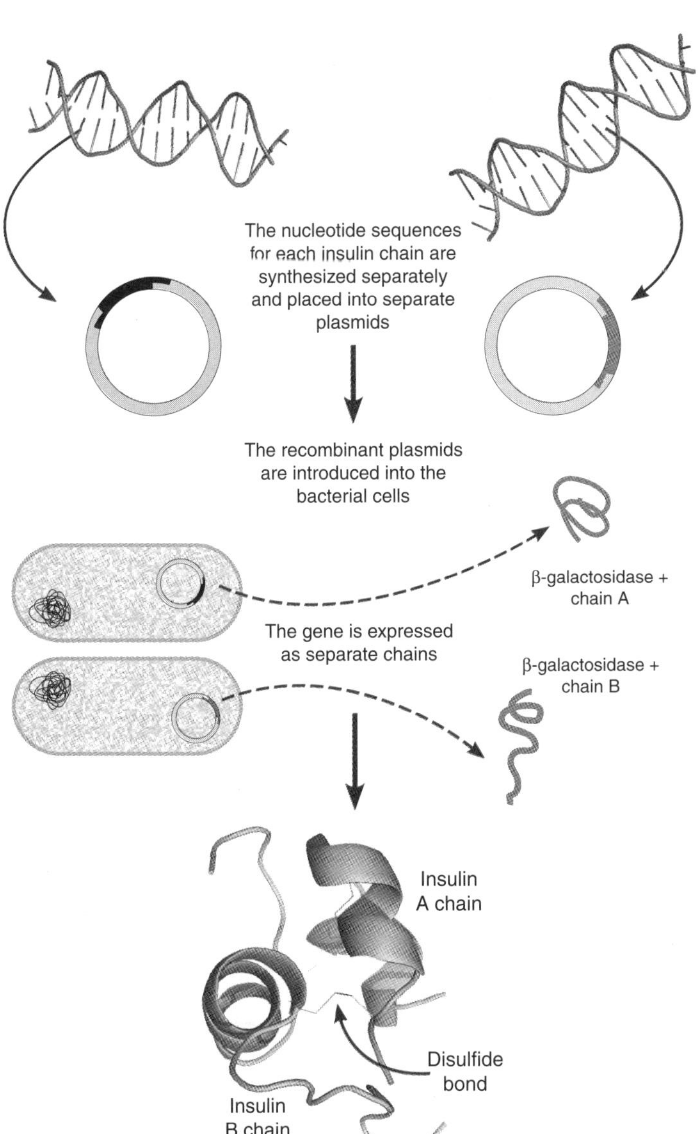
The nucleotide sequences for each insulin chain are synthesized separately and placed into separate plasmids
The recombinant plasmids are introduced into the bacterial cells
β-galactosidase + chain A
The gene is expressed as separate chains
β-galactosidase + chain B
Insulin
A chain
Disulfide
bond
Insulin
B chain

Techniques

The **gene** is **chemically synthesized** as two nucleotide sequences, one for the **insulin A chain** and one for the **insulin B chain**. The two sequences are small enough to be inserted into a plasmid.

Plasmids are extracted from *Escherichia coli*. The gene for the bacterial enzyme **β-galactosidase** is located on the plasmid. To make the bacteria produce insulin, the insulin gene must be linked to the β-galactosidase gene, which carries a promoter for transcription.

Restriction enzymes are used to cut plasmids at the appropriate site and the A and B insulin sequences are inserted. The sequences are joined with the plasmid DNA using **DNA ligase**.

The **recombinant plasmids** are inserted back into the bacteria by placing them together in a culture that favors plasmid uptake by bacteria.

The bacteria are then grown and multiplied in vats under carefully controlled growth conditions.

Outcomes

The product consists partly of β-galactosidase, joined with either the A or B chain of insulin. The chains are extracted, purified, and mixed together. The A and B insulin chains connect via **disulfide cross linkages** to form the functional insulin protein. The insulin can then be made ready for injection in various formulations.

Further Applications

The techniques involved in producing human insulin from genetically modified bacteria can be applied to a range of human proteins and hormones. Proteins currently being produced include human growth hormone, interferon, and factor VIII.

ISBN: 978-1-927173-16-9

Related activities: *Restriction Enzymes, Ligation*

Insulin production in *Saccharomyces*

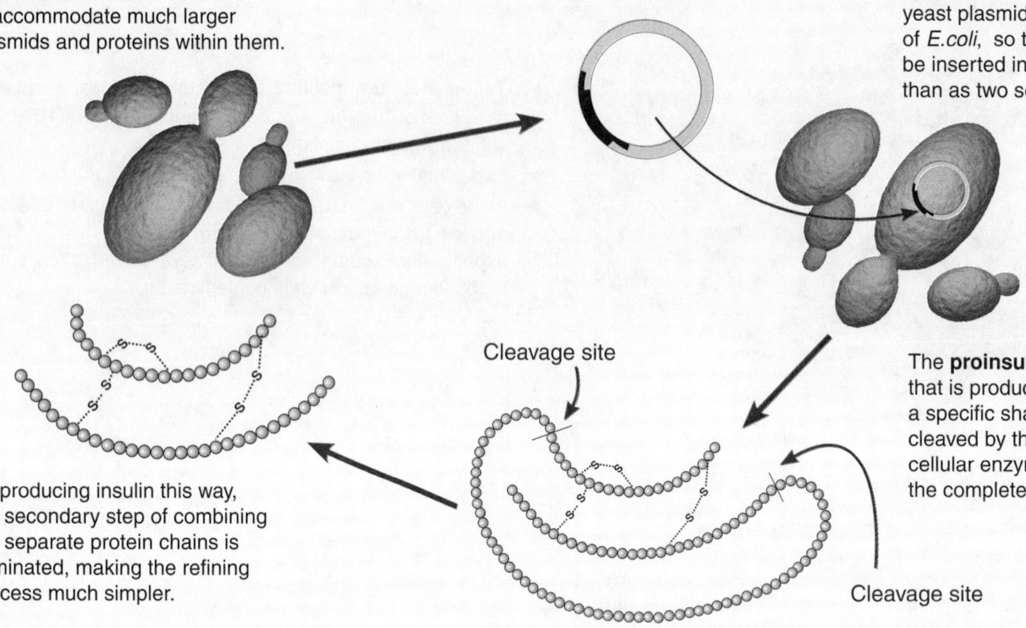

Yeast cells are **eukaryotic** and hence are much larger than bacterial cells. This enables them to accommodate much larger plasmids and proteins within them.

The gene for human insulin is inserted into a plasmid. The yeast plasmid is larger than that of *E.coli*, so the entire gene can be inserted in one piece rather than as two separate pieces.

The **proinsulin** protein that is produced folds into a specific shape and is cleaved by the yeast's own cellular enzymes, producing the completed insulin chain.

Cleavage site

Cleavage site

By producing insulin this way, the secondary step of combining the separate protein chains is eliminated, making the refining process much simpler.

1. Describe the three major problems associated with the traditional method of obtaining insulin to treat diabetes:

   (a) ____________________

   (b) ____________________

   (c) ____________________

2. Explain the reasoning behind using *E. coli* to produce insulin and the benefits that GM technology has brought to diabetics:

3. Explain why, when using *E. coli*, the insulin gene is synthesized as two separate A and B chain nucleotide sequences:

4. Why are the synthetic nucleotide sequences ('genes') 'tied' to the β-galactosidase gene?

5. Yeast (*Saccharomyces cerevisiae*) is also used in the production of human insulin. Discuss the differences in the production of insulin using yeast and *E. coli* with respect to:

   (a) Insertion of the gene into the plasmid:

   (b) Secretion and purification of the protein product:

ISBN: 978-1-927173-16-9

Food for the Masses

It is estimated that by 2050 the world population will reach between 9 and 10 billion people. Currently **1 billion people** (one sixth of the world's population) are **undernourished**. If trends continue, 1.5 billion people will be living under the threat of starvation by 2050, and by 2100 (if global warming is taken into account) nearly half the world's population could be threatened with food shortages. The solution to the problem of food production is complicated. Most of the Earth's arable land has already been developed and currently uses 37% of the Earth's land area, leaving little room to grow more crops or farm more animals. Development of new fast growing and high yield crops appears to be a major part of the solution, but many crops can only be grown under a narrow range of conditions or are susceptible to disease. Moreover, the farming and irrigation of some areas is difficult, costly, and damaging to the environment because vast amounts of water are diverted from their natural courses. **Genetic modification** of plants may help to solve some of these looming problems by producing plants that will require less intensive culture or that will grow in areas previously considered not arable.

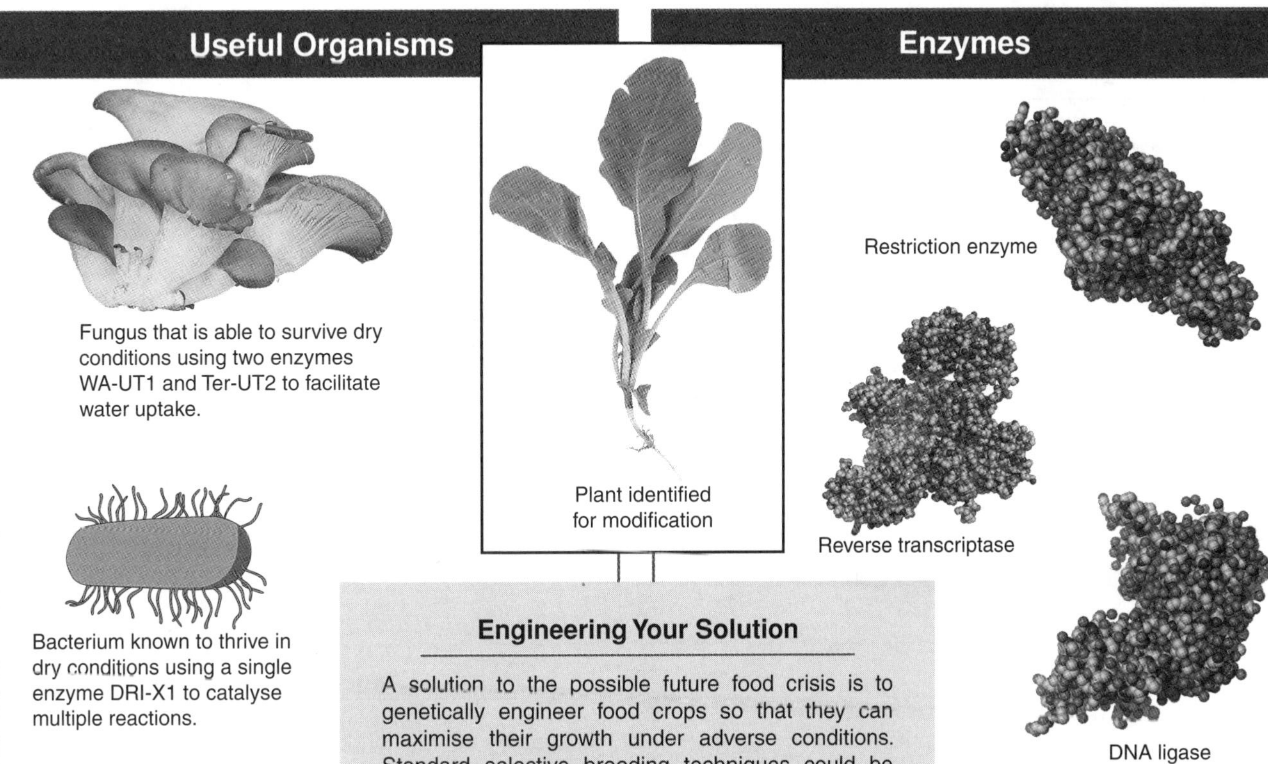
Useful Organisms
Fungus that is able to survive dry conditions using two enzymes WA-UT1 and Ter-UT2 to facilitate water uptake.
Bacterium known to thrive in dry conditions using a single enzyme DRI-X1 to catalyse multiple reactions.
Plant identified for modification
Enzymes
Restriction enzyme
Reverse transcriptase
DNA ligase

Engineering Your Solution

A solution to the possible future food crisis is to genetically engineer food crops so that they can maximise their growth under adverse conditions. Standard selective breeding techniques could be used to do this, but in some plants this may not be possible or feasible and it may require more time than is available. A selection of genetic tools and organisms with useful characteristics are described. **Your task** is to use the items shown to devise a technique to successfully create a plant that could be successfully farmed in semi-desert environments such as sub-Saharan Africa. The following page will take you through the procedure. Not all the items will need to be used.

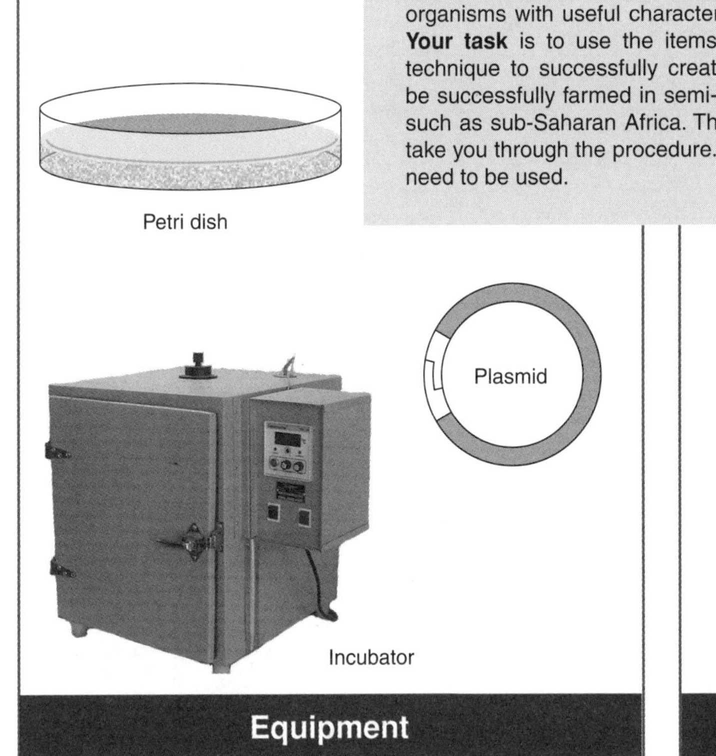
Petri dish
Plasmid
Incubator
Equipment

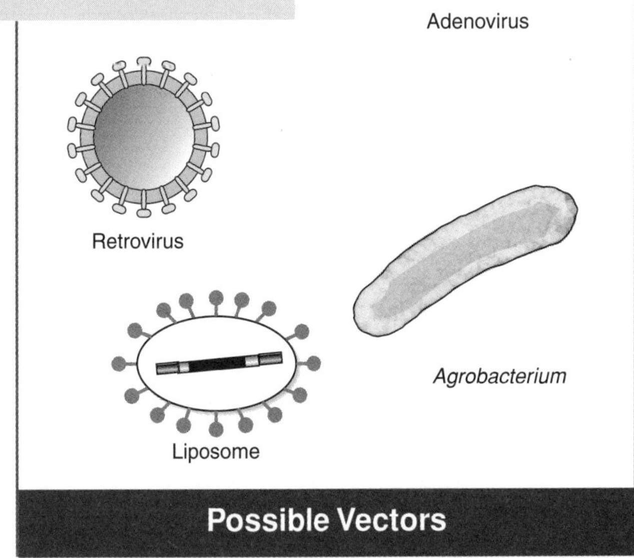
Adenovirus
Retrovirus
Agrobacterium
Liposome
Possible Vectors

ISBN: 978-1-927173-16-9

1. Identify the organism you would chose as a 'donor' of drought survival genes and explain your choice:

2. Describe a process to identify and isolate the required gene(s) and identify the tools to be used:

3. Identify a vector for the transfer of the isolated gene(s) into the crop plant and explain your decision:

4. Explain how the isolated gene(s) would be integrated into the vector's genome:

5. (a) Explain how the vector will transform the identified plant:

(b) Identify the stage of development at which the plant would most easily be transformed. Explain your choice:

6. Explain how the transformed plants could be identified:

7. Explain how a large number of plants can be grown from the few samples that have taken up the new DNA:

ISBN: 978-1-927173-16-9

The Ethics of GMO Technology

The risks associated with using **genetically modified organisms** (GMOs) have been the subject of much debate in recent times. Most experts agree that, provided GMOs are tested properly, the health risks to individuals should be minimal from plant products, although minor problems will occur. Health risks from animal GMOs are potentially more serious, especially when the animals are for human consumption. The potential benefits to be gained from the use of GMOs creates enormous pressure to apply the existing technology. However, there are many concerns, including the environmental and socio-economic effects, and problems of unregulated use. There is also concern about the environmental and economic costs of possible GMO accidents. GMO research is being driven by heavy investment on the part of biotechnology companies seeking new applications for GMOs. Currently a matter of great concern to consumers is the adequacy of government regulations for the labelling of food products with GMO content. This may have important trade implications for countries exporting and importing GMO produce.

Some important points about GMOs

1. The modified DNA is in every cell of the GMO.
2. The mRNA is only expressed in specific tissues.
3. The foreign protein is only expressed in particular tissues but it may circulate in the blood or lymph or be secreted (e.g. milk).
4. In animals, the transgene is only likely to be transmitted from parent to offspring. However, viral vectors may enable accidental transfer of the transgene between unrelated animals.
5. In plants, transmission of the transgene in GMOs is possible by pollen, cuttings, and seeds (even between species).
6. If we eat the animal or plant proper, we will also be eating DNA. The DNA will remain 'intact' if raw, but "degraded" if cooked.
7. Non-transgenic food products may be processed using genetically modified bacteria or yeast, and cells containing their DNA may be in the food product.
8. A transgenic product (e.g. a protein, polypeptide or a carbohydrate) may be in the GMO, but not in the portions sold to the consumer.

Potential effects of GMOs on the world

1. Increase in food production.
2. Decrease in use of pesticides, herbicides and animal remedies.
3. Improvement in the health of the human population and the medicines used to achieve it.
4. Possible development of transgenic products which may be harmful to some (e.g. new proteins causing allergies).
5. May have little real economic benefit to farmers (and the consumer) when increased production (as little as 10%) is weighed against cost, capital, and competition.
6. Possible (uncontrollable) spread of transgenes into other species: plants, indigenous species, animals, and humans.
7. Concerns that the release of GMOs into the environment may be irreversible.
8. Economic sanctions resulting from a consumer backlash against GMO foods and products.
9. Animal welfare and ethical issues: GM animals may suffer poor health and reduced life span.
10. GMOs may cause the emergence of pest, insect, or microbial resistance to traditional control methods.
11. May create a monopoly and dependence of developing countries on companies who are seeking to control the world's commercial seed supply.

POLICE
© Greenpeace

GMO protestors are arrested

Issue: The accidental release of GMOs into the environment.

Problem: Recombinant DNA may be taken up by non-target organisms. These then may have the potential to become pests or cause disease.

Solution: Rigorous controls on the production and release of GMOs. GMOs could have specific genes deleted so that their growth requirements are met only in controlled environments.

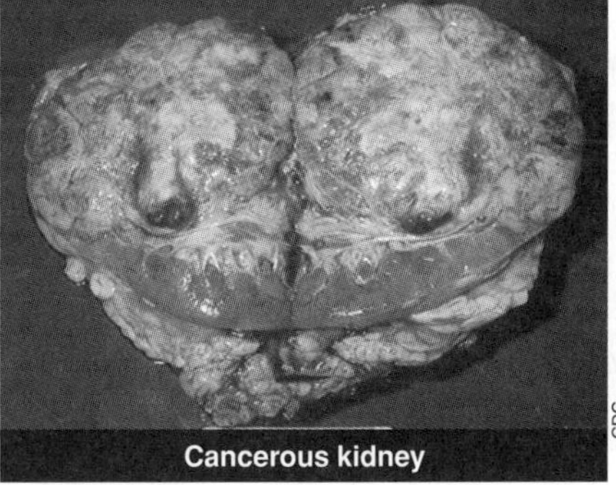
CDC

Cancerous kidney

Issue: A new gene or genes may disrupt normal gene function.

Problem: Gene disruption may trigger cancer. Successful expression of the desired gene is frequently very low.

Solution: A combination of genetic engineering, cloning, and genetic screening so that only those cells that have been successfully transformed are used to produce organisms.

STOP GENETIC POLLUTION!
BIOSAFETY NZ
www.GREENPEACE
BIOSÉC
ARRÊTEZ
POLLUTION
www.GREENPE
Special
© Greenpeace

Protest against GMOs in the environment

Issue: Targeted use of transgenic organisms in the environment.

Problem: Once their desired function, e.g. environmental clean-up, is completed, they may be undesirable invaders in the ecosystem.

Solution: GMOs can be engineered to contain "suicide genes" or metabolic deficiencies so that they do not survive for long in the new environment after completion of their task.

1. Why are genetically modified (GM) plants thought to pose a greater environmental threat than GM animals?

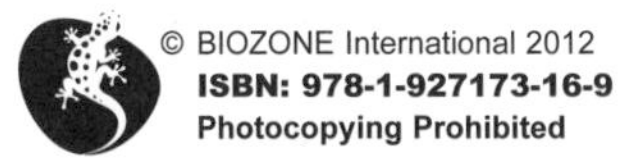

ISBN: 978-1-927173-16-9

2. Describe an advantage and a problem with the use of genetically engineered herbicide resistant crop plants:

(a) Advantage:

(b) Problem:

3. Describe an advantage and a problem with using tropical crops genetically engineered to grow in cold regions:

(a) Advantage:

(b) Problem:

4. Describe an advantage and a problem with using crops that are genetically engineered to grow in marginal habitats (for example, in very saline or poorly aerated soils):

(a) Advantage:

(b) Problem:

5. Describe two uses of transgenic animals within the livestock industry:

(a)

(b)

6. Some years ago, Britain banned the import of a GM, pest resistant corn variety containing marker genes for ampicillin antibiotic resistance. Suggest why the use of antibiotic-resistance genes as markers is no longer common practice:

7. Many agricultural applications of DNA technology make use of transgenic bacteria which infect plants and express a foreign gene. Explain one advantage of each of the following applications of genetic engineering to crop biology:

(a) Development of nitrogen-fixing *Rhizobium* bacteria that can colonise non-legumes such as corn and wheat:

(b) Addition of transgenic *Pseudomonas fluorescens* bacteria into seeds (bacterium produces a pathogen-killing toxin):

8. Some of the public's fears and concerns about genetically modified food stem from moral or religious convictions, while others have a biological basis and are related to the potential biological threat posed by GMOs.
(a) Conduct a class discussion or debate to identify these fears and concerns, and list them below:

(b) Identify which of those you have listed above pose a real biological threat:

ISBN: 978-1-927173-16-9

Cloning by Somatic Cell Nuclear Transfer

Blackface ewe
Dolly
PHOTO: Courtesy Roslin Institute ©

The Issue

- Individuals vary in characteristics, even within specific breeds of animal, such as sheep.
- Clones remove the variability and produce livestock that would develop in a predictable way or produce a consistent quality of product such as wool or milk.
- Clones produced using traditional embryo-splitting are derived from an embryo whose physical characteristics are not completely known. Scientists wanted to speed up the process and produce clones from a proven phenotype.

Concept 1

Somatic cells can be made to return to a dormant or embryonic state so that their genes will not be expressed.

Concept 2

The nucleus of a cell can be removed and replaced with the nucleus of an unrelated cell. Cells can be made to fuse together.

Concept 3

Fertilized egg cells produce embryos. Egg cells that contain the nucleus of a donor cell will produce embryos with DNA identical to the donor cell.

Concept 4

Embryos can be implanted into surrogate mothers and develop to full term with seemingly no ill effects.

Somatic Cell Nuclear Transfer (SCNT)

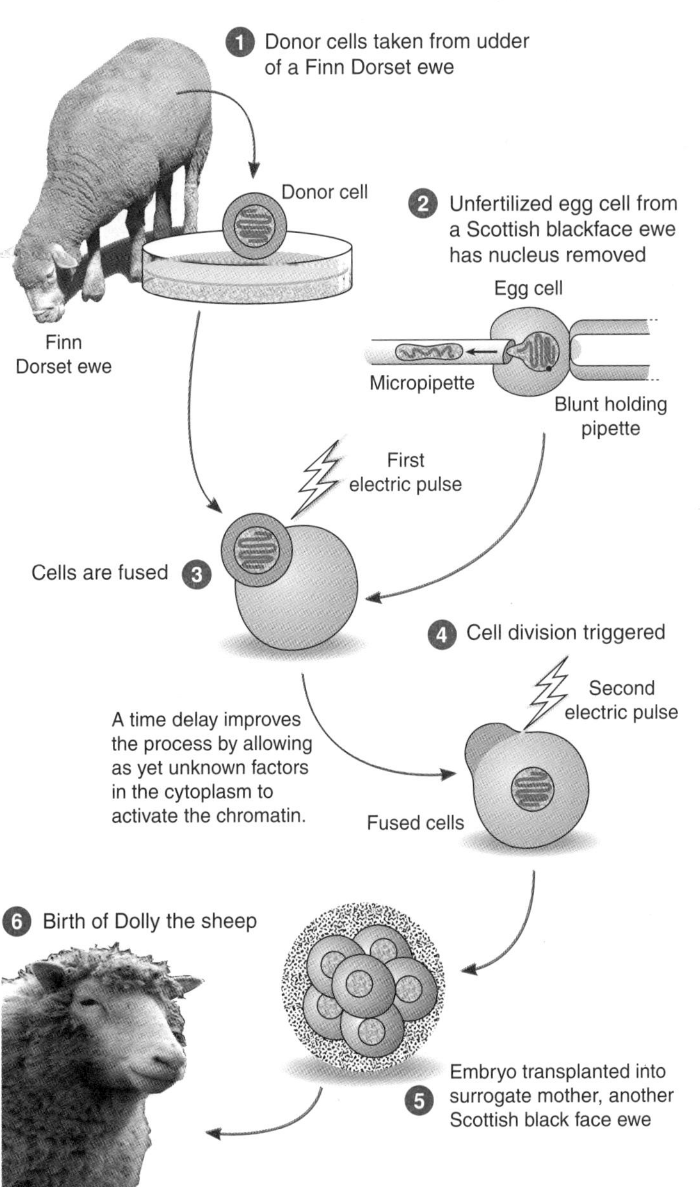
1 Donor cells taken from udder of a Finn Dorset ewe
Donor cell
Finn Dorset ewe
2 Unfertilized egg cell from a Scottish blackface ewe has nucleus removed
Egg cell
Micropipette
Blunt holding pipette
First electric pulse
Cells are fused 3
4 Cell division triggered
Second electric pulse
A time delay improves the process by allowing as yet unknown factors in the cytoplasm to activate the chromatin.
Fused cells
6 Birth of Dolly the sheep
Embryo transplanted into surrogate mother, another Scottish black face ewe 5
PHOTO: Courtesy Roslin Institute ©

Techniques

Donor cells from the udder of a Finn Dorset ewe are taken and cultured in a low nutrient media for a week. The nutrient deprived cells stop dividing and become **dormant**.

An **unfertilized egg** from a Scottish blackface ewe has the nucleus removed using a micropipette. The rest of the cell contents are left intact.

The dormant udder cell and the recipient denucleated egg cell are fused using a mild electric pulse.

A second electric pulse triggers cellular activity and cell division, jump starting the cell into development. This can also be triggered by chemical means.

After six days the embryo is transplanted into a surrogate mother, another Scottish blackface ewe.

After a 148 day gestation 'Dolly' is born. DNA profiling shows she is genetically identical to the original Finn Dorset cell donor.

Outcomes

Dolly, a Finn Dorset lamb, was born at the Roslin Institute (near Edinburgh) in July 1996. She was the first mammal to be cloned from **non-embryonic** cells, i.e. cells that had already differentiated into their final form. Dolly's birth showed that the process leading to cell specialization is not irreversible and that cells can be 'reprogrammed' into an embryonic state. Although cloning seems relatively easy there are many problems that occur. Of the hundreds of eggs that were reconstructed only 29 formed embryos and only Dolly survived to birth.

Further Applications

In animal reproductive technology, cloning has facilitated the rapid production of genetically superior stock. These animals may then be dispersed among commercial herds. The **primary focus** of the new cloning technologies is to provide an economically viable way to rapidly produce transgenic animals with very precise genetic modifications.

Genetic Engineering and Biotechnology

ISBN: 978-1-927173-16-9

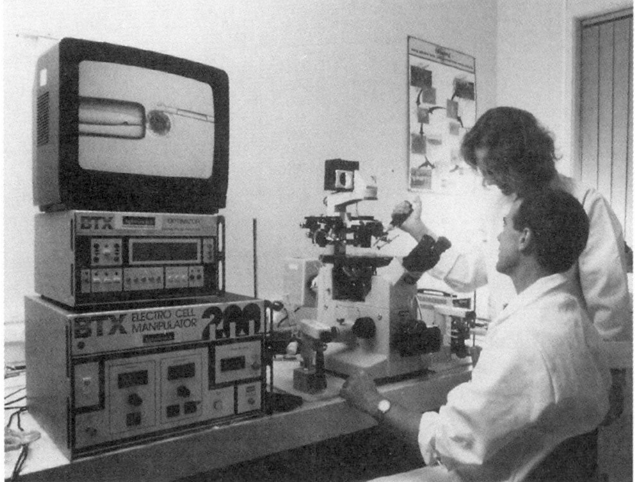
BTX
ELECTRO CELL MANIPULATOR

Dr David Wells and Pavla Misica in the embryo micromanipulation laboratory at AgResearch in Hamilton, New Zealand (monitor's image is enlarged on the right).

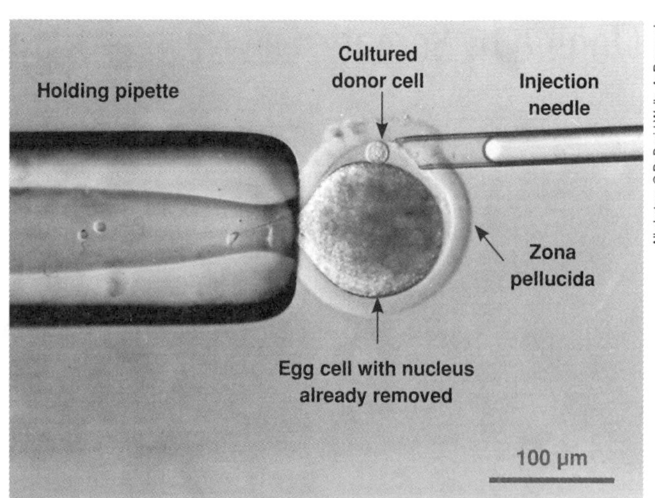
Holding pipette
Cultured donor cell
Injection needle
Zona pellucida
Egg cell with nucleus already removed
100 µm

A single cultured cell is injected underneath the zona pellucida (the outer membrane) and positioned next to the egg cell (step 3 of diagram on previous page).

All photos: © Dr David Wells, AgResearch

Donor cow
10 cloned calves

Adult cloning heralds a new chapter in the breeding of livestock. Traditional breeding methods are slow, unpredictable, and suffer from a time delay in waiting to see what the phenotype is like before breeding the next generation. Adult cloning methods now allow a rapid spread of valuable livestock into commercial use among farmers. It will also allow the livestock industry to respond rapidly to market changes in the demand for certain traits in livestock products. In New Zealand, 10 healthy clones were produced from a single cow (the differences in coat color patterns arise from the random migration of pigment cells in early embryonic development).

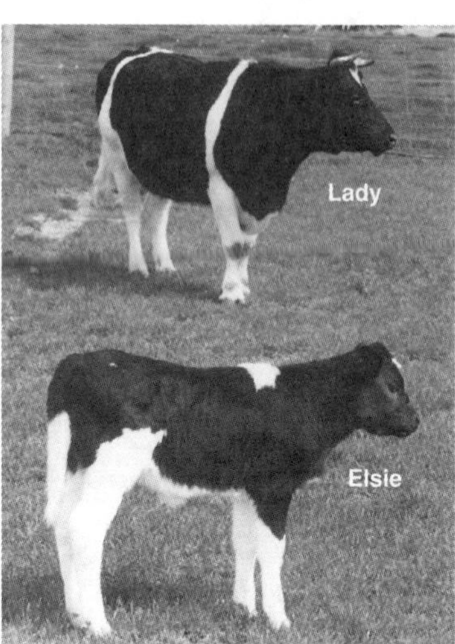
Lady
Elsie

Lady is the last surviving cow of the rare Enderby Island (south of N.Z.) cattle breed. Adult cloning was used to produce her genetic duplicate, Elsie (born 31 July 1998). Elsie represents the first demonstration of the use of adult cloning in animal conservation.

1. What is **adult cloning** (as it relates to somatic cell nuclear transfer or **SCNT**)?:

2. Explain how each of the following events is controlled in the **SCNT** process:

   (a) The switching off of all genes in the donor cell:

   (b) The fusion (combining) of donor cell with enucleated egg cell:

   (c) The activation of the cloned cell into producing an embryo:

3. Describe two potential applications of nuclear transfer technology for the cloning of animals:

   (a)

   (b)

ISBN: 978-1-927173-16-9

Therapeutic Cloning

Stem cells are undifferentiated cells that have the ability to give rise to all the specialized cells of the body. The best source of stem cells is very early embryos. **Embryonic stem cells** (ESC) are **pluripotent** meaning they can give rise to any fetal or adult cell type. The properties of **self renewal** and **potency** shown by stem cells make them suitable for a wide range of applications. These include providing a renewable source of cells for studies of human development and gene regulation, for tests of new drugs and vaccines, for monoclonal antibody production, and for treating diseased and damaged tissue. These technologies require a disease-free and plentiful supply of cells of specific types. **Therapeutic stem cell cloning** is still in its very early stages and, despite its enormous medical potential, research with human embryonic cells is still banned in some countries.

Embryonic Stem Cells

Embryonic stem cells (ESC) are stem cells taken from **blastocysts** (below). Blastocysts are embryos which are about five days old and consist of a hollow ball containing 50-150 cells. ESCs can come from embryos that have been fertilized *in vitro* and then donated for research. They can also come from cloned embryos created using somatic cell nuclear transfer using a donor nucleus.

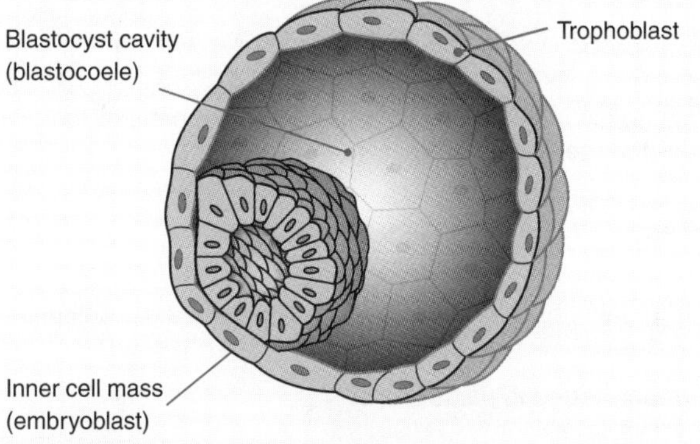
Blastocyst cavity
(blastocoele)
Trophoblast
Inner cell mass
(embryoblast)

Cells derived from the inner cell mass are **pluripotent**. They can become any cells of the body, with the exception of placental cells. When grown *in vitro*, without any stimulation for differentiation, these cells retain their potency through multiple cell divisions. As a consequence of this, **ESC therapies** have potential use in regenerative medicine and tissue replacement. By manipulating the culture conditions, scientists can select and control the type of cells grown (e.g. heart cells). This ability could allow for specific cell types to be grown to treat specific diseases or replace damaged tissue. However, no ESC treatments have been approved to date due to ethical issues.

Ethics of ESC Cloning

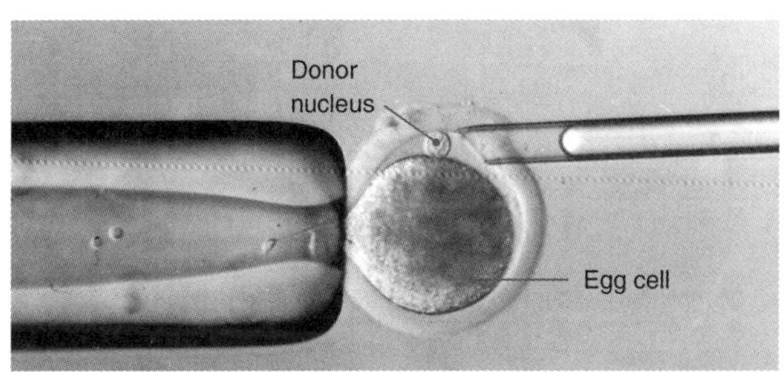
Donor
nucleus
Egg cell

Dr. David Wells, AgResearch

ESC therapy has enormous potential to make life changing improvements to the health of people with diseased or damaged organs. Organs or tissues derived from a patient's ESC could be transplanted back into that patient without fear of tissue rejection or the need for ongoing immunosuppressive drug therapies.

Despite this, many groups oppose the use of therapeutic cloning for many reasons including:

- The technology used to create the embryo could be used for reproductive cloning, i.e. creating a clone of the original human.
- The creation of stem cell line requires the destruction of a human embryo and thus human life.
- Human embryos have the potential to develop into an individual and thus have the same rights of the individual.
- Saving or enhancing the quality of life of an individual does not justify the destruction of the life of another (i.e. the embryo).
- ESC research has not produced any viable long term treatment, while other techniques (e.g. adult stem cells) have.
- There are other stem cell techniques that do not require the creation of an embryo but achieve similar results (e.g. cell lines grown from adult stem cells or umbilical cord blood).

Genetic Engineering and Biotechnology

1. Explain why ESC therapy has such enormous potential in medicine: ____________________

2. Explain why harvesting embryonic stem cells requires the destruction of the embryo: ____________________

3. Outline the ethical debate surrounding ESC, giving two ethical arguments for each side of the debate:

ISBN: 978-1-927173-16-9

Periodicals:
Embryonic stem cells

Related activities: *Stem Cells and Differentiation*
Weblinks: *Stem Cells in the Spotlight*

KEY TERMS: Mix and Match

INSTRUCTIONS: Test your vocabulary by matching each term to its definition, as identified by its preceding letter code.

bioinformatics

clone

DNA amplification

DNA ligase

DNA ligation

DNA polymerase

DNA profiling

embryonic stem cell

gel electrophoresis

genetic modification

human genome project

paternity

plasmid

polymerase chain reaction

primer

recombinant DNA

restriction enzyme

somatic cell nuclear transfer

therapeutic cloning

vector

A An organism or artificial vehicle that is capable of transferring a DNA sequence to another organism.

B The use of computational methods for analyzing biological information. In many cases this involves the use of computers for the analysis of DNA sequences.

C The process of locating regions of a DNA sequence that are variable between individuals in order to distinguish between individuals.

D A short length of DNA used to identify the starting sequence for PCR so that polymerase enzymes can begin amplification.

E An enzyme that is able to cut a length of DNA at a specific sequence or site.

F A kin relationship in which the offspring is directly genetically related to the male of the preceding generation, i.e the father and the offspring.

G The repairing or attaching of fragmented DNA by ligase enzymes.

H A reaction that is used to amplify fragments of DNA using cycles of heating and cooling.

I A small circular piece of DNA commonly found in bacteria.

J Cells taken from a blastocyst which still have the ability to differentiate into any cell in the body and so can be used for therapeutic cloning.

K DNA that has had a new sequence added so that the original sequence has been changed.

L Any organism with genetic information that is identical to the parent organism from which it was created.

M The process of cloning of a organism (usually by SCNT) for the purpose of therapy for a disease, i.e. organs grown from cloned cells can be used to treat the parent organism.

N Enzyme which is able to repair or join DNA fragments.

O An enzyme that is able to replicate DNA and commonly used in PCR to amplify a length of DNA.

P The process of producing more copies of a length of DNA, normally using PCR.

Q A process that is used to separate different lengths of DNA by placing them in a gel matrix placed in a buffered solution through which an electric current is passed.

R The technique of altering the genetic makeup of cells or individuals by the selective removal, insertion, or modification of genes.

S Publicly funded venture to determine the human genome sequence, and map and assign functions to it genes. The first draft became available in 2000. The entire sequence is now available.

T Laboratory technique in which the nucleus of an egg cell is removed and replaced by the nucleus of a donor cell in order to make a genetic clone of the donor organism.

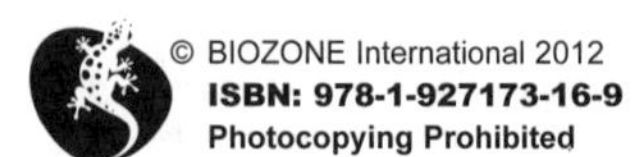

ISBN: 978-1-927173-16-9

Core Topic 5

Ecology

Key concepts

- Energy in ecosystems is captured by autotrophs and is transferred through food chains.
- The decline in the energy available to each successive trophic level limits the number of feeding links in ecosystems.
- Nutrients cycle within and between ecosystems.
- Humans can interfere with nutrient cycles and must manage their impact on the environment.
- Population growth and size is influenced by density dependent and density independent factors.

Key terms

Core

autotroph
carbon cycle
carrying capacity
combustion
community
consumer
decomposer
decomposition
detritivore
ecological pyramid
ecology
ecosystem
emigration
exponential growth
food chain
food web
fossilization
global warming
greenhouse effect
greenhouse gas
habitat
heterotroph
immigration
logistic curve
migration
mortality
natality
nutrient cycle
plateau phase
population
precautionary principle
saprotroph
sigmoid growth
species
transitional phase
trophic efficiency
trophic group
trophic level

Learning Objectives

☐ 1. Use the **KEY TERMS** to compile a glossary for this topic.

Communities and Ecosystems *(5.1)* pages 214-225

☐ 2. Define the terms **ecology**, **species**, **habitat**, **population**, **community**, and **ecosystem** and demonstrate an understanding of the relationship between them.

☐ 3. Distinguish between **autotrophs** and **heterotrophs** and between **consumers**, **detritivores**, and **saprotrophs**. Explain the relationship between these **trophic groups** and the functional role of each group in an ecosystem.

☐ 4. Describe how energy is transferred between **trophic levels** in **food chains** and **food webs**. Describe examples of food chains with at least three linkages (four organisms).

☐ 5. Construct a food web of up to ten organisms. Assign trophic levels to each of the organisms in the web.

☐ 6. Describe how energy enters ecosystems through the activity of autotrophs. Describe energy flow quantitatively using an energy flow diagram. Include reference to trophic levels, direction of energy flow, processes involved in energy transfer, energy sources, and energy sinks. Comment on the efficiency of energy transfers.

☐ 7. Describe food chains quantitatively using **ecological pyramids**. Construct or interpret pyramids of energy, numbers, or biomass for different communities.

☐ 8. State that energy flows through ecosystems but that nutrients are recycled, moving between the atmosphere, the Earth's crust, water, and organisms. Describe the role of saprotrophic bacteria and fungi in nutrient cycles. Use specific examples, e.g. the **carbon cycle** to show how nutrients are exchanged within and between ecosystems

The Greenhouse Effect *(5.2)* pages 225-231

☐ 9. Draw and label a diagram of the carbon cycle to show the processes involved, including photosynthesis, cell respiration, **decomposition**, **fossilization**, and **combustion**.

☐ 10. Describe and explain the causes and consequences of the enhanced **greenhouse effect** (**global warming**). Include an analysis of the changes in concentration of atmospheric CO_2 as documented by historical records.

☐ 11. Outline the **precautionary principle** (proof of no harm) and evaluate its use to justify an immediate strong response to the threats posed by global warming.

☐ 12. Describe the ecological effects of increased global temperatures on Arctic ecosystems. Include reference to the loss of ice habitat, changes in species distribution, spread of pests, and oxidation of permafrost biomass.

Populations *(5.3)* pages 232-235

☐ 13. Outline the effects of **natality**, **mortality**, and migration on population size.

☐ 14. Use an diagram to explain **sigmoid population growth** (the **logistic curve**). Identify the **exponential growth** phase, the **transitional phase,** and the **plateau phase**.

☐ 15. Describe three factors that regulate or set the limit of population increase. Identify these are density dependent or density independent.

Periodicals:

Listings for this chapter are on page 399

Weblinks:

www.thebiozone.com/weblink/IB-3169.html

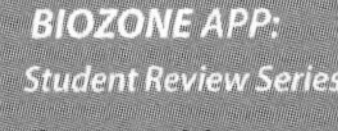

BIOZONE APP:
Student Review Series
Communities

Components of an Ecosystem

The concept of the ecosystem was developed to describe the way groups of organisms are predictably found together in their physical environment. A community comprises all the organisms within an ecosystem. The structure and function of a community is determined by the physical (abiotic) and biotic factors, which determine species distribution and survival.

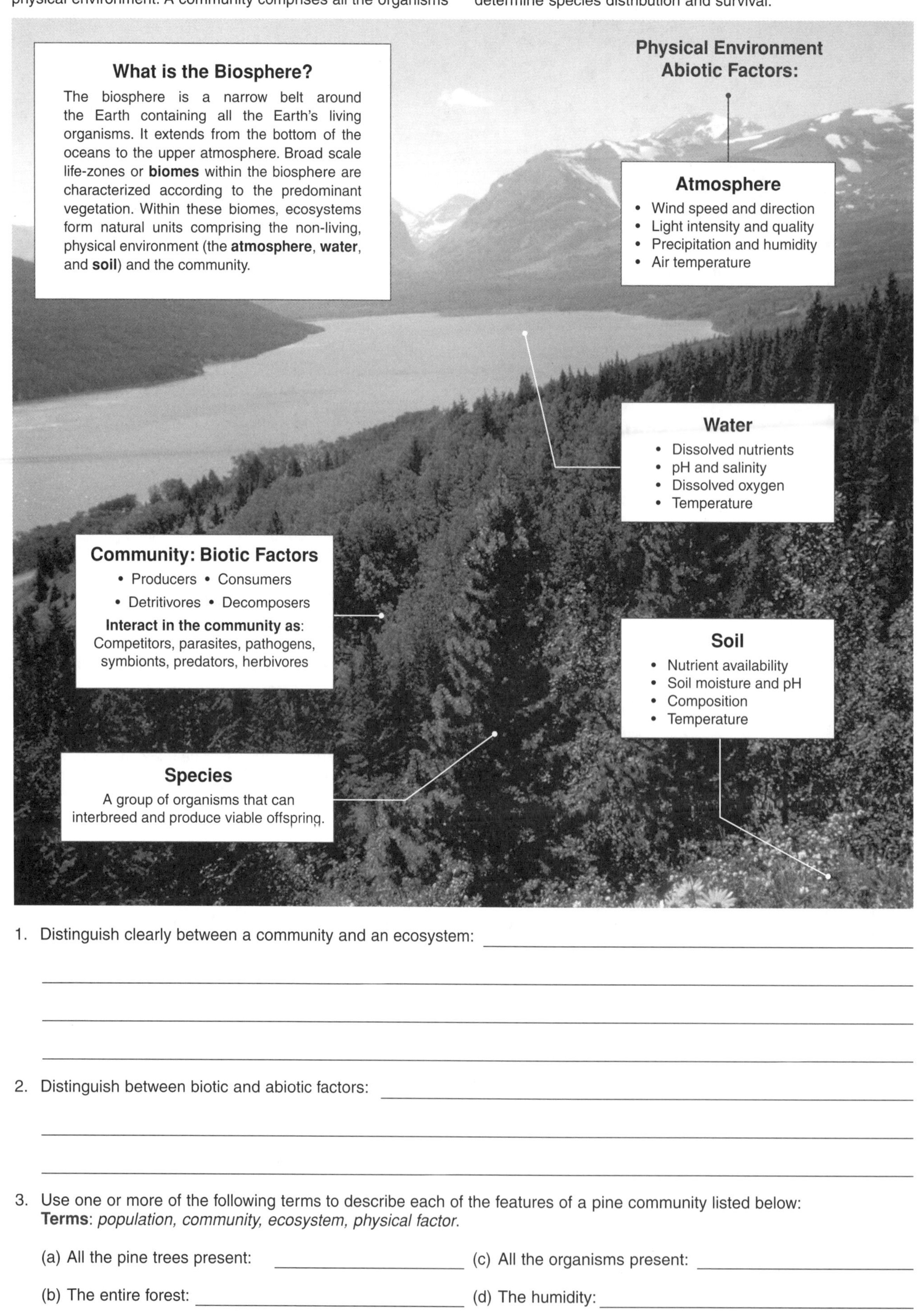
What is the Biosphere?
The biosphere is a narrow belt around the Earth containing all the Earth's living organisms. It extends from the bottom of the oceans to the upper atmosphere. Broad scale life-zones or **biomes** within the biosphere are characterized according to the predominant vegetation. Within these biomes, ecosystems form natural units comprising the non-living, physical environment (the **atmosphere**, **water**, and **soil**) and the community.
Physical Environment Abiotic Factors:
Atmosphere
• Wind speed and direction
• Light intensity and quality
• Precipitation and humidity
• Air temperature
Water
• Dissolved nutrients
• pH and salinity
• Dissolved oxygen
• Temperature
Community: Biotic Factors
• Producers • Consumers
• Detritivores • Decomposers
Interact in the community as:
Competitors, parasites, pathogens, symbionts, predators, herbivores
Soil
• Nutrient availability
• Soil moisture and pH
• Composition
• Temperature
Species
A group of organisms that can interbreed and produce viable offspring.

1. Distinguish clearly between a community and an ecosystem: ______________________

2. Distinguish between biotic and abiotic factors: ______________________

3. Use one or more of the following terms to describe each of the features of a pine community listed below:
Terms: *population, community, ecosystem, physical factor.*

(a) All the pine trees present: ______________ (c) All the organisms present: ______________

(b) The entire forest: ______________ (d) The humidity: ______________

Web links: *racerocks.com*

Periodicals: *Getting to grips with ecology*

ISBN: 978-1-927173-16-9

Food Chains

Every ecosystem has a trophic structure: a hierarchy of feeding relationships that determines the pathways for energy flow and nutrient cycling. Species are assigned to trophic levels on the basis of their sources of nutrition. The first trophic level (**producers**), ultimately supports all other levels. The consumers are those that rely on producers for their energy. Consumers are ranked according to the trophic level they occupy (first order, second order, etc.). The sequence of organisms, each of which is a source of food for the next, is called a **food chain**. Food chains commonly have four links but seldom more than six. Those organisms whose food is obtained through the same number of links belong to the same trophic level. Note that some consumers (particularly top carnivores and omnivores) may feed at several different trophic levels, and many primary consumers eat many plant species. The different food chains in an ecosystem therefore tend to form complex webs of interactions (food webs).

Respiration

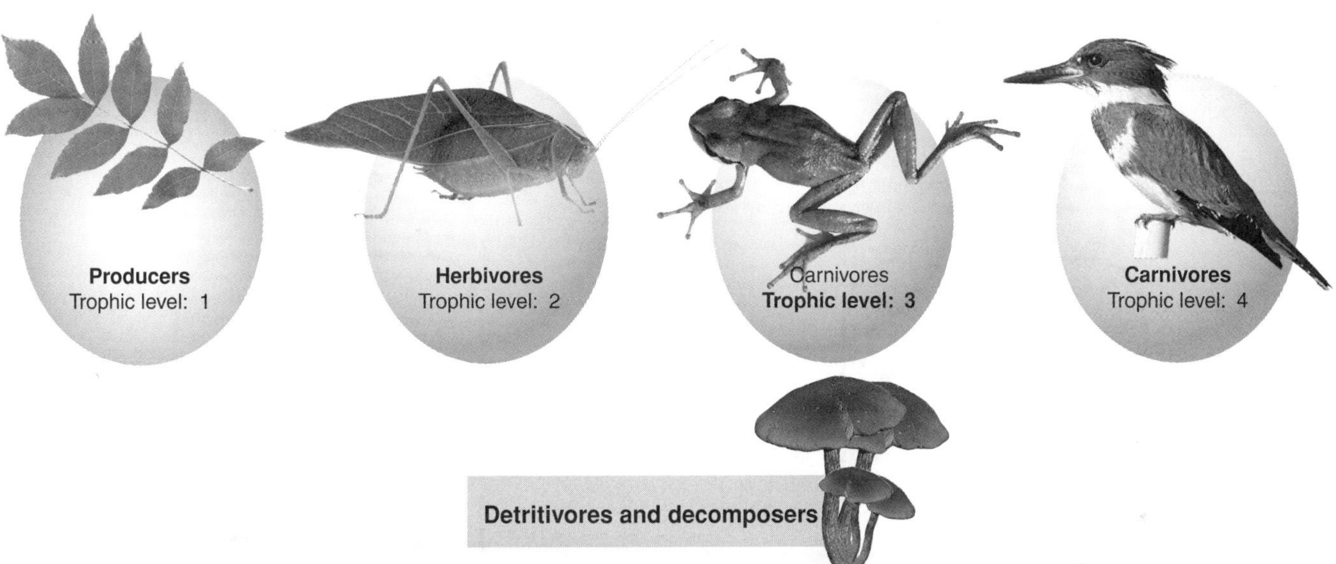
Producers
Trophic level: 1
Herbivores
Trophic level: 2
Carnivores
Trophic level: 3
Carnivores
Trophic level: 4
Detritivores and decomposers

Green plants

Aphids

Ladybug eating aphid
Millipede

Wood-ear fungus

Producers (algae, green plants, and some bacteria) make their own food using simple inorganic carbon sources (e.g. CO_2). Sunlight is the most common energy source for this process.

Consumers (animals, non-photosynthetic protists, and some bacteria) rely on other living organisms or organic particulate matter for their energy and their source of carbon. First order consumers, such as aphids (left), feed on producers. Second (and higher) order consumers, such as ladybugs (center) eat other consumers. **Detritivores** consume (ingest and digest) detritus (decomposing organic material) from every trophic level. In doing so, they contribute to decomposition and the recycling of nutrients. Common detritivores include wood-lice, millipedes (right), and many terrestrial worms.

Decomposers (fungi and some bacteria) obtain their energy and carbon from the extracellular breakdown of dead organic matter (DOM). Decomposers play a central role in nutrient cycling.

The diagram above represents the basic elements of a food chain. In the questions below, you are asked to add to the diagram the features that indicate the flow of energy through the community of organisms.

1. (a) State the original energy source for this food chain: ____________________

   (b) Draw arrows on the diagram above to show how the energy flows through the organisms in the food chain. Label each arrow with the process involved in the energy transfer. Draw arrows to show how energy is lost by respiration.

2. (a) Describe what happens to the **amount** of energy available to each successive trophic level in a food chain:

   ____

   (b) Explain why this is the case: ____________________

   ____

3. Explain what you could infer about the tropic level(s) of the kingfisher, if it was found to eat both katydids and frogs:

   ____

ISBN: 978-1-927173-16-9

Related activities: *Energy Inputs and Outputs, Food Webs, Ecological Pyramids*

Energy Inputs and Outputs

Light is the initial energy source for almost all ecosystems and photosynthesis is the main route by which energy enters most food chains. Energy flows through ecosystems in the high energy chemical bonds within **organic matter** and, in accordance with the second law of thermodynamics, is dissipated as it is transferred through trophic levels. In contrast, nutrients move within and between ecosystems in **biogeochemical cycles** involving exchanges between the atmosphere, the Earth's crust, water, and living organisms. Energy flows through trophic levels rather inefficiently, with only 5-20% of usable energy being transferred to the subsequent level. Energy not used for metabolic processes is lost as heat.

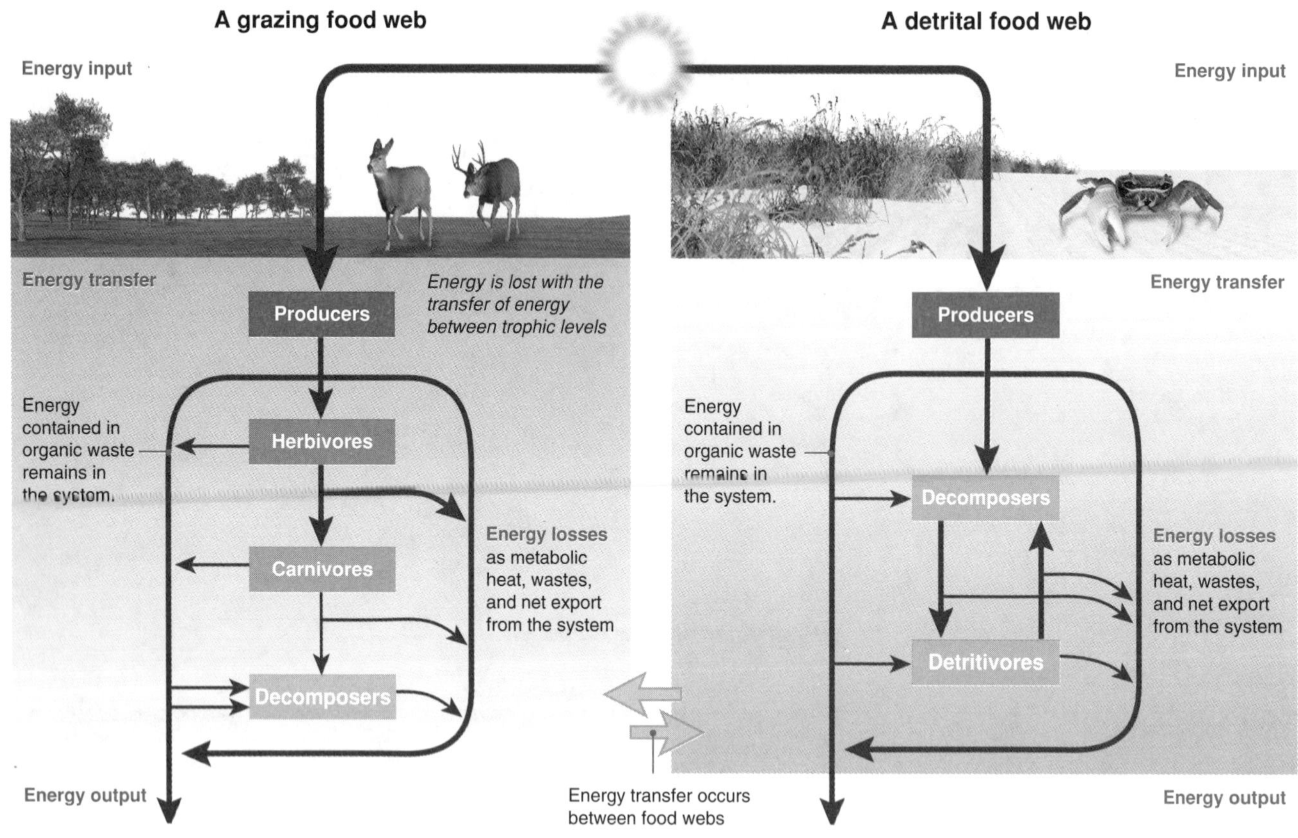
A grazing food web
A detrital food web
Energy input
Energy input
Energy transfer
Producers
Energy is lost with the transfer of energy between trophic levels
Energy contained in organic waste remains in the system.
Herbivores
Carnivores
Decomposers
Energy losses as metabolic heat, wastes, and net export from the system
Energy output
Energy transfer occurs between food webs
Energy transfer
Producers
Energy contained in organic waste remains in the system.
Decomposers
Detritivores
Energy losses as metabolic heat, wastes, and net export from the system
Energy output

1. Discuss the nature of food chains. Include reference to trophic levels, producers, and first and second order consumers:

2. Describe the differences between **producers** and **consumers** with respect to their role in energy transfers:

3. With respect to energy flow, describe a major difference between a detrital and a grazing food web:

4. Distinguish between detritivores and decomposers with respect to how they contribute to nutrient cycling:

***Related activities**: Food Webs, Energy Flow in an Ecosystem*

ISBN: 978-1-927173-16-9

Food Webs

Every ecosystem has a **trophic structure**: a hierarchy of feeding relationships which determines the pathways for energy flow and nutrient cycling. Species are assigned to trophic levels on the basis of their sources of nutrition, with the first trophic level (the **producers**), ultimately supporting all other (consumer) levels. Consumers are ranked according to the trophic level they occupy, although some consumers may feed at several different trophic levels. The sequence of organisms, each of which is a source of food for the next, is called a **food chain**. The different food chains in an ecosystem are interconnected to form a complex web of feeding interactions called a **food web**. In the example of a lake ecosystem below, your task is assemble the organisms into a food web in a way that illustrates their trophic status and their relative trophic position(s).

Feeding Requirements of Lake Organisms

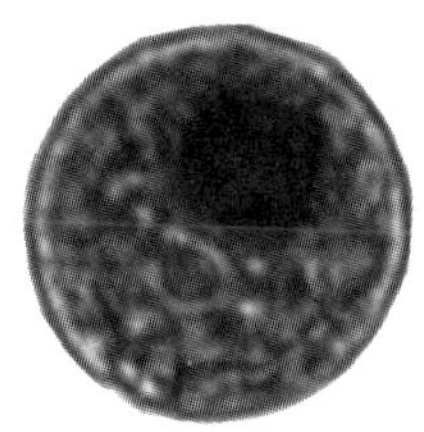

Autotrophic protists
Chlamydomonas (above), *Euglena*
Two of the many genera that form the phytoplankton.

Macrophytes (various species)
A variety of flowering aquatic plants are adapted for being submerged, free-floating, or growing at the lake margin.

Detritus
Decaying organic matter from within the lake itself or it may be washed in from the ake margins.

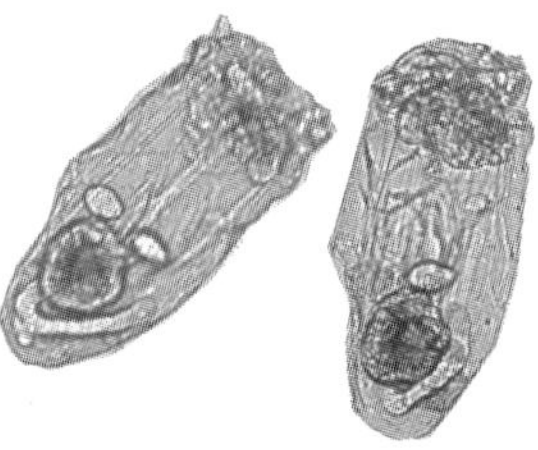

Asplanchna (planktonic rotifer)
A large, carnivorous rotifer that feeds on protozoa and young zooplankton (e.g. small *Daphnia*).

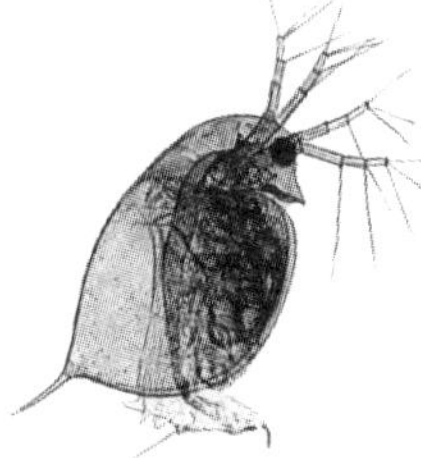

Daphnia
Small freshwater crustacean that forms part of the zooplankton. It feeds on planktonic algae by filtering them from the water with its limbs.

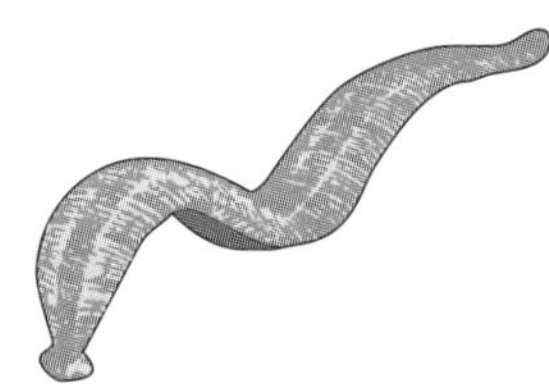

Leech (*Glossiphonia*)
Leeches are fluid feeding predators of smaller invertebrates, including rotifers, small pond snails and worms.

NYSDEC

Three-spined stickleback (*Gasterosteus*)
A common fish of freshwater ponds and lakes. It feeds mainly on small invertebrates such as *Daphnia* and insect larvae.

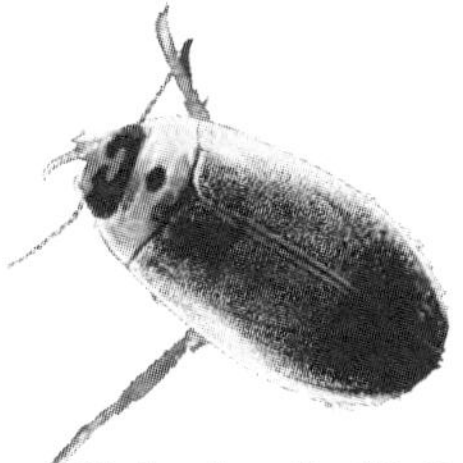

Diving beetle (*Dytiscus*)
Diving beetles feed on aquatic insect larvae and adult insects blown into the lake community. The will also eat organic detritus collected from the bottom mud.

Carp (*Cyprinus*)
A heavy bodied freshwater fish that feeds mainly on bottom living insect larvae and snails, but will also take some plant material (not algae).

Gina Mikel

Dragonfly larva
Large aquatic insect larvae that are voracious predators of small invertebrates including *Hydra*, *Daphnia*, other insect larvae, and leeches.

Great pond snail (*Limnaea*)
Omnivorous pond snail, eating both plant and animal material, living or dead, although the main diet is aquatic macrophytes.

Herbivorous water beetles (e.g. *Hydrophilus*)
Feed on water plants, although the young beetle larvae are carnivorous, feeding primarily on small pond snails.

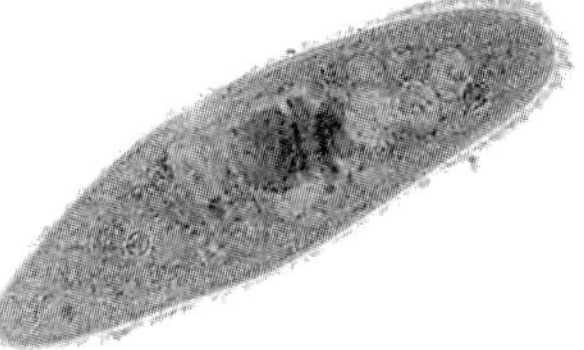

Protozan (e.g. *Paramecium*)
Ciliated protozoa such as *Paramecium* feed primarily on bacteria and microscopic green algae such as *Chlamydomonas*.

Pike (*Esox lucius*)
A top ambush predator of all smaller fish and amphibians. They are also opportunistic predators of rodents and small birds.

Mosquito larva (*Culex* spp.)
The larvae of most mosquito species, e.g. *Culex*, feed on planktonic algae before passing through a pupal stage and undergoing metamorphosis into adult mosquitoes.

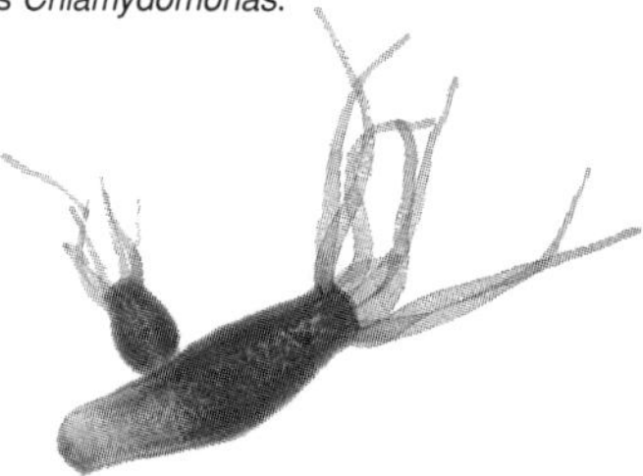

Hydra
A small carnivorous cnidarian that captures small prey items, e.g. small *Daphnia* and insect larvae, using its stinging cells on the tentacles.

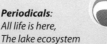

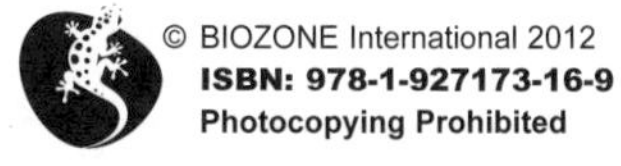

ISBN: 978-1-927173-16-9

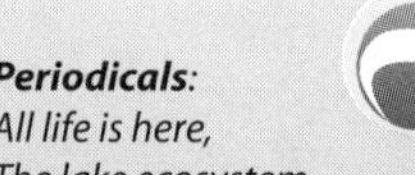

Periodicals:
All life is here,
The lake ecosystem

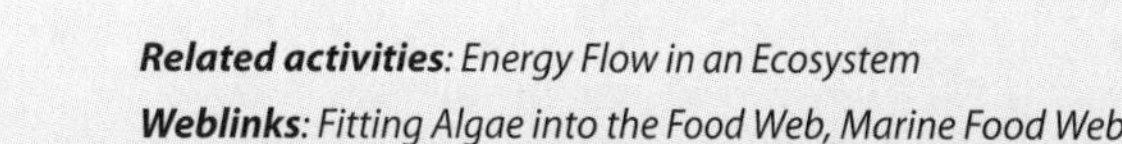

Related activities: *Energy Flow in an Ecosystem*
Weblinks: *Fitting Algae into the Food Web, Marine Food Webs*

1. From the information provided for the lake food web components on the previous page, construct **ten** different **food chains** to show the feeding relationships between the organisms. Some food chains may be shorter than others and most species will appear in more than one food chain. An example has been completed for you.

Example 1: Macrophyte ⟶ Herbivorous water beetle ⟶ Carp ⟶ Pike

(a) ____

(b) ____

(c) ____

(d) ____

(e) ____

(f) ____

(g) ____

(h) ____

(i) ____

(j) ____

2. (a) Use the food chains created above to help you to draw up a **food web** for this community. Use the information supplied to draw arrows showing the flow of **energy** between species (only energy **from** the detritus is required).

   (b) Label each species to indicate its position in the food web, i.e. its trophic level (**T1, T2, T3, T4, T5**). Where a species occupies more than one trophic level, indicate this, e.g. **T2/3**:

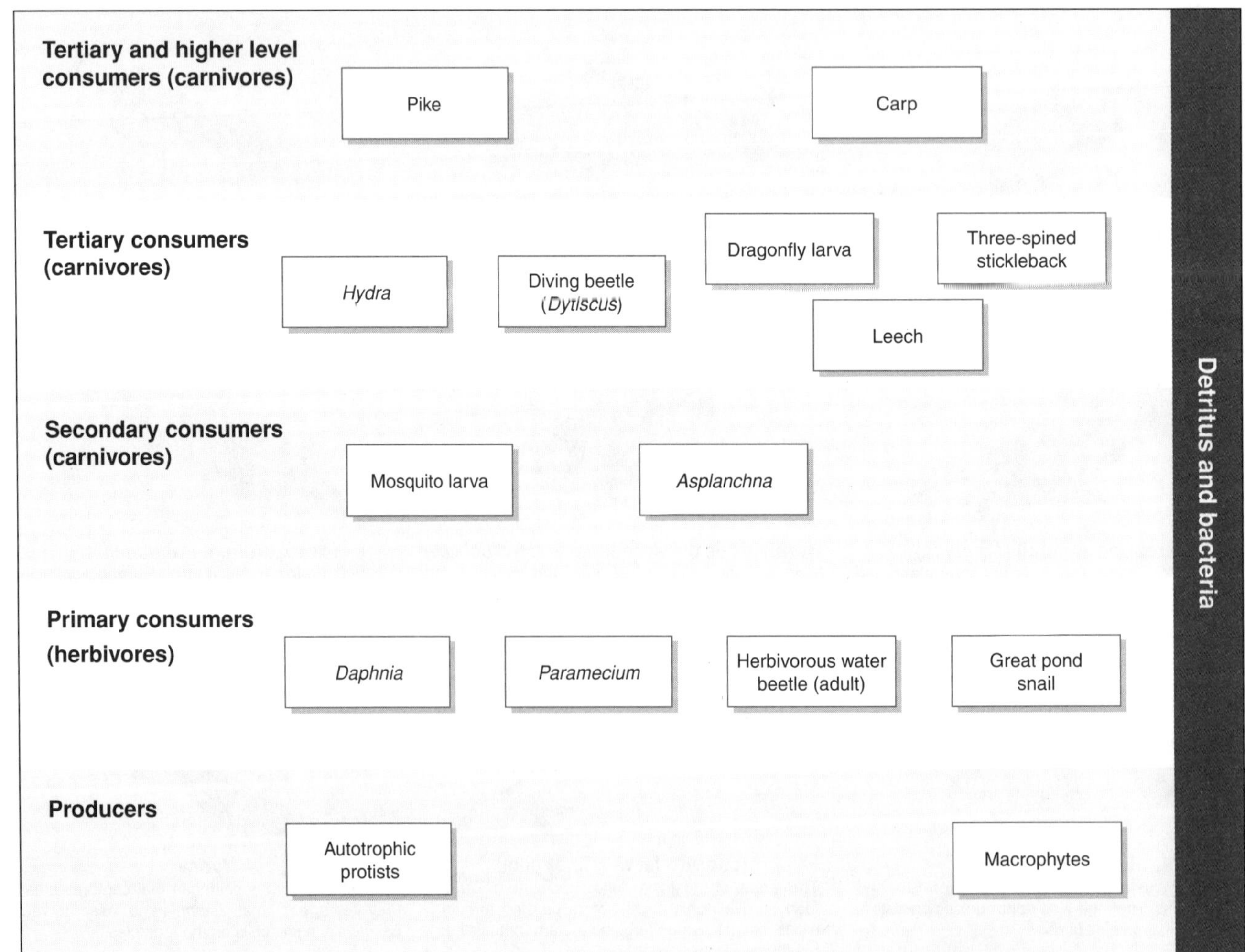
Tertiary and higher level consumers (carnivores)
Pike
Carp
Tertiary consumers (carnivores)
Hydra
Diving beetle (Dytiscus)
Dragonfly larva
Three-spined stickleback
Leech
Secondary consumers (carnivores)
Mosquito larva
Asplanchna
Primary consumers (herbivores)
Daphnia
Paramecium
Herbivorous water beetle (adult)
Great pond snail
Producers
Autotrophic protists
Macrophytes
Detritus and bacteria

ISBN: 978-1-927173-16-9

Energy Flow in an Ecosystem

The flow of energy through an ecosystem can be measured and analyzed. It provides some idea as to the energy trapped and passed on at each trophic level. Each trophic level in a food chain or web contains a certain amount of biomass: the dry weight of all organic matter contained in its organisms. Energy stored in biomass is transferred from one trophic level to another (by eating, defecation etc.), with some being lost as low-grade heat energy to the environment in each transfer. Three definitions are useful:

- **Gross primary production**: The total of organic material produced by plants (including that lost to respiration).
- **Net primary production**: The amount of biomass that is available to consumers at subsequent trophic levels.
- **Secondary production**: The amount of biomass at higher trophic levels (consumer production). Production figures are sometimes expressed as rates (productivity).

The percentage of energy transferred from one trophic level to the next varies between 5% and 20% and is called the **ecological efficiency** (efficiency of energy transfer). An average figure of 10% is often used. The path of energy flow in an ecosystem depends on its characteristics. In a tropical forest ecosystem, most of the primary production enters the detrital and decomposer food chains. However, in an ocean ecosystem or an intensively grazed pasture more than half the primary production may enter the grazing food chain.

Energy Flow Through an Ecosystem

NOTE

Numbers represent **kilojoules** of energy per square metre per year (kJ m^{-2} yr^{-1})

Sunlight falling on plant surfaces 7,000,000

Light absorbed by plants 1,700,000

Energy absorbed from the previous trophic level

100

Trophic level

Energy lost as heat 65

Energy lost to detritus 10

25

Energy passed on to the next trophic level

The energy available to each trophic level will always equal the amount entering that trophic level, minus total losses to that level (due to metabolic activity, death, excretion etc). Energy lost as heat will be lost from the ecosystem. Other losses become part of the detritus and may be utilized by other organisms in the ecosystem

A

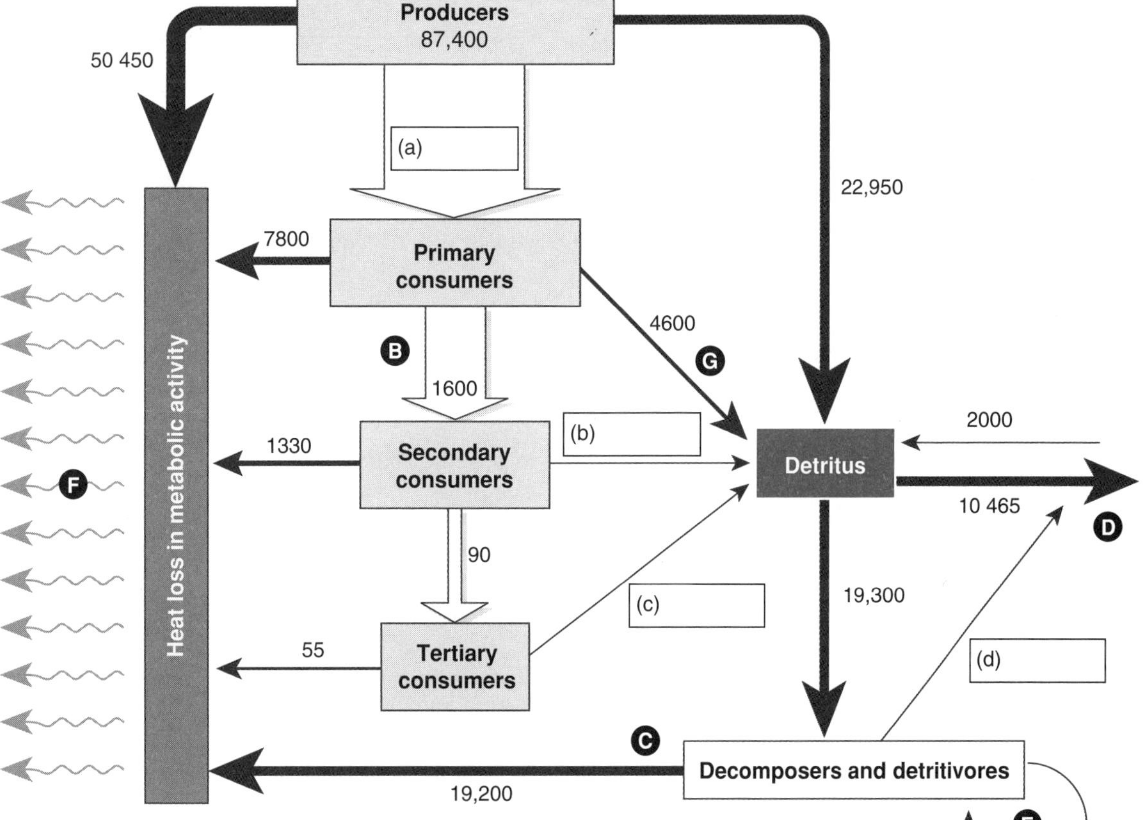
Producers
87,400
50 450
(a)
22,950
Heat loss in metabolic activity
7800
Primary consumers
4600
B
G
1600
2000
Secondary consumers
(b)
1330
Detritus
F
10 465
D
90
(c)
19,300
55
Tertiary consumers
(d)
C
Decomposers and detritivores
19,200
E

ISBN: 978-1-927173-16-9

1. Study the diagram on the previous page illustrating energy flow through a hypothetical ecosystem. Use the example at the top of the page as a guide to calculate the missing values (a)–(d) in the diagram. Note that the sum of the energy inputs always equals the sum of the energy outputs. Place your answers in the spaces provided on the diagram.

2. What is the original source of energy for this ecosystem? __________

3. Identify the processes occurring at the points labelled **A – G** on the diagram:

   A. __________ E. __________

   B. __________ F. __________

   C. __________ G. __________

   D. __________

4. (a) Calculate the percentage of light energy falling on the plants that is absorbed at point **A**:

   Light absorbed by plants ÷ sunlight falling on plant surfaces x 100 = __________

   (b) What happens to the light energy that is not absorbed? __________

5. (a) Calculate the percentage of light energy absorbed that is actually converted (fixed) into producer energy:

   Producers ÷ light absorbed by plants x 100 = __________

   (b) How much light energy is absorbed but not fixed: __________

   (c) Account for the difference between the amount of energy absorbed and the amount actually fixed by producers:

6. Of the total amount of energy **fixed** by producers in this ecosystem (at point **A**) calculate:

   (a) The total amount that ended up as metabolic waste heat (in kJ): __________

   (b) The percentage of the energy fixed that ended up as waste heat: __________

7. (a) State the groups for which detritus is an energy source: __________

   (b) How could detritus be removed or added to an ecosystem? __________

8. Under certain conditions, decomposition rates can be very low or even zero, allowing detritus to accumulate:

   (a) From your knowledge of biological processes, what conditions might slow decomposition rates?

   (b) What are the consequences of this lack of decomposer activity to the energy flow? __________

   (c) Add an additional arrow to the diagram on the previous page to illustrate your answer.

   (d) Describe three examples of materials that have resulted from a lack of decomposer activity on detrital material:

9. The **ten percent rule** states that the total energy content of a trophic level in an ecosystem is only about one-tenth (or 10%) that of the preceding level. For each of the trophic levels in the diagram on the preceding page, determine the amount of energy passed on to the next trophic level as a percentage:

   (a) Producer to primary consumer: __________

   (b) Primary consumer to secondary consumer: __________

   (c) Secondary consumer to tertiary consumer: __________

ISBN: 978-1-927173-16-9

Ecological Pyramids

The trophic levels of any ecosystem can be arranged in a pyramid shape. The first trophic level is placed at the bottom and subsequent trophic levels are stacked on top in their 'feeding sequence'. Ecological pyramids can illustrate changes in the numbers, biomass (weight), or energy content of organisms at each level. Each of these three kinds of pyramids tell us something different about the flow of energy and materials between one trophic level and the next. The type of pyramid you choose in order to express information about the ecosystem will depend on what particular features of the ecosystem you are interested in and, of course, the type of data you have collected.

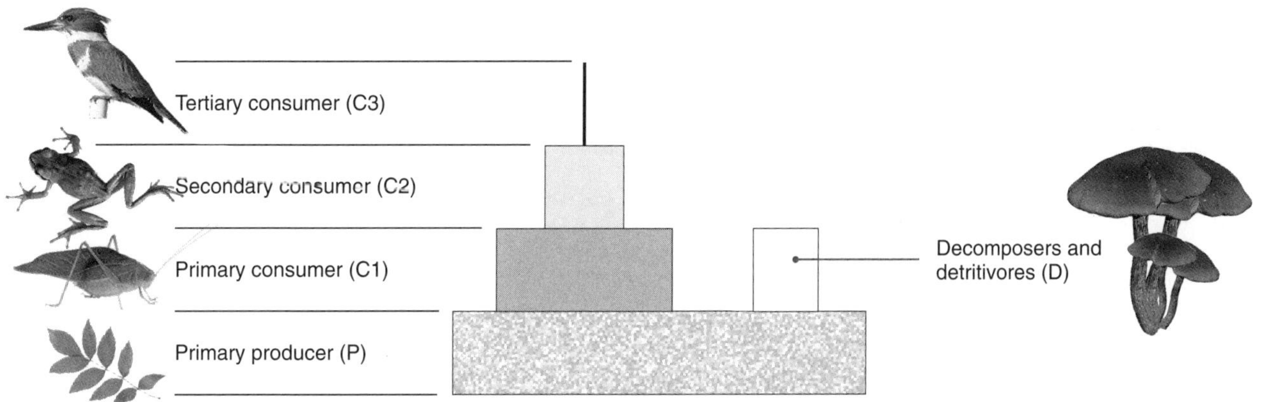

Tertiary consumer (C3)
Secondary consumer (C2)
Primary consumer (C1)
Primary producer (P)
Decomposers and detritivores (D)

The generalized ecological pyramid pictured above shows a conventional pyramid shape, with a large number (or biomass) of producers forming the base for an increasingly small number (or biomass) of consumers. Decomposers are placed at the level of the primary consumers and off to the side. They may obtain energy from many different trophic levels and so do not fit into the conventional pyramid structure. For any particular ecosystem at any one time (e.g. the forest ecosystem below), the shape of this typical pyramid can vary greatly depending on whether the trophic relationships are expressed as numbers, biomass or energy

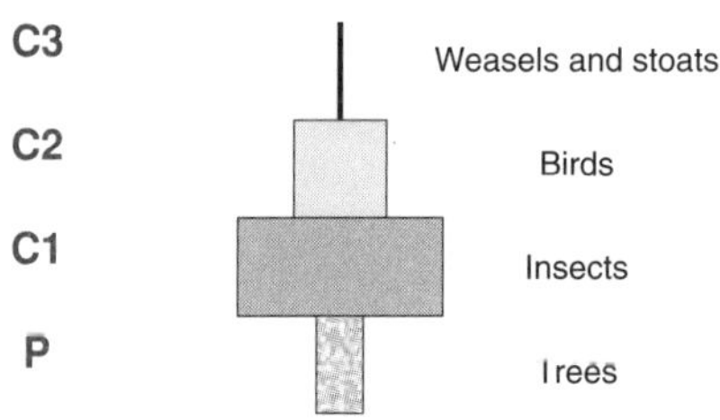

C3
C2
C1
P
Weasels and stoats
Birds
Insects
Trees

Numbers in a forest community

Pyramids of numbers display the number of individual organisms at each trophic level. The pyramid above has few producers, but they may be of a very large size (e.g. trees). This gives an 'inverted pyramid', although not all pyramids of numbers are like this.

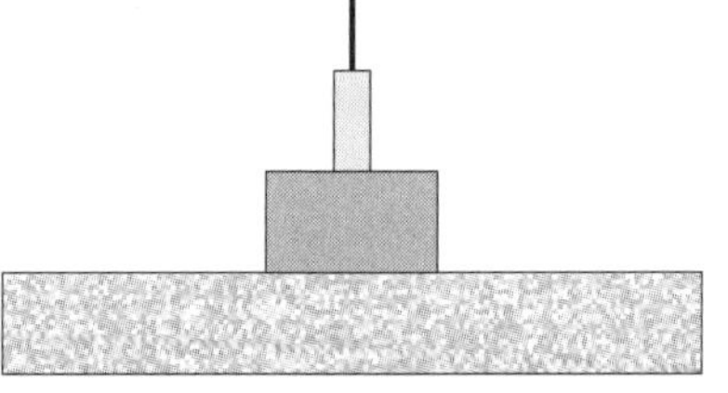

Biomass in a forest community

Biomass pyramids measure the 'weight' of biological material at each trophic level. Water content of organisms varies, so 'dry weight' is often used. Organism size is taken into account, allowing meaningful comparisons of different trophic levels.

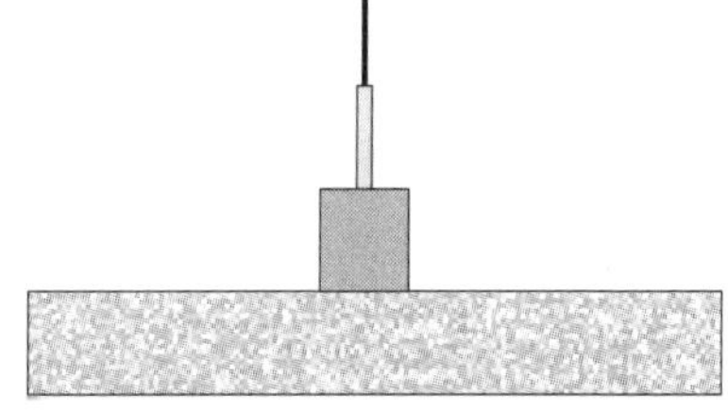

Energy in a forest community

Pyramids of energy are often very similar to biomass pyramids. The energy content at each trophic level is generally comparable to the biomass (i.e. similar amounts of dry biomass tend to have about the same energy content).

1. What do each of the following types of ecological pyramids measure?

   (a) Number pyramid: ______________________

   (b) Biomass pyramid: ______________________

   (c) Energy pyramid: ______________________

2. What is the advantage of using a biomass or energy pyramid rather than a pyramid of numbers to express the relationship between different trophic levels?

____

____

____

____

3. How can a forest community with relatively few producers (see next page) support a large number of consumers?

____

____

ISBN: 978-1-927173-16-9

Related activities: *Food Chains, Energy Flow in an Ecosystem*

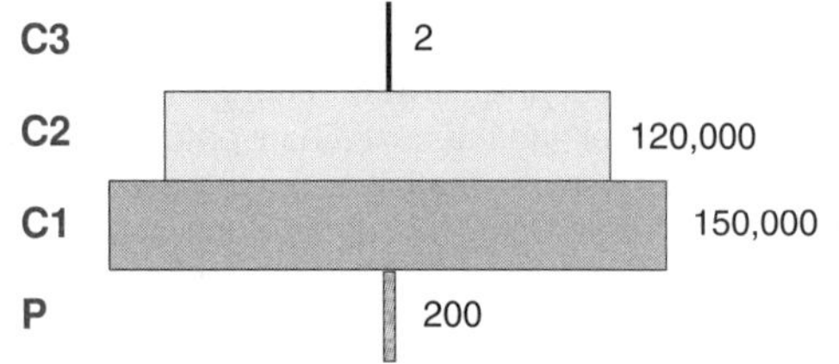
C3
2
C2
120,000
C1
150,000
P
200

C3 1
C2 90,000
C1 200,000
P

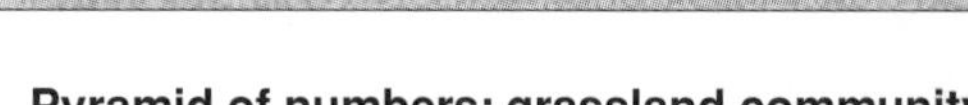
1,500,000

Pyramid of numbers: forest community

In a forest community a few producers may support a large number of consumers. This is due to the large size of the producers; large trees can support many individual consumer organisms. The example above shows the numbers at each trophic level for an oak forest in England, in an area of 10 m^2.

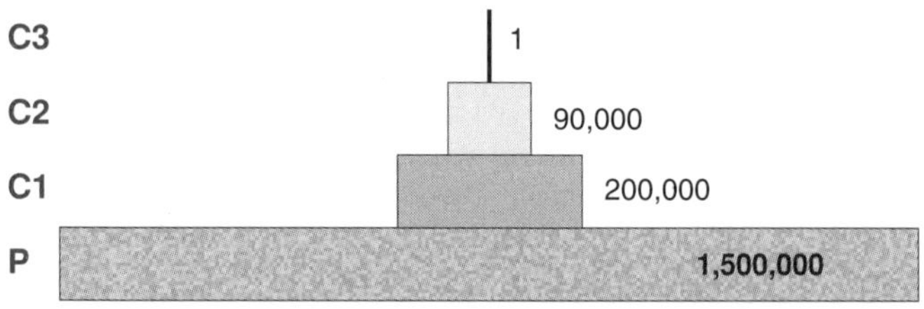

Pyramid of numbers: grassland community

In a grassland community a large number of producers are required to support a much smaller number of consumers. This is due to the small size of the producers. Grass plants can support only a few individual consumer organisms and take time to recover from grazing pressure. The example above shows the numbers at each trophic level for a derelict grassland area (10 m^2) in Michigan, United States.

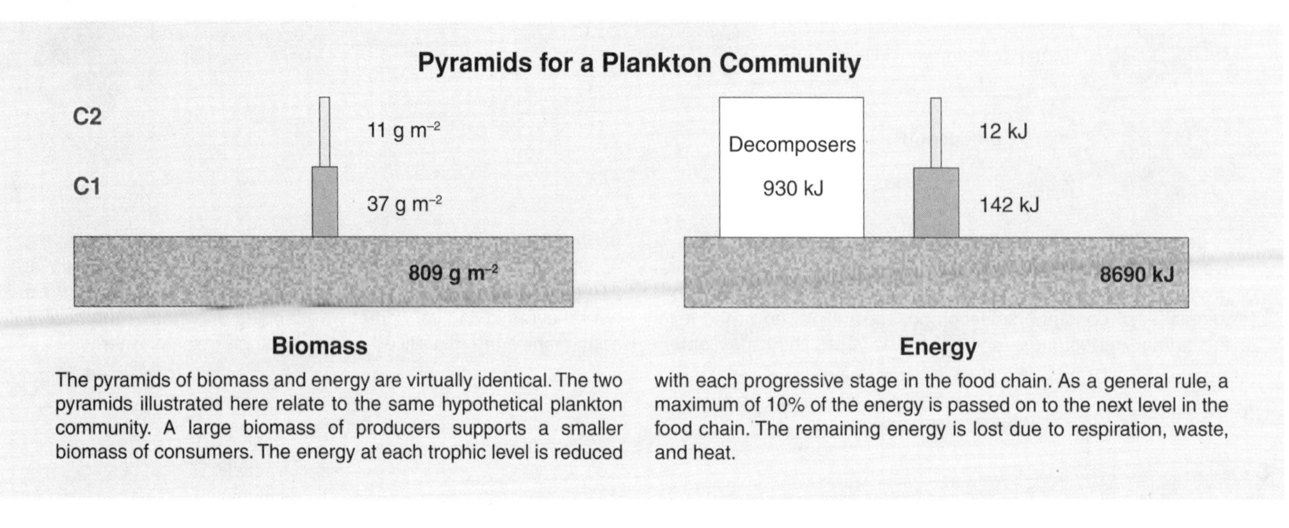
Pyramids for a Plankton Community
C2
11 g m-2
C1
37 g m-2
809 g m-2
Biomass
Decomposers
930 kJ
12 kJ
142 kJ
8690 kJ
Energy

The pyramids of biomass and energy are virtually identical. The two pyramids illustrated here relate to the same hypothetical plankton community. A large biomass of producers supports a smaller biomass of consumers. The energy at each trophic level is reduced with each progressive stage in the food chain. As a general rule, a maximum of 10% of the energy is passed on to the next level in the food chain. The remaining energy is lost due to respiration, waste, and heat.

4. Determine the **energy transfer** between trophic levels in the plankton community example in the above diagram:

(a) Between producers and the primary consumers: ______________________

(b) Between the primary consumers and the secondary consumers: ______________________

(c) Why is the amount of energy transferred from the producer level to primary consumers considerably less than the expected 10% that occurs in many other communities?

____

____

(d) After the producers, which trophic group has the greatest energy content? ______________________

(e) Give a likely explanation why this is the case: ______________________

____

____

An unusual biomass pyramid

The biomass pyramids of some ecosystems appear rather unusual with an inverted shape. The first trophic level has a lower biomass than the second level. What this pyramid does not show is the rate at which the producers (algae) are reproducing in order to support the larger biomass of consumers.

Zooplankton and bottom fauna — 21 g m^{-2}

Algae — 4 g m^{-2}

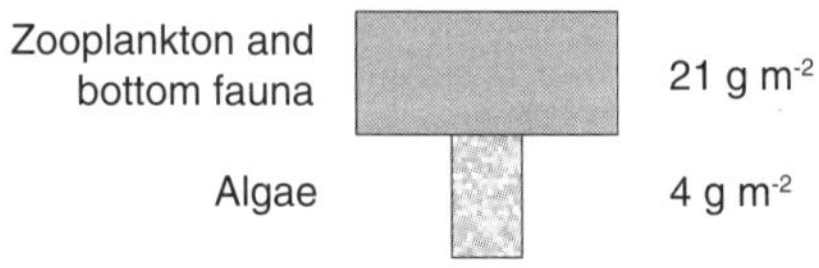

Biomass

5. Give a possible explanation of how a small biomass of producers (algae) can support a larger biomass of consumers (zooplankton):

____

____

____

ISBN: 978-1-927173-16-9

Nutrient Cycles

Nutrient cycling is an important part of every ecosystem. Elements essential for the efficient operation of living systems move through the environment through the processes of uptake and deposition. Commonly, nutrients must be in an ionic (rather than elemental) form in order for plants and animals to have access to them. Some bacteria have the ability to convert elemental forms of nutrients, such as sulfur, into ionic forms and so play an important role in making nutrients available to plants and animals.

Essential Nutrients

Macronutrient	Common form	Function
Carbon (C)	CO_2	Organic molecules
Oxygen (O)	O_2	Respiration
Hydrogen (H)	H_2O	Cellular hydration
Nitrogen (N)	N_2, NO_3^-, NH_4^+	Proteins, nucleic acids
Potassium (K)	K^+	Principal ion in cells
Phosphorus (P)	$H_2PO_4^-$, HPO_4^{2-}	Nucleic acids, lipids
Calcium (Ca)	Ca^{2+}	Membrane permeability
Magnesium (Mg)	Mg^{2+}	Chlorophyll
Sulfur (S)	SO_4^{2-}	Proteins
Micronutrient	**Common form**	**Function**
Iron (Fe)	Fe^{2+}, Fe^{3+}	Chlorophyll, blood
Manganese (Mn)	Mn^{2+}	Enzyme activation
Molybdenum (Mo)	MoO_4^-	Nitrogen metabolism
Copper (Cu)	Cu^{2+}	Enzyme activation
Sodium (Na)	Na^+	Ion in cells
Silicon (Si)	$Si(OH)_4$	Support tissues

Tropical Rainforest

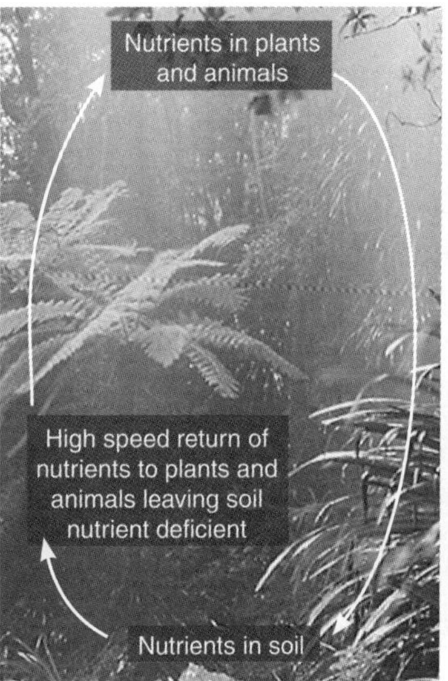
Nutrients in plants and animals
High speed return of nutrients to plants and animals leaving soil nutrient deficient
Nutrients in soil

Temperate Woodland

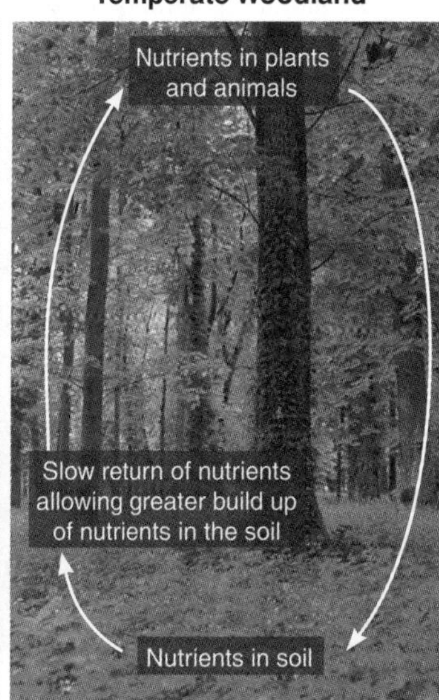
Nutrients in plants and animals
Slow return of nutrients allowing greater build up of nutrients in the soil
Nutrients in soil

The speed of nutrient cycling can vary markedly. Some nutrients are cycled slowly, others quickly. The environment and diversity of an ecosystem can also have a large effect on the speed at which nutrients are recycled.

The Role of Organisms in Nutrient Cycling

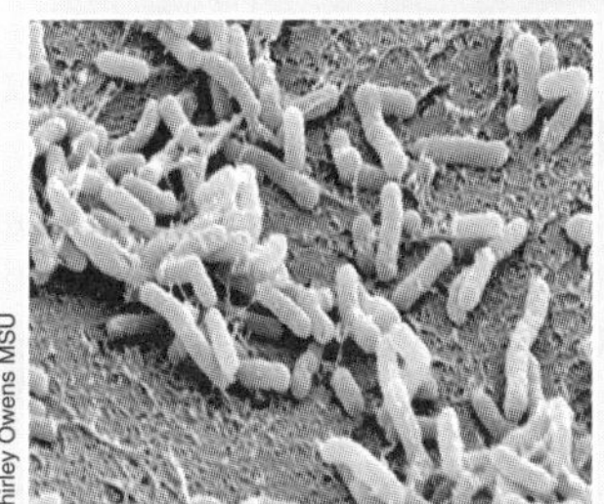

Shirley Owens MSU

Bacteria

Bacteria play an essential role in all nutrient cycles. They have the ability to act as saprophytes, decomposing material, but are also able to convert nutrients from inaccessible to biologically accessible forms.

Fungi

Fungi are saprophytes and play a critical role in decomposing organic material, returning nutrients to the soil or converting them into forms accessible to plants and animals.

Plants

Plants have an important role in absorbing many nutrients from the soil and making them directly available to browsing animals. They also add their own decaying matter to soils.

Animals

Animals utilize and break down materials from bacteria, plants, and fungi and return the nutrients to soils and water via their wastes and when they die.

1. Describe the role of each of the following in nutrient cycling:

   (a) Bacteria: ______

   (b) Fungi: ______

   (c) Plants: ______

   (d) Animals: ______

2. Why are soils in tropical rainforests nutrient deficient relative to soils in temperate woodlands? ______

3. Distinguish between macronutrients and micronutrients: ______

ISBN: 978-1-927173-16-9

Periodicals: Microbes and nutrient cycling

Related activities: Energy Inputs and Outputs
Weblinks: Nitrogen Cycle Animation

The Carbon Cycle

Carbon is an essential element in living systems, providing the chemical framework to form the molecules that make up living organisms (e.g. proteins, carbohydrates, fats, and nucleic acids). Carbon also makes up approximately 0.03% of the atmosphere as the gas carbon dioxide (CO_2), and it is present in the ocean as carbonate and bicarbonate, and in rocks such as limestone. Carbon cycles between the living (biotic) and non-living (abiotic) environment: it is fixed in the process of photosynthesis and returned to the atmosphere in respiration. Carbon may remain locked up in biotic or abiotic systems for long periods of time as, for example, in the wood of trees or in fossil fuels such as coal or oil. Human activity has disturbed the balance of the carbon cycle (the global carbon budget) through activities such as combustion (e.g. the burning of wood and **fossil fuels**) and deforestation.

Respiration
Atmospheric carbon dioxide
Combustion
Dissolving
Carbon in plant and animal tissues
Photosynthesis
Diatoms A
Fossilization
Death and excretion
Drilling for oil and gas
Fossilization
B Decomposers
Oil and gas
Coal
Limestone

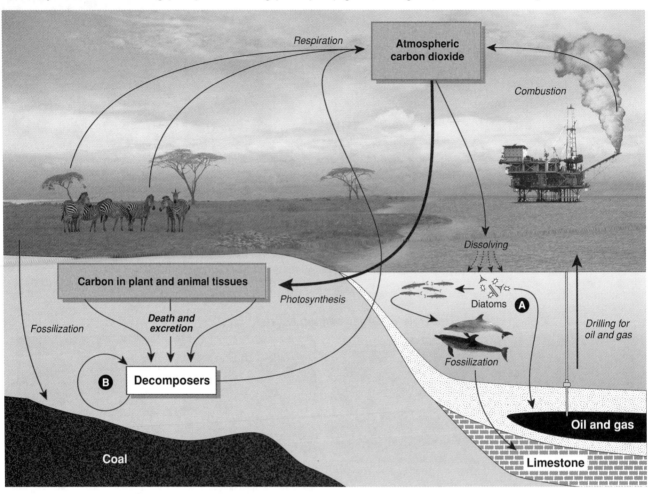

1. Add **arrows** and **labels** to the diagram above to show:
   (a) Dissolving of limestone by acid rain
   (b) Release of carbon from the marine food chain
   (c) Mining and burning of coal
   (d) Burning of plant material.

2. Describe the **biological origin** of the following geological deposits:
   (a) Coal: ____________________
   (b) Oil: ____________________
   (c) Limestone: ____________________

3. (a) What two processes release carbon into the atmosphere? ____________________
   ____
   (b) In what form is the carbon released? ____________________

4. Name the four geological reservoirs (sinks), in the diagram above, that can act as a source of carbon:
   (a) ____________________ (c) ____________________
   (b) ____________________ (d) ____________________

5. (a) Identify the process carried out by diatoms at point [**A**]: ____________________
   (b) Identify the process carried out by decomposers at [**B**]: ____________________

Related activities: Nutrient Cycles
Weblinks: The Carbon Cycle

Periodicals: The case of the missing carbon

ISBN: 978-1-927173-16-9

Termite mound in rainforest

Dung beetle on cow pat

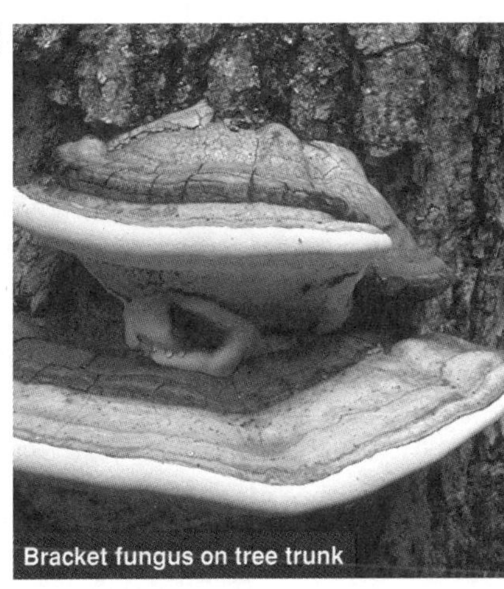
Bracket fungus on tree trunk

Termites: These insects play an important role in nutrient recycling. With the aid of symbiotic protozoans and bacteria in their guts, they can digest the tough cellulose of woody tissues in trees. Termites fulfill a vital function in breaking down the endless rain of debris in tropical rainforests.

Dung beetles: Beetles play a major role in the decomposition of animal dung. Some beetles merely eat the dung, but true dung beetles, such as the scarabs and *Geotrupes*, bury the dung and lay their eggs in it to provide food for the beetle grubs during their development..

Fungi: Together with decomposing bacteria, fungi perform an important role in breaking down dead plant matter in the leaf litter of forests. Some mycorrhizal fungi have been found to link up to the root systems of trees where an exchange of nutrients occurs (a mutualistic relationship).

6. What would be the effect on carbon cycling if there were **no decomposers** present in an ecosystem?

7. Explain the role of each of the following organisms in the carbon cycle:

   (a) Dung beetles:

   (b) Termites:

   (c) Fungi:

8. Bushfires play an important role in some ecosystems. Describe the role of fire in nutrient recycling:

9. In natural circumstances, accumulated reserves of carbon such as peat, coal and oil represent a sink or natural diversion from the cycle. Eventually, the carbon in these sinks returns to the cycle through the action of geological processes which return deposits to the surface for oxidation.

   (a) What is the effect of human activity on the amount of carbon stored in sinks?

   (b) Describe two **global effects** resulting from this activity:

   (c) What could be done to prevent or alleviate these effects?

ISBN: 978-1-927173-16-9

Global Warming

The Earth's atmosphere comprises a mix of gases including nitrogen, oxygen, and water vapor. Small quantities of carbon dioxide (CO_2), methane, and a number of other trace gases are also present. The term 'greenhouse effect' describes the natural process by which heat is retained within the atmosphere by these greenhouse gases, which act as a thermal blanket around the Earth, letting in sunlight, but trapping the heat that would normally radiate back into space. The greenhouse effect results in the Earth having a mean surface temperature of about 15°C, 33°C warmer than it would have without an atmosphere. About 75% of the natural greenhouse effect is due to water vapor. The next most significant agent is CO_2. Fluctuations in the Earth's surface temperature as a result of climate shifts are normal, and the current period of warming climate is partly explained by the recovery after the most recent ice age that finished 10,000 years ago. However since the mid 20th century, the Earth's surface temperature has been increasing. This phenomenon is called global warming and the majority of researchers attribute it to the increase in atmospheric levels of CO_2 and other greenhouse gases emitted into the atmosphere as a result of human activity (i.e. it is anthropogenic). Nine of the ten warmest years on record were in the 2000s (1998 being the third warmest on record). Global surface temperatures in 2005 set a new record but are now tied with 2010 as being the hottest years on record.

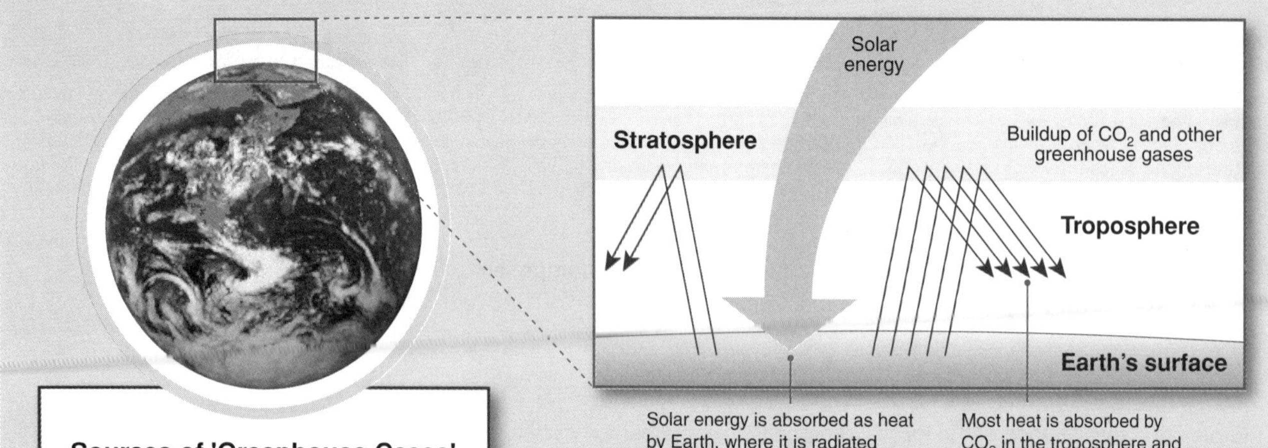
Solar energy
Stratosphere
Buildup of CO_2 and other greenhouse gases
Troposphere
Earth's surface

Solar energy is absorbed as heat by Earth, where it is radiated back into the atmosphere

Most heat is absorbed by CO_2 in the troposphere and radiated back to Earth

Sources of 'Greenhouse Gases'

Carbon dioxide
- Exhaust from cars
- Combustion of coal, wood, oil
- Burning rainforests

Methane
- Plant debris and growing vegetation
- Belching and flatus of cows

Chloro-fluoro-carbons (CFCs)
- Leaking coolant from refrigerators
- Leaking coolant from air conditioners

Nitrous oxide
- Car exhaust

Tropospheric ozone*
- Triggered by car exhaust (smog)

*Tropospheric ozone is found in the lower atmosphere (not to be confused with ozone in the stratosphere)

Greenhouse gas	Tropospheric conc.		Global warming potential *(compared to CO_2)¶*	Atmospheric lifetime *(years)§*
	Pre-industrial 1750	Present day (2008*)		
Carbon dioxide	280 ppm	383.9 ppm	1	120
Methane	700 ppb	1796 ppb	25	12
Nitrous oxide	270 ppb	320.5 ppb	310	120
CFCs	0 ppb	0.39 ppbb	4000+	50-100
HFCs‡	0 ppb	0.045 ppb	1430	14
Tropospheric ozone	25 ppb	34 ppb	17	hours

ppm = parts per million; **ppb** = parts per billion; ‡Hydrofluorcarbons were introduced in the last decade to replace CFCs as refrigerants; * Data from July 2007-June 2008. ¶ Figures contrast the radiative effect of different greenhouse gases relative to CO_2 over 100 years, e.g. over 100 years, methane is 25 times more potent as a greenhouse gas than CO_2 § How long the gas persists in the atmosphere. *Source: CO_2 Information Analysis Centre, Oak Ridge National Laboratory, USA.*

This graph shows how the mean temperature for each year from 1860-2010 (bars) compares with the average temperature between 1961 and 1990. The blue line represents the fitted curve and shows the general trend indicated by the annual data. Most anomalies since 1977 have been above normal; warmer than the long term mean, indicating that global temperatures are tracking upwards. The decade 2001-2010 has been the warmest on record.

Source: Hadley Center for Prediction and Research

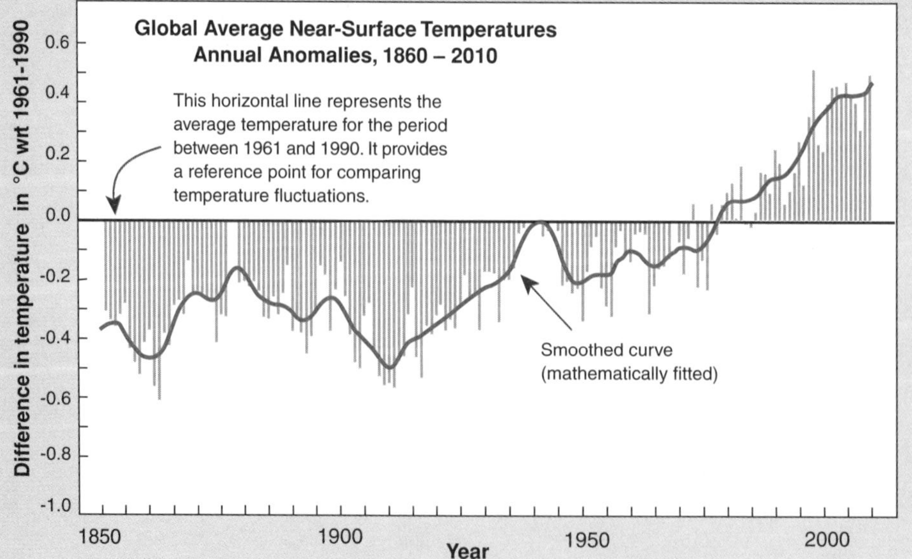
Global Average Near-Surface Temperatures
Annual Anomalies, 1860 – 2010
This horizontal line represents the average temperature for the period between 1961 and 1990. It provides a reference point for comparing temperature fluctuations.
Smoothed curve (mathematically fitted)
Difference in temperature in °C wrt 1961-1990
0.6
0.4
0.2
0.0
-0.2
-0.4
-0.6
-0.8
-1.0
1850
1900
1950
2000
Year

Related activities: *The Carbon Cycle*
Weblinks: *The Greenhouse Effect*

Periodicals:
Global warming

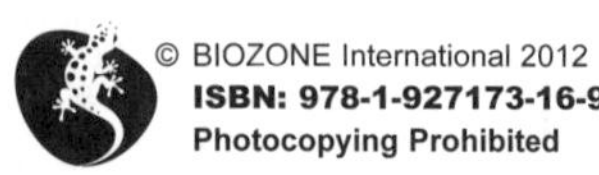

ISBN: 978-1-927173-16-9

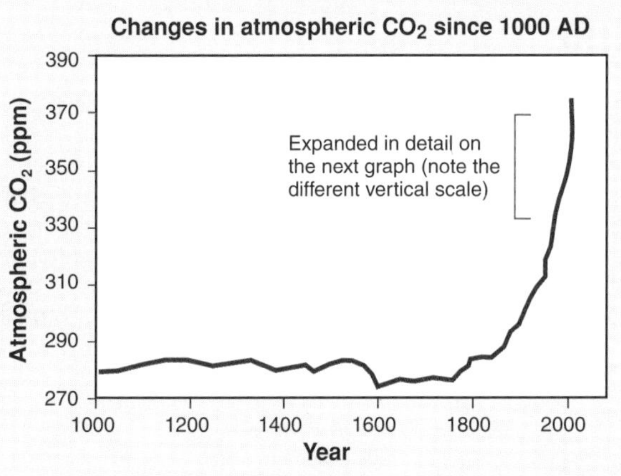
Changes in atmospheric CO_2 since 1000 AD
Atmospheric CO_2 (ppm)
390
370
350
330
310
290
270
Expanded in detail on the next graph (note the different vertical scale)
1000
1200
1400
1600
1800
2000
Year

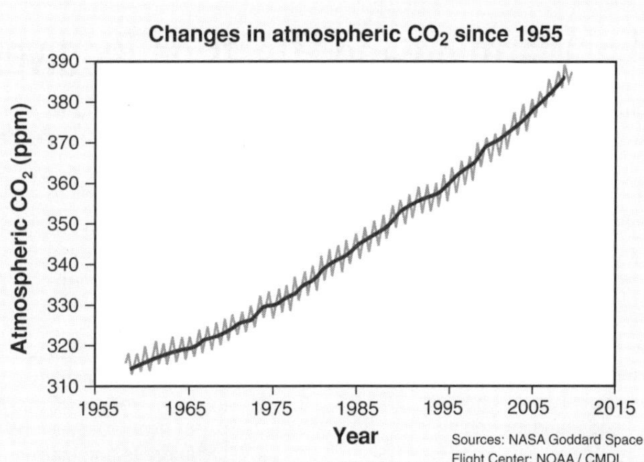
Changes in atmospheric CO_2 since 1955
Atmospheric CO_2 (ppm)
390
380
370
360
350
340
330
320
310
1955
1965
1975
1985
1995
2005
2015
Year
Sources: NASA Goddard Space Flight Center; NOAA / CMDL

Potential Effects of Global Warming

Sea levels are expected to rise by 50 cm by the year 2100. This is the result of the thermal expansion of ocean water and melting of glaciers and ice shelves. Warming may also expand the habitat for many pests, e.g. mosquitoes, shifting the range of infectious diseases.

Forests: Higher temperatures and precipitation changes could increase forest susceptibility to fire, disease, and insect damage. Forest fires release more carbon into the atmosphere and reduces the size of carbon sinks. A richer CO_2 atmosphere will reduce transpiration in plants.

Weather patterns: Global warming may cause regional changes in weather patterns such as El Niño and La Nina, as well as affecting the intensity and frequency of storms. Driven by higher ocean surface temperatures, high intensity hurricanes now occur more frequently.

Water resources: Changes in precipitation and increased evaporation will affect water availability for irrigation, industrial use, drinking, and electricity generation.

Agriculture: Climate change may threaten the viability of important crop-growing regions. Paradoxically, climate change can cause both too much and too little rain.

The ice-albedo effect: Ice has a stabilizing effect on global climate, reflecting nearly all the sun's energy that hits it. As polar ice melts, more of that energy is absorbed by the Earth.

1. Calculate the increase (as a %) in the 'greenhouse gases' between the pre-industrial era and the 2008 measurements (use the data from the table, see facing page). **HINT**: The calculation for carbon dioxide is: (383.9 - 280) ÷ 280 x 100 =

   (a) Carbon dioxide: ________________ (b) Methane: ________________ (c) Nitrous oxide: ________________

2. Explain the zig-zag nature of the atmospheric CO_2 graph to the above right. ________________

3. Explain the greenhouse effect and why it is an important process: ________________

4. Discuss some of the effects global warming will have on human lifestyles: ________________

ISBN: 978-1-927173-16-9

Global Warming and Effects on Biodiversity

Since the last significant period of climate change at the end of the ice age 10,000 years ago, plants and animals have adapted to survive in their current habitats. Accelerated global warming is again changing the habitats that plants and animals live in and this could have significant effects on the biodiversity of specific regions as well as on the planet overall. As temperatures rise, organisms will be forced to move to new areas where temperatures are similar to their current level. Those that cannot move face extinction, as temperatures move outside their limits of tolerance. Changes in precipitation as a result of climate change also affect where organisms can live. Long term changes in climate could see the contraction of many organisms' habitats while at the same time the expansion of others. Habitat migration, the movement of a habitat from its current region into another, will also become more frequent. Already there are a number of cases showing the effects of climate change on a range of organisms.

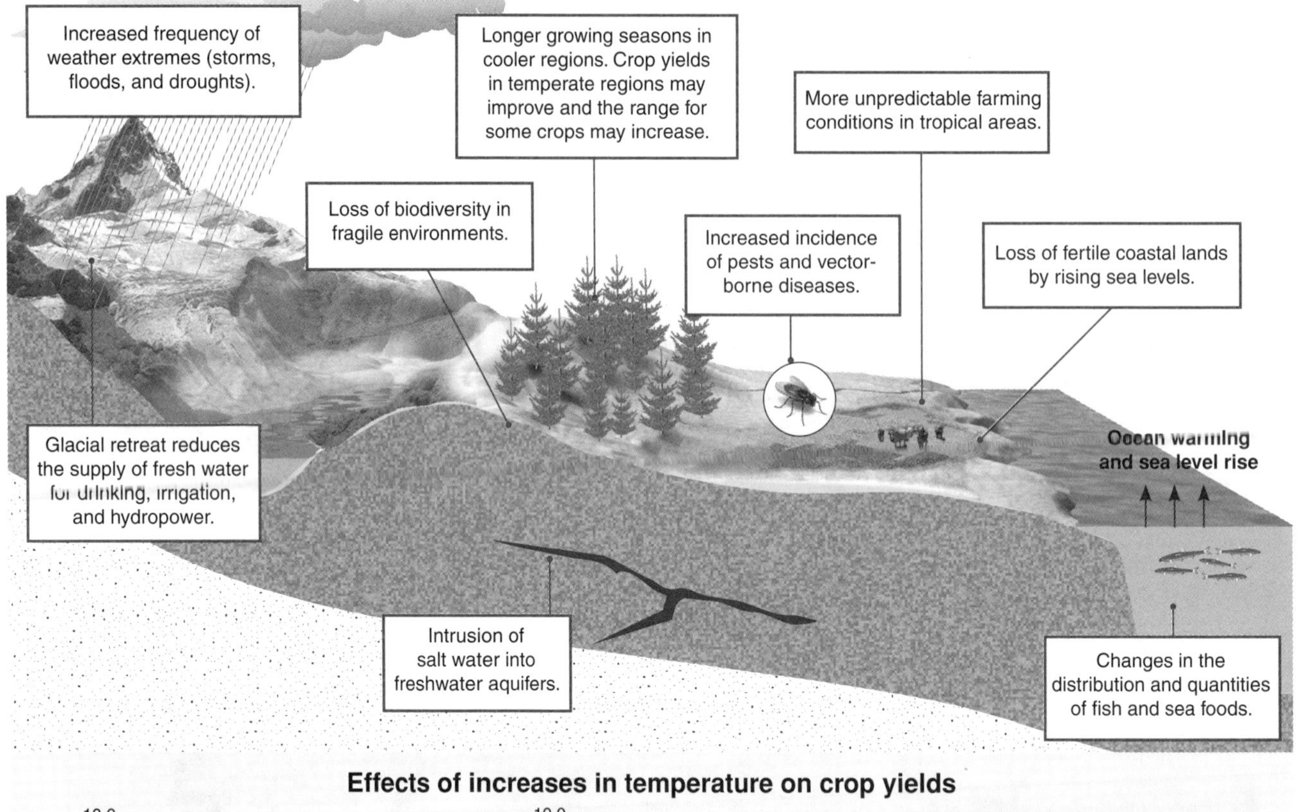

Increased frequency of weather extremes (storms, floods, and droughts).
Longer growing seasons in cooler regions. Crop yields in temperate regions may improve and the range for some crops may increase.
More unpredictable farming conditions in tropical areas.
Loss of biodiversity in fragile environments.
Increased incidence of pests and vector-borne diseases.
Loss of fertile coastal lands by rising sea levels.
Glacial retreat reduces the supply of fresh water for drinking, irrigation, and hydropower.
Ocean warming and sea level rise
Intrusion of salt water into freshwater aquifers.
Changes in the distribution and quantities of fish and sea foods.

Effects of increases in temperature on crop yields

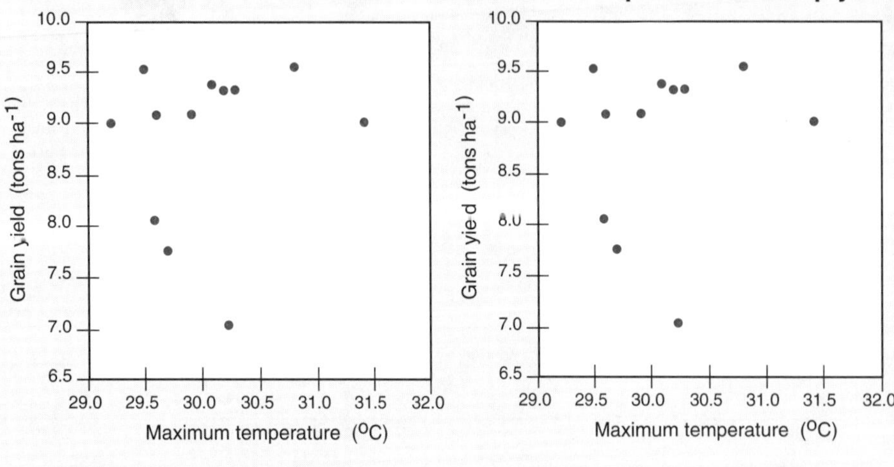

Grain yield (tons ha^{-1})
10.0
9.5
9.0
8.5
8.0
7.5
7.0
6.5
29.0
29.5
30.0
30.5
31.0
31.5
32.0
Maximum temperature (°C)

Studies on the grain production of rice have shown that maximum daytime temperatures have little effect on crop yield. However minimum night time temperatures lower crop yield by as much as 5% for every 0.5°C increase in temperature.

Source: Peng S. et.al. PNAS 2004

Possible effects of increases in temperature on crop damage

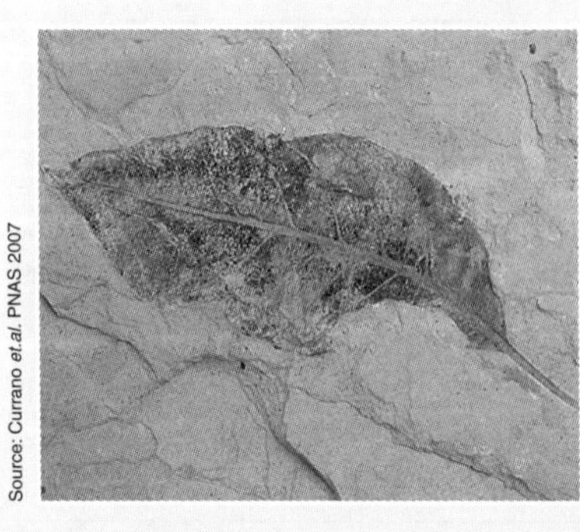

Source: Currano et.al. PNAS 2007

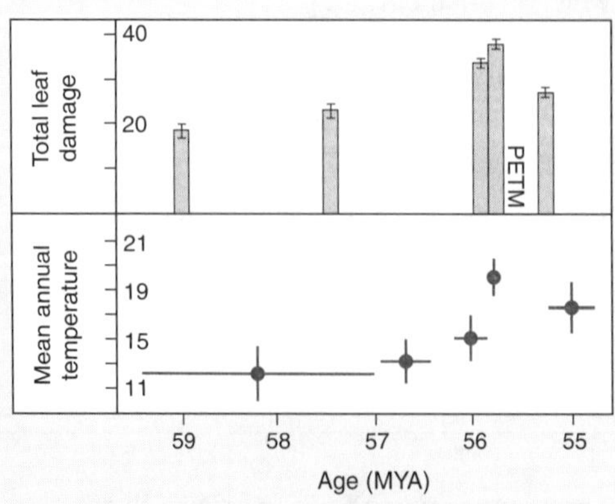

Total leaf damage
40
20
PETM
Mean annual temperature
21
19
15
11
59
58
57
56
55
Age (MYA)

The fossil record shows that global temperatures rose sharply around 56 million years ago. Studies of fossil leaves with insect browse damage indicate that leaf damage peaked at the same time as the Paleocene Eocene Thermal Maximum (PETM). This gives some historical evidence that as temperatures increase, plant damage caused by insects also rises. This could have implications for agricultural crops.

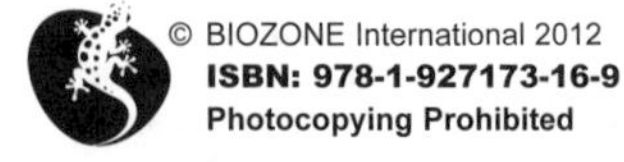

ISBN: 978-1-927173-16-9

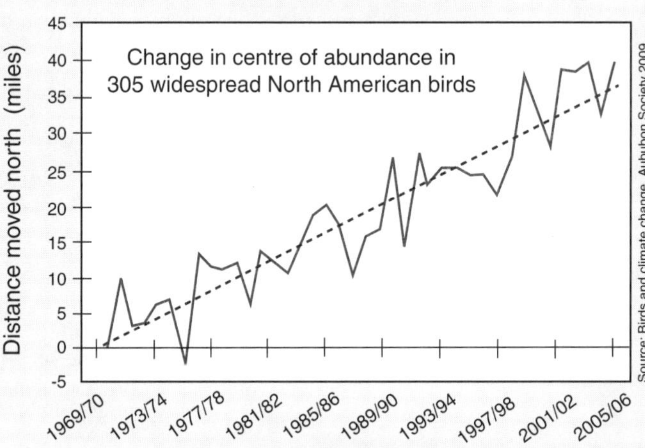
Effects of increases in temperature on animal populations
Change in centre of abundance in 305 widespread North American birds
Distance moved north (miles)
45
40
35
30
25
20
15
10
5
0
-5
1969/70
1973/74
1977/78
1981/82
1985/86
1989/90
1993/94
1997/98
2001/02
2005/06
Source: Birds and climate change, Aububon Society 2009

A number of studies indicate that animals are beginning to be affected by increases in global temperatures. Data sets from around the world show that birds are migrating up to two weeks earlier to summer feeding grounds and are often not migrating as far south in winter.

Animals living at altitude are also affected by warming climates and are being forced to shift their normal range. As temperatures increase, the snow line increases in altitude pushing alpine animals to higher altitudes. In some areas of North America this has resulting the local extinction of the North American pika (*Ochotona princeps*).

Wiki Commons

1. Describe some of the likely effects of global warming on physical aspects of the environment: ____________________

2. (a) Using the information on this and the previous activity, discuss the probable effects of global warming on plant crops:

(b) Suggest how farmers might be able to adjust to these changes: ____________________

3. Discuss the evidence that insect populations are affected by global temperature: ____________________

4. (a) Describe how increases in global temperatures have affected some migratory birds: ____________________

(b) Explain how these changes in migratory patterns might affect food availability for these populations: ____________________

5. Explain how global warming could lead to the local extinction of some alpine species: ____________________

ISBN: 978-1-927173-16-9

Global Warming and Effects on the Arctic

The surface temperature of the Earth is in part regulated by the amount of ice on its surface, which reflects a large amount of heat into space. However, the area and thickness of the polar sea-ice is rapidly decreasing. From 1980 to 2008 the Arctic summer sea-ice minimum almost halved, decreasing by more than 3 million square kilometers. This melting of sea-ice can trigger a cycle where less heat is reflected into space during summer, warming seawater and reducing the area and thickness of ice forming in the winter. At the current rate of reduction, it is estimated that there may be no summer sea-ice left in the Arctic by 2050.

Arctic sea-ice summer minimum 1980: 7.8 million km^2

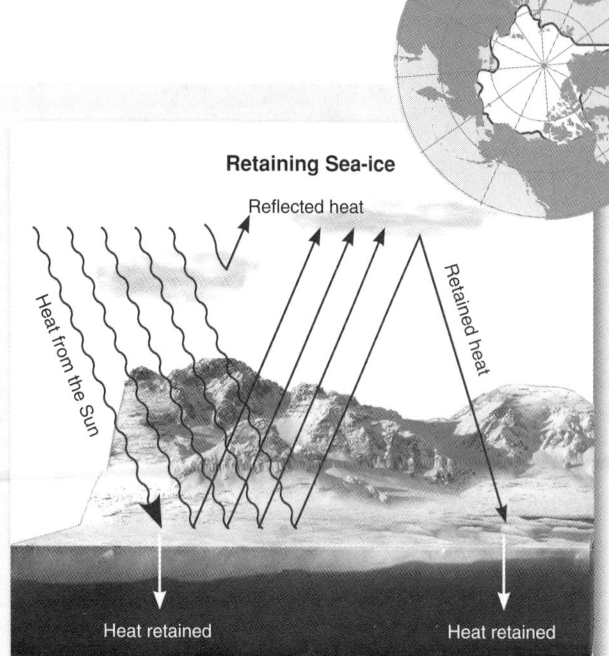
Retaining Sea-ice
Reflected heat
Heat from the Sun
Retained heat
Heat retained
Heat retained

The **albedo** (reflectivity of sea-ice) helps to maintain its presence. Thin sea-ice has a lower albedo than thick sea-ice. More heat is reflected when sea-ice is thick and covers a greater area. This helps to regulate the temperature of the sea, keeping it cool.

Arctic sea-ice summer minimum 2007: Record low, 4.33 million km^2

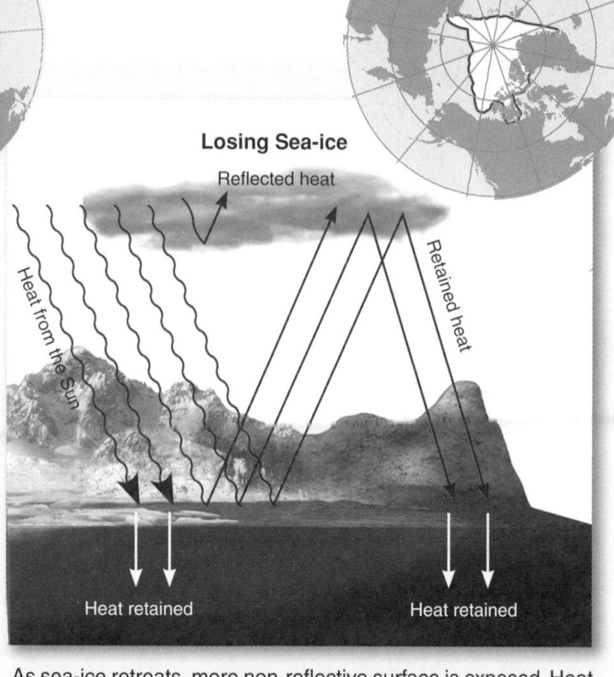
Losing Sea-ice
Reflected heat
Heat from the Sun
Retained heat
Heat retained
Heat retained

As sea-ice retreats, more non-reflective surface is exposed. Heat is retained instead of being reflected, warming both the air and water and causing sea-ice to form later in the autumn than usual. Thinner and less reflective ice forms and perpetuates the cycle.

The temperature in the Arctic has been above average every year since 1988. Coupled with the reduction in summer sea-ice, this is having dire effects on Arctic wildlife such as polar bears, which hunt out on the ice. The reduction in sea-ice reduces their hunting range and forces them to swim longer distances to firm ice. Studies have already shown an increase in drowning deaths of polar bears.

Average* Arctic Air Temperature Fluctuations

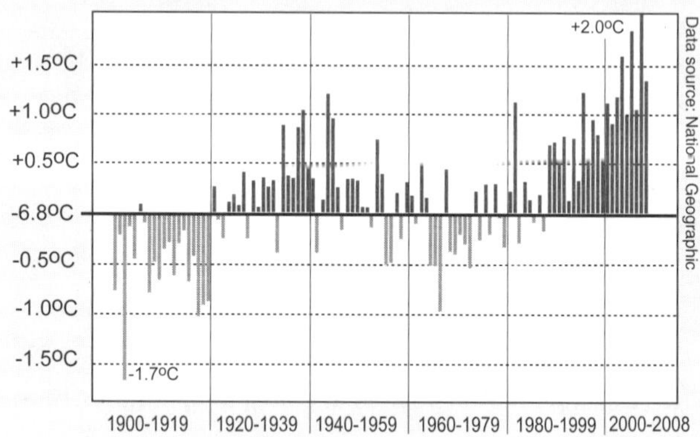
+2.0ºC
+1.5ºC
+1.0ºC
+0.5ºC
-6.8ºC
-0.5ºC
-1.0ºC
-1.5ºC
-1.7ºC
1900-1919
1920-1939
1940-1959
1960-1979
1980-1999
2000-2008
Data source: National Geographic

*Figure shows deviation from the average annual surface air temperature over land. Average calculated on the years 1961-2000.

1. Explain how low sea-ice albedo and volume affects the next year's sea-ice cover: ____________________

2. Discuss the effects of decreasing summer sea-ice on polar wildlife: ____________________

ISBN: 978-1-927173-16-9

Applying the Precautionary Principle

The **precautionary principle** is an analytical tool used to decide if human activity will harm human health or the environment. It aims to *prevent harm occurring* rather than having to manage a problem once it has occurred (i.e. it is better to be safe than sorry).The precautionary principle requires those wanting to carry out an activity to prove that their action will not be harmful. This is different from many current situations where those who believe an action may be harmful have to prove it is harmful it to stop further activity taking place. The precautionary principle places a social responsibility on an action or change maker to protect the public or environment from harm (i.e. they must show their actions are safe before they are allowed to proceed).

When should the precautionary principle be applied?

Global warming provides a good illustrative scenario for how the precautionary principle might be applied.

- Change could occur:
  In places such as Greenland, rising global temperatures would cause ice to melt and glaciers to reduce in size. Polar ice caps would shrink. The Larson B ice shelf in Antarctica has already disintegrated.
- Persistent and irreversible harm:
  The melting of ice caps would destroy and reduce habitat for polar organisms, such as polar bears, possibly driving them to extinction. Sea level rises (caused by the thermal expansion of water and melting of the ice sheets) may flood low lying costal areas.
- Chain reactions and flow on effects:
  A single change caused by global warming could have multiple effects. For example, melting of the glaciers changes habitat structure, causes water shortages in areas reliant on glacial melt water, and can result in localized effects on patterns of freeze and thaw.
- Difficulty in control or repair:
  Because global warming is influenced by global CO_2 production, controlling it is difficult. Unpredicted events (e.g. cyclonic events) that may be difficult (or impossible) to control could occur.
- Uncertainty:
  There is a large amount of evidence to support the fact that the climate is warming. However, there is some debate about the extent to which human activity has contributed to this. Much of this debate is created by those with vested interests in the status quo.
- Current activities can be linked to global warming:
  Human activities, such as burning fossil fuels, produce CO_2, which is a greenhouse gas, and contributes to increased global warming.

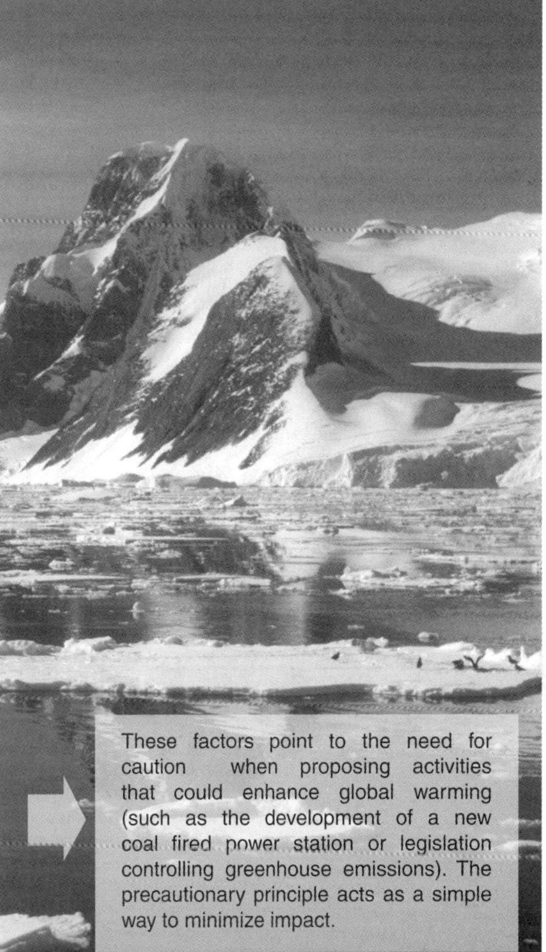

These factors point to the need for caution when proposing activities that could enhance global warming (such as the development of a new coal fired power station or legislation controlling greenhouse emissions). The precautionary principle acts as a simple way to minimize impact.

RADIOACTIVE RELEASE IN OUR WATER
NO MORE

Although the precautionary principle is designed to prevent harm caused by the use of a technology or change to a system, it could also be the cause of damage. The precautionary principle may be applied to solve a problem in one area, and thereby cause another elsewhere. The use of a potentially damaging technology may be forbidden, but not using that technology may be equally damaging.

An example is using nuclear power for electricity generation. Nuclear power generators carry significant environmental risks, yet not using nuclear power may mean greater reliance on fossil fuels, which contributes to accelerated global warming.

1. Explain how the precautionary principle could be applied to prevent exploitation of a resource in an environmentally sensitive area:

2. Give an example, other than the one described, where using the precautionary principle may in fact cause greater harm:

ISBN: 978-1-927173-16-9

Related activities: *Global Warming and Effects on the Arctic*

Features of Populations

Populations have a number of attributes that may be of interest. Usually, biologists wish to determine **population size** (the total number of organisms in the population). It is also useful to know the **population density** (the number of organisms per unit area). The density of a population is often a reflection of the **carrying capacity** of the environment, i.e. how many organisms a particular environment can support. Populations also have structure; particular ratios of different ages and sexes. These data enable us to determine whether the population is declining or increasing in size. We can also look at the **distribution** of organisms within their environment and so determine what particular aspects of the habitat are favoured over others. One way to retrieve information from populations is to **sample** them. Sampling involves collecting data about features of the population from samples of that population (since populations are usually too large to examine in total). Sampling can be done directly through a number of sampling methods or indirectly (e.g. monitoring calls, looking for droppings or other signs). Some of the population attributes that we can measure or calculate are illustrated on the diagram below.

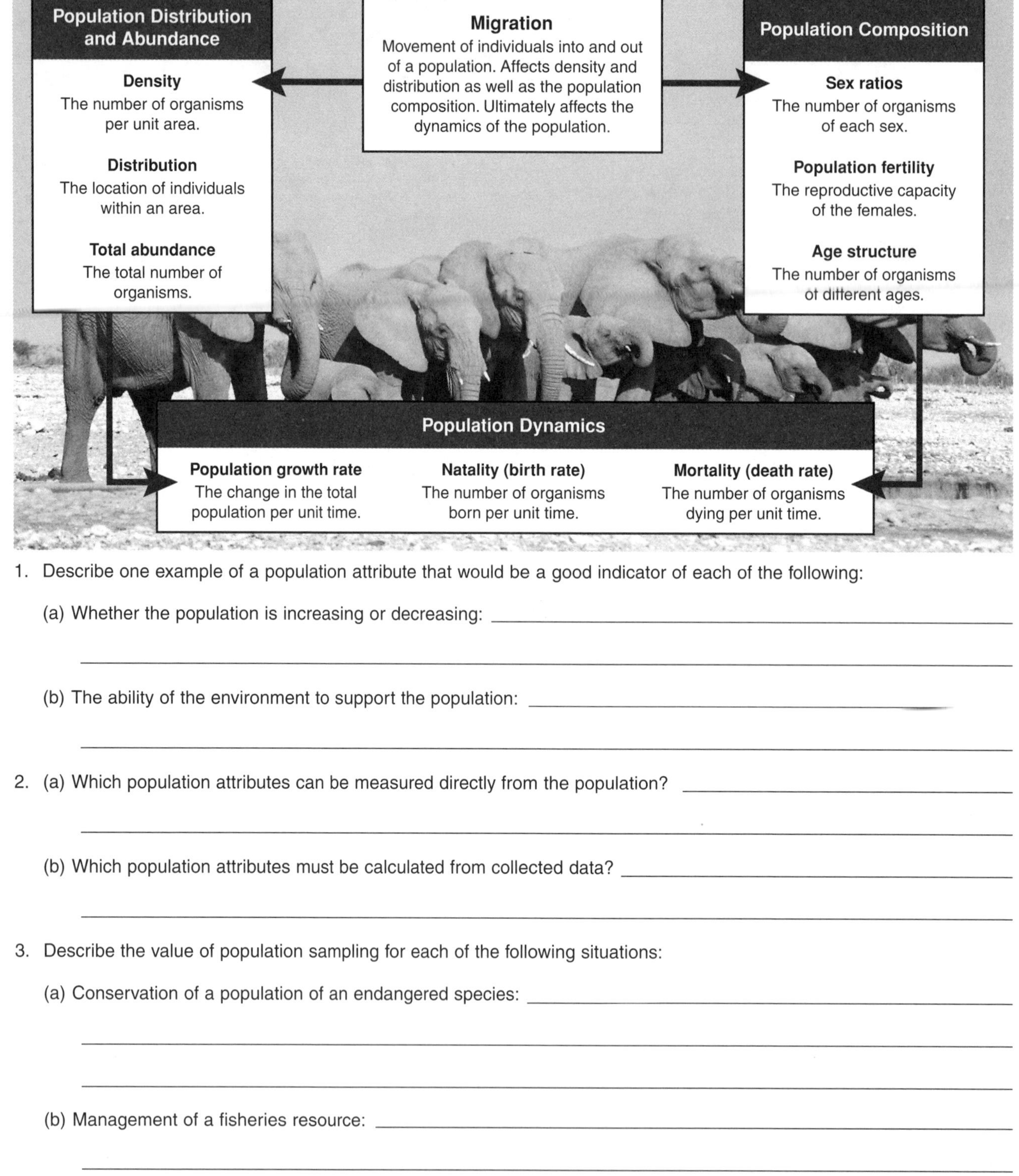
Population Distribution and Abundance
Density
The number of organisms per unit area.
Distribution
The location of individuals within an area.
Total abundance
The total number of organisms.
Migration
Movement of individuals into and out of a population. Affects density and distribution as well as the population composition. Ultimately affects the dynamics of the population.
Population Composition
Sex ratios
The number of organisms of each sex.
Population fertility
The reproductive capacity of the females.
Age structure
The number of organisms of different ages.
Population Dynamics
Population growth rate
The change in the total population per unit time.
Natality (birth rate)
The number of organisms born per unit time.
Mortality (death rate)
The number of organisms dying per unit time.

1. Describe one example of a population attribute that would be a good indicator of each of the following:

   (a) Whether the population is increasing or decreasing: ______

   (b) The ability of the environment to support the population: ______

2. (a) Which population attributes can be measured directly from the population? ______

   (b) Which population attributes must be calculated from collected data? ______

3. Describe the value of population sampling for each of the following situations:

   (a) Conservation of a population of an endangered species: ______

   (b) Management of a fisheries resource: ______

Related activities: Population Regulation

ISBN: 978-1-927173-16-9

Population Regulation

Very few species show continued exponential growth. Population size is regulated by factors that limit population growth. The diagram below illustrates how population size can be regulated by environmental factors. **Density independent factors** may affect all individuals in a population equally. Some, however, may be better able to adjust to them. **Density dependent factors** have a greater affect when the population density is higher. They become less important when the population density is low.

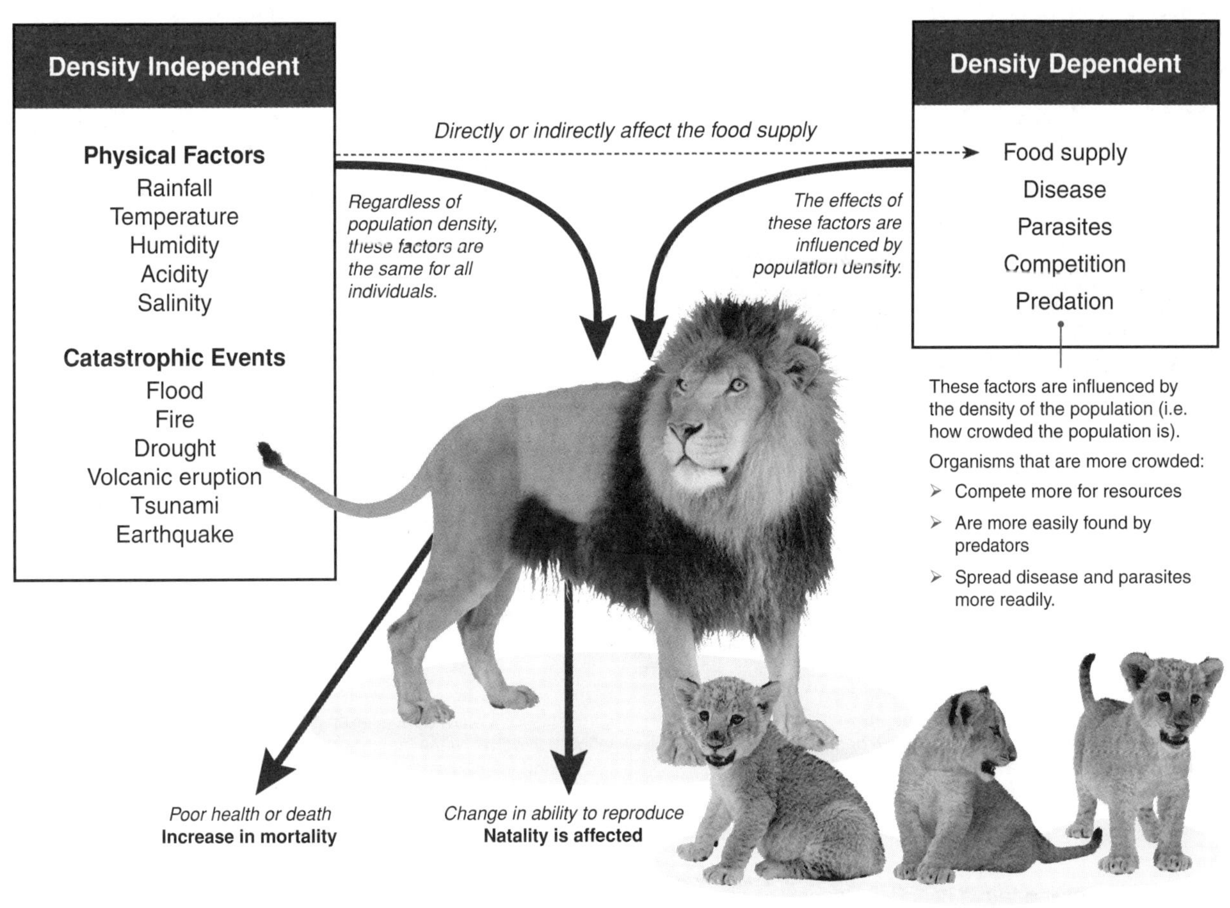

Density Independent
Physical Factors
Rainfall
Temperature
Humidity
Acidity
Salinity
Catastrophic Events
Flood
Fire
Drought
Volcanic eruption
Tsunami
Earthquake
Directly or indirectly affect the food supply
Regardless of population density, these factors are the same for all individuals.
The effects of these factors are influenced by population density.
Density Dependent
Food supply
Disease
Parasites
Competition
Predation
These factors are influenced by the density of the population (i.e. how crowded the population is).
Organisms that are more crowded:
Compete more for resources
Are more easily found by predators
Spread disease and parasites more readily.
Poor health or death
Increase in mortality
Change in ability to reproduce
Natality is affected

1. Discuss the role of **density dependent factors** and **density independent factors** in population regulation. In your discussion, make it clear that you understand the meaning of each of these terms:

2. Why does disease have a greater influence in regulating population size when population density is higher?

3. In cooler climates, aphids go through a huge population increase during the summer months. In autumn, population numbers decline steeply. Describe a density dependent and a density independent factor regulating the population:

   (a) Density dependent:

   (b) Density independent:

ISBN: 978-1-927173-16-9

Population Growth

Organisms do not generally live alone. A **population** is a group of organisms of the same species living together in one geographical area. This area may be difficult to define as populations may comprise widely dispersed individuals that come together only infrequently (e.g. for mating). The number of individuals comprising a population may also fluctuate considerably over time. These changes make populations dynamic: populations gain individuals through births or immigration, and lose individuals through deaths and emigration. For a population in **equilibrium**, these factors balance out and there is no net change in the population abundance. When losses exceed gains, the population declines.

Births, deaths, immigrations (movements into the population) and emigrations (movements out of the population) are events that determine the numbers of individuals in a population. Population growth depends on the number of individuals added to the population from births and immigration, minus the number lost through deaths and emigration. This is expressed as:

Population growth =
Births – Deaths + Immigration – Emigration
(B) (D) (I) (E)

The difference between immigration and emigration gives net migration. Ecologists usually measure the **rate** of these events. These rates are influenced by environmental factors (see below) and by the characteristics of the organisms themselves. Rates in population studies are commonly expressed in one of two ways:

- Numbers per unit time, e.g. 20,150 live births per year.
- Per capita rate (number per head of population), e.g. 122 live births per 1000 individuals per year (12.2%).

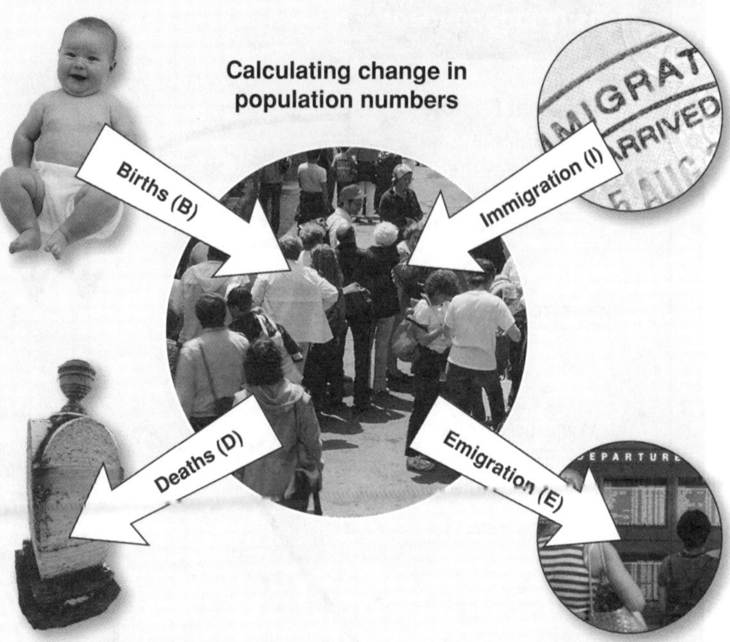

Calculating change in population numbers
Births (B)
Immigration (I)
Deaths (D)
Emigration (E)

Limiting Factors

Population size is also affected by limiting factors; factors or resources that control a process such as organism growth, or population growth or distribution. Examples include availability of food, predation pressure, or available habitat.

Human populations often appear exempt from limiting factors as technology and efficiency solve many food and shelter problems. However, as the last arable land is used and agriculture reaches its limits of efficiency, it is estimated that the human population may peak at around 10 billion by 2050.

1. Define the following terms used to describe changes in population numbers:

   (a) Death rate (mortality): ________________

   (b) Birth rate (natality): ________________

   (c) Net migration rate: ________________

2. Explain how the concept of limiting factors applies to population biology: ________________

3. Using the terms, B, D, I, and E (above), construct equations to express the following (the first is completed for you):

   (a) A population in equilibrium: B + I = D + E

   (b) A declining population: ________________

   (c) An increasing population: ________________

4. A population started with a total number of 100 individuals. Over the following year, population data were collected. Calculate birth rates, death rates, net migration rate, and rate of population change for the data below (as percentages):

   (a) Births = 14: Birth rate = ________ (b) Net migration = +2: Net migration rate = ________

   (c) Deaths = 20: Death rate = ________ (d) Rate of population change = ________

   (e) State whether the population is increasing or declining: ________________

5. The human population is around 6.7 billion. Describe and explain two limiting factors for population growth in humans:

Related activities: Features of Populations

ISBN: 978-1-927173-16-9

Population Growth Curves

Populations becoming established in a new area for the first time frequently undergo a rapid **exponential** (logarithmic) increase in numbers (below, left). As they colonize an area, there are plentiful resources, birth rates are high, and death rates are often low. Exponential growth produces a J-shaped growth curve that rises steeply as more and more individuals contribute to the population increase. If the resources of the new habitat were endless (inexhaustible) then the population would continue to increase at an exponential rate. However, this rarely happens in natural populations. Initially, growth may be exponential (or nearly so) but, as the population grows, its increase will slow and it will stabilize at a level that can be supported by the environment (called the **carrying capacity** or K). This type of growth is called sigmoidal and produces the **logistic growth curve** (below, right). **Established populations** will often fluctuate about K, often in quite a regular way, because there is always a slight lag in the response of the population to the constraints imposed by available resources.

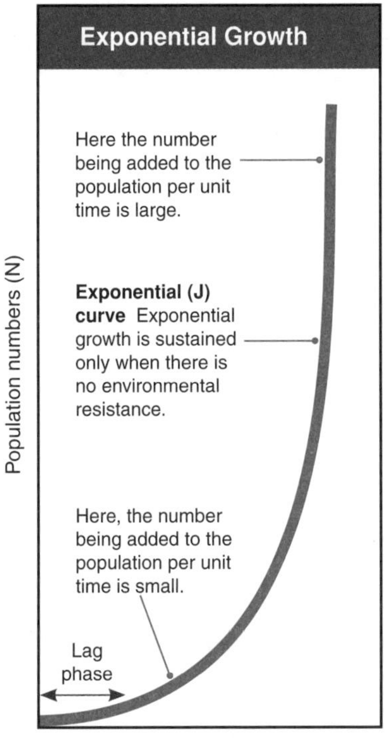
Exponential Growth
Here the number being added to the population per unit time is large.
Exponential (J) curve Exponential growth is sustained only when there is no environmental resistance.
Here, the number being added to the population per unit time is small.
Lag phase
Population numbers (N)
Time

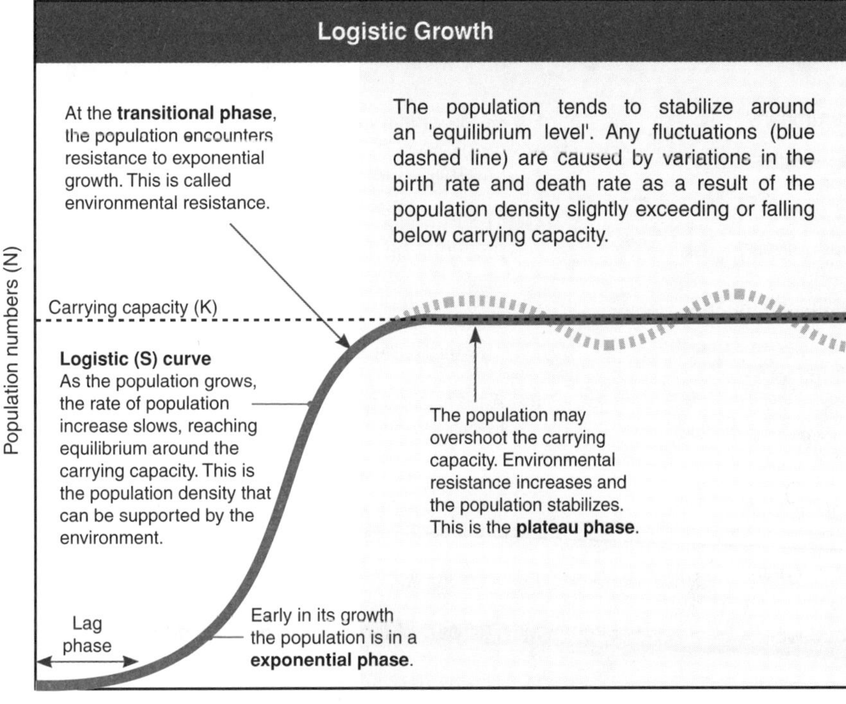
Logistic Growth
At the transitional phase, the population encounters resistance to exponential growth. This is called environmental resistance.
The population tends to stabilize around an 'equilibrium level'. Any fluctuations (blue dashed line) are caused by variations in the birth rate and death rate as a result of the population density slightly exceeding or falling below carrying capacity.
Carrying capacity (K)
Logistic (S) curve As the population grows, the rate of population increase slows, reaching equilibrium around the carrying capacity. This is the population density that can be supported by the environment.
The population may overshoot the carrying capacity. Environmental resistance increases and the population stabilizes. This is the plateau phase.
Lag phase
Early in its growth the population is in a exponential phase.
Population numbers (N)
Time

1. Why don't populations continue to increase exponentially in an environment? ______________________

____

2. What is meant by environmental resistance? ______________________

____

3. (a) What is meant by **carrying capacity**? ______________________

____

(b) Explain the importance of carrying capacity to the growth and maintenance of population numbers: ______________________

____

4. (a) Describe and explain the phases of the logistic growth curve: ______________________

____

____

____

____

____

(b) Explain why a population might overshoot carrying capacity before stabilizing around carrying capacity:

____

____

____

ISBN: 978-1-927173-16-9

KEY TERMS: Mix and Match

INSTRUCTIONS: Test your vocabulary by matching each term to its definition, as identified by its preceding letter code.

autotroph ....

carbon cycle ....

carrying capacity ....

combustion ....

community ....

consumer ....

decomposer ....

density dependent factor ....

detritivore ....

ecosystem ....

exponential growth ....

food chain ....

food web ....

global warming ....

greenhouse effect ....

habitat ....

herbivore ....

heterotroph ....

logistic curve ....

mortality ....

natality ....

producer ....

population ....

precautionary principle ....

saprotroph ....

species ....

transitional phase ....

trophic level ....

A A common sigmoid curve that describes population growth in an environment with limited resources.

B Growth that occurs in multiples based on earlier populations.

C The death rate of a population, (usually expressed as the number of deaths per 1000).

D A factor that regulates the size of a population in proportion to the density of the population.

E An organism that obtains its carbon and energy from other organisms.

F Organism that manufactures its own food from simple inorganic substances.

G The retention of solar energy in the Earth's atmosphere by gases that absorb heat and prevent it being released back into space.

H The maximum population size that can be supported by the environment.

I The birth rate of a population, (usually expressed as the number of births per 1000).

J A taxonomic group denoting organisms that are so genetically alike they can interbreed to produce viable offspring.

K The process of the Earth's surface steadily increasing in temperature (and its projected continuation). Usually attributed the rise in gases produced by fossil fuels and industrial processes.

L A sequence of organisms in which each organism is a source of food for the next.

M The part of the environment which an organism occupies, e.g. stream or grassland.

N The biogeochemical cycle in which carbon is exchanged among the biosphere and the inorganic reservoirs on Earth.

O An organism that feeds on decaying matter (detritus).

P Organism that feeds exclusively on plant material.

Q The total number of individuals of a species within a set habitat or area.

R An exothermic chemical reaction between a fuel and an oxidant. It produces heat and converts the chemical species involved.

S The phase in a logistic growth curve where exponential growth encounters environmental resistance and population growth slows.

T Organism occupying the first trophic level. Usually gains energy via photosynthesis or chemosynthesis.

U An organism that obtains energy from dead material by extracellular digestion.

V A complex series of interactions showing the feeding relationships between organisms in an ecosystem.

W An organism that breaks down dead or decaying matter.

X Any of the feeding levels that energy passes through in an ecosystem.

Y A naturally occurring group of different species living within the same environment and interacting together.

Z The principle that it is better to prevent damage occurring than to manage or repair the damage once it has occurred. It involves proof of no harm by the party involved.

AA Organism that obtains its energy from other living organisms or their dead remains.

BB Community of interacting organisms and the environment (both biotic and abiotic) in which they both live and interact.

ISBN: 978-1-927173-16-9

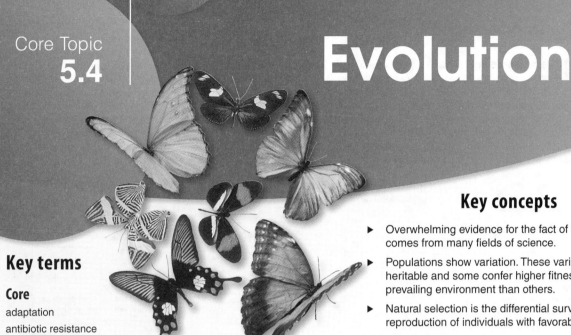

Core Topic
5.4

Evolution

Key concepts

- Overwhelming evidence for the fact of evolution comes from many fields of science.
- Populations show variation. These variations are heritable and some confer higher fitness in the prevailing environment than others.
- Natural selection is the differential survival and reproduction of individuals with favorable variations.
- Evolution is not a thing of the past: it continues and can be observed today.

Key terms

Core

adaptation
antibiotic resistance
differential survival
directional selection
disruptive selection
evolution
fitness
fossil record
homologous structures
industrial melanism
meiosis
natural selection
pentadactyl limb
pesticide resistance
selective breeding
sexual reproduction
stabilizing selection

Learning Objectives

☐ 1. Use the **KEY TERMS** to compile a glossary for this topic.

Evidence for Evolution *(5.4.1-5.4.2)* pages 238-241

☐ 2. Explain what is meant by biological **evolution**.

☐ 3. Outline the evidence for evolution provided by the **fossil record**, **homologous structures** (such as the **pentadactyl limb** of tetrapods), and **selective breeding** of plants and animals.

☐ 4. Understand the term **fitness**. Explain how evolution, through **adaptation**, equips species for survival.

Variation and Evolution *(5.4.3-5.4.7)* pages 238, 242-243

☐ 5. Recall the role of **sexual reproduction** in generating genetic variation.

☐ 6. Outline the fundamental ideas in Darwin's "*Theory of evolution by natural selection*" and their significance. Use this outline as a framework to explain how **natural selection** leads to evolution. Include reference to:
- The tendency of populations to overproduce.
- The fact that overproduction leads to competition.
- The fact that members of a species show variation, that sexual reproduction promotes variation, and that variation is (usually) heritable.
- The differential survival and reproduction of individuals with favorable, heritable variations.

☐ 7. EXTENSION: Recognize three types of natural selection: **stabilizing**, **directional**, and **disruptive selection**. Describe the outcome of each type in a population exhibiting a normal curve in phenotypic variation.

Case Studies in Evolution *(5.4.8)* pages 244-251

☐ 8. Explain two examples of evolution in response to environmental change. Appropriate examples include:
(a) The development of multiple **antibiotic resistance** in bacteria (mandatory). Understand that antibiotic misuse creates the selective environment for resistance to spread, but it does not create the resistance itself.
(b) Adaptation to low nutrient environments in *E. coli*.
(c) Changes to the size and shape of the beaks of Galapagos finches (*Geospiza*).
(d) **Industrial melanism** as an example of **directional selection** in peppered moths (*Biston*). Include reference to changes in the selective environment.
(e) The development of **pesticide resistance** in insects or heavy metal tolerance in plants.
(f) Antigenic shifts in *Influenzavirus*.
(g) Natural selection for skin pigmentation in humans.

Periodicals:
Listings for this chapter are on page 399

Weblinks:
www.thebiozone.com/weblink/IB-3169.html

BIOZONE APP:
Student Review Series
Evolution

Genes, Inheritance, and Selection

Each individual in a population is the carrier of its own particular combination of genetic material. In sexually reproducing organisms, different combinations of genes arise because of the shuffling of the chromosomes during gamete formation. New allele combinations also occur as a result of mate selection and the chance meeting of different gametes from each of the two parents. Some combinations are well suited to the prevailing environment, while others are less so. Those organisms with well suited allele combinations will have greater reproductive success (fitness) than those with less favorable allele combinations and consequently, their genes (alleles) will be represented in greater proportion in subsequent generations. For asexual species, offspring are essentially clones, but new alleles can arise through mutation and some of these may confer a selective advantage. Of course, environments are rarely static, so new allele combinations are always being tested for success.

Genes, Inheritance, and Evolution

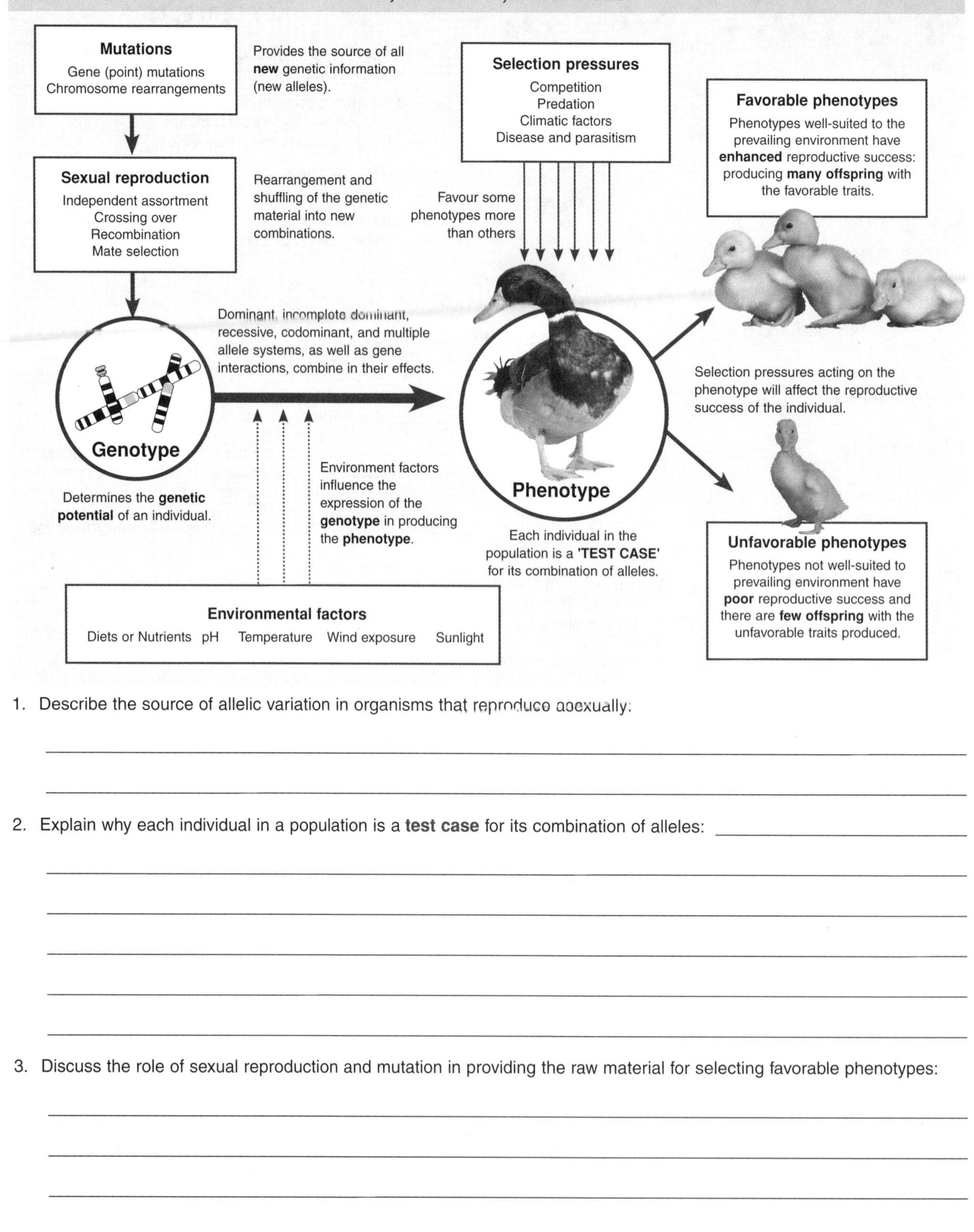
Mutations
Gene (point) mutations
Chromosome rearrangements
Provides the source of all new genetic information (new alleles).
Sexual reproduction
Independent assortment
Crossing over
Recombination
Mate selection
Rearrangement and shuffling of the genetic material into new combinations.
Selection pressures
Competition
Predation
Climatic factors
Disease and parasitism
Favour some phenotypes more than others
Favorable phenotypes
Phenotypes well-suited to the prevailing environment have enhanced reproductive success: producing many offspring with the favorable traits.
Dominant, incomplete dominant, recessive, codominant, and multiple allele systems, as well as gene interactions, combine in their effects.
Genotype
Determines the genetic potential of an individual.
Environment factors influence the expression of the genotype in producing the phenotype.
Phenotype
Each individual in the population is a 'TEST CASE' for its combination of alleles.
Selection pressures acting on the phenotype will affect the reproductive success of the individual.
Unfavorable phenotypes
Phenotypes not well-suited to prevailing environment have poor reproductive success and there are few offspring with the unfavorable traits produced.
Environmental factors
Diets or Nutrients
pH
Temperature
Wind exposure
Sunlight

1. Describe the source of allelic variation in organisms that reproduce asexually:

2. Explain why each individual in a population is a **test case** for its combination of alleles:

3. Discuss the role of sexual reproduction and mutation in providing the raw material for selecting favorable phenotypes:

***Related activities**: Darwin's Theory*

ISBN: 978-1-927173-16-9

The Fossil Record

Fossils are the remains of long-dead organisms that have escaped decay and have, after many years, become part of the Earth's crust. Fossils provide a record of the appearance and extinction of organisms, from species to whole taxonomic groups. Once this record is calibrated against a time scale (by using dating techniques), it is possible to build up a picture of the evolutionary changes that have taken place. The evolution of the horse from the ancestral *Hyracotherium* to modern *Equus* is well documented in the fossil record, and is often used to illustrate the process of evolution. The rich fossil record, which includes numerous **transitional fossils**, has enabled scientists to develop a robust model of horse phylogeny. Horse evolution exhibits a complex tree-like lineage with many divergences (below). It showed no inherent direction, and a diverse array of species coexisted for some time over the 55 million year evolutionary period. The environmental transition from forest to grasslands drove many of the changes observed in the equid fossil record.

Profile with Sedimentary Rocks Containing Fossils

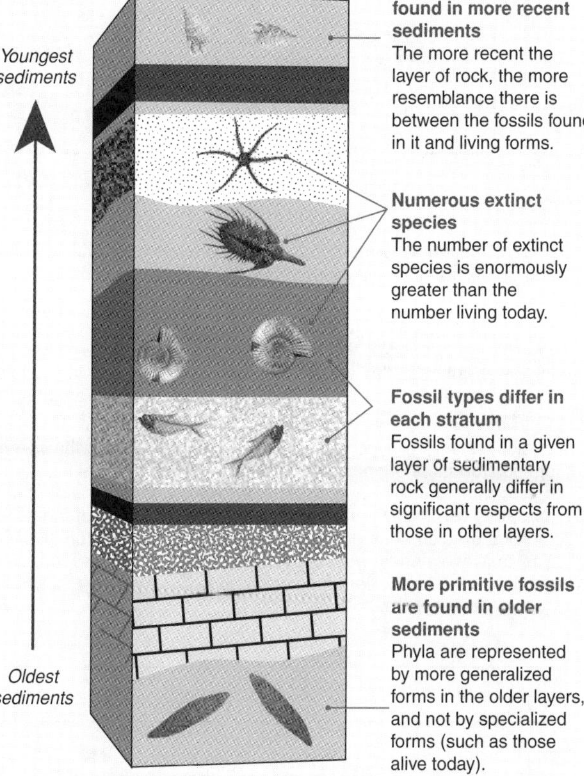
Ground surface
Youngest sediments
Oldest sediments
Recent fossils are found in more recent sediments
The more recent the layer of rock, the more resemblance there is between the fossils found in it and living forms.
Numerous extinct species
The number of extinct species is enormously greater than the number living today.
Fossil types differ in each stratum
Fossils found in a given layer of sedimentary rock generally differ in significant respects from those in other layers.
More primitive fossils are found in older sediments
Phyla are represented by more generalized forms in the older layers, and not by specialized forms (such as those alive today).

Rock strata are layered through time
Rock strata are arranged in the order that they were deposited (unless they have been disturbed by geological events. The most recent layers are near the surface and the oldest are at the bottom.

New fossil types mark changes in environment
In the rocks marking the end of one geological period, it is common to find many new fossils that become dominant in the next. Each geological period had an environment very different from those before and after. Their boundaries coincided with drastic environmental changes and the appearance of new niches. These produced new selection pressures resulting in new adaptive features in the surviving species, as they responded to the changes.

Fossil Evidence of Horse Evolution

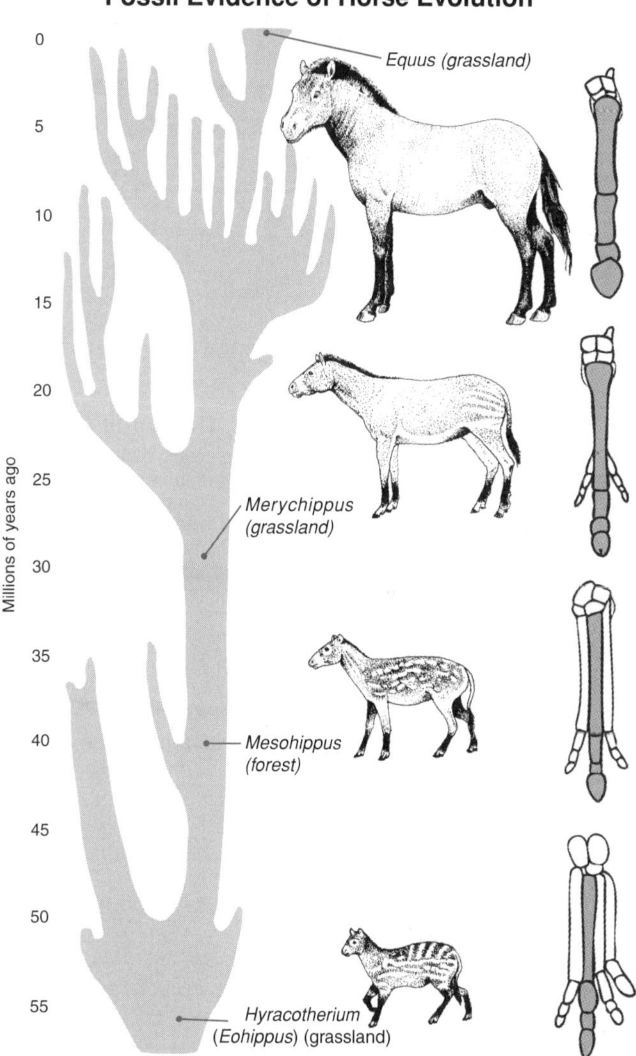
Millions of years ago
0
5
10
15
20
25
30
35
40
45
50
55
Equus (grassland)
Merychippus (grassland)
Mesohippus (forest)
Hyracotherium (*Eohippus*) (grassland)

The fossil record of the horse provides much evidence for evolution (change in body size, limb length, tooth structure, toe reduction). The genus *Equus* (the modern horse) is the only living genus of what was a large and diverse group of animals. Observation of the leg bones of various ancestors of *Equus* show a progressive loss of the outer toe bones, leaving the single middle bone and hoof supporting the animal.

1. Discuss the importance of **fossils** as a record of evolutionary change over time: ______________________

2. In which way does the equid fossil record provide a good example of the evolutionary process? ______________________

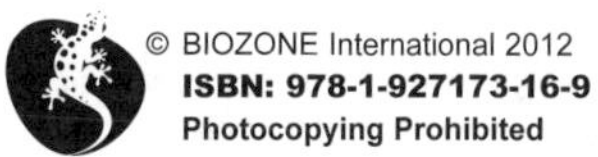

ISBN: 978-1-927173-16-9

Related activities: *Insecticide Resistance*
Weblinks: *Fossil Horse Cybermuseum, Getting into the Fossil Record*

Selection and Population Change

Selective breeding is a method for rapidly producing change in the phenotypic characteristics of a population. Instead of the environment providing the selection pressure for change, humans create it by choosing and breeding together individuals with particular traits.

The example of milk yield (below left) illustrates how humans have directly influenced a trait in cattle. The example of guppy phenotype (below right) shows how humans can indirectly influence change in phenotype by creating an environment that alters selection pressures.

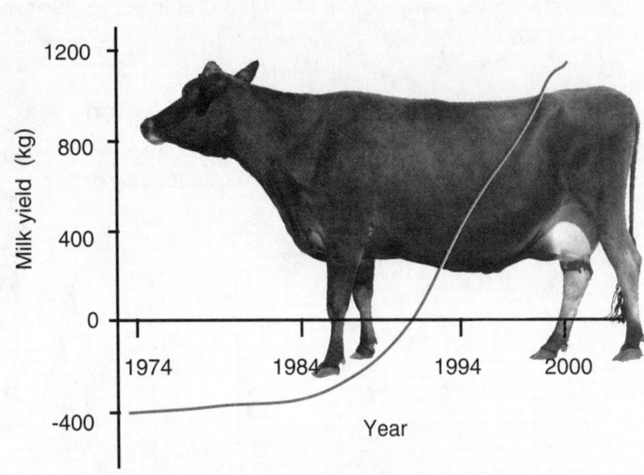
Milk yield per year for cattle in the UK
Milk yield (kg)
1200
800
400
0
-400
1974
1984
1994
2000
Year

Increase in milk yield relative to a baseline figure (0)

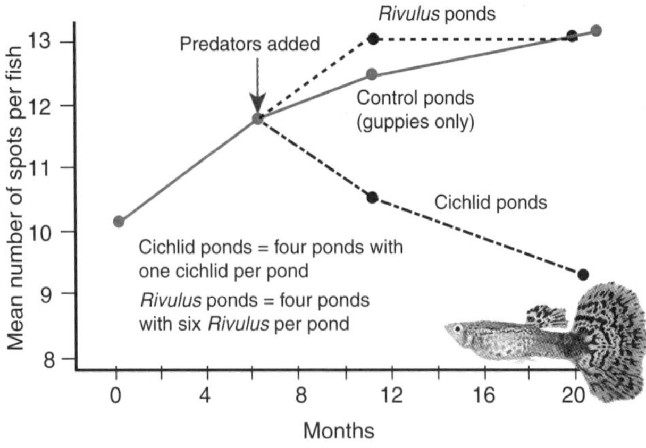
Spots on male guppies vs survival rate
Mean number of spots per fish
13
12
11
10
9
8
Predators added
Rivulus ponds
Control ponds
(guppies only)
Cichlid ponds
Cichlid ponds = four ponds with
one cichlid per pond
Rivulus ponds = four ponds
with six *Rivulus* per pond
0
4
8
12
16
20
Months

The experiment was repeated in the wild using the natural streams and ponds the guppies live in. The same results were obtained.

Increased Milk Production in Domestic Cattle

The domestic livestock of today are the result of thousands of years of **selective breeding**. By choosing and breeding individual animals with particular traits, humans have produced a variety of livestock breeds (and plant varieties) that cater for human needs (e.g. milk, meat, and wool). Since the 1900s, herd breeding records have allowed farmers to systematically improve herd management and increase **genetic gain** (a gain towards the desirable phenotype of a breed). Techniques such as artificial insemination and embryo multiplication have improved milk yield by 1200 kg per lactation over the last 12 years in the UK (above).

Endler's Guppy Experiment

In 1980, John Endler carried out two experiments with guppies. Female guppies prefer to mate with male guppies with bright spots. However, bright spots make a male guppy more obvious to predators.

Endler tested the hypothesis that predation risk limits the number of spots on male guppies. He established ten ponds, mimicking the natural ponds in which guppies live in the wild. After six months, predatory fish (cichlid or *Rivulus*) were added to eight of the ponds. Cichlids prey on adult guppies while *Rivulus* prey only on juveniles. The last two ponds were controls, containing guppies only. The results for the experiment are shown above.

1. Explain how selective breeding provides evidence for evolution: ______________________

____

____

2. (a) Explain why the mean number of spots on male guppies increased in all ponds during the first six months of Endler's guppy experiment:

____

____

(b) Explain why the number of spots on the male guppies in the *Rivulus* pond continued to increase even after the introduction of the predator:

____

____

(c) Explain why the number of spots on male guppies in the cichlid pond decreased after the introduction of the predator:

____

____

3. Using the examples above to help you, compare **selective breeding** and **natural selection**: ______________________

____

____

____

ISBN: 978-1-927173-16-9

Homologous Structures

The evolutionary relationships between groups of organisms is determined mainly by structural similarities called **homologous structures** (homologies), which suggest that they all descended from a common ancestor with that feature. The bones of the forelimb of air-breathing vertebrates are composed of similar bones arranged in a comparable pattern. This is indicative of a common ancestry. The early land vertebrates were amphibians and possessed a limb structure called the **pentadactyl limb**: a limb with five fingers or toes (below left). All vertebrates that descended from these early amphibians, including reptiles, birds and mammals, have limbs that have evolved from this same basic pentadactyl pattern. They also illustrate the phenomenon known as **adaptive radiation**, since the basic limb plan has been adapted to meet the requirements of different niches.

Generalized Pentadactyl Limb

The forelimbs and hind limbs have the same arrangement of bones but they have different names. In many cases bones in different parts of the limb have been highly modified to give it a specialized locomotory function.

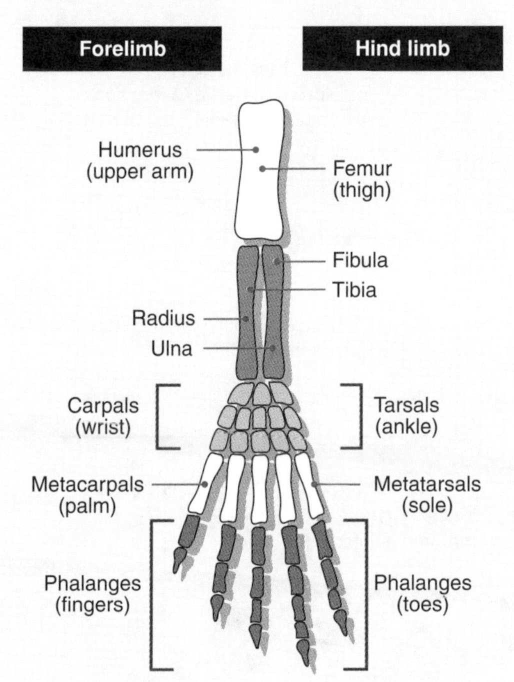
Forelimb
Hind limb
Humerus (upper arm)
Femur (thigh)
Fibula
Tibia
Radius
Ulna
Carpals (wrist)
Tarsals (ankle)
Metacarpals (palm)
Metatarsals (sole)
Phalanges (fingers)
Phalanges (toes)

Specializations of Pentadactyl Limbs

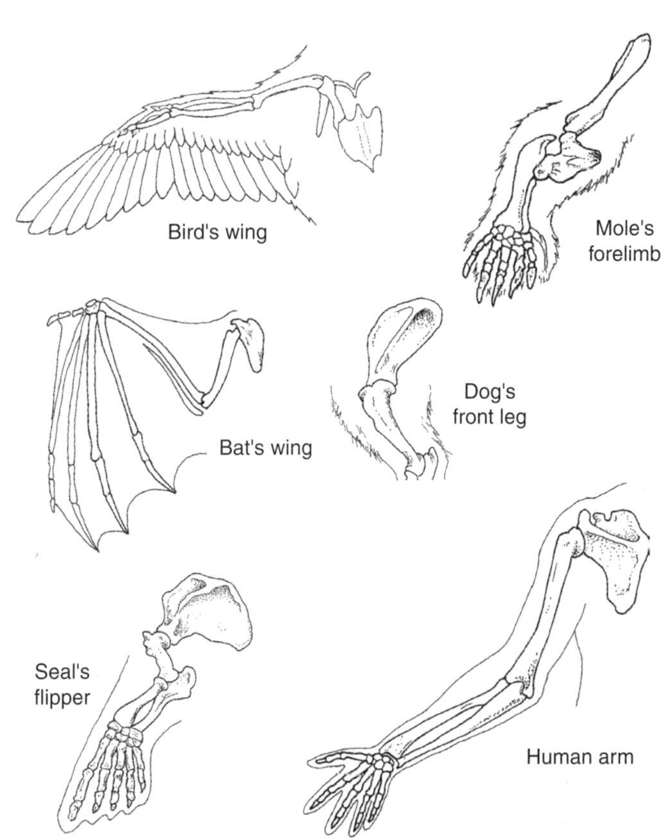
Bird's wing
Mole's forelimb
Bat's wing
Dog's front leg
Seal's flipper
Human arm

1. Briefly describe the purpose of the major anatomical change that has taken place in each of the limb examples above:

(a) Bird wing: Highly modified for flight. Forelimb is shaped for aerodynamic lift and feather attachment.

(b) Human arm: __________

(c) Seal flipper: __________

(d) Dog foot: __________

(e) Mole forelimb: __________

(f) Bat wing: __________

2. Explain how homology in the pentadactyl limb is evidence for adaptive radiation: __________

3. Homology in the behavior of animals (for example, sharing similar courtship or nesting rituals) is sometimes used to indicate the degree of relatedness between groups. How could behavior be used in this way:

ISBN: 978-1-927173-16-9

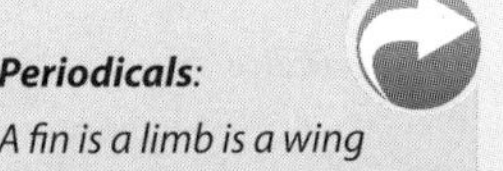

Periodicals:
A fin is a limb is a wing

Weblinks: *Homologous Structures*

Darwin's Theory

In 1859, Darwin and Wallace jointly proposed that new species could develop by a process of natural selection. Natural selection is the term given to the mechanism by which better adapted organisms survive to produce a greater number of viable offspring. This has the effect of increasing their proportion in the population so that they become more common. It is Darwin who is best remembered for the theory of evolution by natural selection through his famous book: '**On the origin of species by means of natural selection**', written 23 years after returning from his voyage on the Beagle, from which much of the evidence for his theory was accumulated. Although Darwin could not explain the origin of variation nor the mechanism of its transmission (this was provided later by Mendel's work), his basic theory of evolution by natural selection (outlined below) is widely accepted today. The study of population genetics has greatly improved our understanding of evolutionary processes, which are now seen largely as a (frequently gradual) change in allele frequencies within a population. Students should be aware that scientific debate on the subject of evolution centres around the relative merits of various alternative hypotheses about the nature of evolutionary processes. The debate is not about the existence of the phenomenon of evolution itself.

Darwin's Theory of Evolution by Natural Selection

Overproduction

Populations produce too many young: many must die

Populations tend to produce more offspring than are needed to replace the parents. Natural populations normally maintain constant numbers. There must therefore be a certain number that die without producing offspring.

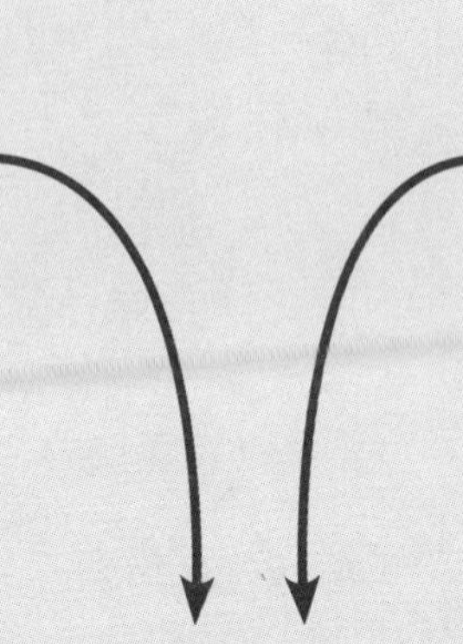

Variation

Individuals show variation: some are more favorable than others

Individuals in a population vary in their phenotype and therefore, their genotype. Some variants are better suited to the prevailing environment and have greater survival and reproductive success.

Andrew Dunn -www.andrewdunnphoto.com

The banded or grove snail, *Cepaea nemoralis*, is famous for the highly variable colors and banding patterns of its shell. These **polymorphisms** are thought to have a role in differential survival in different regions, associated with both the risk of predation and maintenance of body temperature. Dark brown grove snails are more abundant in dark woodlands, whilst snails with light yellow shells and thin banding are more commonly found in grasslands.

Natural Selection

Natural selection favors the best suited at the time

The struggle for survival amongst individuals competing for limited resources will favor those with the most favorable variations. Relatively more of those without favorable variations will die.

Inherited

Variations are inherited: the best suited variants leave more offspring

The variations (both favorable and unfavorable) are passed on to offspring. Each new generation will contain proportionally more descendants of individuals with favorable characters.

1. In your own words, describe how Darwin's theory of evolution by natural selection provides an explanation for the change in the appearance of a species over time:

ISBN: 978-1-927173-16-9

Natural Selection

Natural selection operates on the phenotypes of individuals, produced by their particular combinations of alleles (genotypes). The differential survival of some genotypes over others is called **natural selection** and, as a result of it, organisms with phenotypes most suited to the prevailing environment will be relatively more successful and so leave relatively more offspring. Favorable phenotypes become more numerous while unfavorable phenotypes become less common or disappear. Natural selection is not a static phenomenon; it is always linked to phenotypic suitability in the prevailing environment. It may favor existing phenotypes or shift the phenotypic median one way or another, as is shown below. The top row of diagrams represents the population phenotypic spread before selection, and the bottom row the spread afterwards. Note that balancing selection is similar to disruptive selection, but the balanced polymorphism that results is not associated with phenotypic extremes. Balanced polymorphism can occur as a result of heterozygous advantage or frequency dependent predation.

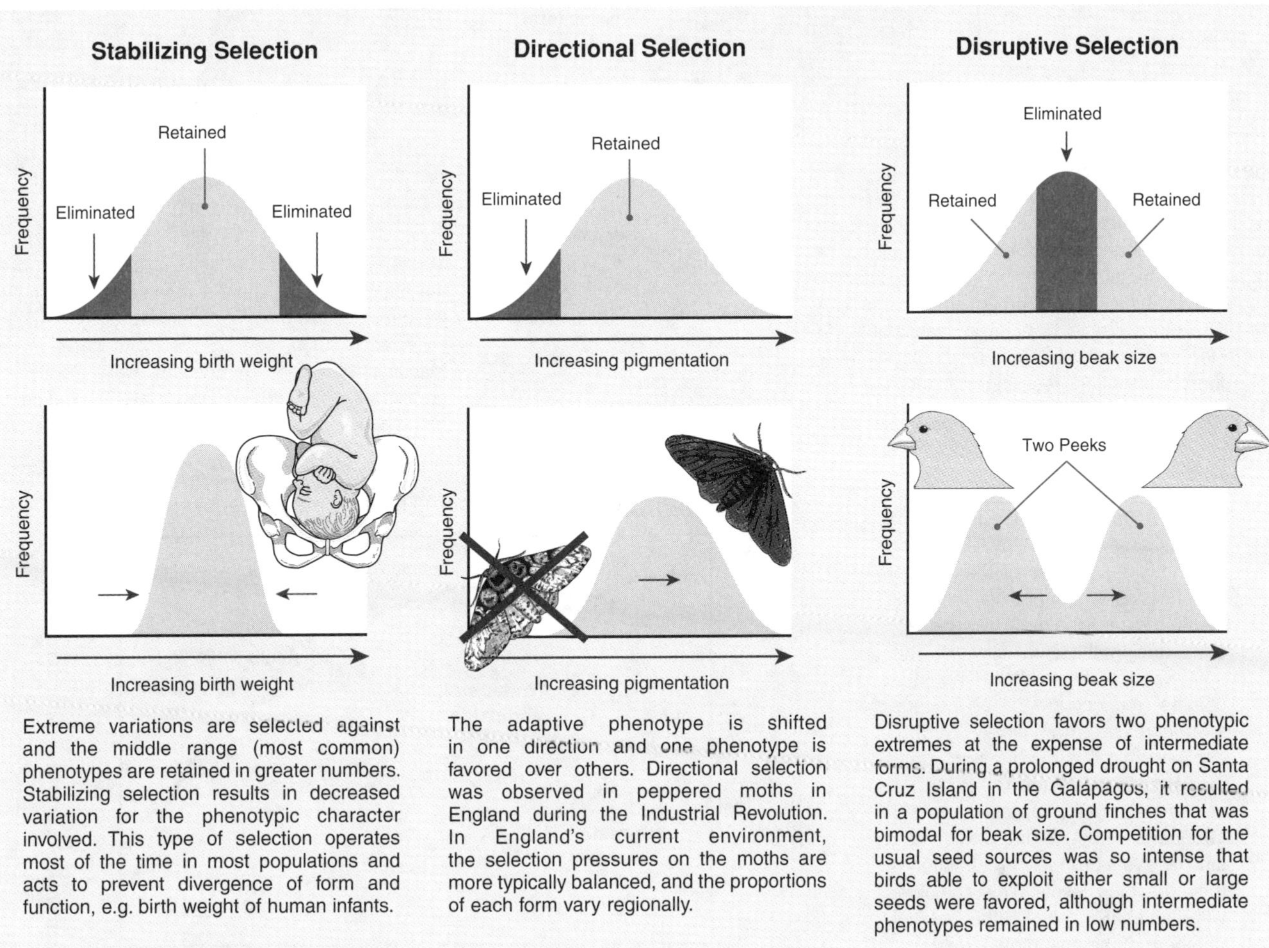
Stabilizing Selection
Directional Selection
Disruptive Selection
Frequency
Retained
Eliminated
Eliminated
Increasing birth weight
Retained
Eliminated
Increasing pigmentation
Eliminated
Retained
Retained
Increasing beak size
Increasing birth weight
Increasing pigmentation
Two Peeks
Increasing beak size

Extreme variations are selected against and the middle range (most common) phenotypes are retained in greater numbers. Stabilizing selection results in decreased variation for the phenotypic character involved. This type of selection operates most of the time in most populations and acts to prevent divergence of form and function, e.g. birth weight of human infants.

The adaptive phenotype is shifted in one direction and one phenotype is favored over others. Directional selection was observed in peppered moths in England during the Industrial Revolution. In England's current environment, the selection pressures on the moths are more typically balanced, and the proportions of each form vary regionally.

Disruptive selection favors two phenotypic extremes at the expense of intermediate forms. During a prolonged drought on Santa Cruz Island in the Galápagos, it resulted in a population of ground finches that was bimodal for beak size. Competition for the usual seed sources was so intense that birds able to exploit either small or large seeds were favored, although intermediate phenotypes remained in low numbers.

1. Explain why fluctuating (as opposed to stable) environments favor disruptive selection:

2. Disruptive selection can be important in the formation of new species:

(a) Describe the evidence from the ground finches on Santa Cruz Island that provides support for this statement:

(b) The ground finches on Santa Cruz Island are one interbreeding population with a strongly bimodal distribution for the character of beak size. Suggest what conditions could lead to the two phenotypic extremes diverging further:

(c) Predict the consequences of the end of the drought and an increased abundance of medium size seeds as food:

ISBN: 978-1-927173-16-9

Related activities: *Industrial Melanism, Selection for Human Birth Weight, Disruptive Selection in Darwin's Finches* ***Weblinks:*** *Natural Selection: A Review*

Directional Selection in Moths

Natural selection may act on the frequencies of phenotypes (and hence genotypes) in populations in one of three different ways (through stabilizing, directional, or disruptive selection). Over time, natural selection may lead to a permanent change in the genetic makeup of a population. Color change in the **peppered moth** (*Biston betularia*) during the Industrial Revolution is often used to show **directional selection** in a polymorphic population. Polymorphic means having more than two forms. Intensive coal burning during this time caused trees to become dark with soot, and the dark form (morph) of peppered moth became dominant.

The gene controlling color in the peppered moth, is located on a single locus. The allele for the melanic (dark) form (**M**) is dominant over the allele for the gray (light) form (**m**).

Olaf Leillinger

Melanic form
Genotype: MM or Mm

Olaf Leillinger

Gray form
Genotype: mm

The peppered moth, *Biston betularia*, has two forms: a gray mottled form, and a dark melanic form. During the Industrial Revolution, the relative abundance of the two forms changed to favour the dark form. The change was thought to be the result of selective predation by birds. It was proposed that the gray form was more visible to birds in industrial areas where the trees were dark. As a result, birds preyed upon them more often, resulting in higher numbers of the dark form surviving.

Museum collections of the peppered moth over the last 150 years show a marked change in the frequency of the melanic form (above right). Moths collected in 1850, prior to the major onset of the Industrial Revolution in England, were mostly the gray form (above left). Fifty years later the frequency of the darker melanic forms had increased.

In the 1940s and 1950s, coal burning was still at intense levels around the industrial centres of Manchester and Liverpool. During this time, the melanic form of the moth was still very dominant. In the rural areas further south and west of these industrial centres, the occurrence of the gray form increased dramatically. With the decline of coal burning factories and the introduction of the Clean Air Act in cities, air quality improved between 1960 and 1980. Sulfur dioxide and smoke levels dropped to a fraction of their previous levels. This coincided with a sharp fall in the relative numbers of melanic moths (right).

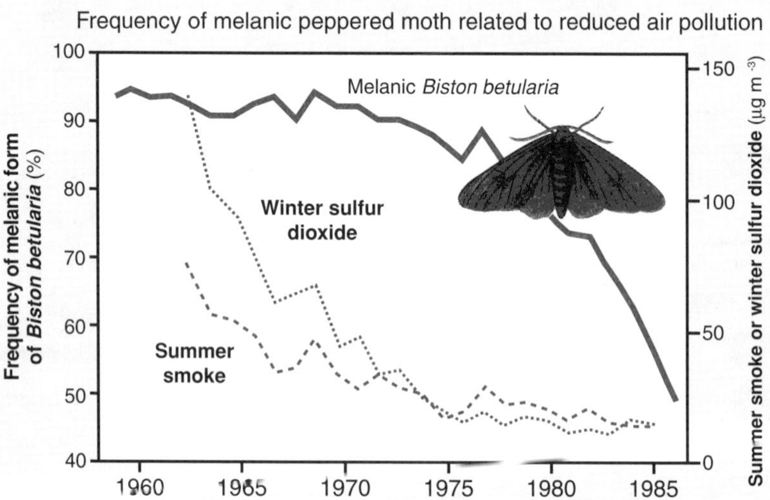
Frequency of melanic peppered moth related to reduced air pollution
Frequency of melanic form of Biston betularia (%)
100
90
80
70
60
50
40
Melanic Biston betularia
Winter sulfur dioxide
Summer smoke
150
100
50
0
Summer smoke or winter sulfur dioxide (µg m⁻³)
1960
1965
1970
1975
1980
1985

1. The populations of peppered moth in England have undergone changes in the frequency of an obvious phenotypic character over the last 150 years. What is the phenotypic character?

____

2. Describe how the selection pressure on the gray form has changed with change in environment over the last 150 years:

____

____

3. Describe the relationship between allele frequency and phenotype frequency: _______________

____

____

4. The level of pollution dropped around Manchester and Liverpool between 1960 and 1985. How did the frequency of the darker melanic form change during this period?

____

____

ISBN: 978-1-927173-16-9

Selection for Skin Color in Humans

Pigmented skin of varying tones is a feature of humans that evolved after early hominins lost the majority of their body hair. However, the distribution of skin color globally is not random; people native to equatorial regions have darker skin tones than people from higher latitudes. For many years, biologists postulated that this was because darker skins had evolved to protect against skin cancer. The problem with this explanation was that skin cancer is not tied to evolutionary fitness because it affects post-reproductive individuals and cannot therefore provide a mechanism for selection. More complex analyses of the physiological and epidemiological evidence has shown a more complex picture in which selection pressures on skin color are finely balanced to produce a skin tone that regulates the effects of the sun's ultraviolet radiation on the nutrients vitamin D and folate, both of which are crucial to successful human reproduction, and therefore evolutionary fitness.

Skin Color in Humans: A Product of Natural Selection

Alaska | France | The Netherlands | Iraq | China | Japan

80° | No data | 80°

Insufficient UV most of year

40° | Insufficient UV one month | 40°

Sufficient UV all year

0° | 0°

Sufficient UV all year

Insufficient UV one month

40° | 40°

Insufficient UV most of year

Adapted from Jablonski & Chaplin, Sci. Am. Oct. 2002

Photo: Lisa Gray

Peru | Liberia | Burundi | Botswana | Southern India | Malaysia

Human skin color is the result of two opposing selection pressures. Skin pigmentation has evolved to protect against destruction of folate by ultraviolet light, but the skin must also be light enough to receive the light required to synthesise vitamin D. Vitamin D synthesis is a process that begins in the skin and is inhibited by dark pigment. Folate is needed for healthy neural development in humans and a deficiency is associated with fatal neural tube defects. Vitamin D is required for the absorption of calcium from the diet and therefore normal skeletal development. Women also have a high requirement for calcium during pregnancy and lactation. Populations that live in the tropics receive enough ultraviolet (UV) radiation to synthesise vitamin D all year long. Those that live in northern or southern latitudes do not. In temperate zones, people lack sufficient UV light to make vitamin D for one month of the year. Those nearer the poles lack enough UV light for vitamin D synthesis most of the year (above). Their lighter skins reflect their need to maximize UV absorption (the photos show skin color in people from different latitudes).

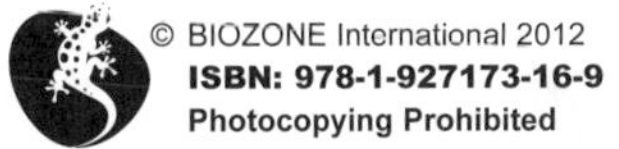

ISBN: 978-1-927173-16-9

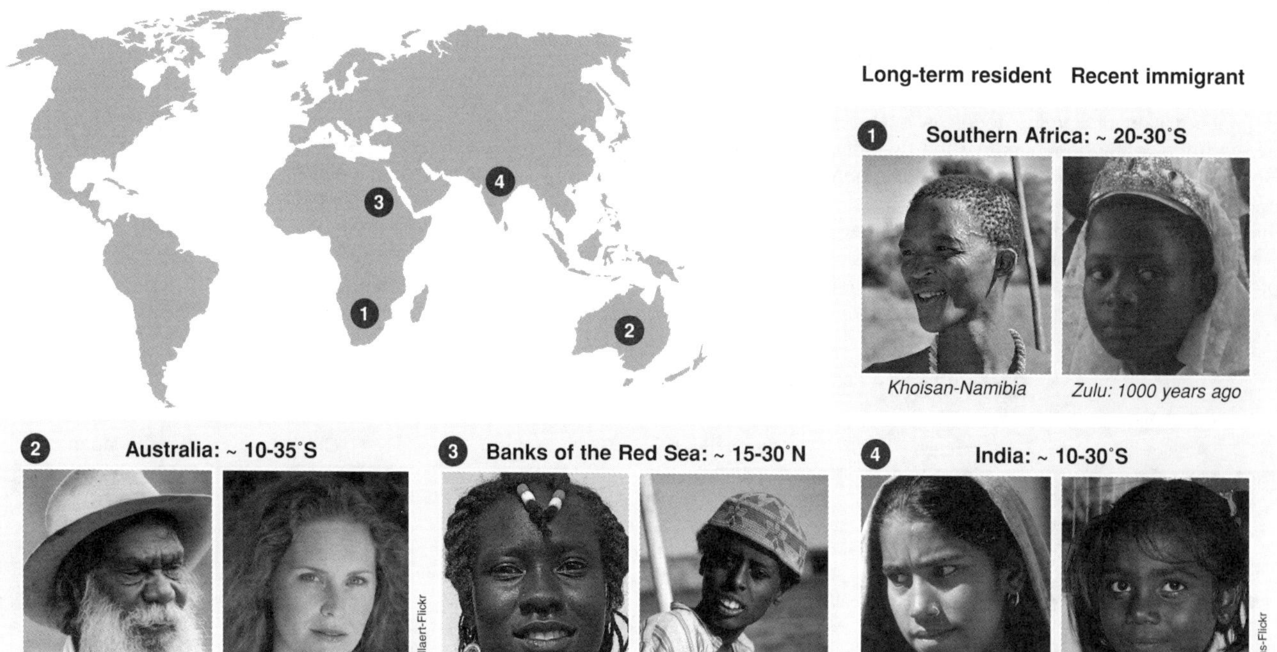

1
2
3
4
Long-term resident
Recent immigrant
1 Southern Africa: ~ 20-30°S
Khoisan-Namibia
Zulu: 1000 years ago
2 Australia: ~ 10-35°S
Aborigine
European: 300 years ago
Rita Willaert-Flickr
3 Banks of the Red Sea: ~ 15-30°N
Nuba-Sudan
Arab: 2000 years ago
4 India: ~ 10-30°S
West Bengal
Tamil: ~100 years ago
Aptychus-Flickr

The skin of people who have inhabited particular regions for millennia has adapted to allow sufficient vitamin D production while still protecting folate stores. In the photos above, some of these original inhabitants are illustrated to the left of each pair and compared with the skin tones of more recent immigrants (to the right of each pair, with the number of years since immigration). The numbered locations are on the map.

1. (a) Describe the role of folate in human physiology: ____________________

   (b) Describe the role of vitamin D in human physiology: ____________________

2. (a) Early hypotheses to explain skin color linked pigmentation level only to the degree of protection it gave from UV-induced skin cancer. Explain why this hypothesis was inadequate in accounting for how skin color evolved:

   ____

   ____

   (b) Explain how the new hypothesis for the evolution of skin color overcomes these deficiencies: ____________________

   ____

   ____

   ____

3. Explain why, in any given geographical region, women tend to have lighter skins (by 3-4% on average) than men:

   ____

   ____

4. The Inuit people of Alaska and northern Canada have a diet rich in vitamin D and their skin color is darker than predicted on the basis of UV intensity at their latitude. Explain this observation:

   ____

   ____

5. (a) What health problems might be expected for people of African origin now living in Canada?

   ____

   ____

   (b) How could these people avoid these problems in their new higher latitude environment? ____________________

   ____

   ____

ISBN: 978-1-927173-16-9

Disruptive Selection in Darwin's Finches

The Galápagos Islands are a group of islands 970 km west of Ecuador. They are home to 13 species of finches descended from a common ancestor. The finches have been closely studied over many years. A study during a prolonged drought on Santa Cruz Island showed how **disruptive selection** can change the distribution of genotypes in a population. During the drought, large and small seed sizes were more abundant than the preferred intermediate seed sizes.

Measurements of the beak length, width, and depth were combined into one **single measure**.

Beak size vs fitness in *Geospiza fortis*

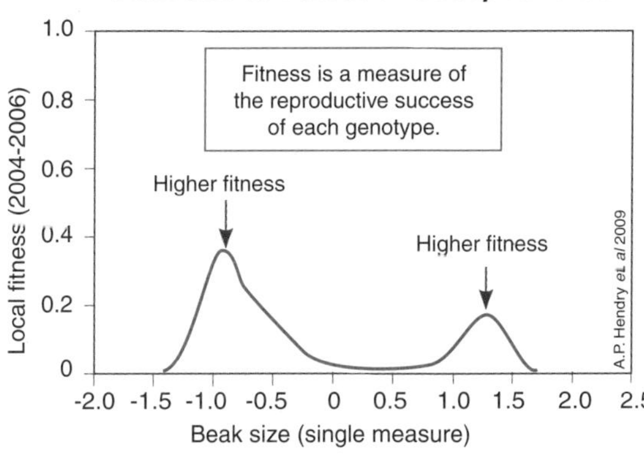
Fitness is a measure of the reproductive success of each genotype.
Higher fitness
Higher fitness
Local fitness (2004-2006)
Beak size (single measure)
A.P. Hendry et al 2009

*Fitness showed a **bimodal distribution** (arrowed) being highest for smaller and larger beak sizes.*

Beak sizes of *G. fortis* were measured over a three year period (2004-2006), at the start and end of each year. At the start of the year, individuals were captured, banded, and their beaks were measured.

The presence or absence of banded individuals was recorded at the end of the year when the birds were recaptured. Recaptured individuals had their beaks measured.

The proportion of banded individuals in the population at the end of the year gave a measure of fitness. Absent individuals were presumed dead (fitness = 0).

Fitness related to beak size showed a bimodal distribution (left) typical of disruptive selection.

Beak size pairing in *Geospiza fortis*

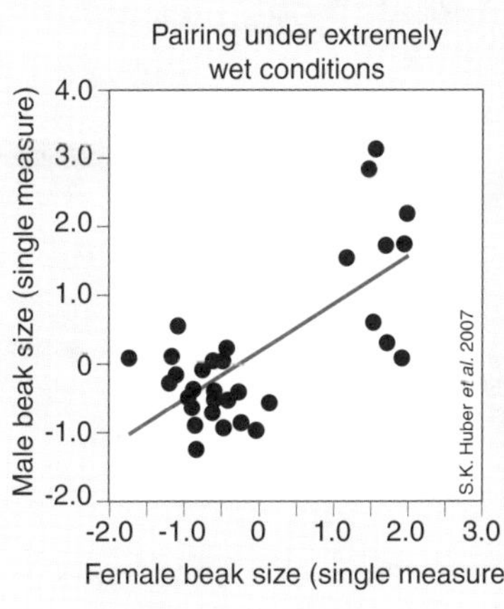
Pairing under extremely wet conditions
Male beak size (single measure)
Female beak size (single measure)
S.K. Huber et al. 2007

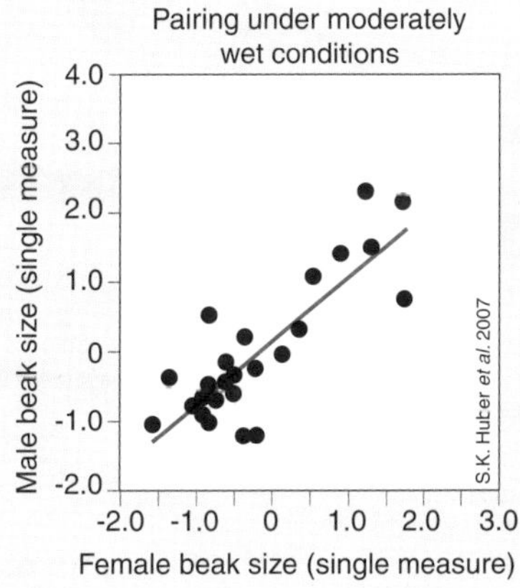
Pairing under moderately wet conditions
Male beak size (single measure)
Female beak size (single measure)
S.K. Huber et al. 2007

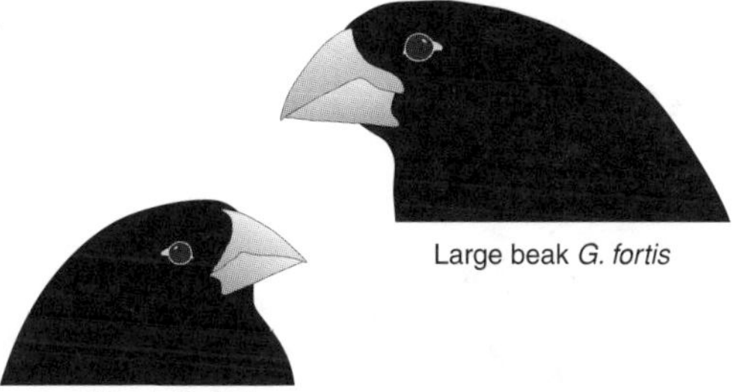
Large beak G. fortis
Small beak G. fortis

A 2007 study found that breeding pairs of birds had similar beak sizes. Male and females with small beaks tended to breed together, and males and females with large beaks tended to breed together. Mate selection maintained the biomodal distribution in the population during extremely wet conditions. If beak size wasn't a factor in mate selection, the beak size would even out.

1. (a) How did the drought affect seed size on Santa Cruz Island? __________

   (b) How did the change in seed size during the drought create a selection pressure for changes in beak size?

2. How does beak size relate to fitness (differential reproductive success) in *G. fortis*? __________

3. (a) Is mate selection in *G. fortis* random / non-random? (delete one)

   (b) Give reasons for your answer: __________

ISBN: 978-1-927173-16-9

Insecticide Resistance

Insecticides are pesticides used to control insects considered harmful to humans, their livelihood, or environment. Insecticides have been used for hundreds of years, but their use has proliferated since the advent of synthetic insecticides (e.g. DDT) in the 1940s. When **insecticide resistance** develops the control agent will no longer control the target species. Insecticide resistance can arise through a combination of behavioral, anatomical, biochemical, and physiological mechanisms, but the underlying process is a form of **natural selection**, in which the most resistant organisms survive to pass on their genes to their offspring. To combat increasing resistance, higher doses of more potent pesticides are sometimes used. This drives the selection process, so that increasingly higher dose rates are required to combat rising resistance. This phenomenon is made worse by the development of multiple resistance in some pest species. High application rates may also kill non-target species, and persistent chemicals may remain in the environment and accumulate in food chains. These concerns have led to some insecticides being banned (DDT has been banned in most developed countries since the 1970s). Insecticides are used in medical, agricultural, and environmental applications, so the development of resistance has serious environmental and economic consequences.

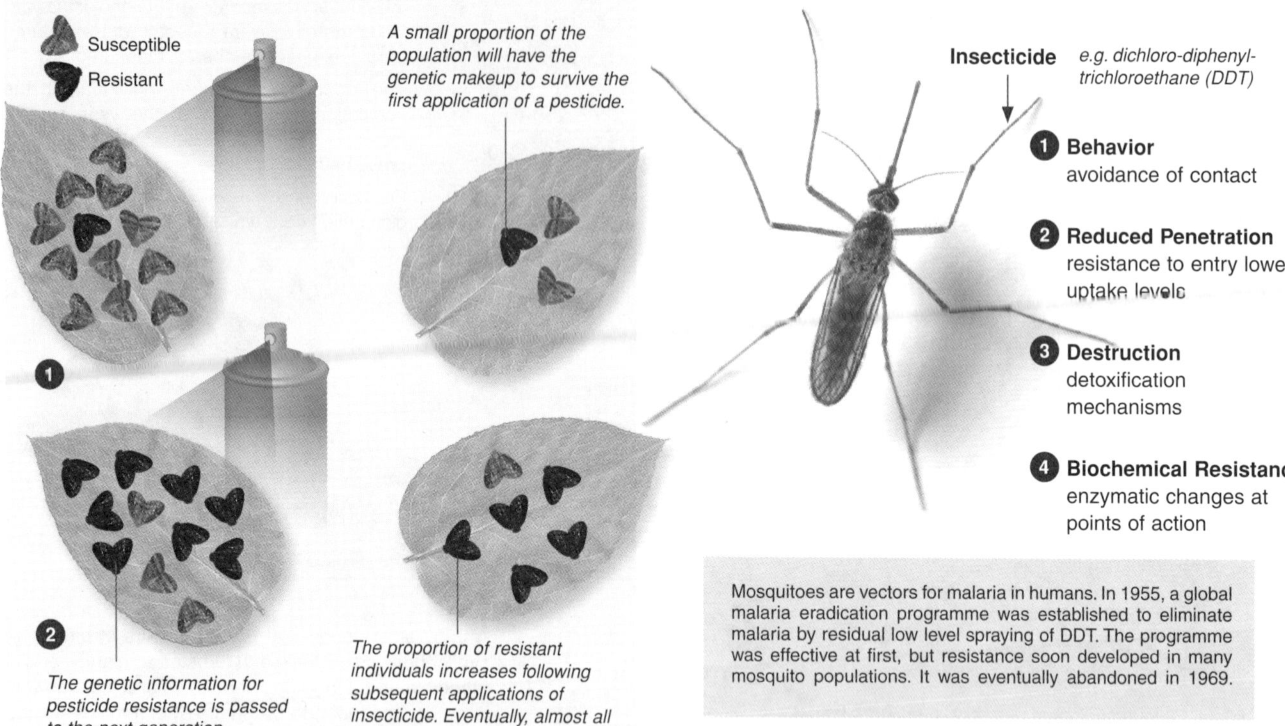

Susceptible
Resistant
A small proportion of the population will have the genetic makeup to survive the first application of a pesticide.
1
2
The genetic information for pesticide resistance is passed to the next generation.
The proportion of resistant individuals increases following subsequent applications of insecticide. Eventually, almost all of the population is resistant.
Insecticide
e.g. dichloro-diphenyl-trichloroethane (DDT)
1 Behavior
avoidance of contact
2 Reduced Penetration
resistance to entry lowers uptake levels
3 Destruction
detoxification mechanisms
4 Biochemical Resistance
enzymatic changes at points of action

Mosquitoes are vectors for malaria in humans. In 1955, a global malaria eradication programme was established to eliminate malaria by residual low level spraying of DDT. The programme was effective at first, but resistance soon developed in many mosquito populations. It was eventually abandoned in 1969.

The Development of Resistance

The application of an insecticide can act as a potent selection pressure for resistance in pest insects. The insecticide acts as a selective agent, and only individuals with greater natural resistance survive the application to pass on their genes to the next generation. These genes (or combination of genes) may spread through all subsequent populations.

Mechanisms of Resistance in Insect Pests

Insecticide resistance in insects can arise through a combination of mechanisms. (1) Increased sensitivity to an insecticide will cause the pest to avoid a treated area. (2) Certain genes (e.g. the *PEN* gene) confer stronger physical barriers, decreasing the rate at which the chemical penetrates the cuticle. (3) Detoxification by enzymes within the insect's body can render the pesticide harmless, and (4) structural changes to the target enzymes make the pesticide ineffective. No single mechanism provides total immunity, but together they transform the effect from potentially lethal to insignificant.

1. Give two reasons why widespread insecticide resistance can develop very rapidly in insect populations:

   (a) ______________________________

   (b) ______________________________

2. Explain how repeated insecticide applications act as a selective agent for evolutionary change in insect populations:

   ____

   ____

3. With reference to synthetic insecticides, discuss the implications of insecticide resistance to human populations:

   ____

   ____

   ____

Related activities: The Evolution of Antibiotic Resistance

ISBN: 978-1-927173-16-9

Evolution in Response to Nutrient Levels

In 1988, Richard Lenski and his research group began the ***E.coli* Long Term Evolution Experiment**. They prepared 12 populations of the bacterium *E. coli* in a minimal growth medium of glucose (i.e. glucose as the limiting growth factor). Each day, 1% of each population was transferred to a flask of fresh medium, while a sample of the previous population was frozen and stored. The experiment has now reached more than 50,000 generations.

The populations were tested at intervals for mutations, but strains were not selected for any particular trait (other than being constantly in a low glucose medium). Measurements made during the experiment have found that cell size has increased, population density has decreased, and **fitness** (contribution to the next generation) in a low glucose medium has increased in all populations compared to the ancestral *E. coli* strain.

The *E. coli* Long Term Evolution Experiment

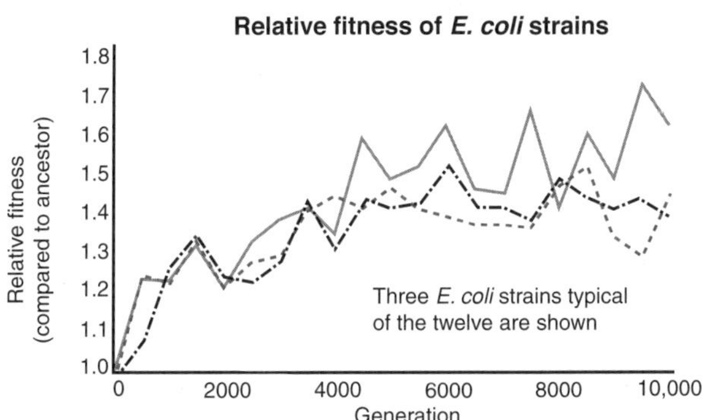
Relative fitness of E. coli strains
Relative fitness (compared to ancestor)
1.8
1.7
1.6
1.5
1.4
1.3
1.2
1.1
1.0
Three E. coli strains typical of the twelve are shown
0
2000
4000
6000
8000
10,000
Generation

Every 500 generations, the fitness of each population was compared to the fitness of the ancestor (denoted as 1). As would be expected, the relative fitness of all the *E. coli* populations increased over time as they adapted to the low glucose environment. However this increased fitness only applies in the low glucose environment. When placed in a different environment, fitness relative to the ancestor actually decreases. Interestingly, although the relative fitness of the populations has followed a similar course, the fitness has varied within and between populations.

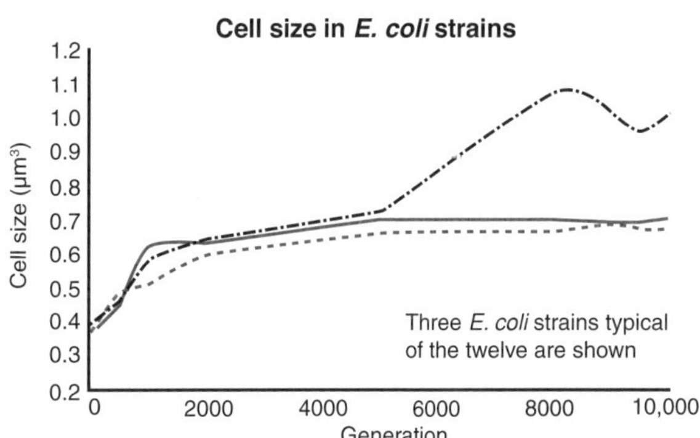
Cell size in E. coli strains
Cell size (μm³)
1.2
1.1
1.0
0.9
0.8
0.7
0.6
0.5
0.4
0.3
0.2
Three E. coli strains typical of the twelve are shown
0
2000
4000
6000
8000
10,000
Generation

The size of the *E. coli* cells was also measured every 500 generations. Cell size increased in all populations and they became rounder. All populations increased their growth rate by about 70% on average. The population density also decreased. The increase in cell size and growth rate is probably an adaptation for acquiring the limited amount of glucose available in the solution. Over the course of the experiment, it has been estimated that each population generated millions of mutations, but only a few have become fixed within the populations.

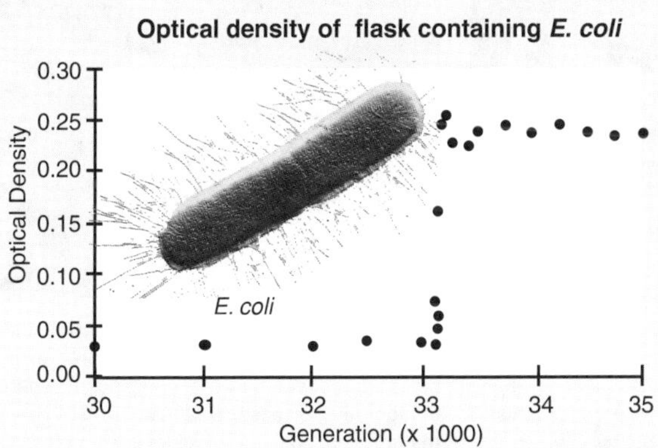
Optical density of flask containing E. coli
Optical Density
0.30
0.25
0.20
0.15
0.10
0.05
0.00
E. coli
30
31
32
33
34
35
Generation (x 1000)

One of the *E. coli* populations obtained a particularly important mutation after 31,000 generations. It became able to metabolize the citrate that was a component of the growth medium (*E. coli* are normally unable to do this). This ability gave the strain increased fitness relative to all the other populations. The mutation for citrate metabolism was noticed when the optical density (cloudiness) of the flask containing the *E. coli* suddenly increased, indicating an increase in the population of bacteria.

Investigations into the citrate mutation found that many other previous mutations were required before the mutation could take effect. Before generation 15,000 it was unlikely for this strain to evolve the ability to metabolize citrate. After generation 15,000 it was more likely to evolve the ability to metabolize citrate.

1. (a) Explain how the fitness of *E. coli* shows evolution does not follow a particular course of adaptations to reach a goal:

   (b) Explain why the increase in the fitness of the *E. coli* populations is only relative to a certain environment:

2. Explain why an increase in growth rate conveys an advantage to the *E. coli* populations:

3. Explain the significance in the mutations after the 15,000 generation mark in the development of citrate metabolism and their meaning in evolutionary development:

ISBN: 978-1-927173-16-9

Related activities: *Evidence for Evolution, The Evolution of Drug Resistance*

The Evolution of Antibiotic Resistance

Antibiotic resistance arises when a genetic change allows bacteria to tolerate levels of antibiotic that would normally inhibit growth. This resistance may arise spontaneously, through mutation or copying error, or by transfer of genetic material between microbes. Genomic analyses from 30,000 year old permafrost sediments show that the genes for antibiotic resistance are not new. They have long been present in the bacterial genome, predating the modern selective pressure of antibiotic use. In the current selective environment, these genes have proliferated and antibiotic resistance has spread. For example, methicillin resistant strains of *Staphylococcus aureus* (MRSA) have acquired genes for resistance to all penicillins. Such strains are called superbugs.

The Evolution of Antibiotic Resistance in Bacteria

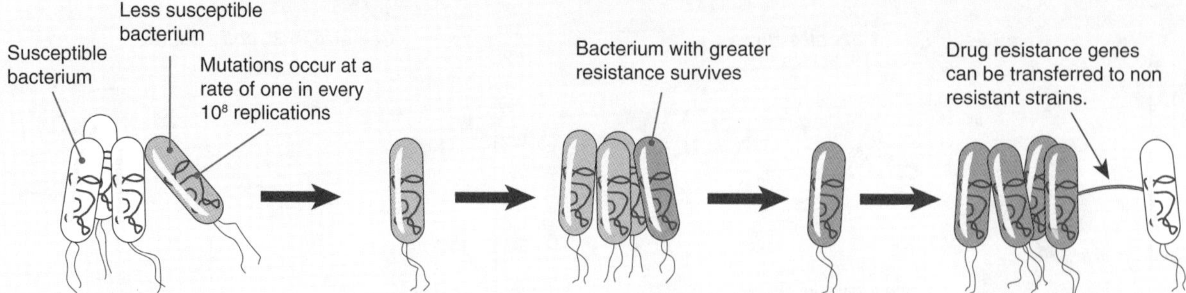
Susceptible bacterium
Less susceptible bacterium
Mutations occur at a rate of one in every 10^8 replications
Bacterium with greater resistance survives
Drug resistance genes can be transferred to non resistant strains.

Any population, including bacterial populations, includes variants with unusual traits, in this case reduced sensitivity to an antibiotic. These variants arise as a result of mutations in the bacterial chromosome. Such mutations are well documented and some are ancient.

When a person takes an antibiotic, only the most susceptible bacteria will die. The more resistant cells remain and continue dividing. Note that the antibiotic does not create the resistance; it provides the environment in which selection for resistance can take place.

If the amount of antibiotic delivered is too low, or the course of antibiotics is not completed, a population of resistant bacteria develops. Within this population too, there will be variation in susceptibility. Some will survive higher antibiotic levels.

A highly resistant population has evolved. The resistant cells can exchange genetic material with other bacteria (via horizontal gene transmission), passing on the genes for resistance. The antibiotic initially used against this bacterial strain will now be ineffective.

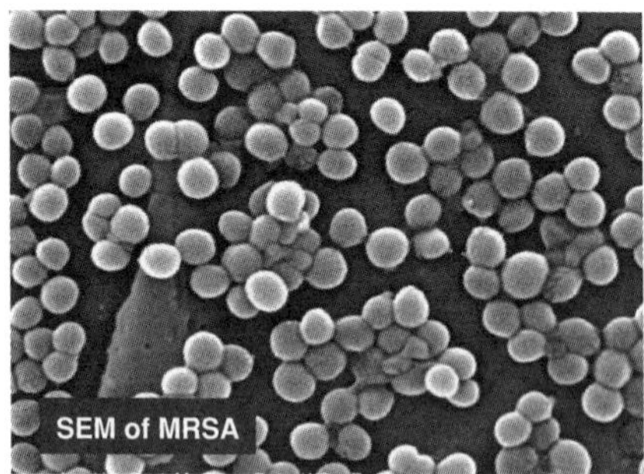
SEM of MRSA

Staphylococcus aureus is a common bacterium responsible for various minor skin infections in humans. MRSA is a variant strain that has evolved resistance to penicillin and related antibiotics. MRSA is troublesome in hospital-associated infections because patients with open wounds, invasive devices (e.g. catheters), or poor immunity are at greater risk for infection than the general public.

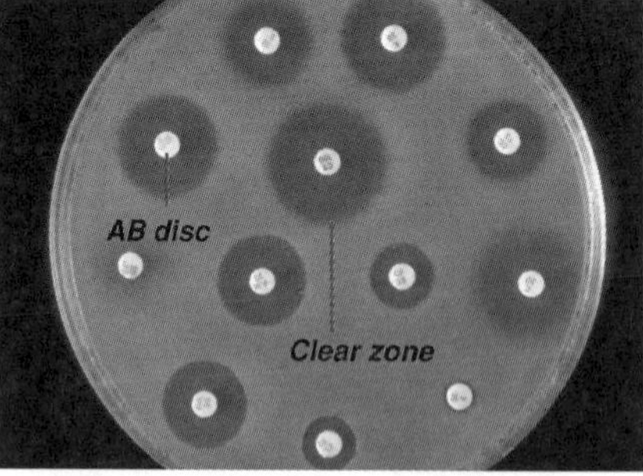
AB disc
Clear zone

The photo above shows an antibiogram plate culture of *Enterobacter sakazakii*, a rare cause of invasive infections in infants. An antibiogram measures the biological resistance of disease-causing organisms to antibiotic agents. The bacterial lawn (growth) on the agar plate is treated with antibiotic discs, and the sensitivity to various antibiotics is measured by the extent of the clearance zone in the bacterial lawn.

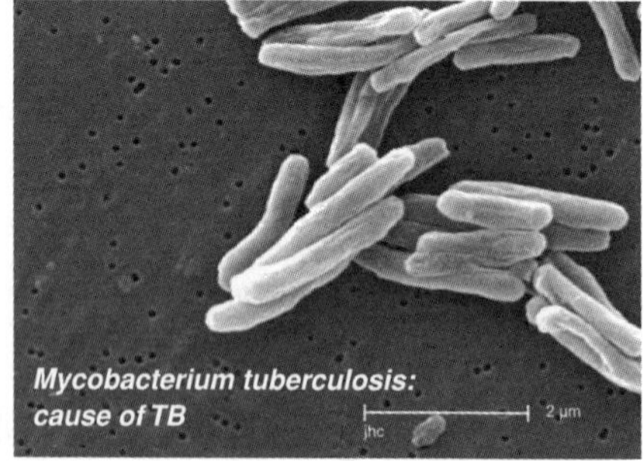
Mycobacterium tuberculosis: cause of TB
2 µm

TB is a disease that has experienced spectacular ups and downs. Drugs were developed to treat it, but then people became complacent when they thought the disease was beaten. TB has since resurged because patients stop their medication too soon and infect others. Today, one in seven new TB cases is resistant to the two drugs most commonly used as treatments, and 5% of these patients die.

1. Describe two ways in which antibiotic resistance can become widespread:

   (a) ______

   (b) ______

2. Genomic evidence indicates that the genes for antibiotic resistance are ancient:

   (a) How could these genes have arisen in the first place? ______

   (b) Why were they not lost from the bacterial genome? ______

   (c) Explain why these genes are proliferating now: ______

Related activities: *Chloroquine Resistance in Protozoa*
Weblinks: *Evolution in E. coli, Why Evolution Matters Now*

Periodicals:
Chasing the superbugs, The enemy within

ISBN: 978-1-927173-16-9

Antigenic Variability in Pathogens

Influenza (flu) is a disease of the upper respiratory tract caused by the viral genus *Influenzavirus*. Globally, up to 500,000 people die from influenza every year. It is estimated that 5-20% of Americans are affected by the flu every year, and a small number of deaths occur as a result. Three types of *Influenzavirus* affect humans. They are simply named *Influenzavirus* A, B, and C. The most common and most virulent of these strains is *Influenzavirus* A, which is discussed in more detail below. Influenza viruses are constantly undergoing genetic changes. **Antigenic drifts** are small changes in the virus which happen continually over time. Such changes mean that the influenza vaccine must be adjusted each year to include the most recently circulating influenza viruses. **Antigenic shift** occurs when two or more different viral strains (or different viruses) combine to form a new subtype. The changes are large and sudden and most people lack immunity to the new subtype. New influenza viruses arising from antigenic shift have caused influenza pandemics that have killed millions people over the last century. *Influenzavirus* A is considered the most dangerous to human health because it is capable of antigenic shift.

Structure of *Influenzavirus*

Viral strains are identified by the variation in their H and N surface antigens. Viruses are able to combine and readily rearrange their RNA segments, which alters the protein composition of their H and N glycoprotein spikes.

The *influenzavirus* is surrounded by an **envelope** containing protein and lipids.

The genetic material is actually closely surrounded by protein capsomeres (these have been omitted here and below right in order to illustrate the changes in the RNA more clearly).

The **neuraminidase (N) spikes** help the virus to detach from the cell after infection.

Hemagglutinin (H) spikes allow the virus to recognize and attach to cells before attacking them.

The viral genome is contained on **eight RNA segments**, which enables the exchange of genes between different viral strains.

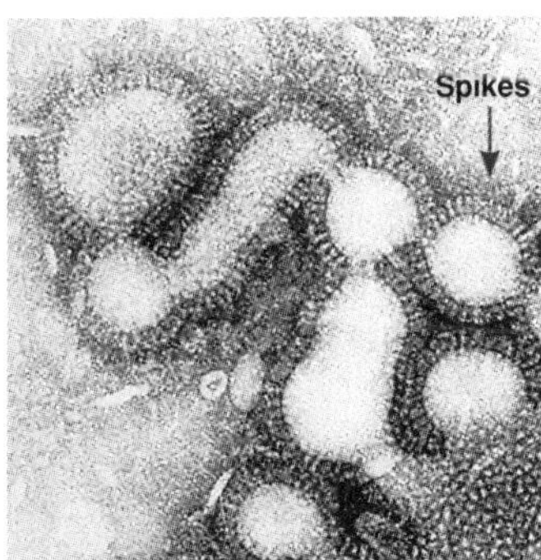
Spikes

Photo right: *Electron micrograph of Influenzavirus showing the glycoprotein spikes projecting from the viral envelope*

Antigenic Shift in *Influenzavirus*

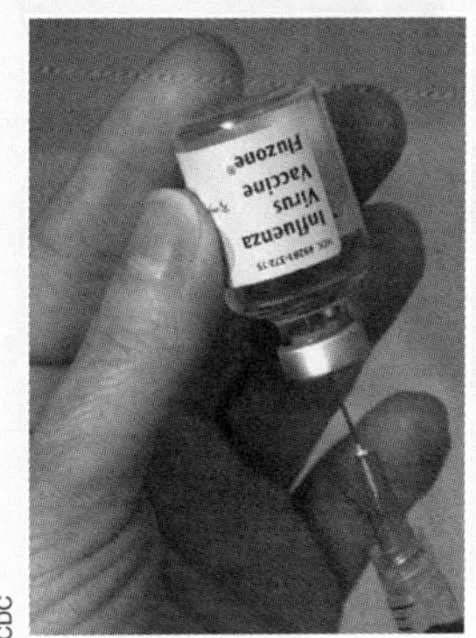

CDC

Influenza vaccination is the primary method for preventing influenza and is 75% effective. The ability of the virus to recombine its RNA enables it to change each year, so that different strains dominate in any one season. The 'flu' vaccination is updated annually to incorporate the antigenic properties of currently circulating strains. Three strains are chosen for each year's vaccination. Selection is based on estimates of which strains will be predominant in the following year.

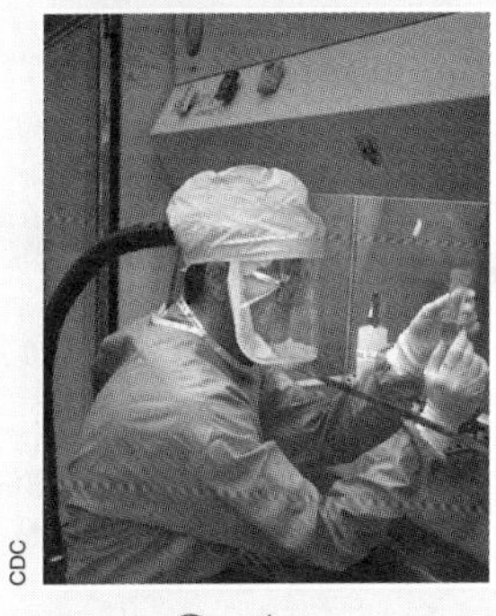

CDC

H1N1, H1N2, and H3N2 (below) are the known *Influenza A* viral subtypes currently circulating among humans. Although the body will have acquired antibodies from previous flu strains, the new combination of N and H spikes is sufficiently different to enable new viral strains to avoid detection by the immune system. The World Health Organization coordinates strain selection for each year's influenza vaccine.

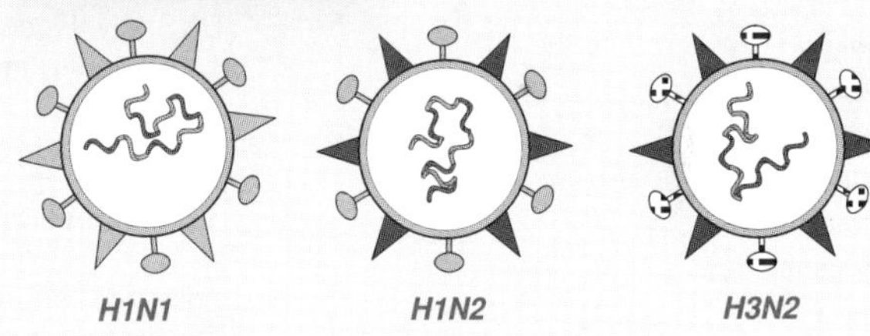
H1N1
H1N2
H3N2

1. The *Influenzavirus* is able to mutate readily and alter the composition of H and N spikes on its surface.

   (a) Why can the virus mutate so readily? ______________________________

   ____

   ____

   (b) How does this affect the ability of the immune system to recognize and respond to the virus? ______________________________

   ____

   ____

   ____

2. Discuss why a virus capable of antigenic shift is more dangerous to humans than a virus undergoing antigenic drift:

   ____

   ____

   ____

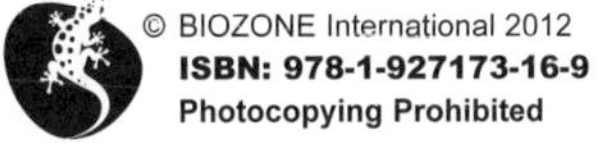

ISBN: 978-1-927173-16-9

Core Topic 5.5

Classification

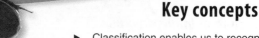

Key concepts

- Classification enables us to recognize and quantify the biological diversity on Earth.
- Organisms are put in taxonomic categories based on shared derived characters.
- Organisms are identified using binomial nomenclature: genus and species.
- Dichotomous classification keys can be used to identify unknown organisms.

Key terms

Core

Angiospermophyta
Animalia
Annelida
Archaea
Arthropoda
Bacteria
binomial nomenclature
Bryophyta
class
Cnidaria
common name
Coniferophyta
dichotomous keys
distinguishing feature
Echinodermata
family,
Filicinophyta
Fungi
genus
kingdom
Linnaeus
Mollusca
order
phylum
Plantae
Platyhelminthes
Porifera
species
taxa
taxonomic category
taxonomy
Tracheophyta
Vertebrata

Periodicals:
Listings for this chapter are on page 399

Weblinks:
www.thebiozone.com/weblink/IB-3169.html

Learning Objectives

☐ 1. Use the **KEY TERMS** to compile a glossary for this topic.

Classification of Life *(5.5.1-5.2.2)* pages 253-259

☐ 2. Describe the principles and importance of biological classification. Recognize **taxonomy** as the study of the theory and practice of classification.

☐ 3. Explain how **binomial nomenclature** is used to classify organisms. Identify the problems associated with using **common names** to describe organisms.

☐ 4. With reference to the work of early taxonomists, particular **Linnaeus**, explain how the social context of scientific work affects the methods and findings of the research. How important is it to view research within the social context of the time? (*TOK*)

☐ 5. EXTENSION: Recognize six kingdoms of life: **Archaea**, **Bacteria**, **Protista, Fungi**, **Plantae**, and **Animalia**.

☐ 6. Recognize seven major taxonomic categories in the five or six kingdom classification system: **kingdom**, **phylum**, **class**, **order**, **family**, **genus**, and **species**. Demonstrate your understanding by describing an example for two different **kingdoms** for each level.

☐ 7. Distinguish taxonomic categories from **taxa**, which are groups of organisms, e.g. 'genus' is a **taxonomic category**, whereas the genus *Drosophila* is a taxon.

Distinguishing Features *(5.5.3-5.2.4)* pages 260-262

☐ 8. Recongize two main Divisions of the Plantae: **Bryophyta** and **Tracheophyta**. Using simple external **distinguishing features**, distinguish between the following plant phyla: **Bryophyta**, **Filicinophyta**, **Coniferophyta**, and **Angiospermophyta**.

☐ 9. Using simple external distinguishing features, distinguish between the following invertebrate phyla: **Porifera**, **Cnidaria**, **Platyhelminthes**, **Annelida**, **Mollusca**, **Echinodermata**, and **Arthropoda**.

☐ 10. EXTENSION: Using simple external distinguishing features, distinguish between the following classes in the phylum **Vertebrata**: Chondrichthyes, Osteichthyes, Amphibia, Reptilia, Aves, Mammalia. If required, recognize three sub-classes of mammals.

Classification Keys *(5.5.5)* pages 263-265

☐ 11. Explain the principles by which **dichotomous keys** are used to identify organisms. Use simple dichotomous keys to recognize and classify some common organisms. Apply and design a key for a group of up to eight organisms.

Classification System

The classification of organisms is designed to reflect how they are related to each other. The fundamental unit of classification of living things is the **species**. Its members are so alike genetically that they can interbreed. This genetic similarity also means that they are almost identical in their physical and other characteristics. Species are classified further into larger, more comprehensive categories (higher taxa). It must be emphasized that all such higher classifications are human inventions to suit a particular purpose.

1. The table below shows part of the classification for humans using the seven major levels of classification. For this question, use the example of the classification of the Ethiopian hedgehog on the next page, as a guide.

   (a) Complete the list of the taxonomic groupings on the left hand side of the table below:

	Taxonomic Group	Human Classification
1.		
2.		
3.		
4.		
5.	Family	Hominidae
6.		
7.		

   (b) Complete the classification for humans (*Homo sapiens*) on the table above.

2. Construct your own acronym or mnemonic to help you remember the principal taxonomic groupings in biology:

3. Describe the two-part scientific naming system (**binomial nomenclature**) which is used to name organisms:

4. Give two reasons why the classification of organisms is important:

   (a)

   (b)

5. Classification has traditionally been based on similarities in morphology but new biochemical methods are now widely used to determine species relatedness. Explain how these are being used to clarify the relationships between species:

6. Mammals have been divided into three major taxa: the monotremes, marsupials, and placentals. Describe the main morphological feature you would use to distinguish each taxon:

   (a) Monotreme (Prototheria):

   (b) Marsupial (Metatheria):

   (c) Placental (Eutheria):

ISBN: 978-1-927173-16-9

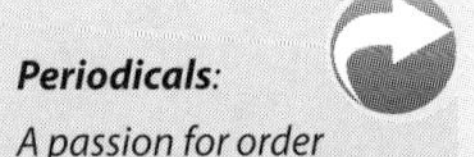

Periodicals:
A passion for order

Related activities: *Classification Keys*

Classification of the Ethiopian Hedgehog

Below is the classification for the **Ethiopian hedgehog**. Only one of each group is subdivided in this chart showing the levels that can be used in classifying an organism. Not all possible subdivisions have been shown here. For example, it is possible to indicate such categories as **super-class** and **sub-family**. The only natural category is the **species**, often separated into geographical **races**, or **sub-species**, which generally differ in appearance.

Kingdom: **Animalia**
Animals; one of five kingdoms

Phylum: **Chordata**
Animals with a notochord (supporting rod of cells along the upper surface)
tunicates, salps, lancelets, and vertebrates

23 other phyla

Sub-phylum: **Vertebrata**
Animals with backbones
fish, amphibians, reptiles, birds, mammals

Class: **Mammalia**
Animals that suckle their young on milk from mammary glands
placentals, marsupials, monotremes

Sub-class: **Eutheria or Placentals**
Mammals whose young develop for some time in the female's reproductive tract gaining nourishment from a placenta
placental mammals

Order: **Insectivora**
Insect eating mammals
An order of over 300 species of primitive, small mammals that feed mainly on insects and other small invertebrates.

17 other orders

Sub-order: **Erinaceomorpha**
The hedgehog-type insectivores. One of the three suborders of insectivores. The other suborders include the tenrec-like insectivores (*tenrecs and golden moles*) and the shrew-like insectivores (*shrews, moles, desmans, and solenodons*).

Family: **Erinaceidae**
The only family within this suborder. Comprises two subfamilies: the true or spiny hodgehogs and the moonrats (gymnures). Representatives in the family include the common European hedgehog, desert hedgehog, and the moonrats.

Genus: ***Paraechinus***
One of eight genera in this family. The genus *Paraechinus* includes three species which are distinguishable by a wide and prominent naked area on the scalp.

7 other genera

Species: ***aethiopicus***
The Ethiopian hedgehog inhabits arid coastal areas. Their diet consists mainly of insects, but includes small vertebrates and the eggs of ground nesting birds.

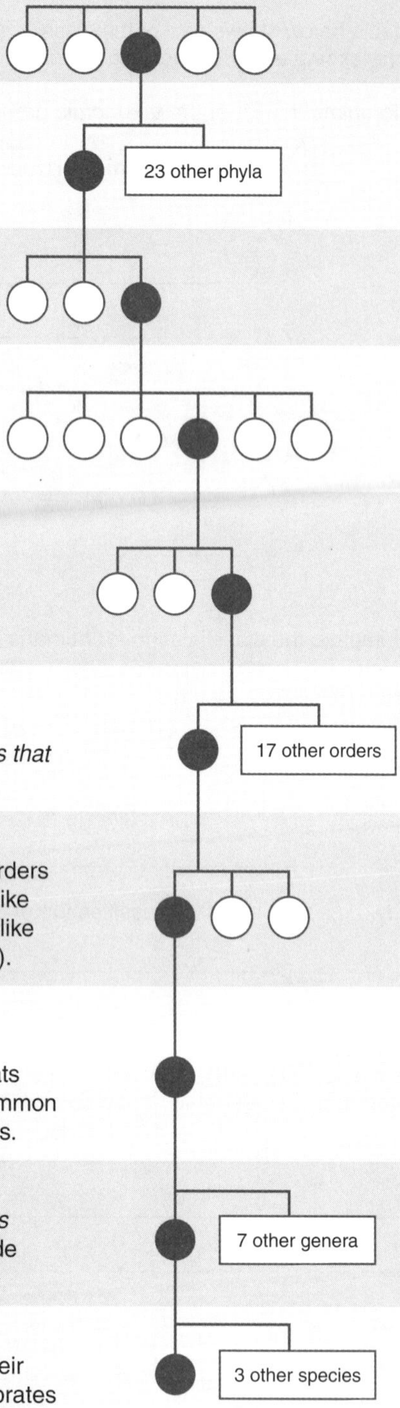

3 other species

The order *Insectivora* was first introduced to group together shrews, moles, and hedgehogs. It was later extended to include tenrecs, golden moles, desmans, tree shrews, and elephant shrews and the taxonomy of the group became very confused. Recent reclassification of the elephant shrews and tree shrews into their own separate orders has made the Insectivora a more cohesive group taxonomically.

Ethiopian hedgehog
Paraechinus aethiopicus

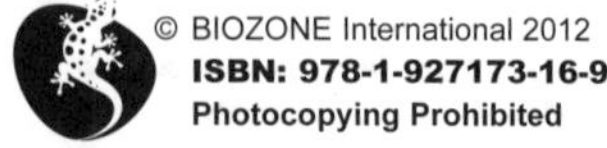

ISBN: 978-1-927173-16-9

Features of Taxonomic Groups

In order to distinguish organisms, it is desirable to classify and name them (a science known as **taxonomy**). An effective classification system requires features that are distinctive to a particular group of organisms. Revised classification systems, recognizing three domains (rather than five or six kingdoms) are now recognized as better representations of the true diversity of life. However, for the purposes of describing the groups with which we are most familiar, the five kingdom system (used here) is still appropriate. The distinguishing features of some major **taxa** are provided in the following pages by means of diagrams and brief summaries. Note that most animals show **bilateral symmetry** (body divisible into two halves that are mirror images). **Radial symmetry** (body divisible into equal halves through various planes) is a characteristic of cnidarians and ctenophores.

SUPERKINGDOM: PROKARYOTAE (Bacteria)

- Also known as prokaryotes. The term moneran is no longer in use.
- Two major bacterial lineages are recognized: the **Archaebacteria** (Archaea) and the more derived **Eubacteria** (Bacteria).
- All have a prokaryotic cell structure: they lack the nuclei and chromosomes of eukaryotic cells, and have smaller (70S) ribosomes.
- Have a tendency to spread genetic elements across species barriers by conjugation, viral transduction, and other processes.
- Asexual. Can reproduce rapidly by binary fission.
- Have evolved a wider variety of metabolism types than eukaryotes.
- Bacteria grow and divide or aggregate into filaments or colonies of various shapes. Colony type is often diagnostic.
- They are taxonomically identified by their appearance (form) and through biochemical differences.

Species diversity: 10,000+ Bacteria are rather difficult to classify to species level because of their relatively rampant genetic exchange, and because their reproduction is asexual.

Eubacteria

- Also known as 'true bacteria', they probably evolved from the more ancient Archaebacteria.
- Distinguished from Archaebacteria by differences in cell wall composition, nucleotide structure, and ribosome shape.
- Diverse group includes most bacteria.
- The **gram stain** is the basis for distinguishing two broad groups of bacteria. It relies on the presence of peptidoglycan in the cell wall. The stain is easily washed from the thin peptidoglycan layer of gram negative walls but is retained by the thick peptidoglycan layer of gram positive cells, staining them a dark violet color.

Gram Positive Bacteria

The walls of gram positive bacteria consist of many layers of peptidoglycan forming a thick, single-layered structure that holds the gram stain.

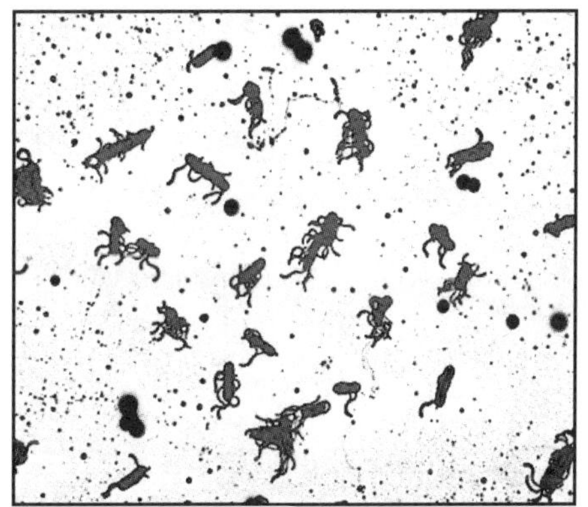

Bacillus alvei: a gram positive, flagellated bacterium. Note how the cells appear dark.

Photos: CDC

Gram Negative Bacteria

The cell walls of gram negative bacteria contain only a small proportion of peptidoglycan, so the dark violet stain is not retained by the organisms.

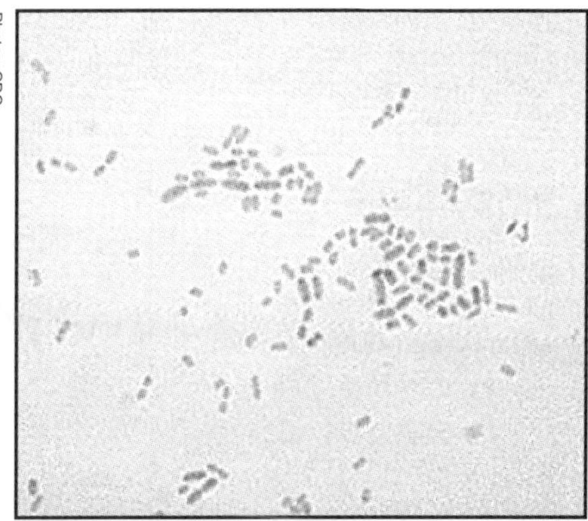

Alcaligenes odorans: a gram negative bacterium. Note how the cells appear pale.

SUPERKINGDOM: EUKARYOTAE
Kingdom: FUNGI

- Heterotrophic.
- Rigid cell wall made of chitin.
- Vary from single celled to large multicellular organisms.
- Mostly saprotrophic (ie. feeding on dead or decaying material).
- Terrestrial and immobile.

Examples:
Mushrooms/toadstools, yeasts, truffles, morels, molds, and lichens.

Species diversity: 80,000 +

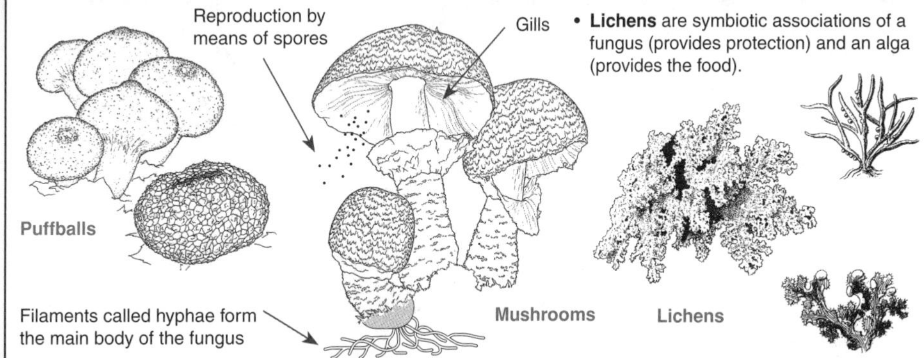
Reproduction by means of spores
Gills
Puffballs
Filaments called hyphae form the main body of the fungus
Mushrooms
Lichens

- **Lichens** are symbiotic associations of a fungus (provides protection) and an alga (provides the food).

Kingdom: PROTISTA

- A diverse group of organisms. They are polyphyletic and so better represented in the 3 domain system.
- Unicellular or simple multicellular.
- Widespread in moist or aquatic environments.

Examples of algae: green, red, and brown algae, dinoflagellates, diatoms.

Examples of protozoa: amoebas, foraminiferans, radiolarians, ciliates.

Species diversity: 55,000 +

Algae 'plant-like' protists

- Autotrophic (photosynthesis)
- Characterized by the type of chlorophyll present

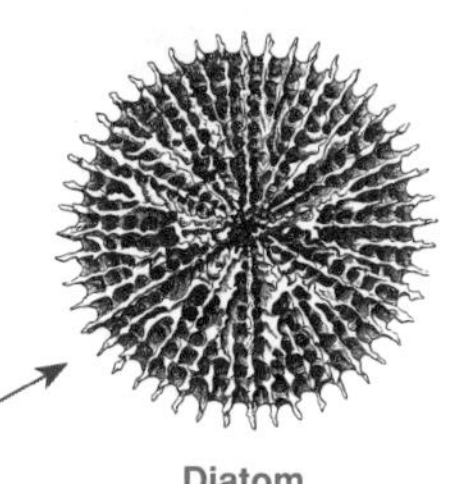
Cell walls of cellulose, sometimes with silica
Diatom

Protozoa 'animal-like' protists

- Heterotrophic nutrition and feed via ingestion
- Most are microscopic (5 µm - 250 µm)

Move via projections called pseudopodia

Lack cell walls

Amoeba

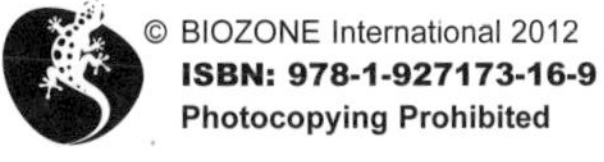

ISBN: 978-1-927173-16-9

Kingdom: PLANTAE

- Multicellular organisms (the majority are photosynthetic and contain chlorophyll).
- Cell walls made of cellulose; food is stored as starch.
- Subdivided into two major divisions based on tissue structure: ***Bryophytes*** (non-vascular plants) and ***Tracheophytes*** (vascular plants).

Non-Vascular Plants:

- Non-vascular, lacking transport tissues (no xylem or phloem).
- Small and restricted to moist, terrestrial environments.
- Do not possess 'true' roots, stems, or leaves.

Phylum Bryophyta: Mosses, liverworts, and hornworts.

Species diversity: 18,600 +

Phylum: Bryophyta

Sexual reproductive structures

Flattened thallus (leaf like structure)

Sporophyte: reproduce by spores

Rhizoids anchor the plant into the ground

Liverworts

Mosses

Vascular Plants:

- Vascular: possess transport tissues.
- Possess true roots, stems, and leaves, as well as stomata.
- Reproduce via spores, not seeds.
- Clearly defined alternation of sporophyte and gametophyte generations.

Seedless Plants:

Spore producing plants, includes:
Phylum Filicinophyta: Ferns
Phylum Sphenophyta: Horsetails
Phylum Lycophyta: Club mosses
Species diversity: 13,000 +

Phylum: Lycophyta

Leaves

Club moss

Phylum: Sphenophyta

Leaves

Horsetail

Phylum: Filicinophyta

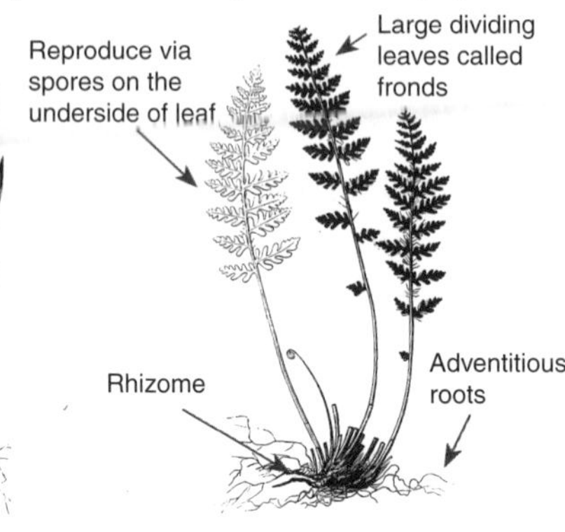

Reproduce via spores on the underside of leaf
Large dividing leaves called fronds
Rhizome
Adventitious roots

Fern

Seed Plants:

Also called Spermatophyta. Produce seeds housing an embryo. Includes:

Gymnosperms

- Lack enclosed chambers in which seeds develop.
- Produce seeds in cones which are exposed to the environment

Phylum Cycadophyta: Cycads
Phylum Ginkgophyta: Ginkgoes
Phylum Coniferophyta: Conifers
Species diversity: 730 +

Phylum: Cycadophyta

Palm-like leaves
Cone

Cycad

Phylum: Ginkgophyta

Flat leaves

Ginkgo

Phylum: Coniferophyta

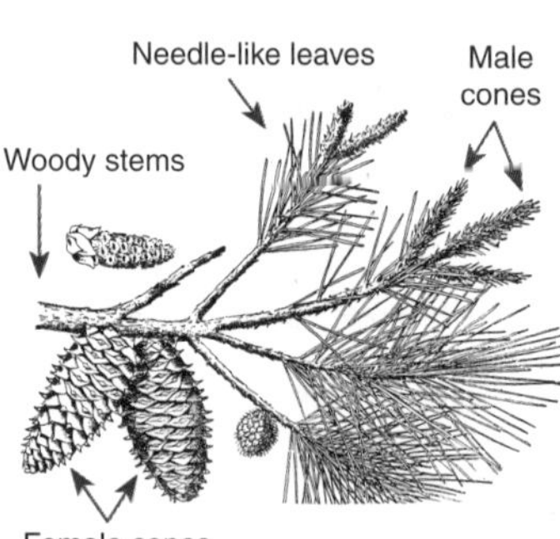

Needle-like leaves
Male cones
Woody stems
Female cones

Conifer

Angiosperms

Phylum: Angiospermophyta

- Seeds in specialized reproductive structures called flowers.
- Female reproductive ovary develops into a fruit.
- Pollination usually via wind or animals.

Species diversity: 260,000 +

The phylum Angiospermophyta may be subdivided into two classes:
Class Monocotyledoneae (Monocots)
Class Dicotyledoneae (Dicots)

Angiosperms: **Monocotyledons**

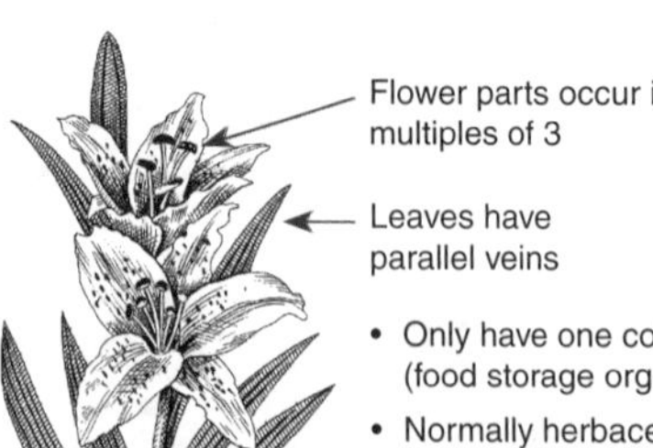

Flower parts occur in multiples of 3
Leaves have parallel veins
Lily

- Only have one cotyledon (food storage organ)
- Normally herbaceous (non-woody) with no secondary growth

Examples: cereals, lilies, daffodils, palms, grasses.

Angiosperms: **Dicotyledons**

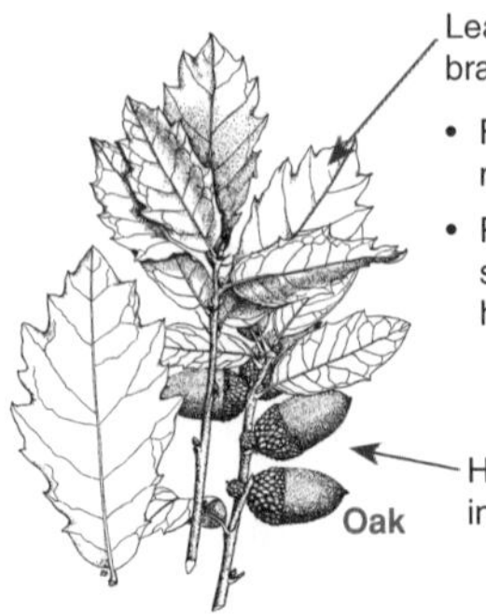

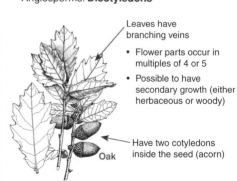

Leaves have branching veins
Have two cotyledons inside the seed (acorn)
Oak

- Flower parts occur in multiples of 4 or 5
- Possible to have secondary growth (either herbaceous or woody)

Examples: many annual plants, trees and shrubs.

ISBN: 978-1-927173-16-9

Kingdom: ANIMALIA

- Over 800,000 species described in 33 existing phyla.
- Multicellular, heterotrophic organisms.
- Animal cells lack cell walls.
- Further subdivided into major phyla on the basis of body symmetry, development of the coelom (protostome or deuterostome), and external and internal structures.

Phylum: Porifera

- Lack organs.
- All are aquatic (mostly marine).
- Asexual reproduction by budding.
- Lack a nervous system.

Examples: sponges.

Species diversity: 8000 +

Body wall perforated by pores through which water enters
• Capable of regeneration (the replacement of lost parts)
• Possess spicules (needle-like internal structures) for support and protection
Tube sponge
Water leaves by a larger opening - the osculum
Sponge
Sessile (attach to ocean floor)

Phylum: Cnidaria

- Diploblastic with two basic body forms:
  Medusa: umbrella shaped and free swimming by pulsating bell.
  Polyp: cylindrical, some are sedentary, others can glide, or somersault or use tentacles as legs.
- Some species have a life cycle that alternates between a polyp stage and a medusa stage.
- All are aquatic (most are marine).

Examples: Jellyfish, sea anemones, hydras, and corals.

Species diversity: 11,000 +

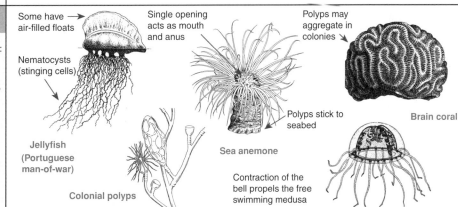

Some have air-filled floats
Nematocysts (stinging cells)
Jellyfish (Portuguese man-of-war)
Single opening acts as mouth and anus
Colonial polyps
Polyps stick to seabed
Sea anemone
Polyps may aggregate in colonies
Brain coral
Contraction of the bell propels the free swimming medusa

Phylum: Rotifera

- A diverse group of small, pseudocoelomates with sessile, colonial, and planktonic forms.
- Most freshwater, a few marine.
- Typically reproduce via cyclic parthenogenesis.
- Characterized by a wheel of cilia on the head used for feeding and locomotion, a large muscular pharynx (mastax) with jaw like trophi, and a foot with sticky toes.

Species diversity: 1500 +

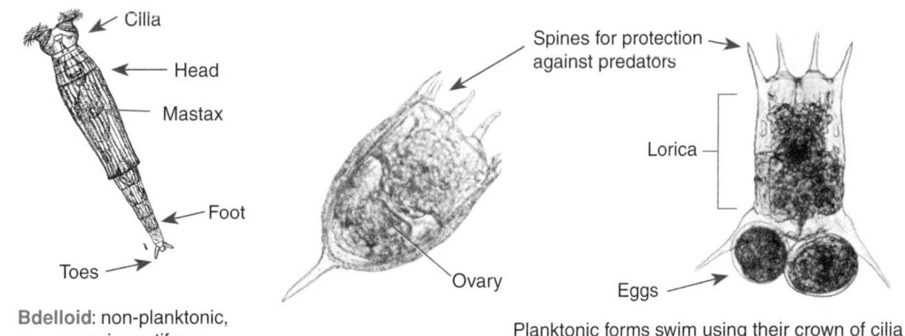
Cilia
Head
Mastax
Foot
Toes
Bdelloid: non-planktonic, creeping rotifer
Spines for protection against predators
Ovary
Lorica
Eggs
Planktonic forms swim using their crown of cilia

Phylum: Platyhelminthes

- Unsegmented. Coelom has been lost.
- Flattened body shape.
- Mouth, but no anus.
- Many are parasitic.

Examples: Tapeworms, planarians, flukes.

Species diversity: 20,000 +

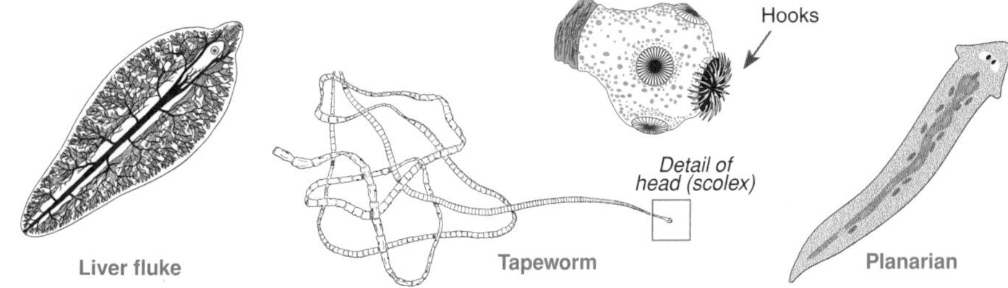
Liver fluke
Hooks
Detail of head (scolex)
Tapeworm
Planarian

Phylum: Nematoda

- Tiny, unsegmented roundworms.
- Many are plant/animal parasites

Examples: Hookworms, stomach worms, lung worms, filarial worms

Species diversity: 80,000 - 1 million

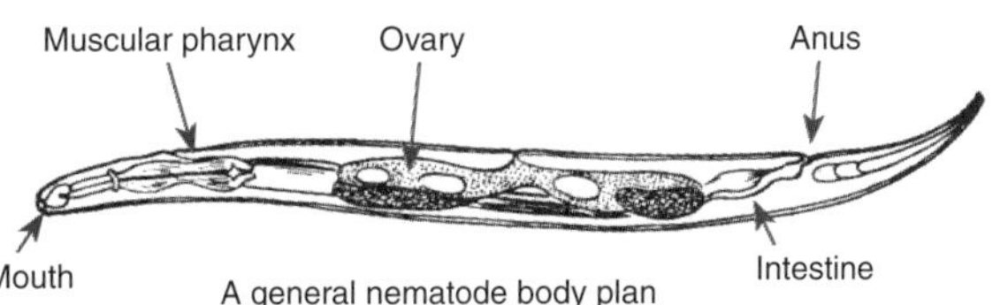
Muscular pharynx
Ovary
Anus
Mouth
Intestine
A general nematode body plan

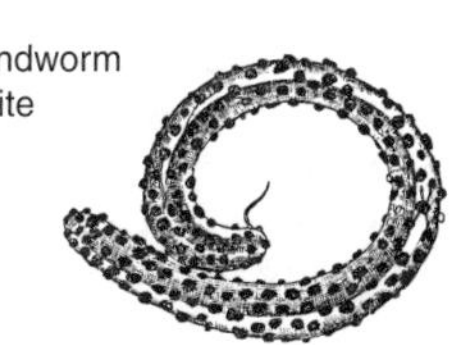
A roundworm parasite

Phylum: Annelida

- Cylindrical, segmented body with chaetae (bristles).
- Move using hydrostatic skeleton and/or parapodia (appendages).

Examples: Earthworms, leeches, polychaetes (including tubeworms).

Species diversity: 15,000 +

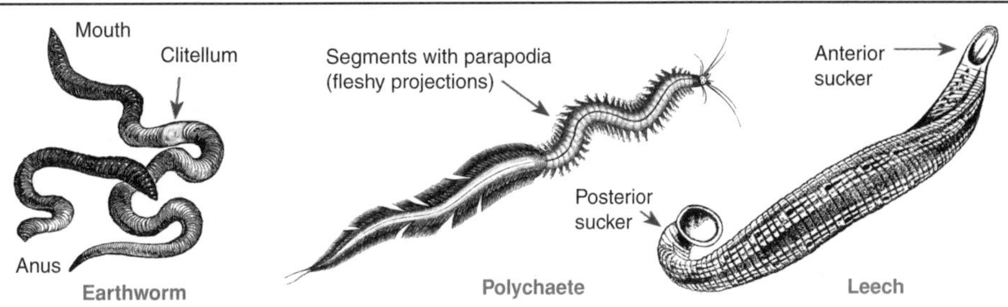
Mouth
Clitellum
Anus
Earthworm
Segments with parapodia (fleshy projections)
Polychaete
Anterior sucker
Posterior sucker
Leech

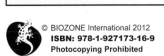

ISBN: 978-1-927173-16-9

Kingdom: ANIMALIA (continued)

Phylum: Mollusca

- Soft bodied and unsegmented.
- Body comprises head, muscular foot, and visceral mass (organs).
- Most have radula (rasping tongue).
- Aquatic and terrestrial species.
- Aquatic species possess gills.

Examples: Snails, mussels, squid.
Species diversity: 110,000 +

Class: Bivalvia Class: Gastropoda Class: Cephalopoda

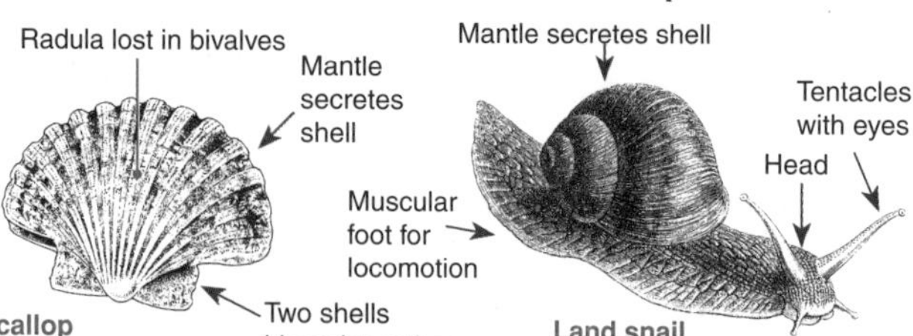

Radula lost in bivalves
Mantle secretes shell
Scallop
Two shells hinged together
Mantle secretes shell
Muscular foot for locomotion
Head
Tentacles with eyes
Land snail
Well developed eyes
Squid
Foot divided into tentacles

Phylum: Arthropoda

- Exoskeleton made of chitin.
- Grow in stages after molting (ecdysis).
- Jointed appendages.
- Segmented bodies.
- Heart found on dorsal side of body.
- Open circulation system.
- Most have compound eyes.

Species diversity: 1 million +
Make up 75% of all living animals.

Arthropods are subdivided into the following classes:

Class: Crustacea (crustaceans)
- Mainly marine.
- Exoskeleton impregnated with mineral salts.
- Gills often present.
- Includes: Lobsters, crabs, barnacles, prawns, shrimps, isopods, amphipods
- **Species diversity**: 35,000 +

Class: Arachnida (chelicerates)
- Almost all are terrestrial.
- 2 body parts: cephalothorax and abdomen (except horseshoe crabs).
- Includes: spiders, scorpions, ticks, mites, horseshoe crabs.
- **Species diversity**: 57,000 +

Class: Insecta (insects)
- Mostly terrestrial.
- Most are capable of flight.
- 3 body parts: head, thorax, abdomen.
- Include: Locusts, dragonflies, cockroaches, butterflies, bees, ants, beetles, bugs, flies, and more
- **Species diversity**: 800,000 +

Myriapoda (=many legs)
Class Diplopoda (millipedes)
- Terrestrial.
- Have a rounded body.
- Eat dead or living plants.
- **Species diversity**: 2000 +

Class Chilopoda (centipedes)
- Terrestrial.
- Have a flattened body.
- Poison claws for catching prey.
- Feed on insects, worms, and snails.
- **Species diversity**: 7000 +

Class: Crustacea

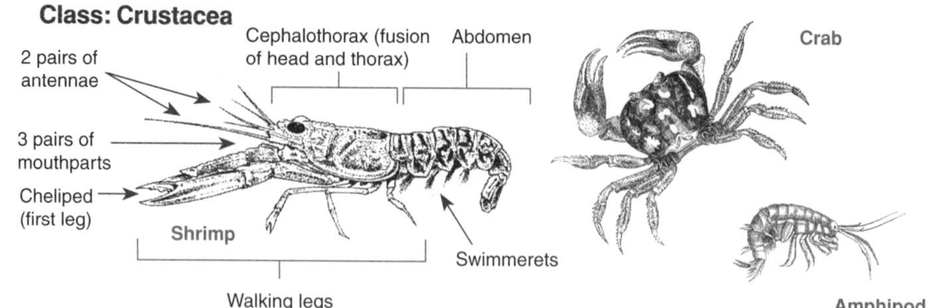

2 pairs of antennae
Cephalothorax (fusion of head and thorax)
Abdomen
3 pairs of mouthparts
Cheliped (first leg)
Shrimp
Swimmerets
Walking legs
Crab
Amphipod

Class: Arachnida

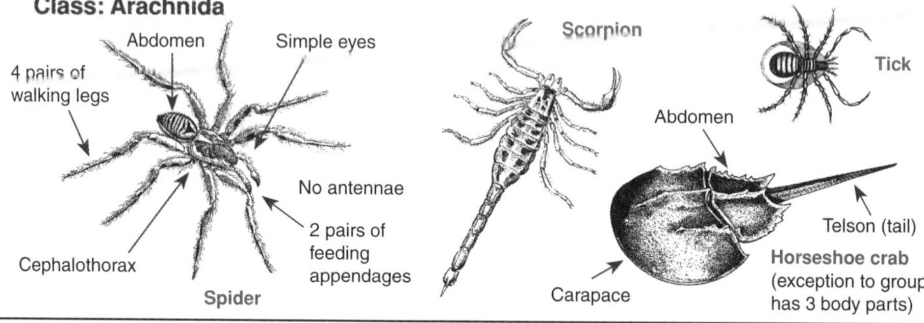

Abdomen
Simple eyes
4 pairs of walking legs
No antennae
2 pairs of feeding appendages
Cephalothorax
Spider
Scorpion
Tick
Abdomen
Telson (tail)
Carapace
Horseshoe crab
(exception to group: has 3 body parts)

Class: Insecta

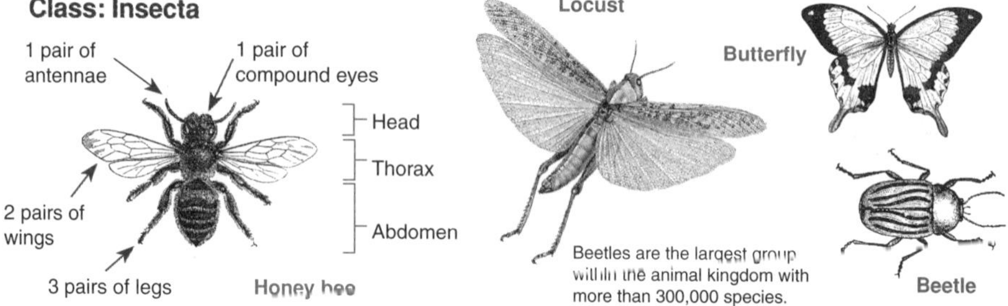

1 pair of antennae
1 pair of compound eyes
Head
Thorax
Abdomen
2 pairs of wings
3 pairs of legs
Honey bee
Locust
Butterfly
Beetles are the largest group within the animal kingdom with more than 300,000 species.
Beetle

Class: Diplopoda

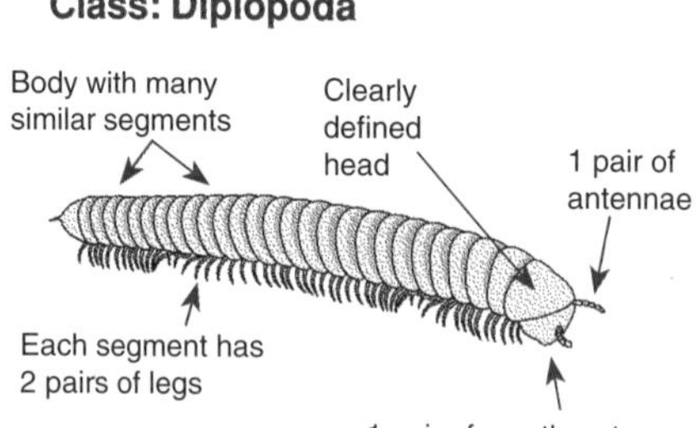

Body with many similar segments
Clearly defined head
1 pair of antennae
Each segment has 2 pairs of legs
1 pair of mouthparts

Class: Chilopoda

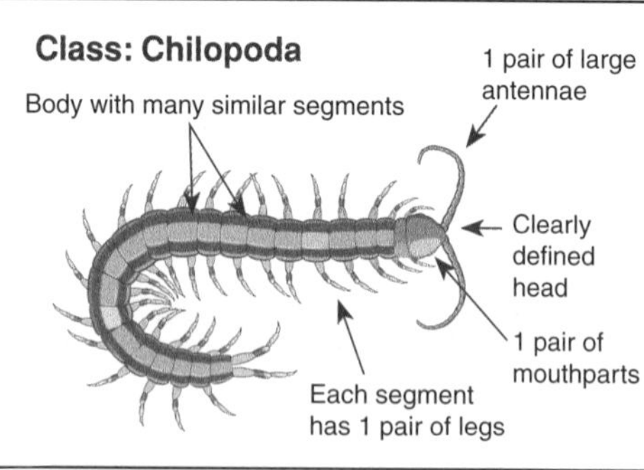

Body with many similar segments
1 pair of large antennae
Clearly defined head
1 pair of mouthparts
Each segment has 1 pair of legs

Phylum: Echinodermata

- Rigid body wall, internal skeleton made of calcareous plates.
- Many possess spines.
- Ventral mouth, dorsal anus.
- External fertilization.
- Unsegmented, marine organisms.
- Tube feet for locomotion.
- Water vascular system.

Examples: Starfish, brittlestars, feather stars, sea urchins, sea lilies.
Species diversity: 6000 +

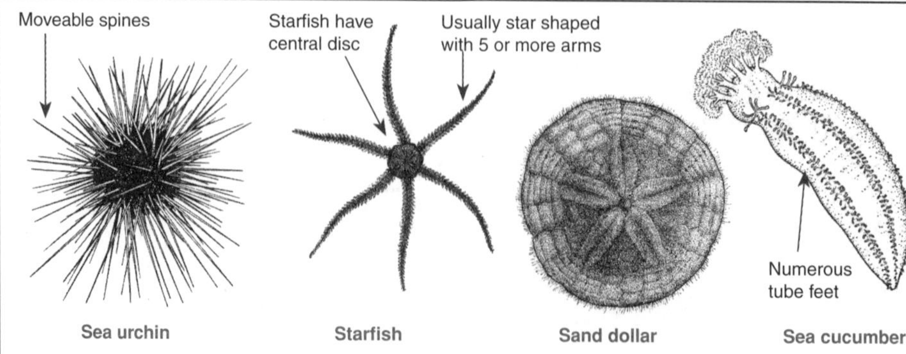

Moveable spines
Starfish have central disc
Usually star shaped with 5 or more arms
Numerous tube feet
Sea urchin
Starfish
Sand dollar
Sea cucumber

ISBN: 978-1-927173-16-9

Kingdom: ANIMALIA (continued)

Phylum: Chordata

- Dorsal notochord (flexible, supporting rod) present at some stage in the life history.
- Post-anal tail present at some stage in their development.
- Dorsal, tubular nerve cord.
- Pharyngeal slits present.
- Circulation system closed in most.
- Heart positioned on ventral side.

Species diversity: 48,000 +

- A very diverse group with several sub-phyla:
  - Urochordata (sea squirts, salps)
  - Cephalochordata (lancelet)
  - Craniata (vertebrates)

Sub-Phylum Craniata (vertebrates)

- Internal skeleton of cartilage or bone.
- Well developed nervous system.
- Vertebral column replaces notochord.
- Two pairs of appendages (fins or limbs) attached to girdles.

Further subdivided into:

Class: Chondrichthyes (cartilaginous fish)

- Skeleton of cartilage (not bone).
- No swim bladder.
- All aquatic (mostly marine).
- Include: Sharks, rays, and skates.

Species diversity: 850 +

Class: Osteichthyes (bony fish)

- Swim bladder present.
- All aquatic (marine and fresh water).

Species diversity: 21,000 +

Class: Amphibia (amphibians)

- Lungs in adult, juveniles may have gills (retained in some adults).
- Gas exchange also through skin.
- Aquatic and terrestrial (limited to damp environments).
- Include: Frogs, toads, salamanders, and newts.

Species diversity: 3900 +

Class Reptilia (reptiles)

- Ectotherms with no larval stages.
- Teeth are all the same type.
- Eggs with soft leathery shell.
- Mostly terrestrial.
- Include: Snakes, lizards, crocodiles, turtles, and tortoises.

Species diversity: 7000 +

Class: Aves (birds)

- Terrestrial endotherms.
- Eggs with hard, calcareous shell.
- Strong, light skeleton.
- High metabolic rate.
- Gas exchange assisted by air sacs.

Species diversity: 8600 +

Class: Mammalia (mammals)

- Endotherms with hair or fur.
- Mammary glands produce milk.
- Glandular skin with hair or fur.
- External ear present.
- Teeth are of different types.
- Diaphragm between thorax/abdomen.

Species diversity: 4500 +

Subdivided into three subclasses: Monotremes, marsupials, placentals.

Class: Chondrichthyes (cartilaginous fish)

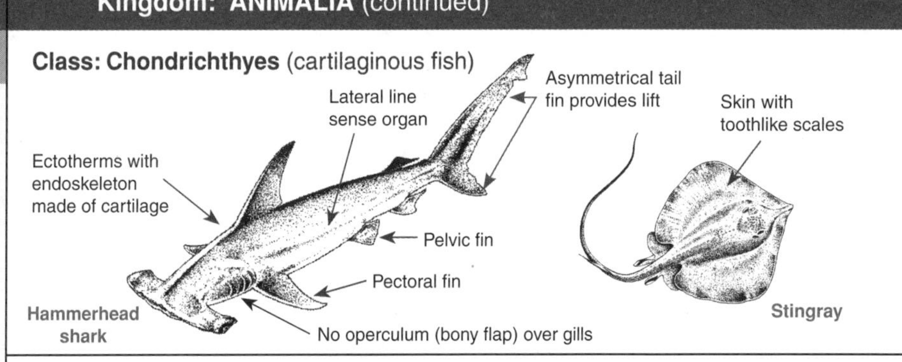
Lateral line sense organ
Asymmetrical tail fin provides lift
Skin with toothlike scales
Ectotherms with endoskeleton made of cartilage
Pelvic fin
Pectoral fin
Hammerhead shark
No operculum (bony flap) over gills
Stingray

Class: Osteichthyes (bony fish)

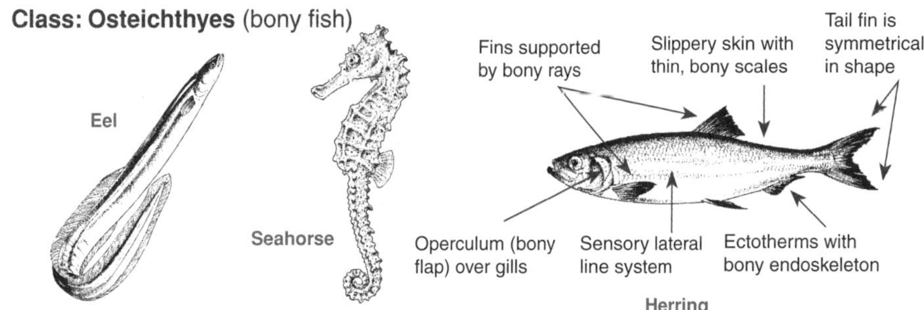
Eel
Seahorse
Fins supported by bony rays
Slippery skin with thin, bony scales
Tail fin is symmetrical in shape
Operculum (bony flap) over gills
Sensory lateral line system
Ectotherms with bony endoskeleton
Herring

Class: Amphibia

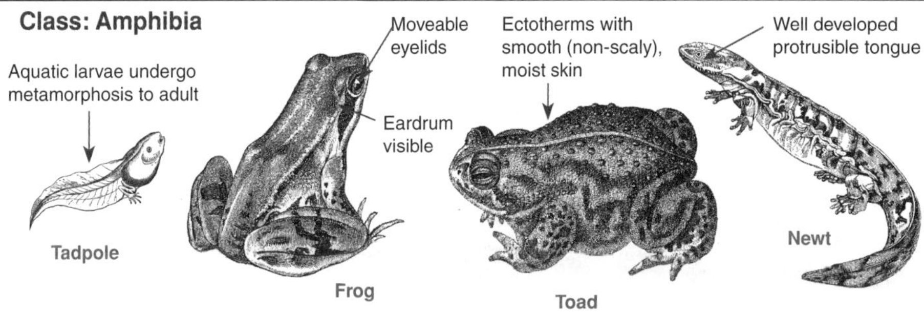
Aquatic larvae undergo metamorphosis to adult
Moveable eyelids
Eardrum visible
Ectotherms with smooth (non-scaly), moist skin
Well developed protrusible tongue
Tadpole
Frog
Toad
Newt

Class: Reptilia

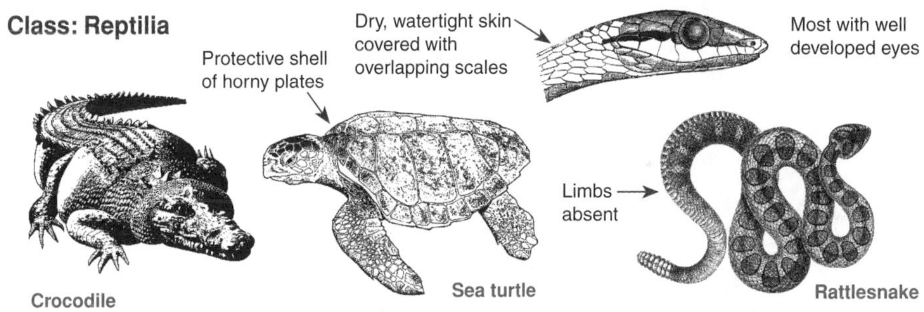
Protective shell of horny plates
Dry, watertight skin covered with overlapping scales
Most with well developed eyes
Limbs absent
Crocodile
Sea turtle
Rattlesnake

Class: Aves

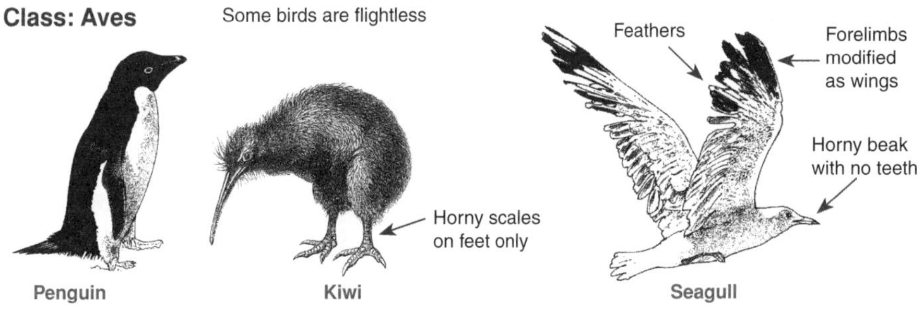
Some birds are flightless
Feathers
Forelimbs modified as wings
Horny beak with no teeth
Horny scales on feet only
Penguin
Kiwi
Seagull

Class: Mammalia

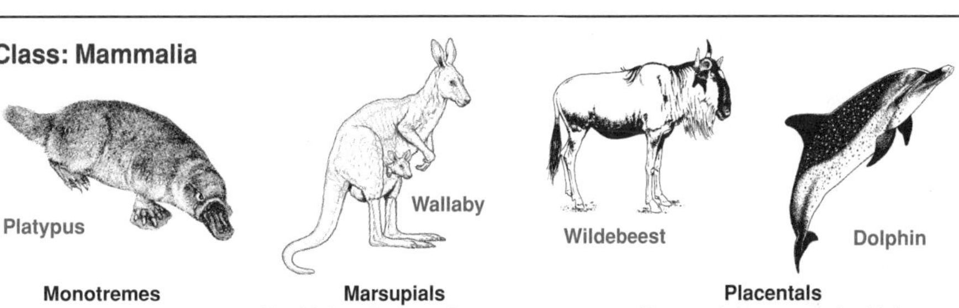
Platypus
Wallaby
Wildebeest
Dolphin

Monotremes
Egg laying mammals

Marsupials
Give birth to live, very immature young which then develop in a pouch

Placentals
Have a placenta and give birth to live, well developed young

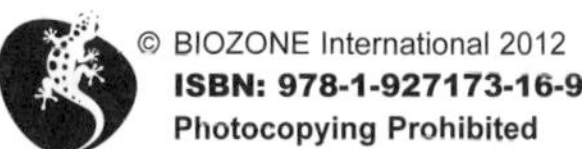

ISBN: 978-1-927173-16-9

Features of Fungi and Plants

Although plants are some of the most familiar organisms in our environment, their classification has not always been straightforward. They were once classified with fungi, although we know now that fungi are unique organisms that differ from other eukaryotes in their mode of nutrition, structural organization, growth, and reproduction. The plant kingdom is monophyletic, meaning that it is derived from a common ancestor. The variety we see in plant taxa today is a result of their enormous diversification from the first plants. Here we recognize four major taxa of the plant kingdom: Bryophyta (mosses and liverworts), Filicinophyta (ferns), Coniferophyta (conifers), and Angiospermophyta (angiosperms).

BOB
BOB
RA
RA

Lichen; Bracket fungus; Liverwort; Moss; Fern frond; Ground fern; Pine tree with female cones; Male pine cone; Coconut palms; Wheat plants; Deciduous tree; Flowering plant

1. Features of Kingdom **Fungi**: ____________________

2. Features of division **Bryophyta**: ____________________

3. Features of phylum **Filicinophyta**: ____________________

4. Features of phylum **Coniferophyta**: ____________________

5. Features of **Angiospermophyta**: ____________________

   (a) Features of monocots: ____________________

   (b) Features of dicots: ____________________

Related activities: *Features of Taxonomic Groups*

Periodicals: *World flowers bloom after recount*

ISBN: 978-1-927173-16-9

Features of Animal Taxa

The animal kingdom is classified into about 35 major **phyla**. Representatives of the more familiar taxa are illustrated below: **cnidarians** (jellyfish, sea anemones, corals), **annelids** (segmented worms), **arthropods** (insects, crustaceans, spiders, scorpions, centipedes, millipedes), **mollusks** (snails, bivalve shellfish, squid, octopus), **echinoderms** (starfish, sea urchins), **vertebrates** from the phylum **chordates** (fish, amphibians, reptiles, birds, mammals). The **arthropods** and the **vertebrates** have been represented in more detail, giving the **classes** for these **phyla**. This activity asks you to describe the **distinguishing features** of each of the taxa below. Underline the feature you think is most important in distinguishing the taxon from others.

Sea anemones

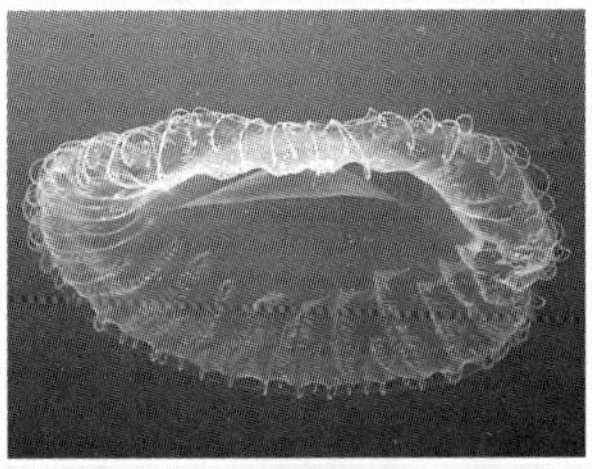

Jellyfish

Tubeworms

Earthworm

BOB

Long-horned beetle

Butterfly

Crab

Woodlouse

Scorpion

Spider

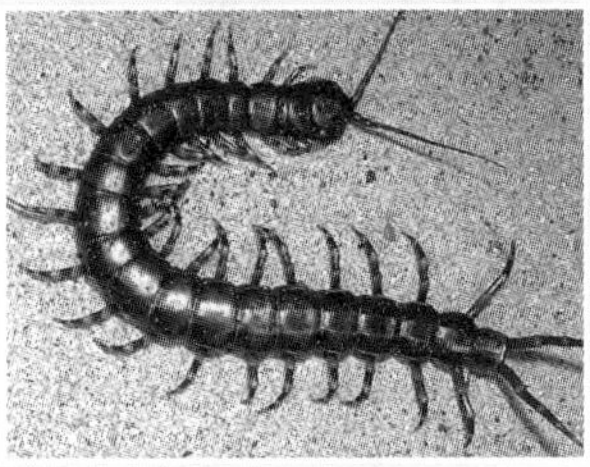

Centipede (Chilopoda)

Millipede (Diplopoda)

1. Features of phylum **Cnidaria**:

2. Features of phylum **Annelida**:

3. Features of phylum **Arthropoda**:

   (a) Features of class **Insecta**:

   (b) Features of class **Crustacea**:

   (c) Features of class **Arachnida**:

   (d) Features of classes **Chilopoda** and **Diplopoda**

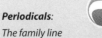

ISBN: 978-1-927173-16-9

Periodicals:
The family line

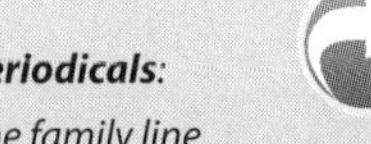

Related activities: *Features of Taxonomic Groups*

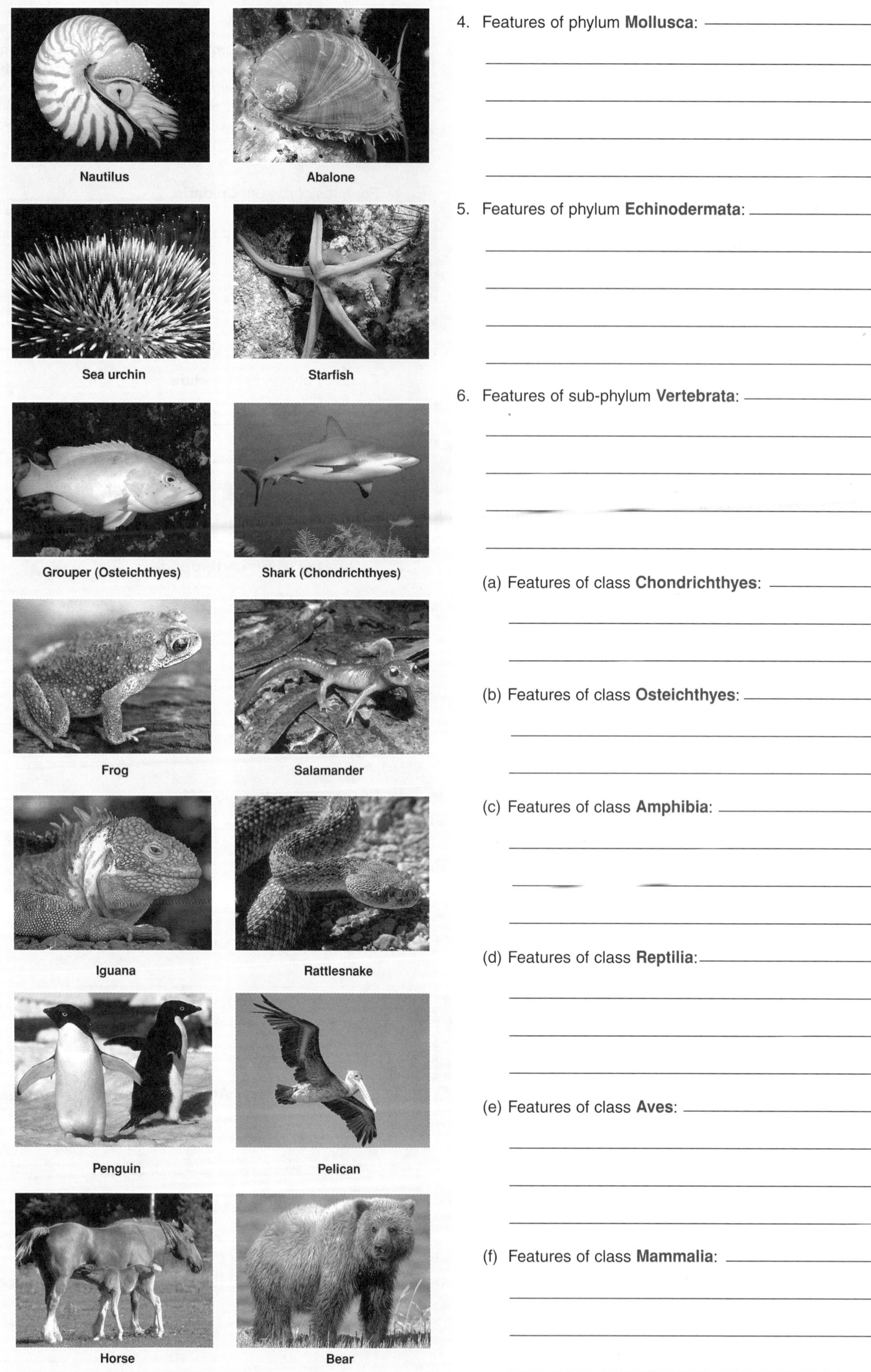

Nautilus | Abalone | Sea urchin | Starfish | Grouper (Osteichthyes) | Shark (Chondrichthyes) | Frog | Salamander | Iguana | Rattlesnake | Penguin | Pelican | Horse | Bear

4. Features of phylum **Mollusca**: ______________________________

5. Features of phylum **Echinodermata**: ______________________________

6. Features of sub-phylum **Vertebrata**: ______________________________

   (a) Features of class **Chondrichthyes**: ______________________________

   (b) Features of class **Osteichthyes**: ______________________________

   (c) Features of class **Amphibia**: ______________________________

   (d) Features of class **Reptilia**: ______________________________

   (e) Features of class **Aves**: ______________________________

   (f) Features of class **Mammalia**: ______________________________

ISBN: 978-1-927173-16-9

Classification Keys

Classification systems provide biologists with a way in which to identify species. They also indicate how closely related, in an evolutionary sense, each species is to others. An organism's classification should include a clear, unambiguous **description**, an accurate **diagram**, and its unique name, denoted by the **genus** and **species**. Classification keys are used to identify an organism and assign it to the correct species (assuming that the organism has already been formally classified and is included in the key). Typically, keys are **dichotomous** and involve a series of linked steps. At each step, a choice is made between two features; each alternative leads to another question until an identification is made. If the organism cannot be identified, it may be a new species or the key may need revision. Two examples of **dichotomous keys** are provided here. The first (below) describes features for identifying the larvae of various genera within the order Trichoptera (caddisflies). From this key you should be able to assign a generic name to each of the caddisfly larvae pictured. The key on the next page identifies aquatic insect orders.

Classification Key for Caddisfly Larvae

The key shown here is a simplified version of one commonly used to identify caddisfly larvae. It identifies the organisms to genus level only. To use the key to identify the larvae pictured below, start at the top and branch at each feature until you reach the bottom.

Caddisfly Larvae
Larvae with portable case
Larvae not in transparent case
Straight case, not spirally coiled
Case made of plant or mineral fragments
Larvae without portable case
Abdominal gill tufts
Abdominal gill tufts absent
Small larvae in transparent case
Case spirally coiled
Case of mineral fragments
Case of plant fragments
Case of smooth secreted material
Genus: Aoteapsyche
Genus: Hydrobiosis
Genus: Oxyethira
Genus: Helicopsyche
Genus: Hudsonema
Genus: Triplectides
Genus: Olinga

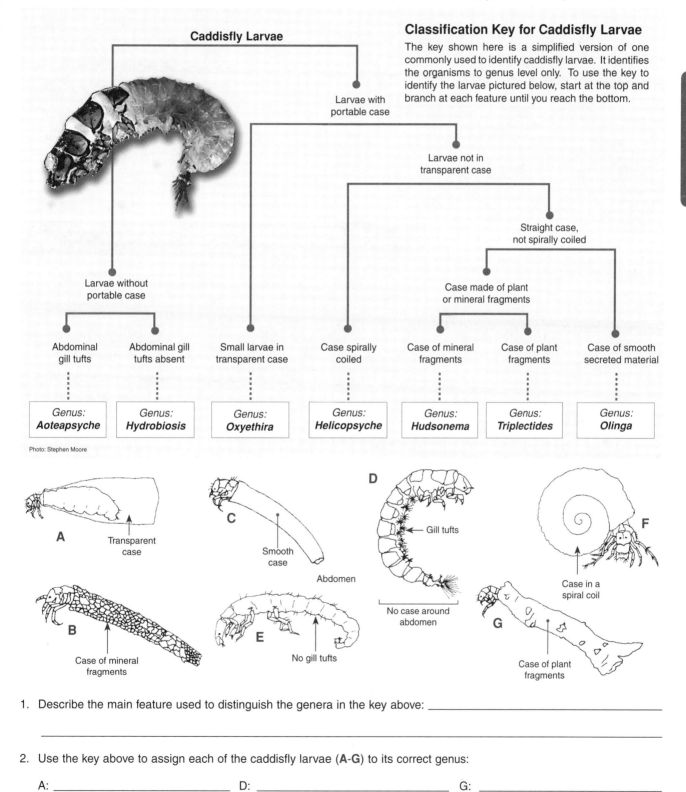

Photo: Stephen Moore

A
Transparent case
C
Smooth case
D
Gill tufts
No case around abdomen
F
Case in a spiral coil
B
Case of mineral fragments
Abdomen
E
No gill tufts
G
Case of plant fragments

1. Describe the main feature used to distinguish the genera in the key above: ______________________

____

2. Use the key above to assign each of the caddisfly larvae (**A-G**) to its correct genus:

A: ____________ D: ____________ G: ____________

B: ____________ E: ____________

C: F:

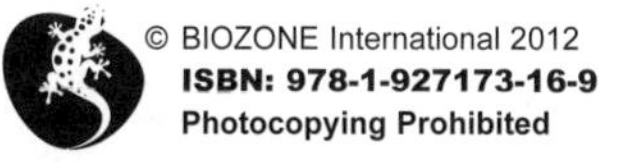

ISBN: 978-1-927173-16-9

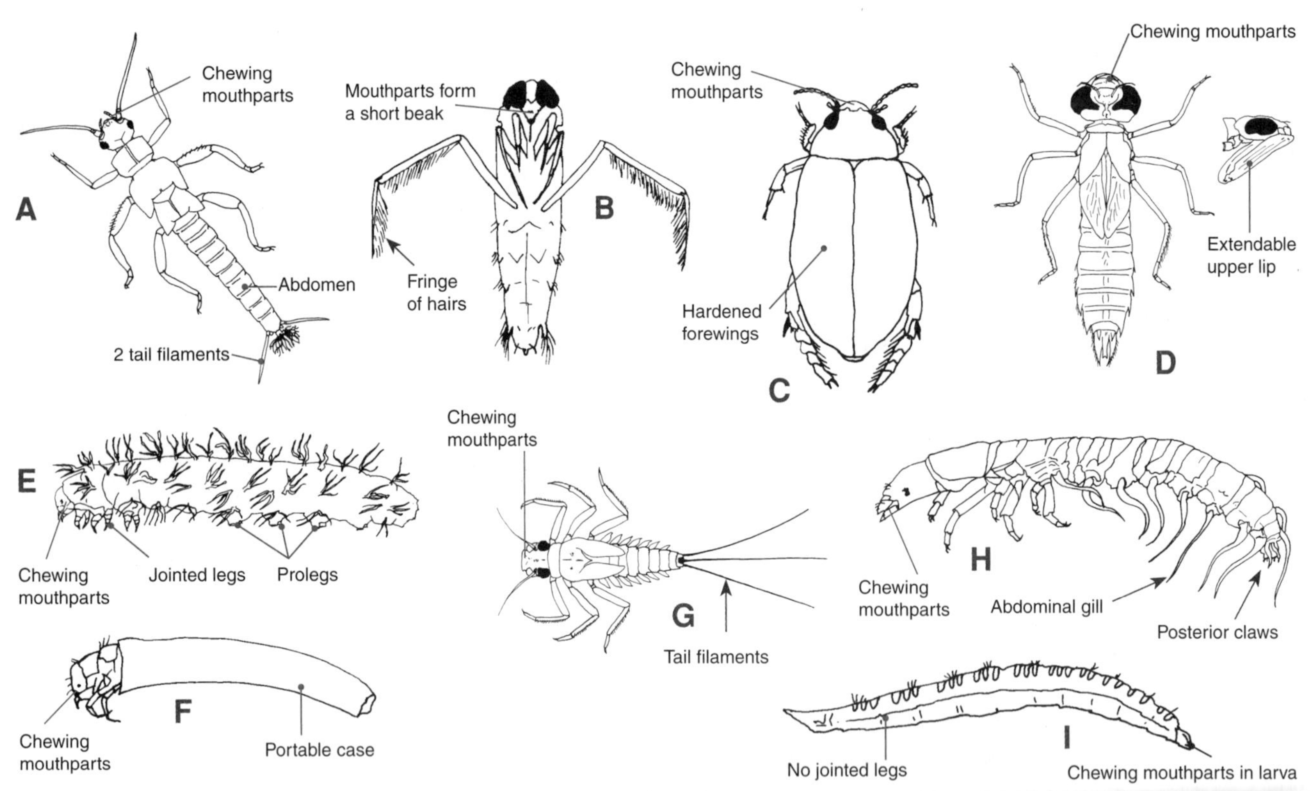

Chewing mouthparts
A
Abdomen
2 tail filaments
Mouthparts form a short beak
B
Fringe of hairs
Chewing mouthparts
Hardened forewings
C
Chewing mouthparts
Extendable upper lip
D
E
Chewing mouthparts
Jointed legs
Prolegs
Chewing mouthparts
G
Tail filaments
H
Chewing mouthparts
Abdominal gill
Posterior claws
F
Chewing mouthparts
Portable case
I
No jointed legs
Chewing mouthparts in larva

2. Use the simplified key to identify each of the orders (by order or common name) of aquatic insects (**A-I**) pictured above:

(a) Order of insect A: ____________________

(b) Order of insect B: ____________________

(c) Order of insect C: ____________________

(d) Order of insect D: ____________________

(e) Order of insect E: ____________________

(f) Order of insect F: ____________________

(g) Order of insect G: ____________________

(h) Order of insect H: ____________________

(i) Order of insect I: ____________________

Key to Orders of Aquatic Insects		
1	Insects with chewing mouthparts; forewings are hardened and meet along the midline of the body when at rest (they may cover the entire abdomen or be reduced in length).	**Coleoptera** (beetles)
	Mouthparts piercing or sucking and form a pointed cone With chewing mouthparts, but without hardened forewings	*Go to 2* *Go to 3*
2	Mouthparts form a short, pointed beak; legs fringed for swimming or long and spaced for suspension on water.	**Hemiptera** (bugs)
	Mouthparts do not form a beak; legs (if present) not fringed or long, or spaced apart.	*Go to 3*
3	Prominent upper lip (labium) extendable, forming a food capturing structure longer than the head.	**Odonata** (dragonflies & damselflies)
	Without a prominent, extendable labium	*Go to 4*
4	Abdomen terminating in three tail filaments which may be long and thin, or with fringes of hairs.	**Ephemeroptera** (mayflies)
	Without three tail filaments	*Go to 5*
5	Abdomen terminating in two tail filaments	**Plecoptera** (stoneflies)
	Without long tail filaments	*Go to 6*
6	With three pairs of jointed legs on thorax	*Go to 7*
	Without jointed, thoracic legs (although non-segmented prolegs or false legs may be present).	**Diptera** (true flies)
7	Abdomen with pairs of non-segmented prolegs bearing rows of fine hooks.	**Lepidoptera** (moths and butterflies)
	Without pairs of abdominal prolegs	*Go to 8*
8	With eight pairs of finger-like abdominal gills; abdomen with two pairs of posterior claws.	**Megaloptera** (dobsonflies)
	Either, without paired, abdominal gills, or, if such gills are present, without posterior claws.	*Go to 9*
9	Abdomen with a pair of posterior prolegs bearing claws with subsidiary hooks; sometimes a portable case.	**Trichoptera** (caddisflies)

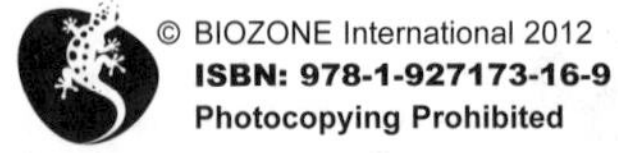

ISBN: 978-1-927173-16-9

Keying Out Plant Species

Dichotomous keys are a useful tool in biology and can enable identification to the species level provided the characteristics chosen are appropriate for separating species. Keys are extensively used by botanists as they are quick and easy to use in the field, although they sometimes rely on the presence of particular plant parts such as fruits or flowers. Some also require some specialist knowledge of plant biology. The following simple activity requires you to identify five species of the genus *Acer* from illustrations of the leaves. It provides valuable practice in using characteristic features to identify plants to species level.

A Dichotomous Key to Some Common Maple Species

1a Adult leaves with five lobes 2

1b Adult leaves with three lobes 4

2a Leaves 7.5-13 cm wide, with smooth edges, lacking serrations along the margin. U shaped sinuses between lobes.

Sugar maple, *Acer saccharum*

2b Leaves with serrations (fine teeth) along the margin 3

3a Leaves 5-13 cm wide and deeply lobed.

Japanese maple, *Acer palmatum*

3b Leaves 13-18 cm wide and deeply lobed.

Silver maple, *Acer saccharinum*

4a Leaves 5-15 cm wide with small sharp serrations on the margins. Distinctive V shaped sinuses between the lobes.

Red maple, *Acer rubrum*

4b Leaves 7.5-13 cm wide without serrations on the margins. Shallow sinuses between the lobes.

Black maple, *Acer nigrum*

1. Use the dichotomous key to the common species of *Acer* to identify the species illustrated by the leaves (drawn to scale). Begin at the top of the key and make a choice as to which of the illustrations best fits the description:

   (a) Species A: ____________________

   (b) Species B: ____________________

   (c) Species C: ____________________

   (d) Species D: ____________________

   (e) Species E: ____________________

2. Identify a feature that could be used to identify maple species when leaves are absent: ____________________

3. Suggest why it is usually necessary to consider a number of different features in order to classify plants to species level:

   ____

4. When identifying a plant, suggest what you should be sure of before using a key to classify it to species level:

   ____

ISBN: 978-1-927173-16-9

KEY TERMS Crossword

Complete the crossword below, which will test your understanding of key terms in this chapter and their meanings

1 2 3 4 5 6 7 8 9 10 11 12 13 14 15 16 17 18 19 20

Clues Across

2. The largest, least specific unit of classification within a kingdom.
4. The largest most general division of living things.
9. One of the larger units of classification, e.g mammals.
11. Phylum of primitive animals. Bodies have a gastrovascular cavity and are composed mainly of a jelly-like substance.
13. Kingdom of multicellular organisms that use chlorophyll to harness light to produce carbohydrates.
16. A unit of classification. A group of related species.
17. Sub-phylum of the kingdom Animalia. Possess backbones, normally made of bone, that protect a dorsal nerve cord.
19. The plant phylum comprising the flowering plants.
20. Group of multicellular organisms that feed off other living organisms. Cells do not possess cell walls.

Clues Down

1. Phylum characterized by the presence of a fleshy, muscular foot and mantle. Most have shells, although in some this may be internal.
2. Phylum with name meaning "pore bearer". Simple multicellular organisms which lack any organs, nervous tissue, digestive system, or circulatory system.
3. Flatworms. Unsegmented, soft bodied bilaterian invertebrates that lack body cavities.
5. Key that gives two options at each step of the identification process. (2 words: 11, 3)
6. Unit of classification used to group together related families.
7. One of the more specific units of classification. A group of related genera.
8. Phylum of cone bearing vascular seed plants.
10. The smallest, most precise unit of classification used. A closely related group of organisms able to breed together.
12. Phylum whose name literally means "jointed foot".
14. Plant division of non-vascular plants.
15. Common name for members of the plant phylum Filiciniophyta.
18. Segmented worms. Includes earthworms, ragworms, and leeches.

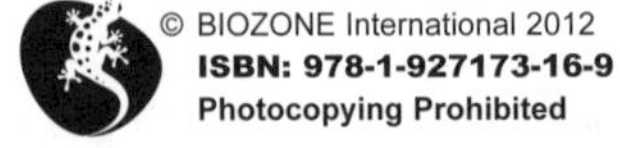

ISBN: 978-1-927173-16-9

Human Health and Physiology (Core)

Digestion	• The role of digestion • The human digestive tract • Digestion, absorption, and transport
The Transport System	• The human circulatory system • The heart and blood vessels • The function of blood
Defense Against Disease	• Pathogens and antibiotics • The first line of defence • Antigens and antibodies • HIV and AIDS
Gas Exchange	• Principles of gas exchange • The human ventilation system
Nerves, Hormones, and Homeostasis	• Structure of the nervous system • Neurons and nerve impulses • Homeostatic mechanisms
Reproduction	• Human reproductive systems • The menstrual cycle • *In vitro* fertilization

Human Health and Physiology (AHL)

Defense Against Disease	• Blood clotting • Clonal selection • Antibody production • Monoclonal antibodies • Vaccination
TOK	*How do we assess the risks associated with vaccination? How are facts communicated?*
Muscles and Movement	• The mechanics of movement • Joint structure and function • Structure and function of muscle
Nerves, Hormones, and Homeostasis	• Gross structure of the kidney • Nephron structure and function • Regulation of urine volume
Reproduction	• Gametogenesis • Gamete structure and fertilization • Pregnancy and the placenta • Fetal development and birth

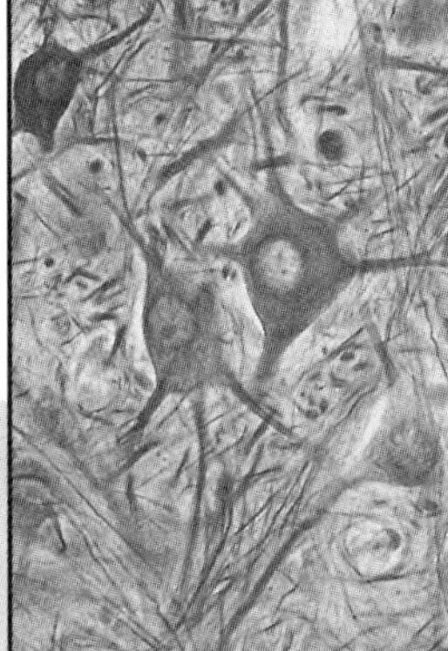

Human body systems interact and exchange materials with the environment to maintain dynamic homeostasis, defend the body against pathogens, and enable reproduction.

Excretion, internal defence, and movement are essential to homeostasis and response to stimuli. Reproduction involves a sequence of events ending in birth.

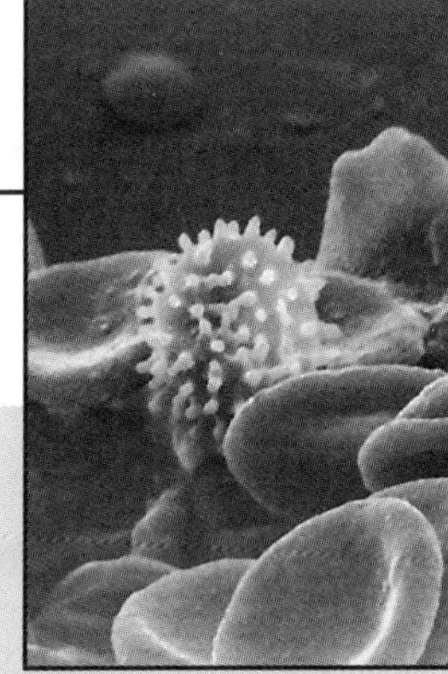

Human Health and Physiology and Plant Science

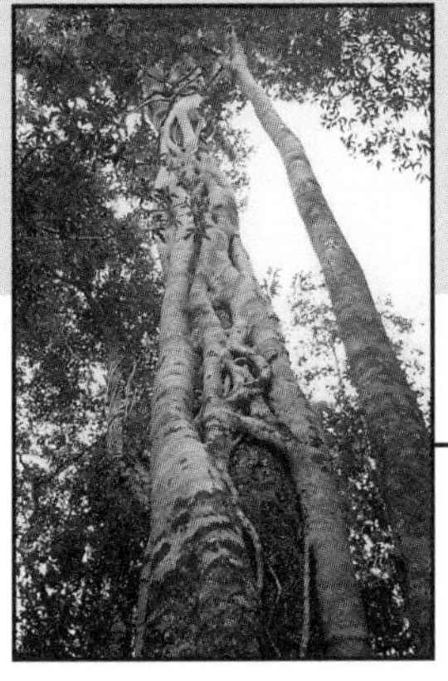

Plants have systems that exchange materials with, and respond to, the environment. Dicot reproduction involves production and germination of seeds.

IB Options CD-ROM for separate purchase include learning objectives and activities to support all the IB options

BIOZONE
IB
BIOLOGY
IB Options
CD-ROM

Plant Science (AHL)

Plant structure and growth	• The structure of plants • Monocots vs dicots • Dicot leaf structure and function • Leaf, stem, and root modifications • Meristems and plant growth • Auxins and phototropism
Transport in angiosperms	• Roots and water and mineral uptake • Plant support • Transpiration and xylem tissue • Translocation and phloem • Adaptations of xerophytes
Reproduction in angiosperms	• Dicot flower structure • Pollination and fertilization • Seed structure and dispersal • Seed germination • Phytochrome and control of flowering

Options on IB CROM

A: Human Nutrition and Health (SL)
B: Physiology of Exercise (SL)
C: Cells and Energy (SL)
D: Evolution (SL/HL)
E: Neurobiology and Behavior (SL/HL)
F: Microbes and Biotechnology (SL/HL)
G: Ecology and Conservation (SL/HL)
H: Further Human Physiology (HL)

Important in this section...

- *Develop understanding of human structure & function*
- *Develop understanding of plant structure & function*
- *Explain how physiological systems are regulated*
- *Explain how physiological systems are interrelated*

Core Topic
6.1

Digestion

Key concepts

- Digestion of food molecules is essential to make nutrient available for assimilation.
- Digestive secretions contain enzymes responsible for breaking down specific food groups.
- The digestive tract is regionally specialized to maximize the efficiency of digestion.
- Nutrients must be absorbed before they can be assimilated. The structure of intestinal villi maximizes absorption across the gut wall.

Key terms

Core

absorption
amylase
assimilation
digestion
digestive tract
egestion
enzyme
esophagus
extracellular digestion
gall bladder
gut
intestinal villi
intracellular digestion
large intestine
liver
lipase
optimum pH
protease
small intestine
stomach
substrate

Learning Objectives

☐ 1. Use the **KEY TERMS** to compile a glossary for this topic.

Role of Digestion *(6.1.1-6.1.3)* page 268

☐ 2. Define the term digestion and explain why large food molecules kust be digested.

☐ 3. Distinguish between **intracellular** and **extracellular** digestion. Explain the need for **enzymes** in digestion and appreciate that digestive enzymes are released in digestive secretions at specific regions in the **gut**.

☐ 4. Distinguish between **amylase**, **protease**, and **lipase** enzymes. State the source, **substrate**, and **optimum pH** for one amylase, one protease, and one lipase enzyme.

Structure of the Digestive Tract *(6.1.4-6.1.6)* pages 270-274

☐ 5. Draw and label a diagram of the human digestive system. Identify **mouth**, **esophagus**, **stomach**, **small intestine**, **large intestine**, **anus**, **liver**, **pancreas**, and **gall bladder**. The diagram should indicate the interconnections between these different regions.

☐ 6. Outline the function of the stomach, small intestine, and large intestine. Include reference to the pH environment of each region, digestive secretions, structural features, primary role and any secondary roles.

Absorption and Transport of Nutrients *(6.1.7)* pages 274-275

☐ 7. Distinguish between **absorption** and **assimilation**.

☐ 8. Describe the structure of an **intestinal villus** and explain how the structure is related to its functional role in absorprtion and transport of the products of digestion.

Periodicals:
Listings for this chapter are on page 400

Weblinks:
www.thebiozone.com/weblink/IB-3169.html

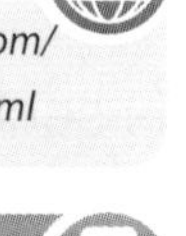

BIOZONE APP:
Student Review Series
Digestive System

The Role of the Digestive System

Nutrients are substances required by the body for energy, metabolism, and tissue growth and repair. Nutrients occur in food, which must be broken down by mechanical and chemical processes before the nutrients can be absorbed into the blood and utilized by the body. Digestion in most animals, including humans, is **extracellular**, meaning it occurs outside the cells themselves and it is the breakdown products that are absorbed by the cells. Digestion is the role of the digestive system.

The Digestive System and Stages of Digestion

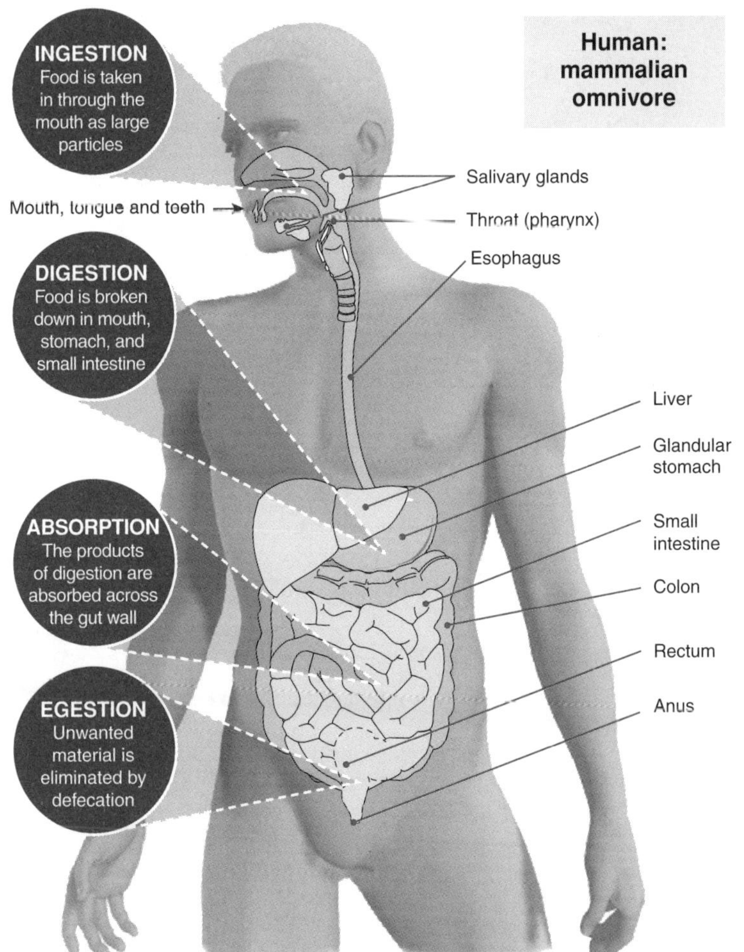
INGESTION
Food is taken in through the mouth as large particles
Mouth, tongue and teeth
DIGESTION
Food is broken down in mouth, stomach, and small intestine
ABSORPTION
The products of digestion are absorbed across the gut wall
EGESTION
Unwanted material is eliminated by defecation
Human: mammalian omnivore
Salivary glands
Throat (pharynx)
Esophagus
Liver
Glandular stomach
Small intestine
Colon
Rectum
Anus

Before the body can use the food we eat, it must be broken down into smaller components that can be absorbed across the intestinal wall. The breakdown of proteins, fats, and carbohydrates is achieved by enzymes and mechanical processes (such as chewing).

ELITE FUEL
SUPPLY
SQUEEZEPACKS
PLUS CAFFEINE
BALANCE
LEMON CHILL FLAVOUR

Some foods are specially designed to be quickly absorbed (e.g. sports gels and drinks). Energy foods (such as the one shown above) contain a mixture of simple monomer molecules (e.g. monosaccharides) for quick absorption and larger polymer molecules (e.g. polysaccharides) for longer lasting energy release.

Enzymes in the Digestive System

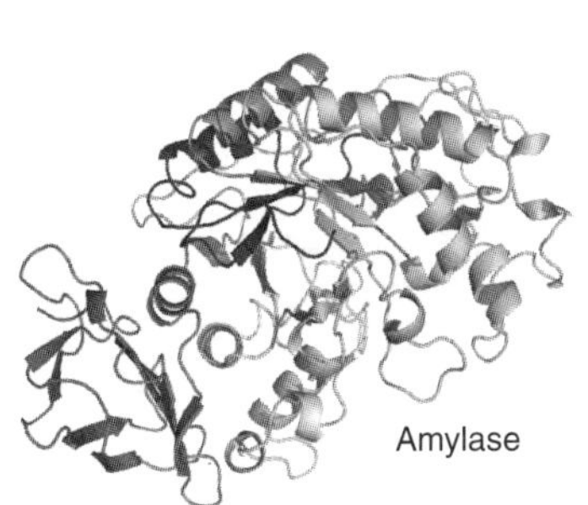
Amylase

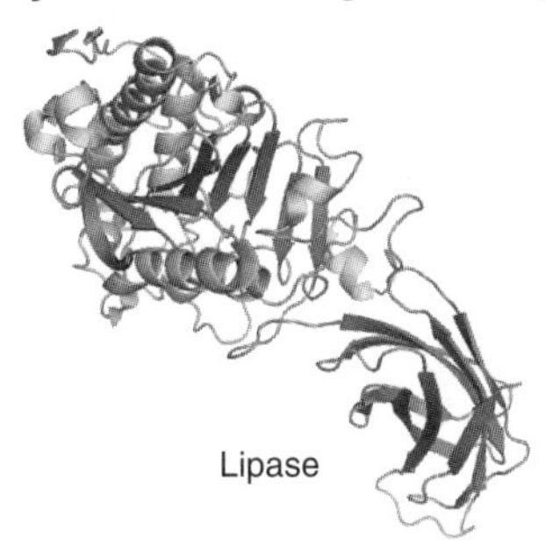
Lipase

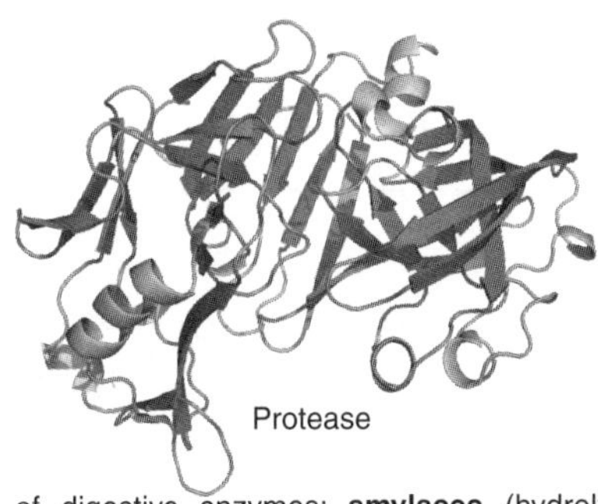
Protease

Enzymes play a key role in the digestive of food. They increase the speed of digestion by catalyzing the breakdown of food polymers (e.g. protein) into smaller monomers (e.g. amino acids) that can be absorbed by the intestinal villi are produced.. There are three main types of digestive enzymes; **amylases** (hydrolyze carbohydrates), **proteases** (hydrolyze protein or peptides), and **lipases** (hydrolyze lipids).

1. (a) How is food broken down during the digestive process? ______________________

____

(b) Why must large food molecules be broken down into smaller molecules? ______________________

____

2. Explain the role of enzymes in the digestive system: ______________________

____

____

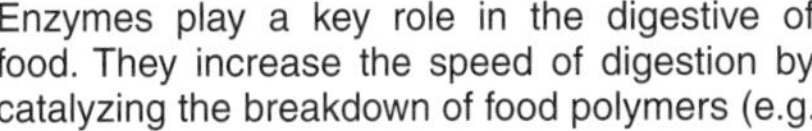

ISBN: 978-1-927173-16-9

Related activities: *Enzymes, Enzyme Reaction Rates, The Human Digestive Tract*

The Human Digestive Tract

An adult consumes an estimated metric tonne of food a year. Food provides the source of the energy required to maintain **metabolism**. The human digestive tract, like most tube-like guts, is regionally specialized to maximize the efficiency of physical and chemical breakdown (**digestion**), **absorption**, and **elimination**. The gut is essentially a hollow, open-ended, muscular tube, and the food within it is essentially outside the body, having contact only with the cells lining the tract. Food is physically moved through the gut tube by waves of muscular contraction, and subjected to chemical breakdown by enzymes contained within digestive secretions. The products of this breakdown are then absorbed across the gut wall. A number of organs are associated with the gut along its length and contribute, through their secretions, to the digestive process at various stages.

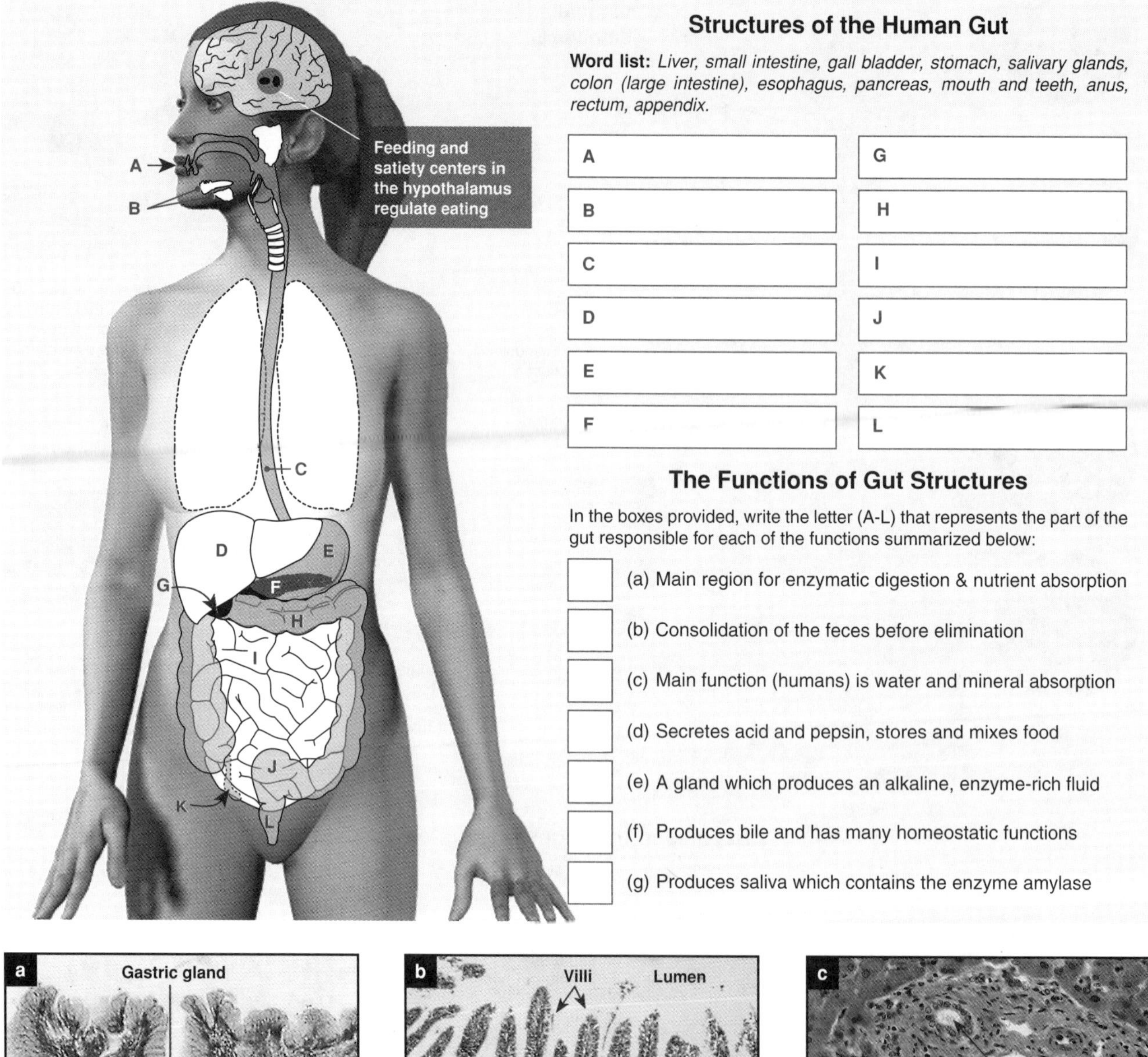
Feeding and satiety centers in the hypothalamus regulate eating
A
B
C
D
E
F
G
H
I
J
K
L

Structures of the Human Gut

Word list: *Liver, small intestine, gall bladder, stomach, salivary glands, colon (large intestine), esophagus, pancreas, mouth and teeth, anus, rectum, appendix.*

A	G
B	H
C	I
D	J
E	K
F	L

The Functions of Gut Structures

In the boxes provided, write the letter (A-L) that represents the part of the gut responsible for each of the functions summarized below:

- (a) Main region for enzymatic digestion & nutrient absorption
- (b) Consolidation of the feces before elimination
- (c) Main function (humans) is water and mineral absorption
- (d) Secretes acid and pepsin, stores and mixes food
- (e) A gland which produces an alkaline, enzyme-rich fluid
- (f) Produces bile and has many homeostatic functions
- (g) Produces saliva which contains the enzyme amylase

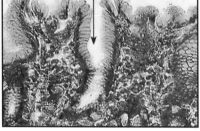

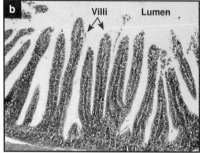

a
Gastric gland
EII
b
Villi
Lumen
EII
c
Bile ducts
EII

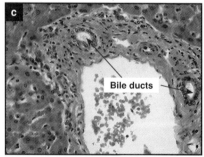

1. In the spaces provided on the diagram above, identify the parts labeled **A-L** (choose from the word list provided). Match each of the **functions** described (a)-(g) with the letter representing the corresponding structure on the diagram.
2. On the same diagram, mark with lines and labels: anal sphincter (**AS**), pyloric sphincter (**PS**), cardiac sphincter (**CS**).
3. Identify the region of the gut illustrated by the photographs (**a**)-(**c**) above. For each one, explain the identifying features:

(a) ______________________________

(b) ______________________________

(c) ______________________________

Related activities: *Digestion, Absorption, and Transport*
Web links: *Digestion Animation, Acid Secretion in the Stomach*

Periodicals: *The anatomy of digestion*

ISBN: 978-1-927173-16-9

The Stomach, Duodenum, and Pancreas

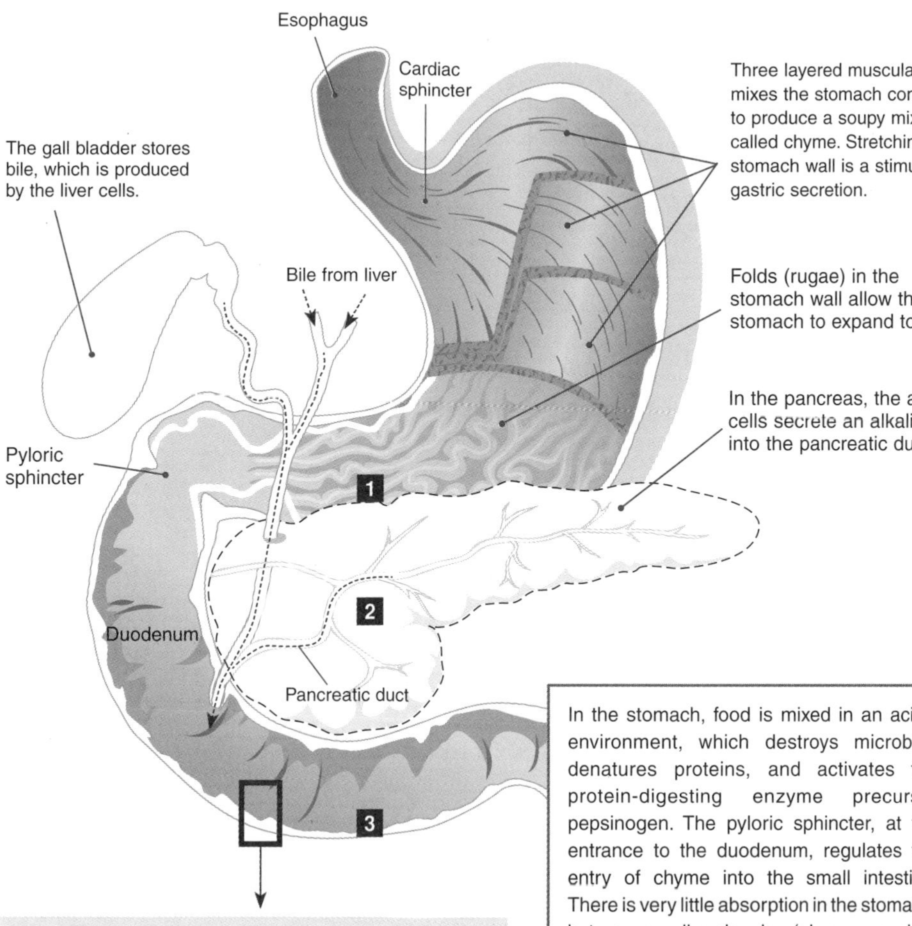
Esophagus
Cardiac sphincter
The gall bladder stores bile, which is produced by the liver cells.
Bile from liver
Pyloric sphincter
Duodenum
Pancreatic duct
1
2
3
Three layered muscular wall mixes the stomach contents to produce a soupy mixture called chyme. Stretching of the stomach wall is a stimulus for gastric secretion.
Folds (rugae) in the stomach wall allow the stomach to expand to 1L.
In the pancreas, the acinar cells secrete an alkaline fluid into the pancreatic duct.

Enzymes and their actions

1 Gastric juice

Acts in stomach (optimal pH)

Pepsin (1.5-2.0)

Protein → peptides

2 Pancreatic juice

Acts in duodenum (optimal pH)

1. Pancreatic amylase (6.7-7.0)
2. Trypsin (7.8-8.7)
3. Chymotrypsin (7.8)
4. Pancreatic lipase (8.0)

1. Starch → maltose
2. Protein → peptides
3. Protein → peptides
4. Fats → fatty acids & glycerol

3 Intestinal juice

Acts in small intestine (optimal pH)

1. Maltase (6.0-6.5)
2. Peptidases (~ 8.0)

1. Maltose → glucose
2. Polypeptides → amino acids

In the stomach, food is mixed in an acidic environment, which destroys microbes, denatures proteins, and activates the protein-digesting enzyme precursor pepsinogen. The pyloric sphincter, at the entrance to the duodenum, regulates the entry of chyme into the small intestine. There is very little absorption in the stomach, but very small molecules (glucose, aspirin, alcohol) are absorbed directly across the stomach wall into the gastric blood vessels surrounding the stomach.

Detail of a Villus (Small Intestine)

The **intestinal villi** project into the gut lumen and provide an immense surface area for nutrient absorption. The villi are lined with **epithelial cells** and each has a brush border of many **microvilli** which further increase the surface area.

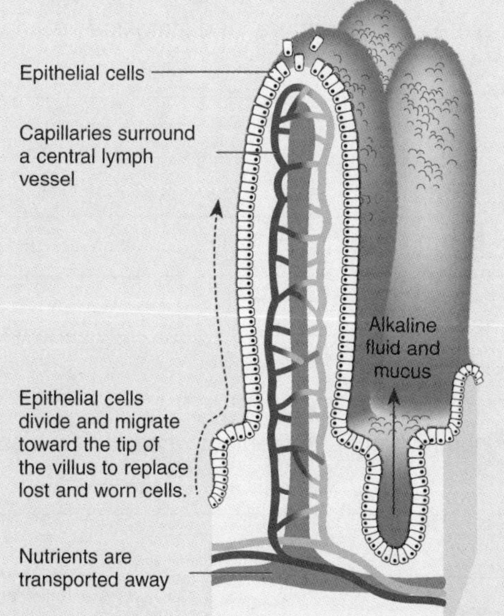
Epithelial cells
Capillaries surround a central lymph vessel
Epithelial cells divide and migrate toward the tip of the villus to replace lost and worn cells.
Alkaline fluid and mucus
Nutrients are transported away

Enzymes bound to the surfaces of the epithelial cells break down peptides and carbohydrate molecules. The breakdown products are then absorbed into the underlying blood and lymph vessels. Tubular exocrine glands and goblet cells secrete alkaline fluid and mucus into the lumen.

Detail of a Gastric Gland (Stomach Wall)

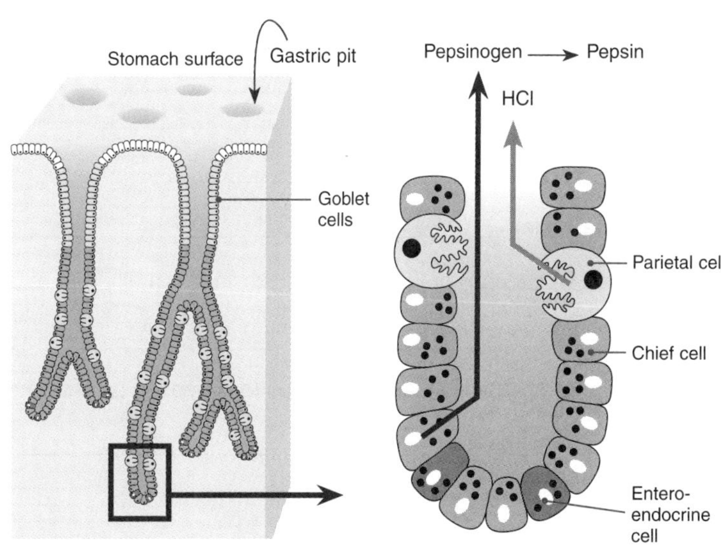
Stomach surface
Gastric pit
Goblet cells
Pepsinogen → Pepsin
HCl
Parietal cell
Chief cell
Entero-endocrine cell

Gastric secretions are produced by **gastric glands**, which pit the lining of the stomach. Chief cells in the gland secrete pepsinogen, a precursor of the enzyme pepsin. Parietal cells produce hydrochloric acid, which activates the pepsinogen. Goblet cells at the neck of the gastric gland secrete mucus to protect the stomach mucosa from the acid. Enteroendocrine cells in the gastric gland secrete the hormone gastrin which acts on the stomach to increase gastric secretion.

Digestion

ISBN: 978-1-927173-16-9

Web links: *Peristalsis and Gastric Emptying, Peristalsis Animation*

The Large Intestine

After most of the nutrients have been absorbed in the small intestine, the remaining fluid contents pass into the large intestine (appendix, cecum, and colon). The fluid comprises undigested or undigestible food, bacteria, dead cells, mucus, bile, ions, and water. In humans and other omnivores, the large intestine's main role is to reabsorb water and electrolytes.

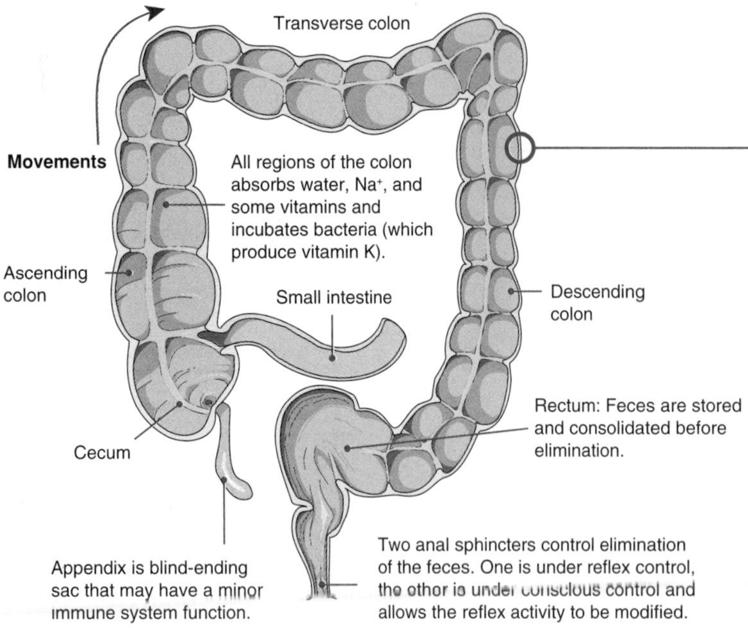

Transverse colon

Movements

All regions of the colon absorbs water, Na^+, and some vitamins and incubates bacteria (which produce vitamin K).

Ascending colon

Small intestine

Descending colon

Rectum: Feces are stored and consolidated before elimination.

Cecum

Appendix is blind-ending sac that may have a minor immune system function.

Two anal sphincters control elimination of the feces. One is under reflex control, the other is under conscious control and allows the reflex activity to be modified.

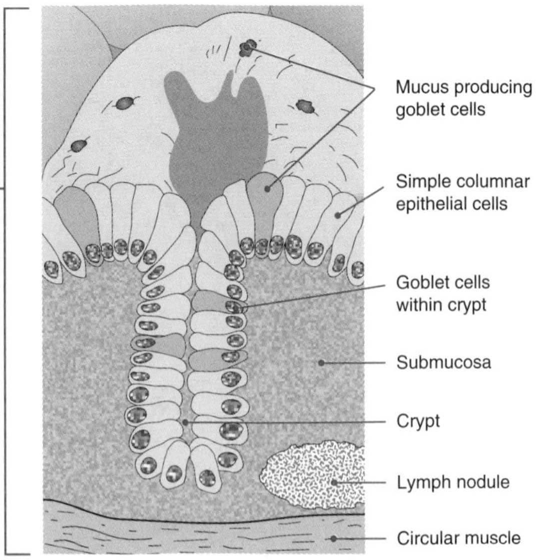

Mucus producing goblet cells

Simple columnar epithelial cells

Goblet cells within crypt

Submucosa

Crypt

Lymph nodule

Circular muscle

The wall of the large intestine is lined with simple columnar epithelium. The epithelium is contains tubular glands called crypts containing mucus-secreting goblet cells. The mucus lubricates the colon wall and aids formation of the feces.

4. Summarize the structural and functional specializations in each of the following regions of the gut:

(a) Stomach: ______

(b) Small intestine: ______

(c) Large intestine: ______

5. Identify two sites for enzyme secretion in the gut, give an example of an enzyme produced there, and state its role:

(a) Site: ______ Enzyme: ______

Enzyme's role: ______

(b) Site: ______ Enzyme: ______

Enzyme's role: ______

6. (a) Suggest why the pH of the gut secretions varies at different regions in the gut: ______

(b) Explain why protein-digesting enzymes (e.g. pepsin) are secreted in an inactive form and then activated after release: ______

7. (a) Describe how food is moved through the digestive tract: ______

(b) Explain how the passage of food through the tract is regulated: ______

8. (a) Predict the consequence of food moving too rapidly through the gut: ______

(b) Predict the consequence of food moving too slowly through the gut: ______

ISBN: 978-1-927173-16-9

The Digestive Role of the Liver

The liver is a large organ, weighing about 1.4 kg, and is well supplied with blood. It carries out several hundred different functions and has a pivotal role in the maintenance of homeostasis. Its role in the digestion of food centers around the production of the alkaline fluid, **bile**, which is secreted at a rate of 0.8-1.0 liter per day. It is also responsible for processing absorbed nutrients, which arrive at the liver via the hepatic portal system. These functions are summarized below.

Digestive Functions of the Liver

The digestive role of the liver is in the production of **bile**. Bile is a yellow, brown, or olive-green alkaline fluid (pH 7.6–8.6), consisting of water and bile salts, cholesterol, lecithin, bile pigments, and several ions. The bile salts are used in the small intestine to break up (**emulsify**) fatty molecules for easier digestion and absorption.

The high pH neutralizes the acid entering the small intestine from the stomach. Bile is also partly an excretory product; the breakdown of red blood cells in the liver produces the principal bile pigment, **bilirubin**. Bacteria act on the bile pigments, giving the brownish color to feces. The production and secretion of bile is regulated through nervous and hormonal mechanisms. Hormones (secretin and cholecystokinin) are released into the blood from the intestinal mucosa in response to the presence of food (especially fat) in the small intestine.

Internal Gross Structure of the Human Liver

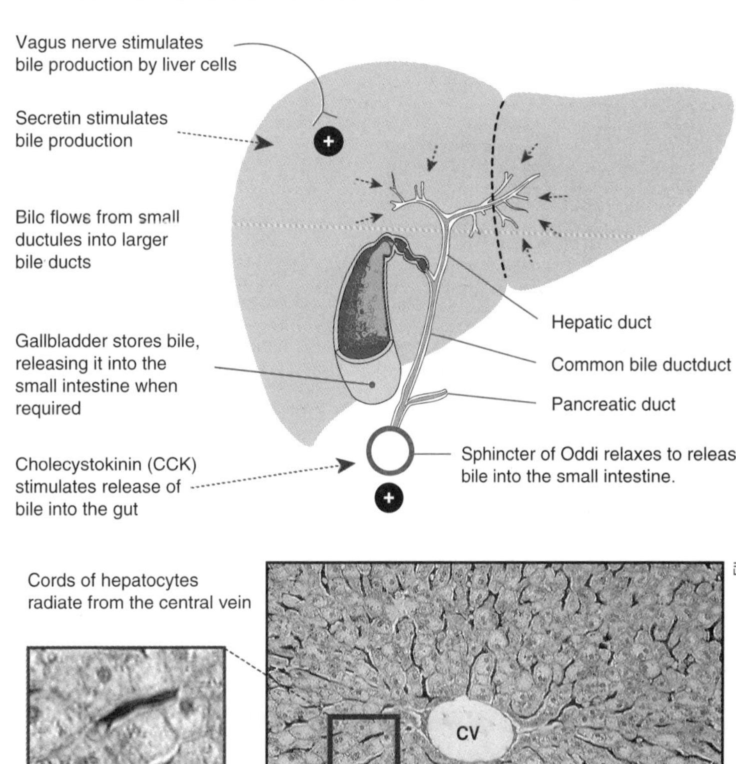

Vagus nerve stimulates bile production by liver cells
Secretin stimulates bile production
Bile flows from small ductules into larger bile ducts
Gallbladder stores bile, releasing it into the small intestine when required
Cholecystokinin (CCK) stimulates release of bile into the gut
Hepatic duct
Common bile ductduct
Pancreatic duct
Sphincter of Oddi relaxes to release bile into the small intestine.
Cords of hepatocytes radiate from the central vein
CV
Individual liver cells
EII

Liver Tissue

The liver tissue is made up of many lobules, each one comprising cords of liver cells (hepatocytes), radiating from a central vein (CV), and surrounded by branches of the hepatic artery, hepatic portal vein, and bile ductule. Bile is produced by the individual liver cells, which secrete it into canaliculi that empty into small bile ducts. The hepatocytes also process the nutrients entering the liver via the hepatic portal system.

1. The liver produces bile. Describe the two main functions of bile in digestion:

   (a) ______

   (b) ______

2. Describe the two primary functions of the liver related to the processing of digestion products arriving from the gut:

   (a) ______

   (b) ______

3. Explain the role of the gall bladder in digestion: ______

4. Describe in what way bile is an excretory product as well as a digestive secretion: ______

5. Name the two principal hormones controlling the production (secretion) and release of bile, and state the effect of each:

   (a) Hormone 1: ______ Effect: ______

   (b) Hormone 2: ______ Effect: ______

6. State the stimulus for hormonal stimulation of bile secretion: ______

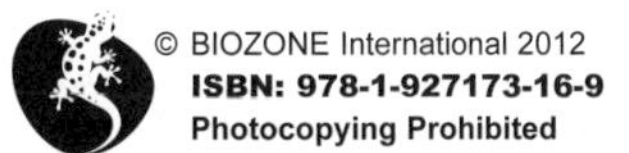

ISBN: 978-1-927173-16-9

Digestion, Absorption, and Transport

All the chemical and physical processes of digestion are aimed at the breakdown of food molecules into forms that can pass through the intestinal lining into the underlying blood and lymph vessels. Starch is broken down first into maltose and short chain carbohydrates such as dextrose, before being hydrolyzed to glucose (below). Breakdown products of other foodstuffs include amino acids (from proteins), and fatty acids, glycerol, and acylglycerols (from fats). The passage of these molecules from the gut into the blood or lymph is called **absorption**. Nutrients are then transported directly or indirectly to the liver for storage or processing. After they have been absorbed, nutrients can be assimilated, i.e incorporated into the substance of the body itself.

Digestion of Starch

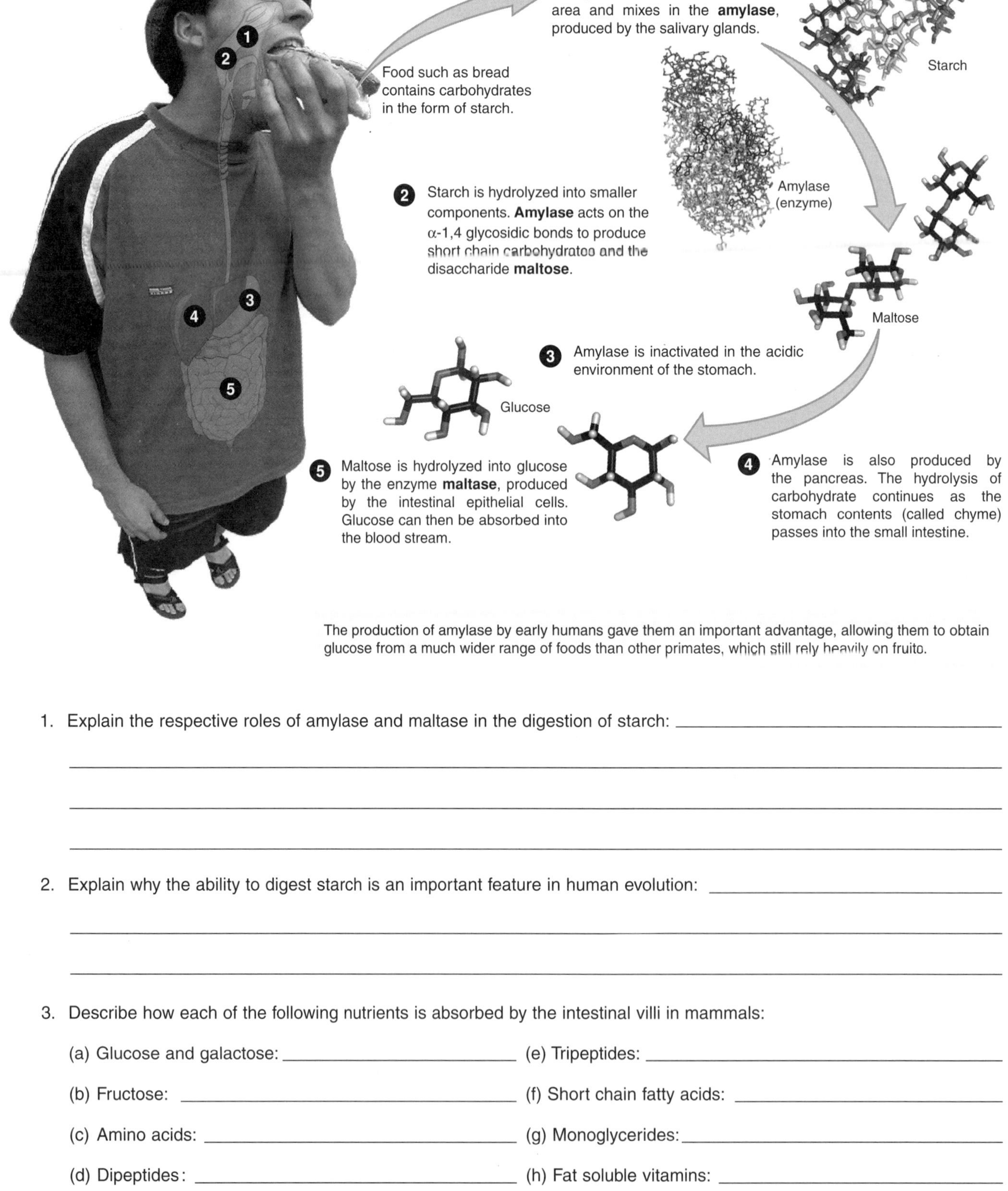

1 Starch digestion begins in the mouth. The **teeth** grind the solid mass, which increases its surface area and mixes in the **amylase**, produced by the salivary glands.
Starch
Food such as bread contains carbohydrates in the form of starch.
Amylase (enzyme)
2 Starch is hydrolyzed into smaller components. **Amylase** acts on the α-1,4 glycosidic bonds to produce short chain carbohydrates and the disaccharide **maltose**.
Maltose
3 Amylase is inactivated in the acidic environment of the stomach.
Glucose
4 Amylase is also produced by the pancreas. The hydrolysis of carbohydrate continues as the stomach contents (called chyme) passes into the small intestine.
5 Maltose is hydrolyzed into glucose by the enzyme **maltase**, produced by the intestinal epithelial cells. Glucose can then be absorbed into the blood stream.

The production of amylase by early humans gave them an important advantage, allowing them to obtain glucose from a much wider range of foods than other primates, which still rely heavily on fruits.

1. Explain the respective roles of amylase and maltase in the digestion of starch: ______

2. Explain why the ability to digest starch is an important feature in human evolution: ______

3. Describe how each of the following nutrients is absorbed by the intestinal villi in mammals:

(a) Glucose and galactose: ______

(b) Fructose: ______

(c) Amino acids: ______

(d) Dipeptides: ______

(e) Tripeptides: ______

(f) Short chain fatty acids: ______

(g) Monoglycerides: ______

(h) Fat soluble vitamins: ______

Related activities: The Human Digestive Tract

Periodicals: The anatomy of digestion

ISBN: 978-1-927173-16-9

Nutrient Absorption by Intestinal Villi

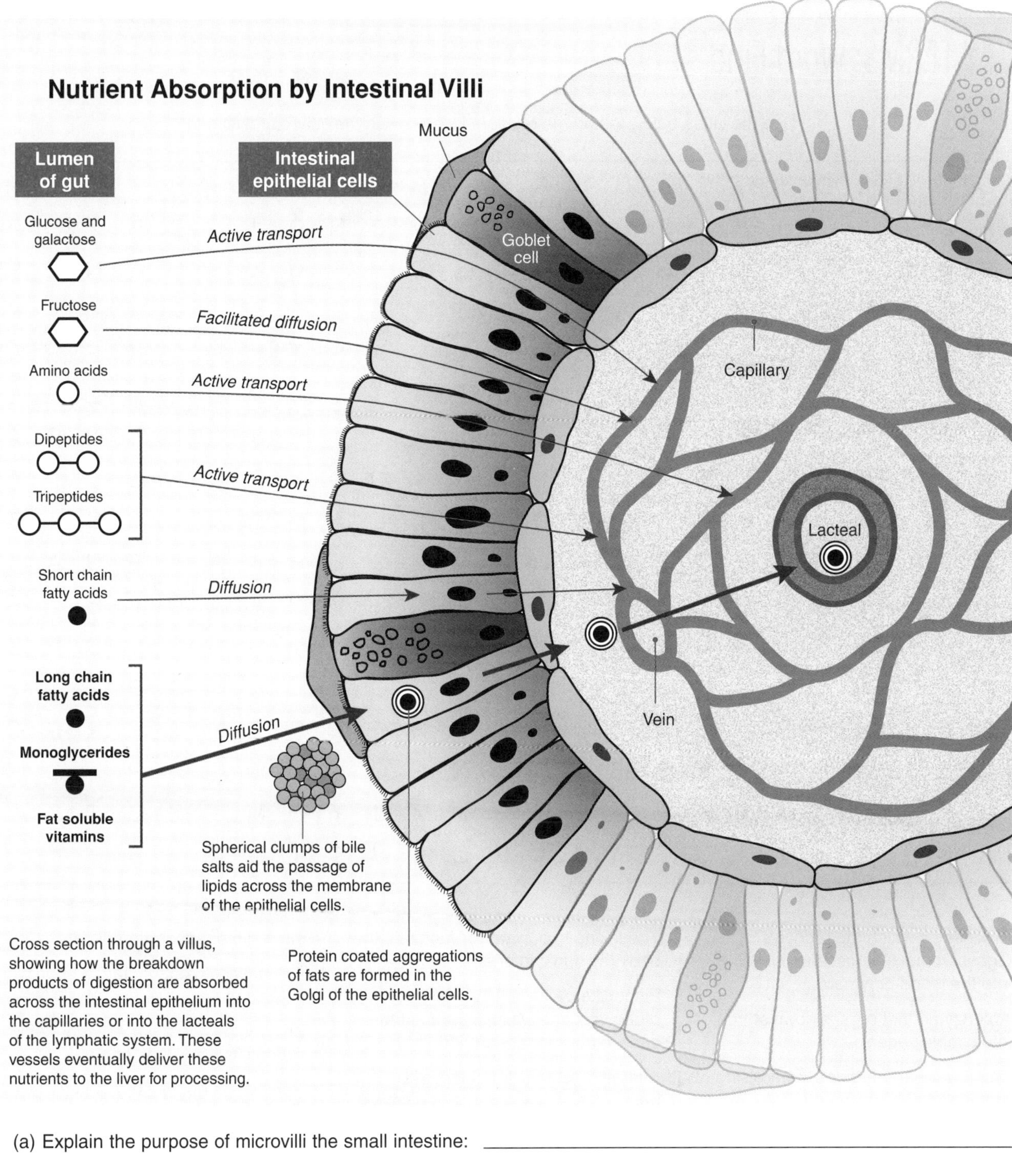

Lumen of gut
Intestinal epithelial cells
Mucus
Goblet cell
Glucose and galactose
Active transport
Fructose
Facilitated diffusion
Amino acids
Active transport
Dipeptides
Tripeptides
Active transport
Short chain fatty acids
Diffusion
Long chain fatty acids
Monoglycerides
Fat soluble vitamins
Diffusion
Capillary
Lacteal
Vein
Spherical clumps of bile salts aid the passage of lipids across the membrane of the epithelial cells.
Protein coated aggregations of fats are formed in the Golgi of the epithelial cells.

Cross section through a villus, showing how the breakdown products of digestion are absorbed across the intestinal epithelium into the capillaries or into the lacteals of the lymphatic system. These vessels eventually deliver these nutrients to the liver for processing.

4. (a) Explain the purpose of microvilli the small intestine: ________________

   (b) What is the advantage of increasing surface area in the small intestine? ________________

   (c) What other features of the small intestine help to increase its surface area? ________________

5. Explain what must happen to long chain fatty acids, monoglycerides and fat soluble vitamins before they can be absorbed into the blood:

ISBN: 978-1-927173-16-9

KEY TERMS: Word Find

Use the clues below to find the relevant key terms in the WORD FIND grid

```
Q N U L D J B Z N U V R K Y B H I N G E S T I O N
A D A A I Q R J Z D E N Z Y M E N D J G D T Z G P
E D B R G D D D S P O O J S S H T U N Y U W N N R
S Q S G E A M Y L A S E X I V G E E I W G J M V O
O Y O E S O Q N Q D B D Y U G F S T K D A K F M T
P R R I T Z G D P P Y H W F J I T R M H L H Z K E
H S P N I A R L G F O J I H K L I S J B L D N D A
A C T T O A S S I M I L A T I O N X P Q B S D F S
G L I E N X H G S P W C N O F E A M R D L A S J E
U P O S F F Q Q P X A R K V H J L H A A A F U A K
S C N T B O N Z X L E S Q Q E I V H A A D F B Z R
E C Q I J F Z L T V Z Y E A N U I T O H D C S R S
T E B N Y C V G I Z K Q P Y J G L L W V D R T I F
H K W E W R J L S T O M A C H E L S F Y E R R D D
X S M A L L I N T E S T I N E O I M Z Q R R A D K
W S B Z M T R K E L W T B H Z J Y D M I Z B T G Y
N K V M O I A U F P N A E G E S T I O N V B E Y L
```

Process by which the products of digestion move across the gut lining into the blood or lymph.

An enzyme that breaks down starch into simple sugars.

The lower part of the gut comprising of the appendix, cecum and colon.

A globular protein which acts as a catalyst to speed up a specific biological reaction.

The muscular tube through which food passes from the mouth to the stomach.

Small organ located near the liver which aids in the digestion of fats by adding bile (produced by the liver) to the food (chime) as it passes through the small intestine.

The taking in of water or food into the body (by drinking or eating).

Fingerlike projections lining the surface of the intestine which increase the surface area for absorption.

A large digestive organ located between the oesophagus and small intestine. It secretes protein digesting enzymes and strong acids to aid the digestion of food.

An enzyme that breaks down fats.

The body's largest internal organ. Carries out important roles in digestion and homeostasis. Its major role in digestion is the production of bile.

The upper part of the gut comprising of the duodenum, jejunum and ileum. The main site of absorption of food.

The molecule or substance upon which an enzyme acts.

Enzyme able to breakdown proteins by hydrolysis of peptide bonds.

Removal of undigested food (in feces) from the gut.

The transformation of molecules contained in food into the molcules used by the body for growth and repair.

The chemical breakdown of food.

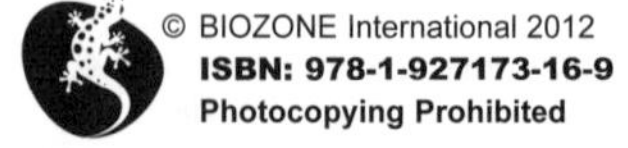

ISBN: 978-1-927173-16-9

Core Topic
6.2

The Transport System

Key terms

Core

antibodies
aorta
arteries (*sing.* artery)
atria (*sing.* atrium)
atrioventricular node
bicuspid valve
blood
blood vessels
capillaries (*sing.* capillary)
coronary arteries
epinephrine (adrenaline)
erythrocytes (= red blood cells)
heart
leukocytes (= white blood cells)
lymphocytes
myogenic contraction
phagocytes
plasma
platelets
pulmonary artery
pulmonary veins
semi-lunar valves
sino-atrial node (pacemaker)
thermoregulation
tricuspid valve
veins
vena cava
ventricle

Key concepts

- An internal transport system is a requirement in most multicellular animals.
- Heart structure reflects function in relation to circulatory and metabolic demands.
- Heart rate is intrinsic but is influenced by extrinsic factors including hormones.
- The structure of blood vessels is related to their functional position in the circulatory system.
- Gas transport in humans is a function of the blood.

Learning Objectives

☐ 1. Use the **KEY TERMS** to compile a glossary for this topic.

Heart Structure and Function *(6.2.1-6.2.4)* pages 278-283

☐ 2. Recall the significance of the surface area: volume relationship to diffusion rates and efficiency of transport of materials into and out of cells. Explain why animals over a certain size require an internal transport system.

☐ 3. Draw and label a diagram of the **heart** to show the four chambers, associated blood vessels, heart valves, and passage of blood through the heart. Include reference to the right and left **atria**, right and left **ventricles**, **aorta**, **vena cava**, **pulmonary artery**, **pulmonary veins**, **coronary arteries**, **semi-lunar valves**, **bicuspid valve**, and **tricuspid valve**.

☐ 4. Describe the function of the **coronary arteries**.

☐ 5. EXTENSION: Describe and explain the cardiac cycle.

☐ 6. Describe and explain the action of the heart in terms of blood flow through the chambers and action of the heart valves.

☐ 7. Outline the regulation of heartbeat in terms of **myogenic muscle contraction**, the role of the **sino-atrial node** (**pacemaker**), nerves (e.g. vagus nerve), the medulla of the brain, and the action of **epinephrine** (adrenaline).
TEACHER'S NOTE: Epinephrine and norepinephrine are now the accepted international terms for adrenaline and noradrenaline respectively. Norepinephrine has similar effects on the heart as epinephrine. Norepinephrine is the neurotransmitter involved in sympathetic nervous stimulation through the cardiac nerve and epinephrine acts similarly as a hormone with general metabolic effects. Both are involved in the fight or flight response.

Structure and Function of Blood Vessels *(6.2.5-6.2.7)* pages 284-288

☐ 8. Describe and explain the structure and function of **blood vessels**, including **arteries**, **capillaries**, and **veins**. Include reference to the functional relationship between these different vessels.

☐ 9. Describe the composition of **blood**, including reference to the **plasma**, **erythrocytes**, **leukocytes** (**phagocytes** and **lymphocytes**), and **platelets**. If required or as extension, outline the role of each of these blood components.

☐ 10. Describe the functional role of blood as a liquid tissue responsible for transporting nutrients, oxygen, carbon dioxide, hormones, **antibodies** (immunoglobulins) and **urea**. Describe the role of blood in **thermoregulation** and the distribution and transfer of heat.

Periodicals:
Listings for this chapter are on page 400

Weblinks:
www.thebiozone.com/weblink/IB-3169.html

BIOZONE APP:
Student Review Series
Cardiovascular System

The Human Circulatory System

The blood vessels of the circulatory system form a vast network of tubes that carry blood away from the heart, transport it to the tissues of the body, and then return it to the heart. The arteries, arterioles, capillaries, venules, and veins are organized into specific routes to circulate the blood throughout the body. The figure below shows a number of the basic **circulatory routes** through which the blood travels. Mammals have a **double circulatory system**: a **pulmonary system** (or circulation), which carries blood between the heart and lungs, and a **systemic system** (circulation), which carries blood between the heart and the rest of the body. The systemic circulation has many subdivisions. Two important subdivisions are the coronary (cardiac) circulation, which supplies the heart muscle, and the **hepatic portal circulation**, which runs from the gut to the liver.

Schematic Overview of the Human Circulatory System

Deoxygenated blood (colored gray below) travels to the right side of the heart via the vena cavae. The heart pumps the deoxygenated blood to the lungs where it releases carbon dioxide and receives oxygen. The oxygenated blood (colored white below) travels via the pulmonary vein back to the heart from where it is pumped to all parts of the body. The **venous system** (figure, left) returns blood from the capillaries to the heart. The **arterial system** (figure right) carries blood from the heart to the capillaries. **Portal systems** carry blood between two capillary beds.

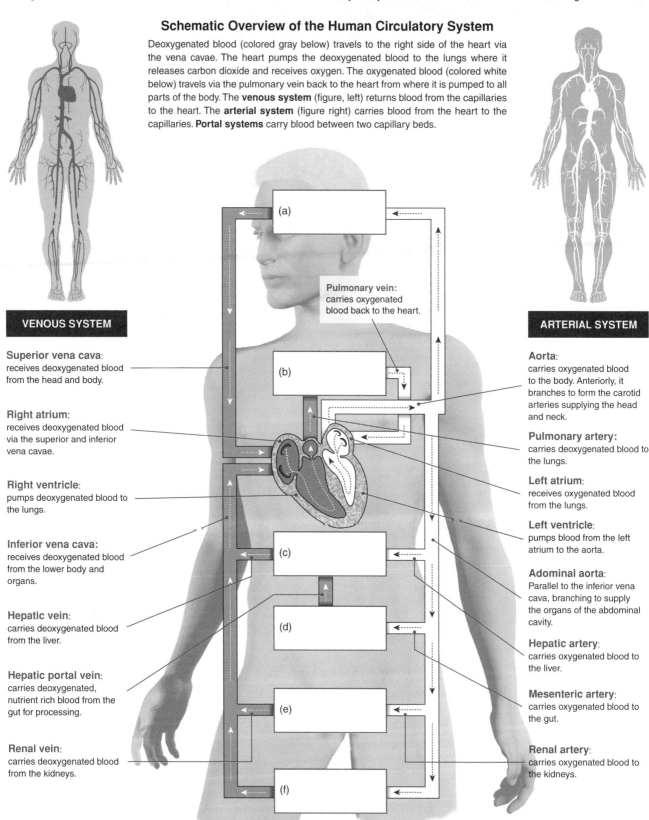

Pulmonary vein: carries oxygenated blood back to the heart.

(a)

(b)

(c)

(d)

(e)

(f)

VENOUS SYSTEM

Superior vena cava: receives deoxygenated blood from the head and body.

Right atrium: receives deoxygenated blood via the superior and inferior vena cavae.

Right ventricle: pumps deoxygenated blood to the lungs.

Inferior vena cava: receives deoxygenated blood from the lower body and organs.

Hepatic vein: carries deoxygenated blood from the liver.

Hepatic portal vein: carries deoxygenated, nutrient rich blood from the gut for processing.

Renal vein: carries deoxygenated blood from the kidneys.

ARTERIAL SYSTEM

Aorta: carries oxygenated blood to the body. Anteriorly, it branches to form the carotid arteries supplying the head and neck.

Pulmonary artery: carries deoxygenated blood to the lungs.

Left atrium: receives oxygenated blood from the lungs.

Left ventricle: pumps blood from the left atrium to the aorta.

Adominal aorta: Parallel to the inferior vena cava, branching to supply the organs of the abdominal cavity.

Hepatic artery: carries oxygenated blood to the liver.

Mesenteric artery: carries oxygenated blood to the gut.

Renal artery: carries oxygenated blood to the kidneys.

1. Complete the diagram above by labelling the boxes with the organs or structures they represent.

ISBN: 978-1-927173-16-9

The Human Heart

The heart is the centre of the human cardiovascular system. It is a hollow, muscular organ, weighing on average 342 grams. Each day it beats over 100,000 times to pump 3780 liters of blood through 100,000 kilometers of blood vessels. It comprises a system of four muscular chambers (two **atria** and two **ventricles**) that alternately fill and empty of blood, acting as a double pump. The left side pumps blood to the body tissues and the right side pumps blood to the lungs. The heart lies between the lungs, to the left of the body's midline, and it is surrounded by a double layered **pericardium** of tough fibrous connective tissue. The pericardium prevents overdistension of the heart and anchors the heart within the **mediastinum**.

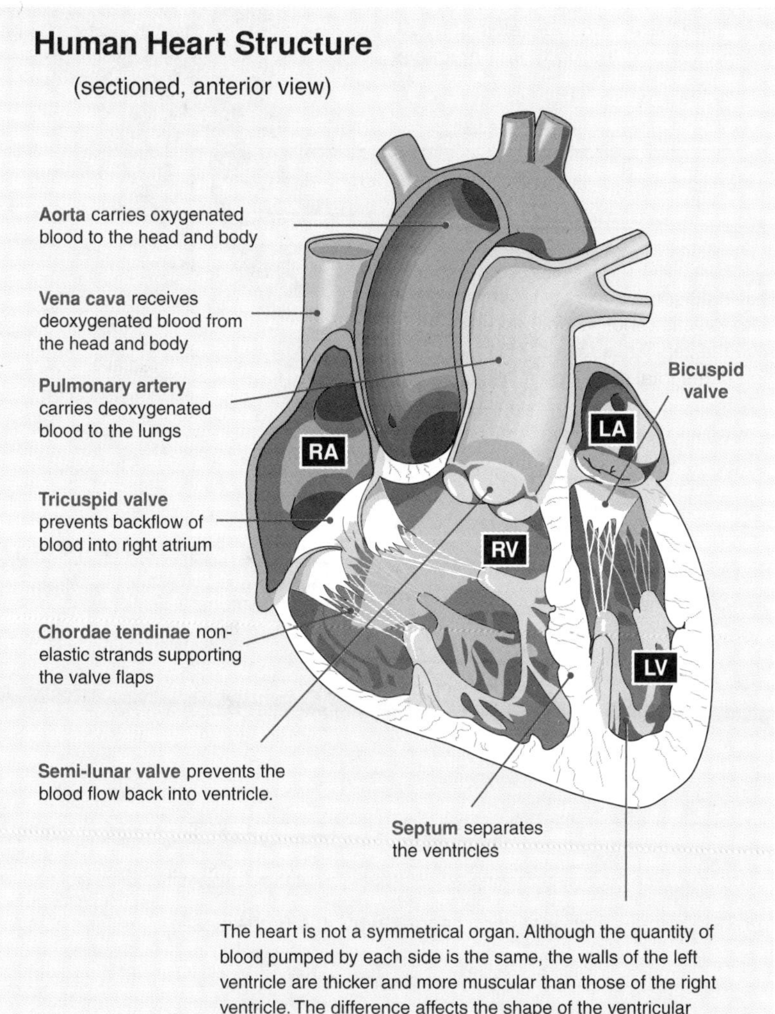
Human Heart Structure
(sectioned, anterior view)
Aorta carries oxygenated blood to the head and body
Vena cava receives deoxygenated blood from the head and body
Pulmonary artery carries deoxygenated blood to the lungs
RA
Tricuspid valve prevents backflow of blood into right atrium
Chordae tendinae non-elastic strands supporting the valve flaps
Semi-lunar valve prevents the blood flow back into ventricle.
Septum separates the ventricles
RV
LA
LV
Bicuspid valve

The heart is not a symmetrical organ. Although the quantity of blood pumped by each side is the same, the walls of the left ventricle are thicker and more muscular than those of the right ventricle. The difference affects the shape of the ventricular cavities, so the right ventricle is twisted over the left.

Key to abbreviations

- **RA** Right atrium: receives deoxygenated blood via the anterior and posterior vena cava
- **RV** Right ventricle: pumps deoxygenated blood to the lungs via the pulmonary artery
- **LA** Left atrium: receives blood returning to the heart from the lungs via the pulmonary veins
- **LV** Left ventricle: pumps oxygenated blood to the head and body via the aorta

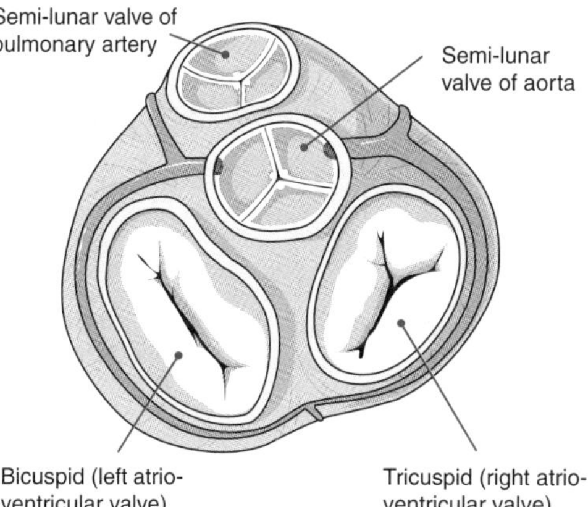
Top view of a heart in section, showing valves
Semi-lunar valve of pulmonary artery
Semi-lunar valve of aorta
Bicuspid (left atrio-ventricular valve)
Tricuspid (right atrio-ventricular valve)

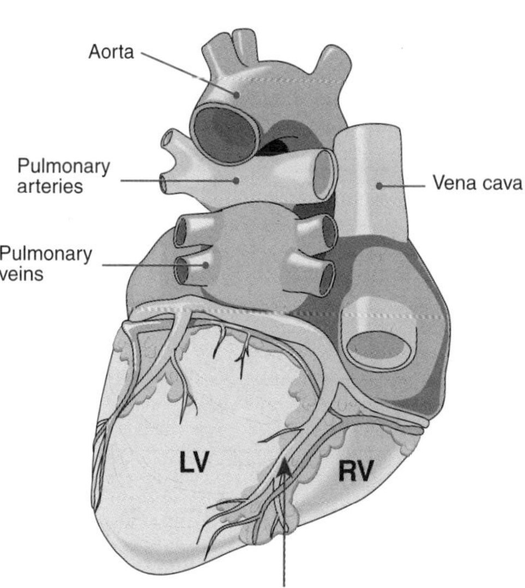
Posterior view of heart
Aorta
Pulmonary arteries
Vena cava
Pulmonary veins
LV
RV

Coronary arteries: The high oxygen demands of the heart muscle are met by a dense capillary network. Coronary arteries arise from the aorta and spread over the surface of the heart supplying the cardiac muscle with oxygenated blood. Deoxygenated blood is collected by cardiac veins and returned to the right atrium via a large coronary sinus.

1. In the schematic diagram of the heart, below, label the four chambers and the main vessels entering and leaving them. The arrows indicate the direction of blood flow. Use large colored circles to mark the position of each of the four valves.

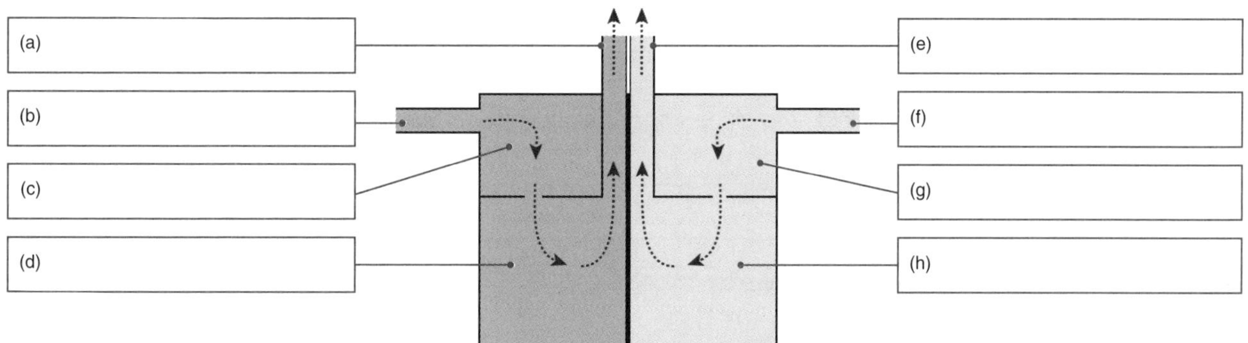
(a)
(b)
(c)
(d)
(e)
(f)
(g)
(h)

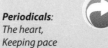

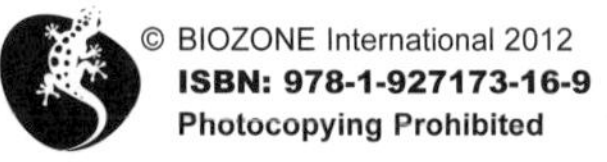

ISBN: 978-1-927173-16-9

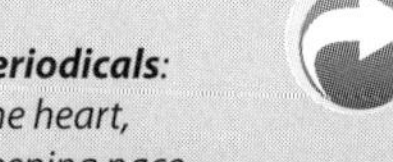
Periodicals: *The heart, Keeping pace*

Related activities: *Cardiac Cycle*
Weblinks: *Anatomy of the Heart, How the Heart Works*

Pressure Changes and the Asymmetry of the Heart

The heart is not a symmetrical organ. The left ventricle and its associated arteries are thicker and more muscular than the corresponding structures on the right side. This asymmetry is related to the necessary pressure differences between the pulmonary (lung) and systemic (body) circulations (not to the distance over which the blood is pumped *per se*). The graph below shows changes in blood pressure in each of the major blood vessel types in the systemic and pulmonary circuits (the horizontal distance not to scale). The pulmonary circuit must operate at a much lower pressure than the systemic circuit to prevent fluid from accumulating in the alveoli of the lungs. The left side of the heart must develop enough "spare" pressure to enable increased blood flow to the muscles of the body and maintain kidney filtration rates without decreasing the blood supply to the brain.

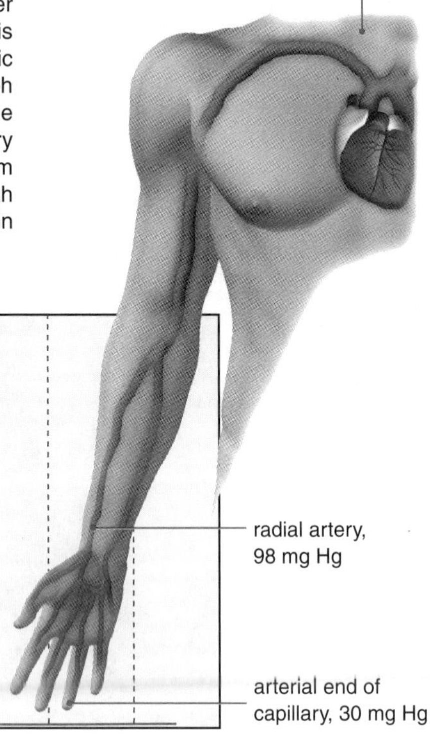

aorta, 100 mg Hg

Pressure (mm Hg)

120
100
80
60
40
20
0

Blood pressure during contraction (systole)

Blood pressure during relaxation (diastole)

The greatest fall in pressure occurs when the blood moves into the capillaries, even though the distance through the capillaries represents only a tiny proportion of the total distance travelled.

radial artery, 98 mg Hg

arterial end of capillary, 30 mg Hg

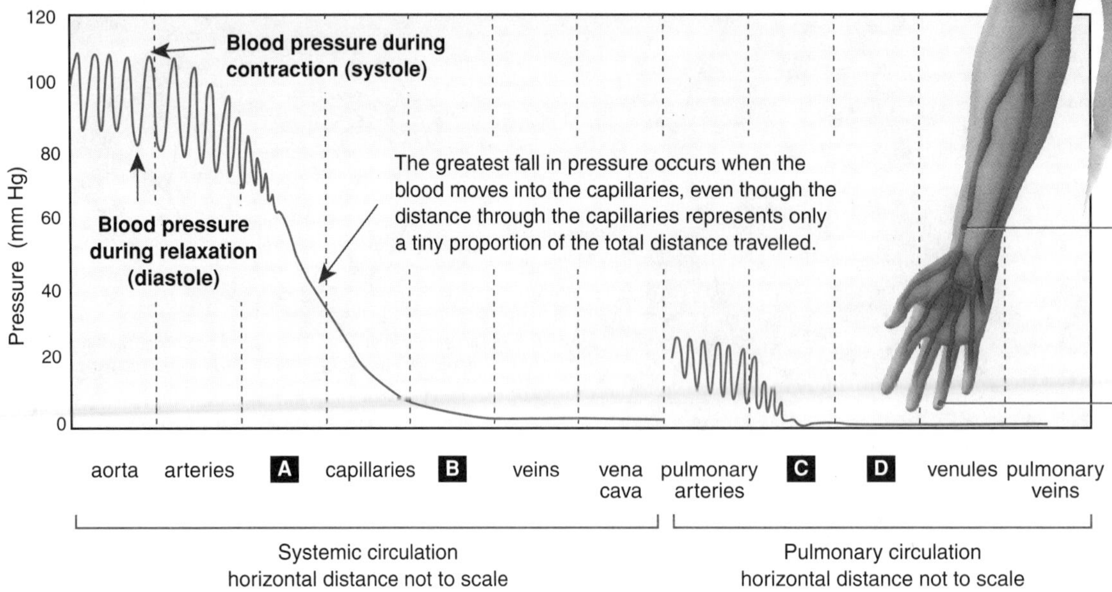

aorta arteries **A** capillaries **B** veins vena cava pulmonary arteries **C** **D** venules pulmonary veins

Systemic circulation
horizontal distance not to scale

Pulmonary circulation
horizontal distance not to scale

2. What is the purpose of the valves in the heart? ______________________________

3. The heart is full of blood, yet it requires its own blood supply. Suggest two reasons why this is the case:

   (a) ______________________________

   (b) ______________________________

4. Predict the effect on the heart if blood flow through a coronary artery is restricted or blocked: ______________________________

5. Identify the vessels corresponding to the letters **A**-**D** on the graph above:

   A: ____________ B: ____________ C: ____________ D: ____________

6. (a) Why must the pulmonary circuit operate at a lower pressure than the systemic system? ______________________________

   (b) Relate this to differences in the thickness of the wall of the left and right ventricles of the heart: ______________________________

7. What are you recording when you take a pulse? ______________________________

ISBN: 978-1-927173-16-9

The Cardiac Cycle

The heart pumps with alternate contractions (**systole**) and relaxations (**diastole**). The **cardiac cycle** refers to the sequence of events of a heartbeat and involves three main stages: atrial systole, ventricular systole, and complete cardiac diastole. Pressure changes within the heart's chambers generated by the cycle of contraction and relaxation are responsible for blood movement and cause the heart valves to open and close, preventing the backflow of blood. The noise of the blood when the valves open and close produces the heartbeat sound (**lubb-dupp**). The heart beat occurs in response to electrical impulses, which can be recorded as a trace, called an **electrocardiogram** or **ECG**. The ECG pattern is the result of the different impulses produced at each phase of the cardiac cycle, and each part is identified with a letter code. An ECG provides a useful method of monitoring changes in heart rate and activity and detection of heart disorders. The electrical trace is accompanied by volume and pressure changes (below).

The Cardiac Cycle

The **pulse** results from the rhythmic expansion of the arteries as the blood spurts from the left ventricle. Pulse rate therefore corresponds to heart rate.

Stage 1: Atrial contraction and ventricular filling
The ventricles relax and blood flows into them from the atria. Note that 70% of the blood from the atria flows passively into the ventricles. It is during the last third of ventricular filling that the atria contract.

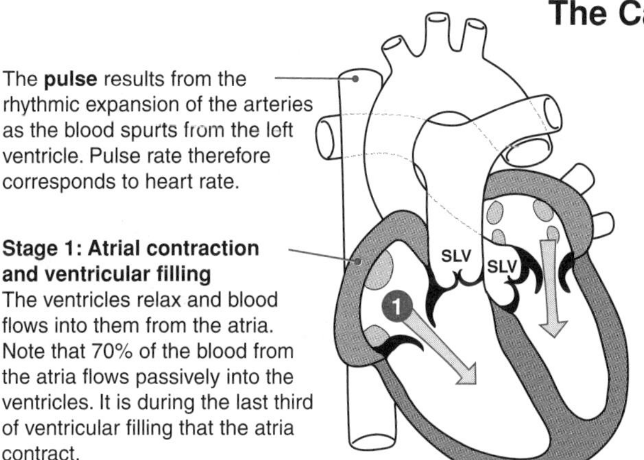
SLV
SLV
1

Heart during ventricular filling

Stage 2: Ventricular contraction
The atria relax, the ventricles contract, and blood is pumped from the ventricles into the aorta and the pulmonary artery. The start of ventricular contraction coincides with the first heart sound.

Stage 3: (not shown) There is a short period of atrial and ventricular relaxation. Semilunar valves (**SLV**) close to prevent backflow into the ventricles (see diagram, left). The cycle begins again. For a heart beating at 75 beats per minute, one cardiac cycle lasts about 0.8 seconds.

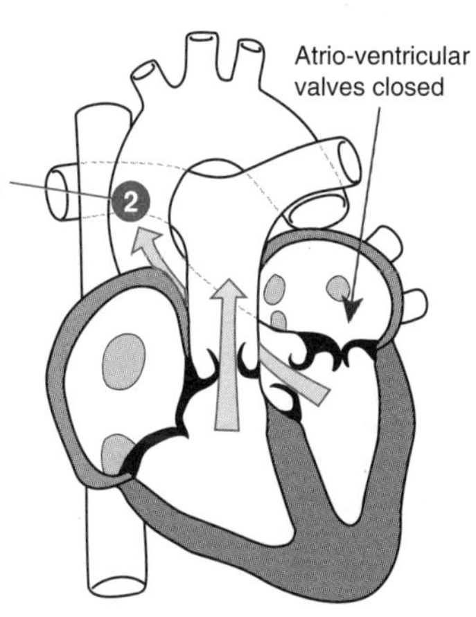
Atrio-ventricular valves closed
2

Heart during ventricular contraction

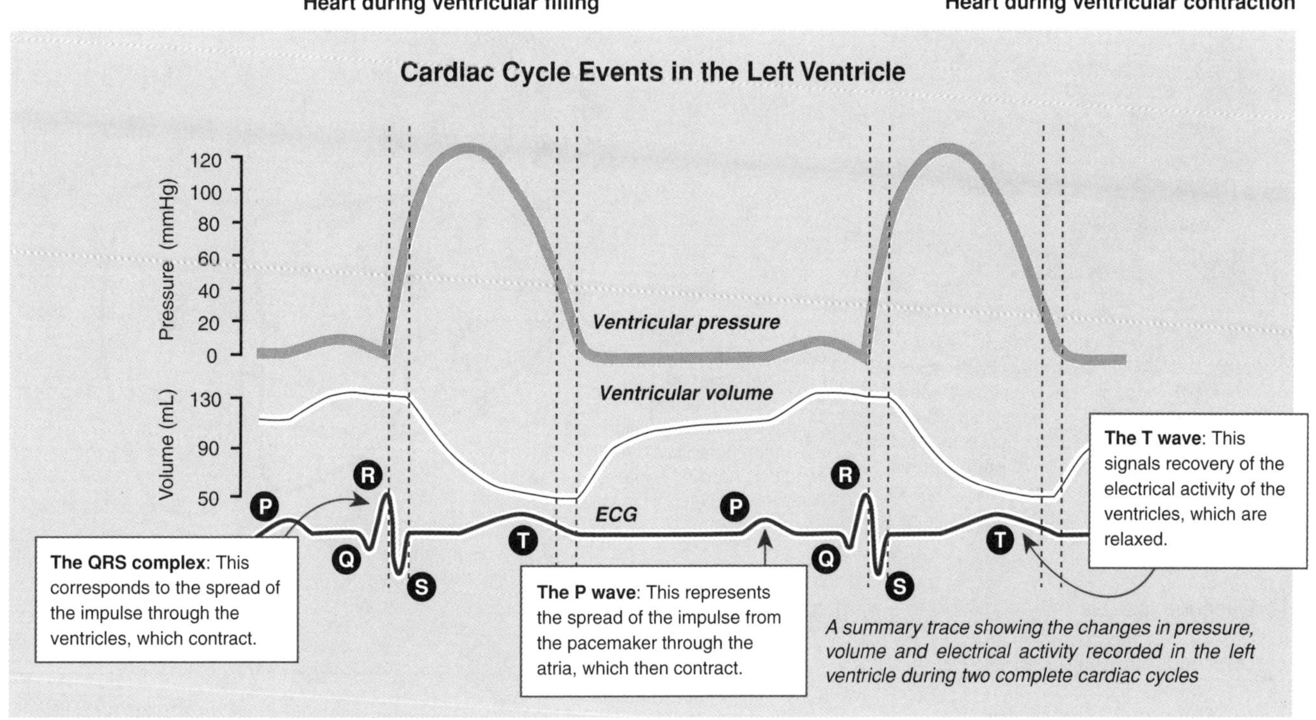
Cardiac Cycle Events in the Left Ventricle
Pressure (mmHg)
120
100
80
60
40
20
0
Ventricular pressure
Volume (mL)
130
90
50
Ventricular volume
ECG
P
Q
R
S
T
The QRS complex: This corresponds to the spread of the impulse through the ventricles, which contract.
The P wave: This represents the spread of the impulse from the pacemaker through the atria, which then contract.
The T wave: This signals recovery of the electrical activity of the ventricles, which are relaxed.

A summary trace showing the changes in pressure, volume and electrical activity recorded in the left ventricle during two complete cardiac cycles

1. Identify each of the following phases of an ECG by its international code:

   (a) Excitation of the ventricles and ventricular systole: ______

   (b) Electrical recovery of the ventricles and ventricular diastole: ______

   (c) Excitation of the atria and atrial systole: ______

2. Suggest the physiological reason for the period of electrical recovery experienced each cycle (the T wave):

   ______

3. Using the letters indicated, mark the points on trace above corresponding to each of the following:

   (a) E: Ejection of blood from the ventricle
   (b) AVC: Closing of the atrioventricular valve
   (c) FV: Filling of the ventricle
   (d) AVO: Opening of the atrioventricular valve

ISBN: 978-1-927173-16-9

Periodicals: Keeping pace: cardiac muscle and heart beat

Weblinks: Cardiac Cycle Animation, The Cardiac Cycle, Cardiac Cycle Movie

Control of Heart Activity

Given adequate supplies of oxygen, the heart will continue to beat when removed from the body. Thus, the origin of the heart-beat is **myogenic**, i.e. contraction is an intrinsic property of the cardiac muscle itself. The heartbeat is regulated by a conduction system consisting of the pacemaker (**sinoatrial node**) and a specialized conduction system of **Purkyne tissue**. The pacemaker sets the basic heart rhythm, but this rate is influenced by the cardiovascular control center in the brainstem, which alters heart rate via parasympathetic (acetylcholine) and sympathetic (noradrenaline) nerves. The hormone **adrenaline** (epinephrine) also influences cardiac output, increasing heart rate in preparation for a **fight or flight response**. Changing the rate and force of heart contraction is the main mechanism for controlling cardiac output in order to meet changing demands.

Generation of the Heartbeat

The basic rhythmic heartbeat is **myogenic**. The nodal cells (SAN and atrioventricular node) spontaneously generate rhythmic action potentials without neural stimulation. The normal resting rate of self-excitation of the SAN is about 50 beats per minute.

The amount of blood ejected from the left ventricle per minute is called the **cardiac output**. It is determined by the **stroke volume** (the volume of blood ejected with each contraction) and the **heart rate** (number of heart beats per minute).

Cardiac muscle responds to stretching by contracting more strongly. The greater the blood volume entering the ventricle, the greater the force of contraction. This relationship is known as **Starling's Law.**

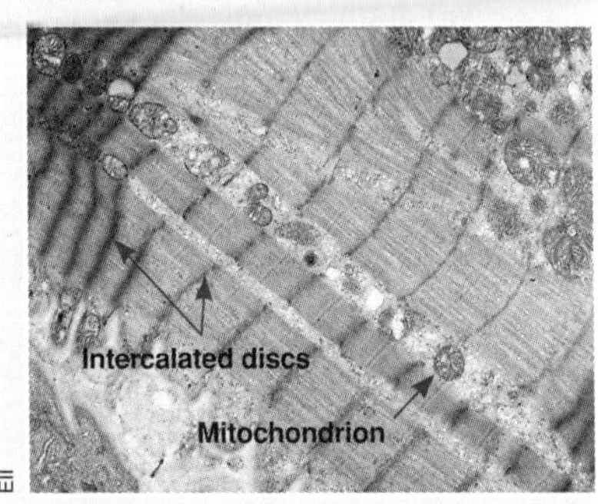
Intercalated discs
Mitochondrion
EII

A TEM photo of cardiac muscle showing branched fibers (muscle cells). Each muscle fiber has one or two nuclei and many large mitochondria. **Intercalated discs** are specialized electrical junctions that separate the cells and allow the rapid spread of impulses through the heart muscle.

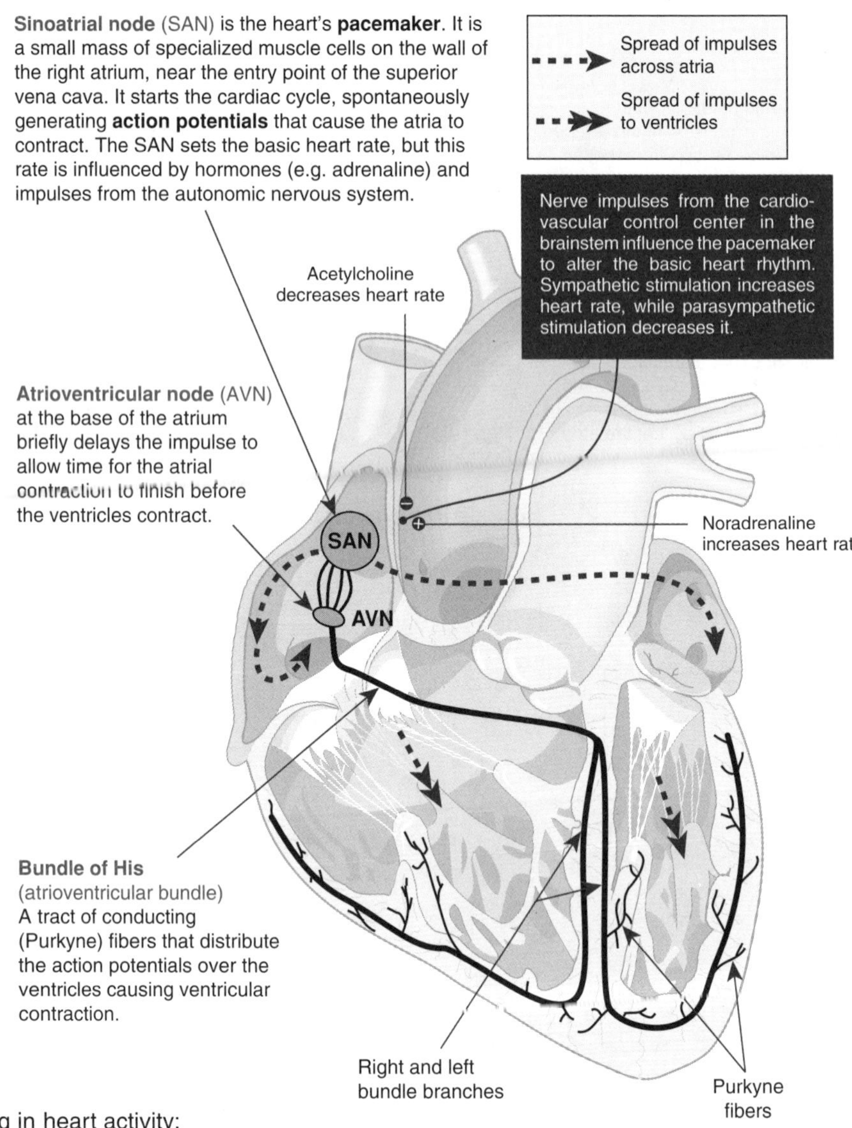
Sinoatrial node (SAN) is the heart's **pacemaker**. It is a small mass of specialized muscle cells on the wall of the right atrium, near the entry point of the superior vena cava. It starts the cardiac cycle, spontaneously generating **action potentials** that cause the atria to contract. The SAN sets the basic heart rate, but this rate is influenced by hormones (e.g. adrenaline) and impulses from the autonomic nervous system.
Spread of impulses across atria
Spread of impulses to ventricles
Nerve impulses from the cardiovascular control center in the brainstem influence the pacemaker to alter the basic heart rhythm. Sympathetic stimulation increases heart rate, while parasympathetic stimulation decreases it.
Acetylcholine decreases heart rate
Atrioventricular node (AVN) at the base of the atrium briefly delays the impulse to allow time for the atrial contraction to finish before the ventricles contract.
SAN
AVN
Noradrenaline increases heart rate
Bundle of His (atrioventricular bundle) A tract of conducting (Purkyne) fibers that distribute the action potentials over the ventricles causing ventricular contraction.
Right and left bundle branches
Purkyne fibers

1. Describe the role of each of the following in heart activity:

   (a) The sinoatrial node: ___

   (b) The atrioventricular node: ___

   (c) The bundle of His: ___

   (d) Intercalated discs: ___

2. What is the significance of delaying the impulse at the AVN? ___

3. What is the advantage of the physiological response of cardiac muscle to stretching? ___

4. The heart-beat is intrinsic. Why is it important to be able to influence the basic rhythm via the central nervous system?

***Related activities:** The Human Heart, The Cardiac Cycle*
***Weblinks:** Your Heart's Electrical System*

ISBN: 978-1-927173-16-9

Review of the Human Heart

A circulatory system is required to transport materials because diffusion is too inefficient and slow to supply all the cells of the body adequately. The circulatory system in humans transports nutrients, respiratory gases, wastes, and hormones, aids in regulating body temperature and maintaining fluid balance, and has a role in internal defense. The circulatory system comprises a network of vessels, a circulatory fluid (blood), and a heart. This activity summarizes key features of the structure and function of the human heart. The necessary information can be found in earlier activities in this topic.

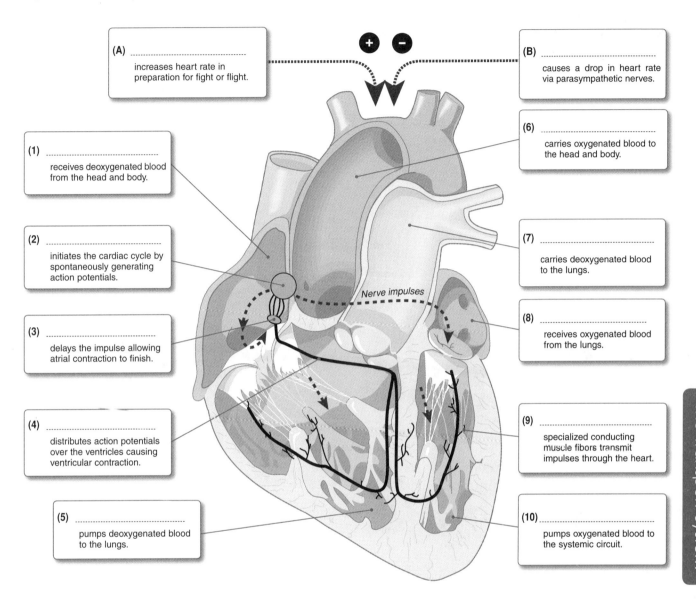

(A) ....
increases heart rate in preparation for fight or flight.

(B) ....
causes a drop in heart rate via parasympathetic nerves.

+ −

(1) ....
receives deoxygenated blood from the head and body.

(2) ....
initiates the cardiac cycle by spontaneously generating action potentials.

(3) ....
delays the impulse allowing atrial contraction to finish.

(4) ....
distributes action potentials over the ventricles causing ventricular contraction.

(5) ....
pumps deoxygenated blood to the lungs.

(6) ....
carries oxygenated blood to the head and body.

(7) ....
carries deoxygenated blood to the lungs.

(8) ....
receives oxygenated blood from the lungs.

(9) ....
specialized conducting muscle fibors transmit impulses through the heart.

(10) ....
pumps oxygenated blood to the systemic circuit.

Nerve impulses

1. On the diagram above, label the identified components of heart structure and intrinsic control (**1-10**), and some of the components of extrinsic control of heart rate (**A-B**).

2. An **ECG** is the result of different impulses produced at each phase of the **cardiac cycle** (the sequence of events in a heartbeat). For each electrical event indicated in the ECG below, describe the corresponding event in the cardiac cycle:

A ....
The spread of the impulse from the pacemaker (sinoatrial node) through the atria.

B ....
The spread of the impulse through the ventricles.

C ....
Recovery of the electrical activity of the ventricles.

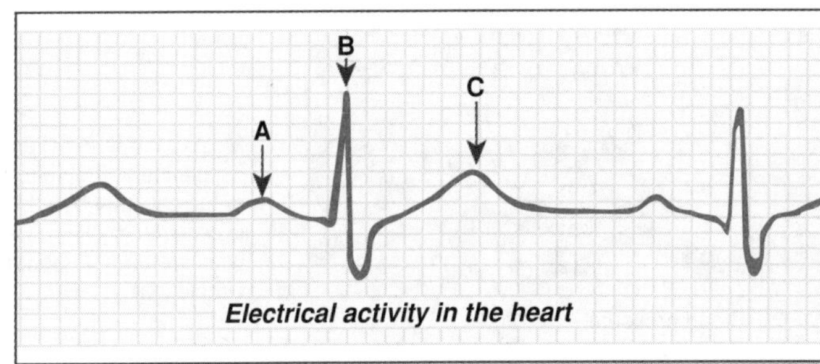
B
C
A
Electrical activity in the heart

3. Describe one treatment that may be indicated when heart rhythm is erratic or too slow: ______________________

____

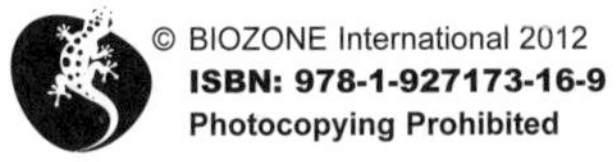

ISBN: 978-1-927173-16-9

Blood Vessels

The blood vessels of the circulatory system connect the fluid environment of the body's cells to the organs that exchange gases, absorb nutrients, and dispose of wastes. In vertebrates, arteries carry blood away from the heart to the capillaries within the tissues. The large arteries leaving the heart branch repeatedly to form distributing arteries, which themselves divide to form small **arterioles** within the tissues and organs. Arterioles deliver blood to the capillaries connecting the arterial and venous systems. Capillaries enable the exchange of nutrients and wastes between the blood and tissues, and they form large networks, especially in tissues and organs with high metabolic rates. The structural differences between blood vessels are related to their functional roles. While vessels close to the heart exhibit all the layers typical of the vessel's type, one or more layers may be absent in vessels more distant from the heart. Capillaries, whose functional role is exchange, consist only of a thin endothelium.

Artery structure

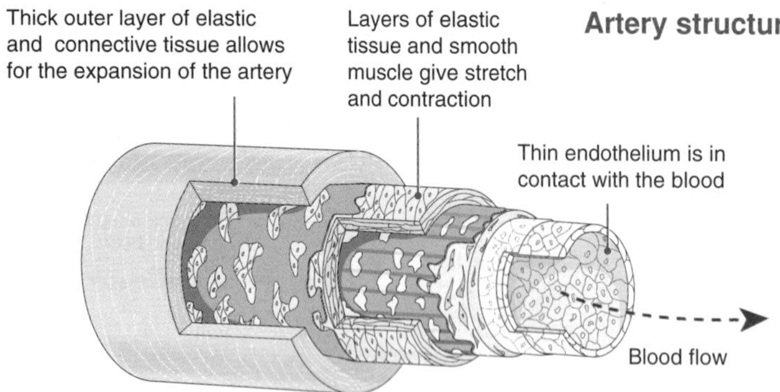

Thick outer layer of elastic and connective tissue allows for the expansion of the artery
Layers of elastic tissue and smooth muscle give stretch and contraction
Thin endothelium is in contact with the blood
Blood flow

Vein structure

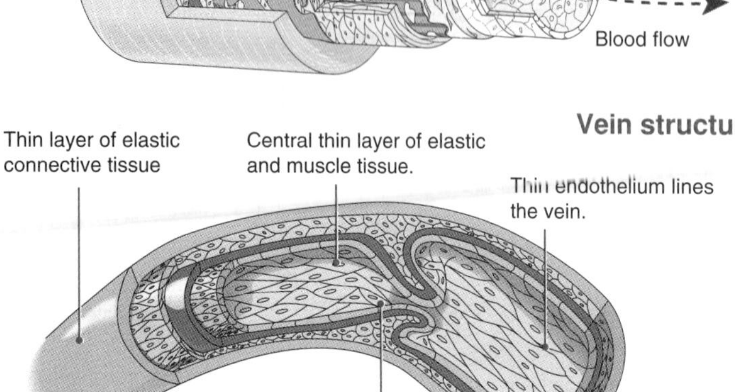

Thin layer of elastic connective tissue
Central thin layer of elastic and muscle tissue.
Thin endothelium lines the vein.
One-way valves are located along the length of veins to prevent the blood from flowing backwards.
Blood flow

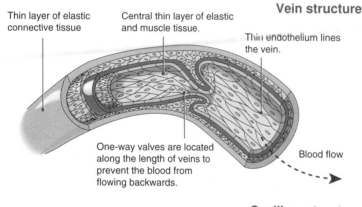

Capillary structure

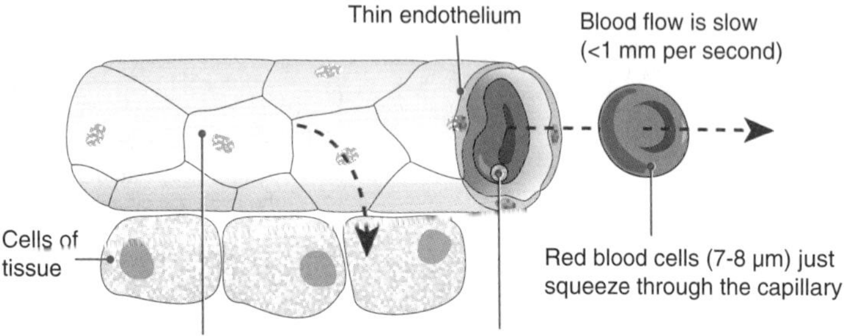

Thin endothelium
Blood flow is slow (<1 mm per second)
Cells of tissue
Red blood cells (7-8 µm) just squeeze through the capillary
Fluid leaks from capillaries to form the interstitial fluid that bathes the tissues.
Large proteins remain in the capillary in solution.

Formation of Tissue Fluid

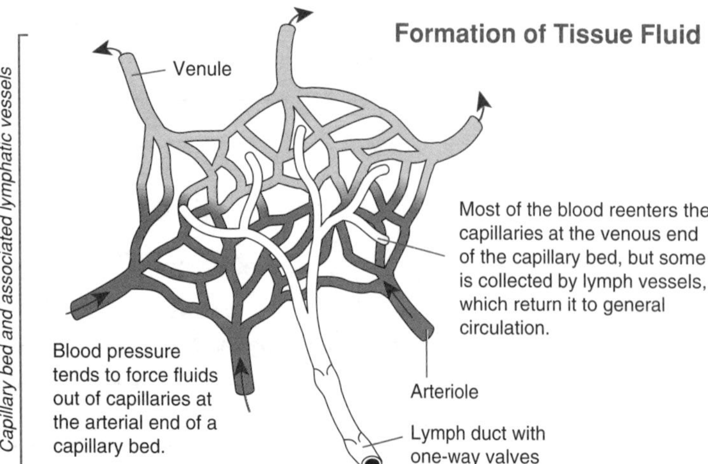

Capillary bed and associated lymphatic vessels
Venule
Most of the blood reenters the capillaries at the venous end of the capillary bed, but some is collected by lymph vessels, which return it to general circulation.
Blood pressure tends to force fluids out of capillaries at the arterial end of a capillary bed.
Arteriole
Lymph duct with one-way valves

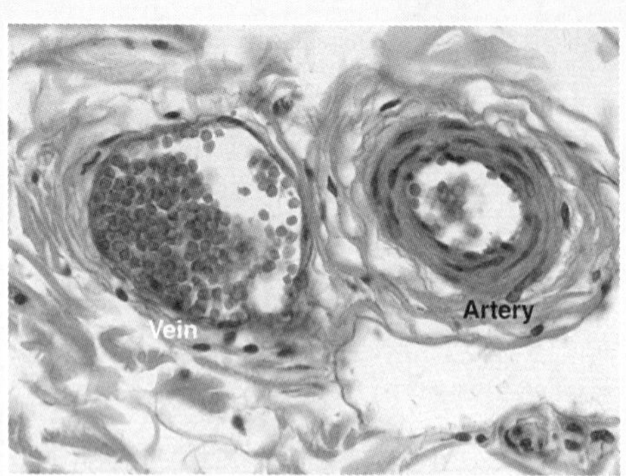

Vein
Artery

Arteries are made up of three layers; an inner layer of thin epthelium called the endothelium, a stretchy middle layer, and a thick outer layer. This structure enables arteries to withstand and maintain high blood pressure. **Veins** are similar in structure to arteries, but have less elastic and muscle tissue. Although veins are less elastic than arteries, they can still expand enough to adapt to changes in the pressure and volume of the blood passing through them.

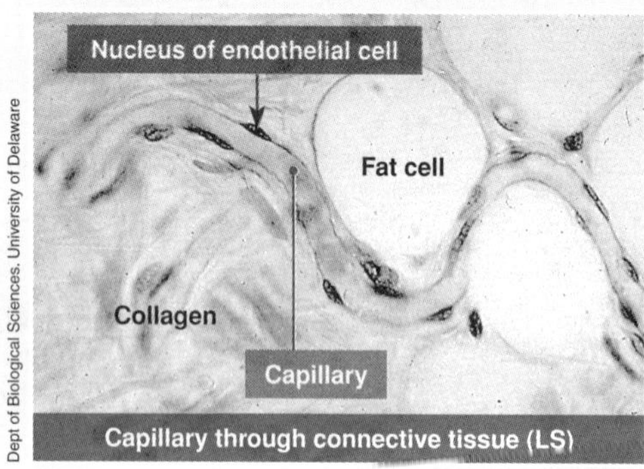

Nucleus of endothelial cell
Fat cell
Collagen
Capillary
Capillary through connective tissue (LS)

Dept of Biological Sciences, University of Delaware

Capillaries are very small blood vessels (4-10 µm diameter) made up of only a single layer of flattened (squamous) epithelial cells. Capillaries form a vast network of vessels that penetrate all parts of the body and are so numerous that no cell is more than 25 µm from any capillary. It is in the capillaries that the exchange of materials between the body cells and the blood takes place.

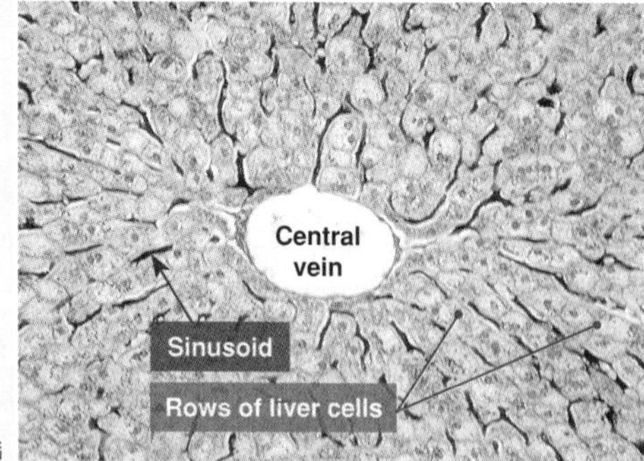

Central vein
Sinusoid
Rows of liver cells

EII

Tiny blood vessels in dense organs, such as the liver (above), are called **sinusoids**. They are wider than capillaries and follow a more convoluted path through the tissue. Instead of the usual endothelial lining, they are lined with phagocytic cells. Sinusoids, like capillaries, transport blood from arterioles to venules.

ISBN: 978-1-927173-16-9

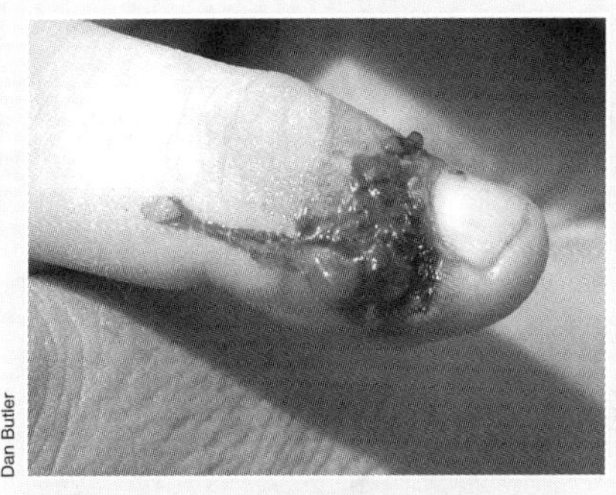
Dan Butler

If a vein is cut, as in the severe finger wound shown above, the blood oozes out slowly in an even flow, and usually clots quickly as it leaves. In contrast, if a cut is made into an artery, the arterial blood spurts rapidly and requires pressure to staunch the flow.

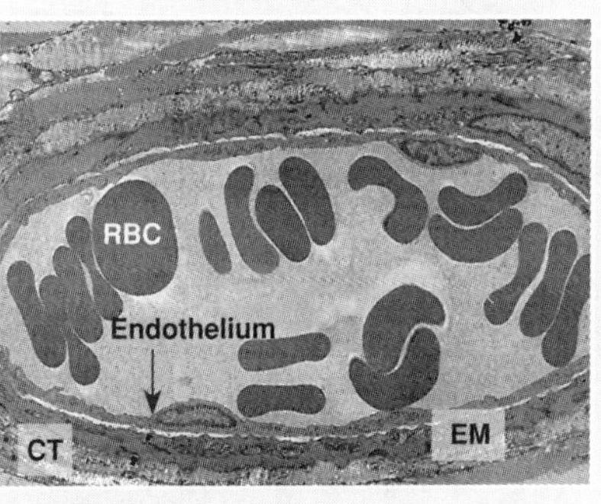
RBC
Endothelium
CT
EM

This TEM shows the structure of a typical vein. Note the red blood cells (RBC) in the lumen of the vessel, the inner layer of of epithelial cells (the endothelium), the central layer of elastic and muscle tissue (EM), and the outer connective tissue (CT) layer.

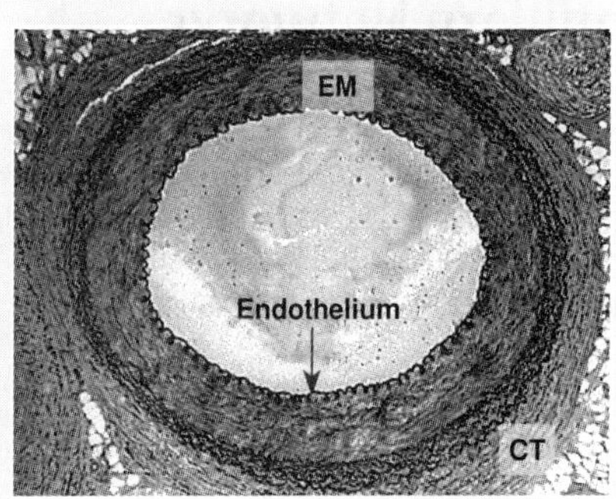
EM
Endothelium
CT

Arteries have a thick central layer of elastic and smooth muscle tissue (EM). Near the heart, arteries have more elastic tissue. This enables them to withstand high blood pressure. Arteries further from the heart have more smooth muscle; this helps them to maintain blood pressure.

1. Describe the contrasting structure of veins and arteries for each of the following properties:

   (a) Thickness of muscle and elastic tissue: ___

   (b) Size of the lumen (inside of the vessel): ___

2. Explain the reasons for the differences you have described above: ___

3. (a) Describe the structure of capillaries, explaining how it differs from that of veins and arteries: ___

   (b) Explain the reasons for these differences: ___

4. Compare the rate and force of blood flow in arteries, veins, and capillaries, explaining reasons for the differences: ___

5. Describe the role of the valves in assisting the veins to return blood back to the heart: ___

6. Explain why blood oozes from a venous wound, rather than spurting as it does from an arterial wound: ___

7. Explain why capillaries form dense networks in tissues with a high metabolic rate: ___

The Transport System

ISBN: 978-1-927173-16-9

Capillary Networks

Capillaries form branching networks where exchanges between the blood and tissues take place. The flow of blood through a capillary bed is called **microcirculation**. In most parts of the body, there are two types of vessels in a capillary bed: the **true capillaries**, where exchanges take place, and a vessel called a **vascular shunt**, which connects the arteriole and venule at either end of the bed. The shunt diverts blood past the true capillaries when the metabolic demands of the tissue are low (e.g. vasoconstriction in the skin when conserving body heat). When tissue activity increases, the entire network fills with blood.

1. Describe the structure of a capillary network:

2. Explain the role of the smooth muscle sphincters and the vascular shunt in a capillary network:

3. (a) Describe a situation where the capillary bed would be in the condition labelled A:

   (b) Describe a situation where the capillary bed would be in the condition labelled B:

4. How does a portal venous system differ from other capillary systems?

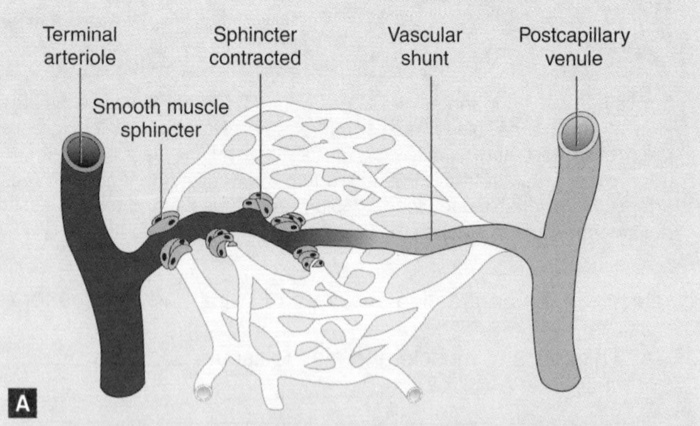

Terminal arteriole
Sphincter contracted
Vascular shunt
Postcapillary venule
Smooth muscle sphincter
A

When the sphincters contract (close), blood is diverted via the vascular shunt to the postcapillary venule, bypassing the exchange capillaries.

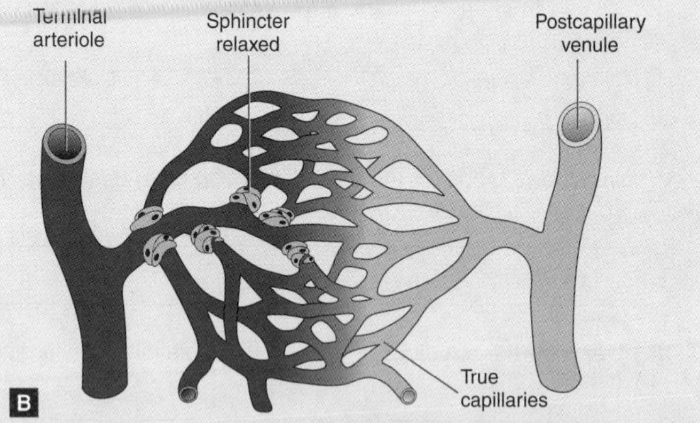

Terminal arteriole
Sphincter relaxed
Postcapillary venule
True capillaries
B

When the sphincters are relaxed (open), blood flows through the entire capillary bed allowing exchanges with the cells of the surrounding tissue.

Connecting Capillary Beds

the role of portal venous systems

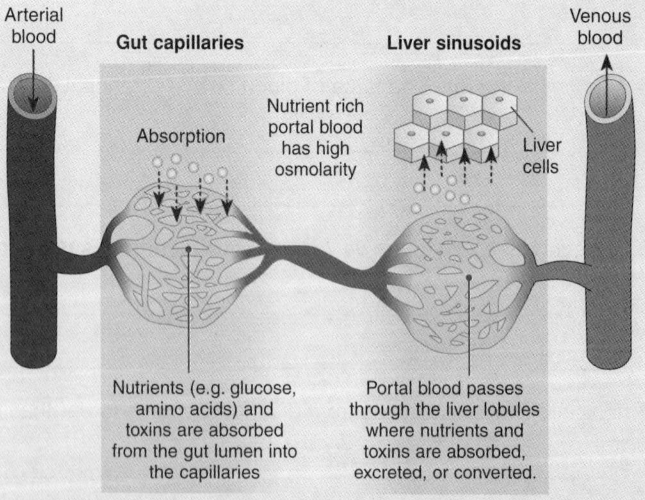

Arterial blood
Gut capillaries
Liver sinusoids
Venous blood
Absorption
Nutrient rich portal blood has high osmolarity
Liver cells
Nutrients (e.g. glucose, amino acids) and toxins are absorbed from the gut lumen into the capillaries
Portal blood passes through the liver lobules where nutrients and toxins are absorbed, excreted, or converted.

A portal venous system occurs when a capillary bed drains into another capillary bed through veins, without first going through the heart. Portal systems are relatively uncommon. Most capillary beds drain into veins which then drain into the heart, not into another capillary bed. The diagram above depicts the hepatic portal system, which includes both capillary beds and the blood vessels connecting them.

ISBN: 978-1-927173-16-9

Blood

Blood makes up about 8% of body weight in humans. Blood is a complex liquid tissue comprising cellular components suspended in plasma. If a blood sample is taken, the cells can be separated from the plasma by centrifugation. The cells (formed elements) settle as a dense red pellet below the transparent, straw-colored plasma. Blood performs many functions: it transports nutrients, respiratory gases, hormones, and wastes; it has a role in thermoregulation through the distribution of heat; it defends against infection; and its ability to clot protects against blood loss. The examination of blood is also useful in diagnosing disease. The cellular components of blood are normally present in particular specified ratios. A change in the morphology, type, or proportion of different blood cells can therefore be used to indicate a specific disorder or infection (see the next page).

Non-Cellular Blood Components

The non-cellular blood components form the plasma. Plasma is a watery matrix of ions and proteins and makes up 50-60% of the total blood volume.

Water

The main constituent of blood and lymph.
Role: Transports dissolved substances. Provides body cells with water. Distributes heat and has a central role in thermoregulation. Regulation of water content helps to regulate blood pressure and volume.

Mineral ions

Sodium, bicarbonate, magnesium, potassium, calcium, chloride.
Role: Osmotic balance, pH buffering, and regulation of membrane permeability. They also have a variety of other functions, e.g. Ca^{2+} is involved in blood clotting.

Plasma proteins

7-9% of the plasma volume.
Serum albumin
Role: Osmotic balance and pH buffering, Ca^{2+} transport.
Fibrinogen and prothrombin
Role: Take part in blood clotting.
Immunoglobulins
Role: Antibodies involved in the immune response.
α-globulins
Role: Bind/transport hormones, lipids, fat soluble vitamins.
β-globulins
Role: Bind/transport iron, cholesterol, fat soluble vitamins.
Enzymes
Role: Take part in and regulate metabolic activities.

Substances transported by non-cellular components

Products of digestion
Examples: sugars, fatty acids, glycerol, and amino acids.
Excretory products
Example: urea
Hormones and vitamins
Examples: insulin, sex hormones, vitamins A and B12.
Importance: These substances occur at varying levels in the blood. They are transported to and from the cells dissolved in the plasma or bound to plasma proteins.

Cellular Blood Components

The cellular components of the blood (also called the formed elements) float in the plasma and make up 40-50% of the total blood volume.

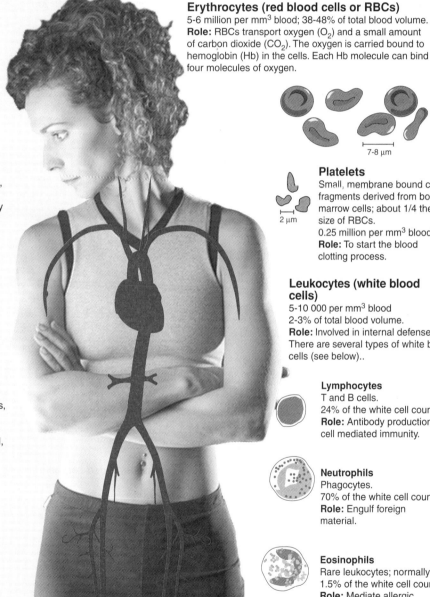

Erythrocytes (red blood cells or RBCs)

5-6 million per mm^3 blood; 38-48% of total blood volume.
Role: RBCs transport oxygen (O_2) and a small amount of carbon dioxide (CO_2). The oxygen is carried bound to hemoglobin (Hb) in the cells. Each Hb molecule can bind four molecules of oxygen.

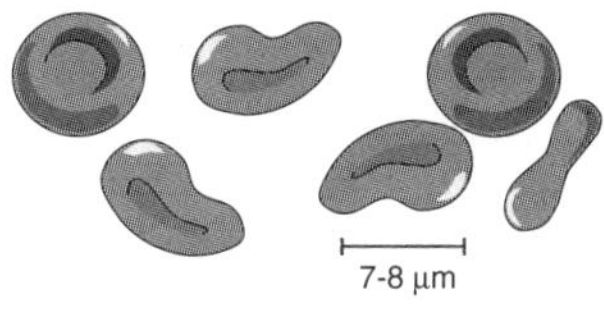
7-8 μm

Platelets

2 μm

Small, membrane bound cell fragments derived from bone marrow cells; about 1/4 the size of RBCs.
0.25 million per mm^3 blood.
Role: To start the blood clotting process.

Leukocytes (white blood cells)

5-10 000 per mm^3 blood
2-3% of total blood volume.
Role: Involved in internal defense. There are several types of white blood cells (see below)..

Lymphocytes
T and B cells.
24% of the white cell count.
Role: Antibody production and cell mediated immunity.

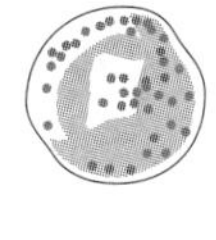

Neutrophils
Phagocytes.
70% of the white cell count.
Role: Engulf foreign material.

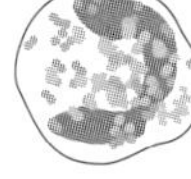

Eosinophils
Rare leukocytes; normally 1.5% of the white cell count.
Role: Mediate allergic responses such as hayfever and asthma.

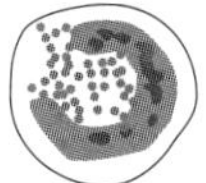

Basophils
Rare leukocytes; normally 0.5% of the white cell count.
Role: Produce heparin (an anti-clotting protein), and histamine. Involved in inflammation.

The Transport System

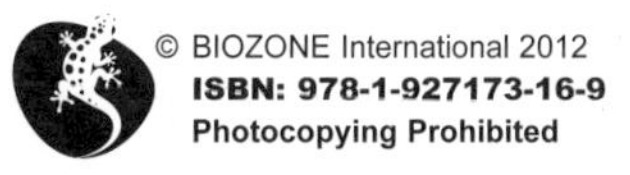

ISBN: 978-1-927173-16-9

Periodicals: *Red blood cells*

Related activities: *Gas Transport in Humans, The Body's Defenses, Thermoregulation in Humans*

The Examination of Blood

Different types of microscopy give different information about blood. A SEM (right) shows the detailed external morphology of the blood cells. A fixed smear of a blood sample viewed with a light microscope (far right) can be used to identify the different blood cell types present, and their ratio to each other. Determining the types and proportions of different white blood cells in blood is called a **differential white blood cell count.** Elevated counts of particular cell types indicate allergy or infection.

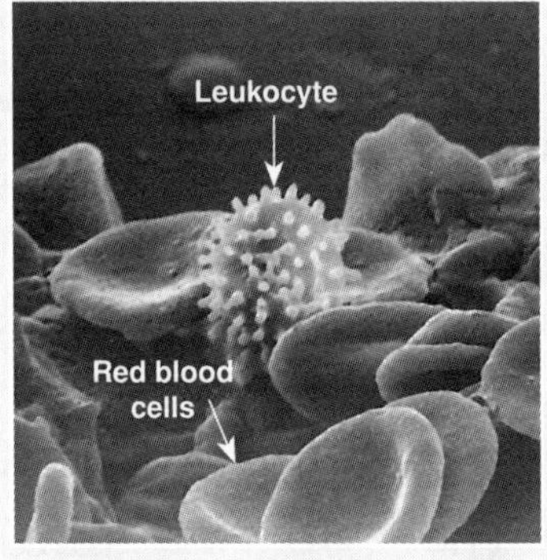
Leukocyte
Red blood cells

SEM of red blood cells and a leukocytes.

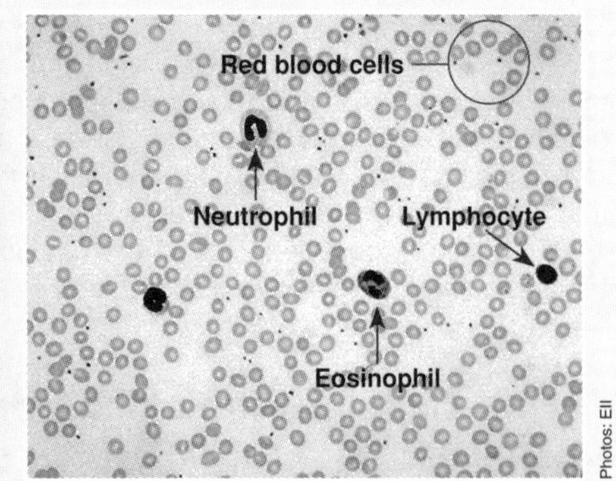
Red blood cells
Neutrophil
Lymphocyte
Eosinophil
Photos: Ell

Light microscope view of a fixed blood smear.

1. For each of the following blood functions, identify the component (or components) of the blood responsible and state how the function is carried out (the mode of action). The first one is done for you:

   (a) **Temperature regulation**. *Blood component*: Water component of the plasma

   Mode of action: Water absorbs heat and dissipates it from sites of production (e.g. organs)

   (b) **Protection against disease.** *Blood component*: __________

   Mode of action: __________

   (c) **Communication between cells, tissues, and organs**. *Blood component*: __________

   Mode of action: __________

   (d) **Oxygen transport.** *Blood component*: __________

   Mode of action: __________

   (e) **CO_2 transport.** *Blood components*: __________

   Mode of action: __________

   (f) **Buffer against pH changes**. *Blood components*: __________

   Mode of action: __________

   (g) **Nutrient supply**. *Blood component*: __________

   Mode of action: __________

   (h) **Tissue repair**. *Blood components*: __________

   Mode of action: __________

   (i) **Transport of hormones, lipids, and fat soluble vitamins**. *Blood component*: __________

   Mode of action: __________

2. Identify a feature that distinguishes red and white blood cells: __________

3. Explain two physiological advantages of red blood cell structure (lacking nucleus and mitochondria):

   (a) __________

   (b) __________

4. Suggest what each of the following results from a differential white blood cell count would suggest:

   (a) Elevated levels of eosinophils (above the normal range): __________

   (b) Elevated levels of neutrophils (above the normal range): __________

   (c) Elevated levels of basophils (above the normal range): __________

   (d) Elevated levels of lymphocytes (above the normal range): __________

ISBN: 978-1-927173-16-9

KEY TERMS: Mix and Match

INSTRUCTIONS: Test your vocabulary by matching each term to its definition, as identified by its preceding letter code.

aorta ....

arteries ....

atria ....

atrioventricular node ....

bicuspid valve ....

blood ....

blood vessels ....

capillaries ....

coronary arteries ....

epinephrine (adrenaline) ....

erythrocytes ....

heart ....

leukocytes ....

lymphocytes ....

myogenic contraction ....

phagocytes ....

plasma ....

platelets ....

pulmonary artery ....

pulmonary vein ....

semi-lunar valves ....

sino-atrial node ....

thermoregulation ....

tricuspid valve ....

veins ....

vena cava ....

ventricle ....

A Contractions initiated by the muscle cells spontaneously and independently of nervous stimulation, e.g. the heartbeat.

B A hormone which increases heart rate.

C The chambers of the heart that receive blood from the body or lungs.

D Small, membrane bound cell fragments derived from bone marrow cells.

E The specialized cardiac cells that initiate the cardiac cycle, and set the basic heart rate. Also call the pacemaker.

F Blood vessels that supply the heart muscle with oxygenated blood.

G Blood cells involved in the body's internal defense. Also call white blood cells.

H Artery which carries oxygenated blood away from the heart to the head and body.

I Large blood vessels with a thick, muscled wall which carries blood away from the heart.

J A blood vessel which carries oxygenated blood from the lungs back to the heart.

K The non-cellular portion of the blood.

L Specialized tissue between the right atrium and the right ventricle of the heart. It delays the delivery of impulse between the atria and ventricles.

M These valves prevent blood flowing back into the ventricles after contraction.

N Specialized leukocytes involved in antibody production and cell mediated immunity.

O The muscular organ responsible for pumping blood around the body.

P The process by which animals regulate body temperature.

Q A chamber of the heart that pumps blood into arteries.

R Circulatory fluid comprising numerous cell types, which moves respiratory gases and nutrients around the body.

S An artery which carries deoxygenated blood to the lungs from the heart.

T The smallest blood vessel. May have a wall only one or two cells thick.

U A large vein which receives deoxygenated blood from the head and body.

V The collective name given to vessels that transport blood around the body.

W Valve which prevents blood flowing back into the left atrium from the left ventricle.

X Blood cells that carries oxygen around the body. Also called red blood cells.

Y Blood vessels that return blood to the heart.

Z Valve preventing blood flowing back from the right ventricle to the right atrium.

AA White blood cells that destroy pathogens by engulfing and digesting them.

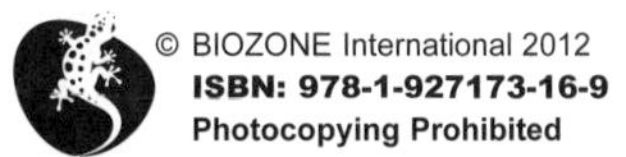

ISBN: 978-1-927173-16-9

Core Topic
6.3

AHL Topic
11.1

Defense Against **Disease**

Key concepts

- The body can distinguish self from non-self and defend itself against pathogens.
- Non-specific defenses target any foreign material.
- The immune response targets specific antigens and has a memory for antigens previously encountered.
- Some pathogens, such as HIV, target immune cells.
- The principles of specific immunity can be applied in the treatment and prevention of disease.

Key terms

Core
AIDS
antibiotics
antibodies (sing. antibody)
antigens
B-cell
HIV
mucous membranes
pathogen
phagocytic leukocytes (= phagocytes)
prevalence
sebum
skin

AHL only
active immunity
blood clotting
clonal selection
helper T-cell
hemostasis
immunological memory
memory cells
monoclonal antibodies
passive immunity
plasma cells
T-cell

Learning Objectives

☐ 1. Use the **KEY TERMS** to compile a glossary for this topic.

Pathogens and Non-Specific Defense *(6.3.1-6.3.4 & 11.1.1)* pages 291-297

☐ 2. Explain what is meant by a **pathogen** and identify examples.

☐ 3. Explain how **antibiotics** work and why they are ineffective against viral pathogens.

☐ 4. Describe **non-specific defenses** in humans, describing the nature and role of each of the following in protecting against pathogens:
(a) **Skin** (including sweat and sebum production) and **mucous membranes**.
(b) Natural anti-bacterial and anti-viral proteins, e.g. interferon.
(c) The inflammatory response and phagocytosis by **phagocytic leukocytes**.

☐ 5. **AHL**: Describe the process of **blood clotting**, and explain its role in **hemostasis** and in restricting entry of pathogens after injury.

Specific Defense *(6.3.5-6.3.6 & 11.1.2-11.1.4)* pages 298-302

☐ 6. Distinguish between **antigens** and **antibodies**.

☐ 7. Explain **antibody** production by **B-cell** clones. Recognize features of antibody structure that form the basis of their specificity to antigens.

☐ 8. **AHL**: Outline the principle of challenge and response, **clonal selection**, and the basis of **immunological memory**. Explain how the immune system is able to respond to the large and unpredictable range of potential antigens.

☐ 9. **AHL**: Giving examples, distinguish between **active immunity** and **passive immunity**.

☐ 10. **AHL**: Explain antigen presentation by **macrophages** and activation of **helper T-cells** leading to activation of B-cells. Describe the formation of antibody-producing **plasma cells** and **memory cells**.

Treating and Preventing Disease *(11.1.5-11.1.7)* pages 303-306

☐ 11. **AHL**: Describe the production and applications of **monoclonal antibodies**.

☐ 12. **AHL**: Explain the principles of **vaccination**, including reference to the **primary** and **secondary response** to infection and the role of memory cells.

☐ 13. **AHL**: Discuss the benefits and dangers of **vaccination**. Discuss the importance of accurate estimation of risk using good scientific data (*TOK*).

HIV and AIDS *(6.3.7-6.3.8)* pages 307-309

☐ 14. Outline the effects of **HIV** on the immune system, with specific reference to the infection of T-cells (lymphocytes) and loss of immune function.

☐ 15. Discuss the cause, transmission, and social implications of **AIDS**. Include reference to the severe impact of AIDS in parts of Africa and the cultural and economic reasons for differences in the prevalence of AIDS globally. Discuss the extent to which individuals in different societies can minimize or eliminate risks associated with specific methods of transmission (TOK).

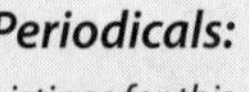

Periodicals:
Listings for this chapter are on page 400

Weblinks:
www.thebiozone.com/weblink/IB-3169.html

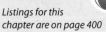

BIOZONE APP:
Student Review Series
Lymphatic System

Pathogens and Disease

Infectious disease refers to disease caused by a **pathogen** (an infectious agent). In 1861, **Louis Pasteur** demonstrated that microorganisms can be present in non-living matter and can contaminate seemingly sterile solutions. He also showed conclusively that microbes can be destroyed by heat. This discovery formed the basis of modern-day **aseptic technique**. Subsequently, scientists developed a better understanding of how pathogens cause disease. The discovery and development of antibiotics last century marked a new era in disease control. **Antibiotics** are chemotherapeutic agents that inhibit or prevent bacterial growth. They are produced naturally by bacteria and fungi, but are now grown or synthesized for use against bacterial infection.

Types of Pathogen

CDC

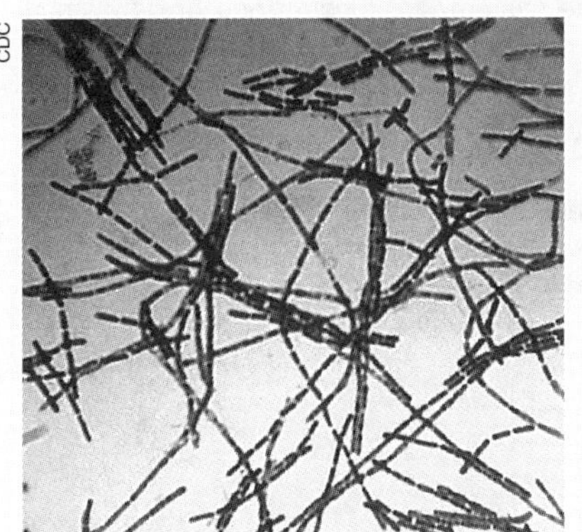

Bacillus anthracis bacterium causes anthrax. The anthrax bacillum can form long-lived spores.

CDC

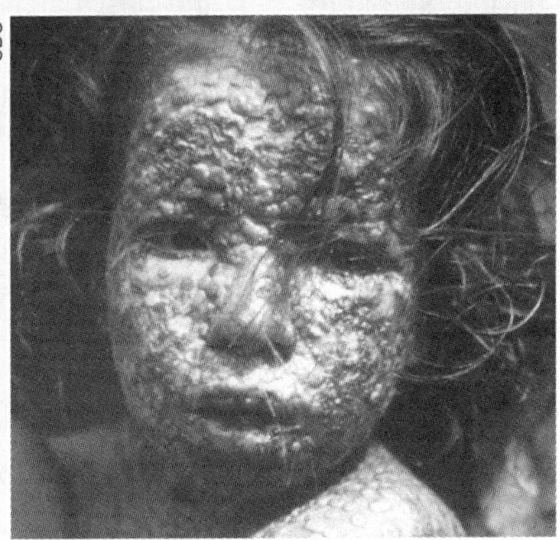

Photo: Bangladeshi girl with smallpox (1973). Smallpox was eradicated from the country in 1977.

Photo: Ute Frevert

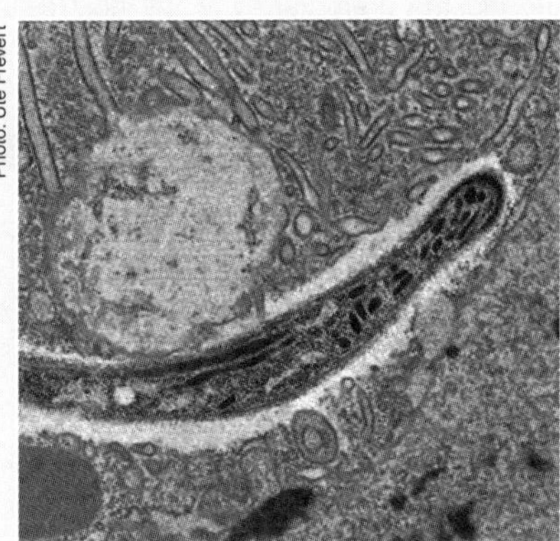

Malaria sporozoite moving through gut epithelia. The parasite is carried by a mosquito vector.

Bacteria: All bacteria are prokaryotes, and are categorized according to the properties of their cell walls and features such as cell shape and arrangement, oxygen requirement, and motility. Many bacteria are useful, but the relatively few species that are pathogenic are responsible for enormous social and economic cost. This is especially so since the rise in incidence of antibiotic resistance.

Viruses cause many common diseases (e.g. the common cold), as well as more serious diseases, such as West Nile fever, and some diseases that have been eradicated as a result of vaccination programmes (e.g. smallpox). Viruses are obligate intracellular parasites and need living host cells in order to multiply. Outside the host cell, they are inert, so antibiotics are ineffective against them.

Eukaryotic pathogens (fungi, algae, protozoa, and parasitic worms) include those responsible for malaria and schistosomiasis. Many are highly specialized parasites with a number of hosts, e.g. the malaria parasite *Plasmodium*, has a mosquito and a human host. Certain stages in the life cycle of *Plasmodium* reside within the host's cells, making them (like viruses) hard to treat.

How Antibiotics Work

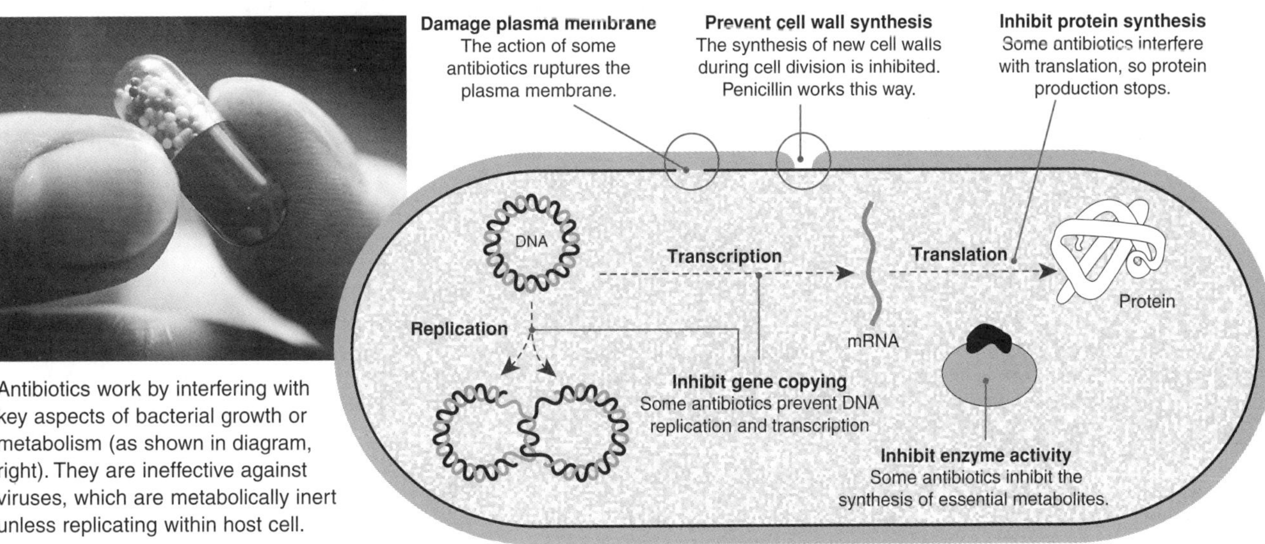
Damage plasma membrane
The action of some antibiotics ruptures the plasma membrane.
Prevent cell wall synthesis
The synthesis of new cell walls during cell division is inhibited. Penicillin works this way.
Inhibit protein synthesis
Some antibiotics interfere with translation, so protein production stops.
DNA
Transcription
Translation
Protein
mRNA
Replication
Inhibit gene copying
Some antibiotics prevent DNA replication and transcription
Inhibit enzyme activity
Some antibiotics inhibit the synthesis of essential metabolites.

Antibiotics work by interfering with key aspects of bacterial growth or metabolism (as shown in diagram, right). They are ineffective against viruses, which are metabolically inert unless replicating within host cell.

A composite diagram of a bacterial cell

1. Explain how an antibiotic's ability to interfere with cell wall formation would prevent growth of the bacterial population:

2. Explain why antibiotics are effective against bacteria, but not against viruses:

3. Why is it generally difficult to control and treat diseases caused by **intracellular parasites**?

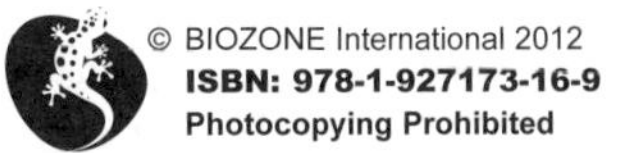

ISBN: 978-1-927173-16-9

Related activities: *Viral Diseases in Humans*
Weblinks: *Microbe Library: How Antibiotics Work*

Viral Diseases in Humans

Viruses are highly specialized intracellular parasites and they are responsible for a wide range of diseases in plants and animals (including humans). Viruses operate by utilizing the host's cellular machinery to replicate new viral particles. Most are able to infect specific types of cells of only one host species. The particular **host range** is determined by the presence of specific receptors on the host cell and the availability of the cellular factors needed for viral multiplication. For animal viruses, the receptor sites are on the plasma membranes of the host cells. Antiviral drugs are difficult to design because they must kill the virus without killing the host cells. Moreover, viruses cannot be attacked when in an inert state. Antiviral drugs work by preventing entry of the virus into the host cell or by interfering with viral replication. Vaccination is still regarded as the most effective way in which to control viral disease. However, vaccination against viruses does not necessarily provide lifelong immunity. New viral strains develop as preexisting strains acquire mutations. These mutations allow the viruses to change their surface proteins and thus evade detection by the host's immune system. The occurrence of new strains of seasonal flu is a good example of this.

Types of Viruses Affecting Humans

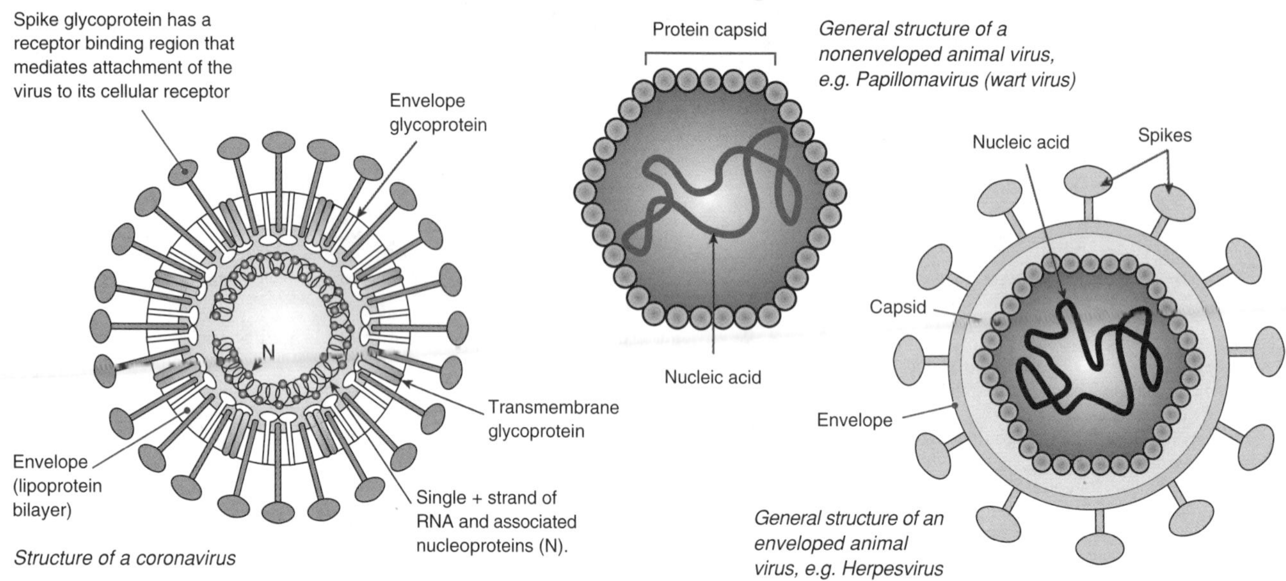
Spike glycoprotein has a receptor binding region that mediates attachment of the virus to its cellular receptor
Envelope glycoprotein
N
Transmembrane glycoprotein
Envelope (lipoprotein bilayer)
Single + strand of RNA and associated nucleoproteins (N).
Protein capsid
Nucleic acid
Nucleic acid
Spikes
Capsid
Envelope

Structure of a coronavirus

General structure of a nonenveloped animal virus, e.g. Papillomavirus (wart virus)

General structure of an enveloped animal virus, e.g. Herpesvirus

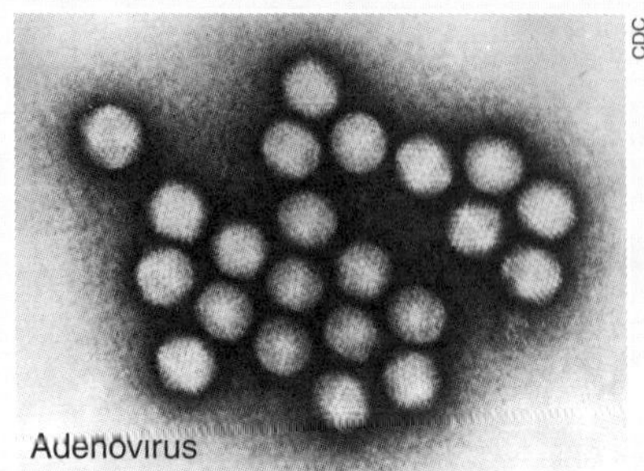
CDC
Adenovirus

Adenoviruses are medium-sized (90-100 nm), nonenveloped viruses containing double-stranded DNA. They commonly cause respiratory illness and are unusually stable to chemical or physical agents, allowing for prolonged survival outside of the body.

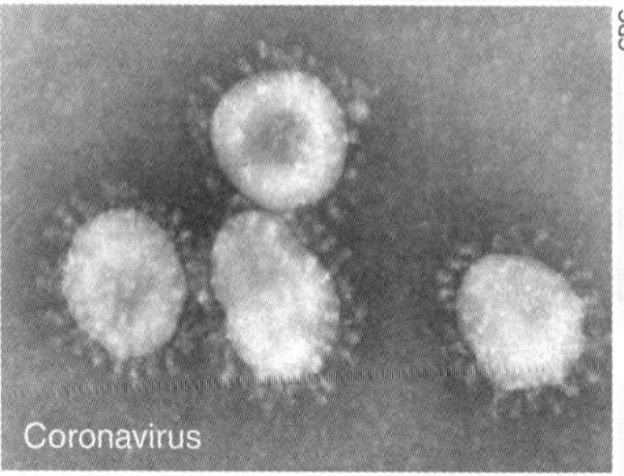
CDC
Coronavirus

Coronaviruses primarily infect the upper respiratory and gastrointestinal tracts of birds and mammals, including humans. Their name derives from the crown or corona of spikes and they have the largest genome of any of the single stranded RNA viruses.

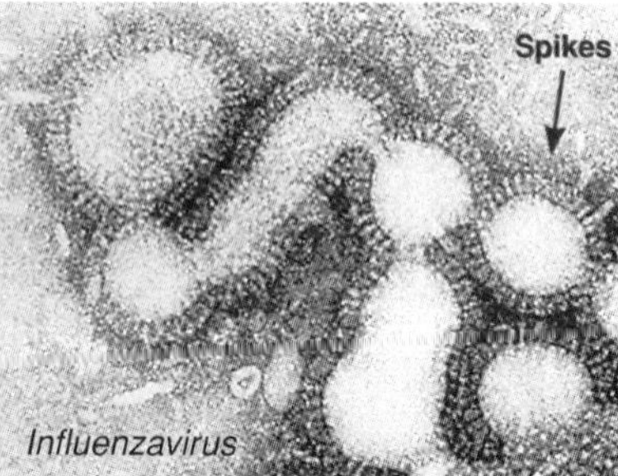
Spikes
Influenzavirus

In some viruses, the capsid is covered by an envelope, which protects the virus from the host's nuclease enzymes. Spikes on the envelope provide a binding site for attachment to the host. ***Influenzavirus*** is an enveloped virus with many glycoprotein spikes.

1. Explain the basis of host specificity in viruses: ______

2. (a) What is the most effective method of controlling viral diseases? ______

   (b) Explain your answer: ______

   (c) Discuss limitations to this approach: ______

ISBN: 978-1-927173-16-9

The Body's Defenses

The human body has a tiered system of defenses to prevent or limit infection by pathogens. The first line of defense has a role in keeping microorganisms from entering the body. If this fails, a second line of defense targets any foreign bodies (including microbes) that manage to get inside. If these defenses fail, the body's immune system provides a third line of (specific) defense. The ability to ward off disease through the various defense mechanisms is called **resistance**. **Non-specific resistance** is provided by the first and second lines of defense. **Specific resistance** (the immune response) is specific to particular pathogens. Part of the immune response involves the production of **antibodies**, which are large proteins that identify and neutralize foreign material such as microorganisms. Antibodies recognize and respond to specific parts of the microbes called **antigens**. Antigens are often proteins or carbohydrates such as fragments of cell wall. The name comes from **anti**body **gen**erator.

1st Line of Defense

The skin provides a physical barrier to the entry of pathogens. Healthy skin is rarely penetrated by microorganisms. Its low pH is unfavorable to the growth of many bacteria and its chemical secretions (e.g. sebum, antimicrobial peptides) inhibit growth of bacteria and fungi. Tears, mucus, and saliva also help to wash bacteria away.

2nd Line of Defense

A range of defense mechanisms operate inside the body to inhibit or destroy pathogens. These responses react to the presence of any pathogen, regardless of which species it is. White blood cells are involved in most of these responses.

It includes the **complement system** whereby plasma proteins work together to bind pathogens and induce an inflammatory responses to help fight infection.

3rd Line of Defense

Once the pathogen has been identified by the immune system, **lymphocytes** launch a range of specific responses to the pathogen, including the production of **antibodies**. Each type of antibody is produced by a B-cell clone and is specific against a particular antigen.

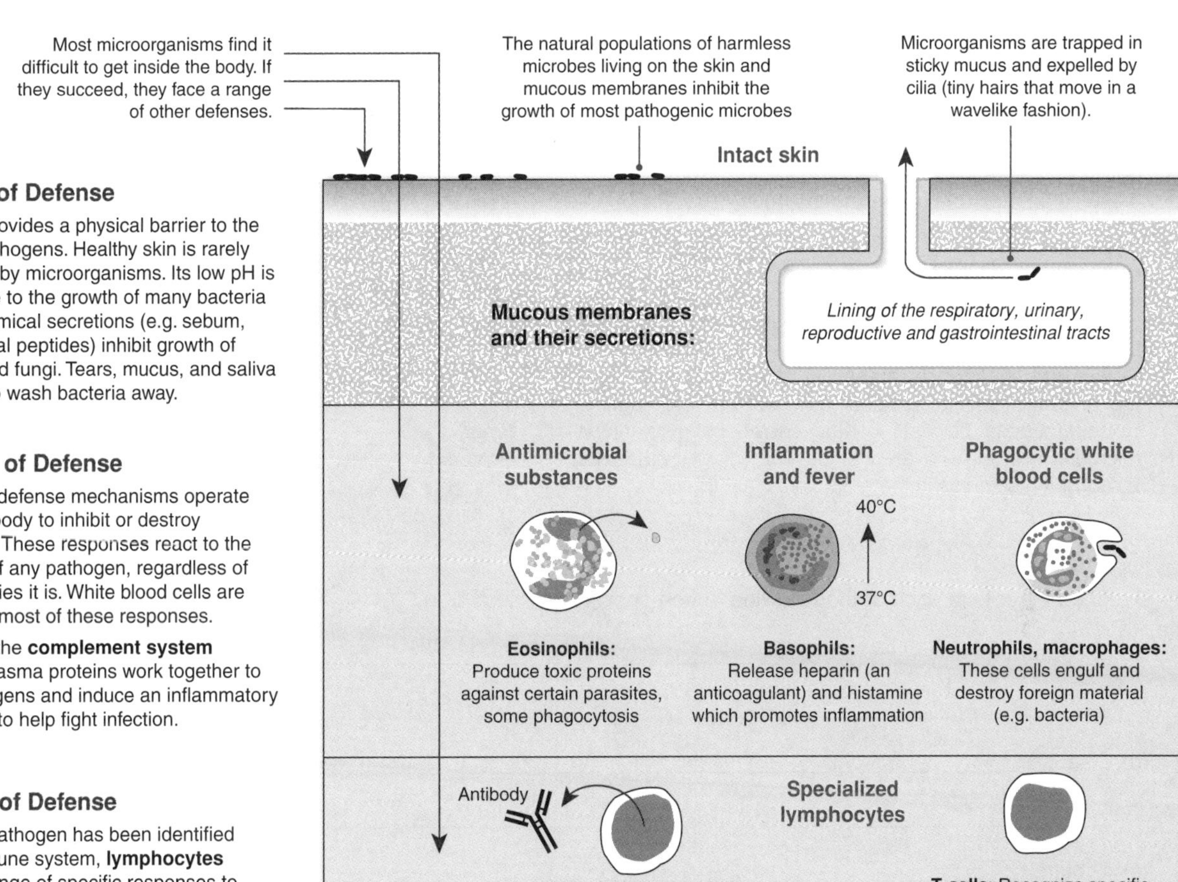
Most microorganisms find it difficult to get inside the body. If they succeed, they face a range of other defenses.
The natural populations of harmless microbes living on the skin and mucous membranes inhibit the growth of most pathogenic microbes
Microorganisms are trapped in sticky mucus and expelled by cilia (tiny hairs that move in a wavelike fashion).
Intact skin
Mucous membranes and their secretions:
Lining of the respiratory, urinary, reproductive and gastrointestinal tracts
Antimicrobial substances
Inflammation and fever
Phagocytic white blood cells
40°C
37°C
Eosinophils: Produce toxic proteins against certain parasites, some phagocytosis
Basophils: Release heparin (an anticoagulant) and histamine which promotes inflammation
Neutrophils, macrophages: These cells engulf and destroy foreign material (e.g. bacteria)
Antibody
Specialized lymphocytes
B-cells: Recognize specific antigens and divide to form antibody-producing clones.
T-cells: Recognize specific antigens and activate specific defensive cells.

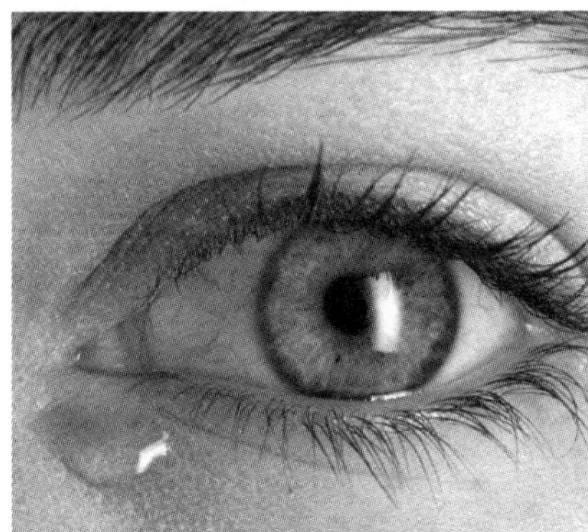

Tears contain antimicrobial substances as well as washing contaminants from the eyes.

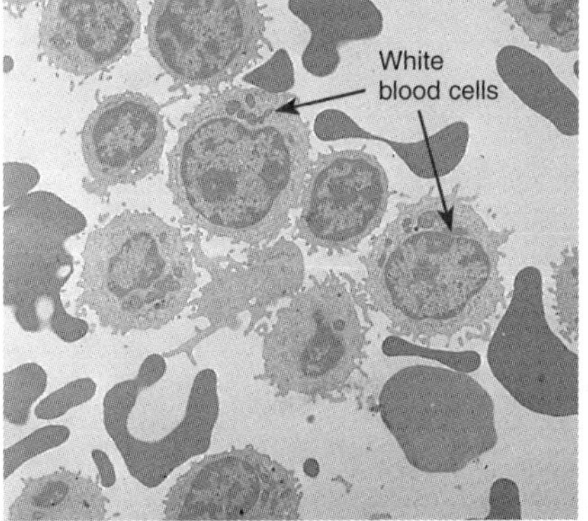
White blood cells
EII

A range of white blood cells (the larger cells in the photograph) form the second line of defense.

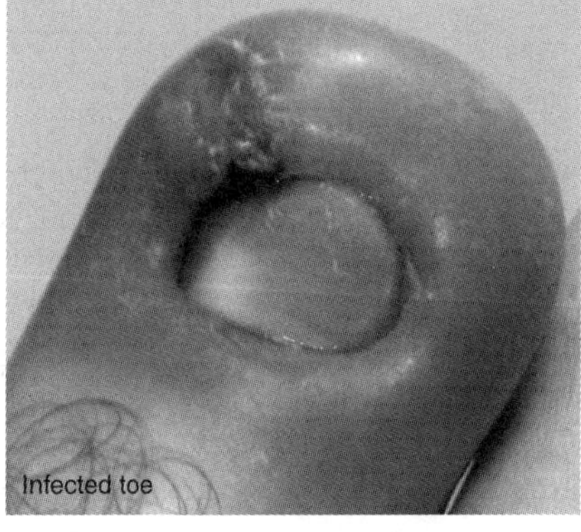
Infected toe
Wikimedia Commons

Inflammation is a localized response to infection characterized by swelling, pain, and redness.

1. How does the skin act as a barrier to prevent pathogens entering the body? ______

ISBN: 978-1-927173-16-9

The Importance of the First Line of Defense

The skin is the largest organ of the body. It forms an important physical barrier against the entry of pathogens into the body. A natural population of harmless microbes live on the skin, but most other microbes find the skin inhospitable. The continual shedding of old skin cells (arrow, right) physically removes bacteria from the surface of the skin. Sebaceous glands in the skin (labelled right) produce sebum, which has antimicrobial properties, and the slightly acidic secretions of sweat inhibit microbial growth.

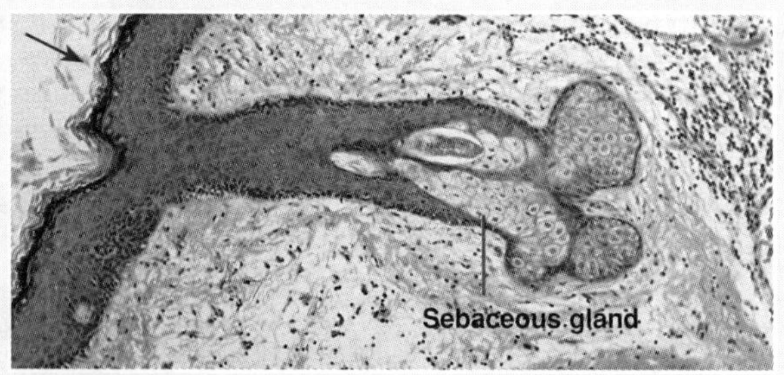
Sebaceous gland

Cilia line the epithelium of the **nasal passage** (below right). Their wave-like movement sweeps foreign material out and keeps the passage free of microorganisms, preventing them from colonizing the body.

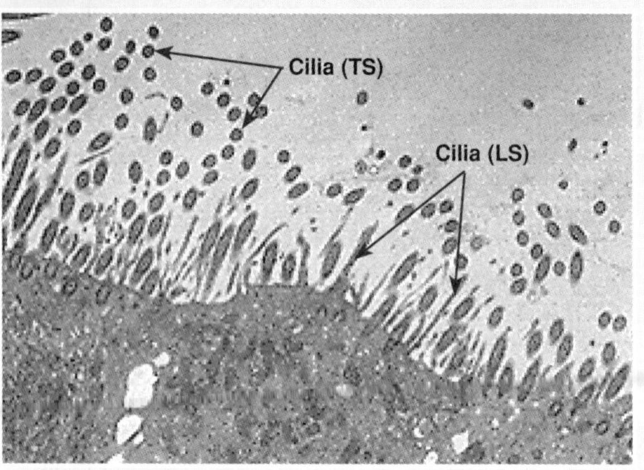
Cilia (TS)
Cilia (LS)
EII

Antimicrobial chemicals are present in many bodily secretions. Tears, saliva, nasal secretions, and human breast milk all contain **lysozymes** and **phospholipases**. Lysozymes kill bacterial cells by catalyzing the hydrolysis of cell wall linkages, whereas phospholipases hydrolyze the phospholipids in bacterial cell membranes, causing bacterial death. Low pH gastric secretions also inhibit microbial growth, and reduce the number of pathogens establishing colonies in the gastrointestinal tract.

2. Describe the role of each of the following in non-specific defense:

(a) Phospholipases: ______________________________

____

(b) Cilia: ______________________________

____

(c) Sebum: ______________________________

____

3. Distinguish between an **antibody** and an **antigen**: ______________________________

____

____

4. Describe the functional role of each of the following defense mechanisms:

(a) Phagocytosis by white blood cells: ______________________________

____

(b) Antimicrobial substances: ______________________________

____

(c) Antibody production: ______________________________

____

5. Explain the value of a three tiered system of defense against microbial invasion: ______________________________

____

____

____

ISBN: 978-1-927173-16-9

Blood Clotting and Defense

Apart from its transport role, **blood** has a role in the body's defense against infection and **hemostasis** (the prevention of bleeding and maintenance of blood volume). The tearing or puncturing of a blood vessel initiates **clotting**. Clotting is normally a rapid process that seals off the tear, preventing blood loss and the invasion of bacteria into the site. Clot formation is triggered by the release of clotting factors from the damaged cells at the site of the tear or puncture. A hardened clot forms a scab, which acts to prevent further blood loss and acts as a mechanical barrier to the entry of pathogens.

Blood Clotting

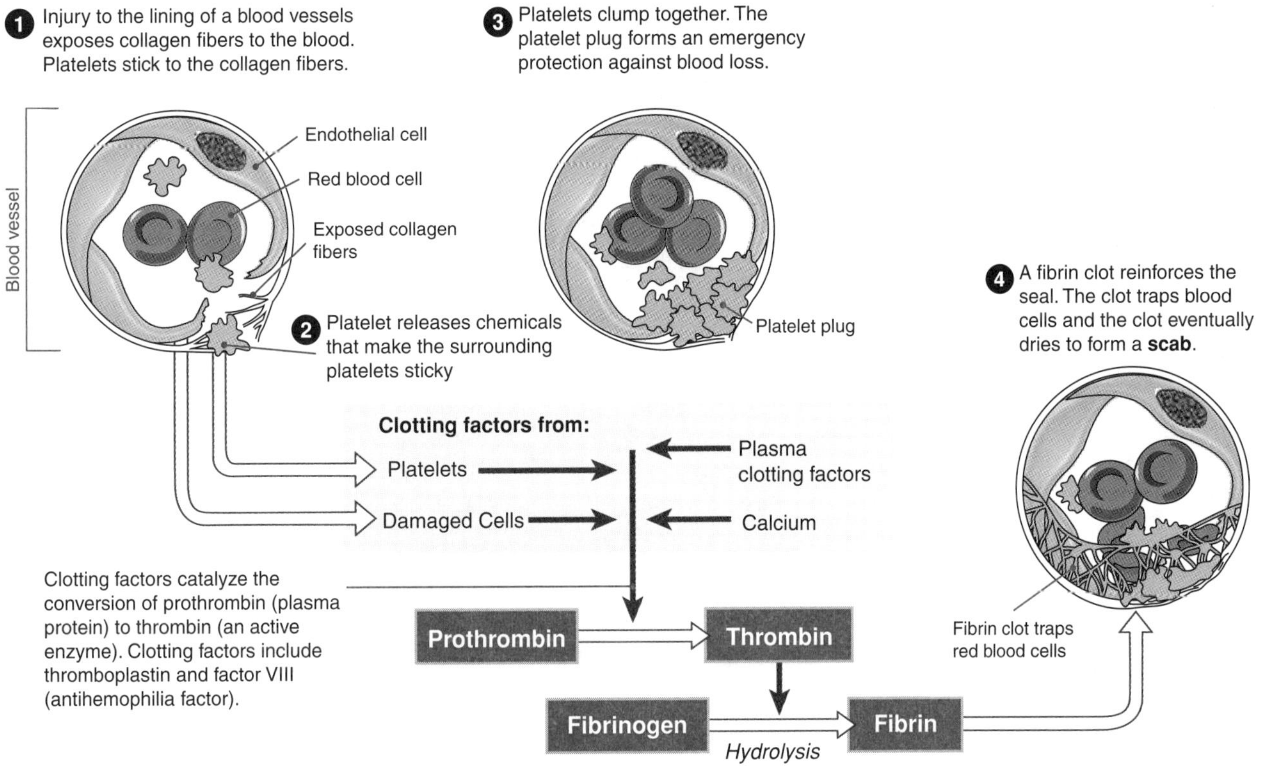

1 Injury to the lining of a blood vessels exposes collagen fibers to the blood. Platelets stick to the collagen fibers.

3 Platelets clump together. The platelet plug forms an emergency protection against blood loss.

Blood vessel

Endothelial cell

Red blood cell

Exposed collagen fibers

2 Platelet releases chemicals that make the surrounding platelets sticky

Platelet plug

4 A fibrin clot reinforces the seal. The clot traps blood cells and the clot eventually dries to form a **scab**.

Clotting factors from:

Platelets

Damaged Cells

Plasma clotting factors

Calcium

Clotting factors catalyze the conversion of prothrombin (plasma protein) to thrombin (an active enzyme). Clotting factors include thromboplastin and factor VIII (antihemophilia factor).

Prothrombin → **Thrombin**

Fibrinogen → *Hydrolysis* → **Fibrin**

Fibrin clot traps red blood cells

1. Explain two roles of the blood clotting system in internal defense and haemostasis:

   (a) ______

   (b) ______

2. Explain the role of each of the following in the sequence of events leading to a blood clot:

   (a) Injury: ______

   (b) Release of chemicals from platelets: ______

   (c) Clumping of platelets at the wound site: ______

   (d) Formation of a fibrin clot: ______

3. (a) What is the role of clotting factors in the blood in formation of the clot? ______

   (b) Why are these clotting factors not normally present in the plasma? ______

4. (a) Name one inherited disease caused by the absence of a clotting factor: ______

   (b) Name the clotting factor involved: ______

Defense Against Disease

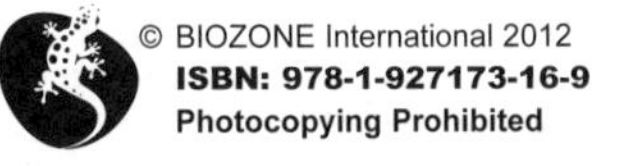

ISBN: 978-1-927173-16-9

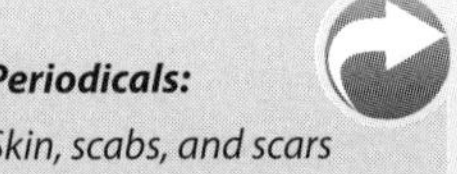

Periodicals:
Skin, scabs, and scars

Weblinks: *Hemostasis*

The Action of Phagocytes

Human cells that ingest microbes and digest them by the process of **phagocytosis** are called **phagocytes**. All are types of white blood cells. During many kinds of infections, especially bacterial infections, the total number of white blood cells increases by two to four times the normal number. The ratio of various white blood cell types changes during the course of an infection.

How a Phagocyte Destroys Microbes

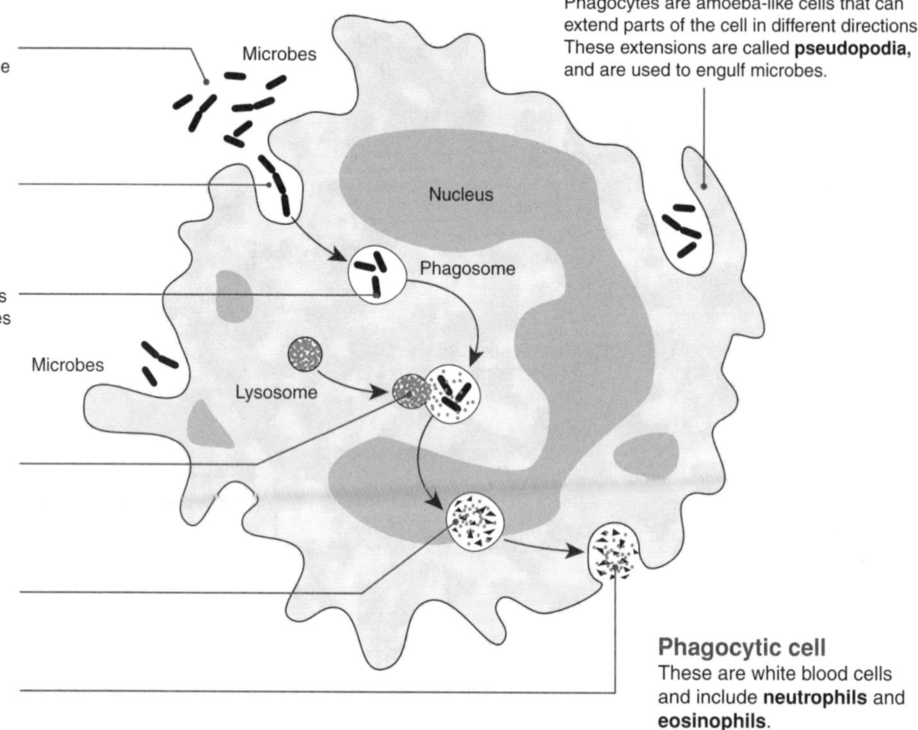

Microbes
Nucleus
Phagosome
Microbes
Lysosome
Phagocytes are amoeba-like cells that can extend parts of the cell in different directions. These extensions are called **pseudopodia,** and are used to engulf microbes.

1 Detection
Phagocyte detects microbes by the chemicals they give off and sticks the microbes to its surface.

2 Ingestion
The microbe is engulfed by the phagocyte wrapping pseudopodia around it to form a vesicle.

3 Phagosome forms
A phagosome (phagocytic vesicle) is formed, which encloses the microbes in a membrane.

4 Fusion with lysosome
Phagosome fuses with a lysosome (which contains powerful enzymes that can digest the microbe).

5 Digestion
The microbes are broken down by enzymes into their chemical constituents.

6 Discharge
Indigestible material is discharged from the phagocyte cell.

Phagocytic cell
These are white blood cells and include **neutrophils** and **eosinophils**.

The Interaction of Microbes and Phagocytes

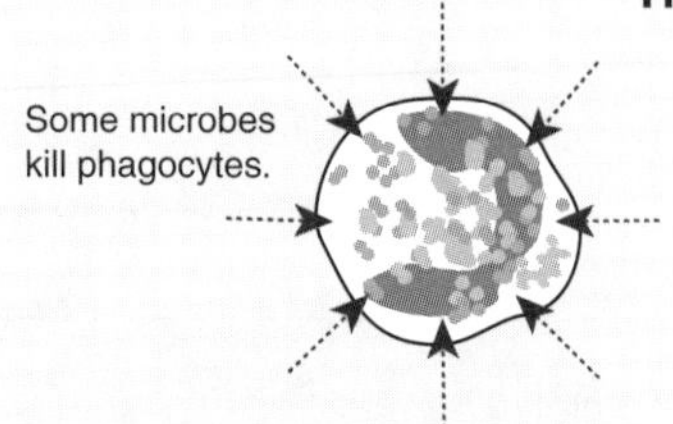

Some microbes kill phagocytes.

Some microbes kill phagocytes

Some microbes produce toxins that can actually kill phagocytes, e.g. toxin-producing staphylococci and the dental plaque-forming bacteria *Actinobacillus*.

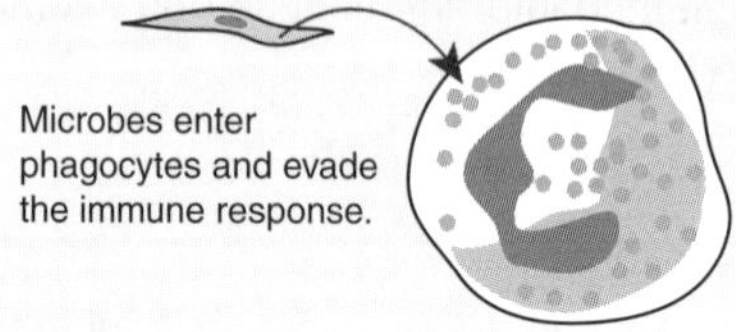

Microbes enter phagocytes and evade the immune response.

Microbes evade immune system

Some microbes can evade the immune system by entering phagocytes. The microbes prevent fusion of the lysosome with the phagosome and multiply inside the phagocyte, almost filling it. Examples include *Chlamydia*, *Mycobacterium tuberculosis*, *Shigella*, and malarial parasites.

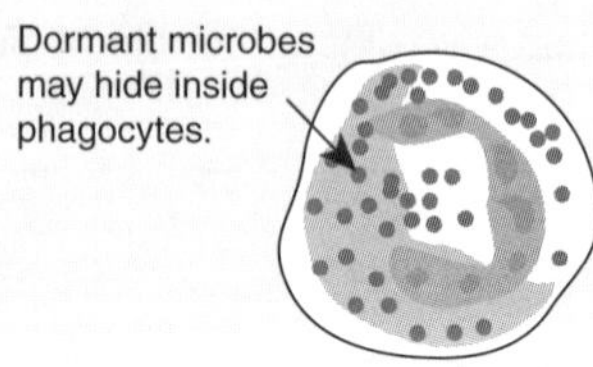

Dormant microbes may hide inside phagocytes.

Dormant microbes hide inside

Some microbes can remain dormant inside the phagocyte for months or years at a time. Examples include the microbes that cause brucellosis and tularemia.

1. Identify the white blood cells capable of phagocytosis: ______________________

____

2. Describe how a blood sample from a patient may be used to determine whether they have a microbial infection (without looking for the microbes themselves):

____

____

3. Explain how some microbes are able to overcome phagocytic cells and use them to their advantage:

____

____

ISBN: 978-1-927173-16-9

Inflammation

Damage to the body's tissues can be caused by physical agents (e.g. sharp objects, heat, radiant energy, or electricity), microbial infection, or chemical agents (e.g. gases, acids and bases). The damage triggers a defensive response called **inflammation**. It is usually characterized by four symptoms: pain, redness, heat, and swelling. The inflammatory response is beneficial and has the following functions: (1) to destroy the cause of the infection and remove it and its products from the body; (2) if this fails, to limit the effects on the body by confining the infection to a small area; (3) replacing or repairing tissue damaged by the infection. The process of inflammation can be divided into three distinct stages. These are described below.

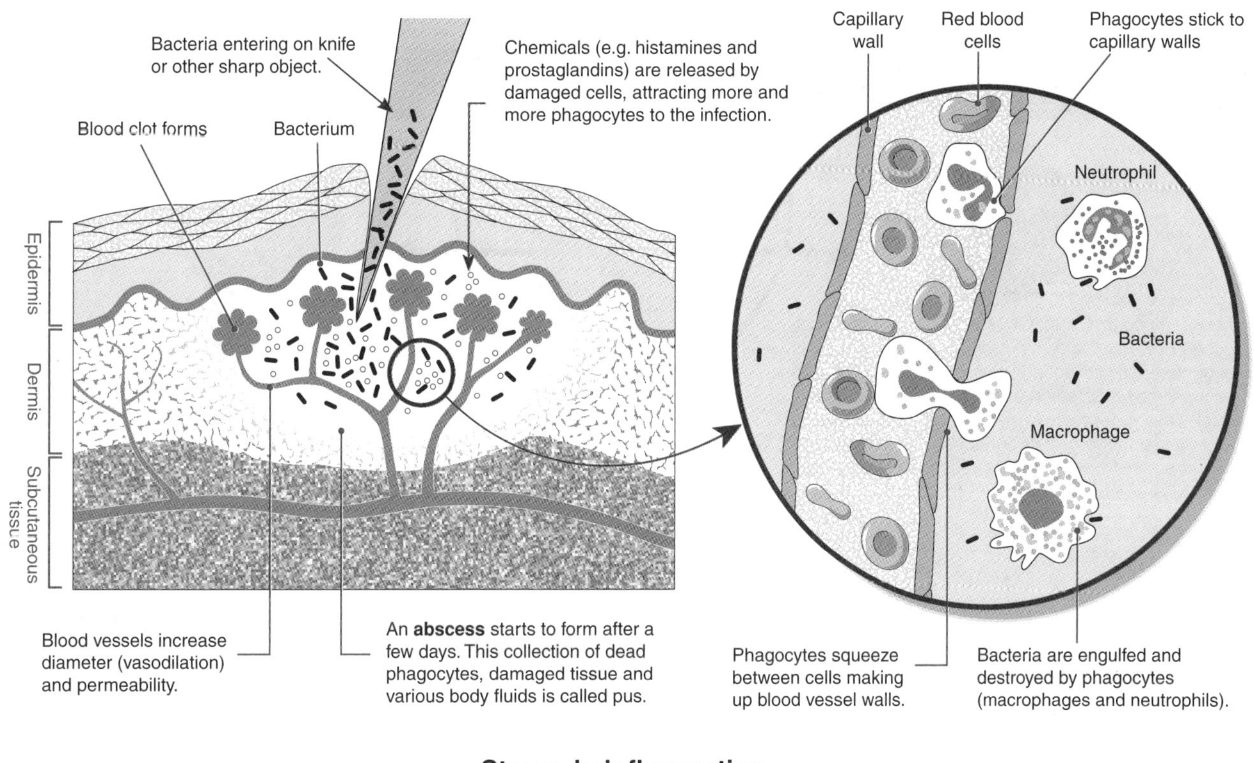
Bacteria entering on knife or other sharp object.
Chemicals (e.g. histamines and prostaglandins) are released by damaged cells, attracting more and more phagocytes to the infection.
Capillary wall
Red blood cells
Phagocytes stick to capillary walls
Blood clot forms
Bacterium
Neutrophil
Epidermis
Dermis
Subcutaneous tissue
Bacteria
Macrophage
Blood vessels increase diameter (vasodilation) and permeability.
An **abscess** starts to form after a few days. This collection of dead phagocytes, damaged tissue and various body fluids is called pus.
Phagocytes squeeze between cells making up blood vessel walls.
Bacteria are engulfed and destroyed by phagocytes (macrophages and neutrophils).

Stages in inflammation

Increased diameter and permeability of blood vessels	Phagocyte migration and phagocytosis	Tissue repair
Blood vessels increase their diameter and permeability in the area of damage. This increases blood flow to the area and allows defensive substances to leak into tissue spaces.	Within one hour of injury, phagocytes appear on the scene. They squeeze between cells of blood vessel walls to reach the damaged area where they destroy invading microbes.	Functioning cells or supporting connective cells create new tissue to replace dead or damaged cells. Some tissue regenerates easily (skin) while others do not at all (cardiac muscle).

1. Outline the three stages of inflammation and identify the beneficial role of each stage:

(a) ______________________________

(b) ______________________________

(c) ______________________________

2. Identify two features of phagocytes important in the response to microbial invasion: ______________________________

3. State the role of histamines and prostaglandins in inflammation: ______________________________

4. Explain why pus forms at the site of infection: ______________________________

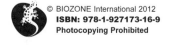

Defense Against Disease

The Immune System

The efficient internal defense provided by the immune system is based on its ability to respond specifically against foreign substances and hold a memory of this response. There are two main components of the immune system: the humoral and the cell-mediated responses. They work separately and together to provide protection against disease. The **humoral immune response** is associated with the serum (the non-cellular part of the blood) and involves the action of **antibodies** secreted by **B-cell lymphocytes**. Antibodies are found in extracellular fluids including lymph, plasma, and mucus secretions. The humoral response protects the body against circulating viruses, and bacteria and their toxins. The **cell-mediated immune response** is associated with the production of specialized lymphocytes called **T-cells**. It is most effective against bacteria and viruses located within host cells, as well as against parasitic protozoa, fungi, and worms.

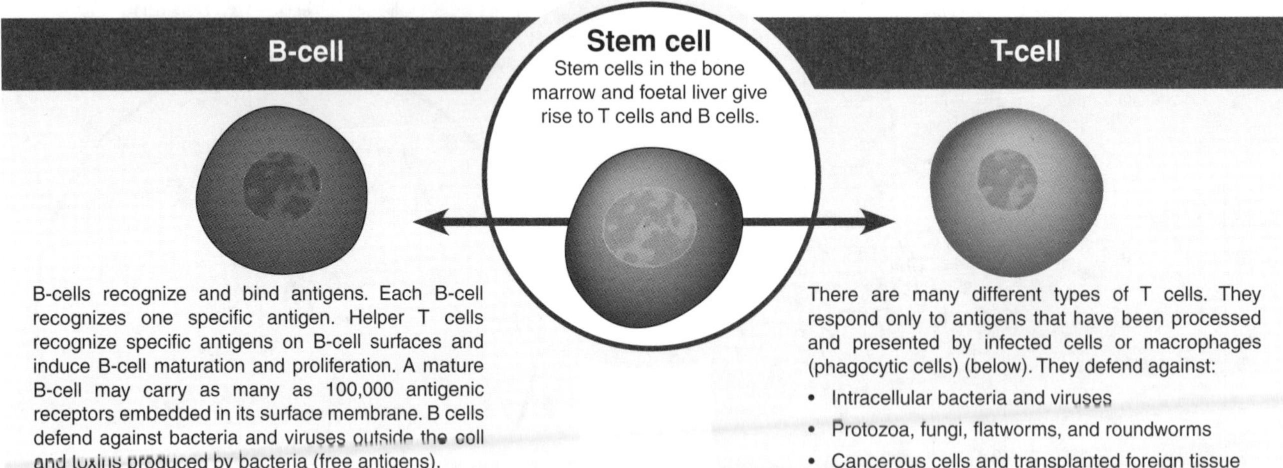
B-cell
Stem cell
Stem cells in the bone marrow and foetal liver give rise to T cells and B cells.
T-cell

B-cells recognize and bind antigens. Each B-cell recognizes one specific antigen. Helper T cells recognize specific antigens on B-cell surfaces and induce B-cell maturation and proliferation. A mature B-cell may carry as many as 100,000 antigenic receptors embedded in its surface membrane. B cells defend against bacteria and viruses outside the cell and toxins produced by bacteria (free antigens).

There are many different types of T cells. They respond only to antigens that have been processed and presented by infected cells or macrophages (phagocytic cells) (below). They defend against:

- Intracellular bacteria and viruses
- Protozoa, fungi, flatworms, and roundworms
- Cancerous cells and transplanted foreign tissue

B Cell and T Cell Activation

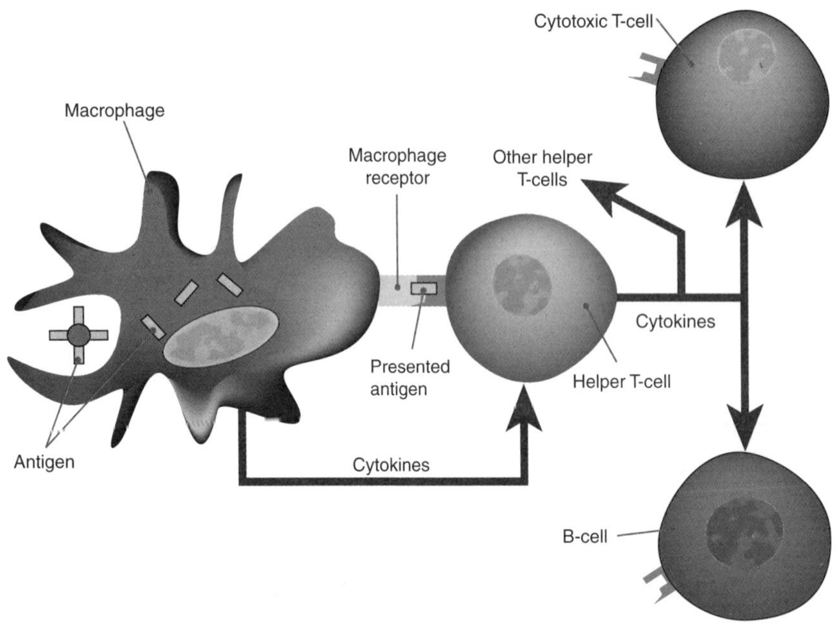
Cytotoxic T-cell
Macrophage
Macrophage receptor
Other helper T-cells
Cytokines
Presented antigen
Helper T-cell
Antigen
Cytokines
B-cell

Helper T-cells are activated by direct cell-to-cell signaling and by signaling to nearby cells using **cytokines** from macrophages.

Macrophages ingest antigens, process them, and present them on the cell surface where they are recognized by helper T-cells. The helper T-cell binds to the antigen and to the macrophage receptor, which leads to activation of the helper T-cell.

The macrophage also produces and releases cytokines, which enhance T-cell activation. The activated T-cell then releases more cytokines which causes the proliferation of other helper T-cells (positive feedback) and helps to activate cytotoxic T-cells and antibody-producing B-cells.

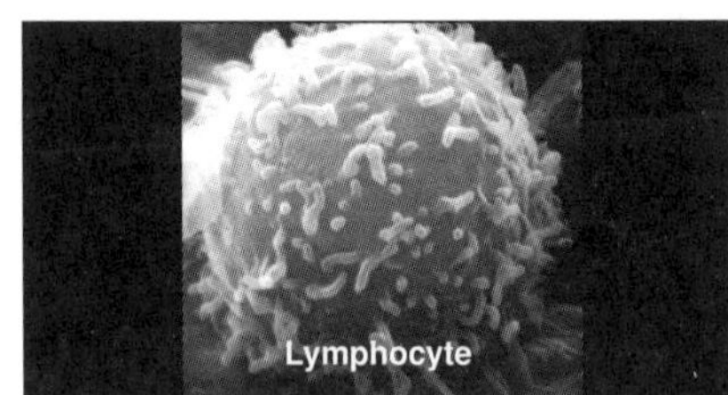
Lymphocyte

1. Describe the general action of the two major divisions in the immune system:

   (a) Humoral immune system: ______________________________

   (b) Cell-mediated immune system: ______________________________

2. Explain how an antigen causes the activation and proliferation of T-cells and B-cells: ______________________________

ISBN: 978-1-927173-16-9

Antibodies

Antibodies and antigens play key roles in the response of the immune system. Antigens are foreign molecules that are able to bind to antibodies (or T-cell receptors) and provoke a specific immune response. Antigens include potentially damaging microbes and their toxins as well as substances such as pollen grains, blood cell surface molecules, and the surface proteins on transplanted tissues. **Antibodies** (also called immunoglobulins) are proteins that are made in response to antigens. They are secreted into the plasma where they circulate and can recognize, bind to, and help to destroy antigens. There are five classes of **immunoglobulins**. Each plays a different role in the immune response (e.g. destroying protozoan parasites, enhancing phagocytosis, and neutralizing toxins and viruses). The human body can produce 100 million antibodies, recognizing many different antigens, including those it has never encountered. Each type of antibody is specific to only one particular antigen.

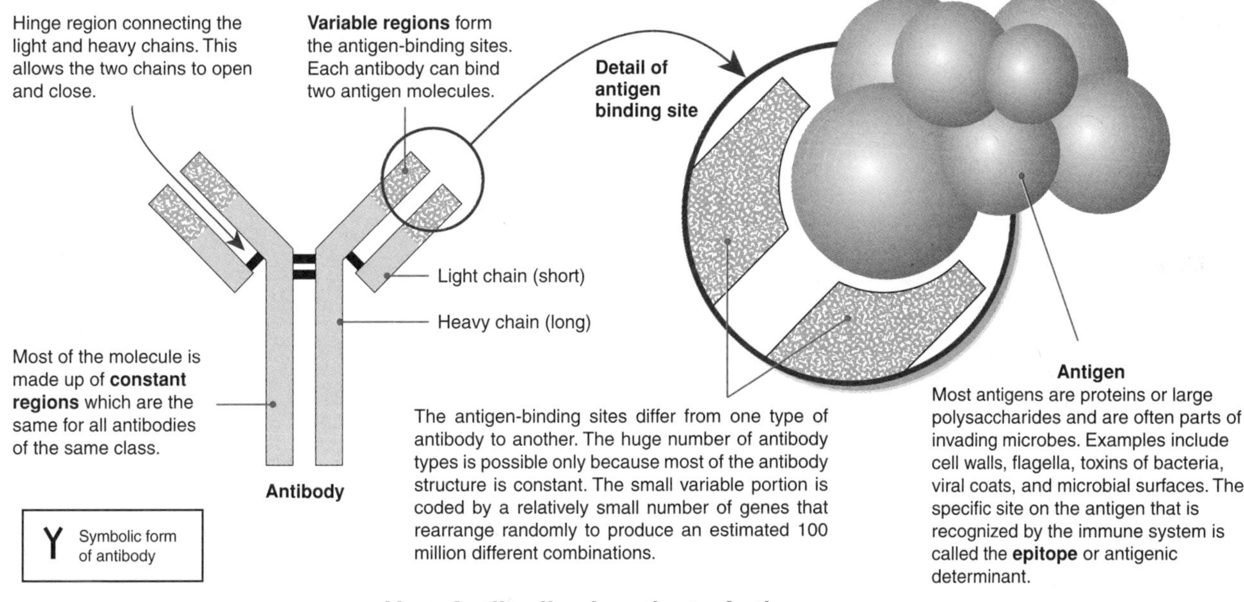
Hinge region connecting the light and heavy chains. This allows the two chains to open and close.
Variable regions form the antigen-binding sites. Each antibody can bind two antigen molecules.
Detail of antigen binding site
Light chain (short)
Heavy chain (long)
Most of the molecule is made up of constant regions which are the same for all antibodies of the same class.
Antibody
Y Symbolic form of antibody
The antigen-binding sites differ from one type of antibody to another. The huge number of antibody types is possible only because most of the antibody structure is constant. The small variable portion is coded by a relatively small number of genes that rearrange randomly to produce an estimated 100 million different combinations.
Antigen
Most antigens are proteins or large polysaccharides and are often parts of invading microbes. Examples include cell walls, flagella, toxins of bacteria, viral coats, and microbial surfaces. The specific site on the antigen that is recognized by the immune system is called the epitope or antigenic determinant.

How Antibodies Inactivate Antigens

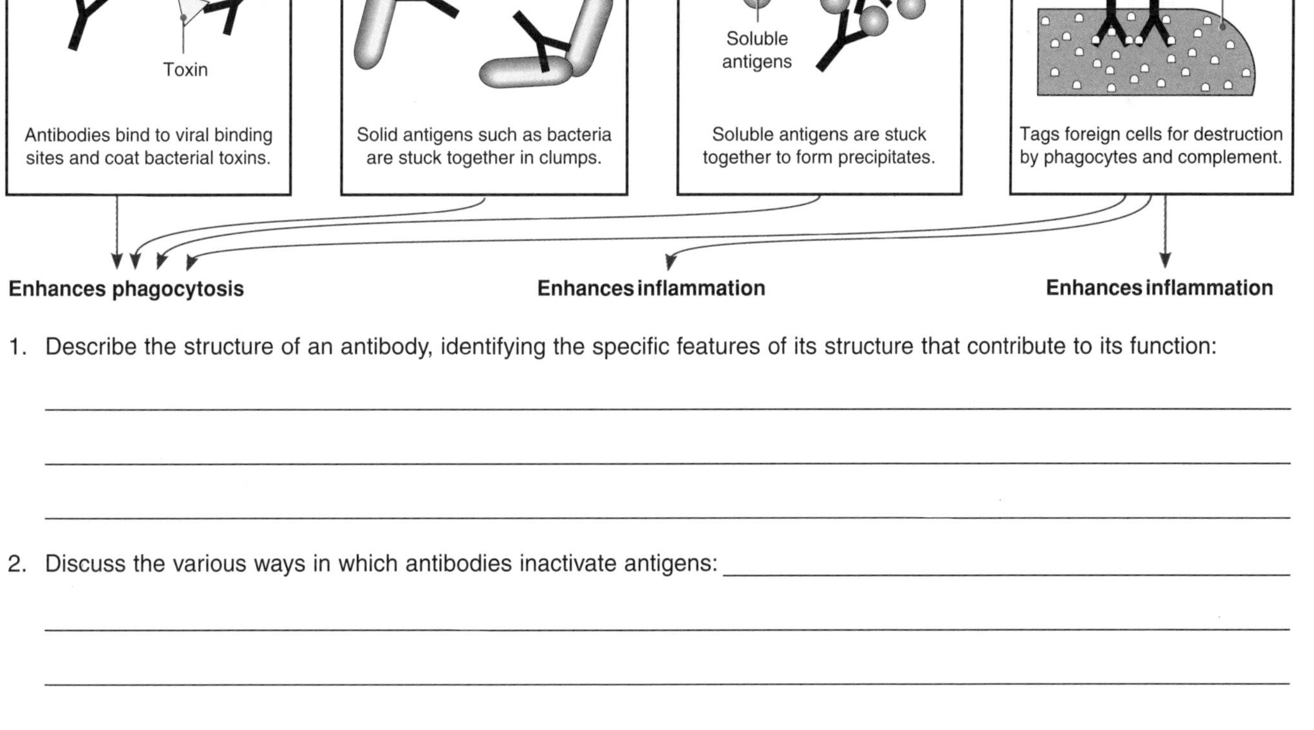
Neutralization
Virus
Toxin
Antibodies bind to viral binding sites and coat bacterial toxins.
Sticking together particulate antigens
Bacterial cell
Solid antigens such as bacteria are stuck together in clumps.
Precipitation of soluble antigens
Soluble antigens
Soluble antigens are stuck together to form precipitates.
Activation of complement
Complement
Bacterial cell
Tags foreign cells for destruction by phagocytes and complement.
Enhances phagocytosis
Enhances inflammation
Enhances inflammation

1. Describe the structure of an antibody, identifying the specific features of its structure that contribute to its function:

2. Discuss the various ways in which antibodies inactivate antigens:

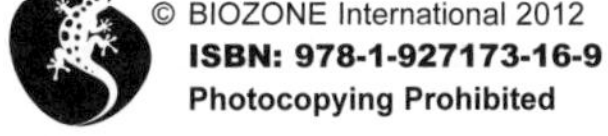

ISBN: 978-1-927173-16-9

Related activities: *The Immune System*
Weblinks: *The Humoral Response*

Clonal Selection

In 1955, the Australian, **Sir Frank Macfarlane Burnet** proposed the **clonal selection theory** to explain how the immune system is able to respond to the large and unpredictable range of potential antigens in the environment. The diagram below describes **clonal selection** after antigen exposure for B-cells. In the same way, a T-cell stimulated by a specific antigen will multiply and develop into different types of T-cells. Clonal selection and differentiation of lymphocytes provide the basis for **immunological memory**.

Five (a-e) of the many B-cells generated during development. Each one can recognize only one specific antigen.

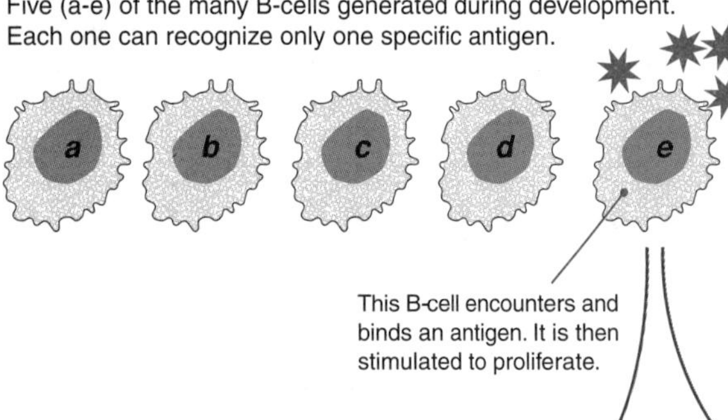
a
b
c
d
e

This B-cell encounters and binds an antigen. It is then stimulated to proliferate.

Clonal Selection Theory

Millions of B-cells form during development. Antigen recognition is randomly generated, so collectively they can recognize many antigens, including those that have never been encountered. Each B-cell makes antibodies corresponding to the specific antigenic receptor on its surface. The receptor reacts only to that specific antigen. When a B-cell encounters its antigen, it responds by proliferating and producing many clones all with the same kind of antibody. This is called clonal selection because the antigen selects the B cells that will proliferate.

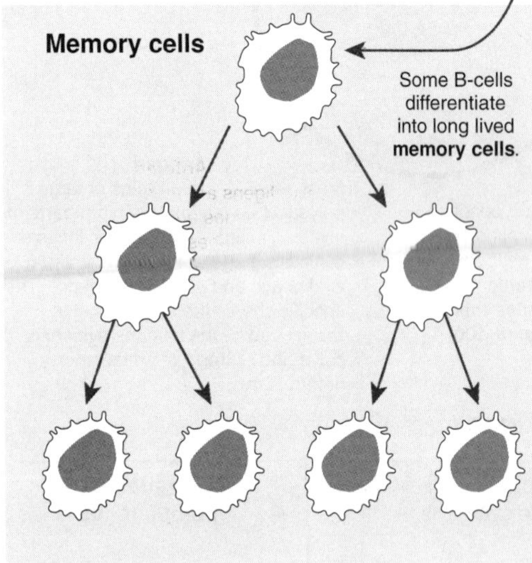
Memory cells
Some B-cells differentiate into long lived **memory cells**.

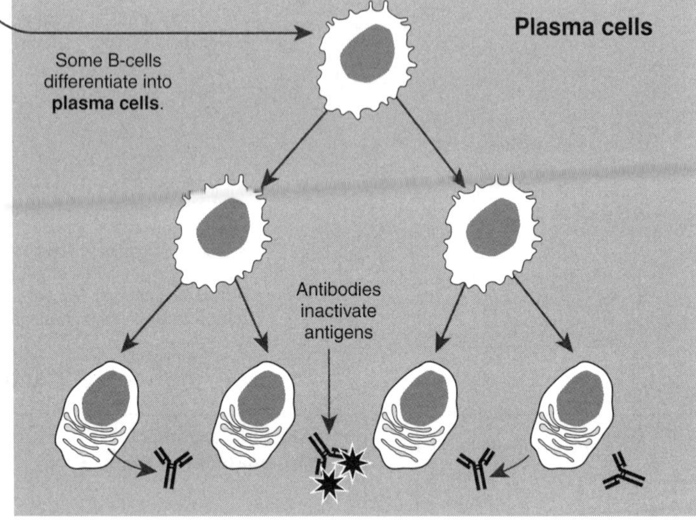
Plasma cells
Some B-cells differentiate into **plasma cells**.
Antibodies inactivate antigens

Some B-cells differentiate into long lived **memory cells**. These are retained in the lymph nodes to provide future immunity (**immunological memory**). In the event of a second infection, B-memory cells react more quickly and vigorously than the initial B-cell reaction to the first infection.

Plasma cells secrete antibodies specific to the antigen that stimulated their development. Each plasma cell lives for only a few days, but can produce about 2000 antibody molecules per second. Note that during development, any B-cells that react to the body's own antigens are selectively destroyed in a process that leads to **self tolerance** (acceptance of the body's own tissues).

1. Describe how clonal selection results in the proliferation of one particular B-cell: ____________________

____

____

____

2. Describe the function of each of the following cells in the immune system response:

   (a) Memory cells: ____________________

   ____

   (b) Plasma cells: ____________________

   ____

3. Explain the basis of **immunological memory**: ____________________

____

____

4. Describe how **self tolerance** develops and explain why it is important: ____________________

____

____

ISBN: 978-1-927173-16-9

Acquired Immunity

We have natural or **innate resistance** to certain illnesses; examples include most diseases of other animal species. **Acquired immunity** refers to the protection an animal develops against certain types of microbes or foreign substances. Immunity can be acquired either passively or actively and is developed during an individual's lifetime. **Active immunity** develops when a person is exposed to microorganisms or foreign substances and the immune system responds. **Passive immunity** is acquired when antibodies are transferred from one person to another. Recipients do not make the antibodies themselves and the effect lasts only as long as the antibodies are present, usually several weeks or months. Immunity may also be **naturally acquired**, through natural exposure to microbes, or **artificially acquired** as a result of medical treatment.

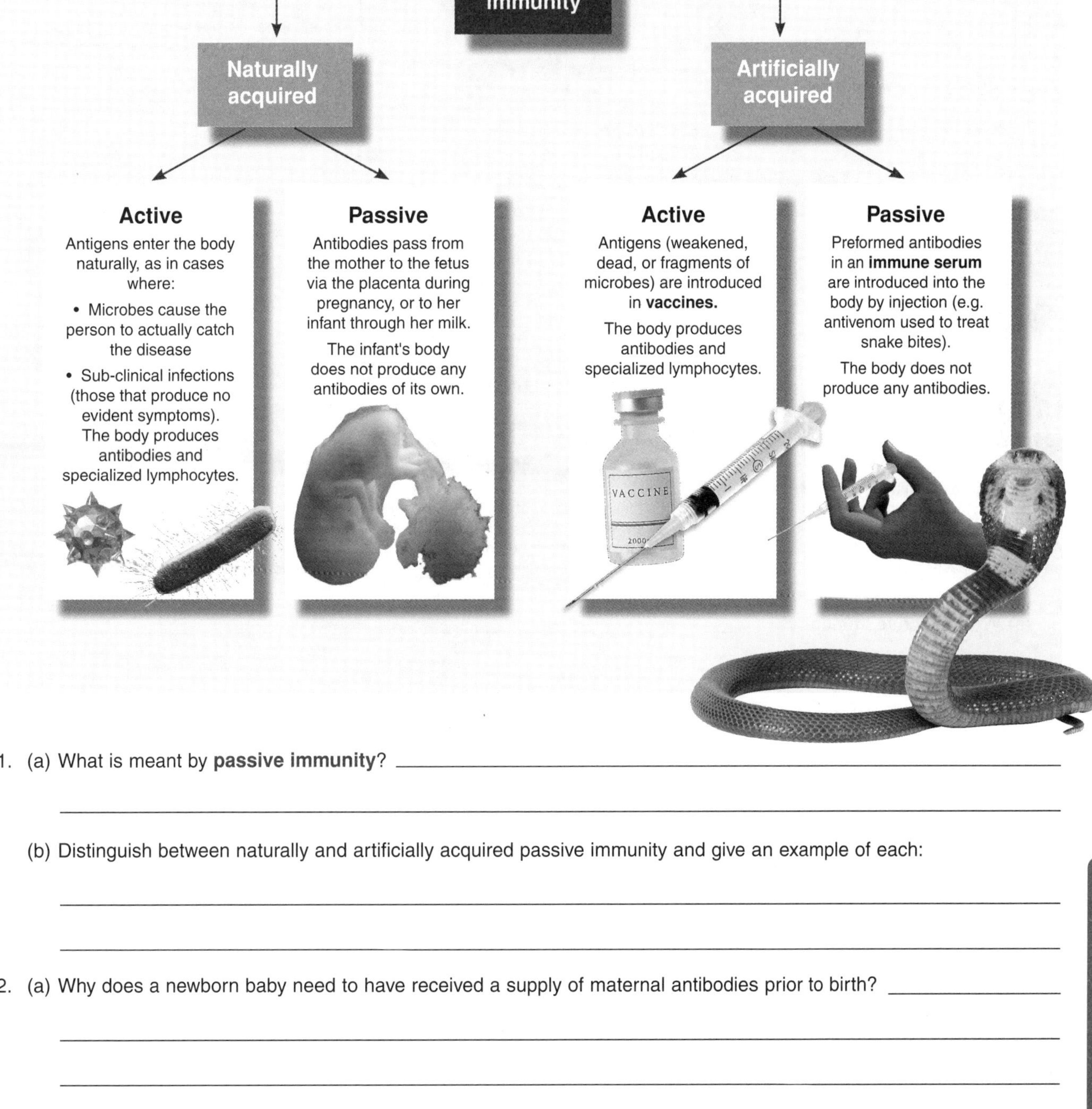

Acquired immunity
Naturally acquired
Artificially acquired
Active
Antigens enter the body naturally, as in cases where:
• Microbes cause the person to actually catch the disease
• Sub-clinical infections (those that produce no evident symptoms).
The body produces antibodies and specialized lymphocytes.
Passive
Antibodies pass from the mother to the fetus via the placenta during pregnancy, or to her infant through her milk.
The infant's body does not produce any antibodies of its own.
Active
Antigens (weakened, dead, or fragments of microbes) are introduced in vaccines.
The body produces antibodies and specialized lymphocytes.
VACCINE
Passive
Preformed antibodies in an immune serum are introduced into the body by injection (e.g. antivenom used to treat snake bites).
The body does not produce any antibodies.

1. (a) What is meant by **passive immunity**? ____________________

   ____

   (b) Distinguish between naturally and artificially acquired passive immunity and give an example of each:

   ____

   ____

2. (a) Why does a newborn baby need to have received a supply of maternal antibodies prior to birth? ____________________

   ____

   ____

   (b) Why is this supply supplemented by antibodies in breast milk? ____________________

   ____

   (c) Would you recommend breast feeding to a new mother? Explain your answer: ____________________

   ____

   ____

   ____

ISBN: 978-1-927173-16-9

Primary and Secondary Responses to Antigens

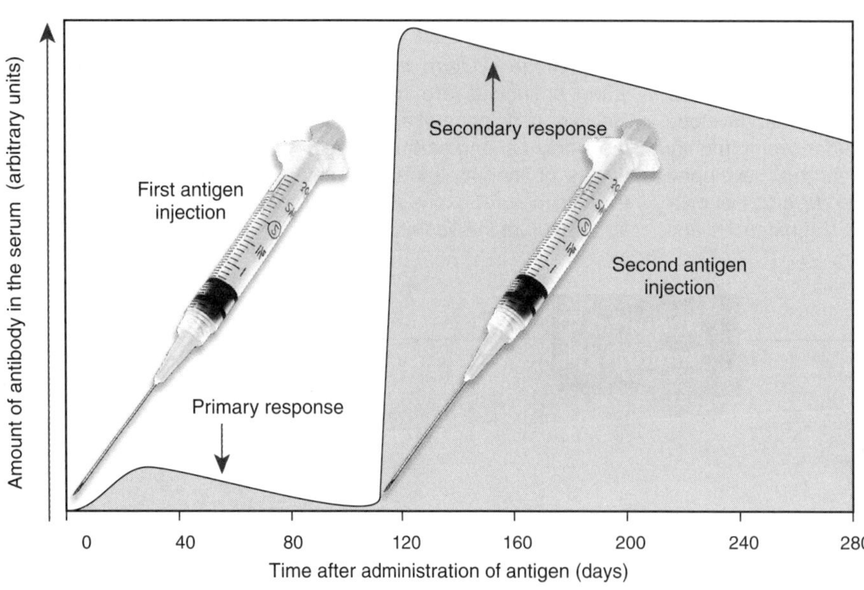

Amount of antibody in the serum (arbitrary units)
First antigen injection
Primary response
Secondary response
Second antigen injection
0
40
80
120
160
200
240
280
Time after administration of antigen (days)

When the B-cells encounter antigens and produce antibodies, the body develops **active immunity** against that antigen.

The initial response to antigenic stimulation, caused by the sudden increase in B-cell clones, is called the **primary response**. Antibody levels as a result of the primary response peak a few weeks after the response begins and then decline. However, because the immune system develops an immunological memory of that antigen, it responds much more quickly and strongly when presented with the same antigen subsequently (the **secondary response**).

This forms the basis of immunization programmes where one or more booster shots are provided following the initial vaccination.

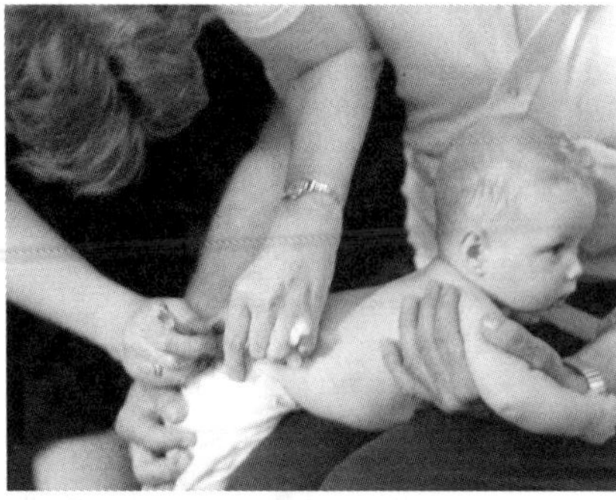

Vaccines against common diseases are given at various stages during childhood according to an immunization schedule. Vaccination has resulted in the decline of some once-common childhood diseases, such as mumps.

Many childhood diseases for which vaccination programmes exist are kept at a low level because of **herd immunity**. If most of the population is immune, those that are not immunized may be protected because the disease is uncommon.

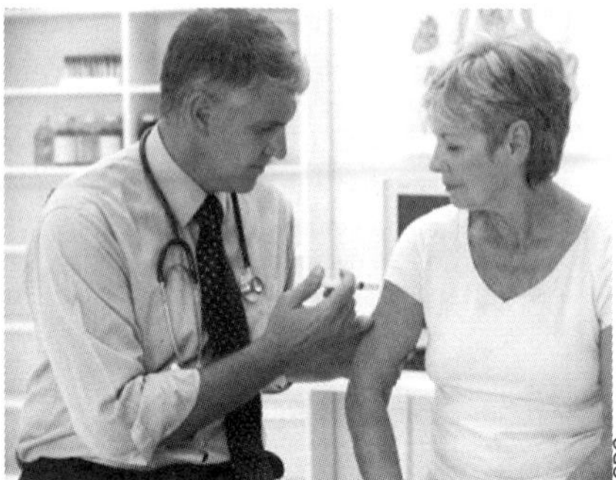

CDC

Most vaccinations are given in childhood, but adults may be vaccinated against a disease (e.g. TB, influenza) if they are in a high risk group (e.g. the elderly) or if they are travelling to a region in the world where a certain disease is prevalent.

3. (a) What is **active immunity**? ______________________________

   (b) Distinguish between naturally and artificially acquired active immunity and give an example of each: ______________________________

4. (a) Describe two differences between the primary and secondary responses to presentation of an antigen: ______________________________

   (b) Why is the secondary response so different from the primary response? ______________________________

5. (a) Explain the principle of **herd immunity**: ______________________________

   (b) Why are health authorities concerned when the vaccination rates for an infectious disease fall?

ISBN: 978-1-927173-16-9

Vaccines and Vaccination

Vaccines operate on the principle that they alert the immune system to the presence of a pathogen by introducing harmless but recognizably foreign antigens against which the body can form antibodies. There are two basic types of vaccine: subunit vaccines and whole-agent vaccines. **Whole-agent vaccines** contain complete nonvirulent microbes, either **inactivated** (killed), or alive but **attenuated** (weakened). Attenuated viruses make very effective vaccines and often provide life-long immunity without the need for booster immunizations. Killed viruses are less effective and many vaccines of this sort have now been replaced by newer subunit vaccines. **Subunit vaccines** contain only the parts of the pathogen that induce the immune response. They are safer than attenuated vaccines because they cannot reproduce in the recipient, and they produce fewer adverse effects because they contain little or no extra material. There are several ways to make subunit vaccines but, in all cases, the subunit vaccine loses its ability to cause disease while retaining its antigenic properties. Some of the most promising vaccines under development consist of naked DNA which is injected into the body and produces an antigenic protein. The safety of DNA vaccines is uncertain but they show promise against rapidly mutating viruses such as influenza and HIV.

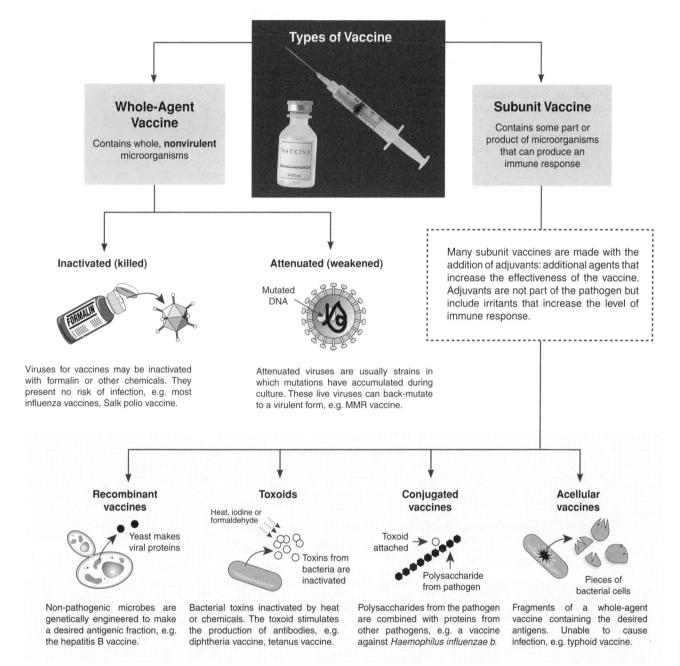

Types of Vaccine

VACCINE
2000ml

Whole-Agent Vaccine

Contains whole, **nonvirulent** microorganisms

Inactivated (killed)

FORMALIN

Viruses for vaccines may be inactivated with formalin or other chemicals. They present no risk of infection, e.g. most influenza vaccines, Salk polio vaccine.

Attenuated (weakened)

Mutated DNA

Attenuated viruses are usually strains in which mutations have accumulated during culture. These live viruses can back-mutate to a virulent form, e.g. MMR vaccine.

Subunit Vaccine

Contains some part or product of microorganisms that can produce an immune response

Many subunit vaccines are made with the addition of adjuvants: additional agents that increase the effectiveness of the vaccine. Adjuvants are not part of the pathogen but include irritants that increase the level of immune response.

Recombinant vaccines

Yeast makes viral proteins

Non-pathogenic microbes are genetically engineered to make a desired antigenic fraction, e.g. the hepatitis B vaccine.

Toxoids

Heat, iodine or formaldehyde

Toxins from bacteria are inactivated

Bacterial toxins inactivated by heat or chemicals. The toxoid stimulates the production of antibodies, e.g. diphtheria vaccine, tetanus vaccine.

Conjugated vaccines

Toxoid attached

Polysaccharide from pathogen

Polysaccharides from the pathogen are combined with proteins from other pathogens, e.g. a vaccine against *Haemophilus influenzae b.*

Acellular vaccines

Pieces of bacterial cells

Fragments of a whole-agent vaccine containing the desired antigens. Unable to cause infection, e.g. typhoid vaccine.

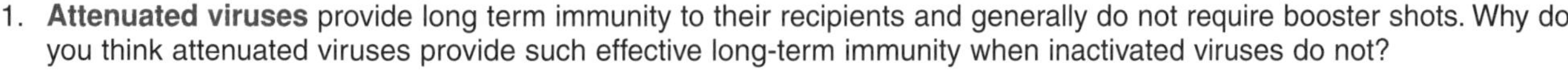

1. **Attenuated viruses** provide long term immunity to their recipients and generally do not require booster shots. Why do you think attenuated viruses provide such effective long-term immunity when inactivated viruses do not?

ISBN: 978-1-927173-16-9

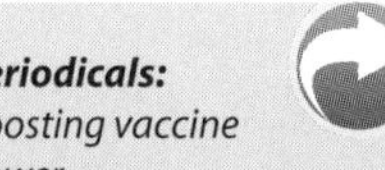

Periodicals: *Boosting vaccine power*

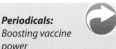

Related activities: *Acquired Immunity*
Weblinks: *Steps in Vaccine Development*

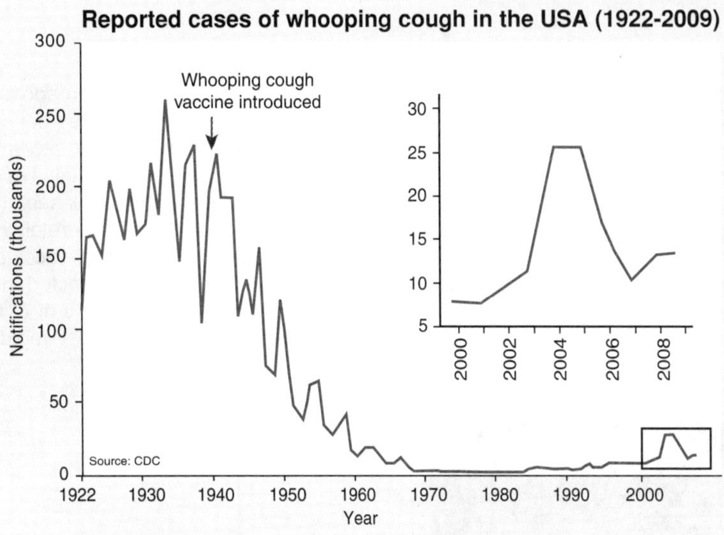
Reported cases of whooping cough in the USA (1922-2009)
Whooping cough vaccine introduced
Notifications (thousands)
Year
Source: CDC

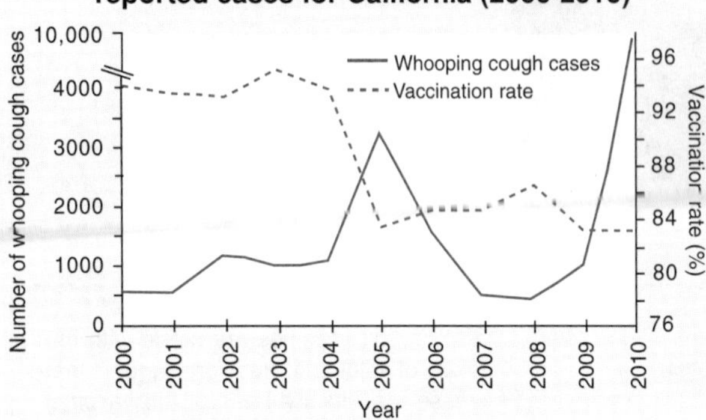
Whooping cough vaccination rates vs reported cases for California (2000-2010)
Whooping cough cases
Vaccination rate
Number of whooping cough cases
Vaccination rate (%)
Year

Case Study: Whooping Cough

Whooping cough is caused by the bacterium *Bordetella pertussis*, and infection may last for two to three months. It is characterized by painful coughing spasms, and a cough that sounds like a "whoop". Severe coughing fits may be followed by periods of vomiting. Inclusion of the whooping cough vaccine into the US immunization schedule in the 1940s has greatly reduced the incidence rates of the disease (left).

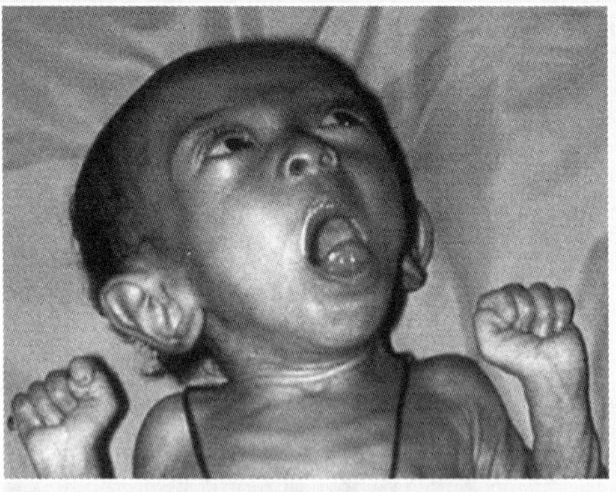
CDC

Above: Infants under six months of age are most at risk of developing complications or dying from whooping cough because they are too young to be fully protected by the vaccine. Ten infants died of whooping cough in California in 2010.

Left: In California, whooping cough vaccination rates have fallen amidst fears that it is responsible for certain health problems such as autism. As a result, rates of whooping cough have increased significantly since 2004. In 2010, over 9000 cases were reported, the highest level in 63 years.

2. How do high vaccination rates help to reduce the incidence of infectious disease?

3. (a) Describe the effect of introducing the whooping cough vaccine into the immunization schedule in the US:

(b) Why do you think whooping cough immunization rates have dropped significantly in California since 2004?

(c) What has been the effect of the lower immunization rates on the number of whooping cough cases?

(d) Suggest why the drop in immunization rates does not perfectly coincide with the increase in disease incidence:

4. Originally the whooping cough vaccine was a whole agent vaccine. In the 1990s it started being manufactured as an acellular vaccine. What advantages does an acellular vaccine have over a whole agent vaccine?

ISBN: 978-1-927173-16-9

Monoclonal Antibodies

A **monoclonal antibody** is an artificially produced antibody that binds to and neutralizes only one specific protein (**antigen**). A monoclonal antibody binds an antigen in the same way that a normally produced antibody does. Monoclonal antibodies are used as diagnostic tools (e.g. detecting pregnancy) or to treat some types of cancer or autoimmune diseases. Therapeutic uses are still limited because the antibodies are produced are from non-human cells and can cause side effects. In the future, production of monoclonal antibodies from human cells will probably result in fewer side effects. Monoclonal antibodies are produced in the laboratory by stimulating the production of B-cells in mice injected with the antigen. These B-cells produce an antibody against a specific antigen. Once isolated, they are made to fuse with immortal tumor cells, and they can be cultured indefinitely in a suitable growing medium (below). Monoclonal antibodies are useful for three reasons: they are all the same (i.e. clones), they can be produced in large quantities, and they are highly specific.

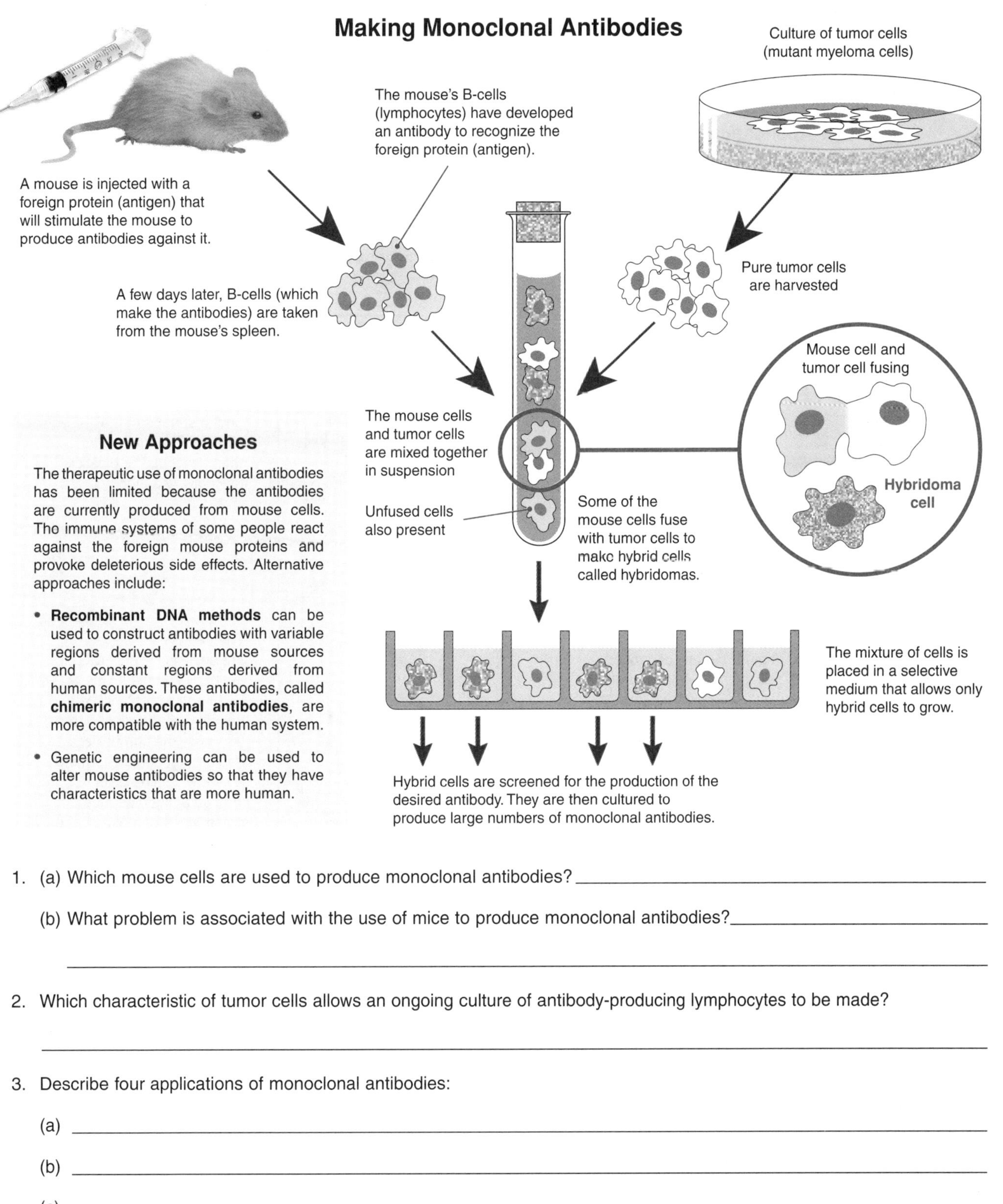
Making Monoclonal Antibodies
A mouse is injected with a foreign protein (antigen) that will stimulate the mouse to produce antibodies against it.
The mouse's B-cells (lymphocytes) have developed an antibody to recognize the foreign protein (antigen).
Culture of tumor cells (mutant myeloma cells)
A few days later, B-cells (which make the antibodies) are taken from the mouse's spleen.
Pure tumor cells are harvested
Mouse cell and tumor cell fusing
Hybridoma cell
The mouse cells and tumor cells are mixed together in suspension
Unfused cells also present
Some of the mouse cells fuse with tumor cells to make hybrid cells called hybridomas.
The mixture of cells is placed in a selective medium that allows only hybrid cells to grow.
Hybrid cells are screened for the production of the desired antibody. They are then cultured to produce large numbers of monoclonal antibodies.

New Approaches

The therapeutic use of monoclonal antibodies has been limited because the antibodies are currently produced from mouse cells. The immune systems of some people react against the foreign mouse proteins and provoke deleterious side effects. Alternative approaches include:

- **Recombinant DNA methods** can be used to construct antibodies with variable regions derived from mouse sources and constant regions derived from human sources. These antibodies, called **chimeric monoclonal antibodies**, are more compatible with the human system.
- Genetic engineering can be used to alter mouse antibodies so that they have characteristics that are more human.

1. (a) Which mouse cells are used to produce monoclonal antibodies? __________

   (b) What problem is associated with the use of mice to produce monoclonal antibodies? __________

2. Which characteristic of tumor cells allows an ongoing culture of antibody-producing lymphocytes to be made?

3. Describe four applications of monoclonal antibodies:

   (a)

   (b)

   (c)

   (d)

ISBN: 978-1-927173-16-9

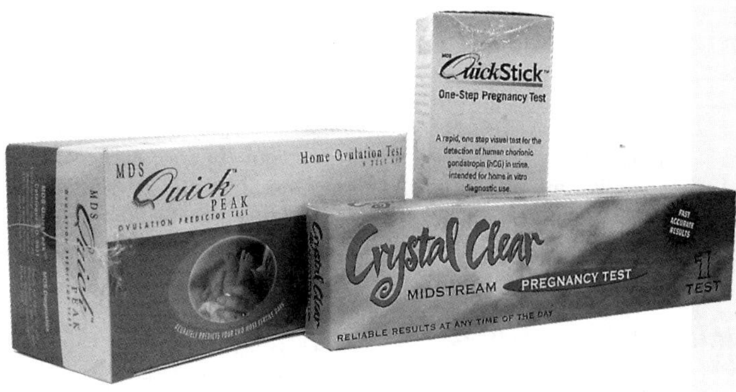
QuickStick
One-Step Pregnancy Test
MDS Quick PEAK
Home Ovulation Test
Crystal Clear
MIDSTREAM PREGNANCY TEST
RELIABLE RESULTS AT ANY TIME OF THE DAY

Detecting Pregnancy using Monoclonal Antibodies

When a woman becomes pregnant, a hormone called **human chorionic gonadotropin** (HCG) is released. HCG accumulates in the bloodstream and is excreted in the urine. Antibodies can be produced against HCG and used in simple test kits (below) to determine if a woman is pregnant. Monoclonal antibodies are also used in other home testing kits, such as those for detecting ovulation time (far left).

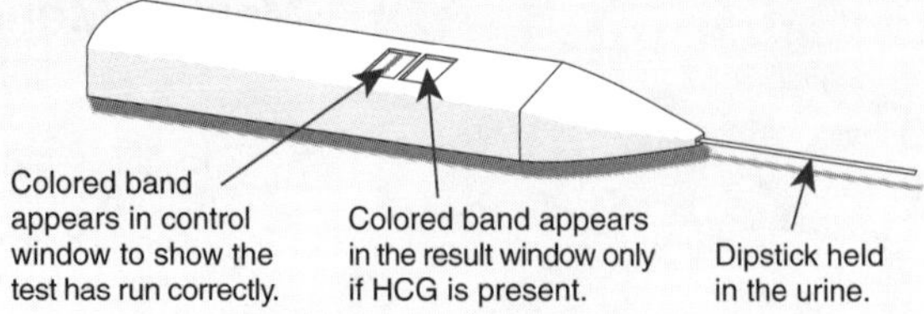
Colored band appears in control window to show the test has run correctly.
Colored band appears in the result window only if HCG is present.
Dipstick held in the urine.

How home pregnancy detection kits work

The test area of the dipstick (below) contains two types of antibodies: free monoclonal antibodies and capture monoclonal antibodies, bound to the substrate in the test window.

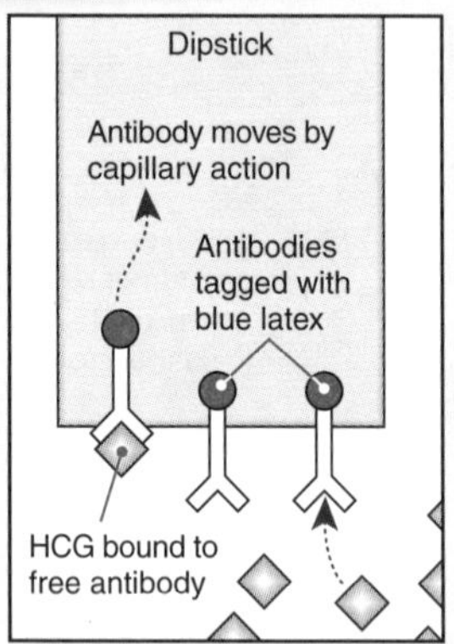
Dipstick
Antibody moves by capillary action
Antibodies tagged with blue latex
HCG bound to free antibody

The free antibodies are specific for HCG and are color-labeled. HCG in the urine of a pregnant woman binds to the free antibodies on the surface of the dipstick. The antibodies then travel up the dipstick by capillary action.

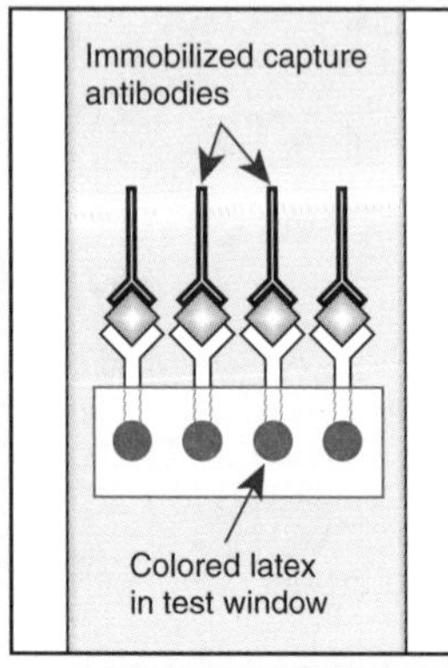
Immobilized capture antibodies
Colored latex in test window

The capture antibodies are specific for the HCG-antibody complex. The HCG-antibody complexes traveling up the dipstick are bound by the immobilized capture antibodies, forming a sandwich. The color labeled antibodies then create a visible color change in the test window.

Other Applications of Monoclonal Antibodies

Diagnostic uses

- Detecting the presence of pathogens such as *Chlamydia* and streptococcal bacteria, distinguishing between *Herpesvirus* I and II, and diagnosing AIDS.
- Measuring protein, toxin, or drug levels in serum.
- Blood and tissue typing.
- Detection of antibiotic residues in milk.

Therapeutic uses

- Neutralizing endotoxins produced by bacteria in blood infections.
- Used to prevent organ rejection, e.g. in kidney transplants, by interfering with the T cells involved with the rejection of transplanted tissue.
- Used in the treatment of some auto-immune disorders such as rheumatoid arthritis and allergic asthma. The monoclonal antibodies bind to and inactivate factors involved in the cascade leading to the inflammatory response.
- Immunodetection and immunotherapy of cancer. Herceptin is a monoclonal antibody for the targeted treatment of breast cancer. Herceptin recognizes receptor proteins on the outside of cancer cells and binds to them. The immune system can then identify the antibodies as foreign and destroy the cell.
- Inhibition of platelet clumping, which is used to prevent reclogging of coronary arteries in patients who have undergone angioplasty. The monoclonal antibodies bind to the receptors on the platelet surface that are normally linked by fibrinogen during the clotting process.

4. For each of the following applications, suggest why an antibody-based test or therapy is so valuable:

(a) Detection of toxins or bacteria in perishable foods: ______________________

____

____

(b) Detection of pregnancy without a doctor's prescription: ______________________

____

(c) Targeted treatment of tumors in cancer patients: ______________________

____

ISBN: 978-1-927173-16-9

HIV and AIDS

AIDS (acquired immune deficiency syndrome) was first reported in the US in 1981. By 1983, the pathogen had been identified as a retrovirus that selectively infects **helper T cells**. It has since been established that HIV arose by the recombination of two simian viruses. It has probably been endemic in some central African regions for decades, as HIV has been found in blood samples from several African nations from as early as 1959. The disease causes a massive deficiency in the immune system due to infection with **HIV** (human immunodeficiency virus). HIV is a **retrovirus** (RNA, not DNA) and is able to splice its genes into the host cell's chromosome. As yet, there is no cure or vaccine, and the disease has taken the form of a **pandemic**, spreading to all parts of the globe and killing more than a million people each year. In southern Africa, AIDS is widespread through the heterosexual community, partly as a result of social resistance to condom use and the high incidence of risky, polygamous behavior.

Capsid
Protein coat that protects the nucleic acids (RNA) within.

Viral envelope
A piece of the cell membrane budded off from the last human host cell.

Nucleic acid
Two identical strands of RNA contain the genetic blueprint for making more HIV viruses.

Reverse transcriptase
Two copies of this important enzyme convert the RNA into DNA once inside a host cell.

Surface proteins
These spikes allow HIV to attach to receptors on the host cells (T cells and macrophages).

The structure of HIV

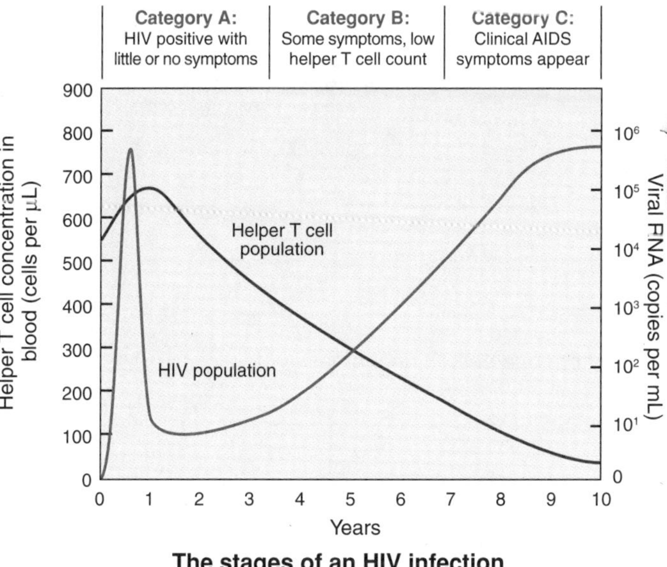
Category A: HIV positive with little or no symptoms
Category B: Some symptoms, low helper T cell count
Category C: Clinical AIDS symptoms appear
Helper T cell concentration in blood (cells per μL)
Viral RNA (copies per mL)
Helper T cell population
HIV population
Years
900
800
700
600
500
400
300
200
100
0
0 1 2 3 4 5 6 7 8 9 10
10^6
10^5
10^4
10^3
10^2
10^1
0

The stages of an HIV infection

AIDS is only the end stage of a HIV infection. Shortly after the initial infection, HIV antibodies appear in the blood. The progress of infection has three clinical categories (shown in the graph above).

HIV/AIDS

Individuals affected by the human immunodeficiency virus (HIV) may have no symptoms, while medical examination may detect swollen lymph glands. Others may experience a short-lived illness when they first become infected (resembling infectious mononucleosis). The range of symptoms resulting from HIV infection is huge, and is not the result of the HIV infection directly. The symptoms arise from an onslaught of secondary infections that gain a foothold in the body due to the suppressed immune system (due to the few helper T cells). These infections are from normally rare fungal, viral, and bacterial sources. Full blown AIDS can also feature some rare forms of cancer. Some symptoms are listed below:

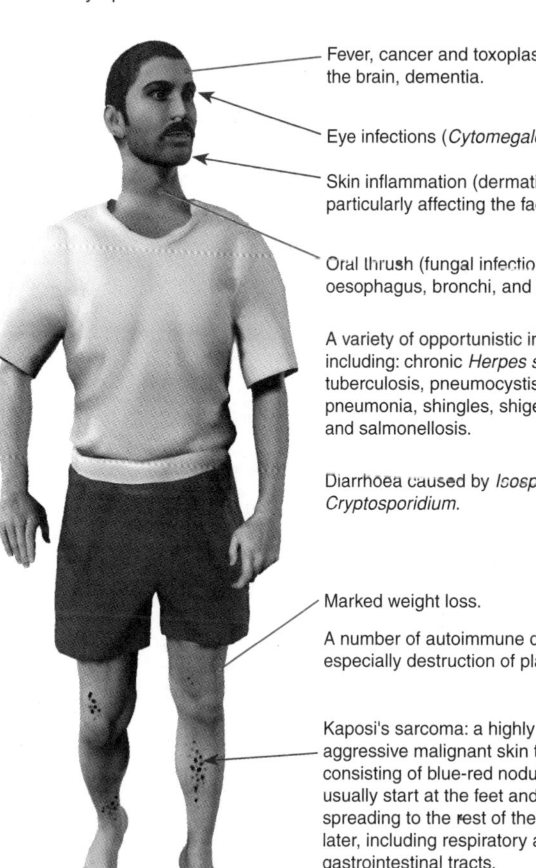
Fever, cancer and toxoplasmosis of the brain, dementia.
Eye infections (*Cytomegalovirus*).
Skin inflammation (dermatitis) particularly affecting the face.
Oral thrush (fungal infection) of the oesophagus, bronchi, and lungs.
A variety of opportunistic infections, including: chronic *Herpes simplex*, tuberculosis, pneumocystis pneumonia, shingles, shigellosis, and salmonellosis.
Diarrhoea caused by *Icospora* or *Cryptosporidium*.
Marked weight loss.
A number of autoimmune diseases, especially destruction of platelets.
Kaposi's sarcoma: a highly aggressive malignant skin tumour consisting of blue-red nodules, usually start at the feet and ankles, spreading to the rest of the body later, including respiratory and gastrointestinal tracts.

1. Explain why the HIV virus has such a devastating effect on the human body's ability to fight disease:

____

____

____

2. Consult the graph above showing the stages of HIV infection (remember, HIV infects and destroys helper T-cells).

   (a) Describe how the virus population changes with the progression of the disease: _______________

   ____

   ____

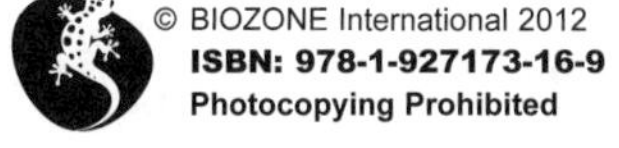

ISBN: 978-1-927173-16-9

Transmission, Diagnosis, Treatment, and Prevention of HIV

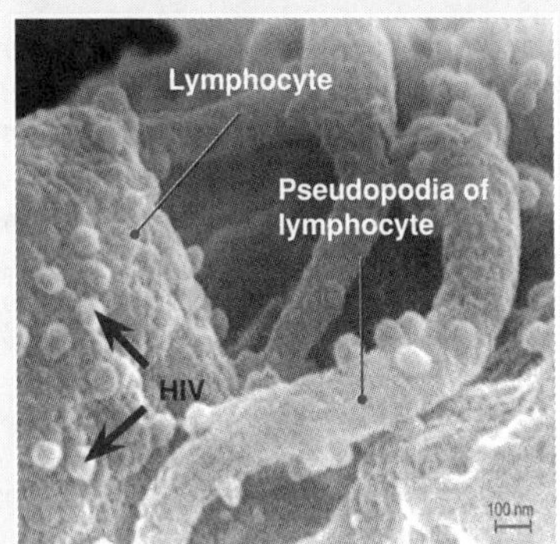
Lymphocyte
Pseudopodia of lymphocyte
HIV
100 nm
CDC

A SEM shows spherical HIV-1 virions on the surface of a human lymphocyte.

Modes of Transmission

1. HIV is transmitted in blood, vaginal secretions, semen, breast milk, and across the placenta.
2. In developed countries, blood transfusions are no longer a likely source of infection because blood is tested for HIV antibodies.
3. Historically, transmission of HIV in developed countries has been primarily through intravenous drug use and homosexual activity, but heterosexual transmission is increasing.
4. Transmission via heterosexual activity has been particularly important to the spread of HIV in Asia and southern Africa, partly because of the high prevalence of risky sexual behavior in these regions.

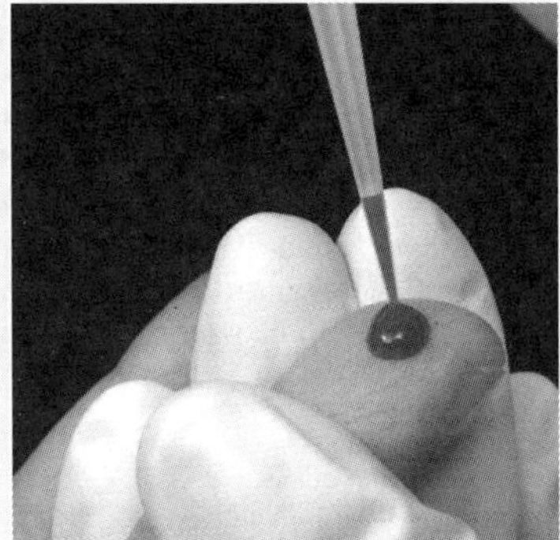
CDC

Diagnosis of HIV is possible using a simple antibody-based test on a blood sample.

HIV is easily transmitted between intravenous drug users who share needles.

Treatment and Prevention

Improving the acceptance and use of safe sex practices and condoms are crucial to reducing HIV infection rates. Condoms are protective irrespective of age, the scope of sexual networks, or the presence of other sexually transmitted infections. HIV's ability to destroy, evade, and hide inside the cells of the human immune system make it difficult to treat. Research into vaccination and chemotherapy is ongoing. The first chemotherapy drug to show promise was the nucleotide analogue AZT, which inhibits reverse transcriptase. Protease inhibitors are also used. These work by blocking the HIV protease so that HIV makes copies of itself that cannot infect other cells. An effective vaccine is still some time away, although recent vaccines based on monoclonal antibody technology appear to be promising.

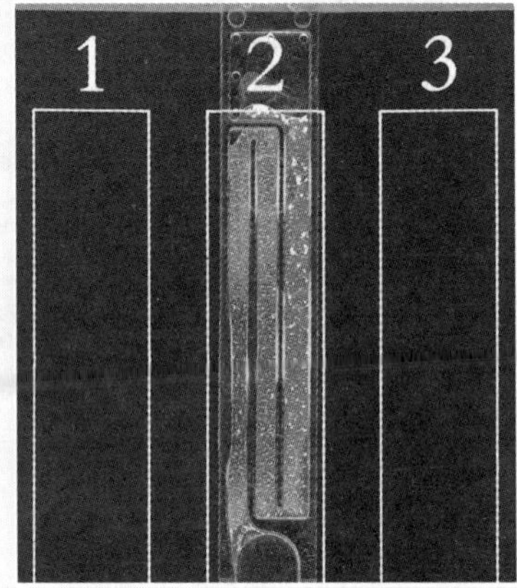
1
2
3
CDC

A positive HIV rapid test result shows clumping (aggregation) where HIV antibodies have reacted with HIV protein-coated latex beads.

(b) Describe how the helper T-cells respond to the infection: ______

3. Describe three common ways in which HIV can be transmitted from one person to another: ______

4. Discuss the social factors that have contributed to a high prevalence of HIV among heterosexuals in southern Africa:

5. Explain why it has been so difficult to develop a **vaccine** for HIV: ______

6. In a rare number of cases, people who have been HIV positive for many years still have no apparent symptoms. comment on the potential importance of this observation and its likely potential in the search for a cure for AIDS:

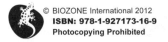

ISBN: 978-1-927173-16-9

The Impact of HIV/AIDS in Africa

Around 10% of the world's population lives in sub-Saharan Africa, yet this region is home to two thirds of HIV-infected people. The impact of HIV-AIDS on Africa's populations, workplaces, and economies is enormous, and is setting back Africa's economic and social progress. The effects of the disease are disproportionate; the vast majority of people living with HIV in Africa are in their working prime. Life expectancies too have fallen; in many African countries they are half what they were 15 years ago (see the graph to the right). This has been detrimental to all aspects of African social and economic structure.

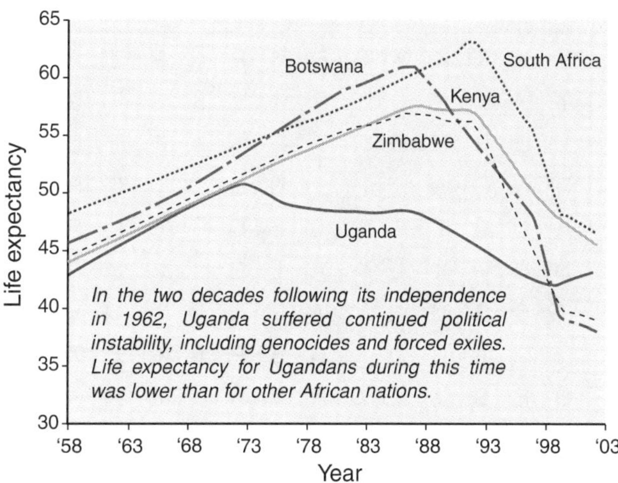

Life expectancy in some southern African countries
Life expectancy
Year
65
60
55
50
45
40
35
30
'58
'63
'68
'73
'78
'83
'88
'93
'98
'03
Botswana
South Africa
Kenya
Zimbabwe
Uganda
In the two decades following its independence in 1962, Uganda suffered continued political instability, including genocides and forced exiles. Life expectancy for Ugandans during this time was lower than for other African nations.

The Impact on the Health Sector

Increasingly, the demand of the AIDS epidemic on health care facilities is not being adequately met. In sub-Saharan Africa, people with HIV-related diseases occupy more than half of all hospital beds. Health care workers are also at high risk of contracting HIV. For example, Botswana lost 17% of its healthcare workforce to AIDS between 1999 and 2005. In some regions, 40% of midwives are HIV positive. Although access to treatment is improving across the continent, most of those needing treatment do not receive it.

Photo courtesy of Karl Mueller; Flickr

This Malawi grandmother is now responsible for the care and support of her four grandchildren, whose parents both died of AIDS. The intergenerational effects of HIV/AIDS are the longest lasting and are related to how the epidemic intensifies poverty and leads to its persistence. The challenge to African countries is in achieving the sustainable development needed to respond effectively to the epidemic.

Information from various sources including UNAIDS, UNDP, and AVERT.org

The Impact on Households

The AIDS epidemic has the greatest impact on poor households. Loss of one or both parents to AIDS results in a loss of income, and leads to a dissolution of family structure. The burden on older people is immense. Parents of adults with AIDS often find themselves supporting a household and caring for their orphaned grandchildren (left). Children (especially girls) may be forced to abandon their educations to work at home or to care for sick relatives. Damage to the education sector has a feedback effect too, as poor education fuels the spread of HIV.

The Impact on Food Security

The AIDS epidemic adds to food insecurity in sub-Saharan Africa. In many African countries, where food is already in short supply, HIV/AIDS is responsible for a severe depletion of the agricultural workforce and a consequent decline in agricultural output. Much of burden of coping rests with women, who are forced to take up roles outside their homes as well as continue as housekeepers, carers, and providers of food.

The Future?

AIDS in Africa is linked to other problems, including poverty, food insecurity, and poor public infrastructure. Efforts to fight the epidemic must work within these constraints. Without control, the AIDS epidemic will continue to present the single greatest barrier to Africa's social and economic development.

1. With reference to the graph of life expectancies in southern African countries:

   (a) Describe and explain the trends in since the late 1980s: __________

   (b) Describe and give a likely explanation of the trends between 1958 and 1988: __________

2. Discuss the barriers sub-Saharan African nations will face in countering the effects of the HIV/AIDS: __________

Defense Against Disease

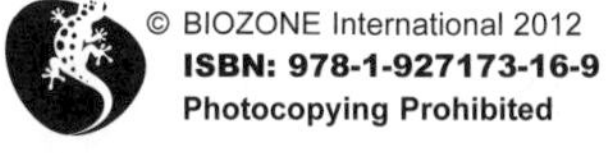

ISBN: 978-1-927173-16-9

KEY TERMS: Word Find

Use the clues below to find the relevant key terms in the WORD FIND grid

```
E Y X K Y P F B E C R Z R E V X S P L I D Y M L S Q S E U Q
Y M H S L K P A V K S F G D I Y E W A I D S W Y G E I Q P L
L U P R O R R K W A I T D P H A G O C Y T E N N I D S Z A E
B C O W J P I H D C U C H Z R K S E B U M Y I D P I O T T V
A O C I F G L E Q T C P R E V A L E N C E T O Z W I Q Y I F
N U F X L A A L E I W Y P C X J T F P T T B C V N N C Z K F
T S B M W Y W P X V L Q V A I J F N D O I E N Z I O L E C O
I M A R O I F E P E P Z X A S X T M L T V T I P H S O M S N
B E B I M D P R A I B H A H M P A C N P R I C P G K N M H Z
I M M Q X W A T Y M K U H I V X D A M H W J M N W I A E E G
O B I K N Y S C U M P E D C E O L Q G D Y Y S R O N L M M X
T R Q P T M S E B U C O M O O A W W B A L X C X P X S O O F
I A P E C W I L U N H H T L N P L A S M A C E L L S E R S N
C N A D E B V L A I Y R B O Q G N T J W P C Y C W Y L Y T Y
S E T T L H E A N T B U L T Z M J J B F T F L X S X E C A Q
N U H C L X Q R H Y W C K G O T H U P B M H B Q B W C E S U
S N O I M M U N O L O G I C A L M E M O R Y Y Q C C T L I Y
I Q G X D M O M C N B P D Q E Y A N T I B O D Y E O I L S Z
E F E I E P M Z O A X M Z A A N T I G E N F A X L J O S Y Y
L A N J M K F M V S B O L B B M Y K T O D P V U L B N K T F
```

CORE CLUES

The acronym of a syndrome caused by a retrovirus that selectively infects helper T cells.

A protein made in response to an antigen.

A foreign molecule that is able to bind to an antibody and provoke a specific immune response.

A lymphocyte that makes antibodies against specific antigens. (2 words; 1, 4)

The abbreviation for the retrovirus which causes AIDS.

A specific white blood cell involved in the adaptive immune response.

A disease-causing organism.

A white blood cell that destroys microbes by digesting them.

Chemotherapeutic agents that inhibit or prevent bacterial growth.

The term describing the number of people in a population affected by a disease.

This is the first line of defense against disease and provides a physical barrier to prevent pathogens entering the body.

A chemical secreted by the skin that inhibits the growth of bacteria and fungi.

A membrane lining body passages that have contact with the air (e.g. nasal passage), these often secrete mucus. (2 words: 6, 8)

AHL ONLY CLUES

The name of an Immunity that is induced in the host itself by the antigen, and is long lasting. (2 words; 6, 8)

A theory for how B-cells and T-cells are selected to target specific antigens invading the body. (2 words; 6, 9)

The ability of the immune system to respond rapidly in the future to antigens encountered in the past. (2 words; 11, 6)

Artificially produced antibodies that specifically bind to a substance; they can then serve to detect or otherwise target that substance. (2 words: 10, 10)

The name of the immunity gained by the receipt of ready-made antibodies.

Type of lymphocyte responsible for the cellular immune response. (2 words; 1, 4)

B-lymphocytes that have been exposed to an antigen and produce large numbers of antibodies. (2 words: 6, 5)

This process stops bleeding and helps to maintain blood volume.

A specialized lymphocyte which promote the activation and function of B-cells and killer T-cells. (3 words: 6, 1, 4)

The process by which a clot is formed to prevent blood loss. (2 words: 5, 8)

These B-cells provide immunological memory against a specific antigen, increasing the reaction to it the next time it is encountered. (2 words: 6, 5)

ISBN: 978-1-927173-16-9

Core Topic
6.4

Gas Exchange

Key concepts

- Organisms exchange respiratory gases with their environment.
- Gas exchange surfaces have specific properties that maximize exchange rates.
- The ventilation system moves respiratory gases in and out of the body and helps to maintain the concentration gradients for diffusion of those gases.
- Respiratory pigments increase the oxygen carrying capacity of the circulatory fluids.

Key terms

Core

alveoli (*sing*. alveolus)
breathing
bronchi
bronchioles
cell respiration
diaphragm
gas exchange
gas exchange surface
intercostal muscles
lungs
spirometry
trachea
ventilation
ventilation system

Learning Objectives

☐ 1. Use the **KEY TERMS** to compile a glossary for this topic.

Principles of Gas Exchange *(6.4.1-6.4.2)* page 312

☐ 2. Explain why organisms must exchange **respiratory gases** with their environment.

☐ 3. Distinguish between **ventilation**, **gas exchange**, and **cell respiration**.

☐ 4. Describe the essential features of **gas exchange surfaces** and their significance in terms of gas exchange rates. Explain the significance of the surface area: volume ratio to the exchange of gases with the environment.

☐ 5. Explain the need for a **ventilation system** and explain how it fulfils its role.

Gas Exchange System *(6.4.3-6.4.5)* pages 313-318

☐ 6. Describe the features of **alveoli** that are adapted to their role in gas exchange.

☐ 7. Draw and label a diagram of the ventilation system, including the **trachea**, **lungs**, **bronchi**, **bronchioles**, and **alveoli**. Draw the detail of an **alveolus** to show its relationship with the surrounding capillaries.

☐ 8. Explain the mechanism of lung ventilation (**breathing**). Include reference to the changes in lung volume and pressure and how these are brought about by the action of the internal and external **intercostal muscles**, the **diaphragm**, and the abdominal muscles. Distinguish between quiet and forced breathing and comment on the involvement of muscles in each of these.

☐ 9. Explain how changes in lung volume during breathing can be measured using **spirometry**. As required, interpret a spirogram showing changes in lung volumes in different situations, including at rest and during exercise.

☐ 10. Understand the relationship between the ventilation system and the internal transport system in humans. Include reference to the transport of oxygen and carbon dioxide.

☐ 11. EXTENSION/CONTEXT: Explain some of the effects of high altitude on the body's ventilation system and ability to obtain oxygen. Describe and explain some of the physiological adjustments that can be made to cope with high altitude. Describe and explain some of the heritable adaptations shown by high altitude populations.

Periodicals:
Listings for this chapter are on page 400

Weblinks:
www.thebiozone.com/weblink/IB-3169.html

BIOZONE APP:
Student Review Series
The Respiratory System

Introduction to Gas Exchange

Living cells require energy for the activities of life. Energy is released in cells by the breakdown of sugars and other substances in the metabolic process called **cell respiration**. As a consequence of this process, gases need to be exchanged (by **diffusion**) between the respiring cells and the environment. In most organisms these gases are carbon dioxide and oxygen. The diagram below illustrates this for an animal (human). Plant cells also respire, but their gas exchange budget is different because they consume CO_2 and produce O_2 in photosynthesis. Effective gas exchange surfaces are thin so that the barrier they present to diffusion is minimized. Diffusion gradients are maintained by transport of gases away from the gas exchange surface.

The Need for Gas Exchange

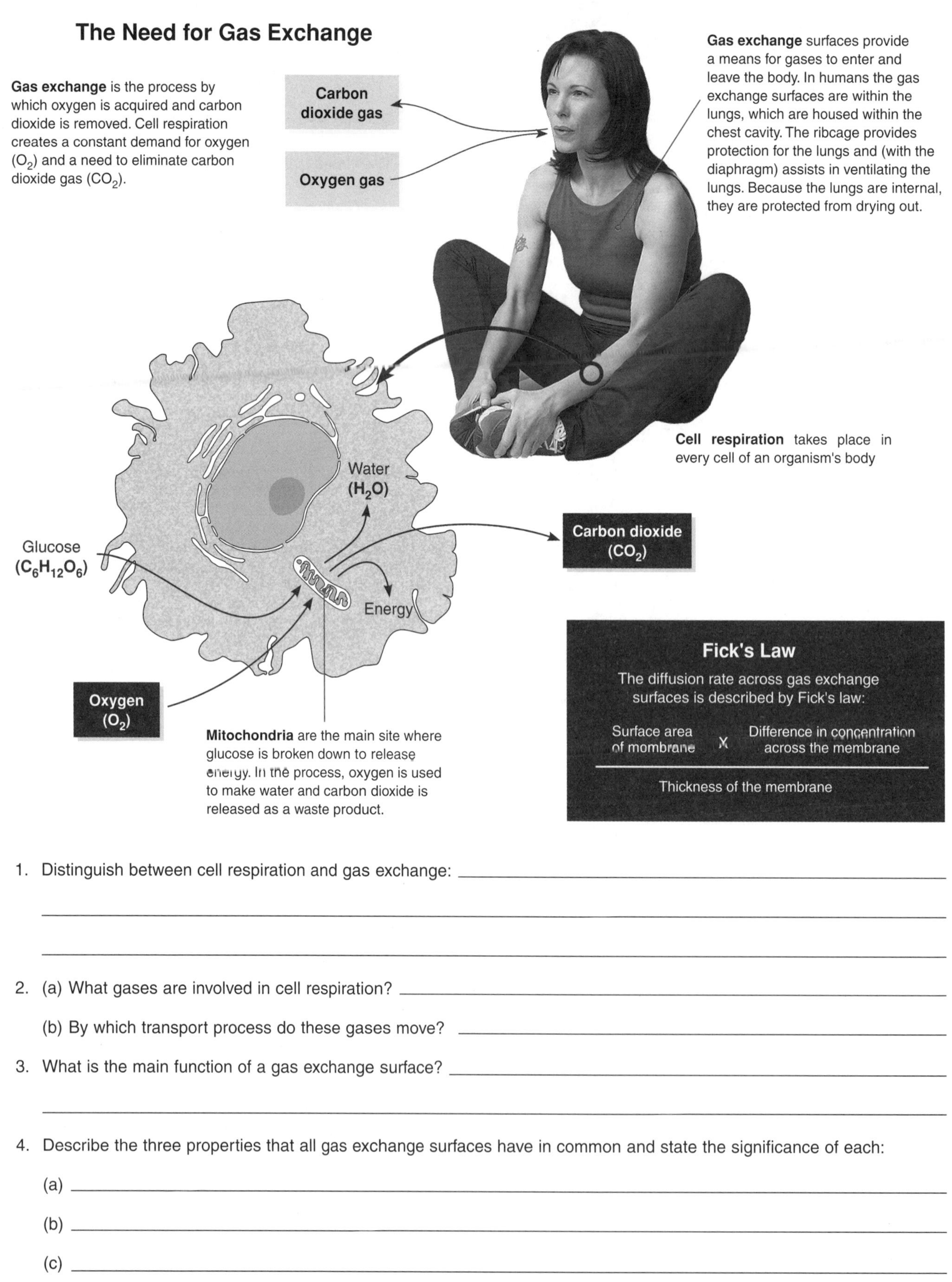
Gas exchange is the process by which oxygen is acquired and carbon dioxide is removed. Cell respiration creates a constant demand for oxygen (O_2) and a need to eliminate carbon dioxide gas (CO_2).
Carbon dioxide gas
Oxygen gas
Gas exchange surfaces provide a means for gases to enter and leave the body. In humans the gas exchange surfaces are within the lungs, which are housed within the chest cavity. The ribcage provides protection for the lungs and (with the diaphragm) assists in ventilating the lungs. Because the lungs are internal, they are protected from drying out.
Cell respiration takes place in every cell of an organism's body
Water (H_2O)
Glucose ($C_6H_{12}O_6$)
Carbon dioxide (CO_2)
Energy
Oxygen (O_2)
Mitochondria are the main site where glucose is broken down to release energy. In the process, oxygen is used to make water and carbon dioxide is released as a waste product.
Fick's Law
The diffusion rate across gas exchange surfaces is described by Fick's law:
Surface area of membrane x Difference in concentration across the membrane
Thickness of the membrane

1. Distinguish between cell respiration and gas exchange: ______________________________

____

____

2. (a) What gases are involved in cell respiration? ______________________________

(b) By which transport process do these gases move? ______________________________

3. What is the main function of a gas exchange surface? ______________________________

____

4. Describe the three properties that all gas exchange surfaces have in common and state the significance of each:

(a) ______________________________

(b) ______________________________

(c) ______________________________

ISBN: 978-1-927173-16-9

The Human Ventilation System

The paired lungs of mammals, including humans, are located within the thorax and are connected to the outside air by way of a system of tubular passageways: the trachea, bronchi, and bronchioles. Ciliated, mucus secreting epithelium lines this system of tubules, trapping and removing dust and pathogens before they reach the gas exchange surfaces. Each lung is divided into a number of lobes, each receiving its own bronchus.

Each bronchus divides many times, terminating in the respiratory bronchioles from which arise 2-11 alveolar ducts and numerous **alveoli** (air sacs). These provide a very large surface area (around 70 m^2) for the exchange of respiratory gases by diffusion between the alveoli and the blood in the capillaries. The details of this exchange across the **gas exchange membrane** (also called the respiratory membrane) are described on the next page.

Morphology of the Ventilation System

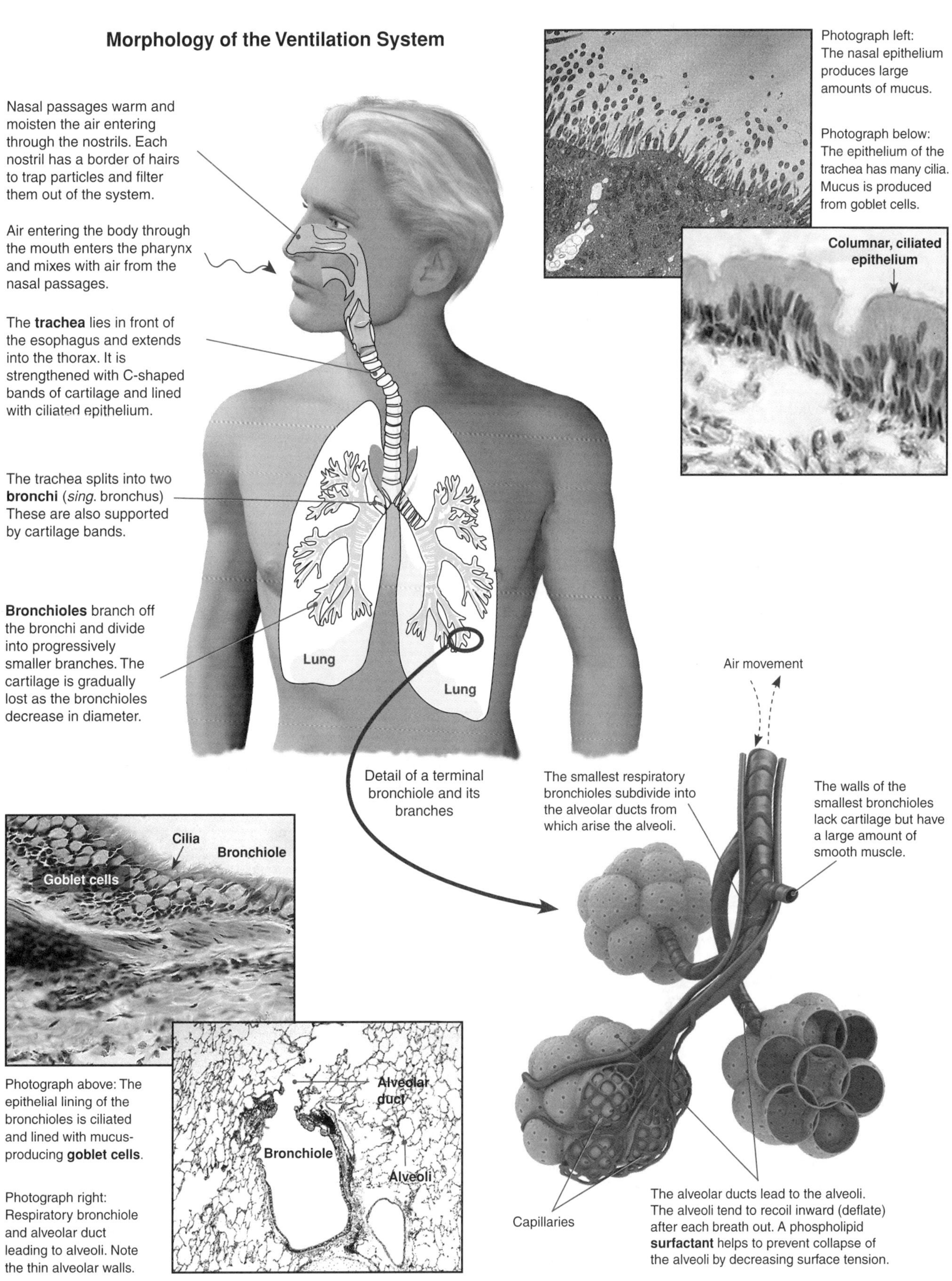
Nasal passages warm and moisten the air entering through the nostrils. Each nostril has a border of hairs to trap particles and filter them out of the system.
Air entering the body through the mouth enters the pharynx and mixes with air from the nasal passages.
The **trachea** lies in front of the esophagus and extends into the thorax. It is strengthened with C-shaped bands of cartilage and lined with ciliated epithelium.
The trachea splits into two **bronchi** (*sing.* bronchus) These are also supported by cartilage bands.
Bronchioles branch off the bronchi and divide into progressively smaller branches. The cartilage is gradually lost as the bronchioles decrease in diameter.
Lung
Lung
Columnar, ciliated epithelium
Detail of a terminal bronchiole and its branches
Air movement
The smallest respiratory bronchioles subdivide into the alveolar ducts from which arise the alveoli.
The walls of the smallest bronchioles lack cartilage but have a large amount of smooth muscle.
Cilia
Bronchiole
Goblet cells
Alveolar duct
Bronchiole
Alveoli
Capillaries
The alveolar ducts lead to the alveoli. The alveoli tend to recoil inward (deflate) after each breath out. A phospholipid **surfactant** helps to prevent collapse of the alveoli by decreasing surface tension.

Photograph left: The nasal epithelium produces large amounts of mucus.

Photograph below: The epithelium of the trachea has many cilia. Mucus is produced from goblet cells.

Photograph above: The epithelial lining of the bronchioles is ciliated and lined with mucus-producing **goblet cells**.

Photograph right: Respiratory bronchiole and alveolar duct leading to alveoli. Note the thin alveolar walls.

All photos on this page: EII

ISBN: 978-1-927173-16-9

An Alveolus

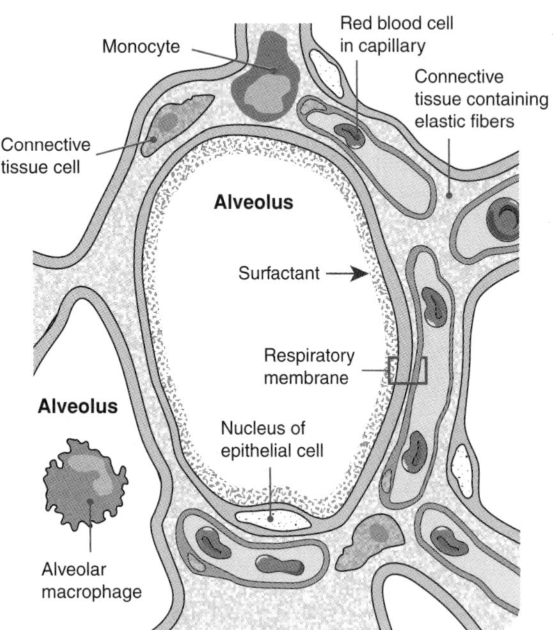

Monocyte
Red blood cell in capillary
Connective tissue containing elastic fibers
Connective tissue cell
Alveolus
Surfactant
Respiratory membrane
Alveolus
Nucleus of epithelial cell
Alveolar macrophage

The diagram above illustrates the physical arrangement of the alveoli to the capillaries through which the blood moves. Phagocytic monocytes and macrophages are also present to protect the lung tissue. Elastic connective tissue gives the alveoli their ability to expand and recoil.

The Respiratory Membrane

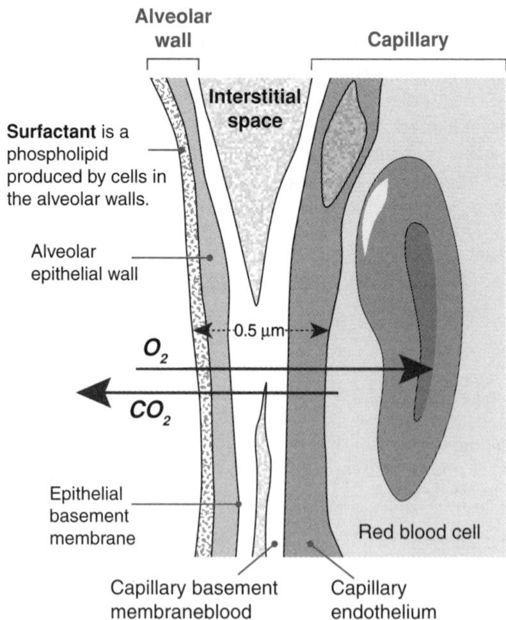

Alveolar wall
Capillary
Interstitial space
Surfactant is a phospholipid produced by cells in the alveolar walls.
Alveolar epithelial wall
0.5 µm
O_2
CO_2
Epithelial basement membrane
Red blood cell
Capillary basement membraneblood
Capillary endothelium

The **respiratory membrane** is the term for the layered junction between the alveolar epithelial cells, the endothelial cells of the capillary, and their associated basement membranes (thin, collagenous layers that underlie the epithelial tissues). Gases move freely across this membrane.

1. (a) Explain how the basic structure of the human respiratory system provides such a large area for gas exchange:

   (b) Identify the general region of the lung where exchange of gases takes place:

2. Describe the structure and purpose of the respiratory membrane:

3. Describe the role of the surfactant in the alveoli:

4. Using the information above and on the previous page, complete the table below summarizing the **histology of the respiratory pathway**. Name each numbered region and use a tick or cross to indicate the presence or absence of particular tissues.

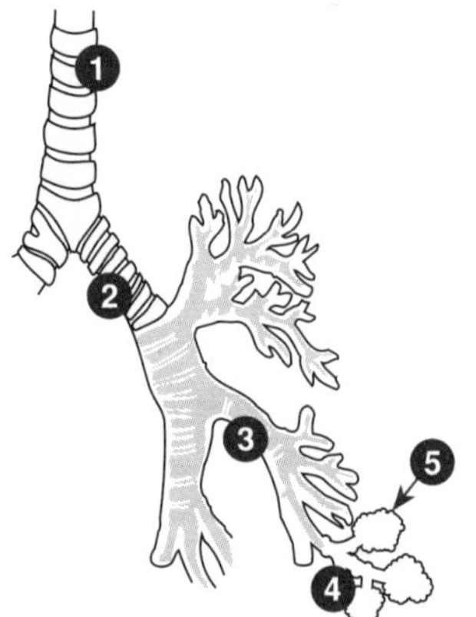

1
2
3
4
5

	Region	Cartilage	Ciliated epithelium	Goblet cells (mucus)	Smooth muscle	Connective tissue
1						✓
2						
3		gradually lost				
4	Alveolar duct		✗	✗		
5					very little	

5. Babies born prematurely are often deficient in surfactant. This causes respiratory distress syndrome; a condition where breathing is very difficult. From what you know about the role of surfactant, explain the symptoms of this syndrome:

ISBN: 978-1-927173-16-9

Breathing in Humans

In mammals, the mechanism of breathing (**ventilation**) provides a continual supply of fresh air to the lungs and helps to maintain a large diffusion gradient for respiratory gases across the gas exchange surface. Oxygen must be delivered regularly to supply the needs of respiring cells. Similarly, carbon dioxide, which is produced as a result of cellular metabolism, must be quickly eliminated from the body. Adequate lung ventilation is essential to these exchanges. The cardiovascular system participates by transporting respiratory gases to and from the cells of the body. The volume of gases exchanged during breathing varies according to the physiological demands placed on the body (e.g. by exercise). These changes can be measured using spirometry.

Inspiration (inhalation or breathing in)

During quiet breathing, inspiration is achieved by increasing the space (therefore decreasing the pressure) inside the lungs. Air then flows into the lungs in response to the decreased pressure inside the lung. Inspiration is always an active process involving muscle contraction.

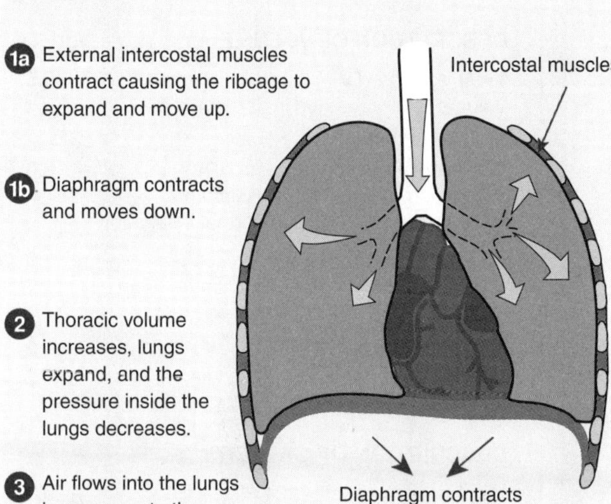
1a External intercostal muscles contract causing the ribcage to expand and move up.
1b Diaphragm contracts and moves down.
2 Thoracic volume increases, lungs expand, and the pressure inside the lungs decreases.
3 Air flows into the lungs in response to the pressure gradient.
Intercostal muscles
Diaphragm contracts and moves down

Expiration (exhalation or breathing out)

During quiet breathing, expiration is achieved passively by decreasing the space (thus increasing the pressure) inside the lungs. Air then flows passively out of the lungs to equalise with the air pressure. In active breathing, muscle contraction is involved in bringing about both inspiration and expiration.

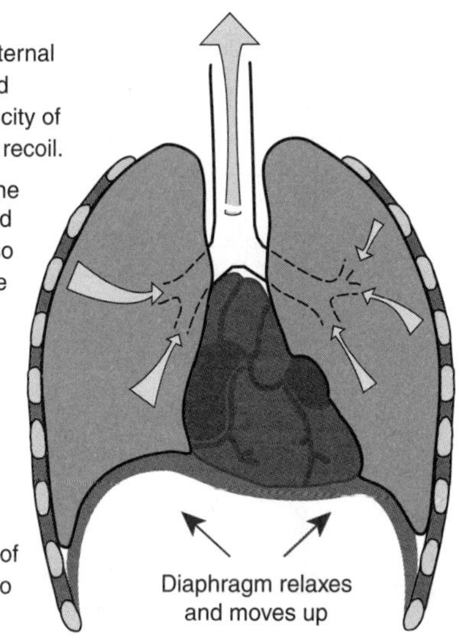
1 In **quiet breathing**, external intercostal muscles and diaphragm relax. Elasticity of the lung tissue causes recoil.
In **forced breathing**, the internal intercostals and abdominal muscles also contract to increase the force of the expiration
2 Thoracic volume decreases and the pressure inside the lungs increases.
3 Air flows passively out of the lungs in response to the pressure gradient.
Diaphragm relaxes and moves up

1. Explain the purpose of breathing:

2. (a) Describe the sequence of events involved in quiet breathing:

   (b) Explain the essential difference between this and the situation during heavy exercise or forced breathing:

3. Identify what other gas is lost from the body in addition to carbon dioxide:

4. Explain the role of the elasticity of the lung tissue in normal, quiet breathing:

5. Breathing rate is regulated through the medullary respiratory center in response to demand for oxygen. The trigger for increased breathing rate is a drop in blood pH. Suggest why this is an appropriate trigger to increase breathing rate:

ISBN: 978-1-927173-16-9

Measuring Lung Function

Changes in lung volume can be measured using a technique called **spirometry**. Total adult lung capacity varies between 4 and 6 liters (L or dm^3) and is greater in males. The **vital capacity**, which describes the volume exhaled after a maximum inspiration, is somewhat less than this because of the residual volume of air that remains in the lungs even after expiration. The exchange between fresh air and the residual volume is a slow process and the composition of gases in the lungs remains relatively constant. Once measured, the tidal volume can be used to calculate the **pulmonary ventilation rate** or PV, which describes the amount of air exchanged with the environment per minute. Measures of respiratory capacity provide one way in which a reduction in lung function can be assessed (for example, as might occur as result of disease or an obstructive lung disorder such asthma).

Determining changes in lung volume using spirometry

The apparatus used to measure the amount of air exchanged during breathing and the rate of breathing is a **spirometer** (also called a respirometer). A simple spirometer consists of a weighted drum, containing oxygen or air, inverted over a chamber of water. A tube connects the air-filled chamber with the subject's mouth, and soda lime in the system absorbs the carbon dioxide breathed out. Breathing results in a trace called a spirogram, from which lung volumes can be measured directly.

During inspiration
Air is removed from the chamber, the drum sinks, and an upward deflection is recorded on the paper on the rotating drum.

During expiration
Air is added to the chamber, the drum rises, and a downward deflection is recorded.

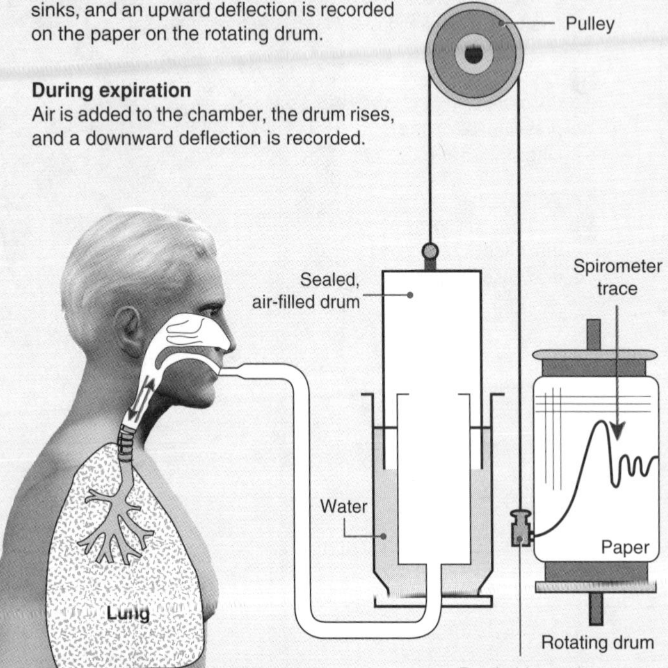
Pulley
Sealed, air-filled drum
Spirometer trace
Water
Paper
Lung
Rotating drum
Pen holder and counter balance

Lung Volumes and Capacities

The air in the lungs can be divided into volumes. Lung capacities are combinations of volumes.

DESCRIPTION OF VOLUME	Vol / L
Tidal volume (TV) Volume of air breathed in and out in a single breath	0.5
Inspiratory reserve volume (IRV) Volume breathed in by a maximum inspiration at the end of a normal inspiration	3.3
Expiratory reserve volume (ERV) Volume breathed out by a maximum effort at the end of a normal expiration	1.0
Residual volume (RV) Volume of air remaining in the lungs at the end of a maximum expiration	1.2
DESCRIPTION OF CAPACITY	
Inspiratory capacity (IC) = TV + IRV Volume breathed in by a maximum inspiration at the end of a normal expiration	3.8
Vital capacity (VC) = IRV + TV + ERV Volume that can be exhaled after a maximum inspiration.	4.8
Total lung capacity (TLC) = VC + RV The total volume of the lungs. Only a fraction of TLC is used in normal breathing	6.0

PRIMARY INDICATORS OF LUNG FUNCTION

Forced expiratory volume in 1 second (FEV_1)
The volume of air that is maximally exhaled in the first second of exhalation.

Forced vital capacity (FVC)
The total volume of air that can be forcibly exhaled after a maximum inspiration.

1. Describe how each of the following might be expected to influence values for lung volumes and capacities obtained using spirometry:

   (a) Height: ______

   (b) Gender: ______

   (c) Age: ______

2. A percentage decline in FEV_1 and FVC (to <80% of normal) are indicators of impaired lung function, e.g in asthma:

   (a) Explain why a forced volume is a more useful indicator of lung function than tidal volume:

   ______

   (b) Asthma is treated with drugs to relax the airways. Suggest how spirometry could be used during asthma treatment:

   ______

***Related activities:** The Human Ventilation System*
***Web links:** Respiratory Basics Learning Activity*

ISBN: 978-1-927173-16-9

Respiratory gas	Approximate percentages of O_2 and CO_2		
	Inhaled air	Air in lungs	Exhaled air
O_2	21.0	13.8	16.4
CO_2	0.04	5.5	3.6

Above: The percentages of respiratory gases in air (by volume) during normal breathing. The percentage volume of oxygen in the alveolar air (in the lung) is lower than that in the exhaled air because of the influence of the **dead air volume** (the air in the spaces of the nose, throat, larynx, trachea and bronchi). This air (about 30% of the air inhaled) is unavailable for gas exchange.

Left: During exercise, the breathing rate, tidal volume, and PV increase up to a maximum (as indicated below).

Spirogram for a male during quiet and forced breathing, and during exercise

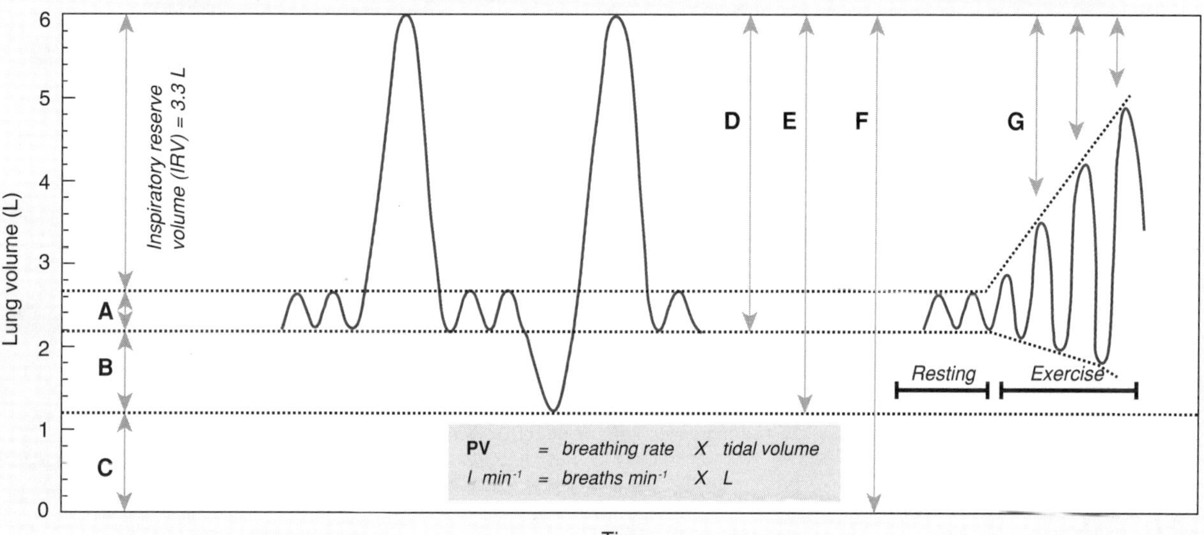
Lung volume (L)
0
1
2
3
4
5
6
Time
Inspiratory reserve volume (IRV) = 3.3 L
A
B
C
D
E
F
G
Resting
Exercise
PV = breathing rate X tidal volume
l min⁻¹ = breaths min⁻¹ X L

3. Using the definitions given on the previous page, identify the volumes and capacities indicated by the letters **A**-**F** on the spirogram above. For each, indicate the volume (vol) in liters (L). The inspiratory reserve volume has been identified:

(a) **A**: ______________ Vol: ________ (d) **D**: ______________ Vol: ________

(b) **B**: ______________ Vol: ________ (e) **E**: ______________ Vol: ________

(c) **C**: ______________ Vol: ________ (f) **F**: ______________ Vol: ________

4. Explain what is happening in the sequence indicated by the letter **G**: ______________

5. Calculate PV when breathing rate is 15 breaths per minute and tidal volume is 0.4 L: ______________

6. (a) Describe what would happen to PV during strenuous exercise: ______________

(b) Explain how this is achieved: ______________

7. The table above gives approximate percentages for respiratory gases during breathing. Study the data and then:

(a) Calculate the difference in CO_2 between inhaled and exhaled air: ______________

(b) Explain where this 'extra' CO_2 comes from: ______________

(c) Explain why the dead air volume raises the oxygen content of exhaled air above that in the lungs: ______________

ISBN: 978-1-927173-16-9

Gas Transport In Humans

The transport of respiratory gases around the body is the role of the blood and its respiratory pigments. Oxygen is transported throughout the body chemically bound to the respiratory pigment **hemoglobin** inside the red blood cells. In the muscles, oxygen from hemoglobin is transferred to and retained by **myoglobin**, a molecule that is chemically similar to hemoglobin except that it consists of only one heme-globin unit. Myoglobin has a greater affinity for oxygen than hemoglobin and acts as an oxygen store within muscles, releasing the oxygen during periods of prolonged or extreme muscular activity. If the myoglobin store is exhausted, the muscles are forced into oxygen debt and must respire anaerobically. The waste product of this, lactic acid, accumulates in the muscle and is transported (as lactate) to the liver where it is metabolized under aerobic conditions.

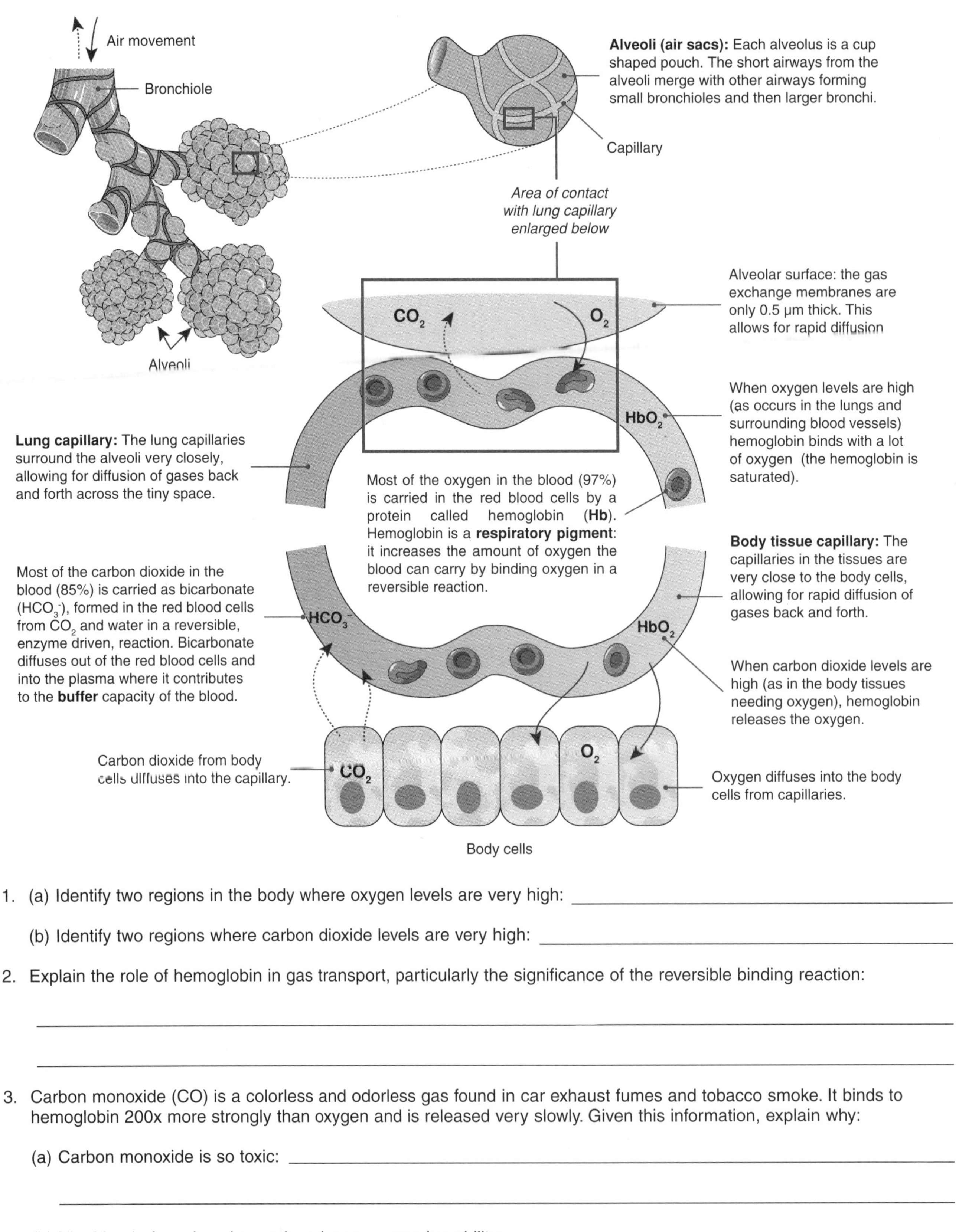

Air movement
Bronchiole
Alveoli
Alveoli (air sacs): Each alveolus is a cup shaped pouch. The short airways from the alveoli merge with other airways forming small bronchioles and then larger bronchi.
Capillary
Area of contact with lung capillary enlarged below
CO_2
O_2
Alveolar surface: the gas exchange membranes are only 0.5 µm thick. This allows for rapid diffusion
HbO_2
When oxygen levels are high (as occurs in the lungs and surrounding blood vessels) hemoglobin binds with a lot of oxygen (the hemoglobin is saturated).
Lung capillary: The lung capillaries surround the alveoli very closely, allowing for diffusion of gases back and forth across the tiny space.
Most of the oxygen in the blood (97%) is carried in the red blood cells by a protein called hemoglobin (**Hb**). Hemoglobin is a **respiratory pigment**: it increases the amount of oxygen the blood can carry by binding oxygen in a reversible reaction.
Body tissue capillary: The capillaries in the tissues are very close to the body cells, allowing for rapid diffusion of gases back and forth.
Most of the carbon dioxide in the blood (85%) is carried as bicarbonate (HCO_3^-), formed in the red blood cells from CO_2 and water in a reversible, enzyme driven, reaction. Bicarbonate diffuses out of the red blood cells and into the plasma where it contributes to the **buffer** capacity of the blood.
HCO_3^-
HbO_2
When carbon dioxide levels are high (as in the body tissues needing oxygen), hemoglobin releases the oxygen.
Carbon dioxide from body cells diffuses into the capillary.
CO_2
O_2
Oxygen diffuses into the body cells from capillaries.
Body cells

1. (a) Identify two regions in the body where oxygen levels are very high: ______

   (b) Identify two regions where carbon dioxide levels are very high: ______

2. Explain the role of hemoglobin in gas transport, particularly the significance of the reversible binding reaction:

______

______

3. Carbon monoxide (CO) is a colorless and odorless gas found in car exhaust fumes and tobacco smoke. It binds to hemoglobin 200x more strongly than oxygen and is released very slowly. Given this information, explain why:

   (a) Carbon monoxide is so toxic: ______

   ______

   (b) The blood of smokers has reduced oxygen carrying ability: ______

   ______

Related activities: *The Human Ventilation System*

ISBN: 978-1-927173-16-9

CONTEXT: The Effects of High Altitude

The air at high altitudes contains less oxygen than the air at sea level. Air pressure decreases with altitude so the pressure (therefore amount) of oxygen in the air also decreases. Sudden exposure to an altitude of 2000 m would make you breathless on exertion and above 7000 m most people would become unconscious. The effects of altitude on physiology are related to this lower oxygen availability. Humans can make some physiological adjustments to life at altitude; this is called **acclimatization**. Populations that have lived at altitude for generations show heritable **adaptations** to high altitude life.

Mountain Sickness

Altitude sickness or mountain sickness is usually a mild illness associated with trekking to altitudes of 5000 meters or so. Common symptoms include headache, insomnia, poor appetite and nausea, vomiting, dizziness, tiredness, coughing and breathlessness. The best way to avoid mountain sickness is to ascend to altitude slowly (no more than 300 m per day above 3000 m). Continuing to ascend with mountain sickness can result in more serious illnesses: accumulation of fluid on the brain (cerebral edema) and accumulation of fluid in the lungs (pulmonary edema). These complications can be fatal if not treated with oxygen and a rapid descent to lower altitude.

Physiological Adjustment to Altitude

Effect	Minutes	Days	Weeks
Increased heart rate	←→ (minutes to early days)		
Increased breathing	←→ (late minutes to days)		
Concentration of blood		←→ (days)	
Increased red blood cell production		←→ (days to weeks)	
Increased capillary density			←→ (weeks)

The human body can make adjustments to life at altitude. Some of these changes take place almost immediately: breathing and heart rates increase. Other adjustments may take weeks (see above). These responses are all aimed at improving the rate of supply of oxygen to the body's tissues. When more permanent adjustments to physiology are made (increased blood cells and capillary networks) heart and breathing rates can return to normal.

Adaptations to High Altitude

Studies on three high altitude human populations (Andean, Tibetan, and Ethiopian highlands) have found three different adaptations for survival to the low levels of oxygen in the air.

Populations living in the Andean Altiplano in South America develop higher concentrations of hemoglobin than populations living at sea level.

Tibetan populations have similar hemoglobin concentrations to populations at sea level, but take more breaths per minute to compensate for the lower oxygen pressure. The average Tibetan man has a ventilation rate of 15 L min^{-1} while an Andean man has a rate of 10.5 L min^{-1}.

Tibetan porter

Concentration of Hemoglobin in the Blood

Group	Hemoglobin concentration (g $100ml^{-1}$)
U.S. sea level mean	15.3
Andean male	19.2
Andean female	17.8
Tibetan male	15.6
Tibetan female	14.2
Ethiopian highlands male	15.9
Ethiopian highlands female	15.0

Recent studies on indigenous populations in the Ethiopian highlands (altitude of >1500 m) have found that they do not experience **hypoxia** (lack of oxygen), but have none of the physiological of the adaptations seen in the Andean or Tibetan populations. Their hemoglobin concentration is normal, and they do not compensate with a higher ventilation rate.

Adaptations in indigenous Ethiopian highlanders may be due to changes in the hypoxia-inducible factor-1 (HIF-1) pathway. This pathway controls the systemic and cellular responses to hypoxic conditions. Several genes are activated at low oxygen concentrations to control the HIF-1 pathway. Positive selection may be acting on these genes in Ethiopian highlanders.

1. (a) Describe the general effects of high altitude on the body: ______________________

   (b) Name the general term given to describe these effects: ______________________

2. (a) Name one short term physiological adjustment that humans make to high altitude: ______________________

   (b) Explain how this adaptation helps to increase the amount of oxygen the body receives: ______________________

3. Describe and explain one adaptation to high altitude that has evolved in some human populations: ______________________

ISBN: 978-1-927173-16-9

Core Topic
6.5

AHL Topic
11.3

Nerves, Hormones, and **Homeostasis**

Key concepts

- The nervous system receives information and coordinates an appropriate response. Neurons are electrically excitable cells that respond to signal input.
- Homeostasis of the internal environment is maintained using hormonal and nervous mechanisms via negative feedback.
- Disruptions to homeostatic regulation can lead to chronic disease.
- The kidney has a central role in removing metabolic wastes and regulating blood volume.

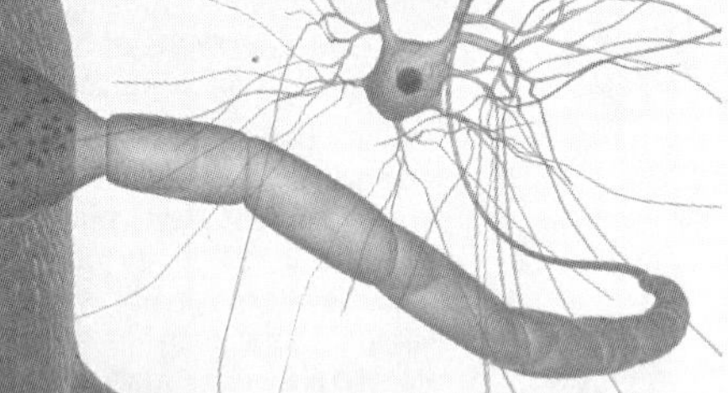

Key terms

Core

action potential
axon
blood glucose
cell body
central nervous system (CNS)
dendrites
depolarization
diabetes mellitus
effector
endocrine system
glucagon
homeostasis
hormone
insulin
motor end plate
motor neuron
myelin sheath
negative feedback
neurotransmitter
nodes of Ranvier
non-myelinated nerve
peripheral nervous system (PNS)
receptor
repolarization
resting potential
target cell
thermoregulation

AHL only

anti-diuretic hormone (ADH)
kidney
cortex
glomerulus
medulla
nephron
osmoregulation
renal pelvis
ultrafiltration
ureter
urine

Periodicals:
Listings for this chapter are on page 400

Weblinks:
www.thebiozone.com/weblink/IB-3169.html

BIOZONE APP:
Student Review Series
Nervous System, Urinary System

Learning Objectives

☐ 1. Use the **KEY TERMS** to compile a glossary for this topic.

Nerves and Nerve Impulses *(6.5.1-6.5.6)* pages 321-327

☐ 2. Describe the organization of the human nervous system, distinguishing between the **central nervous system** (CNS) and the **peripheral nervous system** (PNS). Recognize **neurons** as the basic structural unit of nervous systems. Describe neurons as electrically excitable cells capable of responding to signal input.

☐ 3. Draw and label a diagram of a myelinated **motor neuron**. Include reference to **dendrites cell body**, **axon**, **myelin sheath**, **nodes of Ranvier**, and **motor end plate**.

☐ 4. Using a diagram of a simple reflex arc, describe how nerve impulses are conducted from **receptors** to the CNS, within the CNS, and from the CNS to **effectors**.

☐ 5. Explain how the **resting potential** of a neuron is established and maintained. Describe an **action potential** in terms of **depolarization** and **repolarization**.

☐ 6. Explain how an action potential is propagated along a **non-myelinated nerve**. Include reference to the movement of Na^+ and K^+ and changes in membrane potential.

☐ 7. EXTENSION: Describe how an action potential is propagated in a myelinated nerve.

☐ 8. Explain the principles of synaptic transmission and describe the events occurring at a chemical **synapse** after arrival of an action potential.

Hormones and Homeostasis *(6.5.7-6.5.12)* pages 328-333

☐ 9. Describe the general structure of the **endocrine system** and understand the role of **hormones** as chemical regulators carried by the blood to **target cells**.

☐ 10. Explain the principles of **homeostasis**, including the role of **receptors** in monitoring the levels of variables and **negative feedback** in correcting deviations from the steady state. Outline the role of homeostatic mechanisms in maintaining blood pH, carbon dioxide concentration, blood glucose concentration, body temperature, and fluid levels.

☐ 11. Explain **thermoregulation** in humans including the role of the hypothalamus, skin and associated structures, and muscular activity such as shivering.

☐ 12. Explain the control of **blood glucose**, including the role of **insulin**, **glucagon** and the α and β cells of the pancreatic islets. Distinguish between type I and type II **diabetes mellitus**. Discuss the high prevalence of type II diabetes in some countries (*TOK*).

The Kidney *(11.3)* pages 334-339

☐ 13. **AHL**: Define the term **excretion**. Draw and label a diagram of the **kidney** to show the **cortex**, **medulla**, **renal pelvis**, **ureter** and renal arteries and veins.

☐ 14. **AHL**: Annotate a diagram of a **nephron** to show the function of each part. Explain **ultrafiltration** in the **glomerulus** and reabsorption in the proximal convoluted tubule.

☐ 15. **AHL**: Define **osmoregulation**. Describe the regulation of urine volume and blood volume by the kidney, including the role of **ADH**.

☐ 16. **AHL**: Explain the difference in the concentration of protein, glucose, and urea between blood plasma, glomerular filtrate, and **urine**. Explain the presence of glucose in the urine of untreated diabetics.

Nervous Regulatory Systems

An essential feature of living organisms is their ability to coordinate their activities in response to environmental stimuli. The vertebrate plan is a good model for studying the basics of nervous regulation. Vertebrates detect and respond to environmental change through the nervous and endocrine systems. These two systems are quite different structurally, but interact to coordinate behavior and physiology. The **nervous system** is the body's control and communication center. It has three functions: to detect stimuli, interpret them, and initiate appropriate responses. It comprises millions of **neurons** (nerve cells), which are specialized to transmit information in the form of electrochemical impulses (action potentials). The nervous system forms a signaling network with branches carrying information to and from specific target tissues. Impulses can be transmitted rapidly over considerable distances and although it comprises millions of neural connections, its plan (below) is quite simple.

Coordination by the Nervous System

The vertebrate nervous system consists of the **central nervous system** (brain and spinal cord), and the nerves and receptors outside it (**peripheral nervous system**). Sensory input to receptors comes via stimuli. Information about the effect of a response is provided by feedback mechanisms so that the system can be readjusted. The basic organization of the nervous system can be simplified into a few key components: the sensory receptors, a central nervous system processing point, and the effectors which bring about the response (below):

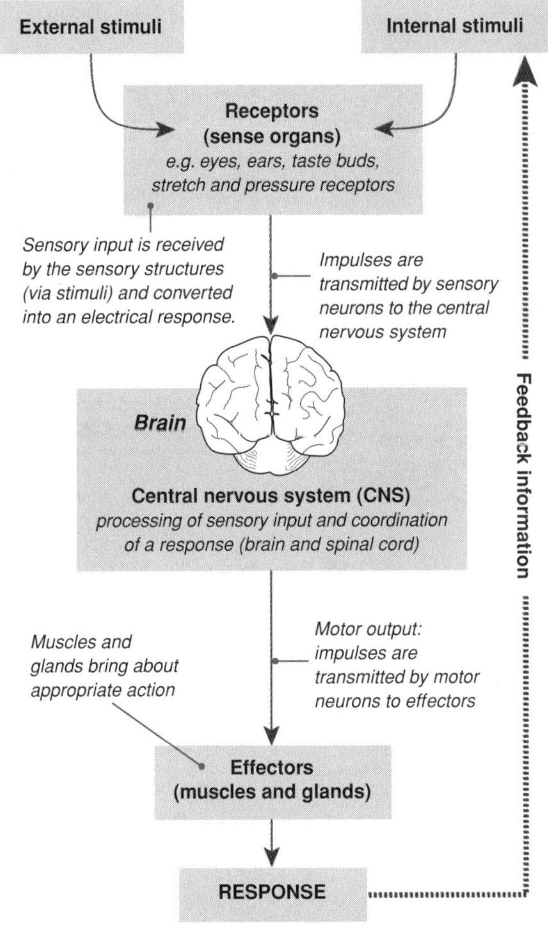
External stimuli
Internal stimuli
Receptors (sense organs)
e.g. eyes, ears, taste buds, stretch and pressure receptors
Sensory input is received by the sensory structures (via stimuli) and converted into an electrical response.
Impulses are transmitted by sensory neurons to the central nervous system
Brain
Central nervous system (CNS)
processing of sensory input and coordination of a response (brain and spinal cord)
Feedback information
Muscles and glands bring about appropriate action
Motor output: impulses are transmitted by motor neurons to effectors
Effectors (muscles and glands)
RESPONSE

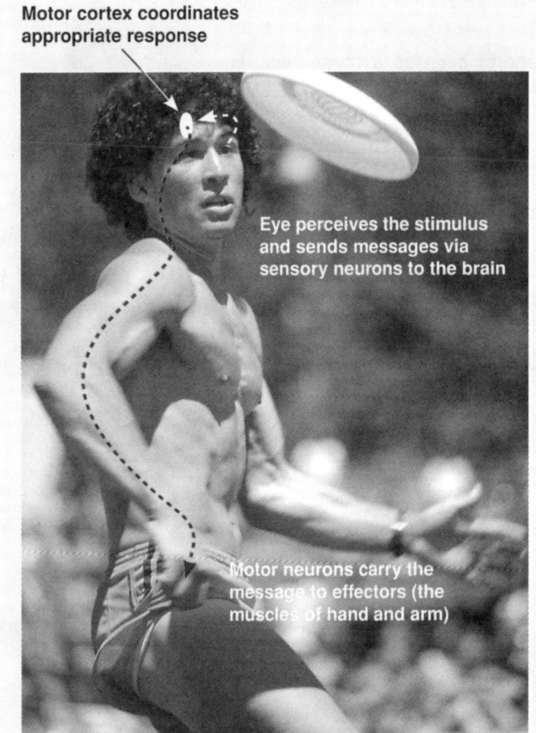
Motor cortex coordinates appropriate response
Eye perceives the stimulus and sends messages via sensory neurons to the brain
Motor neurons carry the message to effectors (the muscles of hand and arm)

In the example above, the frisbee's approach is perceived by the eye. The motor cortex of the brain integrates the sensory message. Coordination of hand and body orientation is brought about through motor neurons to the muscles.

Comparison of nervous and hormonal control

	Nervous control	Hormonal control
Communication	*Impulses across synapses*	*Hormones in the blood*
Speed	*Very rapid (within a few milliseconds)*	*Relatively slow (over minutes, hours, or longer)*
Duration	*Short term and reversible*	*Longer lasting effects*
Target pathway	*Specific (through nerves) to specific cells*	*Hormones broadcast to target cells everywhere*
Action	*Causes glands to secrete or muscles to contract*	*Causes changes in metabolic activity*

1. Identify the three basic components of a nervous system and describe their role:

(a) ______________________________

(b) ______________________________

(c) ______________________________

2. Comment on the significance of the differences between the speed and duration of nervous and hormonal controls:

____

____

____

ISBN: 978-1-927173-16-9

Related activities: *The Human Nervous System*
Weblinks: *Nervous System Animation*

The Human Nervous System

The **nervous system** is the body's control and communication centre. It has three broad functions: detecting stimuli, interpreting them, and initiating appropriate responses. Its basic structure is shown below. Further detail is provided in the following pages.

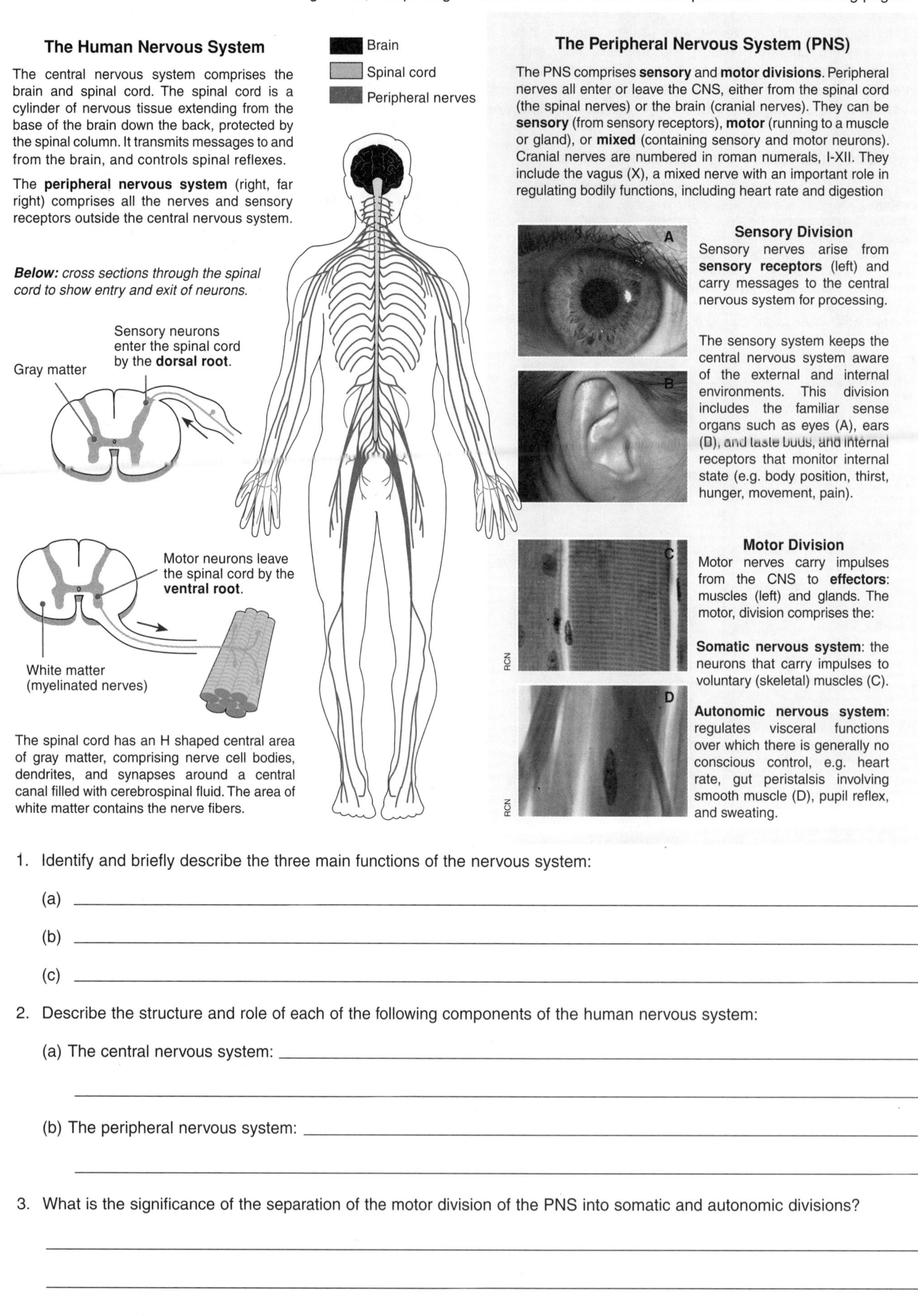

Brain
Spinal cord
Peripheral nerves
Sensory neurons enter the spinal cord by the **dorsal root**.
Gray matter
Motor neurons leave the spinal cord by the **ventral root**.
White matter (myelinated nerves)
A
B
C
D
RCN
RCN

The Human Nervous System

The central nervous system comprises the brain and spinal cord. The spinal cord is a cylinder of nervous tissue extending from the base of the brain down the back, protected by the spinal column. It transmits messages to and from the brain, and controls spinal reflexes.

The **peripheral nervous system** (right, far right) comprises all the nerves and sensory receptors outside the central nervous system.

Below: *cross sections through the spinal cord to show entry and exit of neurons.*

The spinal cord has an H shaped central area of gray matter, comprising nerve cell bodies, dendrites, and synapses around a central canal filled with cerebrospinal fluid. The area of white matter contains the nerve fibers.

The Peripheral Nervous System (PNS)

The PNS comprises **sensory** and **motor divisions**. Peripheral nerves all enter or leave the CNS, either from the spinal cord (the spinal nerves) or the brain (cranial nerves). They can be **sensory** (from sensory receptors), **motor** (running to a muscle or gland), or **mixed** (containing sensory and motor neurons). Cranial nerves are numbered in roman numerals, I-XII. They include the vagus (X), a mixed nerve with an important role in regulating bodily functions, including heart rate and digestion

Sensory Division

Sensory nerves arise from **sensory receptors** (left) and carry messages to the central nervous system for processing.

The sensory system keeps the central nervous system aware of the external and internal environments. This division includes the familiar sense organs such as eyes (A), ears (B), and taste buds, and internal receptors that monitor internal state (e.g. body position, thirst, hunger, movement, pain).

Motor Division

Motor nerves carry impulses from the CNS to **effectors**: muscles (left) and glands. The motor, division comprises the:

Somatic nervous system: the neurons that carry impulses to voluntary (skeletal) muscles (C).

Autonomic nervous system: regulates visceral functions over which there is generally no conscious control, e.g. heart rate, gut peristalsis involving smooth muscle (D), pupil reflex, and sweating.

1. Identify and briefly describe the three main functions of the nervous system:

   (a) ______________________________

   (b) ______________________________

   (c) ______________________________

2. Describe the structure and role of each of the following components of the human nervous system:

   (a) The central nervous system: ______________________________

   ____

   (b) The peripheral nervous system: ______________________________

   ____

3. What is the significance of the separation of the motor division of the PNS into somatic and autonomic divisions?

   ____

   ____

   ____

Related activities: *Nervous Regulatory Systems, Neuron Structure and Function*

ISBN: 978-1-927173-16-9

Neuron Structure and Function

Homeostasis depends on the nervous system detecting, interpreting, and responding appropriately to both internal and external stimuli. Many of these responses are involuntary and are achieved through **reflexes**. Information, in the form of electrochemical impulses, is transmitted along nerve cells (**neurons**) from receptors to effectors. Neurons typically consist of a cell body, dendrites, and an axon (below). The speed of impulse conduction depends primarily on the axon diameter and whether or not the axon is **myelinated**. The speed of impulse conduction is slower in non-myelinated neurons because the impulse must be propagated along the entire length of the axon. Conduction speeds are faster in myelinated neurons because the axon in insulated. The principle behind increasing impulse speed through saltatory conduction is described (as extension) overleaf.

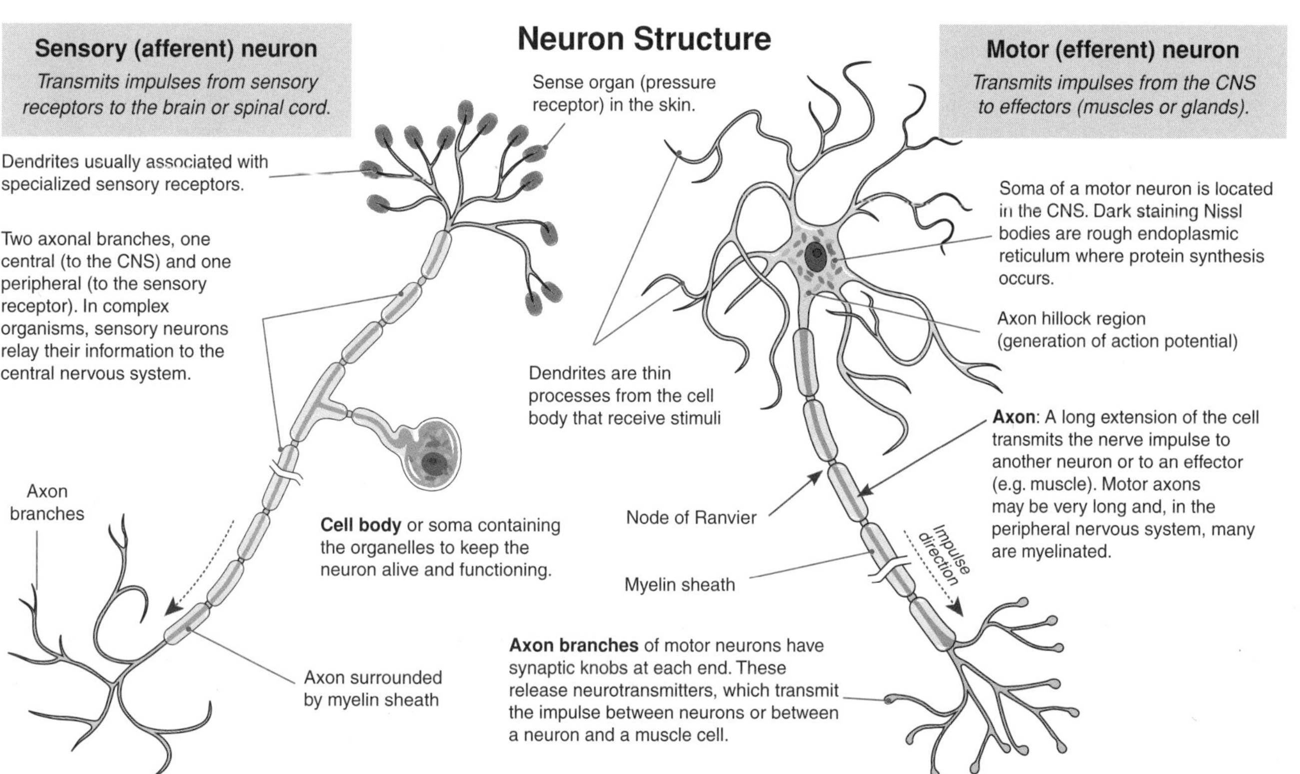
Neuron Structure
Sensory (afferent) neuron
Transmits impulses from sensory receptors to the brain or spinal cord.
Dendrites usually associated with specialized sensory receptors.
Two axonal branches, one central (to the CNS) and one peripheral (to the sensory receptor). In complex organisms, sensory neurons relay their information to the central nervous system.
Axon branches
Sense organ (pressure receptor) in the skin.
Cell body or soma containing the organelles to keep the neuron alive and functioning.
Axon surrounded by myelin sheath
Dendrites are thin processes from the cell body that receive stimuli
Node of Ranvier
Myelin sheath
Impulse direction
Axon branches of motor neurons have synaptic knobs at each end. These release neurotransmitters, which transmit the impulse between neurons or between a neuron and a muscle cell.
Motor (efferent) neuron
Transmits impulses from the CNS to effectors (muscles or glands).
Soma of a motor neuron is located in the CNS. Dark staining Nissl bodies are rough endoplasmic reticulum where protein synthesis occurs.
Axon hillock region (generation of action potential)
Axon: A long extension of the cell transmits the nerve impulse to another neuron or to an effector (e.g. muscle). Motor axons may be very long and, in the peripheral nervous system, many are myelinated.

Nerves, Hormones, and Homeostasis

Where conduction speed is important, the axons of neurons are sheathed within a lipid and protein rich substance called **myelin**. Myelin is produced by **oligodendrocytes** in the central nervous system (CNS) and by **Schwann cells** in the peripheral nervous system (PNS). At intervals along the axons of myelinated neurons, there are gaps between neighboring Schwann cells and their sheaths. These are called **nodes of Ranvier**. Myelin acts as an insulator, increasing the speed at which nerve impulses travel because it prevents ion flow across the neuron membrane and forces the current to "jump" along the axon from node to node.

Myelinated Neurons

Diameter: 1-25 μm

Conduction speed: 6-120 ms^{-1}

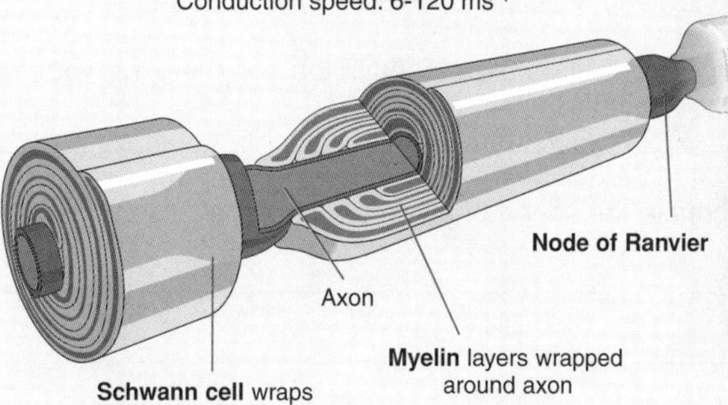
Node of Ranvier
Axon
Myelin layers wrapped around axon
Schwann cell wraps only one axon and produces myelin

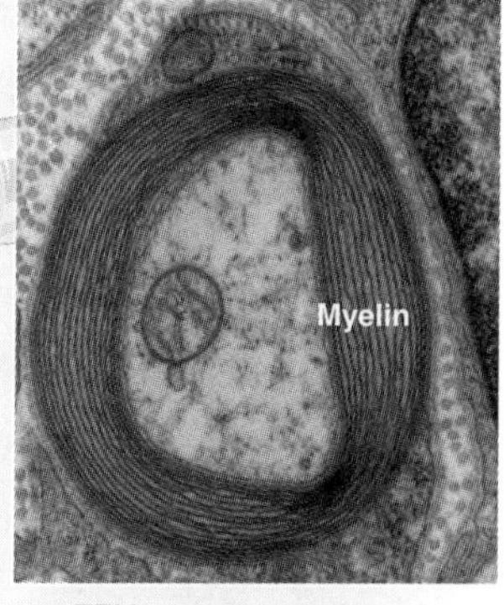
Myelin
WIKI

TEM cross section through a myelinated axon

Non-myelinated axons are relatively more common in the CNS where the distances travelled are less than in the PNS. Here, the axons are encased within the cytoplasmic extensions of oligodendrocytes or Schwann cells, rather than within a myelin sheath. The speed of impulse conduction is slower than in myelinated neurons because the nerve impulse is propagated along the entire axon membrane, rather than jumping from node to node as occurs in myelinated neurons. Conduction speeds are slower than in myelinated neurons, although they are faster in larger neurons (there is less ion leakage from a larger diameter axon).

Non-myelinated Neurons

Diameter: <1 μm in vertebrates

Conduction speed: 0.2-0.5 ms^{-1}

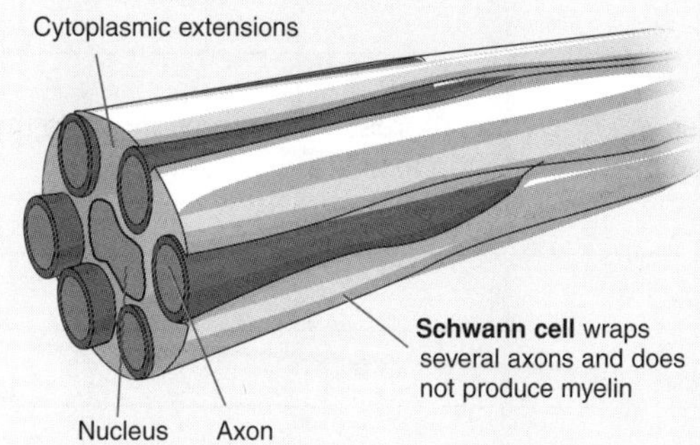
Cytoplasmic extensions
Schwann cell wraps several axons and does not produce myelin
Nucleus
Axon

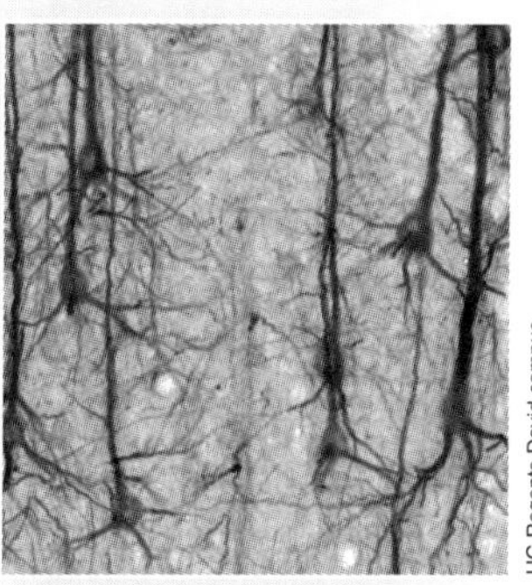
UC Regents David campus

Unmyelinated pyramidal neurons of the cerebral cortex

ISBN: 978-1-927173-16-9

EXTENSION: Saltatory Conduction

Axon myelination is analogous to insulating an electrical wire. It is a characteristic feature of vertebrate nervous systems and it enables them to achieve very rapid speeds of nerve conduction. Myelinated neurons conduct impulses by **saltatory conduction**, a term that describes how the impulse jumps along the fiber. In saltatory conduction, only the nodes of Ranvier are involved in action potential generation. In myelinated (insulated) regions, there is no leakage of ions across the neuron membrane and the action potential at one node is sufficient to trigger an action potential in the next node.

Apart from increasing the speed of the nerve impulse, the myelin sheath helps in reducing energy expenditure because the area of depolarization is decreased (and therefore also the number of sodium and potassium ions that need to be pumped to restore the resting potential).

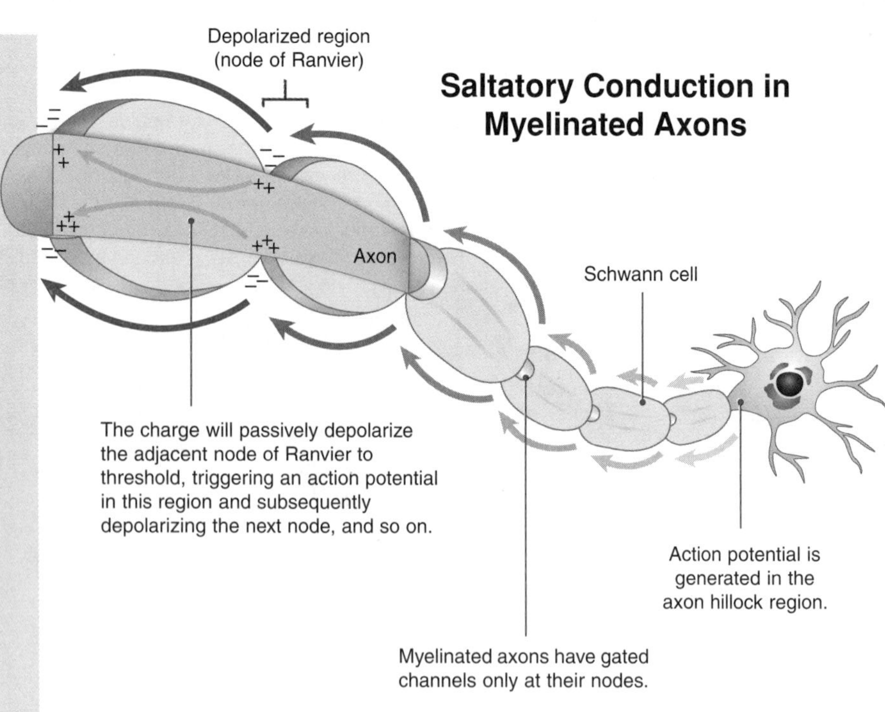
Saltatory Conduction in Myelinated Axons
Depolarized region (node of Ranvier)
Axon
Schwann cell
The charge will passively depolarize the adjacent node of Ranvier to threshold, triggering an action potential in this region and subsequently depolarizing the next node, and so on.
Action potential is generated in the axon hillock region.
Myelinated axons have gated channels only at their nodes.

1. Complete the missing panels of the following table summarizing structural and functional differences between neurons:

	Sensory neuron	Interneuron	Motor neuron
Structure		Short dendrites, long or short axon	
Location	Dendrites outside the spinal cord, cell body in spinal ganglion	Entirely within the CNS	Dendrites and cell body in the spinal cord; axon outside the spinal cord.
Function		Connect sensory & motor neurons	

2. (a) What is the function of myelination in neurons? ___

(b) What cell type is responsible for myelination in the CNS? ___

(c) What cell type is responsible for myelination in the PNS? ___

(d) Why is myelination typically a feature of neurons in the peripheral nervous system? ___

3. How does myelination increase the speed of nerve impulse conduction? ___

4. (a) Describe the adaptive advantage of faster conduction of nerve impulses: ___

(b) Why does increasing the axon diameter also increase the speed of impulse conduction? ___

5. Multiple sclerosis (MS) is a disease involving progressive destruction of the myelin sheaths around axons. Why does MS impair nervous system function even though the axons are still intact?

ISBN: 978-1-927173-16-9

Reflexes

A reflex is an automatic response to a stimulus involving a small number of neurons and a central nervous system (CNS) processing point (usually the spinal cord, but sometimes the brain stem). This type of circuit is called a **reflex arc**. Reflexes permit rapid responses to stimuli. They are classified according to the number of CNS synapses involved; **monosynaptic reflexes** involve only one CNS synapse (e.g. knee jerk reflex), **polysynaptic reflexes** involve two or more (e.g. pain withdrawal reflex). Both are spinal reflexes. The pupil reflex (opening and closure of the pupil) is an example of a cranial reflex.

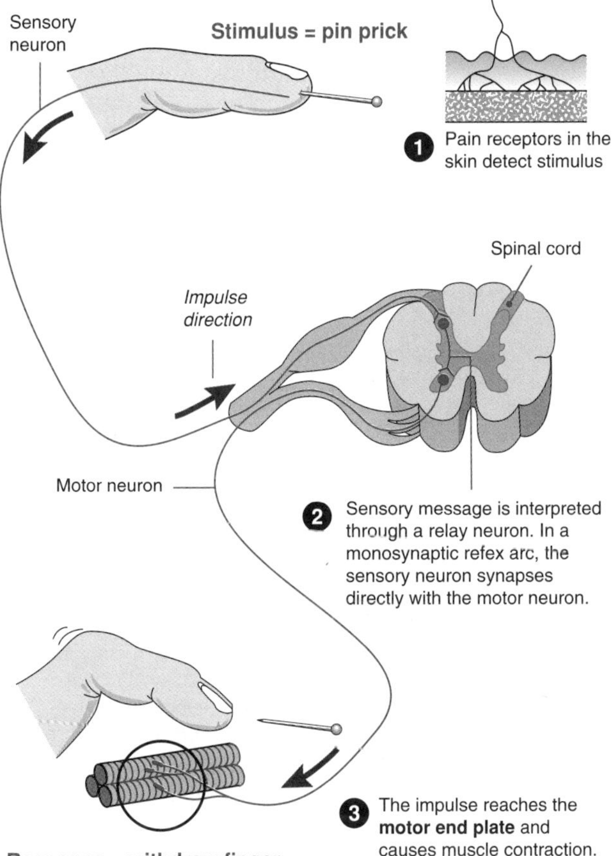
Pain Withdrawal: A Polysynaptic Reflex Arc
Sensory neuron
Stimulus = pin prick
1 Pain receptors in the skin detect stimulus
Spinal cord
Impulse direction
Motor neuron
2 Sensory message is interpreted through a relay neuron. In a monosynaptic refex arc, the sensory neuron synapses directly with the motor neuron.
3 The impulse reaches the **motor end plate** and causes muscle contraction.
Response = withdraw finger

The patella (knee jerk) reflex is a simple deep tendon reflex used to test the function of the femoral nerve and spinal cord segments L2-L4. It helps to maintain posture and balance when walking.

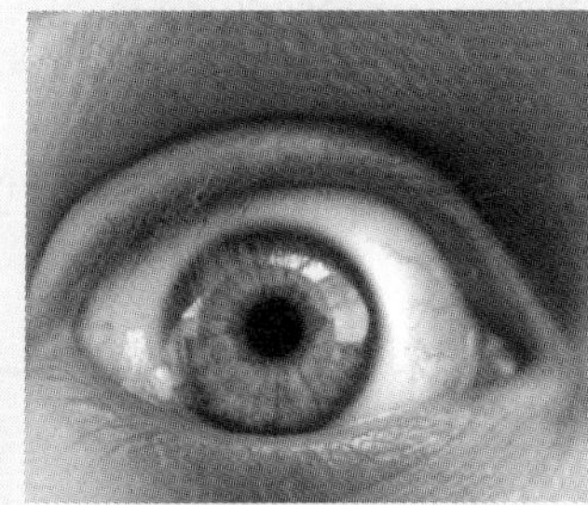

The pupillary light reflex refers to the rapid expansion or contraction of the pupils in response to the intensity of light falling on the retina. It is a polysynaptic cranial reflex and can be used to test for brain death.

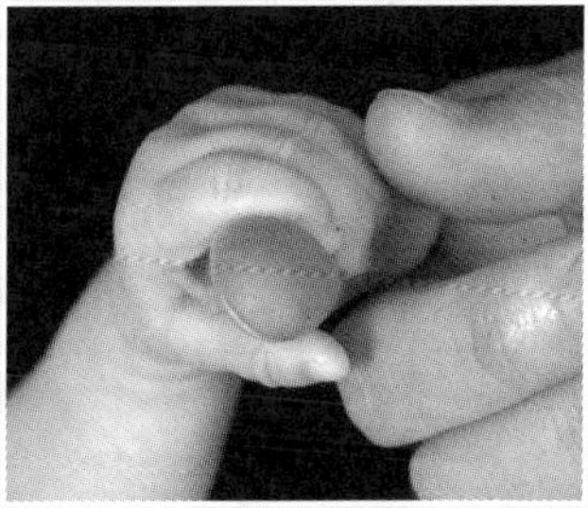

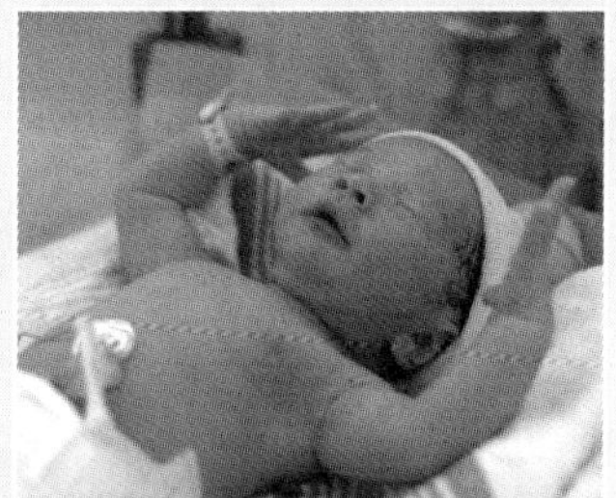

Normal newborns exhibit a number of primitive reflexes in response to particular stimuli. These reflexes disappear within a few months of birth as the child develops. Primitive reflexes include the grasp reflex (above left) and the startle or Moro reflex (above right) in which a sudden noise will cause the infant to throw out its arms, extend the legs and head, and cry. The rooting and sucking reflexes are other examples of primitive reflexes.

1. Why are reasoning and conscious thought not necessary or desirable features of reflex behaviors?

2. Distinguish between a spinal reflex and a cranial reflex and give an example of each:

3. (a) Distinguish between a monosynaptic and a polysynaptic reflex arc and give an example of each:

   (b) Which would produce the most rapid response, given similar length sensory and motor pathways? Explain:

4. (a) With reference to examples, describe the adaptive value of primitive reflexes in newborns:

   (b) Why are newborns tested for the presence of these reflexes?

ISBN: 978-1-927173-16-9

Related activities*: The Human Nervous System, Chemical Synapses*
Weblinks*: Parasympathetic Eye Response, Knee Jerk Reflex*

Transmission of Nerve Impulses

The plasma membranes of cells, including neurons, contain **sodium-potassium ion pumps** which actively pump sodium ions (Na^+) out of the cell and potassium ions (K^+) into the cell. The action of these ion pumps in neurons creates a separation of charge (a potential difference or voltage) either side of the membrane and makes the cells **electrically excitable**. It is this property that enables neurons to transmit electrical impulses. The **resting state** of a neuron, with a net negative charge inside, is maintained by the sodium-potassium pumps, which actively move two K^+ into the neuron for every three Na^+ moved out (below left). When a nerve is stimulated, a brief increase in membrane permeability to Na^+ temporarily reverses the membrane polarity (a depolarization). After the nerve impulse passes, the sodium-potassium pump restores the resting potential. The depolarization is propagated along the axon by local current in non-myelinated fibers and by **saltatory conduction** in myelinated fibers. Impulses pass from neuron to neuron by crossing junctions called **synapses**.

The Resting Neuron

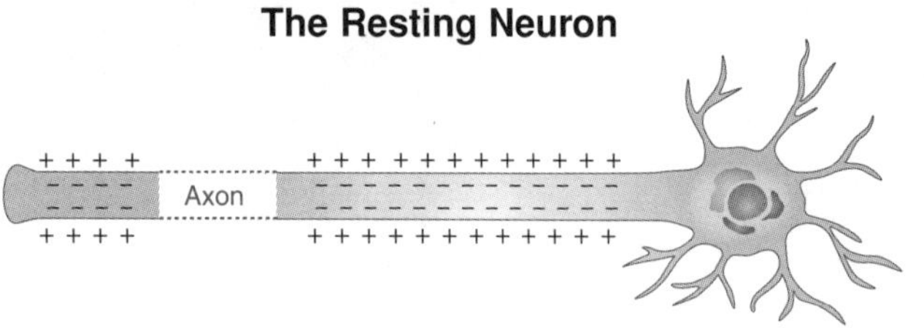

Axon

When a neuron is not transmitting an impulse, the inside of the cell is negatively charged relative to the outside and the cell is said to be electrically polarized. The potential difference (voltage) across the membrane is called the **resting potential**. For most nerve cells this is about -70 mV. Nerve transmission is possible because this membrane potential exists.

The Nerve Impulse

Direction of impulse

Na^+ Na^+

Area of impulse

Next area to be stimulated

Area returning to resting state

When a neuron is stimulated, the distribution of charges on each side of the membrane briefly reverses. This process of **depolarization** causes a burst of electrical activity to pass along the axon of the neuron as an **action potential**. As the charge reversal reaches one region, local currents depolarize the next region and the impulse spreads along the axon.

The Action Potential

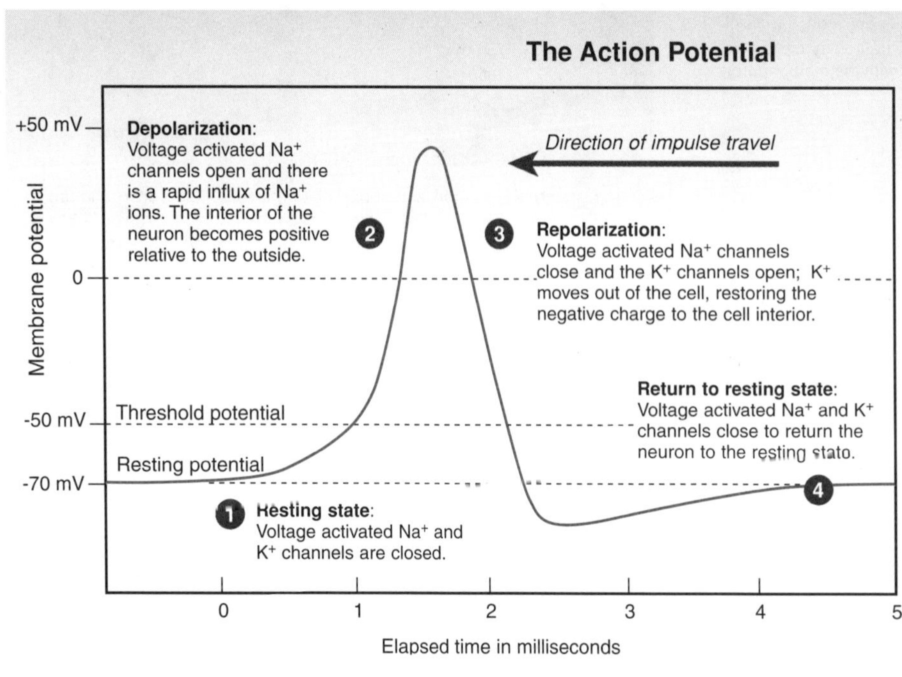

+50 mV
0
-50 mV
-70 mV
Membrane potential
Depolarization: Voltage activated Na+ channels open and there is a rapid influx of Na+ ions. The interior of the neuron becomes positive relative to the outside.
Direction of impulse travel
Repolarization: Voltage activated Na+ channels close and the K+ channels open; K+ moves out of the cell, restoring the negative charge to the cell interior.
Threshold potential
Return to resting state: Voltage activated Na+ and K+ channels close to return the neuron to the resting state.
Resting potential
Resting state: Voltage activated Na+ and K+ channels are closed.
0
1
2
3
4
5
Elapsed time in milliseconds

The depolarization in an axon can be shown as a change in membrane potential (in millivolts). A stimulus must be strong enough to reach the **threshold potential** before an action potential is generated. This is the voltage at which the depolarization of the membrane becomes unstoppable.

The action potential is **all or nothing** in its generation and because of this, impulses (once generated) always reach threshold and move along the axon without attenuation. The resting potential is restored by the movement of potassium ions (K^+) out of the cell. During this **refractory period**, the nerve cannot respond, so nerve impulses are discrete.

1. What is an action potential? ____________________

2. How does an action potential pass along a nerve? ____________________

3. How does the refractory period influence the direction in which an impulse will travel? ____________________

4. Action potentials themselves are indistinguishable from each other. How is the nervous system able to interpret the impulses correctly and bring about an appropriate response?

____

ISBN: 978-1-927173-16-9

Chemical Synapses

Action potentials are transmitted between neurons across junctions, called **synapses**, between the end of one axon and the dendrite or cell body of a receiving neuron. **Chemical synapses** are the most widespread type of synapse in nervous systems. The axon terminal is a swollen knob, and a small gap separates it from the receiving neuron. The synaptic knobs are filled with tiny packets of chemicals called **neurotransmitters**. Transmission involves the diffusion of the neurotransmitter across the gap, where it interacts with the receiving membrane and causes an electrical response. The basic properties of synaptic transmission are shown below. In this example, the neurotransmitter causes a membrane depolarization and the generation of an action potential. Some neurotransmitters have the opposite effect and cause inhibition (e.g. slowing heart rate).

A Cholinergic Synapse

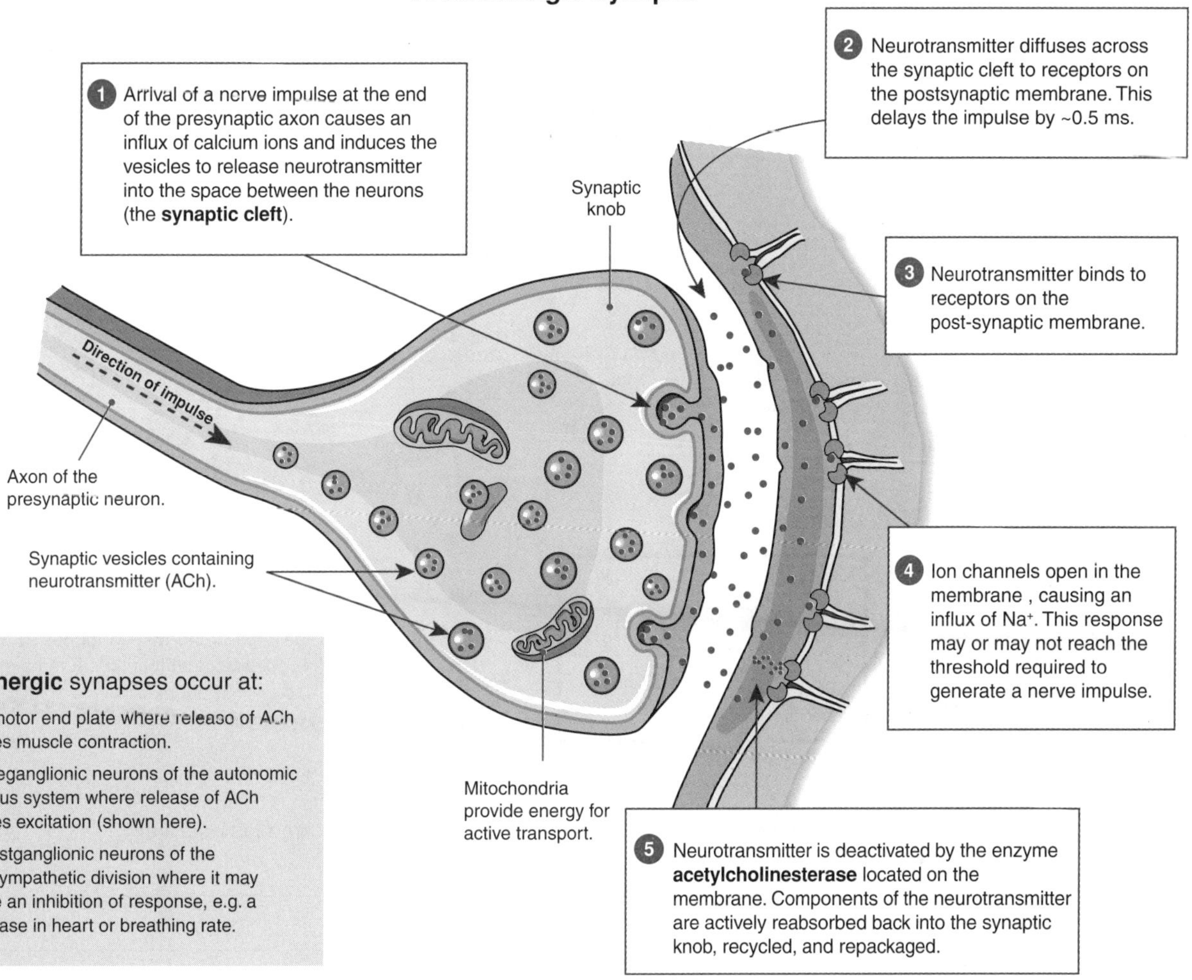

1 Arrival of a nerve impulse at the end of the presynaptic axon causes an influx of calcium ions and induces the vesicles to release neurotransmitter into the space between the neurons (the **synaptic cleft**).
2 Neurotransmitter diffuses across the synaptic cleft to receptors on the postsynaptic membrane. This delays the impulse by ~0.5 ms.
Synaptic knob
3 Neurotransmitter binds to receptors on the post-synaptic membrane.
Direction of impulse
Axon of the presynaptic neuron.
Synaptic vesicles containing neurotransmitter (ACh).
4 Ion channels open in the membrane , causing an influx of Na^+. This response may or may not reach the threshold required to generate a nerve impulse.
Mitochondria provide energy for active transport.
5 Neurotransmitter is deactivated by the enzyme **acetylcholinesterase** located on the membrane. Components of the neurotransmitter are actively reabsorbed back into the synaptic knob, recycled, and repackaged.

Cholinergic synapses occur at:

- The motor end plate where release of ACh causes muscle contraction.
- All preganglionic neurons of the autonomic nervous system where release of ACh causes excitation (shown here).
- All postganglionic neurons of the parasympathetic division where it may cause an inhibition of response, e.g. a decrease in heart or breathing rate.

1. What is a **synapse**? ______________________________

2. What causes the release of neurotransmitter into the synaptic cleft? ______________________________

3. Why is there a brief delay in transmitting an impulse across the synapse? ______________________________

4. (a) How is the neurotransmitter deactivated? ______________________________

   (b) Why is it important for the neurotransmitter substance to be deactivated soon after its release? ______________________________

5. Consult a reference source to identify one function of acetylcholine in the nervous system: ______________________________

6. Suggest one factor that might influence the strength of the response in the receiving cell: ______________________________

ISBN: 978-1-927173-16-9

Periodicals: Bridging the gap

Related activities: *Transmission of Nerve Impulses*
Weblinks: *Nerve Synapse*

Hormonal Regulatory Systems

The endocrine system regulates the body's processes by releasing chemical messengers (hormones) into the bloodstream. Hormones are potent chemical regulators: they are produced in minute quantities yet can have a large effect on metabolism. The endocrine system comprises endocrine cells (organized into endocrine glands), and the hormones they produce. Unlike exocrine glands (e.g. sweat and salivary glands), endocrine glands are ductless glands, secreting hormones directly into the bloodstream rather than through a duct or tube. Some organs (e.g. the pancreas) have both endocrine and exocrine regions, but these are structurally and functionally distinct. The basis of hormonal control and the role of negative feedback mechanisms in regulating hormone levels are described below.

The Mechanism of Hormone Action

Endocrine cells produce hormones and secrete them into the bloodstream where they are distributed throughout the body. Although hormones are broadcast throughout the body, they affect only specific target cells. These target cells have receptors on the plasma membrane which recognize and bind the hormone (see inset, below right). The binding of hormone and receptor triggers the response in the target cell. Cells are unresponsive to a hormone if they do not have the appropriate receptors.

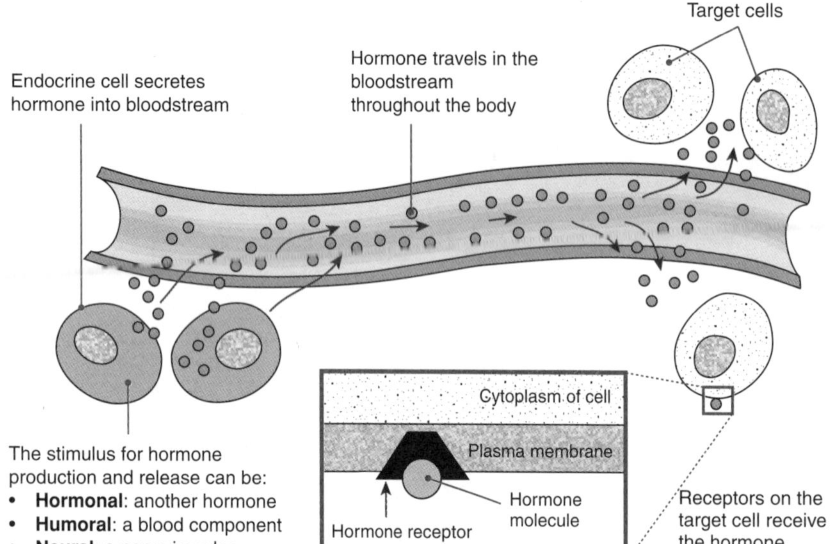
Target cells
Hormone travels in the bloodstream throughout the body
Endocrine cell secretes hormone into bloodstream
Cytoplasm of cell
Plasma membrane
Hormone molecule
Hormone receptor
Receptors on the target cell receive the hormone
The stimulus for hormone production and release can be:
• **Hormonal**: another hormone
• **Humoral**: a blood component
• **Neural**: a nerve impulse

Antagonistic Hormones

Insulin secretion

Blood glucose rises: insulin is released

Raises blood glucose level

Lowers blood glucose level

Blood glucose falls: glucagon is released

Glucagon secretion

The effects of one hormone are often counteracted by an opposing hormone. Feedback mechanisms adjust the balance of the two hormones to maintain a physiological function. Example: insulin acts to decrease blood glucose and glucagon acts to raise it.

1. (a) What are **antagonistic hormones**? Describe an example of how two such hormones operate:

___

___

Example: ___

___

(b) Describe the role of feedback mechanisms in adjusting hormone levels (explain using an example if this is helpful):

___

___

2. How can a hormone influence only the target cells even though all cells may receive the hormone?

___

___

3. Explain why hormonal control differs from nervous system control with respect to the following:

(a) The speed of hormonal responses is slower: ___

___

(b) Hormonal responses are generally longer lasting: ___

___

Related activities: *Nervous Regulatory Systems, Control of Blood Glucose*
Weblinks: *Control of Endocrine Activity*

ISBN: 978-1-927173-16-9

Principles of Homeostasis

Homeostasis refers to the relative physiological constancy of the body, despite external fluctuations. Homeostasis of the internal environment is an essential feature of complex animals and it is the job of the body's **organ systems** to maintain it, even as they make necessary exchanges with the environment. Homeostatic control systems have three functional components: a receptor to detect change, a control centre, and an effector to direct an appropriate response. In **negative feedback** systems, movement away from a steady state triggers a mechanism to counteract further change in that direction. Using negative feedback systems, the body counteracts disturbances and restores the steady state. **Positive feedback** is also used in physiological systems, but to a lesser extent since positive feedback leads to the response escalating in the same direction.

Organ systems maintain a constant internal environment that provides for the needs of all the body's cells, making it possible for animals to move through different and often highly variable external environments. This representation shows how organ systems permit exchanges with the environment. The exchange surfaces are usually internal, but may be connected to the environment via openings on the body surface.

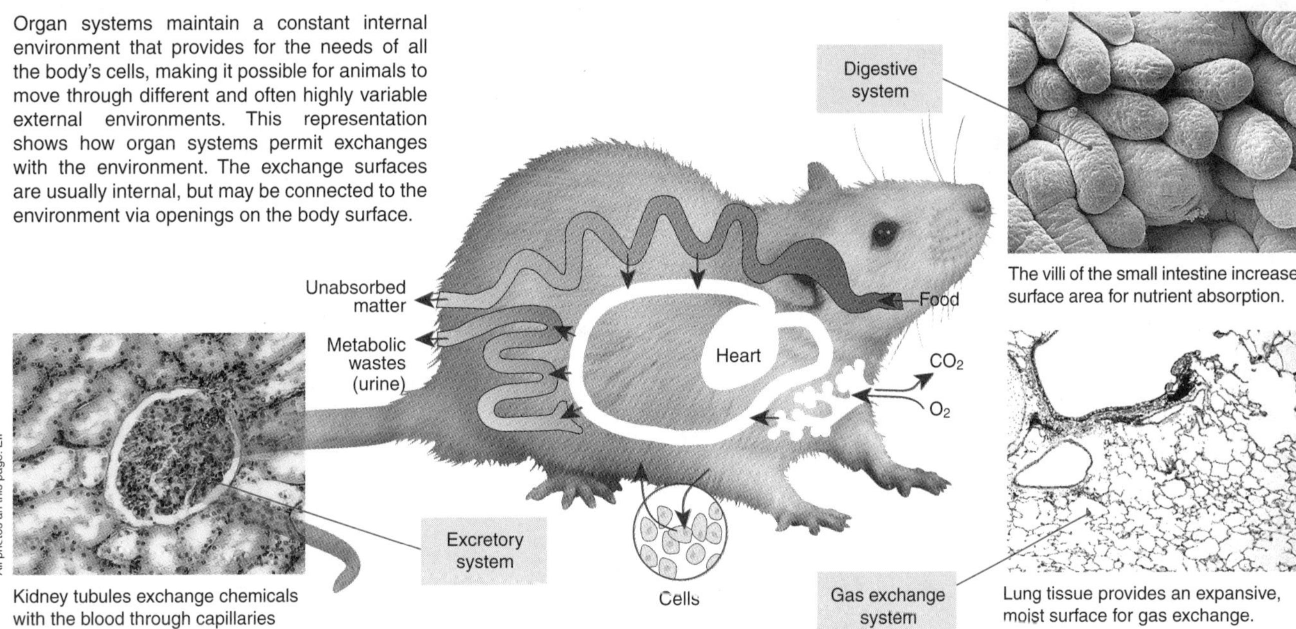
Digestive system
Unabsorbed matter
Metabolic wastes (urine)
Heart
Food
CO_2
O_2
Excretory system
Cells
Gas exchange system

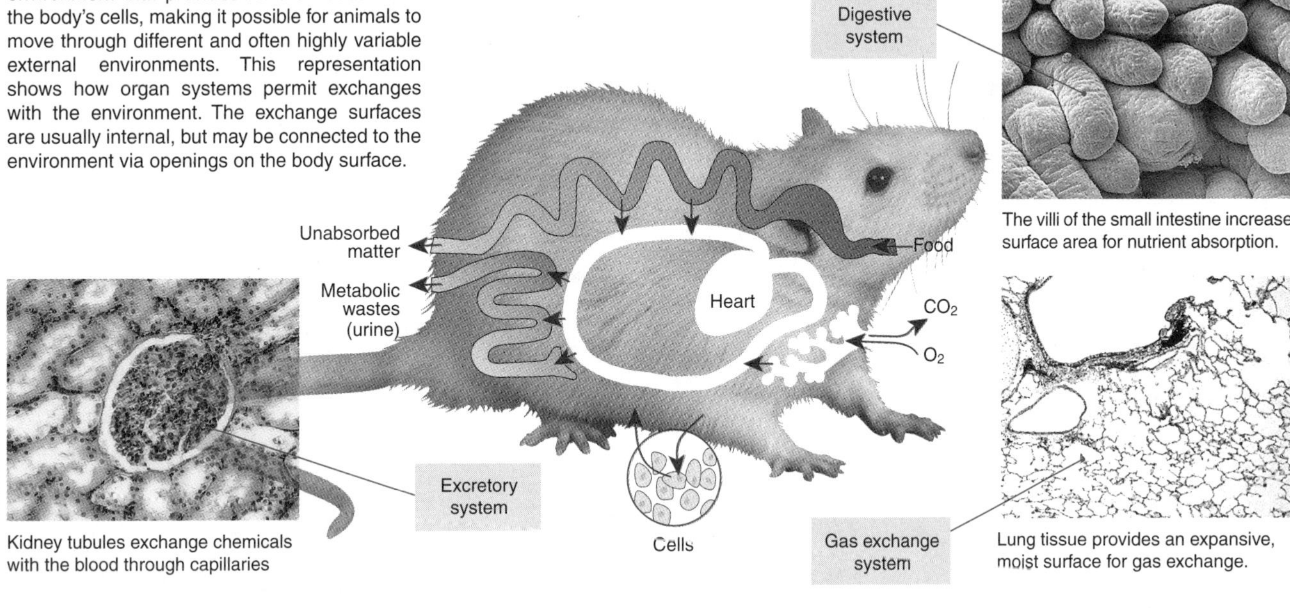

The villi of the small intestine increase surface area for nutrient absorption.

Kidney tubules exchange chemicals with the blood through capillaries

Lung tissue provides an expansive, moist surface for gas exchange.

Negative Feedback and Control Systems

2 Corrective mechanisms activated, e.g. sweating

3 Return to optimum

1 Stress, e.g. exercise generates excessive body heat

Stress, e.g. cold weather causes excessive heat loss

Normal body temperature

Corrective mechanisms activated, e.g. shivering

+

−

1 A stressor, e.g. exercise, takes the internal environment away from optimum.

2 Stress is detected by receptors and corrective mechanisms (e.g. sweating or shivering) are activated.

3 Corrective mechanisms act to restore optimum conditions.

Negative feedback acts to counteract departures from steady state. The diagram shows how stress is counteracted in the case of body temperature.

1. What are the three main components of a regulatory control system in the human body? What is the role of each?

____

____

2. How do negative feedback mechanisms maintain homeostasis in a variable environment? ______________________________

____

____

____

ISBN: 978-1-927173-16-9

Thermoregulation in Humans

In humans and other placental mammals, the temperature regulation center of the body is in the **hypothalamus**. In humans, it has a '**set point**' temperature of 36.7°C. The hypothalamus responds directly to changes in core temperature and to nerve impulses from receptors in the skin. It then coordinates appropriate nervous and hormonal responses to counteract the changes and restore normal body temperature. Like a thermostat, the hypothalamus detects a return to normal temperature and the corrective mechanisms are switched off (**negative feedback**). Toxins produced by pathogens, or substances released from some white blood cells, cause the set point to be set to a higher temperature. This results in **fever** and is an important defense mechanism in the case of infection.

Counteracting Heat Loss

Heat promoting center* in the hypothalamus monitors fall in skin or core temperature below 35.8°C and coordinates responses that generate and conserve heat. These responses are mediated primarily through the **sympathetic nerves** of the autonomic nervous system.

Thyroxine (together with epinephrine) **increases metabolic rate**.

Under conditions of *extreme* cold, epinephrine and thyroxine increase the energy releasing activity of the liver. Under normal conditions, the liver is thermally neutral.

Muscular activity (including *shivering*) produces internal heat.

Erector muscles of hairs contract to raise hairs and increase insulating layer of air. Blood flow to skin decreases (**vasoconstriction**).

Factors causing heat loss

- Wind chill factor accelerates heat loss through conduction.
- Heat loss due to temperature difference between the body and the environment.
- The rate of heat loss from the body is increased by being wet, by inactivity, dehydration, inadequate clothing, or shock.

Factors causing heat gain

- Gain of heat directly from the environment through radiation and conduction.
- Excessive fat deposits make it harder to lose the heat that is generated through activity.
- Heavy exercise, especially with excessive clothing.

Counteracting Heat Gain

Heat losing center* in the hypothalamus monitors any rise in skin or core temperature above 37.5°C and coordinates responses that increase heat loss. These responses are mediated primarily through the **parasympathetic nerves** of the autonomic nervous system.

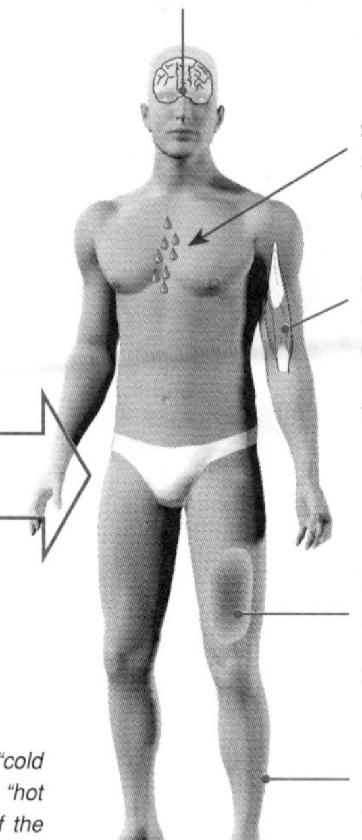

Sweating increases. Sweat cools by evaporation.

Muscle tone and **metabolic rate** are decreased. These mechanisms reduce the body's heat output.

Blood flow to skin (**vasodilation**) increases. This increases heat loss.

Erector muscles of hairs relax to flatten hairs and decrease insulating air layer.

**NOTE: The heat promoting center is also called the "cold centre" and the heat losing center is also called the "hot centre". We have used the terminology descriptive of the activities promoted by the center in each case.*

The Skin and Thermoregulation

Thermoreceptors in the dermis (probably free nerve endings) detect changes in skin temperature outside the normal range and send nerve impulses to the hypothalamus, which mediates a response. Thermoreceptors are of two types: **hot thermoreceptors** detect a rise in skin temperature above 37.5°C while the **cold thermoreceptors** detect a fall below 35.8°C. Temperature regulation by the skin involves **negative feedback** because the output is fed back to the skin receptors and becomes part of a new stimulus-response cycle.

Note that the thermoreceptors detect the temperature change, but the hair erector muscles and blood vessels are the **effectors** for mediating a response.

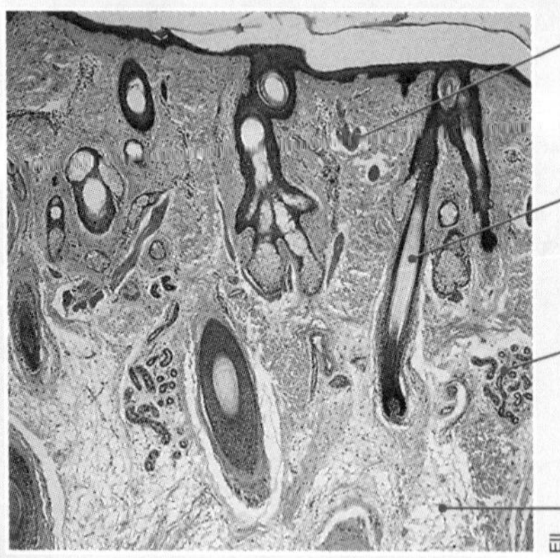

Cross section through the skin of the scalp.

Blood vessels in the dermis dilate (vasodilation) or constrict (vasoconstriction) to respectively promote or restrict heat loss.

Hairs raised or lowered to increase or decrease the thickness of the insulating air layer between the skin and the environment.

Sweat glands produce sweat in response to parasympathetic stimulation from the hypothalamus. Sweat cools through evaporation.

Fat in the subdermal layers insulates the organs against heat loss.

1. State two mechanisms by which body temperature could be reduced after intensive activity (e.g. hard exercise):

   (a) ______________________ (b) ______________________

2. Briefly state the role of the following in regulating internal body temperature:

   (a) The hypothalamus: ______________________

   (b) The skin: ______________________

   (c) Nervous input to effectors: ______________________

   (d) Hormones: ______________________

ISBN: 978-1-927173-16-9

Acid–Base Balance

The pH of the body's fluids must be maintained within a very narrow range (pH 7.35-7.45). The products of metabolic activity are generally acidic and could alter pH considerably without a buffer system to counteract pH changes. The carbonic acid-bicarbonate buffer works throughout the body to maintain the pH of blood plasma close to 7.40. The body maintains the buffer by eliminating either the acid (carbonic acid) or the base (bicarbonate ions). The blood buffers, the lungs, and the kidneys represent the three defense systems against disturbances of pH homeostasis. Changes in carbonic acid concentration can be brought about within seconds via increased or decreased rate of breathing. The renal system, acts more slowly, but can permanently eliminate metabolic acids and regulate the levels of alkaline substances, controlling pH by either excreting or retaining bicarbonate ions.

The Blood Buffer System

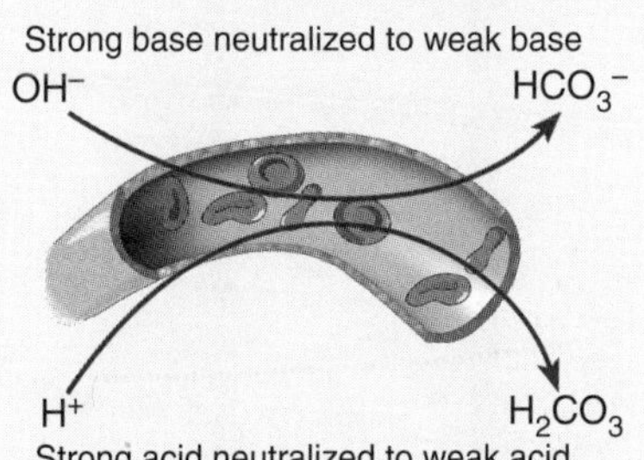
Strong base neutralized to weak base
OH^-
HCO_3^-
H^+
H_2CO_3
Strong acid neutralized to weak acid

A buffer is able to resist changes to the pH of a fluid when either an acid or base is added to it. The bicarbonate ion (HCO_3^-) and its acid, carbonic acid (H_2CO_3), work in the following way:

$$H^+ + HCO_3^- \rightleftharpoons H_2CO_3$$

$$H_2CO_3 \rightleftharpoons H^+ + HCO_3^-$$

If a strong acid (such as HCl) is added to the system a weak acid is formed and thus the pH falls only slightly. Note that the blood also contains proteins, which contain basic and acidic groups that may act either as H^+ acceptors or donors to help maintain blood pH.

The Respiratory System

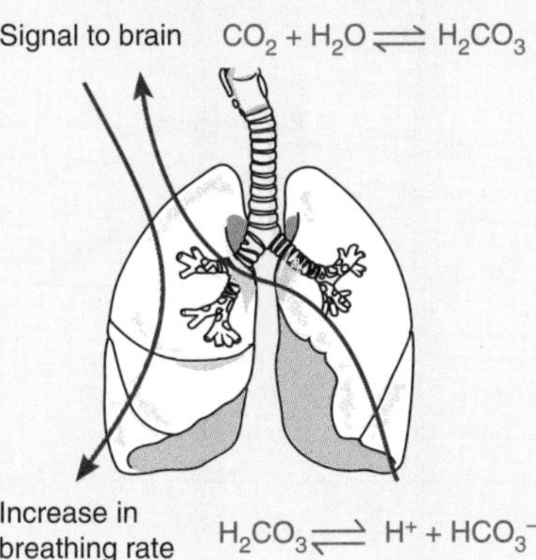
Signal to brain
$CO_2 + H_2O \rightleftharpoons H_2CO_3$
Increase in breathing rate
$H_2CO_3 \rightleftharpoons H^+ + HCO_3^-$

Carbon dioxide (CO_2) in the blood, an end-product of cellular respiration, forms carbonic acid (H_2CO_3) which dissociates to form H^+ and bicarbonate (HCO_3^-). This means that as CO_2 rises in the blood so too does the H^+ concentration. **Chemoreceptors** in the brain detect the rise in H^+ ions and increase the rate of breathing to expel the CO_2. Low levels of CO_2 have the effect of depressing the respiratory system so that H^+ builds up and the pH is once again restored.

The Renal System

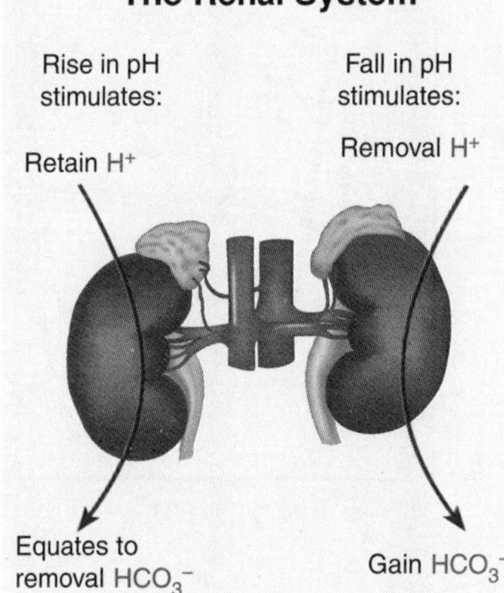
Rise in pH stimulates:
Fall in pH stimulates:
Retain H^+
Removal H^+
Equates to removal HCO_3^-
Gain HCO_3^-

Recall that a net loss of HCO_3^- effectively results in the gain of H^+.

When blood pH rises, bicarbonate is excreted (lost from the body) and H^+ is retained by the tubule cells. Conversely, when blood pH falls, bicarbonate is reabsorbed and H^+ is actively secreted. Urine pH can normally vary from 4.5 to 8.0, reflecting the ability of the renal tubules to excrete or retain ions to maintain the homeostasis of blood pH.

1. Explain why the blood must be kept at a pH between 7.35 and 7.45: ___

2. A drop in the blood pH to below 7.35 is called metabolic acidosis, even though the blood might still be at pH >7 and not strictly acidic. Describe how metabolic acidosis might arise:

3 (a) Describe how the blood buffer system maintains blood pH: ___

(b) Explain the effects of adding a base (e.g. ingestion of alkaline substances) to the system: ___

4. (a) Describe the respiratory response to excess H^+ in the blood: ___

(b) Explain where these H^+ ions come from: ___

(c) Describe how **respiratory acidosis** might arise: ___

5. Explain the role of the renal system in maintaining the pH of the blood: ___

ISBN: 978-1-927173-16-9

Related activities: *Blood, The Physiology of the Kidney, The Human Ventilation System*

Control of Blood Glucose

The endocrine portion of the **pancreas** (the α and β cells of the **islets of Langerhans**) produces two hormones, **insulin** and **glucagon**, which maintain blood glucose at a steady state through **negative feedback**. Insulin promotes a decrease in blood glucose by promoting cellular uptake of glucose and synthesizing glycogen. **Glucagon** promotes an increase in blood glucose through the breakdown of glycogen and the synthesis of glucose from amino acids. When normal blood glucose levels are restored, negative feedback stops hormone secretion. Regulating blood glucose to within narrow limits allows energy to be available to cells as needed. Extra energy is stored as glycogen or fat, and is mobilized to meet energy needs as required. The liver is pivotal in these carbohydrate conversions. One of the consequences of a disruption to this system is the disease **diabetes mellitus**. In type I diabetes, the insulin-producing β cells are destroyed as a result of autoimmune activity and insulin is not produced. In type II diabetes, the pancreatic cells produce insulin, but the body's cells become increasingly resistant to it.

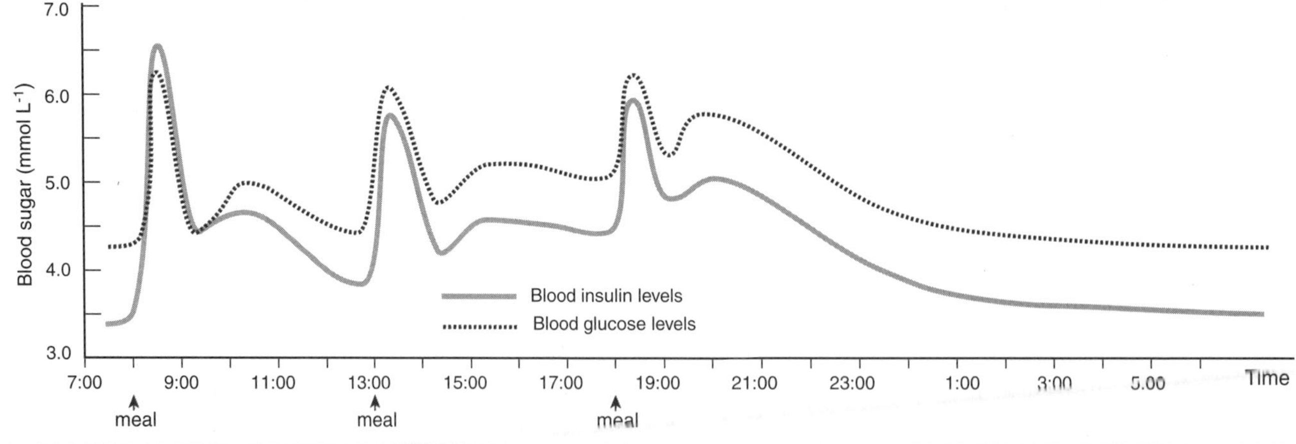
Blood sugar (mmol L⁻¹)
7.0
6.0
5.0
4.0
3.0
Blood insulin levels
Blood glucose levels
7:00
9:00
11:00
13:00
15:00
17:00
19:00
21:00
23:00
1:00
3:00
5.00
Time
meal
meal
meal

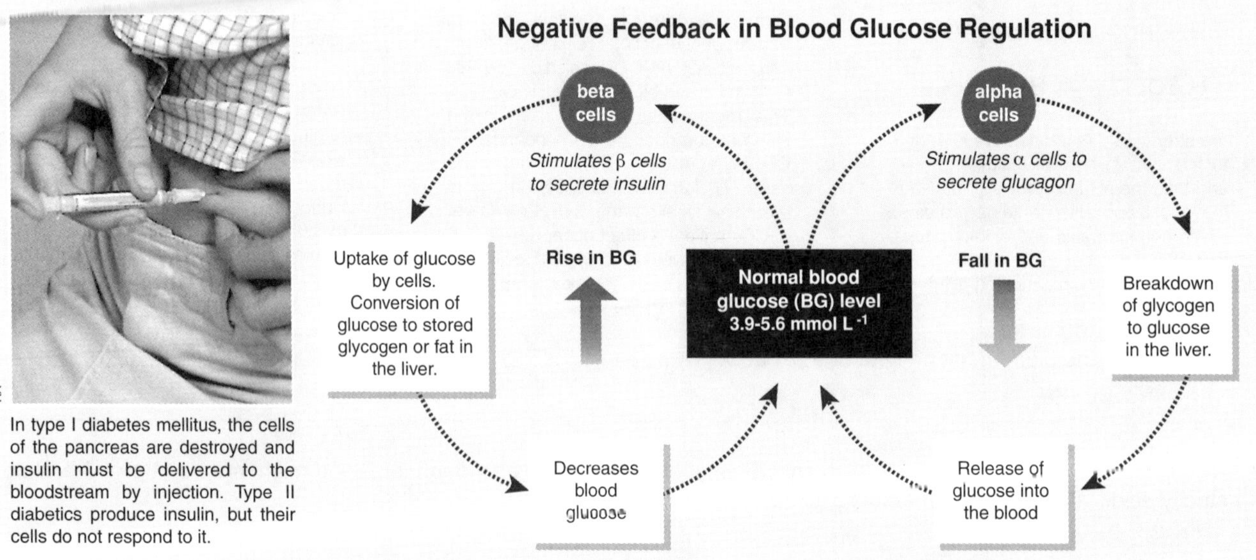
RA
Negative Feedback in Blood Glucose Regulation
beta cells
alpha cells
Stimulates β cells to secrete insulin
Stimulates α cells to secrete glucagon
Uptake of glucose by cells. Conversion of glucose to stored glycogen or fat in the liver.
Rise in BG
Normal blood glucose (BG) level 3.9-5.6 mmol L⁻¹
Fall in BG
Breakdown of glycogen to glucose in the liver.
Decreases blood glucose
Release of glucose into the blood

In type I diabetes mellitus, the cells of the pancreas are destroyed and insulin must be delivered to the bloodstream by injection. Type II diabetics produce insulin, but their cells do not respond to it.

1. (a) What is the stimulus for insulin release? ______________________

   (b) What is the stimulus for glucagon release? ______________________

   (c) How does glucagon bring about an increase in blood glucose level? ______________________

   (d) How does insulin bring about a decrease in blood glucose level? ______________________

2. Explain the pattern of fluctuations in blood glucose and blood insulin levels in the graph above:

3. Identify the mechanism regulating insulin and glucagon secretion (humoral, hormonal, neural): ______________________

Related activities: *Hormonal Regulatory Systems, Diabetes Mellitus*

Periodicals: *Food for thought*

ISBN: 978-1-927173-16-9

Diabetes Mellitus

Diabetes is a general term for a range of disorders sharing two common symptoms: production of large amounts of urine and excessive thirst. **Diabetes mellitus** is the most common form of diabetes and is characterized by **hyperglycemia** (high blood sugar). **Type I** is characterized by a complete lack of insulin production and usually begins in childhood, while **type II** is more typically a disease of older, overweight people whose cells develop a resistance to insulin uptake. Both types are chronic, incurable conditions and are managed differently. Type I is treated primarily with insulin injection, whereas type II sufferers manage their disease through diet and exercise in an attempt to limit the disease's long term detrimental effects.

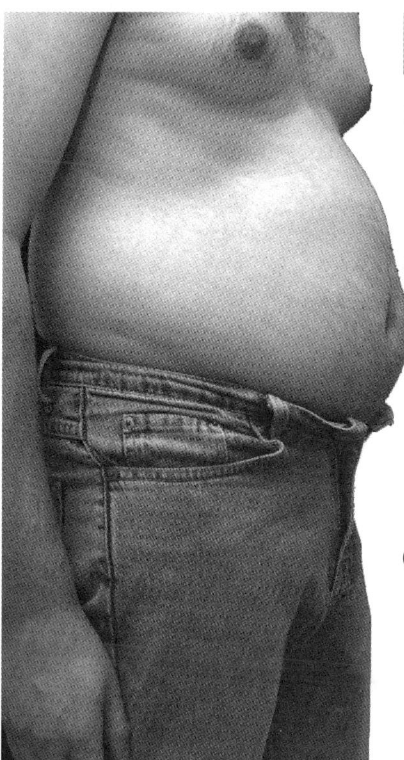

Symptoms of Type II Diabetes Mellitus

1. Symptoms may be mild at first. The body's cells do not respond appropriately to the insulin that is present and blood glucose levels become elevated. Normal blood glucose is 60-110 mg dL^{-1}. In diabetics, fasting blood glucose level is 126 mg dL^{-1} or higher.

2. Symptoms occur with varying degrees of severity:
   - Cells are starved of fuel. This can lead to increased appetite and overeating and may contribute to an existing obesity problem.
   - Urine production increases to rid the body of the excess glucose. Glucose is present in the urine and patients are frequently very thirsty.
   - The body's inability to use glucose properly leads to muscle weakness and fatigue, irritability, frequent infections, and poor wound healing.

3. Uncontrolled elevated blood glucose eventually results in damage to the blood vessels and leads to:
   - coronary artery disease
   - peripheral vascular disease
   - retinal damage, blurred vision and blindness
   - kidney damage and renal failure
   - persistent ulcers and gangrene

Risk Factors

Obesity: BMI greater than 27. Distribution of weight is also important.

Age: Risk increases with age, although the incidence of type 2 diabetes is increasingly reported in obese children.

Sedentary lifestyle: Inactivity increases risk through its effects on bodyweight.

Family history: There is a strong genetic link for type II diabetes. Those with a family history of the disease are at greater risk.

Ethnicity: Certain ethnic groups are at higher risk of developing of type II diabetes.

High blood pressure: Up to 60% of people with undiagnosed diabetes have high blood pressure.

High blood lipids: More than 40% of people with diabetes have abnormally high levels of cholesterol and similar lipids in the blood.

There is no cure for diabetes, but it can be managed to minimize the health effects:

Treating Type II Diabetes

Diabetes is not curable but can be managed to minimize the health effects:

- Regularly check blood glucose level
- Manage diet to reduce fluctuations in blood glucose level
- Take regular exercise
- Reduce weight
- Reduce blood pressure
- Reduce or stop smoking
- Take prescribed anti-diabetic drugs
- In time, insulin therapy may be required

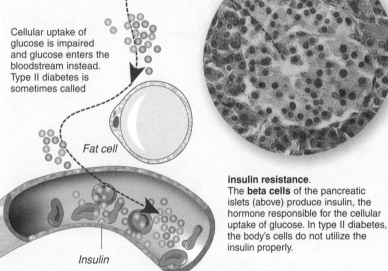

Cellular uptake of glucose is impaired and glucose enters the bloodstream instead. Type II diabetes is sometimes called **insulin resistance.**

Fat cell

Insulin

The **beta cells** of the pancreatic islets (above) produce insulin, the hormone responsible for the cellular uptake of glucose. In type II diabetes, the body's cells do not utilize the insulin properly.

1. Distinguish between type I and type II diabetes, relating the differences to the different methods of treatment:

   ____

   ____

2. What dietary advice would you give to a person diagnosed with type II diabetes? ____________________

   ____

   ____

3. Why is the increase in type II diabetes considered to be epidemic in the developed world? ____________________

   ____

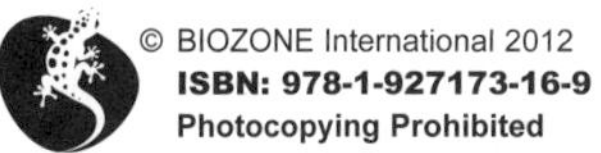

ISBN: 978-1-927173-16-9

Waste Products in Humans

In humans and other mammals, a number of organs are involved in the excretion of the waste products of metabolism: mainly the kidneys, lungs, skin, and gut. The liver is a particularly important organ in the initial treatment of waste products, particularly the breakdown of hemoglobin and the formation of **urea** from ammonia. Excretion should not be confused with the elimination or egestion of undigested and unabsorbed food material from the gut. Note that the breakdown products of hemoglobin (blood pigment) are excreted in bile and pass out with the faeces, but they are not the result of digestion.

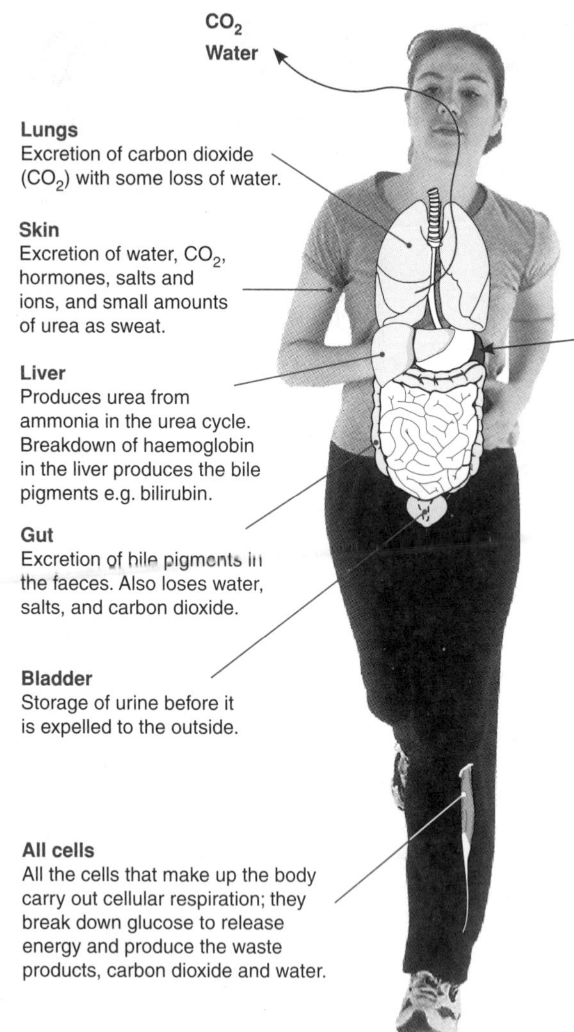

CO_2
Water
Lungs
Excretion of carbon dioxide (CO_2) with some loss of water.
Skin
Excretion of water, CO_2, hormones, salts and ions, and small amounts of urea as sweat.
Liver
Produces urea from ammonia in the urea cycle. Breakdown of haemoglobin in the liver produces the bile pigments e.g. bilirubin.
Gut
Excretion of bile pigments in the faeces. Also loses water, salts, and carbon dioxide.
Bladder
Storage of urine before it is expelled to the outside.
All cells
All the cells that make up the body carry out cellular respiration; they break down glucose to release energy and produce the waste products, carbon dioxide and water.
Kidney
Filtration of the blood to remove urea. Unwanted ions, particularly hydrogen (H^+) and potassium (K^+), and some hormones are also excreted by the kidneys. Some poisons and drugs (e.g. penicillin) are also excreted by active secretion into the urine. Water is lost in excreting these substances and extra water may be excreted if necessary.

Excretion in Humans

In mammals, the kidneys are the main organs of excretion, although the liver, skin, and lungs are also important. As well as ridding the body of nitrogenous wastes, the kidneys also regulate pH and excrete many toxins that are taken in from the environment. Many toxic substances, such as alcohol, are rendered harmless by detoxification in the liver, but the kidneys can also eliminate some by actively secreting them into the urine.

Substance	Origin*	Organ(s) of excretion
Carbon dioxide		
Water		
Bile pigments		
Urea		
Ions (K^+, HCO_3^-, H^+)		
Hormones		
Poisons		
Drugs		

*Origin refers to from where in the body each substance originates

1. Explain the need for excretion:

2. Complete the table above summarizing the origin of excretory products and the main organ(s) of excretion for each.

3. What is the role of the liver in excretion, even though it is not primarily an excretory organ?

4. Why is ammonia converted to urea to be excreted, despite the high energy cost of doing so?

5. In people suffering renal failure, the kidneys cease to produce filtrate. Based on your knowledge of the central role of the kidneys in fluid and electrolyte balance, as well as nitrogen excretion, describe the typical symptoms of kidney failure:

Related activities: The Kidney, The Physiology of the Kidney

ISBN: 978-1-927173-16-9

Water Budget in Humans

We cannot live without water for more than about 100 hours and adequate water is a requirement for physiological function and health. Body water content varies between individuals and through life, from above about 90% of total weight as a fetus to 74% as an infant, 60% as a child, and around 50-59% in adults, depending on gender and age. Gender differences (males usually have a higher water content than females) are the result of differing fat levels. Water intake and output are highly variable but closely matched to less than 0.1% over an extended period. Typical values for water gains and losses, as well as daily water transfers are given below. Men need more water than women due to their higher (on average) fat-free mass and energy expenditure. Infants and young children need more water in proportion to their body weight as they cannot concentrate their urine as efficiently as adults. They also have a greater surface area relative to weight, so water losses from the skin are greater.

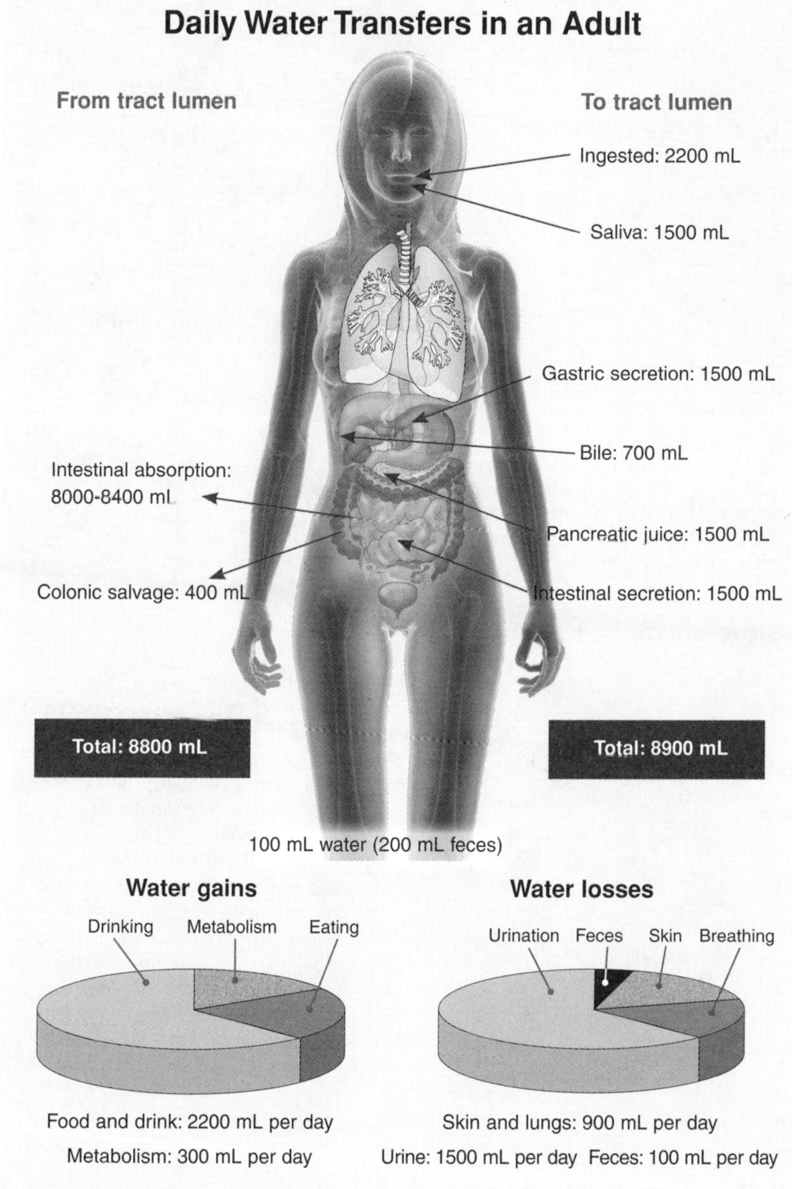
Daily Water Transfers in an Adult
From tract lumen
To tract lumen
Ingested: 2200 mL
Saliva: 1500 mL
Gastric secretion: 1500 mL
Bile: 700 mL
Intestinal absorption: 8000-8400 mL
Pancreatic juice: 1500 mL
Colonic salvage: 400 mL
Intestinal secretion: 1500 mL
Total: 8800 mL
Total: 8900 mL
100 mL water (200 mL feces)
Water gains
Drinking
Metabolism
Eating
Water losses
Urination
Feces
Skin
Breathing
Food and drink: 2200 mL per day
Metabolism: 300 mL per day
Skin and lungs: 900 mL per day
Urine: 1500 mL per day
Feces: 100 mL per day

About 63% of our daily requirement for water is met through drinking fluids, 25% is obtained from food, and the remaining 12% comes from metabolism (the oxidation of glucose to ATP, CO_2, and water).

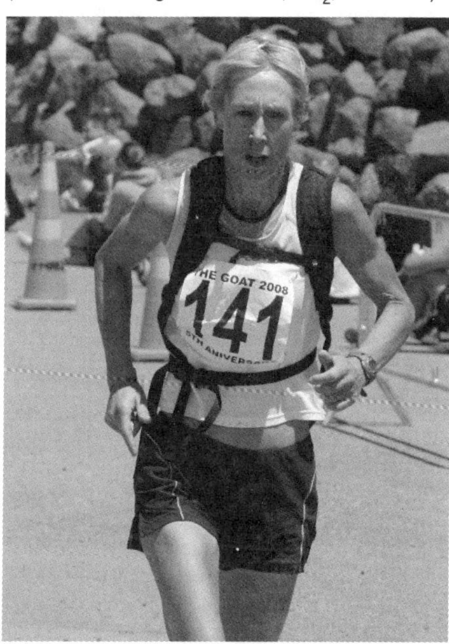
141

Typically, we lose 60% of body water through urination, 36% through the skin and lungs, and 4% in feces. Losses through the skin and from the lungs (breathing) average about 900 mL per day or more during heavy exercise. These are called **insensible losses**.

1. Explain how metabolism provides water for the body's activities: __________

2. Describe four common causes of physiological dehydration:

   (a) __________ (c) __________

   (b) __________ (d) __________

3. Some recent sports events have received media coverage because athletes have collapsed after excessive water intakes. This condition, called **hyponatremia** or water intoxication, causes nausea, confusion, diminished reflex activity, stupor, and eventually coma. From what you know of fluid and electrolyte balances in the body, explain these symptoms:

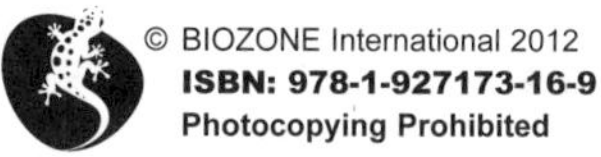

ISBN: 978-1-927173-16-9

The Kidney

The mammalian urinary system consists of the kidneys and bladder, and their associated blood vessels and ducts. The kidneys have a plentiful blood supply from the renal artery. The blood plasma is filtered by the kidneys to form urine. Urine is produced continuously, passing along the ureters to the bladder, a hollow muscular organ lined with smooth muscle and stretchable epithelium. Each day the kidneys filter about 180 L of plasma. Most of this is reabsorbed, leaving a daily urine output of about 1 L. By adjusting the composition of the fluid excreted, the kidneys help to maintain the body's internal chemical balance. Mammalian kidneys are very efficient, producing a urine that is concentrated to varying degrees depending on requirements.

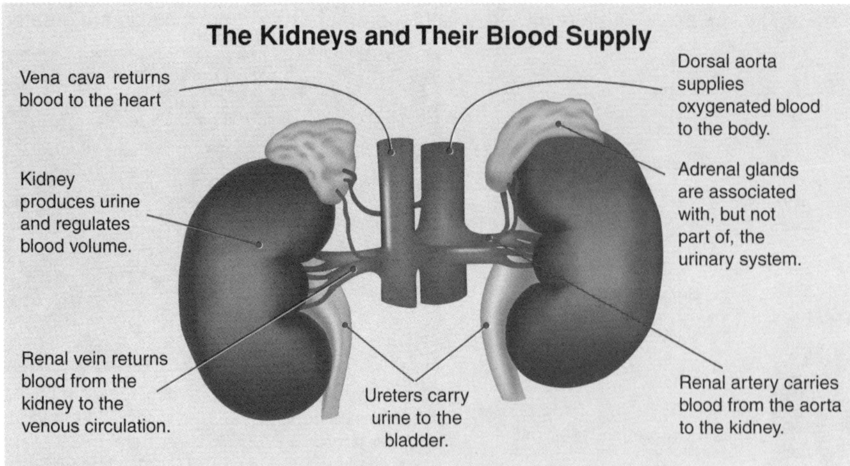
The Kidneys and Their Blood Supply
Vena cava returns blood to the heart
Dorsal aorta supplies oxygenated blood to the body.
Kidney produces urine and regulates blood volume.
Adrenal glands are associated with, but not part of, the urinary system.
Renal vein returns blood from the kidney to the venous circulation.
Ureters carry urine to the bladder.
Renal artery carries blood from the aorta to the kidney.

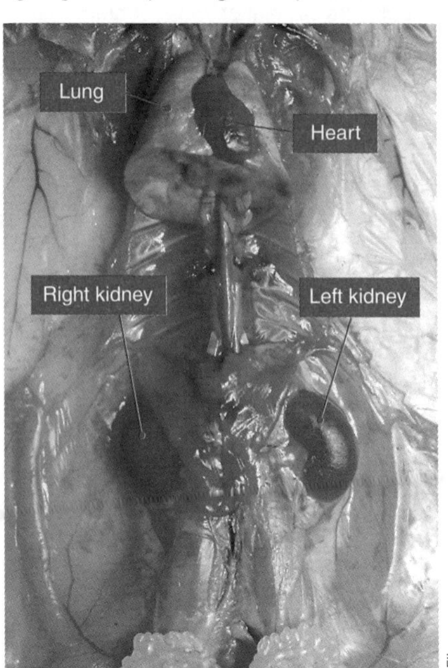
Lung
Heart
Right kidney
Left kidney
EII

The kidneys of humans (above), rats (dissection, right), and other mammals are distinctive, bean shaped organs that lie at the back of the abdominal cavity to either side of the spine. The kidneys lie outside the peritoneum of the abdominal cavity and are partly protected by the lower ribs. Each kidney is surrounded by three layers of tissue. The inner-most renal capsule is a smooth fibrous membrane that acts as a barrier against trauma and infection. The two outer layers comprise fatty tissue and fibrous connective tissue. These protect the kidney and anchor it firmly in place.

Internal Structure of the Human Kidney

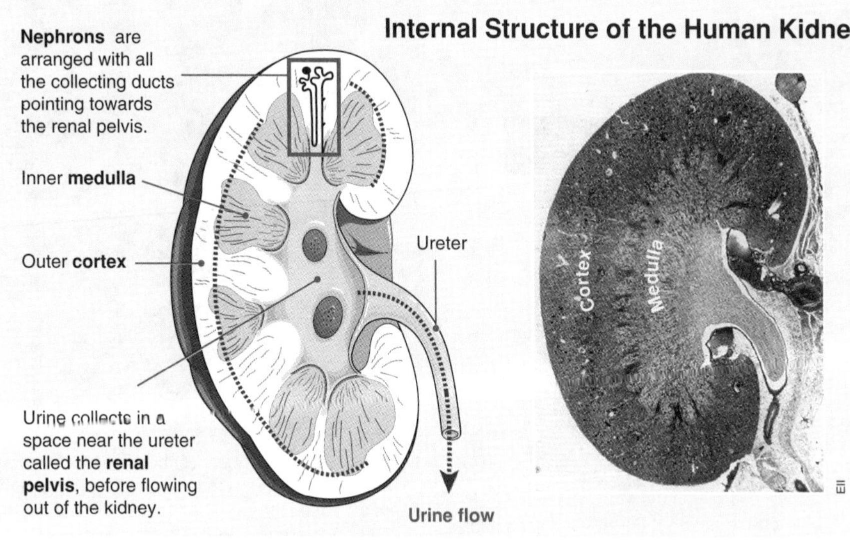
Nephrons are arranged with all the collecting ducts pointing towards the renal pelvis.
Inner **medulla**
Outer **cortex**
Urine collects in a space near the ureter called the **renal pelvis**, before flowing out of the kidney.
Ureter
Urine flow
Cortex
Medulla
EII

Human kidneys are about 100-120 mm long and 25 mm thick. The precise alignment of the nephrons (the filtering elements of the kidney) and their associated blood vessels gives the kidney tissue it characteristic striated (striped) appearance, as shown in the diagram (far left) and LM cross section (left). It is the precise arrangement of nephrons in the kidney that makes it possible to accommodate all the filtering units required. Each kidney contains more than 1 million nephrons. They are selective filter elements, which regulate blood composition and pH, and excrete wastes and toxins.

1. Identify the components of the urinary system and describe their functions: ______

2. Calculate the percentage of the plasma reabsorbed by the kidneys: ______

3. Why does the kidney receive blood at a higher pressure than other organs? ______

4. What is the purpose of the fatty connective tissue surrounding the kidneys? ______

Related activities: *The Physiology of the Kidney, Control of Urine Output*

ISBN: 978-1-927173-16-9

The Physiology of the Kidney

The functional unit of the kidney, the **nephron**, is a selective filter element, comprising a renal corpuscle and its associated tubules and ducts. Ultrafiltration, i.e. forcing fluid and dissolved substances through a membrane by pressure, occurs in the first part of the nephron, across the membranes of the capillaries and the glomerular capsule. The passage of water and solutes into the nephron and the formation of the glomerular filtrate depends on the pressure of the blood entering the afferent arteriole (below). If it increases, filtration rate increases. When it falls, glomerular filtration rate also falls. This process is so precisely regulated that, in spite of fluctuations in arteriolar pressure, glomerular filtration rate per day stays constant. After formation of the initial filtrate, the **urine** is modified through secretion and tubular reabsorption according to physiological needs at the time.

Nephron Structure

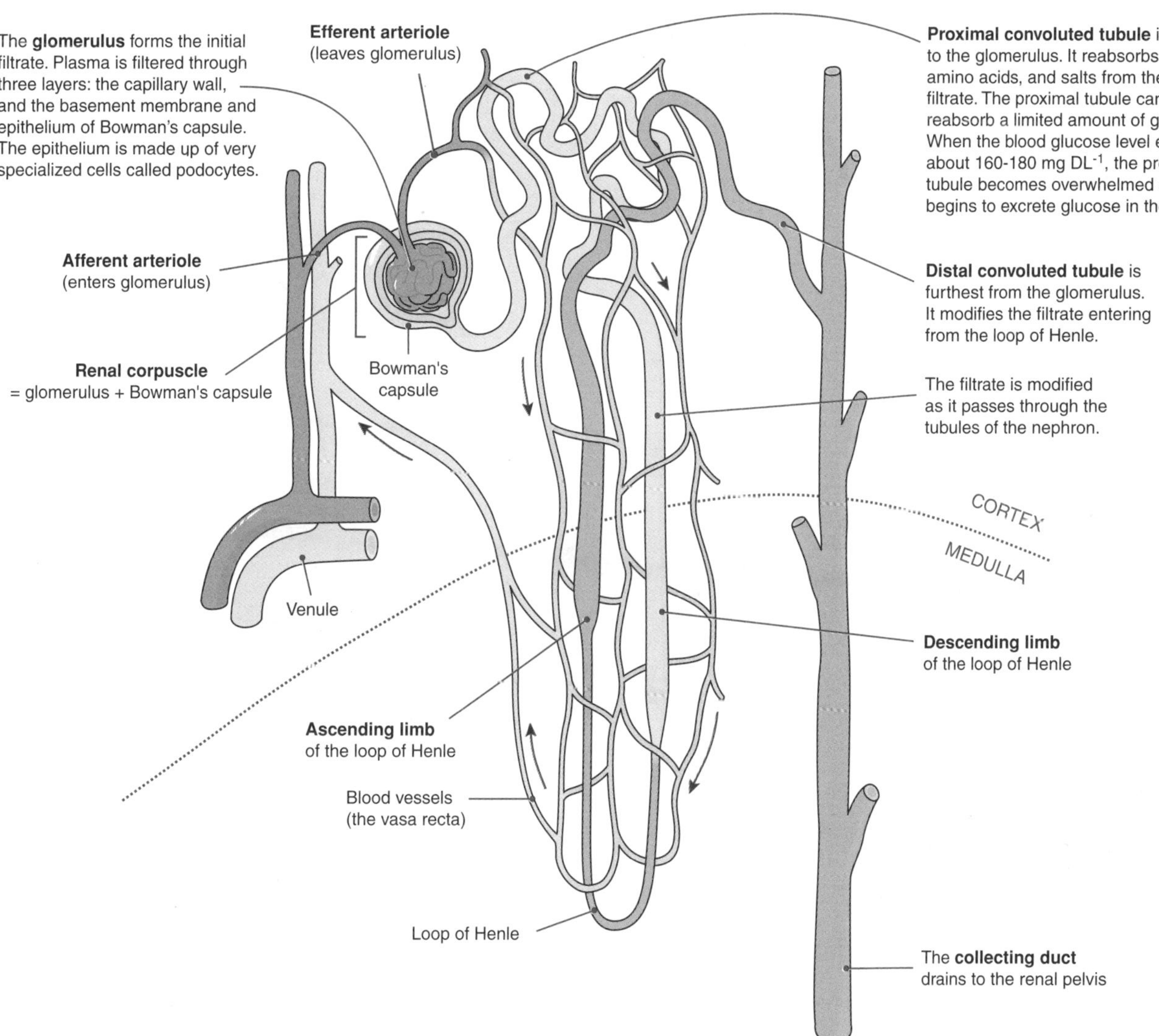
The **glomerulus** forms the initial filtrate. Plasma is filtered through three layers: the capillary wall, and the basement membrane and epithelium of Bowman's capsule. The epithelium is made up of very specialized cells called podocytes.
Efferent arteriole (leaves glomerulus)
Afferent arteriole (enters glomerulus)
Renal corpuscle = glomerulus + Bowman's capsule
Bowman's capsule
Venule
Proximal convoluted tubule is nearest to the glomerulus. It reabsorbs glucose, amino acids, and salts from the initial filtrate. The proximal tubule can only reabsorb a limited amount of glucose. When the blood glucose level exceeds about 160-180 mg DL^{-1}, the proximal tubule becomes overwhelmed and begins to excrete glucose in the urine.
Distal convoluted tubule is furthest from the glomerulus. It modifies the filtrate entering from the loop of Henle.
The filtrate is modified as it passes through the tubules of the nephron.
CORTEX
MEDULLA
Descending limb of the loop of Henle
Ascending limb of the loop of Henle
Blood vessels (the vasa recta)
Loop of Henle
The **collecting duct** drains to the renal pelvis

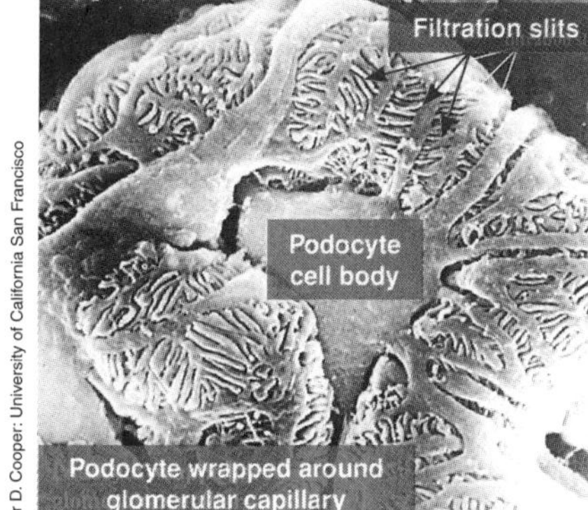
Filtration slits
Podocyte cell body
Podocyte wrapped around glomerular capillary
Dr D. Cooper: University of California San Francisco

The epithelium of Bowman's capsule is made up of specialized cells called **podocytes**. The finger-like cellular processes of the podocytes wrap around the capillaries of the glomerulus, and the plasma filtrate passes through the filtration slits between them.

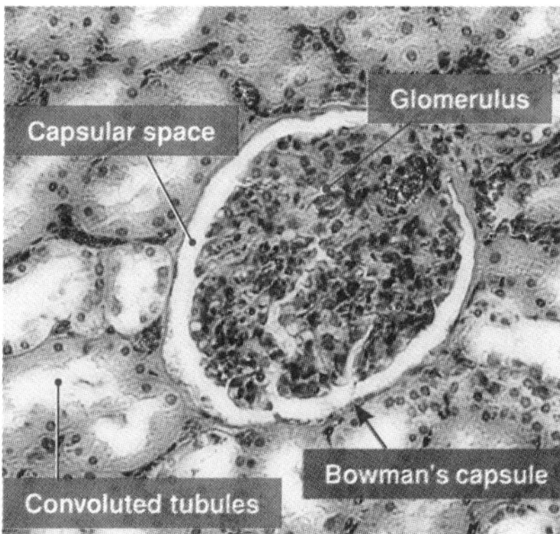
Capsular space
Glomerulus
Convoluted tubules
Bowman's capsule
EII

Bowman's capsule is a double walled cup, lying in the cortex of the kidney. It encloses a dense capillary network called the **glomerulus**. The capsule and its enclosed glomerulus form a **renal corpuscle**. In this section, the convoluted tubules can be seen surrounding the renal corpuscle.

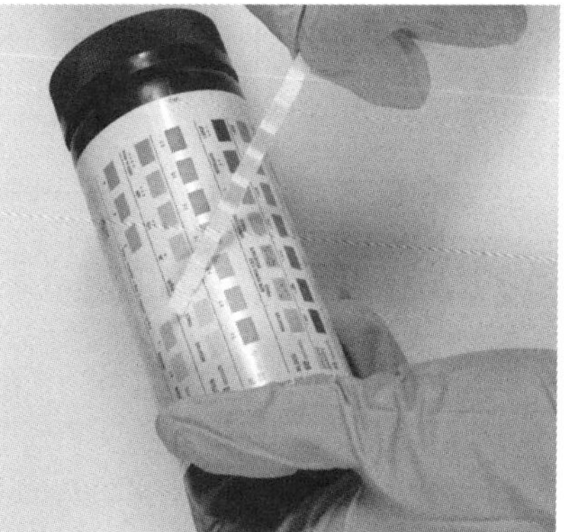

Dipstick urinalysis is commonly used to detect metabolic errors. Less than 0.1% of glucose filtered by the glomerulus normally appears in urine. The presence of glucose in the urine is usually due to untreated diabetes mellitus, which is characterized by high blood glucose levels.

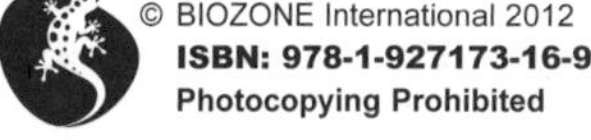

ISBN: 978-1-927173-16-9

Weblinks: *Kidney Vascular System, Interactive Kidney Quiz, The Juxtaglomerular Apparatus*

Summary of Activities in the Kidney Nephron

Urine formation begins by **ultrafiltration** of the blood, as fluid is forced through the capillaries of the glomerulus, forming a filtrate similar to blood but lacking cells and proteins. The filtrate is then modified by secretion and **reabsorption** to add or remove substances (e.g. ions). The processes involved in urine formation are summarized below for each region of the nephron: glomerulus, proximal convoluted tubule, loop of Henle, distal convoluted tubule, and collecting duct.

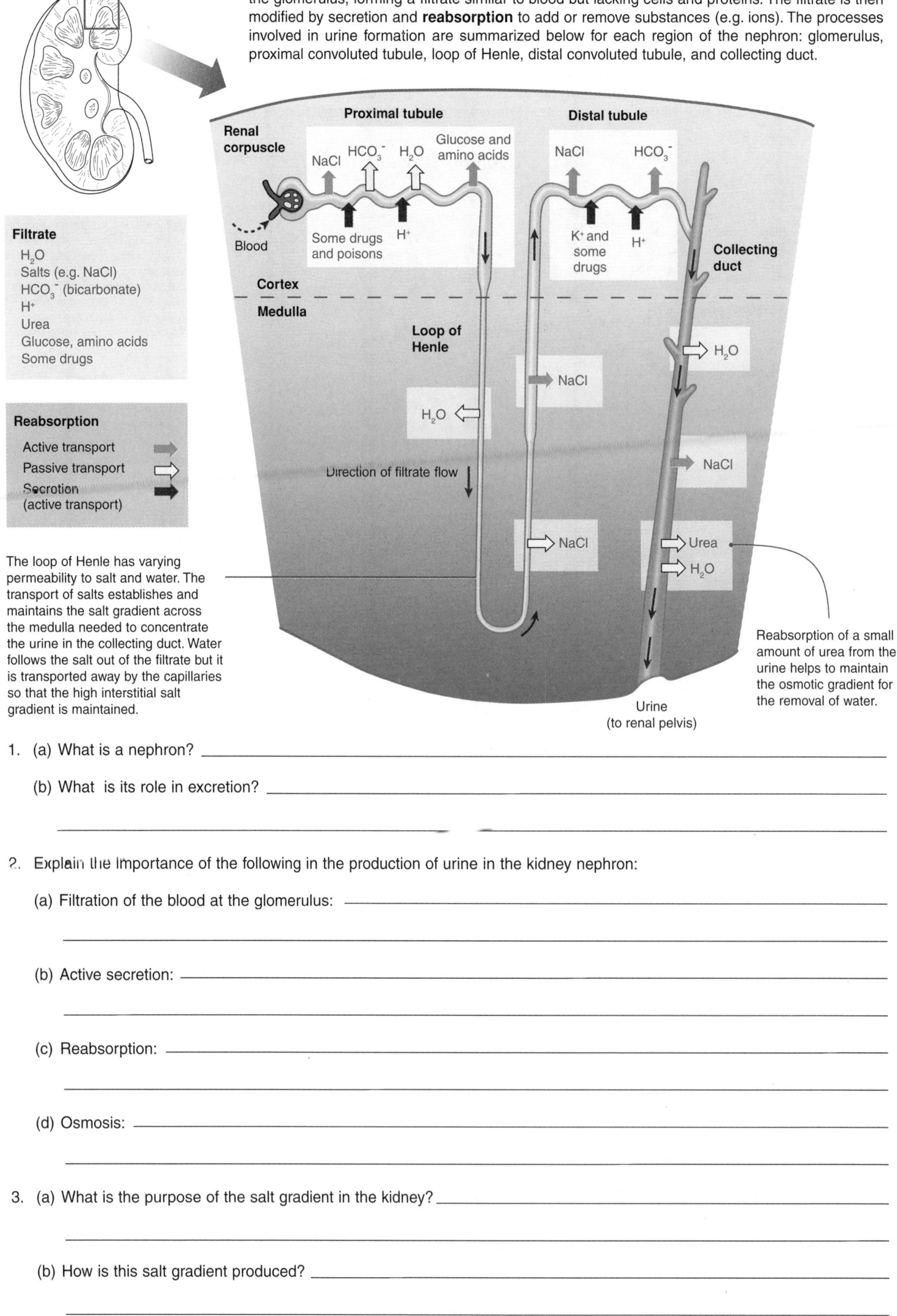

Filtrate
H2O
Salts (e.g. NaCl)
HCO3- (bicarbonate)
H+
Urea
Glucose, amino acids
Some drugs
Reabsorption
Active transport
Passive transport
Secretion
(active transport)
The loop of Henle has varying permeability to salt and water. The transport of salts establishes and maintains the salt gradient across the medulla needed to concentrate the urine in the collecting duct. Water follows the salt out of the filtrate but it is transported away by the capillaries so that the high interstitial salt gradient is maintained.
Proximal tubule
Distal tubule
Renal corpuscle
NaCl
HCO3-
H2O
Glucose and amino acids
Blood
Some drugs and poisons
H+
K+ and some drugs
Collecting duct
Cortex
Medulla
Loop of Henle
NaCl
H2O
Direction of filtrate flow
Urea
Reabsorption of a small amount of urea from the urine helps to maintain the osmotic gradient for the removal of water.
Urine
(to renal pelvis)

1. (a) What is a nephron? ______________________________

   (b) What is its role in excretion? ______________________________

   ____

2. Explain the importance of the following in the production of urine in the kidney nephron:

   (a) Filtration of the blood at the glomerulus: ______________________________

   ____

   (b) Active secretion: ______________________________

   ____

   (c) Reabsorption: ______________________________

   ____

   (d) Osmosis: ______________________________

   ____

3. (a) What is the purpose of the salt gradient in the kidney? ______________________________

   ____

   (b) How is this salt gradient produced? ______________________________

   ____

ISBN: 978-1-927173-16-9

Control of Urine Output

Variations in salt and water intake, and in the environmental conditions to which we are exposed, contribute to fluctuations in blood volume and composition. The primary role of the kidneys is to regulate blood volume and composition (including the removal of nitrogenous wastes), so that homeostasis is maintained. This is achieved through varying the volume and composition of the urine. Two hormones, **antidiuretic hormone** (ADH) and **aldosterone**, are involved in the process.

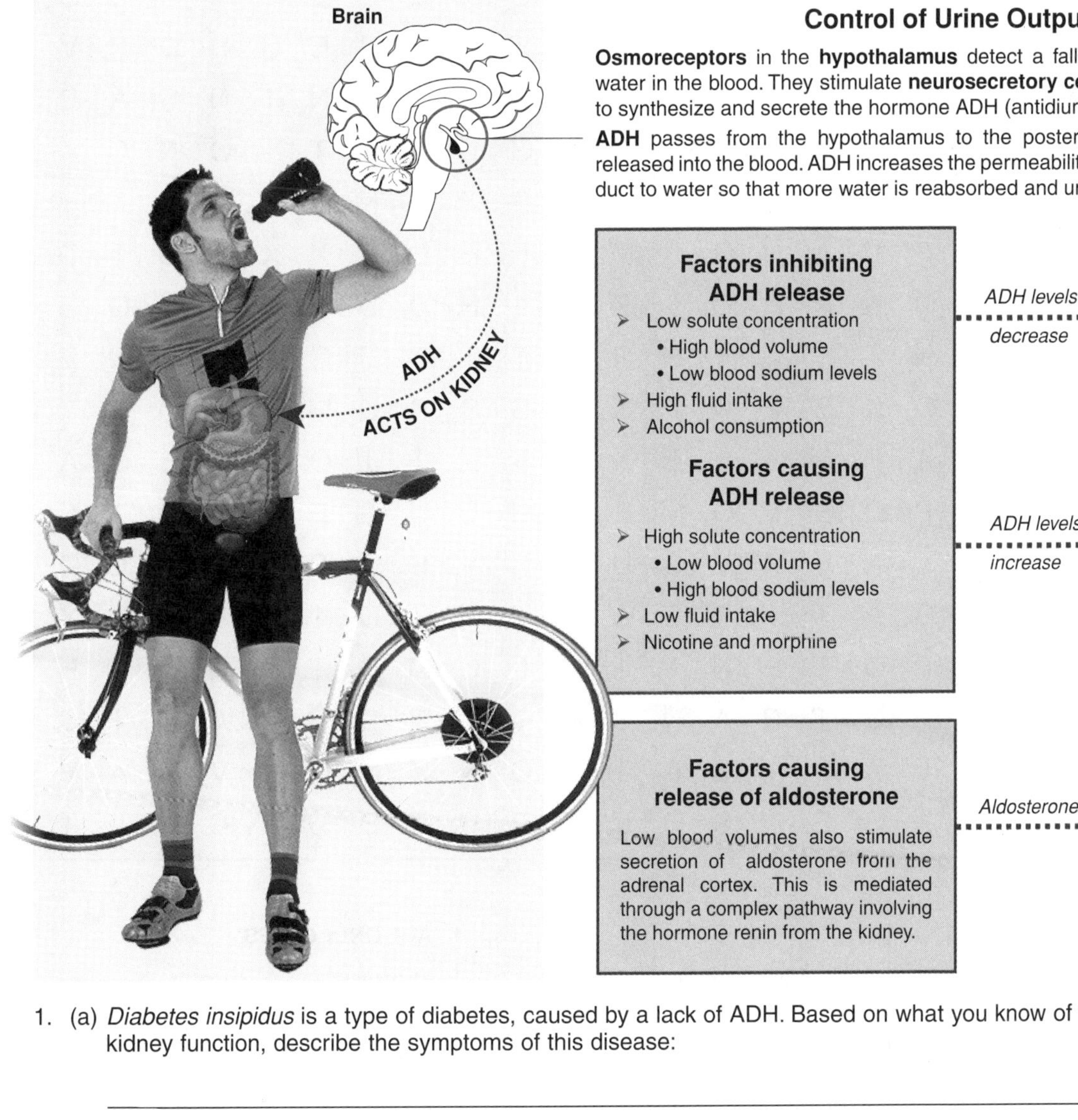
Brain
ADH
ACTS ON KIDNEY

Control of Urine Output

Osmoreceptors in the **hypothalamus** detect a fall in the concentration of water in the blood. They stimulate **neurosecretory cells** in the hypothalamus to synthesize and secrete the hormone ADH (antidiuretic hormone).

ADH passes from the hypothalamus to the posterior pituitary where it is released into the blood. ADH increases the permeability of the kidney collecting duct to water so that more water is reabsorbed and urine volume decreases.

Factors inhibiting ADH release

- Low solute concentration
  - High blood volume
  - Low blood sodium levels
- High fluid intake
- Alcohol consumption

ADH levels decrease → Water reabsorption decreases. Urine output increases.

Factors causing ADH release

- High solute concentration
  - Low blood volume
  - High blood sodium levels
- Low fluid intake
- Nicotine and morphine

ADH levels increase → Water reabsorption increases. Urine output decreases.

Factors causing release of aldosterone

Low blood volumes also stimulate secretion of aldosterone from the adrenal cortex. This is mediated through a complex pathway involving the hormone renin from the kidney.

Aldosterone → Sodium reabsorption increases, water follows, blood volume restored.

1. (a) *Diabetes insipidus* is a type of diabetes, caused by a lack of ADH. Based on what you know of the role of ADH in kidney function, describe the symptoms of this disease:

   (b) How might this disorder be treated?

2. Why does alcohol consumption (especially to excess) cause dehydration and thirst?

3. (a) What is the effect of aldosterone on the kidney nephron?

   (b) What is the net result of its action?

4. How do negative feedback mechanisms operate to regulate blood volume and urine output?

ISBN: 978-1-927173-16-9

Related activities: *Principles of Homeostasis, The Physiology of the Kidney*
Weblinks: *Adrenaline and ADH*

KEY TERMS: Word Find

Use the clues below to find the relevant key terms in the WORD FIND grid

P	A	P	E	R	I	P	H	E	R	A	L	N	E	R	V	O	U	S	S	Y	S	T	E	M
R	N	P	Y	P	N	W	X	D	I	A	B	E	T	E	S	M	E	L	L	I	T	U	S	O
H	X	O	T	H	E	R	M	O	R	E	G	U	L	A	T	I	O	N	E	U	R	O	N	V
O	U	H	F	A	U	L	T	R	A	F	I	L	T	R	A	T	I	O	N	I	N	S	A	D
M	O	U	S	F	P	Q	L	W	O	S	M	O	R	E	G	U	L	A	T	I	O	N	C	K
E	C	Y	B	S	X	R	E	E	Q	E	E	V	T	P	B	X	P	K	K	R	I	Q	T	I
O	P	P	M	K	L	U	F	M	A	H	O	R	M	O	N	E	D	F	X	K	D	C	I	D
S	I	N	E	P	H	R	O	N	N	O	D	L	Z	R	J	H	D	U	A	Y	S	K	O	N
T	I	Y	P	X	M	A	N	T	I	D	I	U	R	E	T	I	C	H	O	R	M	O	N	E
A	W	I	Y	A	K	D	O	K	B	Z	N	F	S	F	S	S	R	D	M	L	I	O	P	Y
S	L	S	N	S	Y	N	A	P	S	E	I	E	X	C	R	E	T	I	O	N	U	B	O	A
I	Q	R	T	E	M	F	P	O	G	L	O	M	E	R	U	L	U	S	E	M	Z	T	T	E
S	C	E	N	T	R	A	L	N	E	R	V	O	U	S	S	Y	S	T	E	M	P	E	E	O
H	Q	T	S	U	N	E	G	A	T	I	V	E	W	Z	V	Y	B	O	I	I	L	M	N	Y
R	E	S	T	I	N	G	P	O	T	E	N	T	I	A	L	C	M	O	T	O	R	D	T	H
T	V	W	W	N	E	U	R	O	T	R	A	N	S	M	I	T	T	E	R	W	W	K	I	C
C	U	W	D	E	P	O	L	A	R	I	Z	A	T	I	O	N	X	Y	O	K	V	N	A	V
H	Q	V	V	H	R	B	X	S	E	N	S	O	R	Y	P	U	Z	B	U	W	C	M	L	I

CORE CLUES

The portion of the nervous system that comprises the brain and spinal cord. (3 words: 7, 7, 6).

All the nerves and sensory receptors outside of the central nervous system comprise this. (3 words: 11, 7, 6).

A nerve cell, typically consisting of a cell body, dendrites and an axon.

The potential difference across the cell membrane of a neuron when there is no impulse passing. (2 words: 7, 9).

Self propagating rapid depolarization of a neuronal membrane, generated by movement of sodium and potassium ions. (2 words: 6, 9).

The change in membrane potential so that it becomes less negative.

The relative physiological constancy of the body despite external fluctuations.

A chemical messenger that is released by axon terminals in response to an action potential.

The junction between two neurons or between a neuron and an effector.

A disease caused by the body's inability to produce or react to insulin. Characterized by the presence of glucose in the urine. (2 words: 8, 8)

Chemical messenger that induces a specific physiological response.

A feedback mechanism by which physiological fluctuations are counteracted is called _ _ _ _ _ _ _ _ feedback.

The process by which animals regulate body temperature.

Neuron that transmits impulses from the central nervous system to effectors (e.g. muscles or glands).

Neuron that transmits sensory impulses from sensory receptors to the brain or spinal cord.

AHL ONLY CLUES

Elimination (by an organism) of waste products of metabolism.

A knot of capillaries in the kidney bound within the Bowman's capsule.

Functional unit of the kidney containing the glomerulus, Bowman's capsule, convoluted tubules, loop of Henle and collecting duct.

The active regulation of osmotic pressure in an organism (through water and ion regulation).

Hormone that increases the permeability of the kidney collecting duct to water so that more water is reabsorbed and urine volume increases. Historically known as vasopressin. (2 words 12, 7).

Bean shaped organ in vertebrates used to remove and concentrate metabolic wastes from the blood.

Process in the kidney by which small molecules and ions are separated from larger ones in the blood to form the renal filtrate.

ISBN: 978-1-927173-16-9

Core Topic **6.6**

AHL Topic **11.4**

Reproduction

Key concepts

- The male reproductive system produces and delivers male gametes to the female reproductive tract.
- The female reproductive system provides the environment for fertilization and embryonic development.
- Hormones control sexual development and reproduction.
- Gametogenesis involves mitotic and meiotic divisions, cell growth and differentiation.
- Fetal growth and development is maintained by the placenta, which acts as a temporary endocrine organ for the maintenance of pregnancy.

Key terms

Core

endometrium
estrogen
FSH
in vitro fertilization (IVF)
LH
menstrual cycle
menstruation
ovulation
progesterone
reproductive system

AHL Only

acrosome reaction
amniotic fluid
amniotic sac
cortical reaction
egg
embryo
epididymis
gamete
gametogenesis
HCG
interstitial (Leydig) cells
germinal epithelium
meiosis
oogenesis
oxytocin
placenta
positive feedback
pregnancy
primary follicles
prostate gland
puberty
secondary oocyte
semen
seminal vesicle
Sertoli cells
spermatogenesis
spermatozoa (sperm)
zygote

Periodicals:
Listings for this chapter are on page 400

Weblinks:
www.thebiozone.com/weblink/IB-3169.html

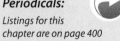

BIOZONE APP:
Student Review Series
Reproduction and Development

Learning Objectives

☐ 1. Use the **KEY TERMS** to compile a glossary for this topic.

The Human Reproductive System *(6.6)* pages 342-344, 353-354

☐ 2. Use labeled diagrams to describe the structure of the human male and female **reproductive systems** and associated structures, noting the relative size and position of the organs.

☐ 3. Describe the role of hormones in the menstrual cycle with reference to **FSH**, **LH**, **estrogen**, and **progesterone**.

☐ 4. Explain the features of the **menstrual cycle** including the development of the ovarian follicles, **ovulation**, development of corpora lutea, the cyclical changes to the **endometrium**, and **menstruation**. Using an annotated diagram, relate the changes in the menstrual cycle to the changes in the hormones regulating the cycle.

☐ 5. List three roles of **testosterone**.

☐ 6. Outline the process of *in vitro* fertilization (IVF). Discuss the ethical issues associated with the use of IVF. Discuss the potential risks in the procedures associated with IVF, including inducement of superovulation and artificial selection of sperm (*TOK*).

Gametogenesis *(11.4.1-11.4.8)* pages 345-347

☐ 7. **AHL**: Annotate a light micrograph of testis tissue to show the location and function of the **interstitial** (**Leydig**) **cells**, **germinal epithelium**, developing **spermatozoa**, and **Sertoli cells**. Describe the processes involved in **spermatogenesis**.

☐ 8. **AHL**: Describe the role of LH, testosterone, and FSH in spermatogenesis.

☐ 9. **AHL**: Annotate a diagram of the ovary to show the location and function of the **germinal epithelium**, **primary follicles**, mature follicles, and **secondary oocyte**. Describe the processes involved in **oogenesis**.

☐ 10. **AHL**: Draw and label diagrams of mature **gametes** (sperm and egg).

☐ 11. **AHL**: Describe the role of the **epididymis**, **seminal vesicle**, and **prostate gland** in the production of **semen**.

☐ 12. **AHL**: Compare **gametogenesis** in males and females, including the number of gametes and the timing of the formation and release of gametes.

Fertilization, Development, and Birth *(11.4.9-11.4.15)* pages 348-352

☐ 13. **AHL**: Describe fertilization, including the **acrosome reaction**, penetration of the egg membrane, and the **cortical reaction**.

☐ 14. **AHL**: Explain the source and the role of **HCG** in early pregnancy. Outline the events in the early development of the **embryo**.

☐ 15. **AHL**: Describe the structure and function of the **placenta**, including its role as an endocrine organ and in the exchange of material between the maternal and fetal blood. Describe the role of the **amniotic sac** and **amniotic fluid** during pregnancy.

☐ 16. **AHL**: Describe the process of birth and its hormonal control, including the role of **positive feedback** in bringing labor to its completion.

The Male Reproductive System

The reproductive role of the male is to produce the sperm and deliver them to the female. When a sperm combines with an egg, it contributes half the genetic material of the offspring and, in humans and other mammals, determines its sex. The reproductive structures of human males (described below) are in many ways typical of other mammals.

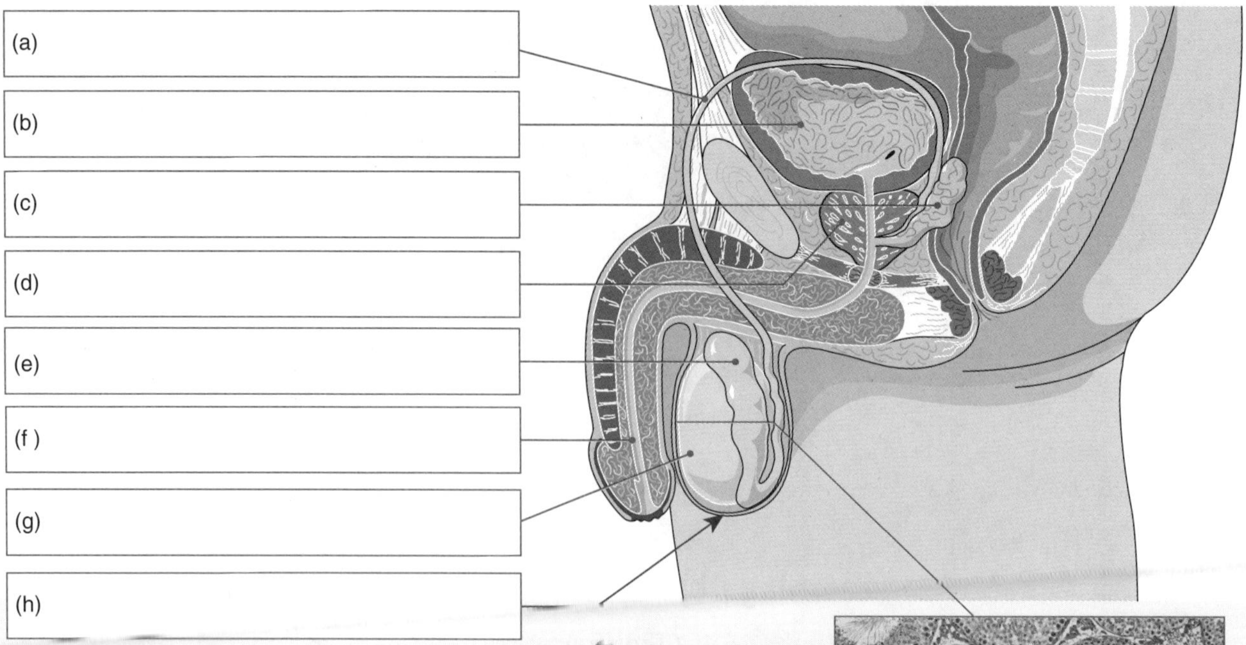

(a)

(b)

(c)

(d)

(e)

(f)

(g)

(h)

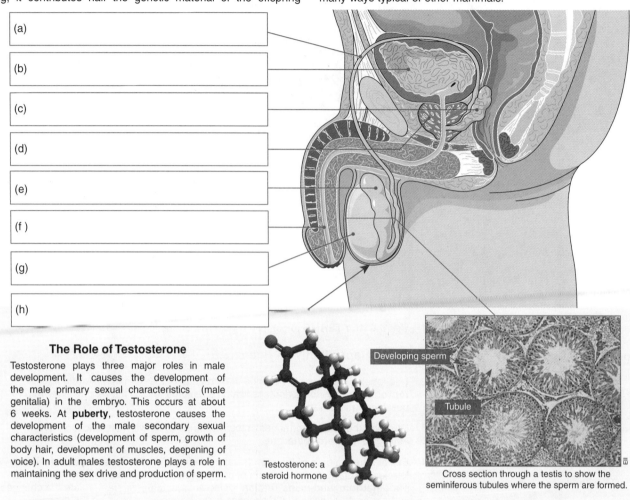

The Role of Testosterone

Testosterone plays three major roles in male development. It causes the development of the male primary sexual characteristics (male genitalia) in the embryo. This occurs at about 6 weeks. At **puberty**, testosterone causes the development of the male secondary sexual characteristics (development of sperm, growth of body hair, development of muscles, deepening of voice). In adult males testosterone plays a role in maintaining the sex drive and production of sperm.

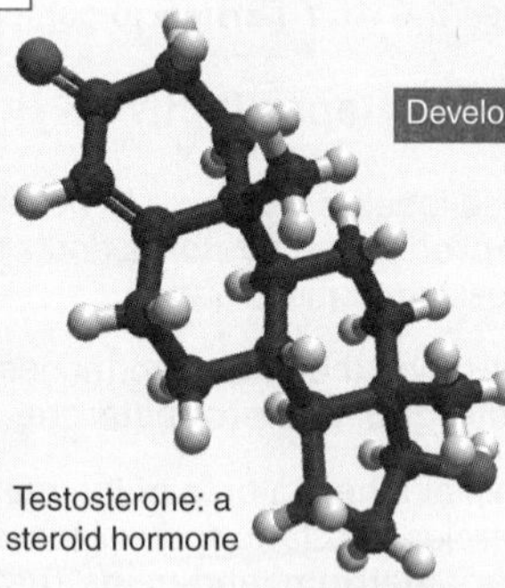

Testosterone: a steroid hormone

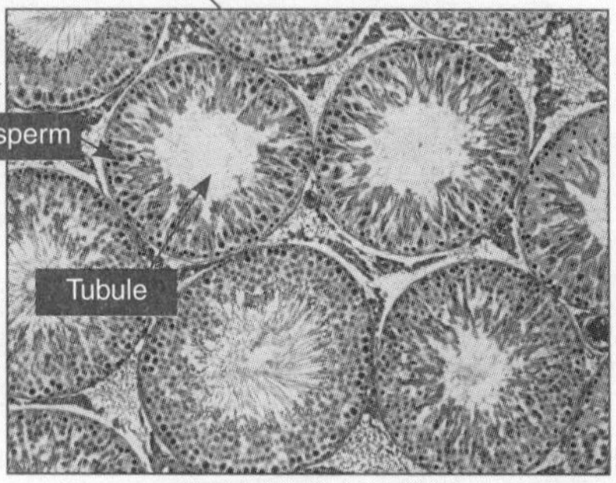

Developing sperm

Tubule

Cross section through a testis to show the seminiferous tubules where the sperm are formed.

1. The male human reproductive system and associated structures are shown above. Using the following word list, and the weblinks provide below, identify the labeled parts (write your answers in the spaces provided on the diagram).
Word list: *bladder, scrotal sac, sperm duct (vas deferens), epididymis, seminal vesicle, testis, urethra, prostate gland*

2. In a short sentence, state the function of each of the structures labeled (a)-(h) in the diagram above:

(a) ______

(b) ______

(c) ______

(d) ______

(e) ______

(f) ______

(g) ______

(h) ______

3. Describe the three roles of testosterone in male development and the male reproductive system: ______

4. State the two main roles of the male reproductive system: ______

Weblinks: *Male Reproductive System*

Periodicals:
Spermatogenesis

ISBN: 978-1-927173-16-9

The Female Reproductive System

The female reproductive system in mammals produces eggs, receives the penis and sperm during sexual intercourse, and houses and nourishes the young. Female reproductive systems in mammals are similar in their basic structure (uterus, ovaries etc.) but the shape of the uterus and the form of the placenta during pregnancy vary. The human system is described below.

The Female Reproductive System

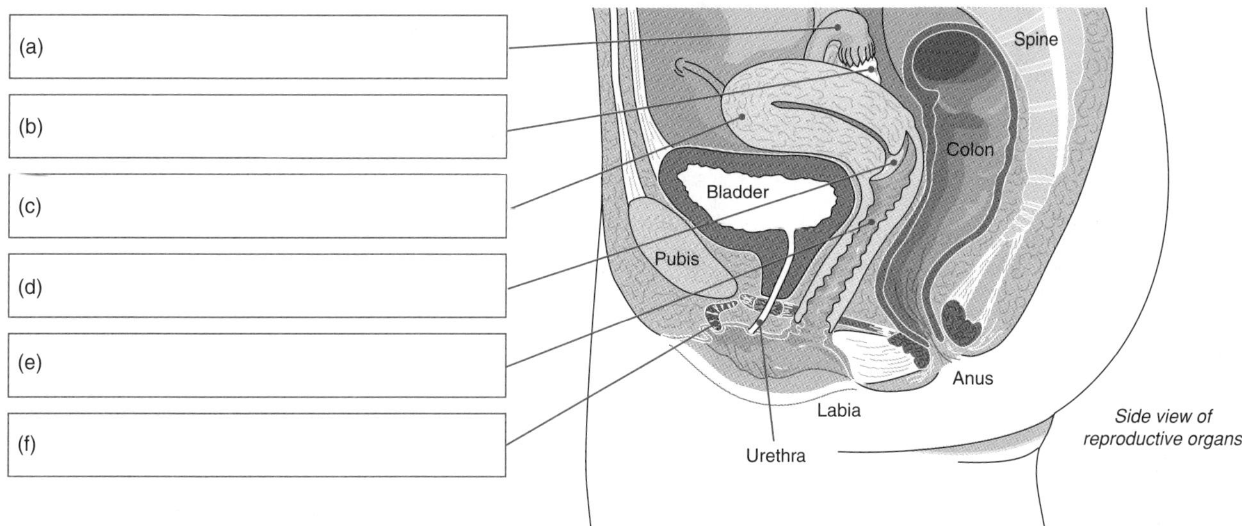
(a)
(b)
(c)
(d)
(e)
(f)
Spine
Colon
Bladder
Pubis
Anus
Labia
Urethra

Side view of reproductive organs

Ovulation and Implantation

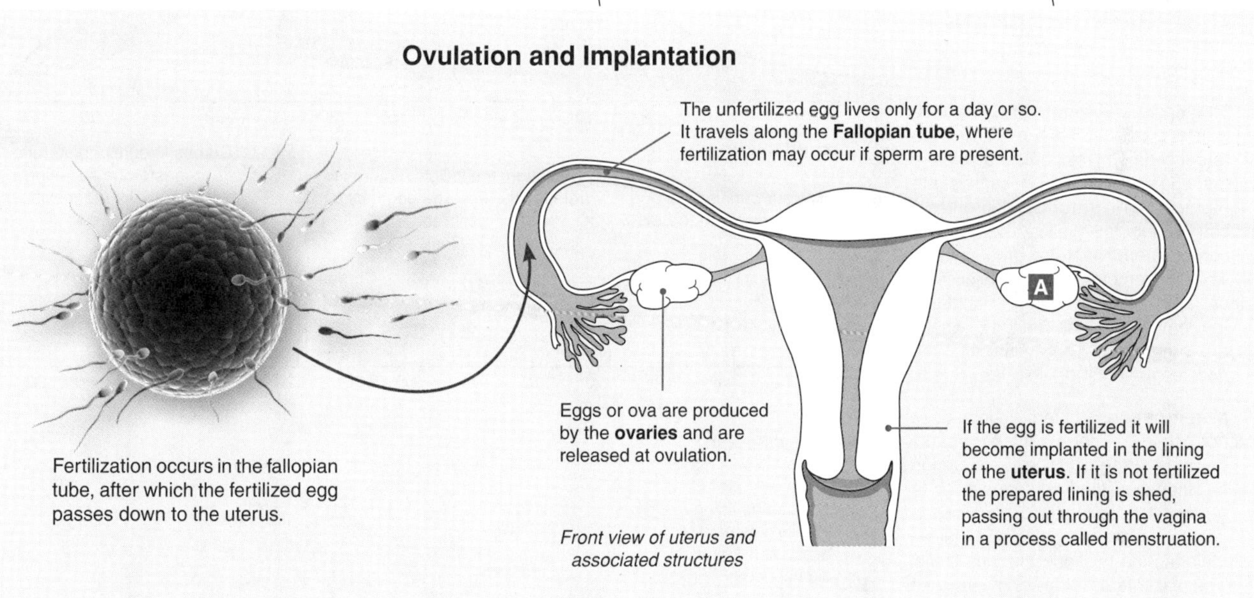
The unfertilized egg lives only for a day or so. It travels along the **Fallopian tube**, where fertilization may occur if sperm are present.
A
Eggs or ova are produced by the **ovaries** and are released at ovulation.
If the egg is fertilized it will become implanted in the lining of the **uterus**. If it is not fertilized the prepared lining is shed, passing out through the vagina in a process called menstruation.
Fertilization occurs in the fallopian tube, after which the fertilized egg passes down to the uterus.

Front view of uterus and associated structures

1. The female human reproductive system and associated structures are illustrated above. Using the word list, and the weblinks below to identify the labeled parts. **Word list**: *ovary, uterus (womb), vagina, fallopian tube (oviduct), cervix, clitoris.*

2. In a few words or a short sentence, state the function of each of the structures labeled (a) - (e) in the above diagram:

   (a) ______

   (b) ______

   (c) ______

   (d) ______

   (e) ______

3. (a) Name the organ labeled (**A**) in the diagram: ______

   (b) Name the event associated with this organ that occurs every month: ______

   (c) Name the process by which mature ova are produced: ______

4. Where does fertilization occur? ______

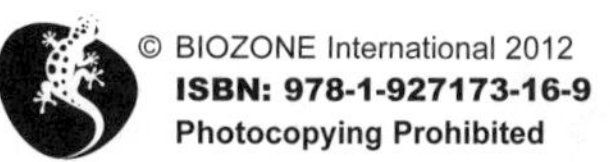

ISBN: 978-1-927173-16-9

Related activities: *The Menstrual Cycle, Oogenesis*
Weblinks: *Female Reproductive System*

The Menstrual Cycle

In non-primate mammals the reproductive cycle is characterized by a **breeding season** and an **estrous cycle** (a period of greater sexual receptivity during which ovulation occurs). In contrast, humans and other primates are sexually receptive throughout the year and may mate at any time. Like all placental mammals, their uterine lining thickens in preparation for pregnancy. However, unlike other mammals, primates shed this lining as a discharge through the vagina if fertilization does not occur. This event, called **menstruation**, characterizes the human reproductive or **menstrual cycle**. In human females, the menstrual cycle starts from the first day of bleeding and lasts for about 28 days. It involves a predictable series of changes that occur in response to hormones. The cycle is divided into three phases (see below), defined by the events in each phase.

The Menstrual Cycle

Luteinizing hormone (LH) and follicle stimulating hormone (FSH): These hormones from the anterior pituitary have numerous effects. FSH stimulates the development of the ovarian follicles resulting in the release of estrogen. Estrogen levels peak, stimulating a surge in LH and triggering ovulation.

Hormone levels: Of the follicles that begin developing in response to FSH, usually only one (the Graafian follicle) becomes dominant. In the first half of the cycle, estrogen is secreted by this developing Graafian follicle. Later, the Graafian follicle develops into the corpus luteum (below right) which secretes large amounts of progesterone (and smaller amounts of estrogen).

The corpus luteum: The Graafian follicle continues to grow and then (around day 14) ruptures to release the egg (ovulation). LH causes the ruptured follicle to develop into a corpus luteum (yellow body). The corpus luteum secretes progesterone which promotes full development of the uterine lining, maintains the embryo in the first 12 weeks of pregnancy, and inhibits the development of more follicles.

Menstruation: If fertilization does not occur, the corpus luteum breaks down. Progesterone secretion declines, causing the uterine lining to be shed (menstruation). If fertilization occurs, high progesterone levels maintain the thickened uterine lining. The placenta develops and nourishes the embryo completely by 12 weeks.

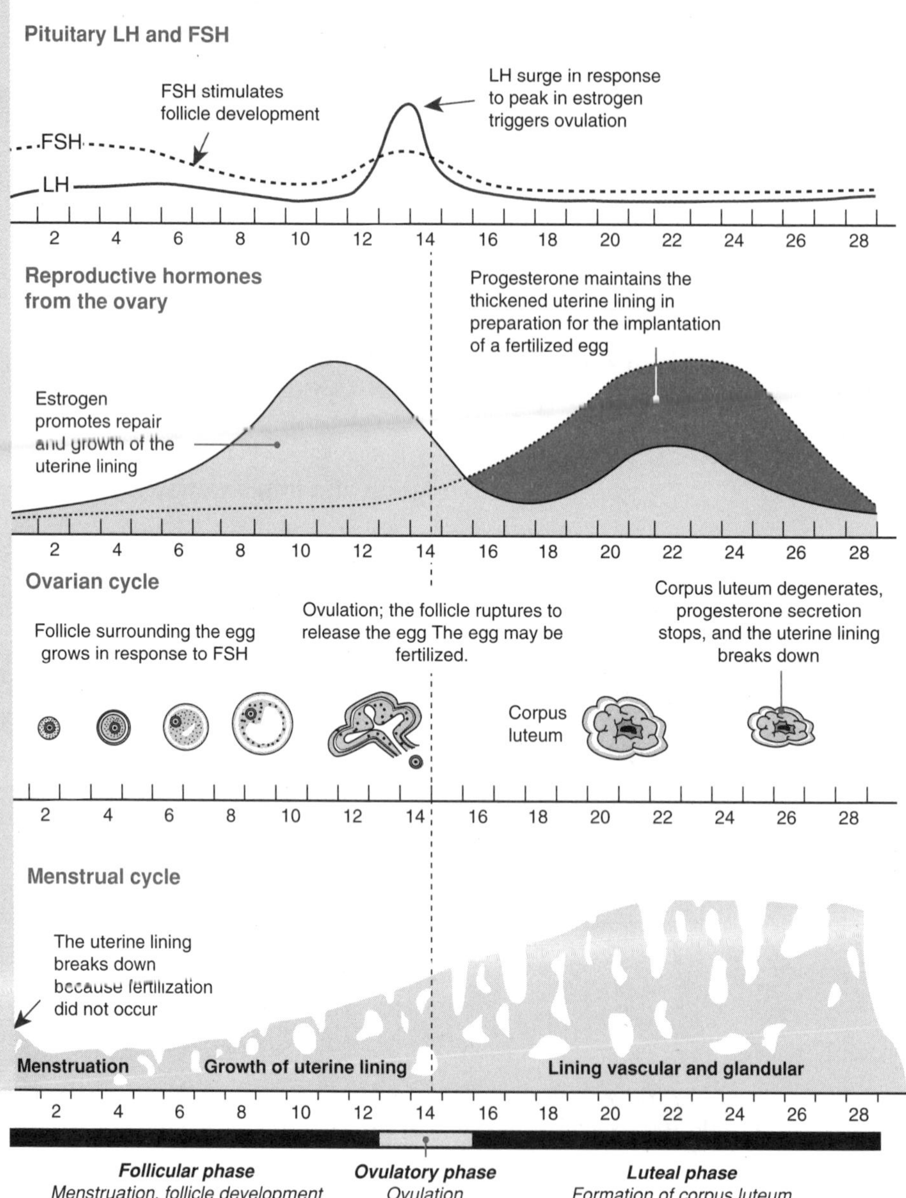
Pituitary LH and FSH
FSH stimulates follicle development
LH surge in response to peak in estrogen triggers ovulation
FSH
LH
2 4 6 8 10 12 14 16 18 20 22 24 26 28
Reproductive hormones from the ovary
Progesterone maintains the thickened uterine lining in preparation for the implantation of a fertilized egg
Estrogen promotes repair and growth of the uterine lining
Ovarian cycle
Follicle surrounding the egg grows in response to FSH
Ovulation; the follicle ruptures to release the egg The egg may be fertilized.
Corpus luteum degenerates, progesterone secretion stops, and the uterine lining breaks down
Corpus luteum
Menstrual cycle
The uterine lining breaks down because fertilization did not occur
Menstruation
Growth of uterine lining
Lining vascular and glandular
Day of the cycle:
Follicular phase
Menstruation, follicle development
Ovulatory phase
Ovulation
Luteal phase
Formation of corpus luteum

1. Name the hormone responsible for:

   (a) Follicle growth: ______________________ (b) Ovulation: ______________________

2. Each month, several ovarian follicles begin development, but only one (the Graafian follicle) develops fully:

   (a) Name the hormone secreted by the developing follicle: ______________________

   (b) State the role of this hormone during the follicular phase: ______________________

   (c) Suggest what happens to the follicles that do not continue developing: ______________________

3. (a) Identify the principal hormone secreted by the corpus luteum: ______________________

   (b) State the purpose of this hormone: ______________________

4. State the hormonal trigger for menstruation: ______________________

ISBN: 978-1-927173-16-9

Spermatogenesis

Gametogenesis involves meiotic division to produce male and female gametes for the purpose of sexual reproduction. Male mammals produce sperm in the testis by a process called **spermatogenesis**. In humans, sperm production begins at puberty and continues throughout a male's lifetime, but does decline with age. Thousands of sperm are produced every second, and take approximately two months to fully mature.

Spermatogenesis

Spermatogenesis is the process by which mature spermatozoa (sperm) are produced in the testis. In humans, they are produced at the rate of about 120 million per day. Spermatogenesis is regulated by the hormones **follicle stimulating hormone** (FSH) (from the anterior pituitary) and testosterone (secreted from the testes in response to **luteinizing hormone** (LH) (from the anterior pituitary). Spermatogonia, in the outer layer of the seminiferous tubules, multiply throughout reproductive life. Some of them divide by meiosis into spermatocytes, which produce spermatids. These are transformed into mature sperm by the process of spermiogenesis in the seminiferous tubules of the testis. Full sperm motility is achieved in the epididymis.

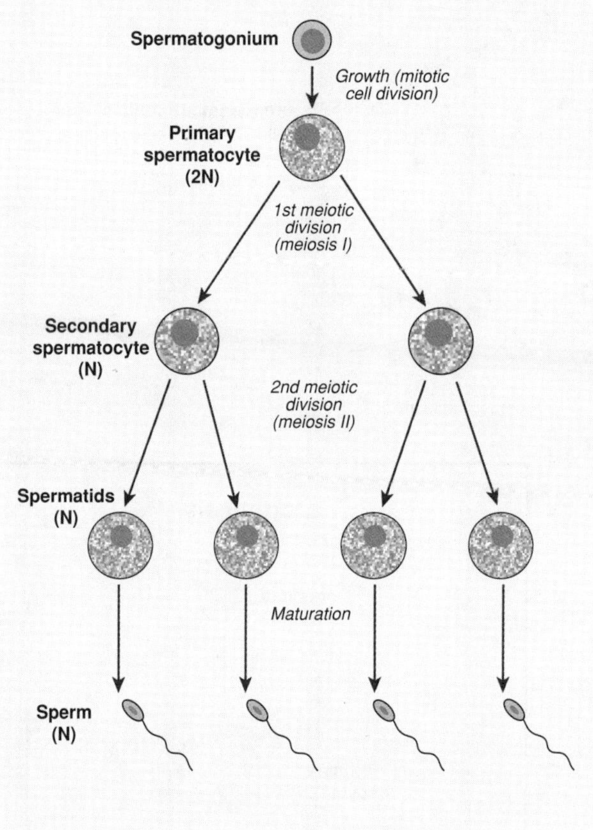
Spermatogonium
Growth (mitotic cell division)
Primary spermatocyte (2N)
1st meiotic division (meiosis I)
Secondary spermatocyte (N)
2nd meiotic division (meiosis II)
Spermatids (N)
Maturation
Sperm (N)

Cross Section Through Seminiferous Tubule

The photograph below shows maturing sperm (arrowed) with tails projecting into the lumen of the seminiferous tubule. Their heads are embedded in the Sertoli cells in the tubule wall and they are ready to break free and move to the epididymis where they complete their maturation. The same cross-section is illustrated diagrammatically (bottom).

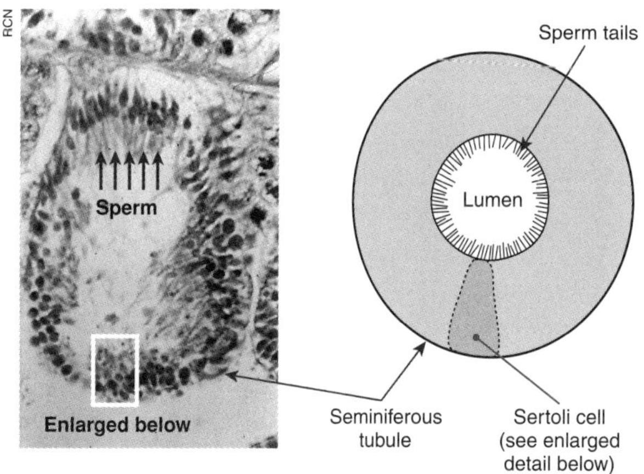
RCN
Sperm
Enlarged below
Sperm tails
Lumen
Seminiferous tubule
Sertoli cell (see enlarged detail below)

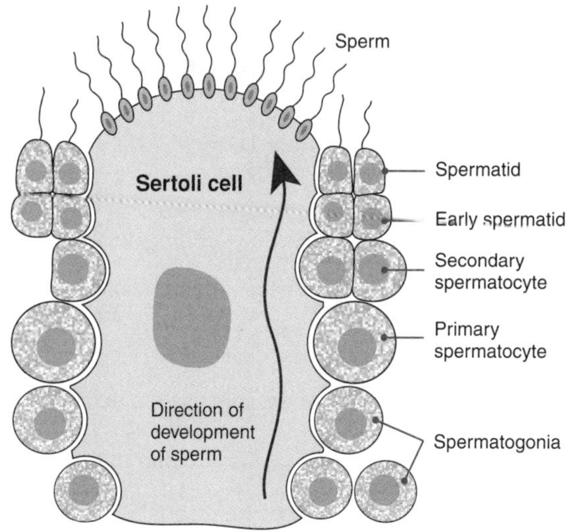
Sperm
Sertoli cell
Spermatid
Early spermatid
Secondary spermatocyte
Primary spermatocyte
Spermatogonia
Direction of development of sperm

1. (a) Name the process by which mature sperm are formed: ____________________

   (b) Identify where this process takes place: ____________________

   (c) State how many mature sperm form from one primary spermatocyte: ____________________

   (d) State the type of cell division which produces mature sperm cells: ____________________

2. Describe the role of FSH and LH in sperm cell production: ____________________

3. Each ejaculation of a healthy, fertile male contains 100-400 million sperm. Suggest why so many sperm are needed:

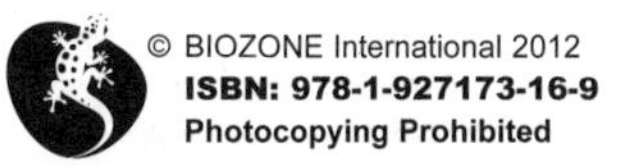

ISBN: 978-1-927173-16-9

Oogenesis

Egg cell (**ovum**, *plural* **ova**) production in females occurs by **oogenesis**. Unlike spermatogenesis, no new eggs are produced after birth. Instead a human female is born with her entire complement of immature eggs. These remain in prophase of meiosis I throughout childhood. After puberty most commonly a single egg cell is released from the ovaries at regular monthly intervals (the menstrual cycle), arrested in metaphase of meiosis II. This second division is only completed upon fertilization. The release of egg cells from the ovaries takes place from the onset of puberty until menopause, when menstruation ceases.

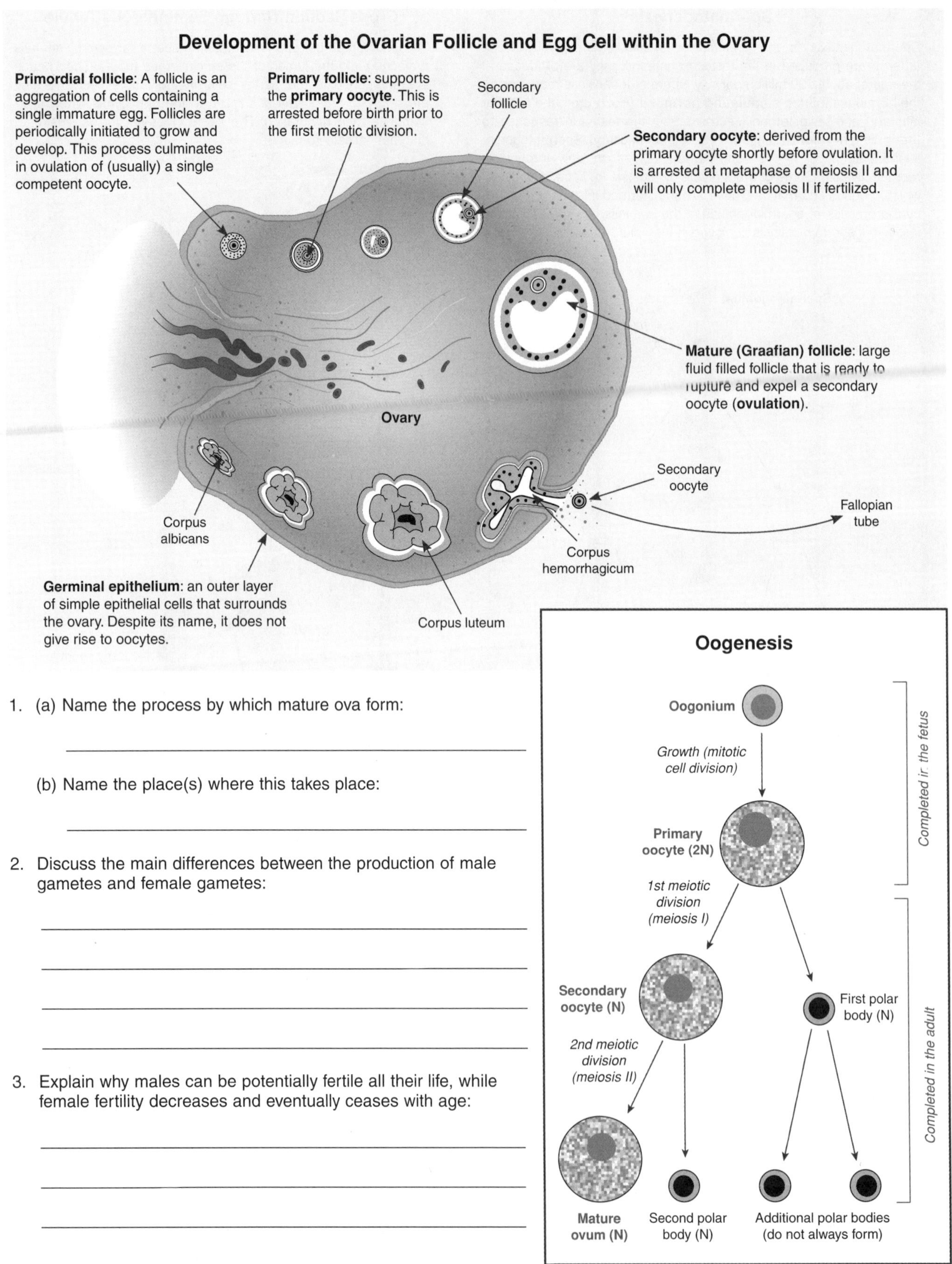
Development of the Ovarian Follicle and Egg Cell within the Ovary
Primordial follicle: A follicle is an aggregation of cells containing a single immature egg. Follicles are periodically initiated to grow and develop. This process culminates in ovulation of (usually) a single competent oocyte.
Primary follicle: supports the primary oocyte. This is arrested before birth prior to the first meiotic division.
Secondary follicle
Secondary oocyte: derived from the primary oocyte shortly before ovulation. It is arrested at metaphase of meiosis II and will only complete meiosis II if fertilized.
Mature (Graafian) follicle: large fluid filled follicle that is ready to rupture and expel a secondary oocyte (ovulation).
Ovary
Secondary oocyte
Fallopian tube
Corpus hemorrhagicum
Corpus albicans
Corpus luteum
Germinal epithelium: an outer layer of simple epithelial cells that surrounds the ovary. Despite its name, it does not give rise to oocytes.
Oogenesis
Oogonium
Growth (mitotic cell division)
Primary oocyte (2N)
1st meiotic division (meiosis I)
Secondary oocyte (N)
First polar body (N)
2nd meiotic division (meiosis II)
Mature ovum (N)
Second polar body (N)
Additional polar bodies (do not always form)
Completed in the fetus
Completed in the adult

1. (a) Name the process by which mature ova form:

   ____

   (b) Name the place(s) where this takes place:

   ____

2. Discuss the main differences between the production of male gametes and female gametes:

   ____

3. Explain why males can be potentially fertile all their life, while female fertility decreases and eventually ceases with age:

   ____

Related activities: *Spermatogenesis, Gametes*
Weblinks: *Oogenesis, Comparison of Spermatogenesis and Oogenesis*

ISBN: 978-1-927173-16-9

Gametes

Gametes are the sex cells of organisms. The gametes of male and female mammals differ greatly in their size, shape, and number. These differences reflect their very different roles in fertilization and reproduction. Male gametes are called **sperm** and female gametes are called eggs or **ovum** (plural ova). Mammalian sperm are highly motile and produced in large numbers. Eggs are large, few in number, and immobile in themselves. They move as a result of the wave-like motion produced by the ciliated cells lining the Fallopian tube. Egg cells contain some food sources to nourish the developing embryo. In mammals, this food source is small because once implantation into the uterus takes place, the fetus derives its nutrient supply from the mother's blood supply.

Egg Structure and Function

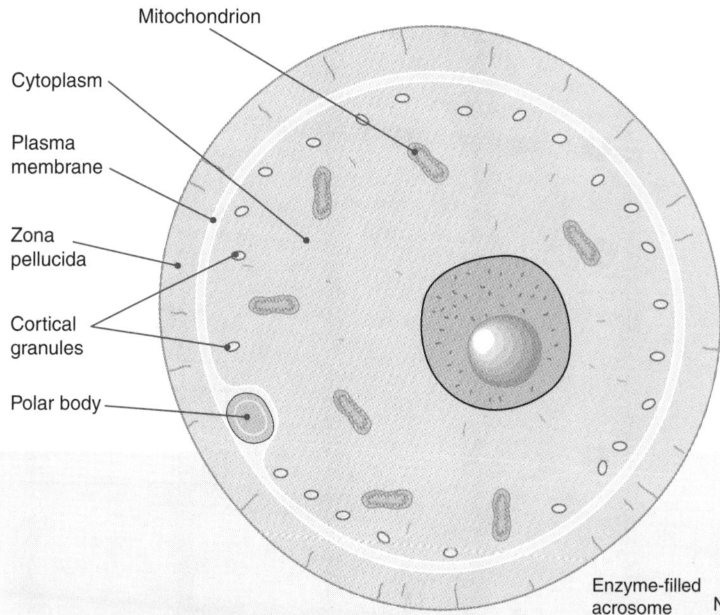
Mitochondrion
Cytoplasm
Plasma membrane
Zona pellucida
Cortical granules
Polar body

The ovum has no propulsion mechanism and is a simpler structure than the sperm cell. It is required to survive for a much longer time than a sperm, so it contains many more nutrients and metabolites and, as a result, it is much larger than a sperm cell (up to 100 µm).

The contents of the ovum are similar to that of a typical mammalian cell, although it is externally surrounded by a jelly-like glycoprotein called the zona pellucida. A small polar body (the remnants of a sister cell) lies between the plasma membrane and zona pellucida. Cortical granules around the inner edge of the plasma membrane contain enzymes that are released once a sperm has penetrated the egg, forming a block to prevent further sperm entry (the cortical reaction).

Sperm Structure and Function

Mature spermatozoa (sperm) are produced by **spermatogenesis** in the testes. Meiotic division of spermatocytes produces spermatids, which then differentiate into mature sperm.

The sperm's structure reflects its purpose, which is to swim along the fluid environment of the female reproductive tract to the ovum, penetrate the ovum's protective barrier, and donate its genetic material. A sperm cell comprises three regions: a headpiece, containing the nucleus and penetrative enzymes, an energy-producing mid-piece, and a tail for propulsion.

Human sperm live only about 48 hours, but they swim quickly and there are so many of them (millions per ejaculation) that usually some are able to reach the egg to fertilize it.

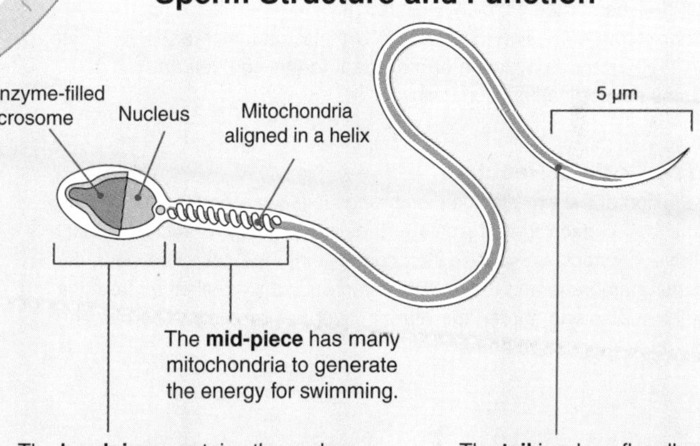
Enzyme-filled acrosome
Nucleus
Mitochondria aligned in a helix
5 µm

The **mid-piece** has many mitochondria to generate the energy for swimming.

The **headpiece** contains the nucleus and the acrosome, which contains the enzymes that help penetrate the egg.

The **tail** is a long flagellum that propels the sperm in its swim to the egg.

1. Explain why sperm need to be motile: ______________________

____

____

2. (a) Explain how the egg cell moves along the Fallopian tube: ______________________

____

(b) Explain why a mature ovum needs to be so many times larger than a sperm: ______________________

____

____

3. Explain why the sperm cell has a large number of mitochondria: ______________________

____

4. Describe the mechanism that ensures that only one sperm cell enters the egg: ______________________

____

____

ISBN: 978-1-927173-16-9

Related activities: *Spermatogenesis, Oogenesis*

Fertilization and Early Growth

When an egg cell is released from the ovary it is arrested in metaphase of meiosis II and is termed a secondary oocyte. **Fertilization** occurs when a sperm penetrates an egg cell at this stage and the sperm and egg nuclei unite to form the zygote. Fertilization is always regarded as time 0 in a period of gestation (pregnancy) and has five distinct stages (below). After fertilization, the zygote begins its **development** i.e. its growth and differentiation into a multicellular organism (see next page).

Fertilization (Time 0)

The stages in fertilization are represented below in a numbered sequence (1-5)

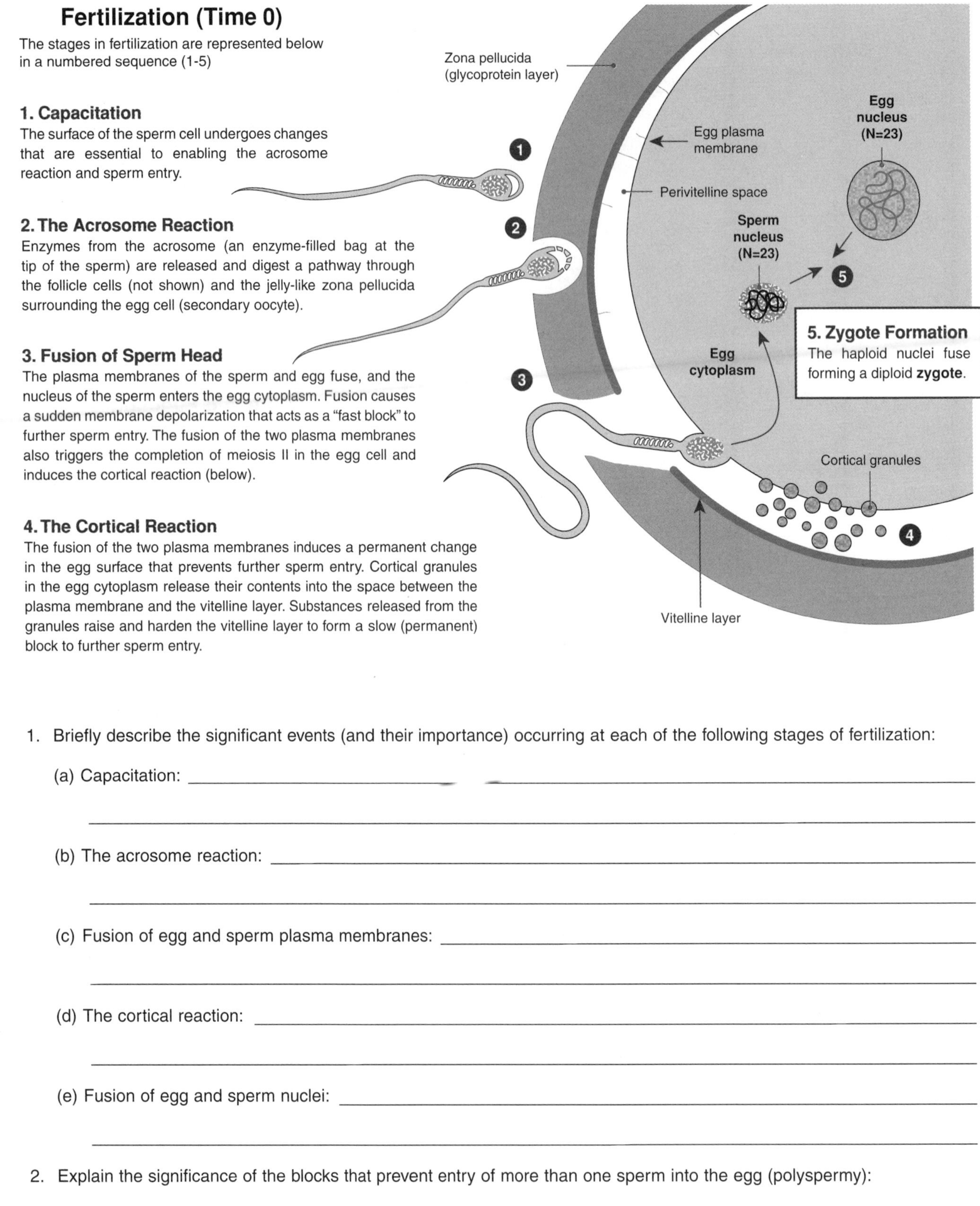
Zona pellucida (glycoprotein layer)
Egg plasma membrane
Egg nucleus (N=23)
Perivitelline space
Sperm nucleus (N=23)
Egg cytoplasm
Cortical granules
Vitelline layer
1
2
3
4
5

1. Capacitation

The surface of the sperm cell undergoes changes that are essential to enabling the acrosome reaction and sperm entry.

2. The Acrosome Reaction

Enzymes from the acrosome (an enzyme-filled bag at the tip of the sperm) are released and digest a pathway through the follicle cells (not shown) and the jelly-like zona pellucida surrounding the egg cell (secondary oocyte).

3. Fusion of Sperm Head

The plasma membranes of the sperm and egg fuse, and the nucleus of the sperm enters the egg cytoplasm. Fusion causes a sudden membrane depolarization that acts as a "fast block" to further sperm entry. The fusion of the two plasma membranes also triggers the completion of meiosis II in the egg cell and induces the cortical reaction (below).

4. The Cortical Reaction

The fusion of the two plasma membranes induces a permanent change in the egg surface that prevents further sperm entry. Cortical granules in the egg cytoplasm release their contents into the space between the plasma membrane and the vitelline layer. Substances released from the granules raise and harden the vitelline layer to form a slow (permanent) block to further sperm entry.

5. Zygote Formation

The haploid nuclei fuse forming a diploid **zygote**.

1. Briefly describe the significant events (and their importance) occurring at each of the following stages of fertilization:

   (a) Capacitation: ______________________

   (b) The acrosome reaction: ______________________

   (c) Fusion of egg and sperm plasma membranes: ______________________

   (d) The cortical reaction: ______________________

   (e) Fusion of egg and sperm nuclei: ______________________

2. Explain the significance of the blocks that prevent entry of more than one sperm into the egg (polyspermy):

____

Related activities: _Female Reproductive System, The Placenta_
Weblinks: _The Acrosome Reaction_

ISBN: 978-1-927173-16-9

Early Growth and Development

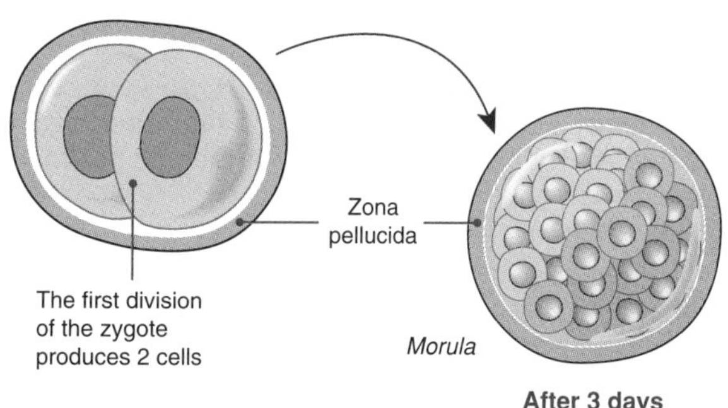
Zona
pellucida
The first division
of the zygote
produces 2 cells
Morula
After 3 days

Cleavage and Development of the Morula

Immediately after fertilization, rapid cell division takes place. These early cell divisions are called **cleavage** and they increase the number of cells, but not the size of the zygote. The first cleavage is completed after 36 hours, and each succeeding division takes less time. After three days, successive cleavages have produced a solid mass of cells called the **morula**, (left) which is still about the same size as the original zygote.

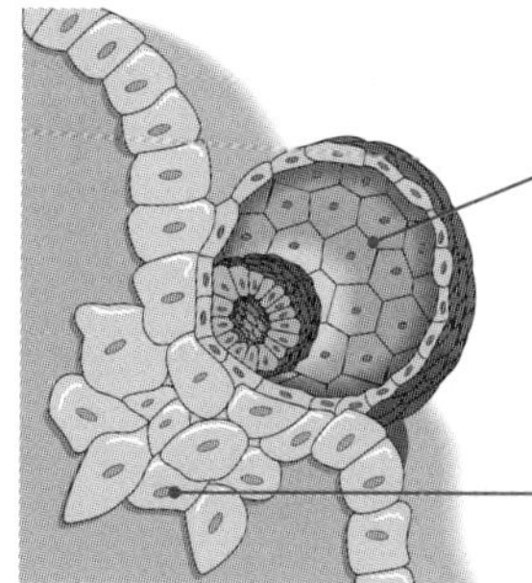
The **blastocyst**, a hollow ball of cells, embeds into the uterine wall using enzymes to digest and penetrate the lining.
The uterine lining provides nourishment for the embryo until the placenta develops.

Implantation of the Blastocyst (after 6-8 days)

After several days in the uterus, the morula develops into the blastocyst. It makes contact with the uterine lining and pushes deeply into it, ensuring a close maternal-fetal contact. Blood vessels provide early nourishment as they are opened up by enzymes secreted by the blastocyst. The embryo produces **HCG** (human chorionic gonadotropin), which prevents degeneration of the corpus luteum and signals that the woman is pregnant.

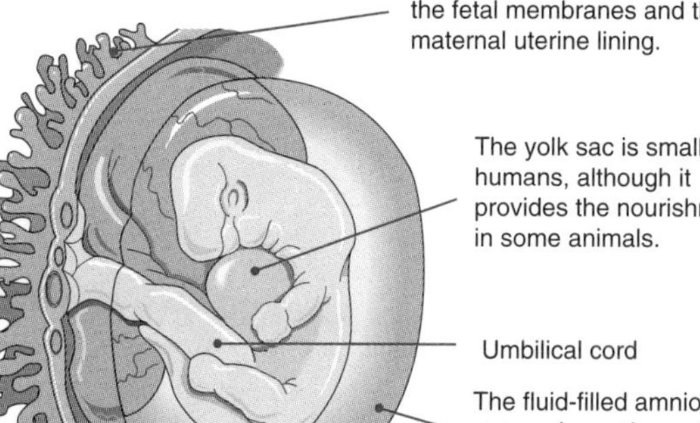
The placenta develops from the fetal membranes and the maternal uterine lining.
The yolk sac is small in humans, although it provides the nourishment in some animals.
Umbilical cord
The fluid-filled amniotic sac encloses the embryo in the amniotic fluid.

5 week old embryo

Embryo at 5-8 Weeks

Five weeks after fertilization, the embryo is only 4-5 mm long, but already the central nervous system has developed and the heart is beating. The embryonic membranes have formed; the amnion encloses the embryo in a fluid-filled space, and the allanto-chorion forms the fetal portion of the placenta. From two months the embryo is called a fetus. It is still small (30-40 mm long), but the limbs are well formed and the bones are beginning to harden. The face has a flat, rather featureless appearance with the eyes far apart. Fetal movements have begun and brain development proceeds rapidly. The placenta is well developed, although not fully functional until 12 weeks. The umbilical cord, containing the fetal umbilical arteries and vein, connects fetus and mother.

3. (a) Explain why the egg cell, when released from the ovary, is termed a secondary oocyte: ______________________

   ____

   (b) At which stage is its meiotic division completed? ______________________

4. What contribution do the sperm and egg cell make to each of the following:

   (a) The nucleus of the zygote? Sperm contribution: ____________ Egg contribution: ____________

   (b) The cytoplasm of the zygote? Sperm contribution: ____________ Egg contribution: ____________

5. What is meant by cleavage? Explain its significance to the early development of the embryo: ______________________

   ____

   ____

6. (a) What is the importance of implantation to the early nourishment of the embryo? ______________________

   ____

   (b) What is the purpose of HCG production by the embryo? ______________________

   ____

   ____

7. State why the fetus is particularly prone to damage from drugs towards the end of the first trimester (2-3 months):

   ____

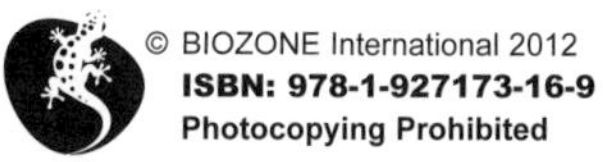

ISBN: 978-1-927173-16-9

The Placenta

As soon as an embryo embeds in the uterine wall it begins to obtain nutrients from its mother and increase in size. At two months, when the major structures of the adult are established, it is called a fetus. It is entirely dependent on its mother for nutrients, oxygen, and elimination of wastes. The placenta is the specialised organ that performs this role, enabling exchange between fetal and maternal tissues, and allowing a prolonged period of fetal growth and development within the protection of the uterus. The placenta also has an endocrine role, producing hormones that enable the pregnancy to be maintained.

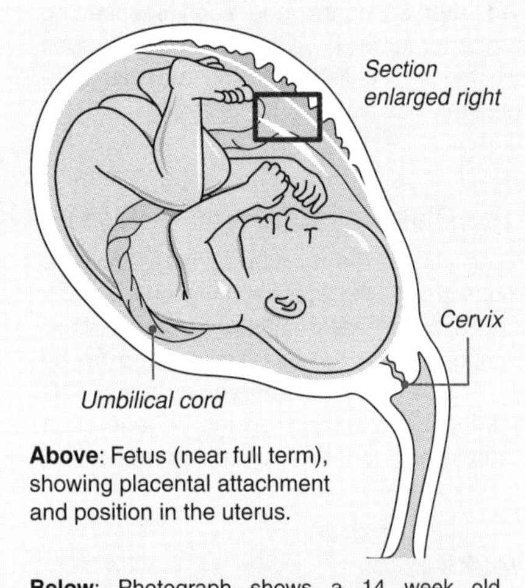
Section enlarged right
Cervix
Umbilical cord

Above: Fetus (near full term), showing placental attachment and position in the uterus.

Below: Photograph shows a 14 week old fetus. Limbs are fully formed, many bones are beginning to ossify, and joints begin to form. Facial features are becoming more fully formed.

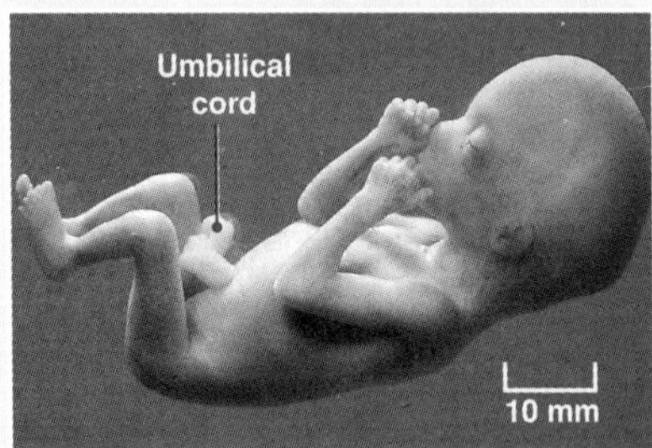
Umbilical cord
10 mm

Schematic diagram showing part of the placenta in section

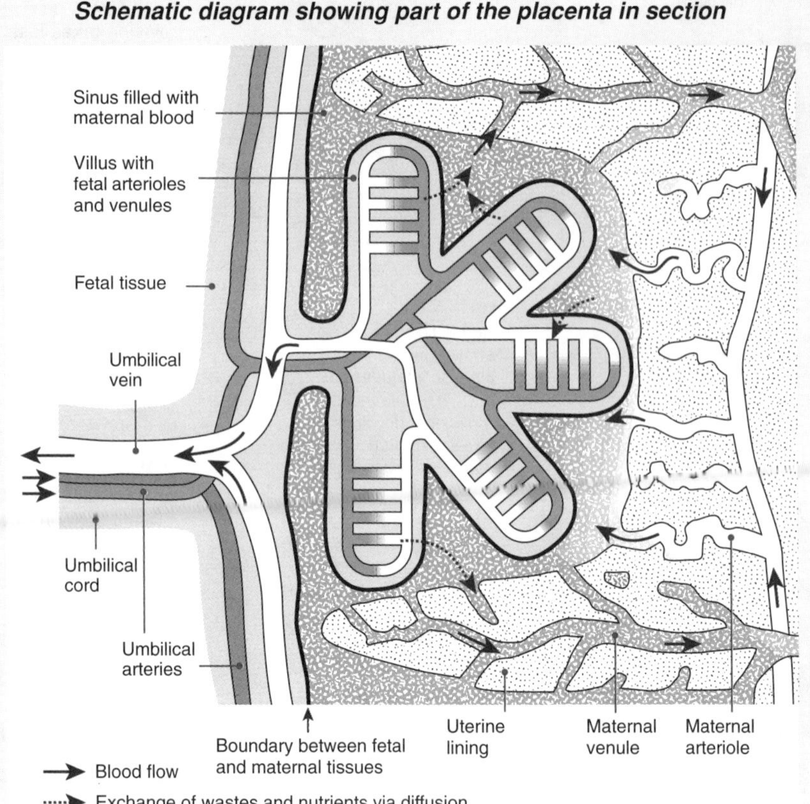
Sinus filled with maternal blood
Villus with fetal arterioles and venules
Fetal tissue
Umbilical vein
Umbilical cord
Umbilical arteries
Boundary between fetal and maternal tissues
Uterine lining
Maternal venule
Maternal arteriole
Blood flow
Exchange of wastes and nutrients via diffusion

The placenta is a disc-like organ, about the size of a dinner plate and weighing about 1 kg. It develops when fingerlike projections (villi) from the fetal membranes grow into the uterine lining. The villi contain the numerous capillaries connecting the fetal arteries and vein. They continue invading the maternal tissue until they are bathed in the maternal blood sinuses. The maternal and fetal blood vessels are in such close proximity that oxygen and nutrients can diffuse from the maternal blood into the capillaries of the villi. From the villi, the nutrients circulate in the umbilical vein, returning to the fetal heart. Carbon dioxide and other wastes leave the fetus through the umbilical arteries, pass into the capillaries of the villi, and diffuse into the maternal blood. Note that fetal blood and maternal blood do not mix: the exchanges occur via diffusion through thin walled capillaries.

1. Describe the structure of the human placenta and explain its function: ______________________

____

____

____

2. The umbilical cord contains the fetal arteries and vein. Describe the status of the blood in each type of fetal vessel:

   (a) Fetal arteries: Oxygenated and containing nutrients / Deoxygenated and containing nitrogenous wastes (delete one)

   (b) Fetal vein: Oxygenated and containing nutrients / Deoxygenated and containing nitrogenous wastes (delete one)

3. **Teratogens** are substances that may cause malformations in embryonic development (e.g. nicotine, alcohol):

   (a) Why do substances ingested by the mother have the potential to be harmful to the fetus?

   ____

   (b) Why is cigarette smoking so harmful to fetal development? ______________________

   ____

ISBN: 978-1-927173-16-9

The Hormones of Pregnancy

Human reproductive physiology occurs in a cycle (the **menstrual cycle**) which follows a set pattern and is regulated by the interplay of several hormones. Control of hormone release is brought about through feedback mechanisms: the levels of the female reproductive hormones, estrogen and progesterone, regulate the secretion of the pituitary hormones that control the ovarian cycle (see earlier pages). Pregnancy interrupts this cycle and maintains the corpus luteum and the placenta as endocrine organs with the specific role of maintaining the developing fetus for the period of its development. During the last month of pregnancy the peptide hormone oxytocin induces the uterine contraction that will expel the baby from the uterus.

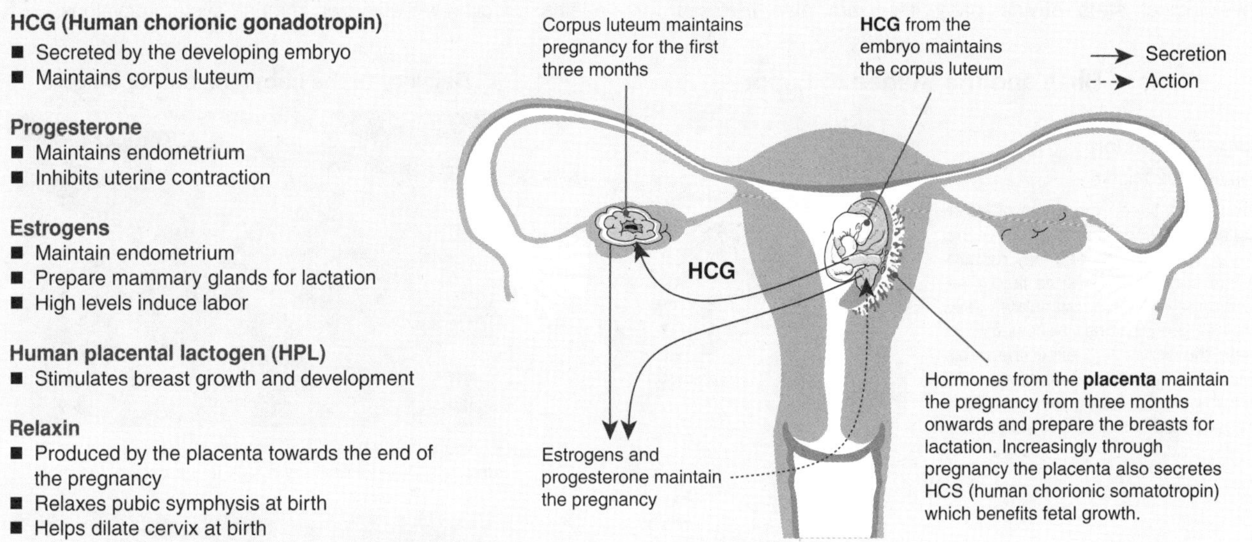
HCG (Human chorionic gonadotropin)
■ Secreted by the developing embryo
■ Maintains corpus luteum
Progesterone
■ Maintains endometrium
■ Inhibits uterine contraction
Estrogens
■ Maintain endometrium
■ Prepare mammary glands for lactation
■ High levels induce labor
Human placental lactogen (HPL)
■ Stimulates breast growth and development
Relaxin
■ Produced by the placenta towards the end of the pregnancy
■ Relaxes pubic symphysis at birth
■ Helps dilate cervix at birth
Corpus luteum maintains pregnancy for the first three months
HCG from the embryo maintains the corpus luteum
Secretion
Action
HCG
Estrogens and progesterone maintain the pregnancy
Hormones from the placenta maintain the pregnancy from three months onwards and prepare the breasts for lactation. Increasingly through pregnancy the placenta also secretes HCS (human chorionic somatotropin) which benefits fetal growth.

Hormonal Changes During Pregnancy, Birth, and Lactation

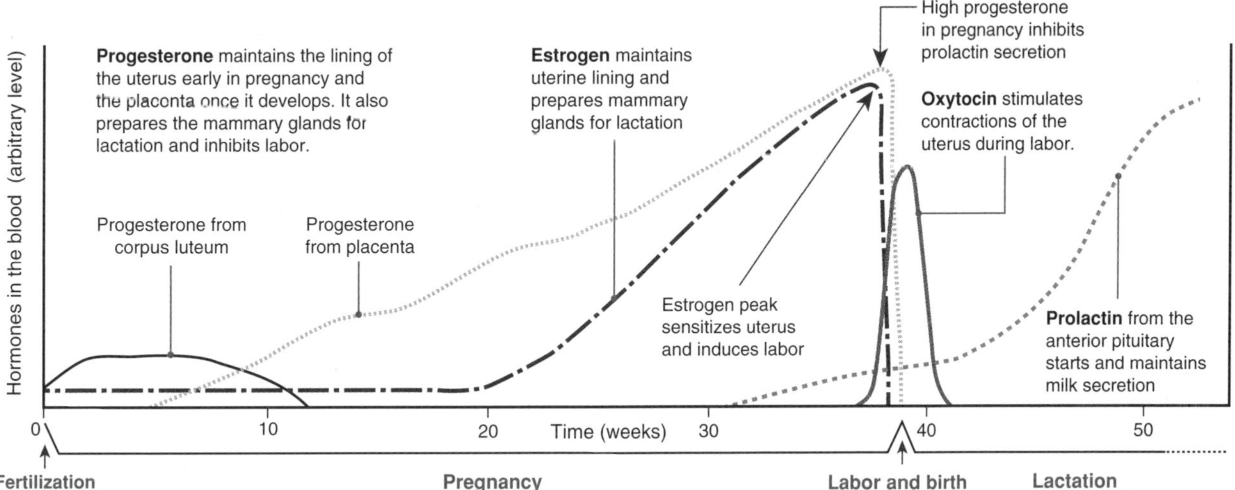
Hormones in the blood (arbitrary level)
Progesterone maintains the lining of the uterus early in pregnancy and the placenta once it develops. It also prepares the mammary glands for lactation and inhibits labor.
Estrogen maintains uterine lining and prepares mammary glands for lactation
High progesterone in pregnancy inhibits prolactin secretion
Oxytocin stimulates contractions of the uterus during labor.
Progesterone from corpus luteum
Progesterone from placenta
Estrogen peak sensitizes uterus and induces labor
Prolactin from the anterior pituitary starts and maintains milk secretion
0
10
20
Time (weeks)
30
40
50
Fertilization
Pregnancy
Labor and birth
Lactation

During the first 12-16 weeks pregnancy, the **corpus luteum** secretes enough progesterone to maintain the uterine lining and sustain the developing embryo. After this, the placenta takes over as the primary endocrine organ of pregnancy. **Progesterone** and **estrogen** from the placenta maintain the uterine lining, inhibit the development of further ova (eggs), and prepare the breast tissue for **lactation** (milk production). At the end of pregnancy, the placenta loses competency, progesterone levels fall, and high estrogen levels trigger the onset of labor. The estrogen peak coincides with an increase in oxytocin, which stimulates uterine contractions in a positive feedback loop: the contractions and the increasing pressure of the cervix from the infant stimulate release of more oxytocin, and more contractions and so on, until the infant exits the birth canal. After birth, the secretion of prolactin increases. Prolactin maintains lactation during the period of infant nursing.

1. (a) Why is the corpus luteum the main source of progesterone in early pregnancy? ____________________

____

(b) What hormones are responsible for maintaining pregnancy? ____________________

2. (a) Name two hormones involved in labor (onset of the birth process): ____________________

(b) Describe two physiological factors in initiating labor: ____________________

____

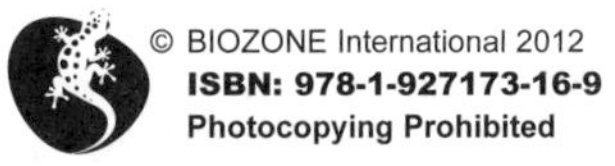

ISBN: 978-1-927173-16-9

Related activities: The Menstrual Cycle
Weblinks: Chemistry of Pregnancy Tests, HCG Pregnancy Test

Birth

A human pregnancy (the period of **gestation**) lasts, on average, about 38 weeks after fertilization. It ends in labor, the birth of the baby, and expulsion of the placenta. During pregnancy, progesterone maintains the placenta and inhibits contraction of the uterus. At the end of a pregnancy, increasing estrogen levels overcome the influence of progesterone and labor begins. Prostaglandins, factors released from the placenta, and the physiological state of the baby itself are also involved in triggering the actual timing of labor onset. Labor itself comprises three stages (below), and ends with the delivery of the placenta. After birth, the mother provides nutrition for the infant through **lactation**: the production and release of milk from mammary glands. Breast milk provides infants with a complete, easily digested food for the first 4-6 months of life. All breast milk contains maternal antibodies, which give the infant protection against infection while its own immune system develops.

Birth and the Stages of Labor

Stage 1: Dilation

Duration: 2-20 hours

The time between the onset of labor and complete opening (dilation) of the cervix. The amniotic sac may rupture at this stage, releasing its fluid. The hormone **oxytocin** stimulates the uterine contractions necessary to dilate the cervix and expel the baby. It is these uterine contractions that give the pain of labor, most of which is associated with this first stage.

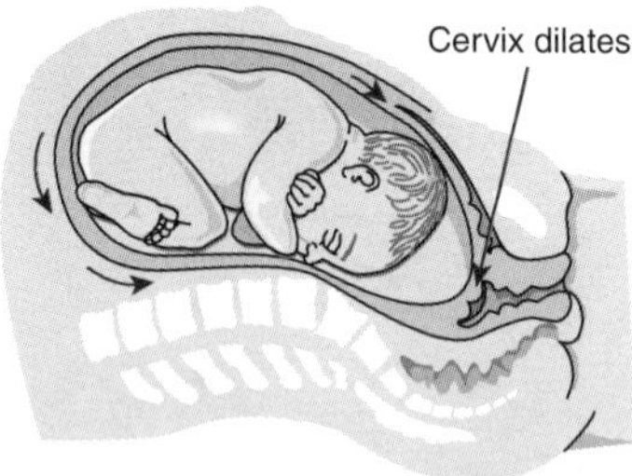
Cervix dilates

Stage 2: Expulsion

Duration: 2-100 minutes

The time from full dilation of the cervix to delivery. Strong, rhythmic contractions of the uterus pass in waves (arrows), and push the baby to the end of the vagina, where the head appears.

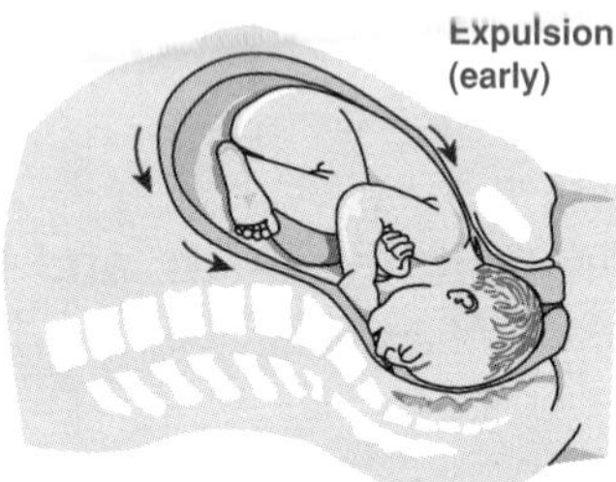
Expulsion (early)

As labor progresses, the time between each contraction shortens. Once the head is delivered, the rest of the body usually follows very rapidly. Delivery completes stage 2.

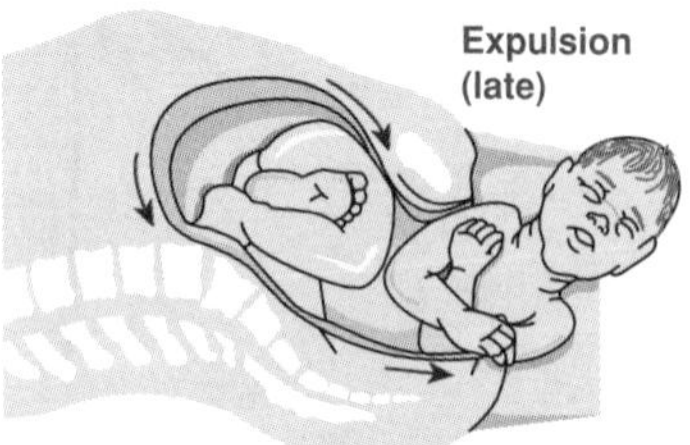
Expulsion (late)

Stage 3: Delivery of placenta

Time: 5-45 minutes after delivery

The third or **placental stage**, refers to the expulsion of the placenta from the uterus. After the placenta is delivered, the placental blood vessels constrict to stop bleeding.

Umbilical cord

Placenta

Delivery of the baby: the end of stage 2

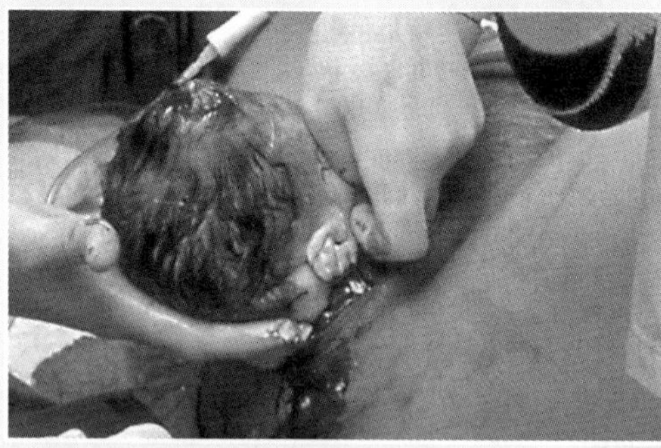

Delivery of the head. This baby is face forward. The more usual position for delivery is face to the back of the mother.

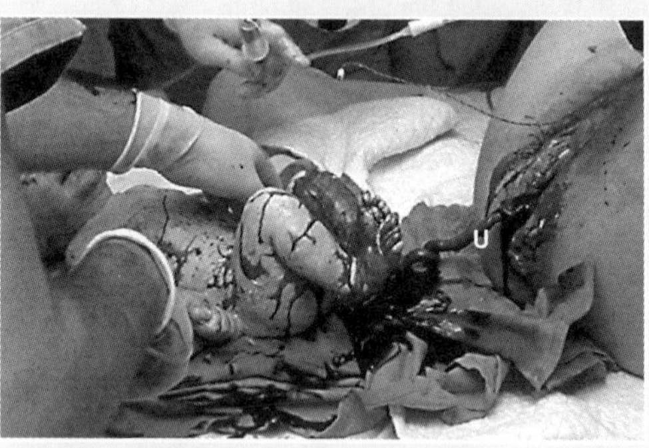
U

Full delivery of the baby. Note the umbilical cord (U), which supplies oxygen until the baby's breathing begins.

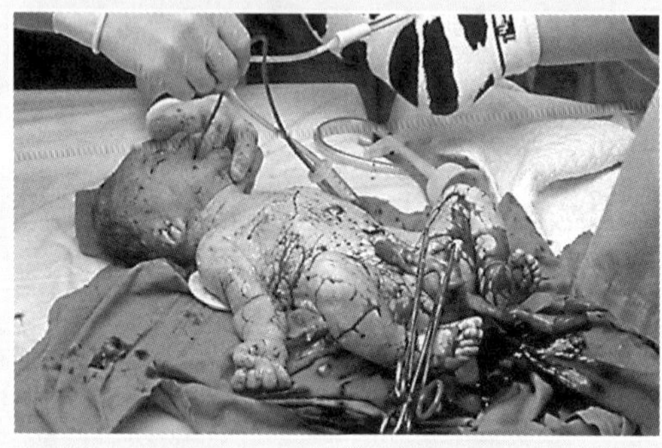

Post-birth check of the baby. The baby is still attached to the placenta and the airways are being cleared of mucus.

1. Name the three stages of birth, and briefly state the main events occurring in each stage:

(a) Stage 1: ____________________

(b) Stage 2: ____________________

(c) Stage 3: ____________________

2. (a) Which hormone is responsible for triggering the onset of labor? ____________________

(b) What two other factors might influence the timing of labor onset? ____________________

____

____

***Related activities**: The Hormones of Pregnancy*

ISBN: 978-1-927173-16-9

In Vitro Fertilization

***In vitro* fertilization** (IVF) may be used to overcome infertility which may result from a disturbance of any of the factors involved in fertilization or embryonic development. Female infertility may be due to a failure to ovulate, requiring stimulation of the ovary, with or without hormone therapy. For couples with one or both partners incapable of providing suitable gametes, it may be possible for them to receive eggs and/or sperm from donors.

Fertility drugs may be used to induce the production of many eggs for use in IVF, although the natural cycle of ovulation can be used to collect the egg. Fertility drugs stimulate the pituitary gland and may induce the simultaneous release of numerous eggs; an event called superovulation. If each egg is allowed to be fertilized, the resulting embryos may then be frozen after 24-72 hours culture.

Causes of Infertility

Infertility is a common problem (as many as one in six couples require help from a specialist). The cause of the infertility may be inherited, due to damage caused by an infectious disease, or psychological.

Causes of male infertility:

- *Penis*: Fails to achieve or maintain erection; abnormal ejaculation.
- *Testes*: Too few sperm produced or sperm are abnormally shaped, have impaired motility, or too short lived.
- *Vas deferens*: A blockage or a structural abnormality may impede the passage of sperm.

Causes of female infertility:

- *Fallopian tubes*: Blockage may prevent sperm from reaching egg; one or both tubes may be damaged (disease) or absent (congenital).
- *Ovaries*: Eggs may fail to mature or may not be released.
- *Uterus*: Abnormality or disorder may prevent implantation of the egg.
- *Cervix*: Antibodies in cervical mucus may damage or destroy the sperm.

In Vitro Fertilization

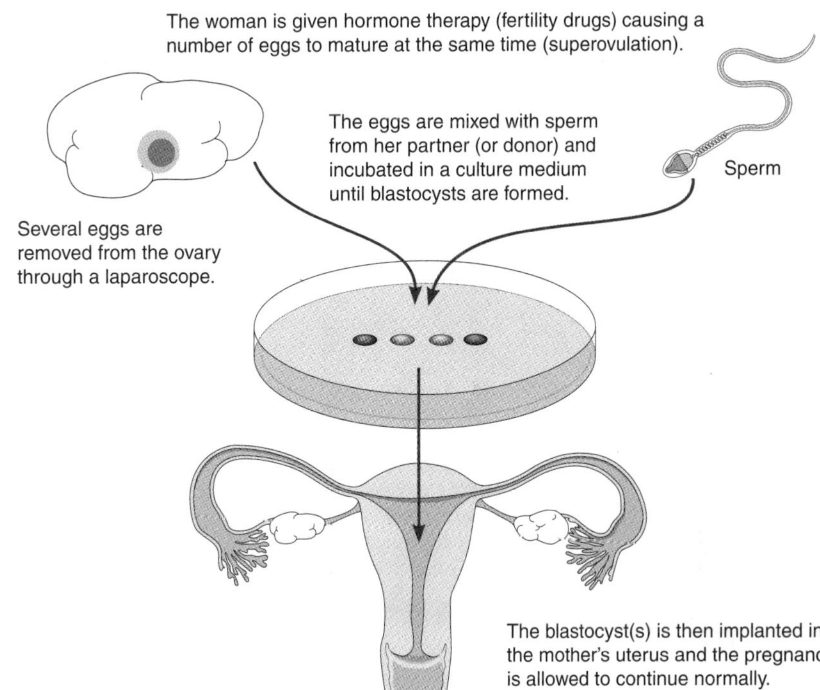
The woman is given hormone therapy (fertility drugs) causing a number of eggs to mature at the same time (superovulation).
Several eggs are removed from the ovary through a laparoscope.
The eggs are mixed with sperm from her partner (or donor) and incubated in a culture medium until blastocysts are formed.
Sperm
The blastocyst(s) is then implanted in the mother's uterus and the pregnancy is allowed to continue normally.

Biological Origins of Gamete Donations

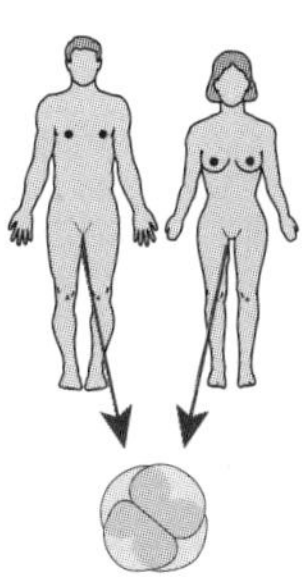

Both partners provide gametes for IVF or GIFT (they donate their own gametes).

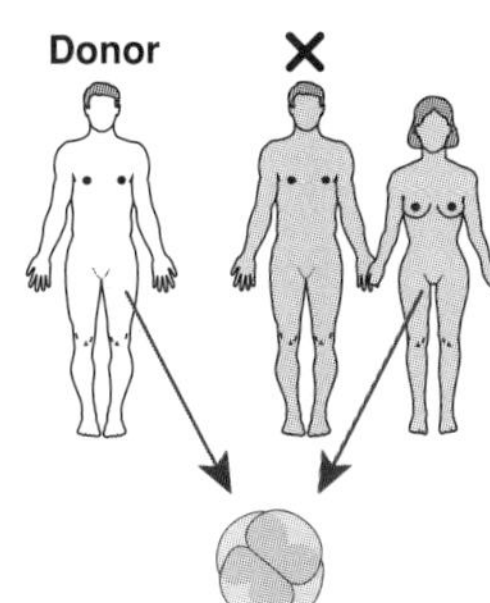
Donor
X

Male partner unable to provide sperm; sperm from male donor.

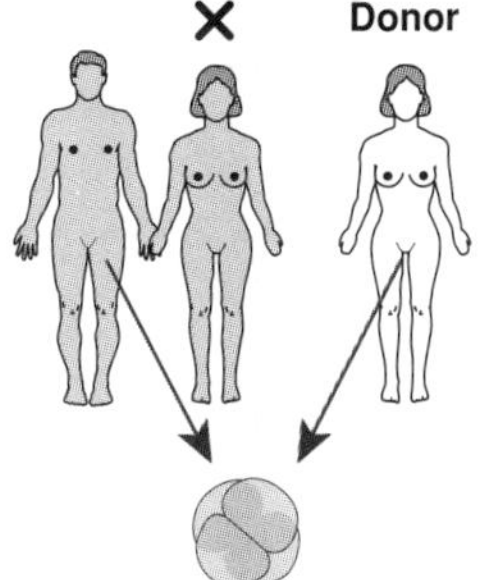
X
Donor

Female partner unable to provide eggs; egg from female donor.

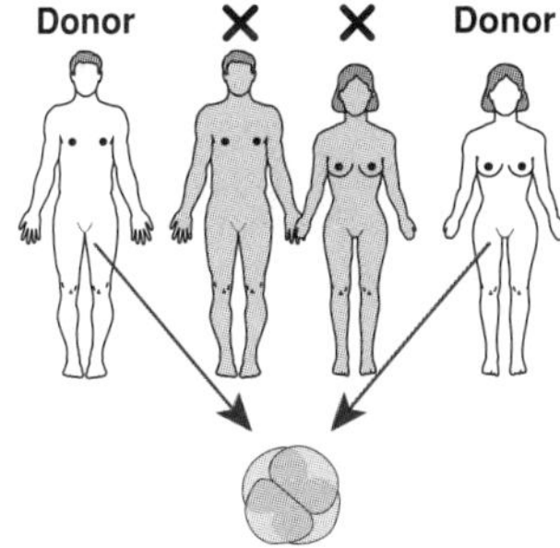
Donor
X
X
Donor

Both partners unable to provide gametes; sperm and egg obtained from donors.

1. Describe three causes of female infertility: ______

2. Describe three causes of male infertility: ______

3. Describe the key stages of **IVF**: ______

Reproduction

ISBN: 978-1-927173-16-9

IVF raises a number of ethical issues including the concept of personhood, religious beliefs, and the rights and responsibilities of the individual, parents, and community as a whole. It also raises issues over the health and psychological effects off the offspring.

Ethical Issue 1:
The rights of the pre-embryo (blastocyst)

Multiple blastocysts are transferred to a woman's uterus to increase the chances of implantation. After implantation, many of these blastocysts are destroyed by selective pregnancy reduction.

When does personhood or the individual begin?

- If it begins at conception, destruction of the extra embryos technically constitutes murder.
- Many different ideas and definitions exist over the start of personhood or individuality.
  - Can the pre-embryo technically be called an individual during the period of totipotency (about 3 weeks)? During this period, any one of its cells could develop into an individual.

Ethical Issue 2:
Possible wrongs to the couple by the use of IVF

Multiple blastocysts are transferred to a woman's uterus to increase the chances of implantation.

- A multiple pregnancy can have psychological and health effects on the parents.
- A multiple pregnancy can have health effects on the embryos.
- The parents may have to bear the cost of IVF, putting financial strain on them (and possible resentment towards others).
  - Cost may be indirect, because the offspring may have health problems (either caused by IVF or naturally occurring).

Ethical Issue 3:
Possible wrongs to the offspring by the use of IVF

There is some medical evidence to suggest IVF babies have a higher chance of medical problems such as pre-term birth, low birth weight, spina bifida, and heart defects.

- Parents with genetic defects preventing them from conceiving naturally could pass these defects to the offspring via IVF.

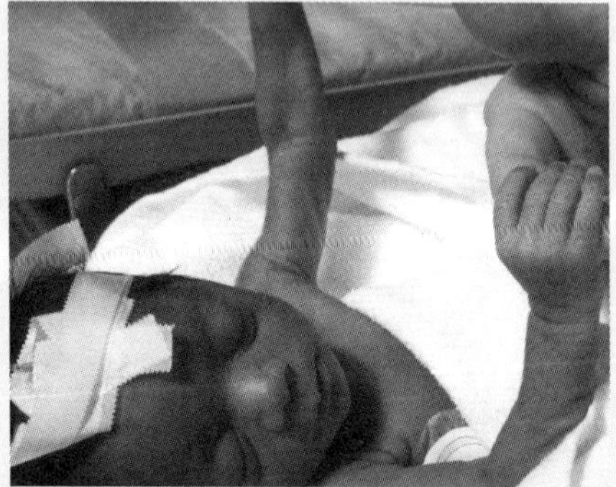

Ethical Issue 4:
Possible wrongs to the community by the use of IVF

IVF is a costly procedure.

- Couples who can afford IVF may be putting money and effort into conception instead of the community.
- The community may have to bare the cost of IVF and welfare for financially struggling parents.
- Offspring with health issues due to IVF may be an ongoing burden to the community.

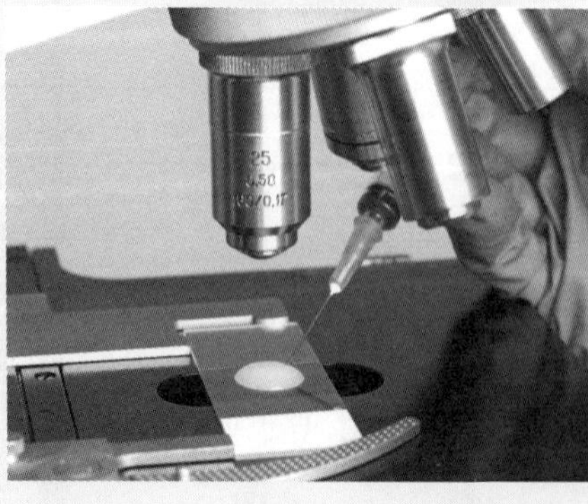

4. Evaluate the ethical and medical issues surrounding IVF treatment. In your opinion what is the best ethical approach to using IVF? Use examples where possible to illustrate your ideas:

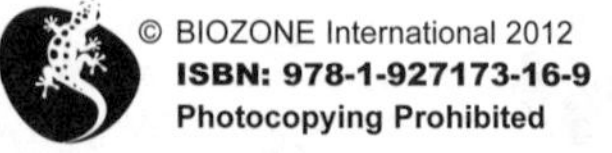

ISBN: 978-1-927173-16-9

KEY TERMS: Mix and Match

INSTRUCTIONS: Test your vocabulary by matching each term to its definition, as identified by its preceding letter code.

CORE

endometrium	**A**	The principal male sex hormone.
estrogen	**B**	The vascularized inner lining of the mammalian uterus.
FSH	**C**	A steroid hormone that maintains the endometrium and placenta (once developed), and inhibits labor.
in vitro fertilization	**D**	The hormone responsible for the development of the ovarian follicles.
LH	**E**	A steroid hormone, which functions as the primary female sex hormone.
menstrual cycle	**F**	The process in a female's menstrual cycle by which a mature ovarian follicle ruptures and releases an ovum.
ovulation	**G**	The fertilization of the ovum (egg cell) by sperm cells outside the body to produce embryos for implantation into the uterus.
progesterone	**H**	A surge in this hormone is responsible for ovulation.
testosterone	**I**	The cycle of changes in reproductive physiology occurring in fertile female humans.

AHL only

acrosome reaction	**A**	A hormone produced made by the developing embryo. It maintains the corpus luteum in early pregnancy.
corpus luteum	**B**	The culmination of a human pregnancy when the infant is expelled from the uterus.
embryo	**C**	The creation of an ovum.
fertilization	**D**	The period of physical changes during which a child's body becomes a reproductively capable adult body.
fetus	**E**	A process in sperm cells that enables the sperm to break through the egg's coating so that fertilization can occur.
gametogenesis	**F**	The earliest stage of development, until about 8 weeks post-fertilization in humans.
HCG	**G**	The initial cell formed from the union of two gametes when a new organism is produced by means of sexual reproduction.
labor	**H**	An egg cell or female gamete.
mature (Graafian) follicle	**I**	Reproductive cell derived from the primary oocyte. It is arrested at meiosis I and will only undergo meiosis II if fertilized.
meiosis	**J**	A reduction division involved in the production of gametes.
oogenesis	**K**	An organ, characteristic of most mammals, that enables exchanges (via the blood supply) of nutrient, gases, and wastes, between mother and fetus.
ovum	**L**	A temporary endocrine structure in the first trimester of pregnancy in mammals, involved in the hormonal maintenance of the endometrium.
oxytocin	**M**	Large fluid filled follicle containing a secondary oocyte ready for ovulation.
placenta	**N**	The fusion of sperm and egg. Also called conception.
puberty	**O**	A hormone with a role in uterine contraction in labor and milk letdown during lactation.
secondary oocyte	**P**	The process by which gametes are produced.
sperm	**Q**	The term for a developing mammal (or other live-bearing vertebrate) after the embryonic stage but before birth.
spermatogenesis	**R**	The creation of mature spermatozoa (sperm cells) in males.
zygote	**S**	The male gamete.

ISBN: 978-1-927173-16-9

AHL Topic
11.2

Muscles and Movement

Key concepts

- Muscles generate movement by contracting against a rigid skeleton.
- Most movement occurs around joints. Synovial joints offer the greatest range of movement.
- The muscular system is organized into discrete muscles, which work as antagonistic pairs.
- In muscle, the movement of actin filaments against myosin filaments creates contraction. ATP and calcium ions are required.

Key terms

AHL only

actin filament
agonist
antagonist
antagonistic muscles
bone
cartilage
contraction
extension
flexion
H zone
insertion (of muscle) joint
joint capsule
ligament
muscle
muscle fiber
myofibril
myosin filament
neuromuscular junction
origin (of muscle)
prime mover
sarcolemma
sarcomere
sarcoplasmic reticulum
sliding filament theory
striated (skeletal) muscle
synergist
synovial fluid
synovial joint
tendon
Z line

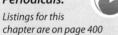

Periodicals:
Listings for this chapter are on page 400

Weblinks:
www.thebiozone.com/weblink/IB-3169.html

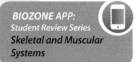

BIOZONE APP: Student Review Series
Skeletal and Muscular Systems

Learning Objectives

☐ 1. Use the **KEY TERMS** to compile a glossary for this topic.

Bones and Joints *(11.2.1-11.2.4)* pages 357-359

☐ 2. **AHL**: State the roles of **bones**, **ligaments**, **muscles**, **tendons**, and **nerves** in generating movement in humans

☐ 3. **AHL**: Label a diagram of a human elbow **joint** to show the **cartilage**, **synovial fluid**, **joint capsule**, named bones, and **antagonistic muscles** (biceps and triceps).

☐ 4. **AHL**: Outline the function of each of the structures identified above. Understand the terms **origin** and **insertion** of muscles as these are important when describing the movement of bones. Take careful note of the insertion of the biceps muscle and think carefully about its role in movement of the forearm.

☐ 5. **AHL**: EXTENSION: Recall that most muscles work in antagonistic pairs and explain why this is the case. Describe the events during **flexion** and **extension** of the elbow. Understand the terms **prime mover**, **synergist**, **agonist**, and **antagonist**.

☐ 6. **AHL**: Compare the movements of the hip joint and knee joint.

Muscles and Movement *(11.2.5-11.2.8)* pages 359-363

☐ 7. **AHL**: Describe the ultrastructure of a **striated** (skeletal) **muscle fiber**, identifying the **myofibrils** with light and dark bands, mitochondria, **sarcoplasmic reticulum**, nuclei, and **sarcolemma**.

☐ 8. **AHL**: Draw and label a diagram show the structure of a sarcomere, including the **Z lines** and **actin** and **myosin filaments**. Describe the composition and arrangement of the **filaments**, explaining how these give rise to the light and dark bands.

☐ 9. **AHL**: Explain the sequence of events in **contraction** of skeletal muscle, from the arrival of an action potential at the motor end plate to shortening of the muscle fiber. Include reference to the role of **actin** and **myosin filaments**, cross-bridge formation, ATP, the **sarcoplasmic reticulum**, and **calcium ions**. Recognize this sequence of events as the **sliding filament theory** of muscle contraction.

☐ 10. **AHL**: Analyze electron micrographs to find the state of contraction of muscle fibers.

The Basis of Human Movement

Bones are too rigid to bend without damage. To allow movement, the skeletal system consists of many bones held together at **joints** by flexible connective tissues called **ligaments**. All movements of the skeleton occur at joints: points of contact between bones, or between cartilage and bones. Joints may be classified structurally as fibrous, cartilaginous, or synovial based on whether fibrous tissue, cartilage, or a joint cavity separates the bones of the joint. Each of these joint types allows a certain degree of movement. **Fibrous joints**, such as the sutures of the skull, generally allow little or no movement. **Cartilaginous joints** (e.g the pubic **symphysis**) generally allow slight movement, while **synovial joints** enable free movement in one or more planes (below). Bones are made to move about a joint by the force of muscles acting upon them.

Cartilaginous Joints

Here, the bone ends are connected by cartilage. Most allow limited movement although some (e.g. between the first ribs and the sternum) are immovable.

Immovable Fibrous Joints

The bones are connected by fibrous tissue. In some (e.g. sutures of the skull), the bones are tightly bound by connective tissue fibers and there is no movement.

Synovial Joints

These allow free movement in one or more planes. The articulating bone ends are separated by a joint cavity containing lubricating synovial fluid (see overleaf).

Hyaline cartilage forms the immovable joint between rib 1 and the sternum

Intervertebral discs of fibrocartilage between vertebrae

Fibrocartilage connecting the pubic bones anteriorly

Ball and socket

Humerus

Hinge joint

Humerus

Radius

Ulna

Saddle joint

Thumb

Condyloid joint

Knuckle

Plane joint

Intertarsal

Fibrous Joints with slight give

In some fibrous joints, the connective tissue fibers joining the bones are long enough to allow very slight give in the joint.

Tibia

Fibula

Fibrous connective tissue strands join the distal ends of the tibia and fibula.

Movement Around Synovial Joints

A B C D E

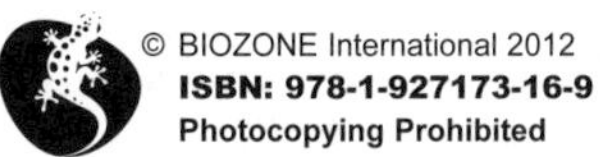

ISBN: 978-1-927173-16-9

***Related activities**: The Mechanics of Movement*
***Web links**: Skeletons and Joints, Movement at Joints*

Structure of a Synovial Joint

Synovial joints (right and below) allow free movement of body parts in varying directions (one, two or three planes). The elbow joint is a hinge joint and typical of a synovial joint. Like most synovial joints, it is reinforced by ligaments (not all shown). In the diagram, the brachialis muscle, which inserts into the ulna and is the prime mover for flexion of the elbow, has been omitted to show the joint structure. Muscles are labeled blue and bones bold.

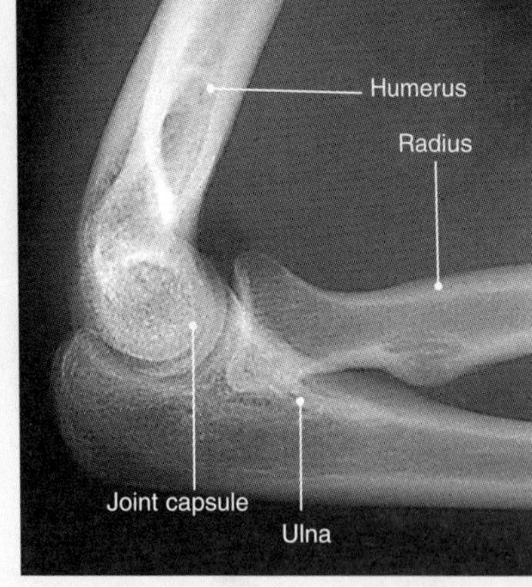
Humerus
Radius
Joint capsule
Ulna

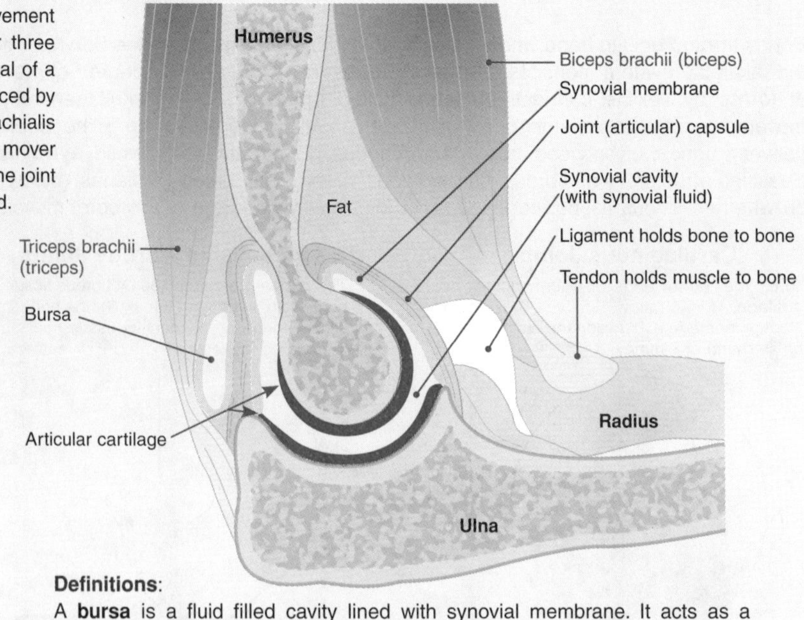
Humerus
Biceps brachii (biceps)
Synovial membrane
Joint (articular) capsule
Synovial cavity
(with synovial fluid)
Fat
Triceps brachii
(triceps)
Ligament holds bone to bone
Tendon holds muscle to bone
Bursa
Radius
Articular cartilage
Ulna

Definitions:
A **bursa** is a fluid filled cavity lined with synovial membrane. It acts as a cushion, e.g. between tendon and bone, or between bones.
Cartilage is a flexible connective tissue. It protects a joint surface against wear.

1. Define the following terms and state the role of each in movement:

   (a) Joint: ______

   (b) Ligament: ______

   (c) Muscle: ______

   (d) Tendon: ______

   (e) Nerve: ______

2. Classify each of the synovial joint models (**A-E**) at the bottom of the previous page, according to the descriptors below:

   (a) Pivot: ______ (b) Hinge: ______ (c) Ball-and-socket: ______ (d) Saddle: ______ (e) Gliding: ______

3. Compare the movements of the hip joint and the knee joint: ______

4. (a) Describe the features common to most synovial joints: ______

   (b) Explain the role that synovial fluid and cartilage play in the structure and function of a synovial joint:

   ______

5. Describe the major difference between a synovial joint and a cartilaginous joint: ______

ISBN: 978-1-927173-16-9

The Mechanics of Movement

We are familiar with the many different bodily movements achievable through the action of muscles. Contractions in which the length of the muscle shortens in the usual way are called **isotonic contractions**: the muscle shortens and movement occurs. When a muscle contracts against something immovable and does not shorten the contraction is called **isometric**. Skeletal muscles are attached to bones by tough connective tissue structures called **tendons**. They always have at least two attachments: the **origin** and the **insertion**. They create movement of body parts when they contract across **joints**. The type and degree of movement achieved depends on how much movement the joint allows and where the muscle is located in relation to the joint. Some common types of body movements are described below (left panel). Because muscles can only pull and not push, most body movements are achieved through the action of opposing sets of muscles (below, right panel).

The Action of Antagonistic Muscles

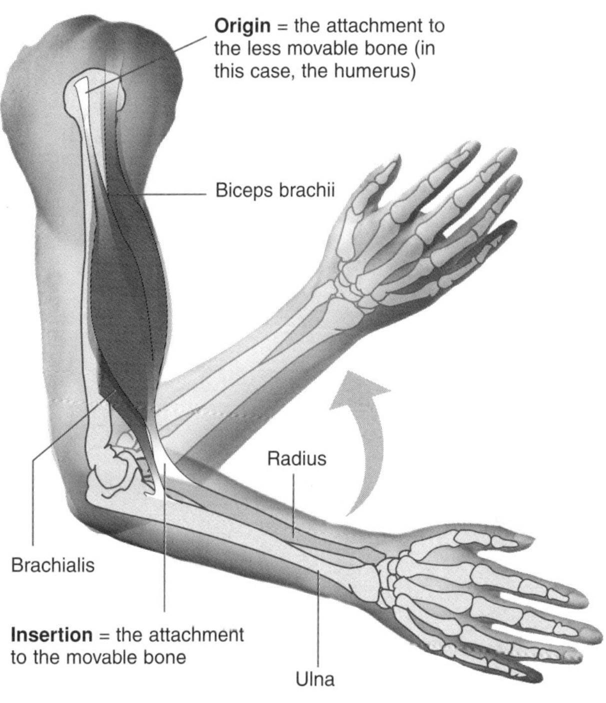
Origin = the attachment to the less movable bone (in this case, the humerus)
Biceps brachii
Radius
Brachialis
Insertion = the attachment to the movable bone
Ulna

Two muscles are involved in flexing the forearm. The **brachialis**, which underlies the biceps brachii and has an origin half way up the humerus, is the **prime mover**. The more obvious **biceps brachii**, which is a two headed muscle with two origins and a common insertion near the elbow joint, acts as the synergist. During contraction, the insertion moves towards the origin.

The skeleton works as a system of levers. The joint acts as a **fulcrum** (or pivot), the muscles exert the **force**, and the weight of the bone being moved represents the **load**. The flexion (bending) and extension (unbending) of limbs is caused by the action of **antagonistic muscles**. Antagonistic muscles work in pairs and their actions oppose each other. During movement of a limb, muscles other than those primarily responsible for the movement may be involved to fine tune the movement.

Every coordinated movement in the body requires the application of muscle force. This is accomplished by the action of agonists, antagonists, and synergists. The opposing action of agonists and antagonists (working constantly at a low level) also produces muscle tone. Note that either muscle in an antagonistic pair can act as the agonist or **prime mover**, depending on the particular movement (for example, flexion or extension).

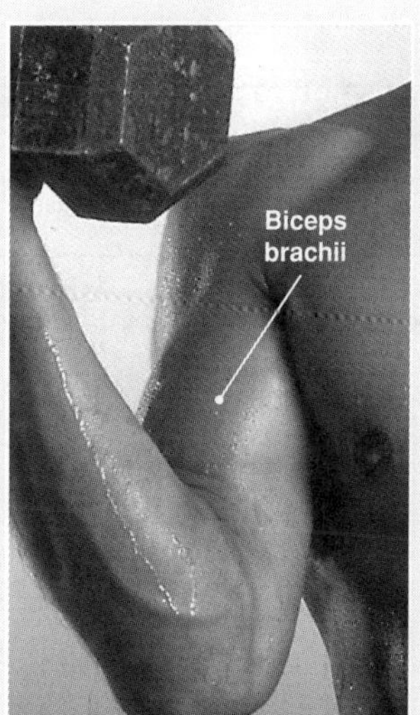
Biceps brachii

Agonists or prime movers: muscles that are primarily responsible for the movement and produce most of the force required.

Antagonists: muscles that oppose the prime mover. They may also play a protective role by preventing over-stretching of the prime mover.

Synergists: muscles that assist the prime movers and may be involved in fine-tuning the direction of the movement.

During flexion of the forearm (left) the **brachialis** muscle acts as the prime mover and the **biceps brachii** is the synergist. The antagonist, the **triceps brachii** at the back of the arm, is relaxed. During extension, their roles are reversed.

Movement at Joints

The synovial joints of the skeleton allow free movement in one or more planes. The articulating bone ends are separated by a joint cavity containing lubricating synovial fluid. Two types of synovial joint, the shoulder ball and socket joint and the hinge joint of the elbow, are illustrated below.

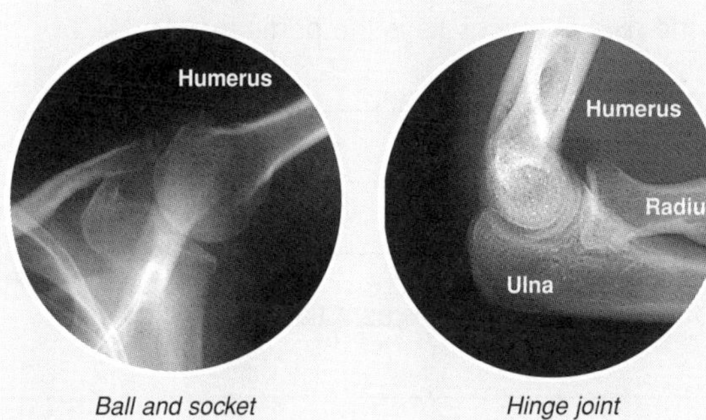
Humerus
Humerus
Radius
Ulna

Ball and socket *Hinge joint*

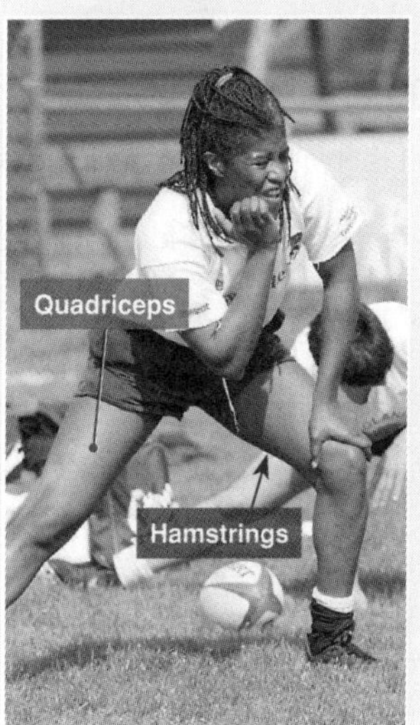
Quadriceps
Hamstrings

Movement of the upper leg is achieved through the action of several large groups of muscles, collectively called the **quadriceps** and the **hamstrings**.

The hamstrings are actually a collection of three muscles, which act together to flex the leg.

The quadriceps at the front of the thigh (a collection of four large muscles) opposes the motion of the hamstrings and extends the leg.

When the prime mover contracts forcefully, the antagonist also contracts very slightly. This stops overstretching and allows greater control over thigh movement.

ISBN: 978-1-927173-16-9

Related activities: *Skeletal Muscle Structure and Function*
Web links: *Muscles in Action*

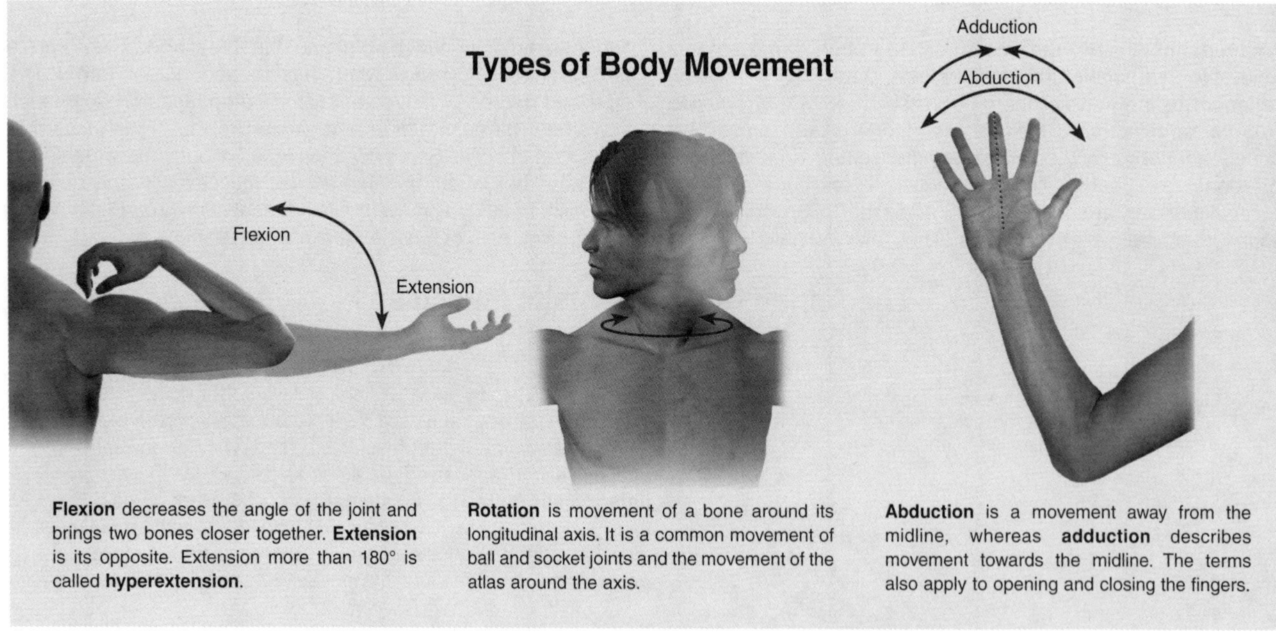
Types of Body Movement
Adduction
Abduction
Flexion
Extension

Flexion decreases the angle of the joint and brings two bones closer together. **Extension** is its opposite. Extension more than 180° is called **hyperextension**.

Rotation is movement of a bone around its longitudinal axis. It is a common movement of ball and socket joints and the movement of the atlas around the axis.

Abduction is a movement away from the midline, whereas **adduction** describes movement towards the midline. The terms also apply to opening and closing the fingers.

1. Describe the role of each of the following muscles in moving a limb:

   (a) Prime mover: ______

   (b) Antagonist: ______

   (c) Synergist: ______

2. Explain why the muscles that cause movement of body parts tend to operate as antagonist pairs: ______

3. Describe the relationship between muscles and joints Using appropriate terminology, explain how antagonistic muscles act together to raise and lower a limb:

4. Explain the role of joints in the movement of body parts: ______

5. (a) Identify the insertion for the biceps brachii during flexion of the forearm: ______

   (b) Identify the insertion of the brachialis muscle during flexion of the forearm: ______

   (c) Identify the antagonist during flexion of the forearm: ______

   (d) Given its insertion, describe the forearm movement during which the biceps brachialis is the prime mover:

6. (a) Describe a forearm movement in which the brachialis is the antagonist: ______

   (b) Identify the prime mover in this movement: ______

7. (a) Describe the actions that take place in the neck when you nod your head up and down as if saying "yes":

   (b) Describe the action being performed when a person sticks out their thumb to hitch a ride: ______

ISBN: 978-1-927173-16-9

Skeletal Muscle Structure and Function

Skeletal muscle (also called striated or voluntary muscle) is organized into bundles of muscle cells or fibers. Each fiber is a single cell with many nuclei and each fiber is itself a bundle of smaller myofibrils arranged lengthwise. Each myofibril is in turn composed of two kinds of myofilaments (thick and thin), which overlap to form light and dark bands. The alternation of these light and dark bands gives skeletal muscle its striated or striped appearance. The sarcomere, bounded by the dark Z lines, forms one complete contractile unit. Muscle fibers are innervated by the branches of motor neurons, each of which terminates in a specialized cholinergic synapse called the **neuromuscular junction** (or motor end plate). A motor neuron and all the fibers it innervates (which may be a few or several hundred) are called a **motor unit**. Graded responses in the muscle as a whole are achieved by varying the number of motor units active at any one time (recruitment of motor units).

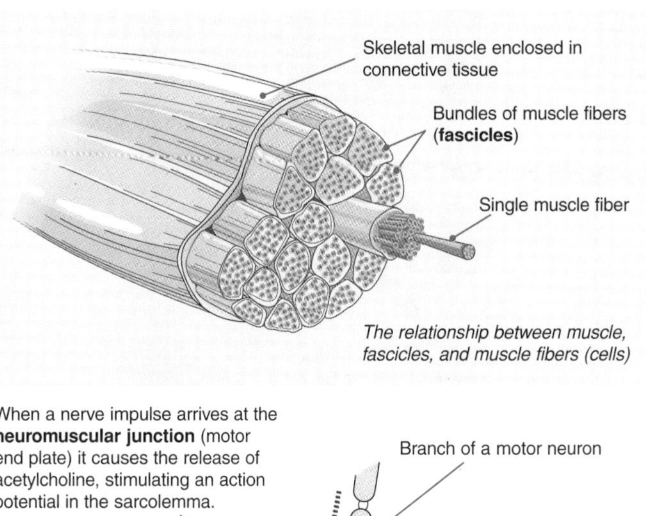
Skeletal muscle enclosed in connective tissue
Bundles of muscle fibers (**fascicles**)
Single muscle fiber

The relationship between muscle, fascicles, and muscle fibers (cells)

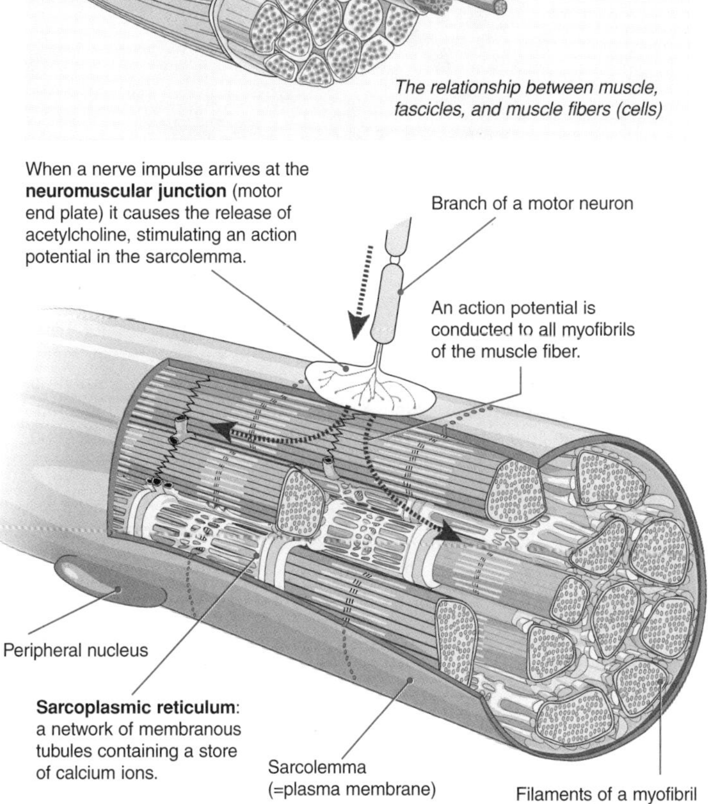
When a nerve impulse arrives at the **neuromuscular junction** (motor end plate) it causes the release of acetylcholine, stimulating an action potential in the sarcolemma.
Branch of a motor neuron
An action potential is conducted to all myofibrils of the muscle fiber.
Peripheral nucleus
Sarcoplasmic reticulum: a network of membranous tubules containing a store of calcium ions.
Sarcolemma (=plasma membrane)
Filaments of a myofibril seen in cross section

Longitudinal section of a sarcomere

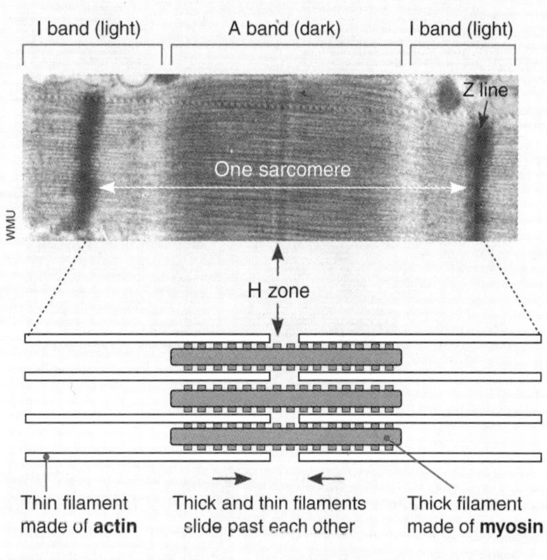
I band (light)
A band (dark)
I band (light)
Z line
One sarcomere
WMU
H zone
Thin filament made of **actin**
Thick and thin filaments slide past each other
Thick filament made of **myosin**

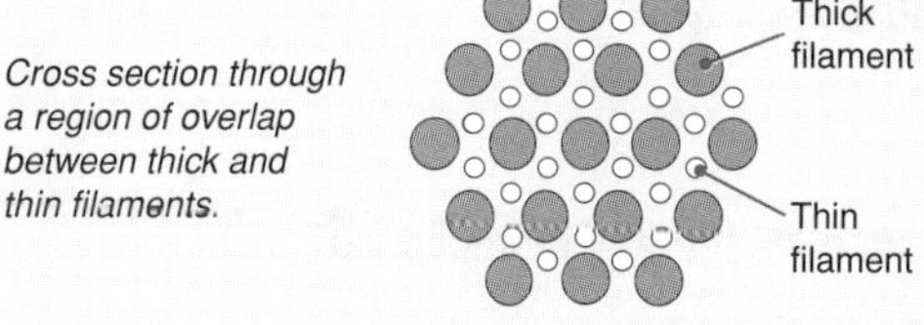
Cross section through a region of overlap between thick and thin filaments.
Thick filament
Thin filament

The photograph of a sarcomere (above) shows the banding pattern arising as a result of the highly organized arrangement of thin and thick filaments. It is represented schematically in longitudinal section and cross section.

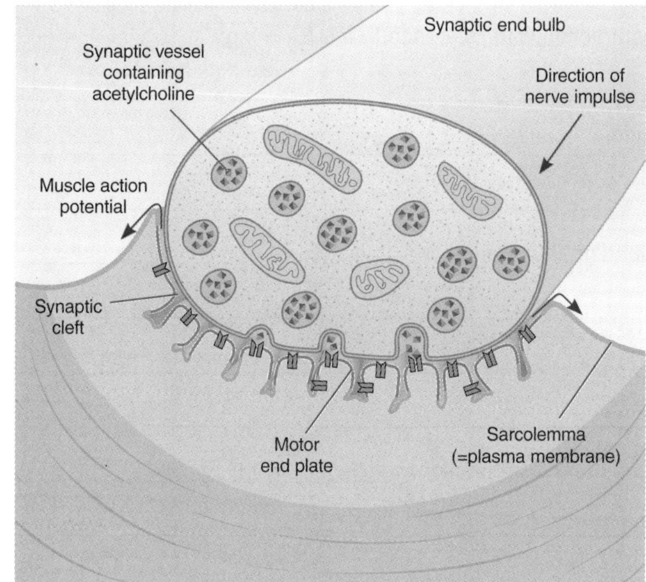
Synaptic end bulb
Synaptic vessel containing acetylcholine
Direction of nerve impulse
Muscle action potential
Synaptic cleft
Motor end plate
Sarcolemma (=plasma membrane)

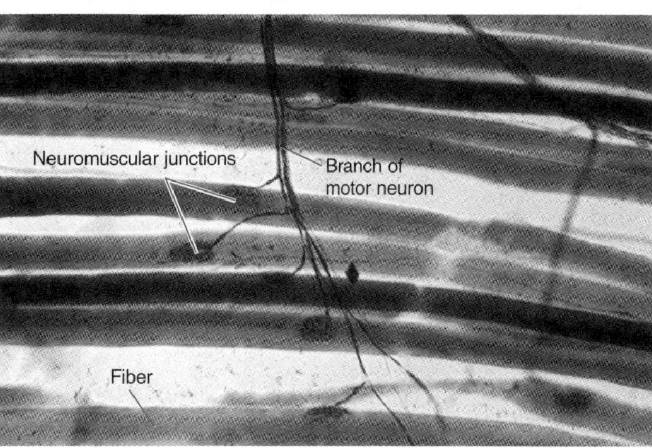
Neuromuscular junctions
Branch of motor neuron
Fiber

Above: Axon terminals of a motor neuron supplying a muscle. The branches of the axon terminate on the sarcolemma of a fiber at regions called the neuromuscular junction. Each fiber receives a branch of an axon, but one axon may supply many muscle fibers.

Left: Diagrammatic representation of the neuromuscular junction.

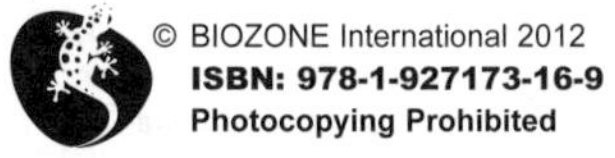

ISBN: 978-1-927173-16-9

Periodicals: *Human muscle: structure and function*

Related activities: *The Sliding Filament Theory, Chemical Synapses*
Weblinks: *Muscle Structure and Function*

The Banding Pattern of Myofibrils

Within a myofibril, the thin filaments, held together by the **Z lines**, project in both directions. The arrival of an action potential sets in motion a series of events that cause the thick and thin filaments to slide past each other. This is called **contraction** and it results in shortening of the muscle fiber and is accompanied by a visible change in the appearance of the myofibril: the I band and the sarcomere shorten and H zone shortens or disappears (below).

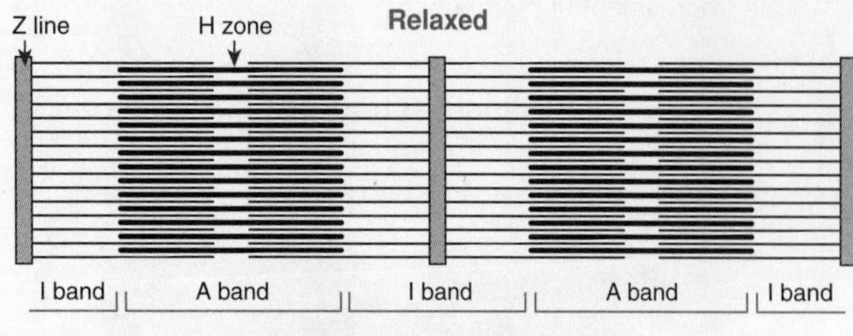
Z line
H zone
Relaxed
I band
A band
I band
A band
I band

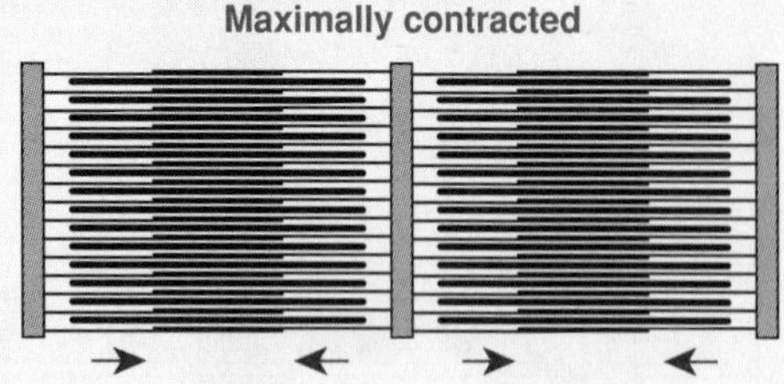
Maximally contracted

The response of a single muscle fiber to stimulation is to contract maximally or not at all; its response is referred to as the **all-or-none law** of muscle contraction. If the stimulus is not strong enough to produce an action potential, the muscle fiber will not respond. However skeletal muscles as a whole are able to produce varying levels of contractile force. These are called **graded responses**.

When Things Go Wrong

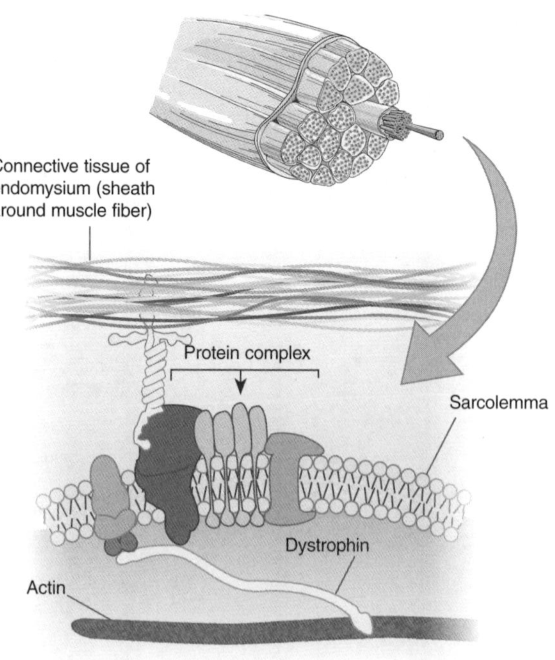
Connective tissue of endomysium (sheath around muscle fiber)
Protein complex
Sarcolemma
Dystrophin
Actin

Duchenne's muscular dystrophy is an X-linked disorder caused by a mutation in the gene DMD, which codes for the protein **dystrophin**. The disease causes a rapid deterioration of muscle, eventually leading to loss of function and death. It is the most prevalent type of muscular dystrophy and affects only males. Dystrophin is an important structural component within muscle tissue and it connects muscles fibers to the extracellular matrix through a protein complex on the sarcolemma. The absence of dystrophin allows excess calcium to penetrate the sarcolemma (the fiber's plasma membrane). This damages the sarcolemma, and eventually results in the death of the cell. Muscle fibers die and are replaced with adipose and connective tissue.

1. Describe what the neuromuscular junction is: ______________________

2. (a) Explain the cause of the banding pattern visible in striated muscle: ______________________

   (b) Explain the change in appearance of a myofibril during contraction with reference to the following:

   The I band: ______________________

   The H zone: ______________________

   The sarcomere: ______________________

3. Describe the purpose of the connective tissue sheaths surrounding the muscle and its fascicles: ______________________

4. Explain what is meant by the all-or-none response of a muscle fiber: ______________________

5. Explain why the inability to produce **dystrophin** leads to a loss of muscle function: ______________________

ISBN: 978-1-927173-16-9

The Sliding Filament Theory

The previous activity described how muscle contraction is achieved by the thick and thin muscle filaments sliding past one another. This sliding is possible because of the structure and arrangement of the thick and thin filaments. The ends of the thick myosin filaments are studded with heads or **cross bridges** that can link to the thin filaments next to them. The thin filaments contain the protein actin, but also a regulatory protein complex. When the cross bridges of the thick filaments connect to the thin filaments, a shape change moves one filament past the other. Two things are necessary for cross bridge formation: calcium ions, which are released from the **sarcoplasmic reticulum** when the muscle receives an action potential, and ATP, which is hydrolyzed by ATPase enzymes on the myosin. When cross bridges attach and detach in sarcomeres throughout the muscle cell, the cell shortens. Although a muscle fiber responds to an action potential by contracting maximally, skeletal muscles as a whole can produce varying levels of contractile force. These **graded responses** are achieved by changing the frequency of stimulation (**frequency summation**) and by changing the number and size of motor units recruited (**multiple fiber summation**). Maximal contractions of a muscle are achieved when nerve impulses arrive at the muscle at a rapid rate and a large number of motor units are active at once.

The Sliding Filament Theory

Muscle contraction requires calcium ions (Ca^{2+}) and energy (in the form of ATP) in order for the thick and thin filaments to slide past each other. The steps are:

1. The binding sites on the **actin** molecule (to which myosin 'heads' will locate) are blocked by a complex of two protein molecules: tropomyosin and troponin.
2. Prior to muscle contraction, ATP binds to the heads of the myosin molecules, priming them in an erect high energy state. Arrival of an action potential causes a release of Ca^{2+} from the sarcoplasmic reticulum. The Ca^{2+} binds to the troponin and causes the blocking complex to move so that the myosin binding sites on the actin filament become exposed.
3. The heads of the cross-bridging myosin molecules attach to the binding sites on the actin filament. Release of energy from the hydrolysis of ATP accompanies the cross bridge formation.
4. The energy released from ATP hydrolysis causes a change in shape of the myosin **cross bridge**, resulting in a bending action (*the power stroke*). This causes the actin filaments to slide past the myosin filaments towards the centre of the sarcomere.
5. (Not illustrated). Fresh ATP attaches to the myosin molecules, releasing them from the binding sites and repriming them for a repeat movement. They become attached further along the actin chain as long as ATP and Ca^{2+} are available.

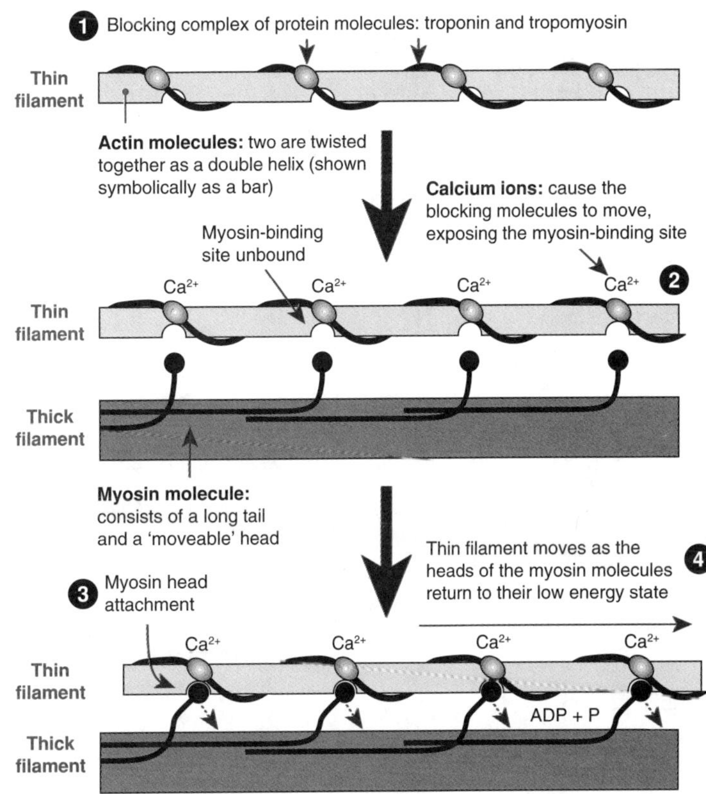

1 Blocking complex of protein molecules: troponin and tropomyosin
Thin filament
Actin molecules: two are twisted together as a double helix (shown symbolically as a bar)
Calcium ions: cause the blocking molecules to move, exposing the myosin-binding site
Myosin-binding site unbound
Ca2+
2
Thin filament
Thick filament
Myosin molecule: consists of a long tail and a 'moveable' head
3 Myosin head attachment
Thin filament moves as the heads of the myosin molecules return to their low energy state 4
ADP + P
Thin filament
Thick filament

1. Match the following chemicals with their functional role in muscle movement (draw a line between matching pairs):

(a) Myosin	• Bind to the actin molecule in a way that prevents myosin head from forming a cross bridge
(b) Actin	• Supplies energy for the flexing of the myosin 'head' (power stroke)
(c) Calcium ions	• Has a moveable head that provides a power stroke when activated
(d) Troponin-tropomyosin	• Two protein molecules twisted in a helix shape that form the thin filament of a myofibril
(e) ATP	• Bind to the blocking molecules, causing them to move and expose the myosin binding site

2. Describe the two ways in which a muscle as a whole can produce contractions of varying force:

(a) ______________________________

____

(b) ______________________________

____

3. (a) Identify the two things necessary for cross bridge formation: ______________________________

(b) Explain where each of these comes from: ______________________________

____

Muscles and Movement

ISBN: 978-1-927173-16-9

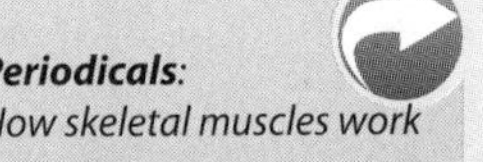

Periodicals: *How skeletal muscles work*

Related activities: *Muscle Structure and Function*
Weblinks: *Muscle Cell Contraction, Sliding Filament*

KEY TERMS: Word Find

Use the clues below to find the relevant key terms in the WORD FIND grid

M M J T X L N U O H C V A V P D W T F J T V R K Y

A U T P S N I F B A G L K U Y Q C E E I D S C T Z

G U S X F L K G Q H H X G V B O N E B R J X J R O

O U G C I J I K A E B H L J L I T C S A H X Q W T

N B L C L F O D J M B O Y A L K Y G M G V A F X Q

I O P S I E R I I F E T N Z J R S M B Y Y C V J B

S U Q Q E J S N N N P N C S A R C O M E R E F D G

T X S T X Q Y L L T G E T O S N B N T T E H F W W

U K F E T Y N H T A C F S E S O B Z Z D E H L K P

N N O N E U O M A J C A I M Y O F I B R I L S E W

J J H D N W V B N T Q H P L S K D G H G X H D P Y

C Z I O S Z I E T U K L T S A C O B H A I S S I U

I V F N I V A O A Z K Z S Q U M F S A Q C D A I E

M D C S O A L P G Q C R F J G L E A Q C V T M M U

A R N L N T E B O F L E X I O N E N V G Z E I S V

A L I F Q B V R N G B C R J O R N P T V Q Y A N S

I I R I I G L O I V R M P Z W J K B E T J B O J N

F F V R C B Q V S S Y A V C G I V S X R H C A I F

J C G K G Q E V T X S I W H H C H F A E U E S K K

F H Z O N E C R S I X P P P V F N R K N O O O O N

Q V L O D X C O N T R A C T I O N C C K Y D H R P

K J L F U D O B B X D R G O W G S N R M I F S Y Y

Material composed mainly of calcium, phosphorus and sodium that forms the endoskeleton of vertebrates.

Part of the muscle structure where filaments of myosin and actin overlap. The contraction of this structure is caused by the movement of these filaments past each other.

Tissue composed of fibers able to contract.

A connective tissue structure that connects bones to bones.

A connective tissue structure that connects muscles to bone.

A muscle cell is also called this.

Cylindrical organelles in muscle cells comprising bundles of actin and myosin filaments.

This protein forms the thin filaments in muscle myofibrils.

This protein forms the thick filaments in muscle myofibrils.

The name describing shortening of a muscle fiber.

Bending movement that decreases the angle between two body parts.

A straightening movement that increases the angle between body parts.

The most common and most movable type of joint in the body of a mammal, in which the articulating surfaces are surrounded by a capsule.

A model for muscle contraction in which filaments slide past one another. (3 words: 7, 8, 6)

The envelope of tissue surrounding the cavity of the synovial joint.

The border that separates and links sacomeres in a muscle myofibril.

The zone of thick filament that is not over lapped by thin filament in a relaxed muscle myofibril.

Muscle that is primarily responsible for a specific movement and produces most of the force required.

A muscle working against the action of another.

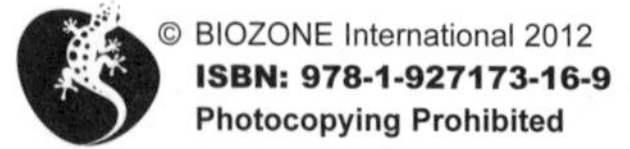

ISBN: 978-1-927173-16-9

AHL Topic
9

Plant **Science**

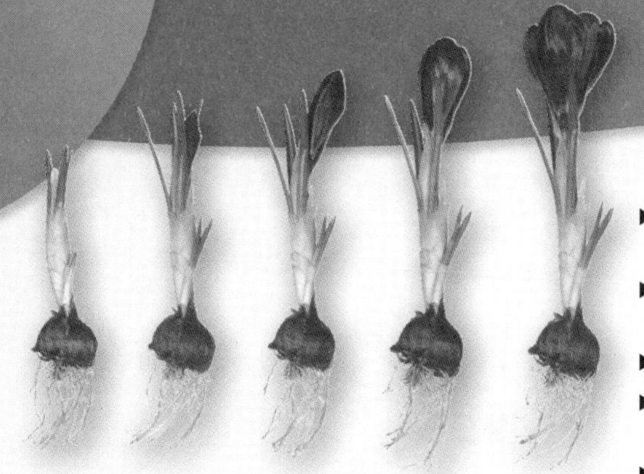

Key concepts

- The plant body is specialized to make exchanges with the environment to maximize photosynthetic rate.
- Water loss is a consequence of gas exchange but plants have adaptations to reduce water loss.
- Transpiration drives water uptake in plants.
- Translocation moves carbohydrate around the plant from sources to sinks.
- The seed houses a dormant embryonic plant. Special conditions are required to start germination.
- Plant hormones regulate responses to environment.

Key terms

AHL only

abscisic acid
active transport
adhesion
amylase
apical dominance
apical meristem
auxin
bulb
cell turgor
cellulose
cohesion-tension
dicotyledonous plant (dicot)
diffusion
flower
gibberellin
guard cells
lateral meristem
leaf
lignin
long-day plant
mesophyll
mineral ion
monotyledonous plant monocot)
phloem
phototropism
phytochrome
plan diagram
root
short-day plant
sink
source
stem
stomata
tendril
translocation
transpiration
transpiration pull
tuber
xerophyte
xylem

Periodicals:
Listings for this chapter are on page 400

Weblinks:
www.thebiozone.com/weblink/IB-3169.html

Learning Objectives

□ 1. Use the **KEY TERMS** to compile a glossary for this topic.

Plant Structure and Growth *(9.1)* pages 366-381

□ 2. **AHL**: Draw and label **plan diagrams** of a dicot **stem** and **leaf**, showing the distribution of tissues.

□ 3. **AHL**: Outline three differences between the structure of monocot and dicot plants.

□ 4. **AHL**: Relate the distribution of tissues in a dicot leaf to their functional roles. (e.g. light absorption, gas exchange, water conservation, transport).

□ 5. **AHL**: Describe modifications of **roots**, stems, and leaves and describe the function of the modification. Examples include **bulbs**, **tubers**, storage roots, and **tendrils**.

□ 6. **AHL**: Identify the location of **apical** and **lateral meristems** in dicot plants. Compare growth at apical and lateral meristems in dicot plants.

□ 7. **AHL**: Explain the control of plant growth using the example of the role of **auxin** in **phototropism**. Describe the evidence for the role of auxin in plant growth responses.

Transport in Angiosperms *(9.2)* pages 382-389

□ 8. **AHL**: Relate the structure and properties of the **root** system to its functional role in the uptake of water and **mineral ions**. List the ways in which mineral ions in the soil move into the root. Explain how the roots absorb mineral ions by **active transport**.

□ 9. **AHL**: Describe how plants support themselves, including reference to thickened **cellulose** cell walls, **cell turgor**, and lignified xylem.

□ 10. **AHL**: Describe **transpiration** in a flowering plant,. Include reference to the roles of xylem, **cohesion-tension**, **adhesion**, **transpiration pull** and **evaporation**.

□ 11. **AHL**: Describe how guard cells regulate transpiration by opening and closing the stomata and explain the role of **abscisic acid** in this.

□ 12. **AHL**: Explain the effect of abiotic factors, specifically humidity, light, air movement, and temperature on transpiration rate.

□ 13. **AHL**: Describe the adaptations of **xerophytes** that help to reduce transpiration.

□ 14. **AHL**: Describe **translocation** in the **phloem**, identifying **sources** and **sinks** in the transport of sucrose and amino acids.

Reproduction in Angiosperms *(9.3)* pages 390-397

□ 15. **AHL**: Draw and label a diagram of a animal-pollinated **flower** (dicot).

□ 16. **AHL**: Distinguish between pollination, fertilization, and seed dispersal. Draw and label a diagram to show the external and internal structure of a named dicot seed.

□ 17. **AHL**: Describe **germination** in a typical, starchy seed. Explain the conditions required to break dormancy and outline the metabolic processes involved, including the role of water absorption, production of **amylase** and formation of **gibberellin**.

□ 18. **AHL**: Explain the control of flowering by **phytochrome** in **long-day** and **short-day plants**. How could this system be used to manipulate flowering in commercial plants?

The General Structure of Plants

The support and transport systems in plants are closely linked; many of the same tissues are involved in both systems. Primitive plants (e.g. mosses and liverworts) are small and low growing, and have no need for support and transport systems. If a plant is to grow to any size, it must have ways to hold itself up against gravity and to move materials around its body. The body of a flowering plant has three parts: **roots** anchor the plant and absorb nutrients from the soil, **leaves** produce sugars by photosynthesis, and **stems** link the roots to the leaves and provide support for the leaves and reproductive structures. Vascular tissues (xylem and phloem) link all plant parts so that water, minerals, and manufactured food can be transported between different regions. All plants rely on fluid pressure within their cells (turgor) to give some support to their structure.

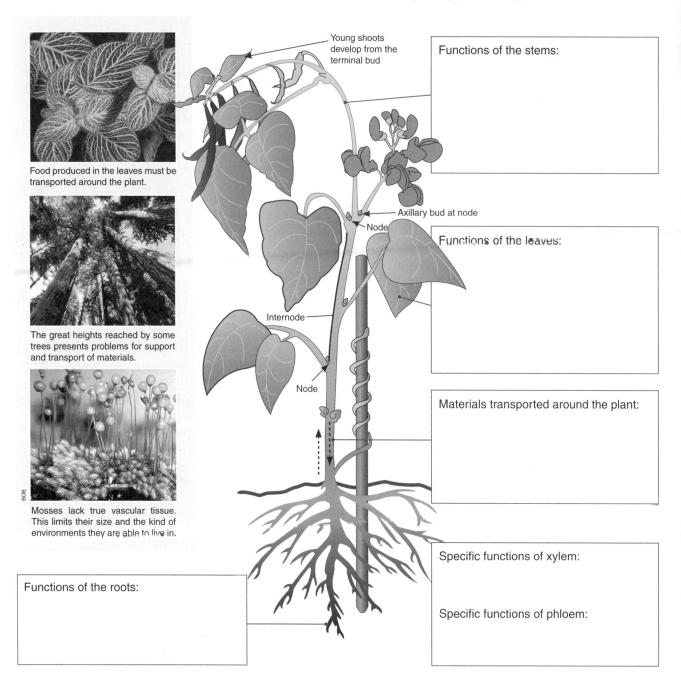

Food produced in the leaves must be transported around the plant.

The great heights reached by some trees presents problems for support and transport of materials.

BOB

Mosses lack true vascular tissue. This limits their size and the kind of environments they are able to live in.

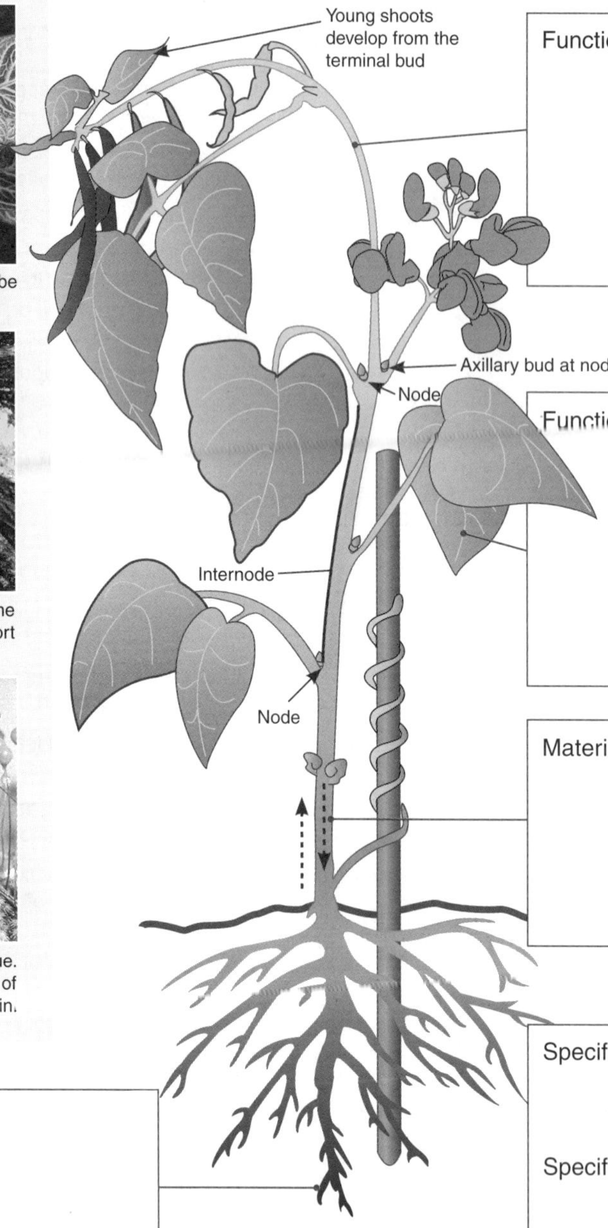

Young shoots develop from the terminal bud

Axillary bud at node

Node

Internode

Node

Functions of the stems:

Functions of the leaves:

Materials transported around the plant:

Specific functions of xylem:

Specific functions of phloem:

Functions of the roots:

1. In the boxes provided in the diagram above:
   (a) List the main functions of the leaves, roots and stems (remember that the leaves themselves have leaf veins).
   (b) List the materials that are transported around the plant body.
   (c) Describe the functions of the transport tissues: xylem and phloem.

2. What is the solvent for all materials transported around the plant? ______________________

3. State what processes are involved in the transport of sap in the following tissues:

   (a) The xylem: ______________________

   (b) The phloem: ______________________

ISBN: 978-1-927173-16-9

Monocots vs Dicots

The distinction between **monocots** and **dicots** only became common in botany in the late 18th century. Until then, plants were commonly classified by form (e.g. shrubs, vines, trees, etc). The names refer to the number of cotyledons (or seed leaves) in the embryo but this is the only feature that is definitive alone. Without an embryo, it can sometimes be difficult to determine if a plant is a monocot or a dicot, as some plants display features of both and some characteristics (e.g. flower parts) can be disguised by specialized adaptations. The features listed below are commonly used to distinguish monocots and dicots, although it is best to use multiple features to determine with confidence into which class a plant should be placed.

Features of Monocots and Dicots

Monocots	Dicots	Description of terms
Embryo has a single **cotyledon**	Embryo has two **cotyledons**	Cotyledons are the embryonic seed leaves
Pollen has a single furrow or pore	**Pollen** has three furrows of pores	Pollen is powderlike material produced by seed plants, containing immature male gametophytes.
Flower parts are in multiples of three	**Flower parts** are in multiples of four or five	Flower parts include the petals, stamens, and carpel.
Major **leaf veins** are in parallel	**Leaf veins** form net-like lattice	There is often a single central major leaf vein in dicots
Stem **vascular bundles** are scattered	Stem **vascular bundles** are in a ring	Vascular bundles (xylem and phloem) form the transport system in a plant
Roots are **adventitious**	Major tap (central) root from which others arise	Adventitious roots form fibrous mats
Secondary growth absent	**Secondary growth** generally present.	Secondary growth produces wood

1. Look carefully at the structures in the photographs below and identify if the plant is a monocot or a dicot. Give reasons for your identification below:

(a)

(b)

(c)

(d)

(e)

(f)

ISBN: 978-1-927173-16-9

Related activities: *Dicot Stems and Roots*

Leaf Structure and Gas Exchange

The epidermis of leaves is covered with tiny pores, called **stomata**. Each stoma has a special **guard cell** on each side. When the guard cells around a stoma are turgid, the stoma opens, and when they are flaccid, the cells collapse together and close the stoma. Stomata allow gas exchange between the air and the photosynthetic cells inside the leaf, but they are also the major routes for water loss. About 90% of water loss from a plant occurs via stomata. About 10% occurs directly through the waxy cuticle. This figure can be much higher if the cuticle is thin (as in ferns), or lower in plants with thick, waxy leaves (succulents).

Gas Exchanges and the Function of Stomata

Gases enter and leave the leaf by way of stomata. Inside the leaf (as illustrated by a dicot, right), the large air spaces and loose arrangement of the spongy mesophyll facilitate the diffusion of gases and provide a large surface area for gas exchanges.

Respiring plant cells use oxygen (O_2) and produce carbon dioxide (CO_2). These gases move in and out of the plant and through the air spaces by diffusion.

When the plant is photosynthesizing, the situation is more complex. Overall there is a net consumption of CO_2 and a net production of oxygen. The fixation of CO_2 maintains a gradient in CO_2 concentration between the inside of the leaf and the atmosphere. Oxygen is produced in excess of respiratory needs and diffuses out of the leaf. These **net** exchanges are indicated by the arrows on the diagram.

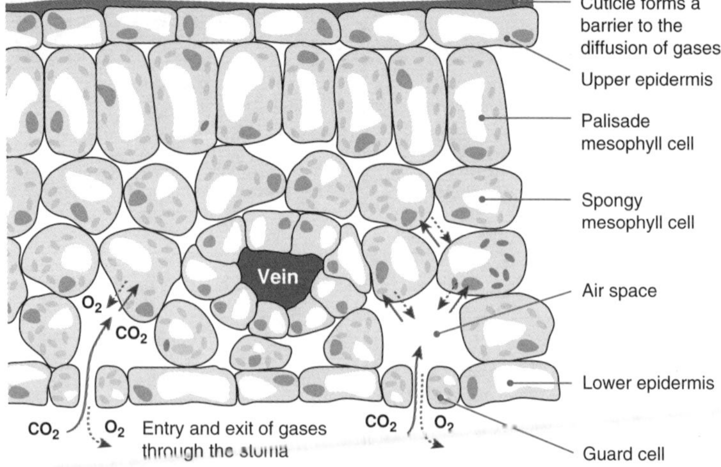

Cuticle forms a barrier to the diffusion of gases
Upper epidermis
Palisade mesophyll cell
Spongy mesophyll cell
Air space
Lower epidermis
Guard cell
Vein
O_2
CO_2
CO_2
O_2
Entry and exit of gases through the stoma
CO_2
O_2

Net gas exchanges in a photosynthesizing dicot leaf

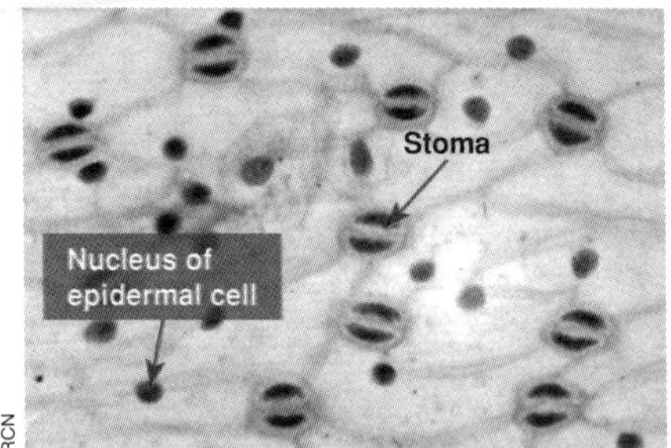

Stoma
Nucleus of epidermal cell
RCN

A surface view of the leaf epidermis of a dicot (above) illustrating the density and scattered arrangement of stomata. In dicots, stomata are usually present only on the lower leaf surface.

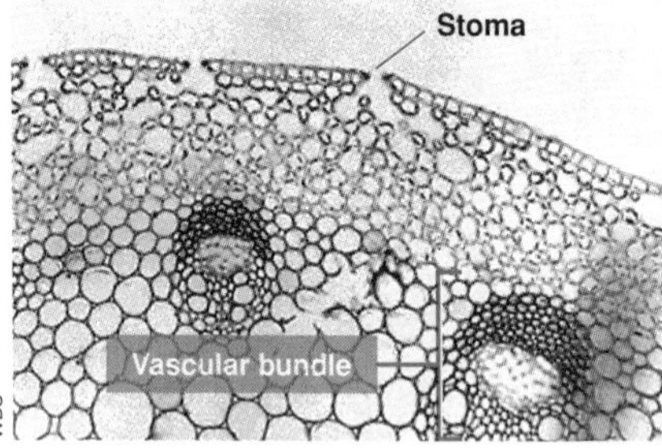

Stoma
Vascular bundle
WBS

The stems of some plants (e.g. the buttercup above) are photosynthetic. Gas exchange between the stem tissues and the environment occurs through stomata in the outer epidermis.

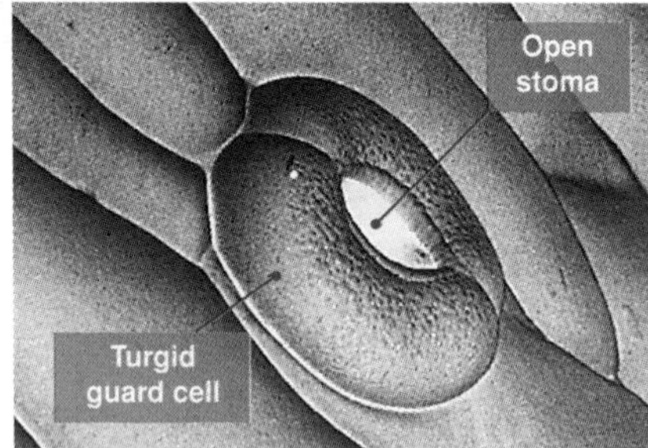

Open stoma
Turgid guard cell

Closure of stomata is regulated by abscisic acid (ABA). ABA causes K^+ ions to leave the guard cells and water to follow (by osmosis). This causes a loss of guard cell turgor and closes the stomatal pore.

1. Identify two ways in which the continuous air spaces through the plant facilitate gas exchange:

   (a) ______________________________

   (b) ______________________________

2. (a) Name the pores in the leaf that facilitate gas exchange in plants: ______________________________

   (b) Briefly outline their role in gas exchange in an angiosperm: ______________________________

3. Describe an adaptation of photosynthetic stems for maintaining gas exchange: ______________________________

4. Plants carrying a defect for the synthesis of ABA show a wilting phenotype. Explain why: ______________________________

5. Explain why plants with a thin cuticle require a moist or humid environment: ______________________________

Related activities: *Transpiration*

ISBN: 978-1-927173-16-9

Dicot Stems and Roots

The stem and root systems of plants are closely linked. Stems are the primary organs for supporting the plant, whereas roots anchor the plant in the ground, absorb water and minerals from the soil, and transport these materials to other parts of the plant body. Roots may also act as storage organs, storing excess carbohydrate reserves until they are required by the plant. Like most parts of the plant, stems and roots contain vascular tissues. These take the form of bundles containing the xylem and phloem and strengthening fibers. The entire plant body, including the roots and stems is covered in an epidermis but, unlike most of the plant, the root epidermis has only a thin cuticle that presents no barrier to water entry. Young roots are also covered with **root hairs**. Compared with stems, roots are relatively simple and uniform in structure, and their features are associated with aeration of the tissue and transport of water and minerals form the soil. Dicot stems and roots are described below.

Dicot Stem Structure

In dicots, the vascular bundles are arranged in an orderly fashion around the stem. Each vascular bundle contains **xylem** (to the inside) and **phloem** (to the outside). Between the phloem and the xylem is the **vascular cambium**. This is a layer of cells that divide to produce the thickening of the stem. The middle of the stem, called the **pith**, is filled with thin-walled parenchyma cells. The vascular bundles in dicots are arranged in an orderly way around the periphery of the stem (below).

Xylem to the inside of the vascular cambium

Phloem to the outside of the vascular cambium

Enlarged below

Cortex

Pith

Vascular cambium is responsible for secondary growth

Fibre cap

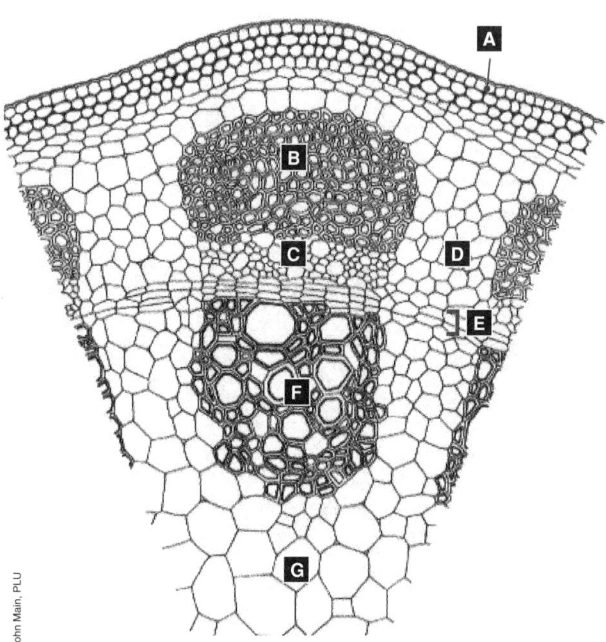
A
B
C
D
E
F
G
John Main, PLU

Cross section through a typical dicot stem

Dicot Root Structure

The primary tissues of a dicot root are simple in structure. The large cortex is made up of parenchyma (packing) cells, which store starch and other substances. The air spaces between the cells are essential for aeration of the root tissue, which is non-photosynthetic. The vascular tissue, xylem (X) and phloem (P) forms a central cylinder through the root and is surrounded by the **pericycle**, a ring of cells from which lateral roots arise.

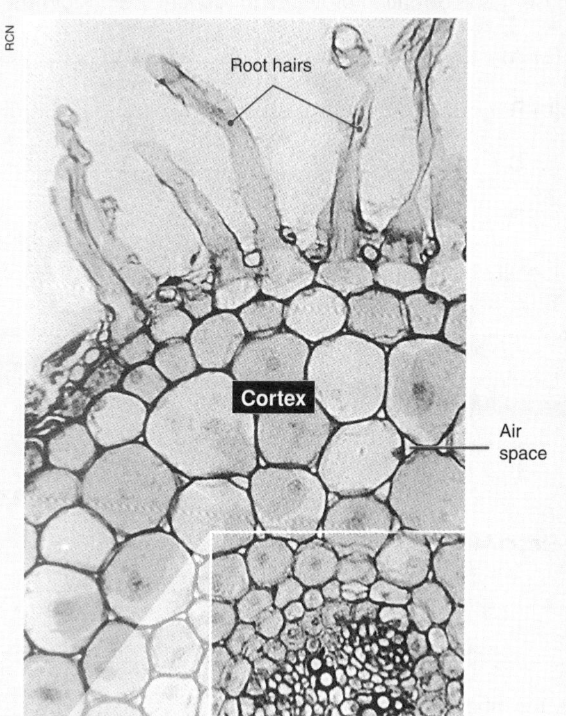
RCN
Root hairs
Cortex
Air space

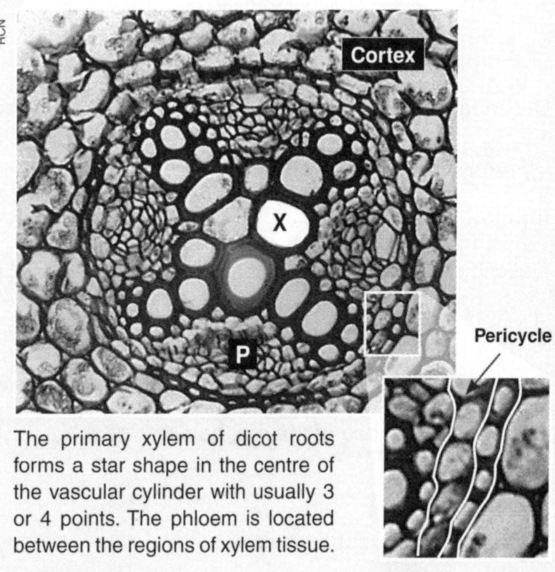
RCN
Cortex
X
P
Pericycle

The primary xylem of dicot roots forms a star shape in the centre of the vascular cylinder with usually 3 or 4 points. The phloem is located between the regions of xylem tissue.

Endodermis regulates the flow of water into the root

ISBN: 978-1-927173-16-9

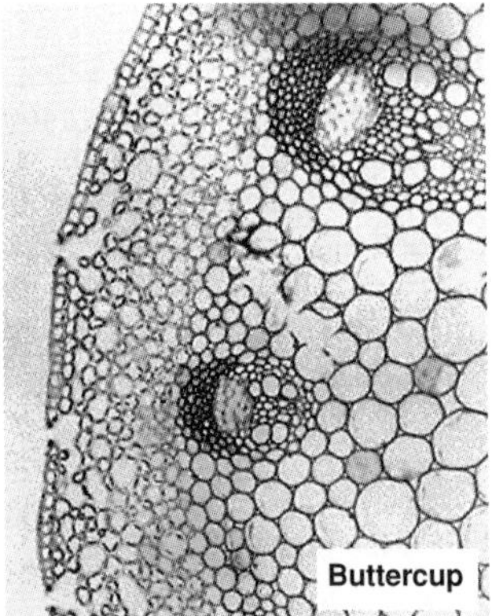
Buttercup

In plants with photosynthetic stems, CO_2 enters the stem through stomata in the epidermis. The air spaces in the cortex are more typical of leaf mesophyll than stem cortex.

University of Florida

Strawberry plants send out runners. These are above-ground, trailing stems that form roots at their nodes. The plant uses this mechanism to spread vegetatively over a wide area.

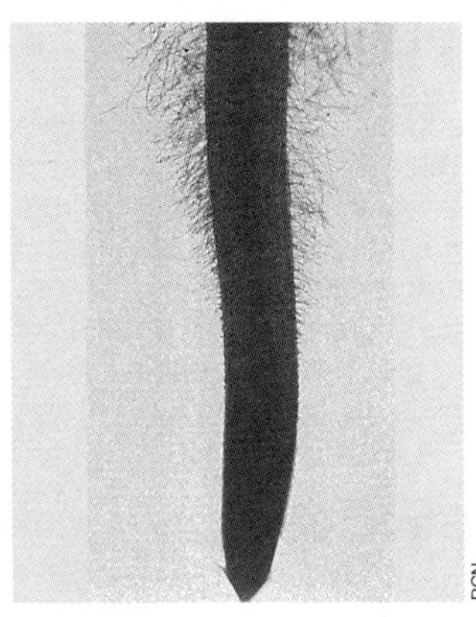
RCN

Root hairs are located just behind the region of cell elongation in the root tip. The root tip is covered by a slimy root cap. This protects the dividing cells of the tip and lubricates root movement.

The roots and their associated root hairs provide a very large surface area for the uptake of water and ions, as shown in this photograph of the roots of a hydroponically grown plant.

1. Use the information provided to identify the structures **A-G** in the photograph of the dicot stem on the previous page:

   (a) **A**: ____________________ (e) **E**: ____________________

   (b) **B**: ____________________ (f) **F**: ____________________

   (c) **C**: ____________________ (g) **G**: ____________________

   (d) **D**: ____________________

2. Identify the feature that distinguishes stems from other parts of the plant: ____________________

3. Describe a distinguishing feature of stem structure in dicots: ____________________

4. Describe the role of the vascular cambium: ____________________

5. Describe three functions of roots: ____________________

6. Describe two distinguishing features of internal anatomy of a primary dicot root:

   (a) ____________________

   (b) ____________________

7. Describe the role of the parenchyma cells of the root cortex: ____________________

8. Explain the purpose of the root hairs: ____________________

9. Explain why the root tip is covered by a cap of cells: ____________________

ISBN: 978-1-927173-16-9

Modifications in Plants

Various parts of the plant body may be modified for a specific role. Some **biennial plants**, e.g. carrots, store carbohydrates in fleshy **storage roots** during their first year of growth and use this store the following year to fuel the development of flowers, fruits, and seeds. The specialized **buttress** roots and **aerial** roots of some large tropical tree species (e.g. the banyan tree) provide support in thin tropical soils. Mangroves also have specialized aerial roots, called pneumatophores, which enable gas exchange in the water-logged substrate. In some **epiphytes**, e.g. the orchids, the aerial roots may be photosynthetic. **Parasitic** plants, such as mistletoe, produce rootlike organs that penetrate and parasitize the host's tissues. The stems of some plants also function as storage or photosynthetic organs. Horizontal underground stems, or **rhizomes**, can become swollen to provide a food store in the same way as root tubers. Similarly, **corms** are upright underground stems that become thickened with stored food. **Bulbs** are large buds with thick non-photosynthetic, food storage leaves clustered on short stems, e.g. onions. Cacti are a familiar example of stem and leaf modification. The fleshy green stems store water and photosynthesize while the leaves are modified into defensive spines. The tendrils of legumes and the traps of insectivorous plants are other familiar leaf modifications.

Modifications of Plant Parts for Storage

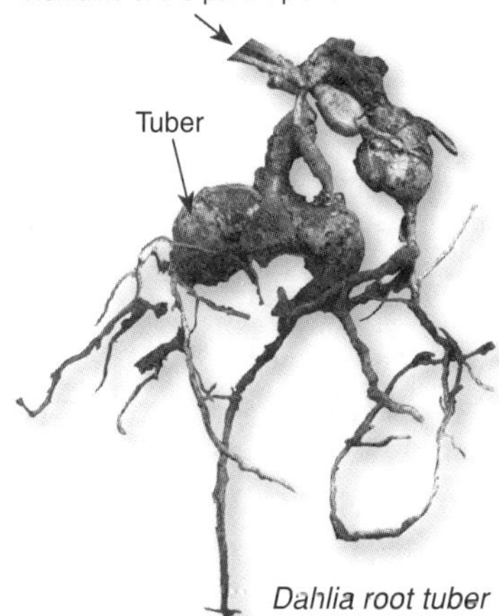
Remains of the parent plant
Tuber
Dahlia root tuber

Root tubers, e.g. dahlias (above), lack terminal and lateral buds. Both stem and root tubers can give rise to new individuals, thereby providing a means of vegetative propagation.

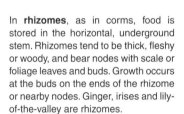

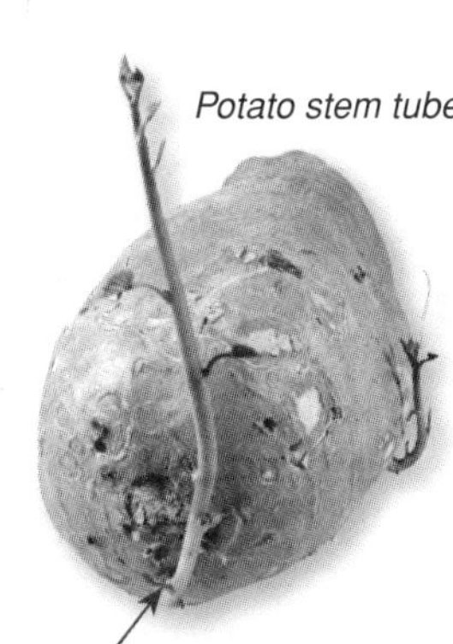
Potato stem tuber
'Eye' (lateral bud)

Tubers are the swollen part of an underground stem or root, usually modified for storing food. The potato is a **stem tuber**, as shown by the presence of terminal and lateral buds.

Food is stored in fleshy "scale" leaves
Stem plate to which the leaves are attached

A true **bulb** is really just a typical shoot compressed into a shortened form. Fleshy storage leaves are attached to a stem plate and form concentric circles around the growing tip. New roots form from the lower part of the stem.

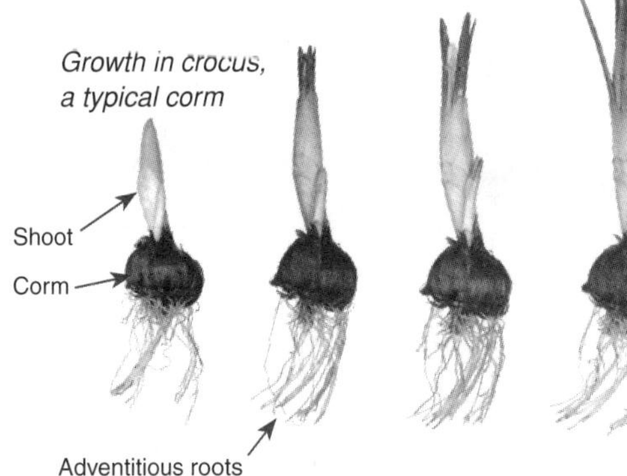
Growth in crocus, a typical corm
Shoot
Corm
Adventitious roots

In a **corm**, food is stored in stem tissue. Corms look like bulbs, but if you cut a corm in half you see a mass of homogenous tissue rather than concentric rings of fleshy leaves as in a bulb. Cyclamen, gladiolus, and crocus (above) are corms.

Underground stem containing stored food
Iris rhizome

In **rhizomes**, as in corms, food is stored in the horizontal, underground stem. Rhizomes tend to be thick, fleshy or woody, and bear nodes with scale or foliage leaves and buds. Growth occurs at the buds on the ends of the rhizome or nearby nodes. Ginger, irises and lily-of-the-valley are rhizomes.

Supportive and Breathing Roots

istock

Many plants have specialized roots growing from the stem into the soil. These **prop roots** are seen in mangroves (above) and corn. They provide stability for the plant in the substrate.

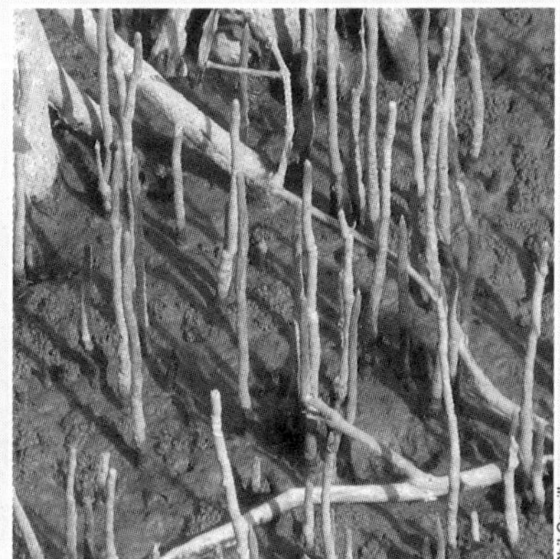
Phil Camill

Pneumatophores are the specialized 'breathing roots' of some types of mangroves. They grow up into the air and absorb oxygen-rich air via surface openings in the wood called lenticels.

istock

The tropical **banyan tree** sends down rope-like aerial roots from its branches. These anchor in the soil and become very thick, forming massive columns that support the heavy branches.

ISBN: 978-1-927173-16-9

Related activities: *Xerophytes*
Web links: *Types of Roots*

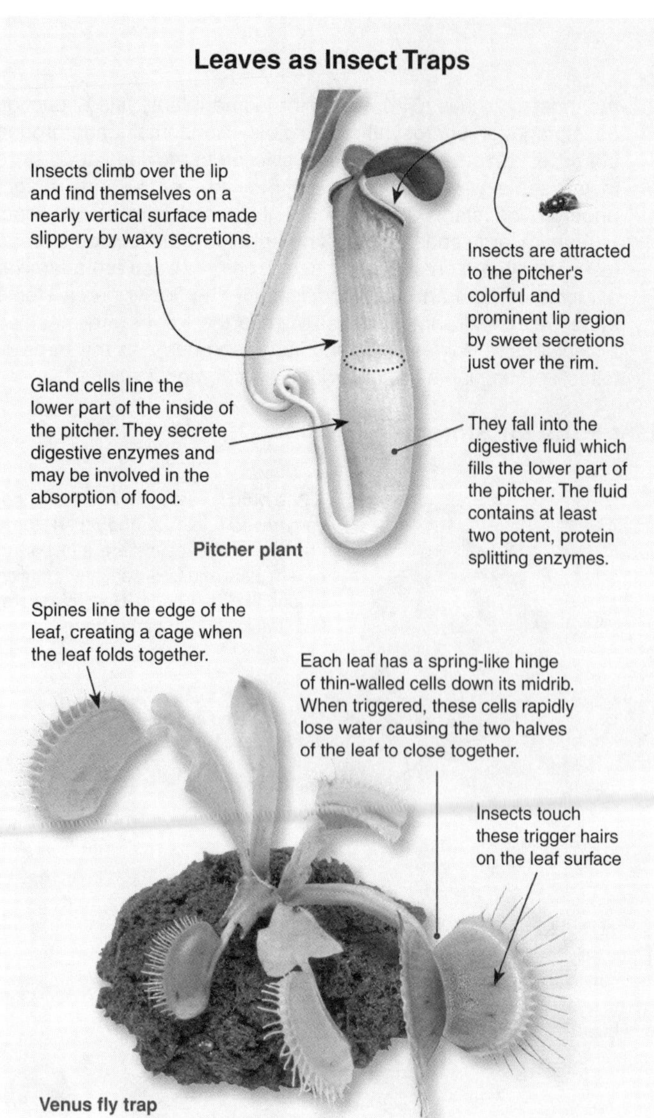
Leaves as Insect Traps
Insects climb over the lip and find themselves on a nearly vertical surface made slippery by waxy secretions.
Insects are attracted to the pitcher's colorful and prominent lip region by sweet secretions just over the rim.
Gland cells line the lower part of the inside of the pitcher. They secrete digestive enzymes and may be involved in the absorption of food.
They fall into the digestive fluid which fills the lower part of the pitcher. The fluid contains at least two potent, protein splitting enzymes.
Pitcher plant
Spines line the edge of the leaf, creating a cage when the leaf folds together.
Each leaf has a spring-like hinge of thin-walled cells down its midrib. When triggered, these cells rapidly lose water causing the two halves of the leaf to close together.
Insects touch these trigger hairs on the leaf surface
Venus fly trap

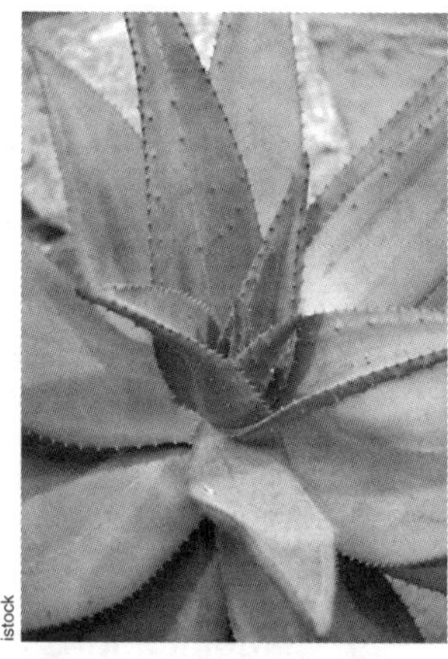
istock

Many arid-adapted plants, like this aloe, have succulent leaves that are photosynthetic and also modified for internal storage of water.

Tendril

A tendril is a thread-like structure leaf modification that helps a plant to climb over other plants or objects to gain access to light.

Opening bud
Bud scale
LJ Grauke

In temperate climates, the buds of woody plants (e.g. **hickory**) are protected over winter by modified leaves called bud scales. The waxy scales prevent desiccation and insulate the bud against the cold.

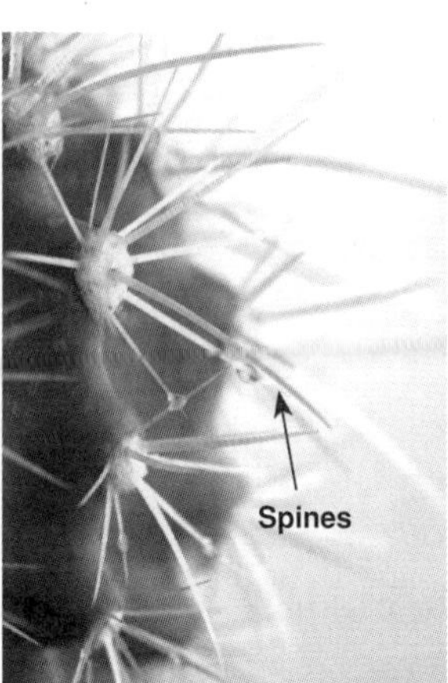
Spines

Many cacti have leaves reduced to sharp non-photosynthetic spines, which act to deter browsers. In the example above, the spines (leaves) grow from shortened shoots that arise from the photosynthetic stem.

1. For each of the following, identify the plant part that has been modified, and describe the modification and its purpose. Give an example in each case. The first one has been completed for you:

   (a) Bulb: Leaves are modified for food storage in the dormant plant. The leaf bases are fleshy and tightly packed together on a shortened stem. Example: onion, garlic, tulip, lily.

   (b) Corm: ______

   (c) Bud scales: ______

   (d) Tendrils: ______

   (e) Venus flytrap 'trap': ______

2. Discuss the role of aerial roots in named examples, explaining how they benefit the plant in each case:

______

ISBN: 978-1-927173-16-9

Qualitative Practical Work

Biological drawings should include as much detail as you need to distinguish different structures and types of tissue, but avoid unnecessary detail. Tissue preparations are rarely neat and tidy and there may be areas where you cannot see detail and where cells will appear to overlie one another. In these cases you will need to infer detail where possible from adjacent cells. Avoid shading as this can smudge and obscure detail. Labeling involves interpretation based on your knowledge and labels should be away from the drawing with label lines pointing to the structures identified. Add a title and any details of the image such as magnification. In this activity, you will practice the skills required to translate what is viewed into a good biological drawing.

Above: *Use relaxed viewing when drawing at the microscope. Use one eye (the left for right handers) to view and the right eye to view and direct your drawing.*

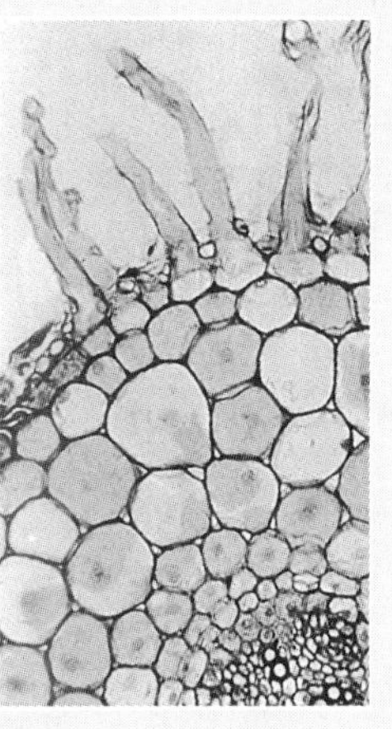

Above: Light micrograph TS through a *Ranunculus* root.
Left: Use one eye (the left for right handers) to view and the right eye to view and direct your drawing.

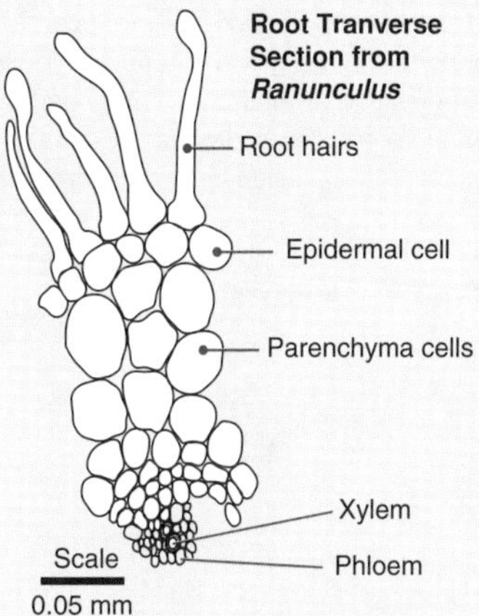
Root Tranverse Section from *Ranunculus*
Root hairs
Epidermal cell
Parenchyma cells
Xylem
Phloem
Scale
0.05 mm

A biological drawing of the same section. A biological drawing is different from a diagram, which is idealized and may contain more structure than can be seen in one section.

TASK

Complete the biological drawing of a cross section through a dicot leaf (below). Use the example above of the *Ranunculus* root as a guide to the detail required in your drawing

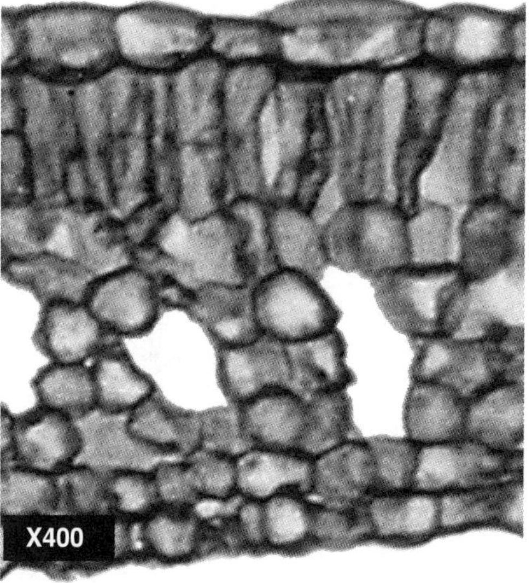
X400
RCN

Light micrograph of a cross section through a leaf.

ISBN: 978-1-927173-16-9

Weblinks: *Scientific Drawings for Biological Courses*

Plant Meristems

The differentiation of plant cells occurs only at specific regions called **meristems**. Two types of growth can contribute to an increase in the size of a plant. **Primary growth**, which occurs in the **apical meristem** of the buds and root tips, increases the length (height) of a plant. **Secondary growth** (not discussed here) increases plant girth and occurs in the lateral meristem in the stem. All plants show primary growth but only some show secondary growth (the growth that produces woody tissues).

Primary Growth

Primary growth occurs at the **apical meristem** (root and shoot tips). Three types of **primary meristem** are produced from the apical meristem: procambium, protoderm, and ground meristem. In dicots, the **procambium** forms vascular bundles that are found in a ring near the epidermis and surrounded by cortex. As the procambium divides, the cells on the inside become primary **xylem** and those on the outside become primary **phloem**.

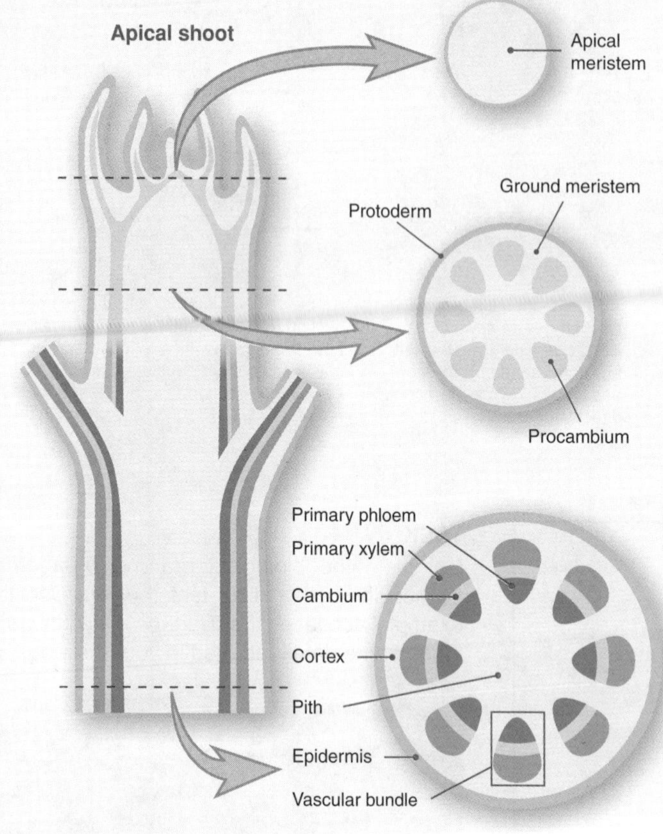

Apical shoot
Apical meristem
Ground meristem
Protoderm
Procambium
Primary phloem
Primary xylem
Cambium
Cortex
Pith
Epidermis
Vascular bundle

Primary Tissues Generated by the Meristem

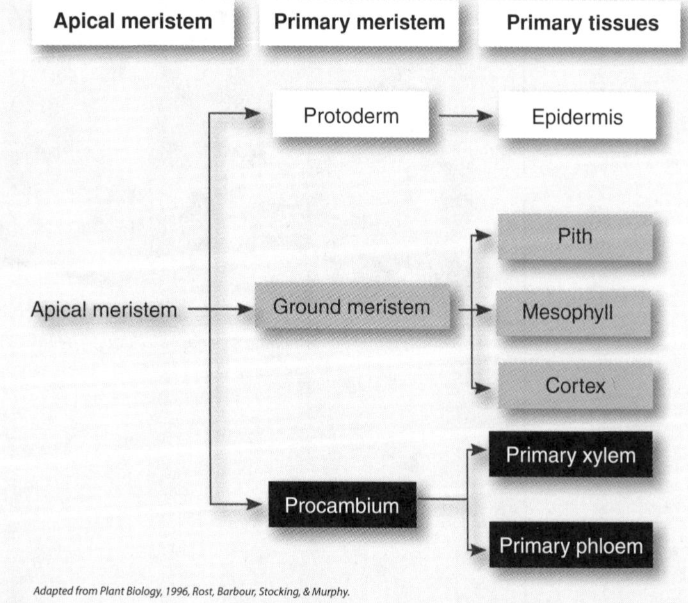

Apical meristem
Primary meristem
Primary tissues
Protoderm
Epidermis
Pith
Apical meristem
Ground meristem
Mesophyll
Cortex
Primary xylem
Procambium
Primary phloem

Adapted from Plant Biology, 1996, Rost, Barbour, Stocking, & Murphy.

1. Describe the role of the meristems in plants:

2. Describe the location of the meristems and relate this to how plants grow:

3. Describe a distinguishing feature of meristematic tissue:

4. Discuss the structure and formation of the primary tissues in dicot plants:

ISBN: 978-1-927173-16-9

Support in Plants

Plants support themselves in their environment and maintain the positions that enable them to carry out essential processes. All plants are provided some support by **cell turgor**. For very small plants, this is sufficient. Terrestrial vascular plants have strengthening tissues that may be hardened with lignin, and many also produce secondary growth (wood). For aquatic plants the water provides support, and adaptations are primarily to maintain the plant in the photic zone and to remain anchored.

Aquatic Environment

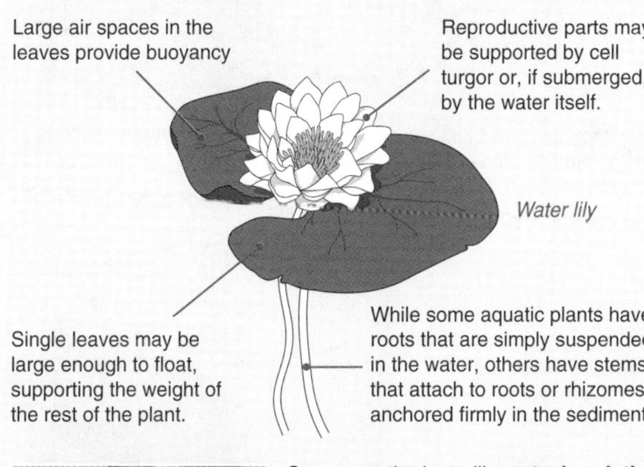
Large air spaces in the leaves provide buoyancy
Reproductive parts may be supported by cell turgor or, if submerged, by the water itself.
Water lily
Single leaves may be large enough to float, supporting the weight of the rest of the plant.
While some aquatic plants have roots that are simply suspended in the water, others have stems that attach to roots or rhizomes anchored firmly in the sediment.

Some aquatic plants, like **water hyacinth**, have swollen petioles that act as floats. Many form floating mats which block water ways and are serious weeds e.g. *Salvinia* and alligator weed.

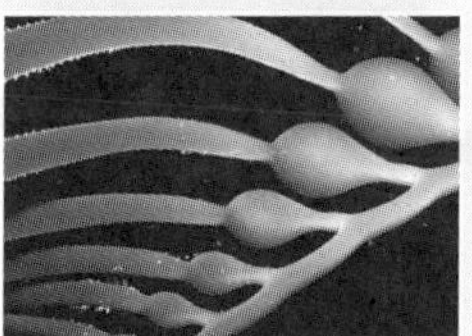

Marine and freshwater **algae** are not plants but have plant like qualities (e.g. chlorophyll). They lack vascular tissue and are supported by the water. Buoyancy may be assisted by air-filled floats or projections of the cell wall, which increase surface area (as in the case of diatoms).

Many floating or semi-aquatic plants, such as **water lilies**, have expanded leaves that provide a large surface area for flotation and photosynthesis. Such floating leaves support the submerged parts of the plant. These plants have roots that are attached to the bottom sediment.

Terrestrial Environment

Vascular plants with secondary thickening (woody tissue) can reach an enormous size. The Australian *Eucalyptus regnans* is reputed to grow to over 140 m in height. Their structural tissues (e.g. **xylem** and **wood**) provide support against gravity. Leaves and reproductive parts are supported by **cell turgor**.

RA

Roots anchor the plant in the ground, forming a stable base for growth. The roots of large hard-wood trees can form **buttresses**, providing extra support in poor soils.

Moss
BOB

Bryophytes (mosses and liverworts) lack any vascular tissue: the plant body is supported by **cell turgor**. As a consequence, they are small and their upward growth is restricted. Although there are no true roots, filamentous rhizoids anchor the plant in the ground.

Liverwort
BOB

Liverworts are simpler in structure than mosses. The gametophyte (the main plant body) is a flattened structure that may be a lobed thallus or leaf-like, depending on the species. The plant lies flat against the substrate, rising just a centimeter or two above the ground.

Fern

Ferns are tracheophytes (vascular plants). They have well developed vascular tissues that provide support and allow transport of nutrients and water around the plant. Because of this, they are able to grow to considerable heights (e.g. tree ferns).

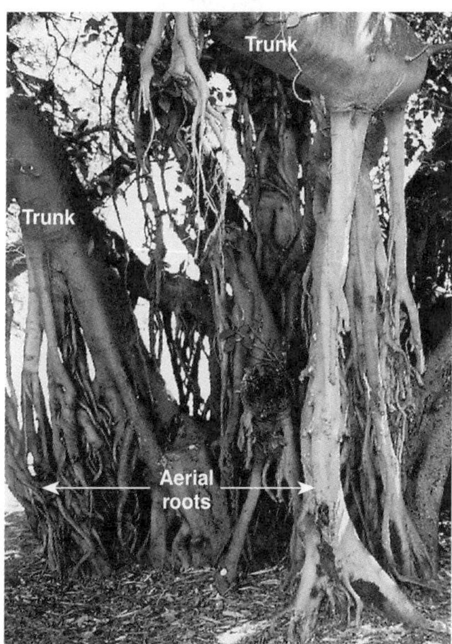
Trunk
Trunk
Aerial roots
RA

The **Lord Howe Island fig** has ten or more trunks that develop and form aerial roots. These grow downwards and once anchored in the ground, they provide extra support for the heavy weight of the trunks, allowing them to cover a wide area. Without such support, the trunks would collapse.

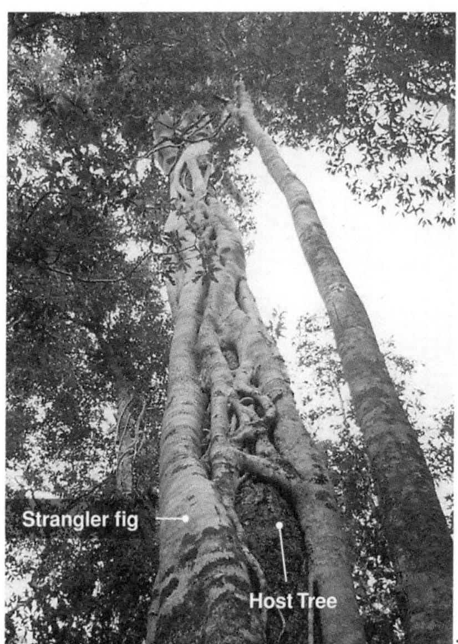
Strangler fig
Host Tree
RA

Strangler figs begin life high in the forest canopy as epiphytes. They develop roots that grow towards the forest floor. Once rooted in the soil they grow rapidly, embracing the trunk of a host tree with roots and shading it. As the host tree increases in girth, the fig cuts off the sap supply to its roots, killing it.

Pneumatophores
RA

Mangroves grow on mudflat shorelines. The root system cannot penetrate far into the mud due to the lack of oxygen. Support is provided by roots that are sent out in all directions just below the surface. Pneumatophores or breathing roots, (seen above) arise from these shallow lateral roots.

ISBN: 978-1-927173-16-9

Related activities: *Modifications in Plants*

Support in Woody Plants

Secondary xylem which consists of massed xylem vessels and fibers, makes up the bulk of the stem as **wood**, providing considerable strength.

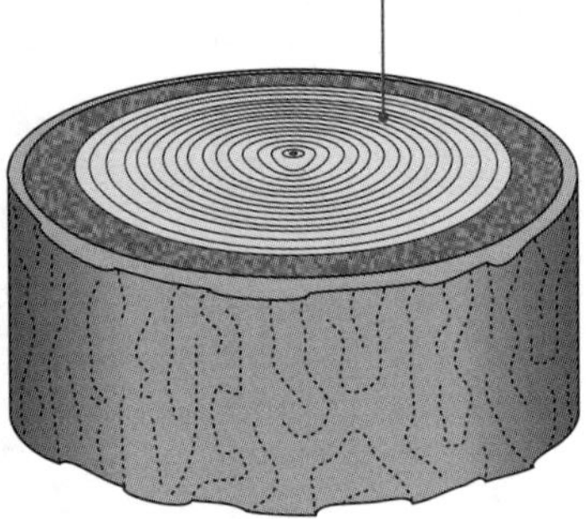

The strengthening of cells with lignin gives support to stems in all vascular plants. Lignin, together with cell turgor, is particularly important in non-woody (herbaceous) plants.

Support in Herbaceous Plants

Turgor pressure inside the parenchyma cells provides a strong inflating force that pushes against the epidermal layer.

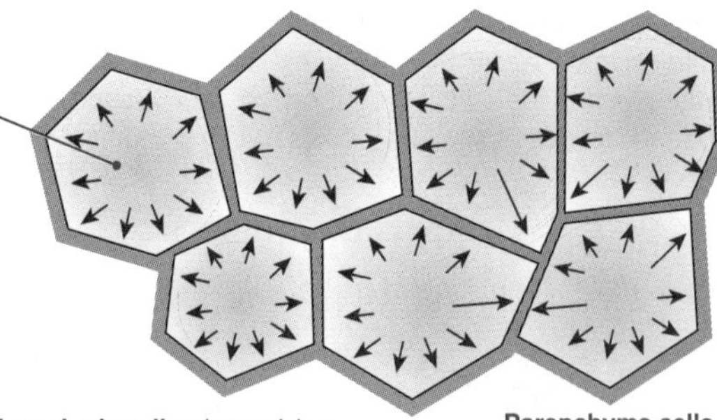
Parenchyma cells

Vascular bundles (comprising xylem and phloem) enhance the ability of herbaceous stems to resist tension and compression.

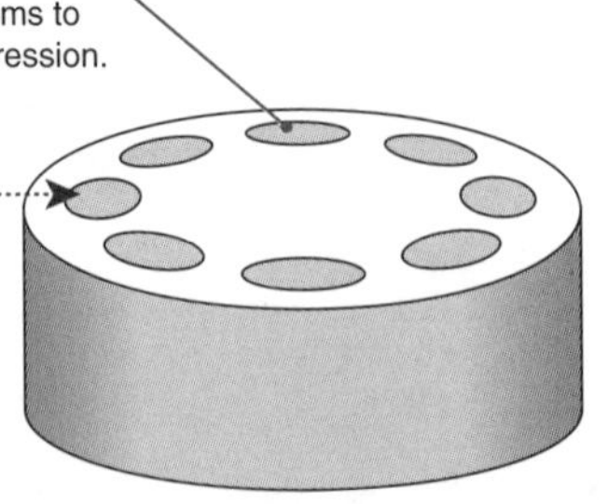

Xylem vessel with spiral thickening produced as a result of **lignin** deposition.

1. Contrast the main problems experienced by aquatic and terrestrial plants in supporting themselves:

2. Describe how the following are achieved in the aquatic protists and plants named below:

   (a) Maintaining a stationary position in seaweeds:

   (b) Keeping the fronds of kelp near the surface:

   (c) Keeping water lily pads floating on the surface:

3. Describe the function of **buttresses** on the trunks of large rainforest hardwood trees:

4. Explain the role of the following in providing support for vascular plants:

   (a) Lignin:

   (b) Turgor pressure:

   (c) Vascular bundles:

   (d) Secondary xylem:

5. Describe how strangler fig trees overcome support problems during the early stage of their development:

ISBN: 978-1-927173-16-9

Xylem

Xylem is the principal **water conducting tissue** in vascular plants. It is also involved in conducting dissolved minerals, in food storage, and in supporting the plant body. As in animals, tissues in plants are groupings of different cell types that work together for a common function. Xylem is a **complex tissue**. In angiosperms, it is composed of five cell types: tracheids, vessels, xylem parenchyma, sclereids (short sclerenchyma cells), and fibers. The tracheids and vessel elements form the bulk of the tissue. They are heavily strengthened and are the conducting cells of the xylem. Parenchyma cells are involved in storage, while fibers and sclereids provide support. When mature, xylem is dead.

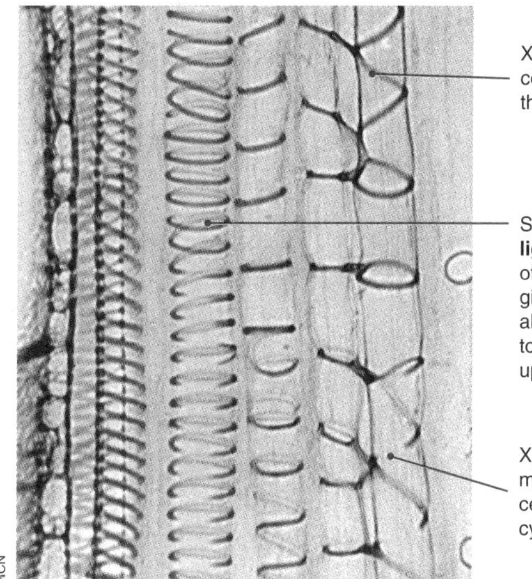
RCN
Xylem vessels form continuous tubes throughout the plant.
Spiral thickening of **lignin** around the walls of the vessel elements give extra strength allowing the vessels to remain rigid and upright.
Xylem is dead when mature. Note how the cells have lost their cytoplasm.

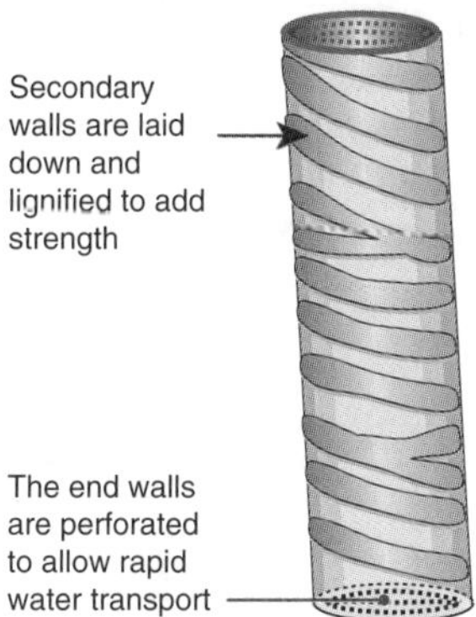
Vessel element
Secondary walls are laid down and lignified to add strength
The end walls are perforated to allow rapid water transport

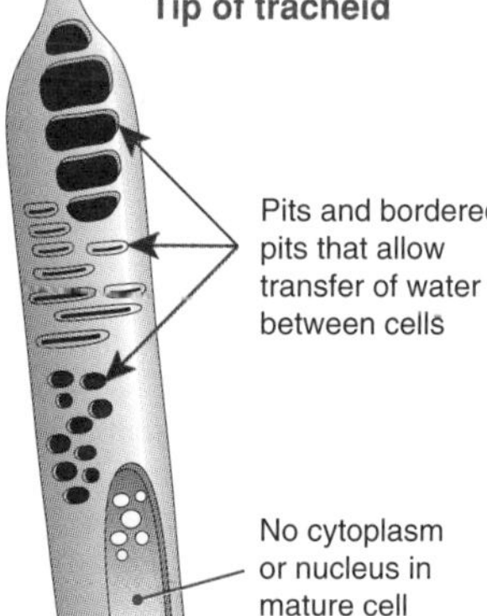
Tip of tracheid
Pits and bordered pits that allow transfer of water between cells
No cytoplasm or nucleus in mature cell

Vessel elements and tracheids are the two conducting cells types in xylem. Tracheids are long, tapering hollow cells. Water passes from one tracheid to another through thin regions in the wall called **pits**. Vessel elements have pits, but the end walls are also perforated and water flows unimpeded through the stacked elements.

The Structure of Xylem Tissue

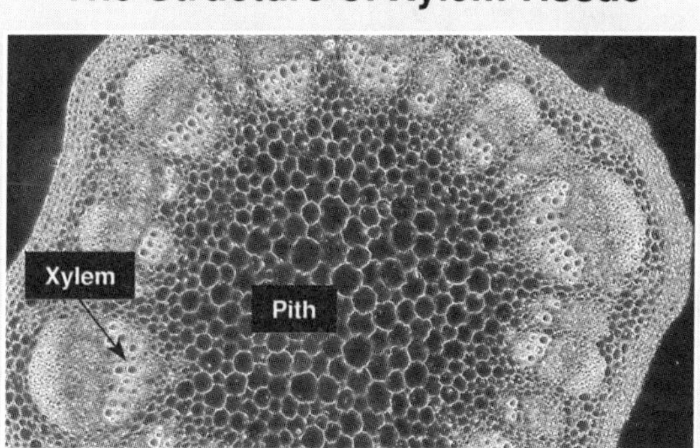
Xylem
Pith

This cross section through the stem, *Helianthus* (sunflower) shows the central pith, surrounded by a peripheral ring of vascular bundles. Note the xylem vessels with their thick walls.

Fibers are a type of sclerenchyma cell. They are associated with vascular tissues and usually occur in groups. The cells are very elongated and taper to a point and the cell walls are heavily thickened. Fibers give mechanical support to tissues, providing both strength and elasticity.

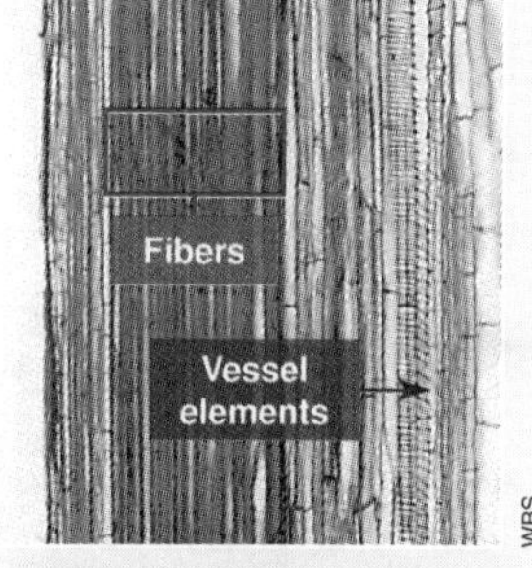
Fibers
Vessel elements
WBS

Vessel elements are found only in the xylem of angiosperms. They are large diameter cells that offer very low resistance to water flow. The possession of vessels (stacks of vessel elements) provides angiosperms with a major advantage over gymnosperms and ferns as they allow for very rapid water uptake and transport.

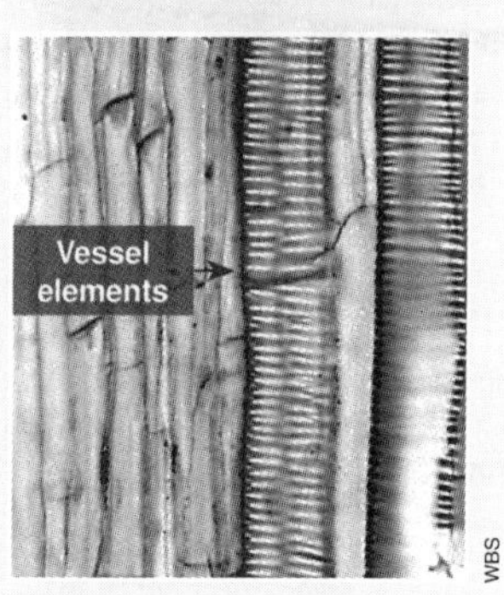
Vessel elements
WBS

1. Describe the function of **xylem**: ______________________________

____

2. Identify the four main cell types in xylem and explain their role in the tissue:

(a) ______________________________

(b) ______________________________

(c) ______________________________

(d) ______________________________

3. Describe one way in which xylem is strengthened in a mature plant: ______________________________

4. Describe a feature of vessel elements that increases their efficiency of function: ______________________________

____

ISBN: 978-1-927173-16-9

Phloem

Like xylem, **phloem** is a complex tissue, comprising a variable number of cell types. Phloem is the principal **food (sugar) conducting tissue** in vascular plants, transporting dissolved sugars around the plant. The bulk of phloem tissue comprises the **sieve tubes** (sieve tube members and sieve cells) and their companion cells. The sieve tubes are the principal conducting cells in phloem and are closely associated with the **companion cells** (modified parenchyma cells) with which they share a mutually dependent relationship. Other parenchyma cells, concerned with storage, occur in phloem, and strengthening fibers and sclereids (short sclerenchyma cells) may also be present. Unlike xylem, phloem is alive when mature.

LS through a sieve tube end plate

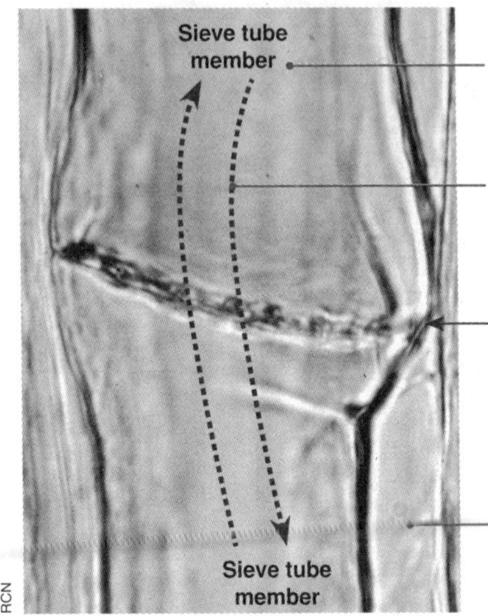
Sieve tube member
Sieve tube member
RCN

The sieve tube members lose most of their organelles but are still alive when mature

Sugar solution flows in both directions

Sieve tube end plate
Tiny holes (arrowed in the photograph below) perforate the sieve tube elements

Companion cell: a cell adjacent to the sieve tube member, responsible for keeping it alive

TS through a sieve tube end plate

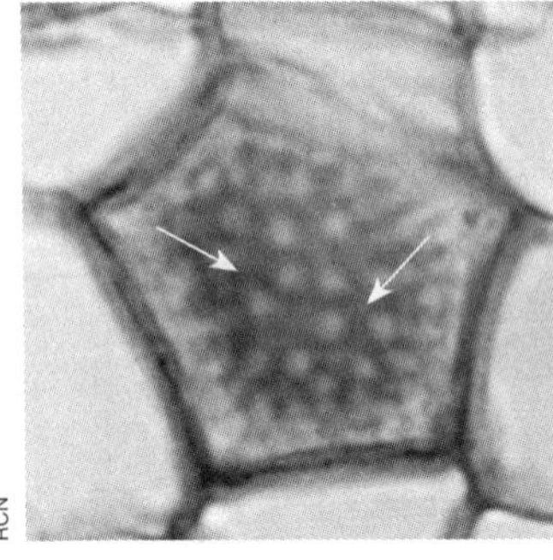
RCN

Adjacent sieve tube members are connected through **sieve plates** through which phloem sap flows.

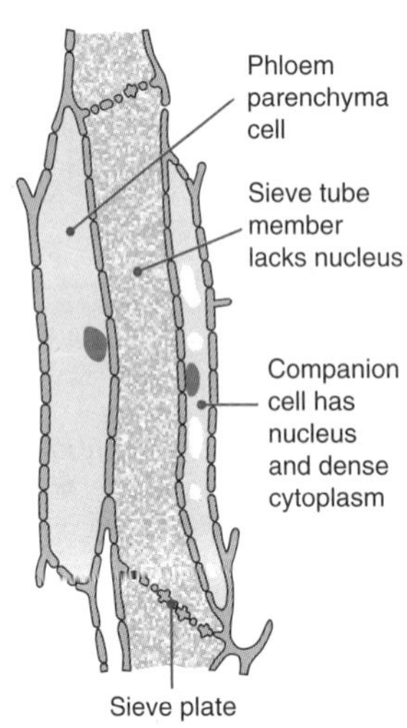
Phloem parenchyma cell
Sieve tube member lacks nucleus
Companion cell has nucleus and dense cytoplasm
Sieve plate

The Structure of Phloem Tissue

Phloem is alive at maturity and functions in the transport of sugars and minerals around the plant. Like xylem, it forms part of the structural vascular tissue of plants.

Fibers are associated with phloem as they are in xylem. Here they are seen in cross section where you can see the extremely thick cell walls and the way the fibers are clustered in groups. See the previous page for a view of fibers in longitudinal section.

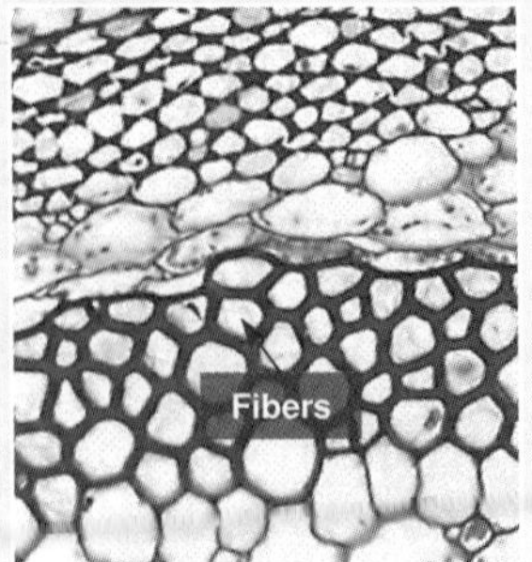
Fibers

In this cross section through a buttercup root, the smaller companion cells can be seen lying alongside the sieve tube members. It is the sieve tube members that, end on end, produce the **sieve tubes**. They are the conducting tissue of phloem.

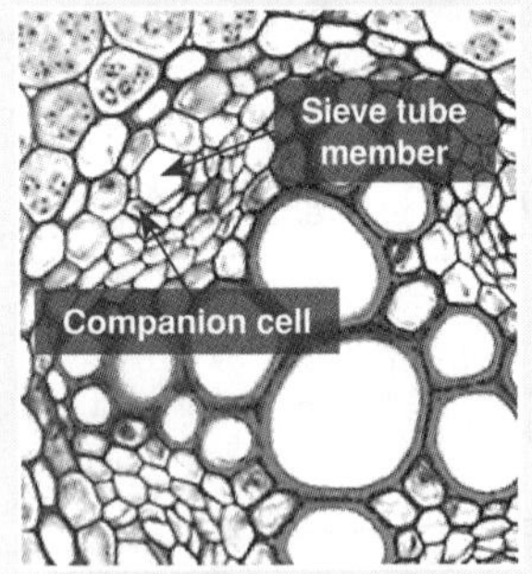
Sieve tube member
Companion cell

In this longitudinal section of a buttercup root, each sieve tube member has a thin **companion cell** associated with it. Companion cells retain their nucleus and control the metabolism of the sieve tube member next to them. They also have a role in the loading and unloading of sugar into the phloem.

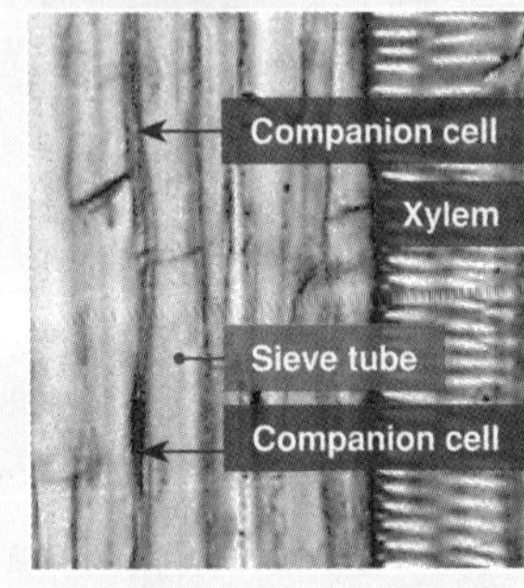
Companion cell
Xylem
Sieve tube
Companion cell

1. Describe the function of **phloem**: __________

2. Describe two differences between xylem and phloem: __________

3. Explain the purpose of the **sieve plate** at the ends of each sieve tube member: __________

4. (a) Name the conducting cell type in phloem: __________

   (b) Explain two roles of the companion cell in phloem: __________

5. State the purpose of the phloem parenchyma cells: __________

6. Identify a type of cell that provides strengthening in phloem: __________

ISBN: 978-1-927173-16-9

Tropisms and Growth Responses

Tropisms are plant growth responses to external stimuli, in which the stimulus direction determines the direction of the growth response. Tropisms are identified according to the stimulus involved, e.g. photo- (light), gravi- (gravity), hydro- (water), and may be positive or negative depending on whether the plant moves towards or away from the stimulus respectively.

(a) ....
A positive growth response to a chemical stimulus. *Example: Pollen tubes grow towards a chemical, possibly calcium ions, released by the ovule of the flower.*

(b) ....
Stems and coleoptiles grow away from the direction of the Earth's gravitational pull.

(c) ....
Growth response to water. Roots are influenced primarily by gravity but will also grow towards water.

(d) ....
Growth responses to light, particularly directional light. Coleoptiles, young stems, and some leaves show a positive response.

(e) ....
Roots respond positively to the Earth's gravitational pull, and curve downward after emerging through the seed coat.

(f) ....
Growth responses to touch or pressure. Tendrils (modified leaves) have a positive coiling response stimulated by touch.

Plant growth responses are adaptive in that they position the plant body in a suitable growing environment, within the limits of the position in which it germinated. They also tend to reinforce each other. For example, shoots grow away from gravity and towards the light.

Root mass in a hydroponically grown plant

Sweet pea tendrils

Germinating pollen

Thale cress bending to the light

Kristian Peters

1. Identify each of the plant tropisms described in (a)-(f) above. State whether the response is positive or negative.

2. Define the term **tropism** and explain the adaptive value of tropic responses to a plant in its environment:

3. Describe the adaptive value of the following tropisms:

(a) Positive gravitropism in roots:

(b) Positive phototropism in coleoptiles:

(c) Positive thigmomorphogenesis in weak stemmed plants:

(d) Positive chemotropism in pollen grains:

ISBN: 978-1-927173-16-9

Related activities: *Investigating Phototropism*
Weblinks: *Plants in Motion*

Transport and Effects of Auxins

Auxins are **phytohormones** (plant growth substances) that have a central role in a wide range of growth and developmental responses in vascular plants. **Indole-3-acetic acid** (IAA) is the most potent native auxin in intact plants. It was the first discovered and is the most studied. Although its actions on various aspects of plant growth are well known, it most commonly acts in concert with (or in opposition to) other phytohormones, especially cytokinins and gibberellins. The response of any particular plant tissue to IAA depends on the tissue itself, the concentration of the hormone, the timing of its release, and the presence of other phytohormones. For example, in undifferentiated tissue, the application of IAA and cytokinin in equal concentrations promotes xylem development, but a relatively higher concentration of auxin promotes rooting. Gradients in auxin concentration during growth prompt differential responses in specific tissues and contribute to the plant's organ development and directional growth.

AUXINS IN PLANTS:

The most important auxin produced by plants is indole-3-acetic acid (IAA). It plays important roles in a number of plant activities, although its action is often influenced by the presence of other plant growth factors. Auxin is important in:

- apical dominance
- positive phototropic response in stems
- positive gravitropic response in roots
- development of the embryo
- promoting cell elongation and growth in stem length
- promoting cell enlargement and differentiation in cambium (growth of secondary vascular tissues)
- leaf formation and fruit development
- root initiation and development
- delaying the onset of senescence. Young leaves and fruit produce auxins and while they do so, they remain attached to the stem and abscission is inhibited.

PRODUCTION OF AUXIN:

Auxin is synthesized in the meristematic tissues, mainly the root tips and shoot tips, but also in young leaves and flowers.

HOW AUXIN MOVES IN PLANTS:

Synthesis of auxin is not always the site of action so it must be moved to other locations. Auxin moves through the plant by two mechanisms:

- through the phloem (in the sap) and the xylem (in the transpiration stream)
- from cell to cell by diffusion but also via membrane transporters

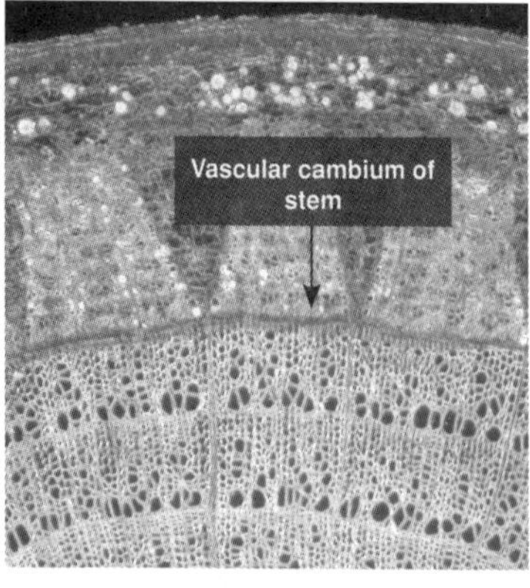
Vascular cambium of stem

Auxin promotes the activity of the vascular cambium (above) and the differentiation of xylem and phloem. Other growth regulators (e.g. cytokinins) influence this by increasing the sensitivity of the tissues to IAA.

Jiron (John O'Neill)

Auxin delays fruit senescence and is required for fruit growth. As the seeds mature, they release auxin, which diffuses to the surrounding flower parts, which develop into the fruit covering the seeds.

NASA

Auxins are responsible for apical dominance in shoots. Auxin is produced in the shoot tip and diffuses down to inhibit the development of the lateral buds.

Left: Auxin promotes growth in stem length. The effect is stronger if gibberellins are also present. This photograph shows a healthy Arabidopsis plant, next to a stunted auxin signal-transduction mutant.

1. Explain the role of auxin (IAA) in the following plant growth processes:

   (a) Apical dominance: ______

   (b) Stem growth: ______

   (c) Secondary growth: ______

2. Why does pruning (removing the central leader) induce bushy growth in plants? ______

3. How is auxin (IAA) able to bring about quite different responses in different plant tissues? ______

ISBN: 978-1-927173-16-9

Investigating Phototropism

Phototropism in plants was linked to a growth promoting substance in the 1920s. A number of classic experiments, investigating phototropic responses in severed coleoptiles, gave evidence for the hypothesis that auxin was responsible for tropic responses in stems. Auxins promote cell elongation. Stem curvature in response to light can therefore result from the differential distribution of auxin either side of a stem. However, the mechanisms of hormone action in plants are still not well understood. Auxins increase cell elongation only over a certain concentration range. At certain levels, auxins stop inducing elongation and begin to inhibit it. There is *some* experimental evidence that contradicts the original auxin hypothesis and the early experiments have been criticized for oversimplifying the real situation. Outlined below are some experiments that investigate plant responses to light, and the role of hormone(s) in controlling these (also see following activities on phytohormones).

1. **Directional light:** A pot plant is exposed to direct sunlight near a window and as it grows, the shoot tip turns in the direction of the sun. If the plant was rotated, it adjusted by growing towards the sun in the new direction.

   (a) What hormone regulates this growth response?

   ____

   (b) What is the name of this growth response?

   ____

   (c) How do the cells behave to bring about this change in shoot direction at:

   Point **A**? ______________________________

   Point **B**? ______________________________

   (d) Which side (A or B) would have the highest concentration of hormone?

   ____

   (e) Draw a diagram of the cells as they appear across the stem from point A to B (in the rectangle on the right).

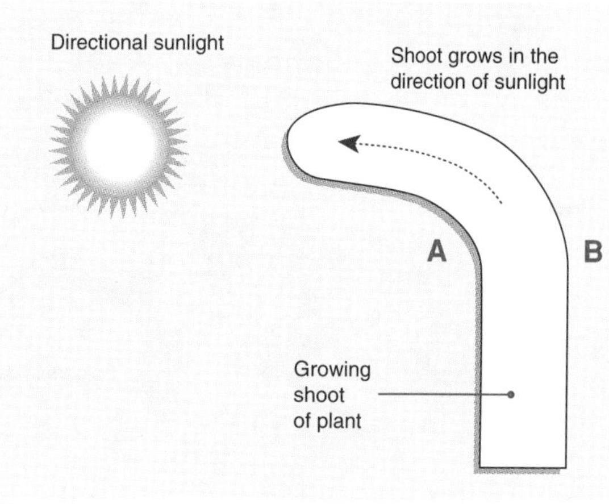
Directional sunlight
Shoot grows in the direction of sunlight
A
B
Growing shoot of plant

Draw your cells here:

2. **Light excluded from shoot tip:** With a tin foil cap placed over the top of the shoot tip, light is prevented from reaching it. When growing under these conditions, the direction of growth does not change towards the light source, but grows straight up. State what conclusion can you come to about the source and activity of the hormone that controls the growth response:

   ____

   ____

   ____

   ____

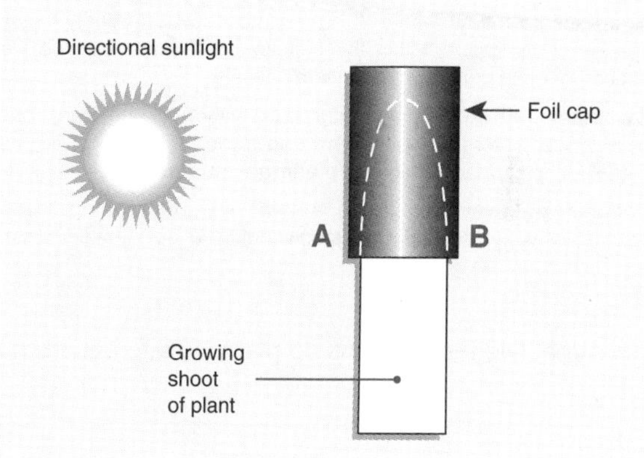
Directional sunlight
Foil cap
A
B
Growing shoot of plant

3. **Cutting into the transport system:** Two identical plants were placed side-by-side and subjected to the same directional light source. Razor blades were cut half-way into the stem, thereby interfering with the transport system of the stem. Plant A had the cut on the same side as the light source, while Plant B was cut on the shaded side. Predict the growth responses of:

   Plant **A**: ______________________________

   ____

   ____

   Plant **B**: ______________________________

   ____

   ____

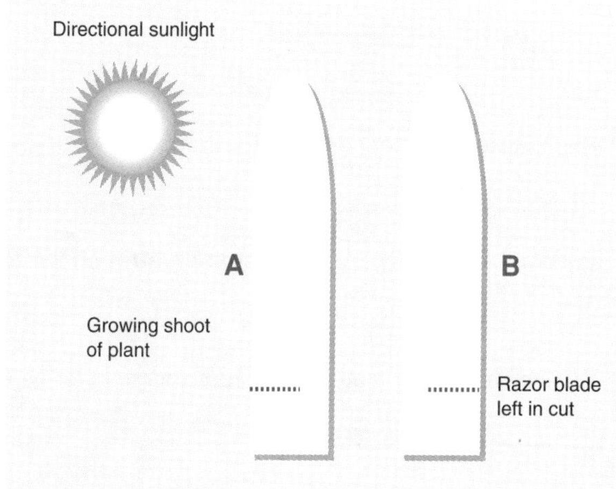
Directional sunlight
A
B
Growing shoot of plant
Razor blade left in cut

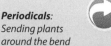

Uptake at the Root

Plants need to take up water and minerals constantly. They must compensate for the loss of water from the leaves and provide the materials they need for the manufacture of food. The uptake of water and minerals is mostly restricted to the younger, most recently formed cells of the roots and the root hairs. Water uptake is a passive process and most occurs through the free spaces outside the plasma membranes. Mineral ions move through the soil to the plant root by mass flow (dissolved in water) and by diffusion. Most plants also have mutualistic relationships with fungi, which are efficient at absorbing mineral ions and make some available to the plant root. Uptake of the mineral ions by the root tissue itself occurs by passive or by active transport.

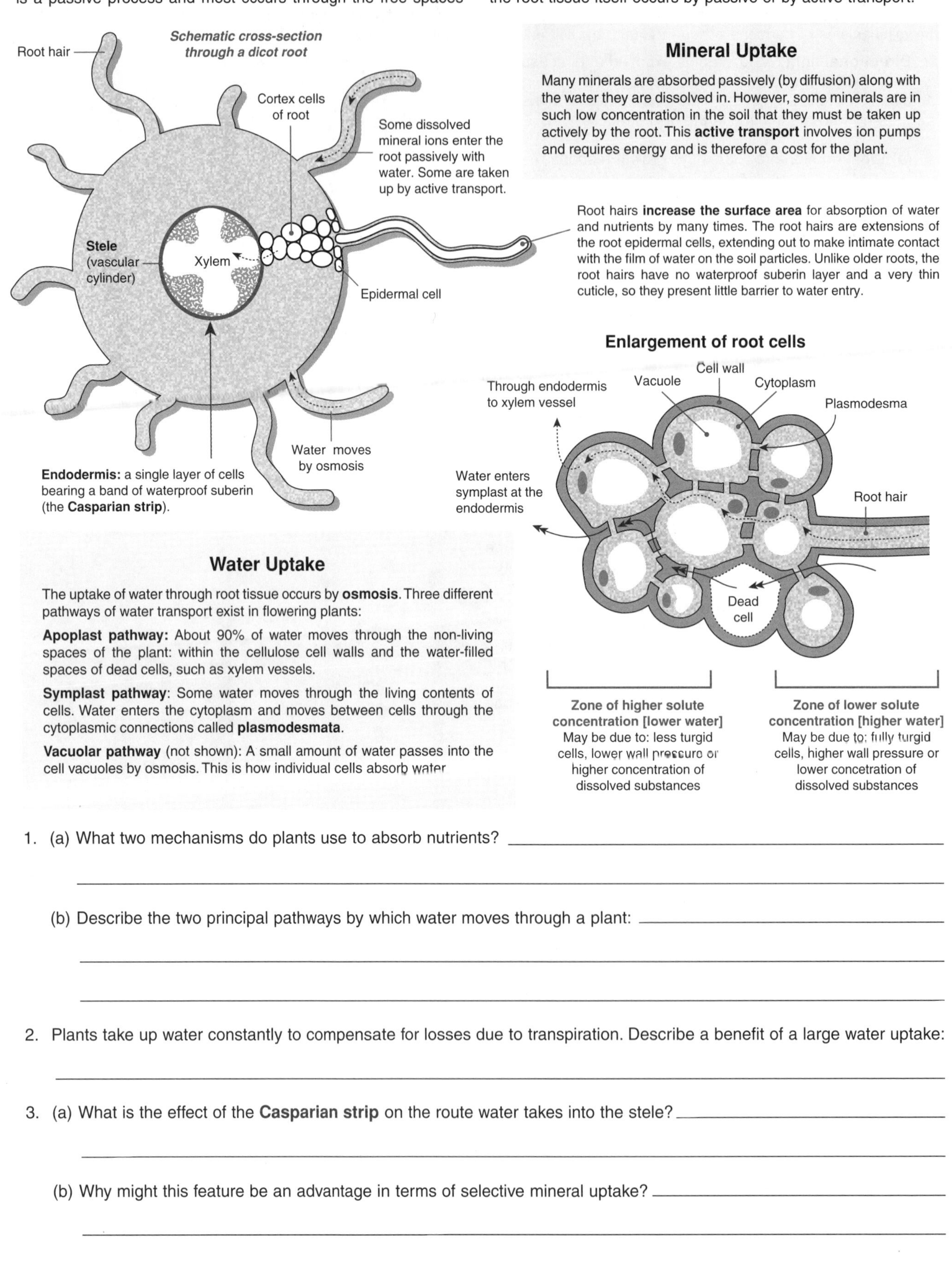

Schematic cross-section through a dicot root
Root hair
Cortex cells of root
Some dissolved mineral ions enter the root passively with water. Some are taken up by active transport.
Stele (vascular cylinder)
Xylem
Epidermal cell
Water moves by osmosis
Endodermis: a single layer of cells bearing a band of waterproof suberin (the Casparian strip).
Root hairs increase the surface area for absorption of water and nutrients by many times. The root hairs are extensions of the root epidermal cells, extending out to make intimate contact with the film of water on the soil particles. Unlike older roots, the root hairs have no waterproof suberin layer and a very thin cuticle, so they present little barrier to water entry.
Enlargement of root cells
Cell wall
Vacuole
Cytoplasm
Plasmodesma
Through endodermis to xylem vessel
Water enters symplast at the endodermis
Root hair
Dead cell
Zone of higher solute concentration [lower water]
May be due to: less turgid cells, lower wall pressure or higher concentration of dissolved substances
Zone of lower solute concentration [higher water]
May be due to: fully turgid cells, higher wall pressure or lower concetration of dissolved substances

Mineral Uptake

Many minerals are absorbed passively (by diffusion) along with the water they are dissolved in. However, some minerals are in such low concentration in the soil that they must be taken up actively by the root. This **active transport** involves ion pumps and requires energy and is therefore a cost for the plant.

Water Uptake

The uptake of water through root tissue occurs by **osmosis**. Three different pathways of water transport exist in flowering plants:

Apoplast pathway: About 90% of water moves through the non-living spaces of the plant: within the cellulose cell walls and the water-filled spaces of dead cells, such as xylem vessels.

Symplast pathway: Some water moves through the living contents of cells. Water enters the cytoplasm and moves between cells through the cytoplasmic connections called **plasmodesmata**.

Vacuolar pathway (not shown): A small amount of water passes into the cell vacuoles by osmosis. This is how individual cells absorb water

1. (a) What two mechanisms do plants use to absorb nutrients? ______

   (b) Describe the two principal pathways by which water moves through a plant: ______

2. Plants take up water constantly to compensate for losses due to transpiration. Describe a benefit of a large water uptake: ______

3. (a) What is the effect of the **Casparian strip** on the route water takes into the stele? ______

   (b) Why might this feature be an advantage in terms of selective mineral uptake? ______

***Related activities:** Dicot Stems and Roots*
***Weblinks:** Water Uptake in Plants, Mineral Uptake in Roots*

ISBN: 978-1-927173-16-9

Transpiration

Plants lose water all the time, despite the adaptations they have to help prevent it (e.g. waxy leaf cuticle). Approximately 99% of the water a plant absorbs from the soil is lost by evaporation from the leaves and stem. This loss, mostly through stomata, is called **transpiration** and the flow of water through the plant is called the **transpiration stream**. Plants rely on a gradient in solute concentration from the roots to the air to move water through their cells. Water flows passively from soil to air along a gradient of increasing solute (decreasing water) concentration. This gradient is the driving force in the ascent of water up a plant. A number of processes contribute to water movement up the plant: transpiration pull, cohesion, and root pressure. Transpiration may seem wasteful, but it has benefits; evaporative water loss cools the plant and the transpiration stream helps the plant to maintain an adequate mineral uptake, as many essential minerals occur in low concentrations in the soil.

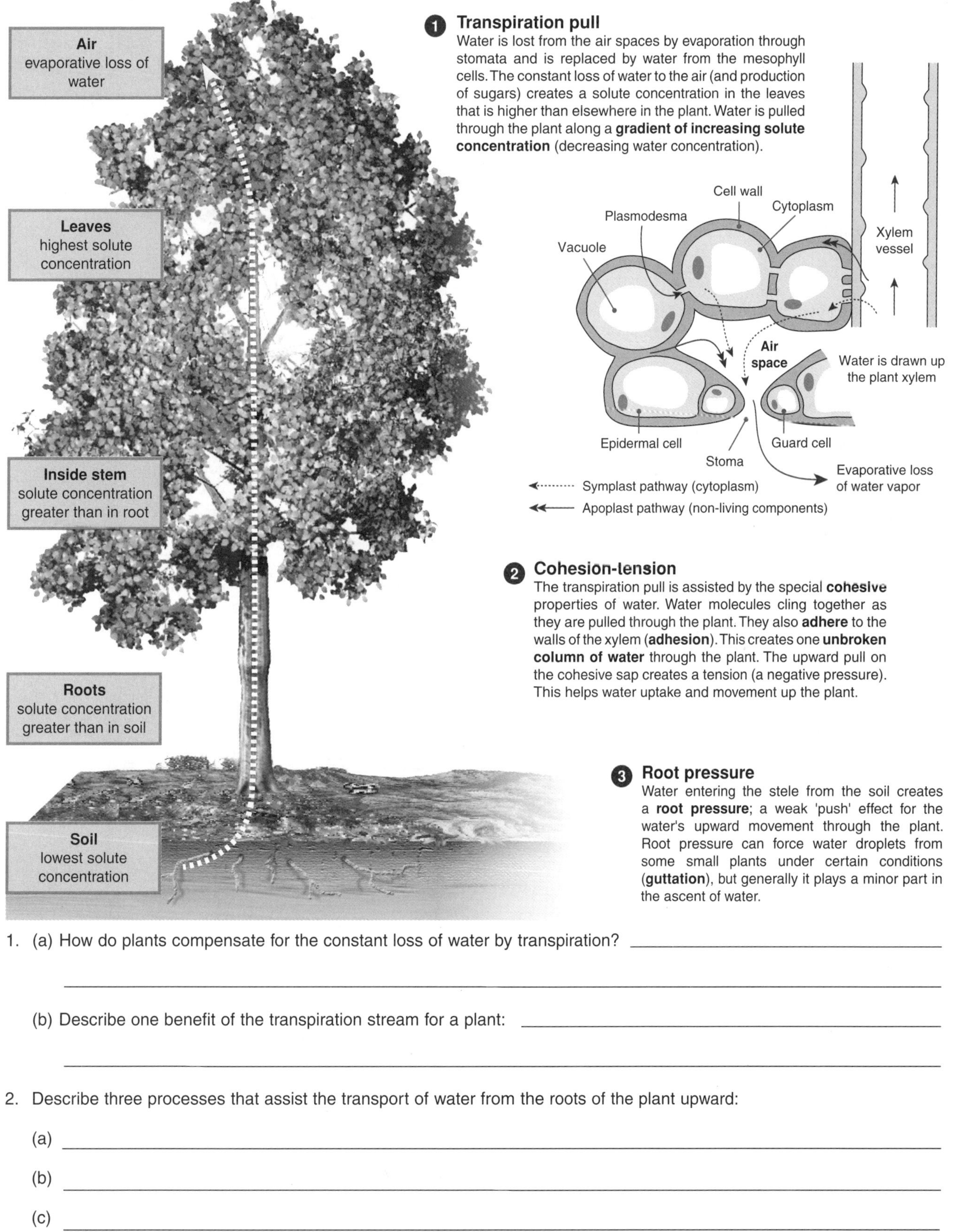
Air
evaporative loss of water
Leaves
highest solute concentration
Inside stem
solute concentration greater than in root
Roots
solute concentration greater than in soil
Soil
lowest solute concentration
Cell wall
Cytoplasm
Plasmodesma
Vacuole
Xylem vessel
Air space
Water is drawn up the plant xylem
Epidermal cell
Guard cell
Stoma
Evaporative loss of water vapor
Symplast pathway (cytoplasm)
Apoplast pathway (non-living components)

1 Transpiration pull

Water is lost from the air spaces by evaporation through stomata and is replaced by water from the mesophyll cells. The constant loss of water to the air (and production of sugars) creates a solute concentration in the leaves that is higher than elsewhere in the plant. Water is pulled through the plant along a **gradient of increasing solute concentration** (decreasing water concentration).

2 Cohesion-tension

The transpiration pull is assisted by the special **cohesive** properties of water. Water molecules cling together as they are pulled through the plant. They also **adhere** to the walls of the xylem (**adhesion**). This creates one **unbroken column of water** through the plant. The upward pull on the cohesive sap creates a tension (a negative pressure). This helps water uptake and movement up the plant.

3 Root pressure

Water entering the stele from the soil creates a **root pressure**; a weak 'push' effect for the water's upward movement through the plant. Root pressure can force water droplets from some small plants under certain conditions (**guttation**), but generally it plays a minor part in the ascent of water.

1. (a) How do plants compensate for the constant loss of water by transpiration? ______________________

____

(b) Describe one benefit of the transpiration stream for a plant: ______________________

____

2. Describe three processes that assist the transport of water from the roots of the plant upward:

(a) ____

(b) ____

(c) ____

ISBN: 978-1-927173-16-9

Periodicals:
How trees lift water, High tension

***Related activities:** Uptake at the Root, Investigating Transpiration*
***Weblinks:** Transpiration Animation*

The Potometer

A potometer is a simple instrument for investigating transpiration rate (water loss per unit time). The equipment is simple and easy to obtain. A basic potometer, such as the one shown right, can easily be moved around so that transpiration rate can be measured under different environmental conditions.

Some of the physical conditions investigated are:

- Humidity or vapor pressure (high or low)
- Temperature (high or low)
- Air movement (still or windy)
- Light level (high or low)
- Water supply

It is also possible to compare the transpiration rates of plants with different adaptations e.g. comparing transpiration rates in plants with rolled leaves vs rates in plants with broad leaves. If possible, experiments like these should be conducted simultaneously using replicate equipment. If conducted sequentially, care should be taken to keep the environmental conditions the same for all plants used.

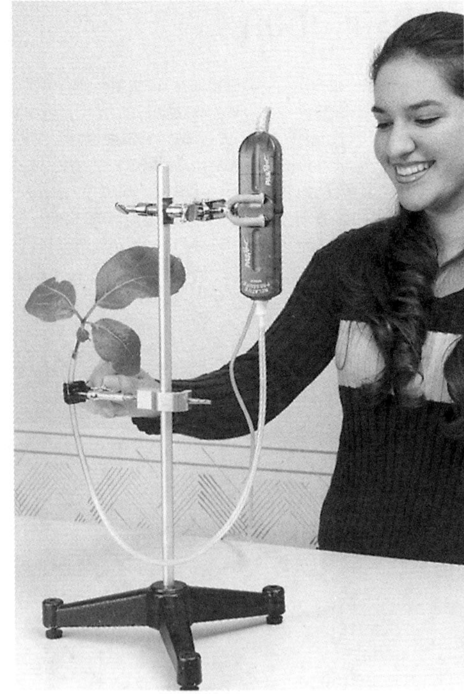

3. Describe three environmental conditions that increase the rate of transpiration in plants, explaining how they operate:

(a) ______________________________

(b) ______________________________

(c) ______________________________

4. The **potometer** (above) is an instrument used to measure transpiration rate. Briefly explain how it works:

____

____

5. An experiment was conducted on transpiration from a hydrangea shoot in a potometer. The experiment was set up and the plant left to stabilize (environmental conditions: still air, light shade, 20°C). The plant was then subjected to different environmental conditions and the water loss was measured each hour. Finally, the plant was returned to original conditions, allowed to stabilize and transpiration rate measured again. The data are presented below:

Experimental conditions	Temperature (°C)	Humidity (%)	Transpiration rate (gh^{-1})
(a) Still air, light shade, room temperature	20°C	70	1.20
(b) Moving air, light shade	20°C	70	1.60
(c) Still air, bright sunlight	23°C	70	3.75
(d) Still air and dark, moist chamber	19.5°C	100	0.05

(a) Identify the control in this experiment: ______________________________

____

(b) State which factors increased transpiration rate, explaining how each has its effect: ______________________________

____

____

(c) Why did the plant have such a low transpiration rate in humid, dark conditions? ______________________________

____

____

____

____

ISBN: 978-1-927173-16-9

Investigating Plant Transpiration

The relationship between the rate of transpiration and the environment can be investigated using a potometer. It is useful to tabulate and graph the data from such an experiment. Graphs and tables display data in a way that makes it easy to see trends or relationships between different variables. This activity describes a plant transpiration experiment and provides data on the effect of four different conditions on transpiration rate. Guidelines for drawing the appropriate graphs are provided.

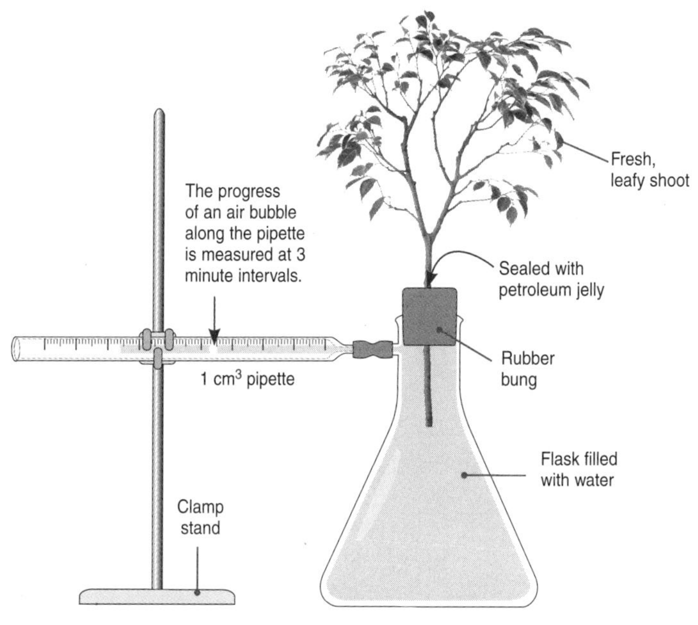
The progress of an air bubble along the pipette is measured at 3 minute intervals.
Fresh, leafy shoot
Sealed with petroleum jelly
Rubber bung
1 cm³ pipette
Flask filled with water
Clamp stand

The Apparatus

This experiment investigated the influence of environmental conditions on plant transpiration rate. Four conditions were studied: room conditions (ambient), wind, bright light, and high humidity. After setting up the potometer, the apparatus was equilibrated for 10 minutes, and the position of the air bubble in the pipette was recorded. This is the time 0 reading. The plant was then exposed to one of the environmental conditions. Students recorded the location of the air bubble every three minutes over a 30 minute period. The potometer readings for each environmental condition are presented in Table 1 (next page).

The Aim

To investigate the effect of environmental conditions on the transpiration rate of plants.

Background

Plants lose water all the time by evaporation from the leaves and stem. This loss, mostly through pores in the leaf surfaces, is called **transpiration**. Despite the adaptations plants have to help prevent water loss (e.g. waxy leaf cuticle), 99% of the water a plant absorbs from the soil is lost by evaporation. Environmental conditions can affect transpiration rate.

A class was divided into four groups to study how four different environmental conditions (ambient, wind, bright light, and high humidity) affected transpiration rate. A **potometer** was used to measure transpiration rate (water loss per unit time). A basic potometer, such as the one shown left, can easily be moved around so that transpiration rate can be measured under different environmental conditions.

Guidelines for Drawing Line Graphs

Line graphs are used when one variable (the independent variable) affects another, the dependent variable.

Fig. 1: Cumulative water loss in μL from a geranium shoot in still and moving air.

Graphs (called figures) should have a concise, explanatory title. If several graphs appear in your report they should be numbered consecutively.

A key identifies symbols. This information sometimes appears in the title.

Still air
Moving air

Volume of water loss (μL)
20 40 60 80 100 120 140 160 180 200 220

Time (s)
0 30 60 90 120 150 180

Plot points accurately. Different responses can be distinguished using different symbols, lines or bar colors.

Label both axes and provide appropriate units of measurement if necessary.

Two or more sets of results can be plotted on the same figure and distinguished by a key. For time series, it is appropriate to join the plotted points with a line.

Place the dependent variable e.g. biological response, on the vertical (Y) axis (if you are drawing a scatter graph it does not matter).

Each axis should have an appropriate scale. Decide on the scale by finding the maximum and minimum values for each variable.

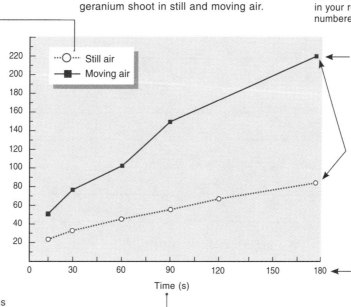

NOTE: The data must be continuous for both variables.

Place the independent variable e.g. time or treatment, on the horizontal (X) axis

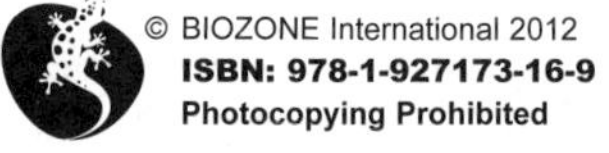

ISBN: 978-1-927173-16-9

Table 1. Potometer readings

Treatment \ Time (min)	0	3	6	9	12	15	18	21	24	27	30
Ambient	0	0.002	0.005	0.008	0.012	0.017	0.022	0.028	0.032	0.036	0.042
Wind	0	0.025	0.054	0.088	0.112	0.142	0.175	0.208	0.246	0.283	0.325
High humidity	0	0.002	0.004	0.006	0.008	0.011	0.014	0.018	0.019	0.021	0.024
Bright light	0	0.021	0.042	0.070	0.091	0.112	0.141	0.158	0.183	0.218	0.239

1. (a) Plot the potometer data from Table 1 on the grid provided. Use the guidelines for drawing line graphs on the previous page as a reference if you need help:

   (b) Identify the independent variable: ______

2. (a) Identify the control: ______

   (b) Explain the purpose of including an experimental control in an experiment: ______

   (c) Which factors increased water loss? ______

   (d) How does each environmental factor influence water loss? ______

   (e) Explain why the plant lost less water in humid conditions: ______

ISBN: 978-1-927173-16-9

Xerophytes

Plants adapted to dry conditions are called **xerophytes** and they show structural (xeromorphic) and physiological adaptations for water conservation. These typically include small, hard leaves, and epidermis with a thick cuticle, sunken stomata, succulence, and permanent or temporary absence of leaves. Xerophytes may live in humid environments, provided that their roots are in dry micro-environments (e.g. the roots of epiphytic plants that grow on tree trunks or branches). The nature of the growing environment is important in many other situations too. **Halophytes** (salt tolerant plants) and alpine species may show xeromorphic features in response to the scarcity of obtainable water and high water losses in these environments.

Leaves modified into spines or hairs to reduce water loss. Light coloured spines reflect solar radiation.

Squat, rounded shape reduces surface area. The surface tissues of many cacti are tolerant of temperatures in excess of 50°C.

Shallow, but extensive fibrous root system.

Stem becomes the major photosynthetic organ, plus a reservoir for water storage.

Water table low

Dry Desert Plant

Desert plants, such as cacti, must cope with low or sporadic rainfall and high transpiration rates. A number of structural adaptations (diagram left) reduce water losses, and enable them to access and store available water. Adaptations such as waxy leaves also reduce water loss and, in many desert plants, germination is triggered only by a certain quantity of rainfall.

Acacia trees have **deep root systems**, allowing them to draw water from lower water table systems.

Hairs

The outer surface of many succulents are coated in fine hairs, which traps air close to the surface reducing transpiration rate.

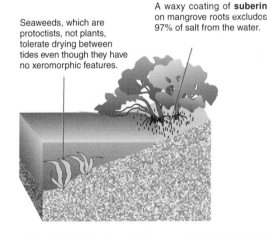

A waxy coating of **suberin** on mangrove roots excludes 97% of salt from the water.

Seaweeds, which are protoctists, not plants, tolerate drying between tides even though they have no xeromorphic features.

Ocean Margin Plant

Land plants that colonize the shoreline must have adaptations to obtain water from their saline environment while maintaining their osmotic balance. In addition, the shoreline is often a windy environment, so they frequently show xeromorphic adaptations that enable them to reduce water losses.

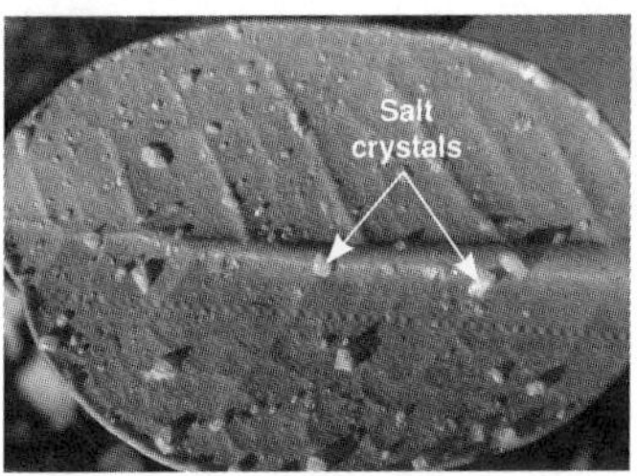
Salt crystals

To maintain osmotic balance, mangroves can secrete absorbed salt as salt crystals (above), or accumulate salt in old leaves which are subsequently shed.

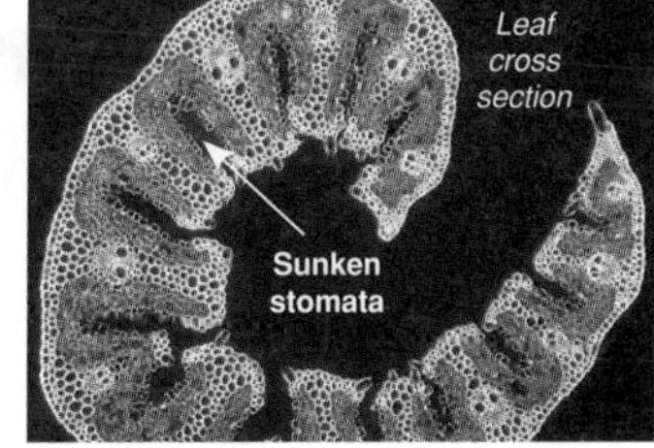
Leaf cross section
Sunken stomata

Grasses found on shoreline coasts (where it is often windy), curl their leaves and have sunken stomata to reduce water loss by transpiration.

Methods of water conservation in various plant species

Adaptation for water conservation	Effect of adaptation	Example
Thick, waxy cuticle to stems and leaves	Reduces water loss through the cuticle.	*Pinus* sp. ivy (*Hedera*), sea holly (*Eryngium*), prickly pear (*Opuntia*).
Reduced number of stomata	Reduces the number of pores through which water loss can occur.	Prickly pear (*Opuntia*), *Nerium* sp.
Stomata sunken in pits, grooves, or depressions Leaf surface covered with fine hairs Massing of leaves into a rosette at ground level	Moist air is trapped close to the area of water loss, reducing the diffusion gradient and therefore the rate of water loss.	**Sunken stomata**: *Pinus* sp., *Hakea* sp. Hairy leaves: lamb's ear. **Leaf rosettes**: dandelion (*Taraxacum*), daisy.
Stomata closed during the light, open at night	CAM metabolism: CO_2 is fixed during the night, water loss in the day is minimized.	**CAM plants**, e.g. American aloe, pineapple, *Kalanchoe, Yucca*.
Leaves reduced to scales, stem photosynthetic Leaves curled, rolled, or folded when flaccid	Reduction in surface area from which transpiration can occur.	**Leaf scales**: broom (*Cytisus*). **Rolled leaf**: marram grass (*Ammophila*), *Erica* sp.
Fleshy or succulent stems Fleshy or succulent leaves	When readily available, water is stored in the tissues for times of low availability.	**Fleshy stems**: *Opuntia*, candle plant (*Kleinia*). **Fleshy leaves**: *Bryophyllum*.
Deep root system below the water table	Roots tap into the lower water table.	Acacias, oleander.
Shallow root system absorbing surface moisture	Roots absorb overnight condensation.	Most cacti.

ISBN: 978-1-927173-16-9

Periodicals:
Cacti

Related activities: *Transpiration, Modifications in Plants*
Weblinks: *Some Adaptations to Habitats, Desert Plant Survival*

Adaptations in halophytes and drought tolerant plants

RA

Ice plant (*Carpobrotus*): The leaves of many desert and beach dwelling plants are fleshy or succulent. The leaves are triangular in cross section and crammed with water storage cells. The water is stored after rain for use in dry periods. The shallow root system is able to take up water from the soil surface, taking advantage of any overnight condensation.

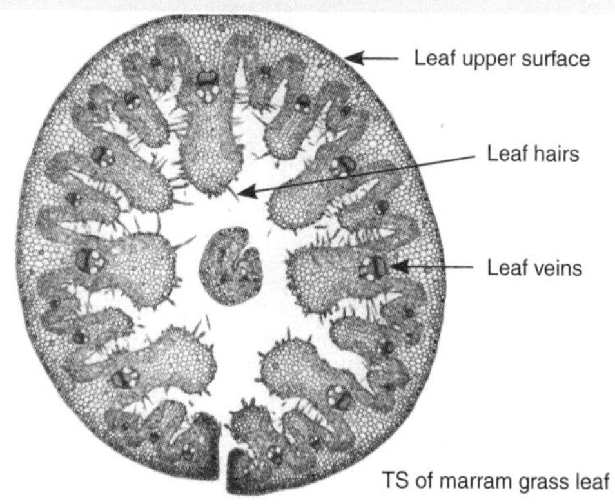
Leaf upper surface
Leaf hairs
Leaf veins
TS of marram grass leaf
RCN

Marram grass (*Ammophila*): The long, wiry leaf blades of this beach grass are curled downwards with the stomata on the inside. This protects them against drying out by providing a moist microclimate around the stomata. Plants adapted to high altitude often have similar adaptations.

RA

Ball cactus (*Echinocactus grusonii*): In many cacti, the leaves are modified into long, thin spines which project outward from the thick fleshy stem. This reduces the surface area over which water loss can occur. The stem stores water and takes over as the photosynthetic organ. As in succulents, a shallow root system enables rapid uptake of surface water.

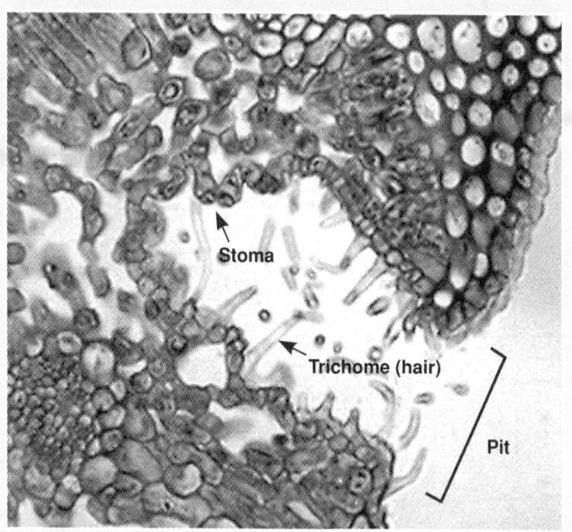
Stoma
Trichome (hair)
Pit
WBS

Oleander is a xerophyte from the Mediterranean region with many water conserving features. It has a thick multi-layered epidermis and the stomata are sunken in trichome-filled pits on the leaf underside. The pits restrict water loss to a greater extent than they reduce uptake of carbon dioxide.

1. Explain the purpose of **xeromorphic** adaptations: ____________________

2. Describe three xeromorphic adaptations of plants:

   (a) ____________________

   (b) ____________________

   (c) ____________________

3. Describe a physiological mechanism by which plants can reduce water loss during the daylight hours:

4. How does creating a moist microenvironment around the areas of water loss reduce transpiration rate?

5. Why do seashore plants (halophytes) exhibit many xeromorphic features? ____________________

ISBN: 978-1-927173-16-9

Translocation

Phloem transports the organic products of photosynthesis (sugars) through the plant in a process called **translocation**. In angiosperms, the sugar moves through the sieve-tube members, which are arranged end-to-end and perforated with sieve plates. Apart from water, phloem sap comprises mainly sucrose (up to 30%). It may also contain minerals, hormones, and amino acids, in transit around the plant. Movement of sap in the phloem is from a **source** (a plant organ where sugar is made or mobilized) to a **sink** (a plant organ where sugar is stored or used). Loading sucrose into the phloem at a source involves energy expenditure; it is slowed or stopped by high temperatures or respiratory inhibitors. In some plants, unloading the sucrose at the sinks also requires energy, although in others, diffusion alone is sufficient to move sucrose from the phloem into the cells of the sink organ.

Phloem sap moves from source to sink at rates as great as 100 m h^{-1}: too fast to be accounted for by cytoplasmic streaming. The most acceptable model for phloem movement is the **pressure-flow** (bulk flow) hypothesis. Phloem sap moves by bulk flow, which creates a pressure (hence the term "pressure-flow"). The key elements in this model are outlined below and in steps 1-4 at right. Note that, for simplicity, the cells that lie between the source (and sink) cells and the phloem sieve-tube have been omitted.

1. Loading sugar into the phloem increases the solute concentration inside the sieve-tube cells. This causes the sieve-tubes to take up water by osmosis.
2. The water uptake creates a hydrostatic pressure that forces the sap to move along the tube, just as pressure pushes water through a hose.
3. The pressure gradient in the sieve tube is reinforced by the active unloading of sugar and consequent loss of water by osmosis at the sink (e.g. root cell).
4. Xylem recycles the water from sink to source.

DH

Measuring Phloem Flow

Experiments investigating flow of phloem often use aphids. Aphids feed on phloem sap (left) and act as natural **phloem probes**. When the mouthparts (stylet) of an aphid penetrate a sieve-tube cell, the pressure in the sieve-tube force-feeds the aphid. While the aphid feeds, it can be severed from its stylet, which remains in place in the phloem. The stylet serves as a tiny tap that exudes sap. Using different aphids, the rate of flow of this sap can be measured at different locations on the plant.

Phloem Transport

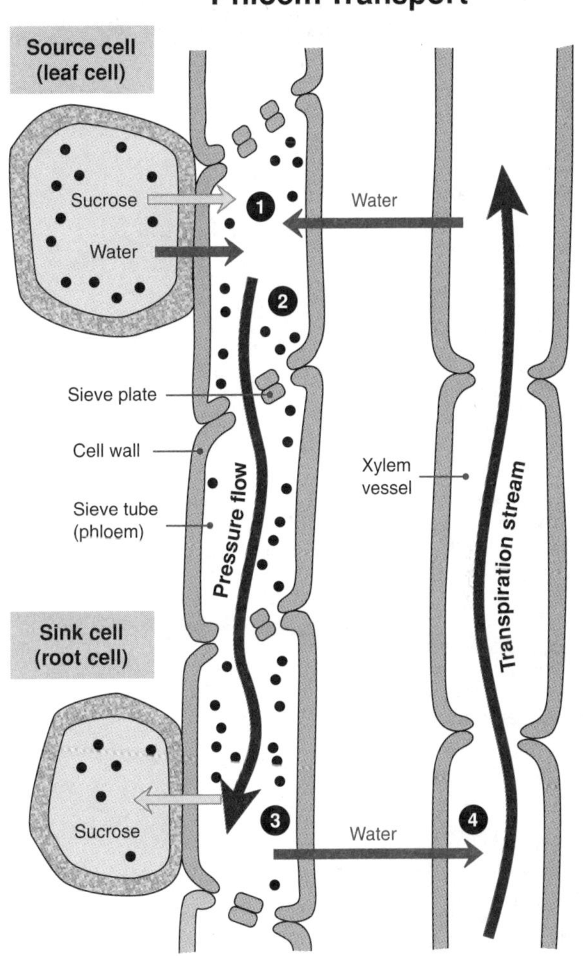
Source cell (leaf cell)
Sucrose
Water
1
Water
2
Sieve plate
Cell wall
Sieve tube (phloem)
Pressure flow
Xylem vessel
Transpiration stream
Sink cell (root cell)
Sucrose
3
Water
4

Source: Modified after Campbell *Biology* 1993

1. (a) From what you know about osmosis, explain why water follows the sugar as it moves through the phloem:

   (b) What is meant by '**source to sink**' flow in phloem transport?

2. Why does a plant need to move food around, particularly from the leaves to other regions?

3. Mature phloem is a live tissue, whereas xylem (the water transporting tissue) is dead when mature. Why is it necessary for phloem to be alive to be functional, whereas xylem can function as a dead tissue?

4. Why do non-vascular plants (mosses and liverworts) manage without specialized tissues to transport sugars?

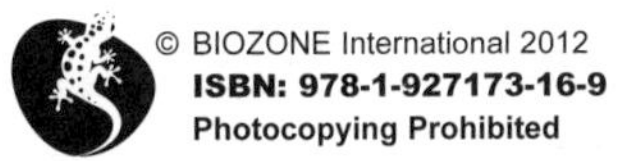

ISBN: 978-1-927173-16-9

Periodicals: *High tension*

Related activities: Xylem, Phloem, Ion Pumps
Weblinks: Sucrose Transport, Sucrose Transport in the Phloem

Insect Pollinated Flowers

Flowering plants (**angiosperms**) are highly successful organisms. The egg cell is retained within the flower of the parent plant and the male gametes (contained in the **pollen**) must be transferred to it by **pollination** in order for fertilization to occur. Most angiosperms are **monoecious**, with male and female parts on the same plant. Some of these plants will self-pollinate, but most have mechanisms that make this difficult or impossible. The female and male parts may be physically separated in the flower, or they may mature at different times. **Dioecious plants** avoid this problem by carrying the male and female flowers on separate plants. Flowers are pollinated in three different ways (animal, wind or water) and their structures differ accordingly. Of the animal pollinators, insects provide the greatest effectiveness of pollination as well as the most specialized pollination. Flowers attract insects with brightly colored petals, scent, and offers of food such as nectar and pollen.

Cross Section of an Insect Pollinated Flower

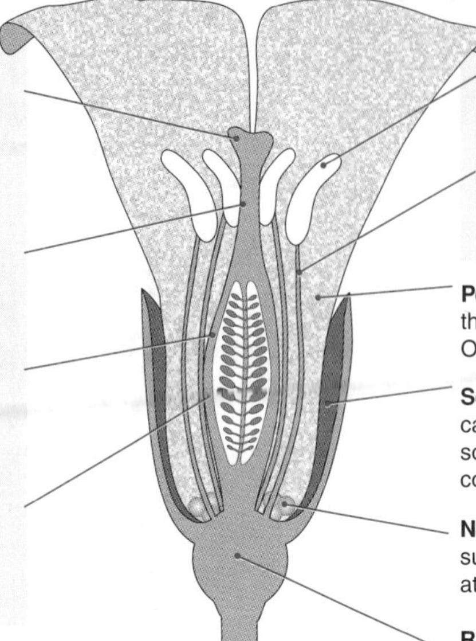

Stigma: The receptive part of the carpel. Pollen grains will germinate only if they land here.

Style: The structure that supports the stigma.

Ovary: The base of the carpel where the ovules develop.

Ovules: These are eggs and once fertilized, become the seeds. The ovule skin becomes the seed coat or testa.

An entire female part is the carpel. There may be one or more carpels per flower.

Anther: Top portion of the stamen, the male organ of reproduction.

Filament: The slender stalk of the stamen that supports the anther.

Petals: Collectively, these form the corolla. Often brightly colored.

Sepals: Together form the calyx. Usually green, but sometimes the same colour as the petals.

Nectary: Plants produce a sugary liquid called nectar to attract insects to the flower.

Receptacle: The swollen base of the flower. Sometimes it forms the succulent tissue of the fruit.

Magnolias are an ancient plant group, with very generalized flowers that are accessible to their beetle pollinators.

Orchids are well known for the many structural variations in their flowers, which are often highly specialized. They frequently have only one specific insect pollinator; a relationship that has arisen through coevolution.

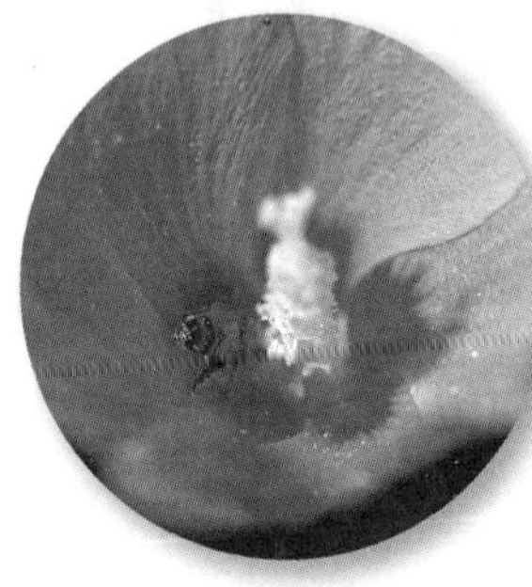

The petals of flowers guide insects towards the pollen or nectar at the centre of the flower using various colors and lines known as nectar guides. In this way, wandering insects are enticed into entering the flower and transfer pollen in the most efficient way.

Bees and many other insects are able to detect ultraviolet light. Many flowers contain pigments that reflect UV producing a specific pattern visible to insects but not to other animals. In this way, plants can use their flowers to specifically attract preferred insect pollinators.

1. What is the difference between monoecious and dioecious plants?

2. What is the difference between the stigma and the anther?

3. How are flowers used to attract specific insect pollinators to a plant?

ISBN: 978-1-927173-16-9

Pollination and Fertilization

Before the egg and sperm can fuse in fertilization, the pollen (containing the male gametes) must be transferred from the male reproductive structures to the female structures in **pollination**. Plants rarely self-pollinate; although they can be made to do so. Adaptations to ensure cross pollination include structural and physiological mechanisms associated with the flowers or cones themselves, and reliance on wind and animal pollinators. Once pollination has occurred the pollen grain can develop to produce the pollen tube allowing the sperm nuclei to enter the ovule and fertilization to take place.

Mechanisms for Ensuring Cross - Pollination

Male willow catkin
Photo: Ernie

An effective way of ensuring cross pollination is to have separate male and female plants. This occurs in about 6% of angiosperms including willows and holly. Other plants produce separate male and female flowers on the same plant. These may develop at different times so that pollen cannot fertilize the same plant.

Tulip anthers and stigma

Some plants produce flowers with both male and female structures. They can ensure cross pollination by developing the anthers at a different time to the stigma.

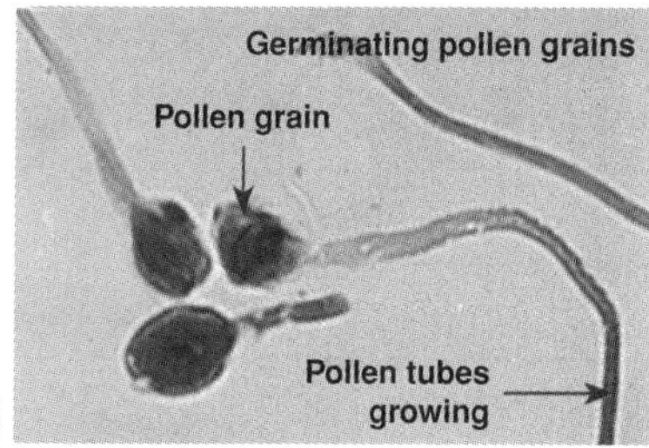
Germinating pollen grains
Pollen grain
Pollen tubes growing
RCN

In many plants, pollen will not germinate if it lands on the stigma of the same plant ensuring that the egg cells are not fertilized by sperm from the same plant.

Growth of the Pollen Tube and Double Fertilization

Pollen grains are immature male gametophytes, formed by mitosis of haploid microspores within the pollen sac. Pollination is the actual transfer of the pollen from the stamens to the stigma. Pollen grains cannot move independently. They are usually carried by wind (**anemophily**) or animals (**entomophily**). After landing on the sticky stigma, the pollen grain is able to complete development, germinating and growing a pollen tube that extends down to the ovary. Directed by chemicals (usually calcium), the pollen tube enters the ovule through the **micropyle**, a small gap in the ovule. A **double fertilization** takes place. One sperm nucleus fuses with the egg to form the zygote. A second sperm nucleus fuses with the two polar nuclei within the embryo sac to produce the endosperm tissue (3N). There are usually many ovules in an ovary, therefore many pollen grains (and fertilizations) are needed before the entire ovary can develop.

Different pollens are variable in shape and pattern, and genera can be easily distinguished on the basis of their distinctive pollen. This feature is exploited in the relatively new field of forensic botany; the tracing of a crime through botanical evidence. The species specific nature of pollen ensures that only genetically compatible plants will be fertilized. Some species, such as *Primula*, produce two pollen types, and this assists in cross pollination between different flower types.

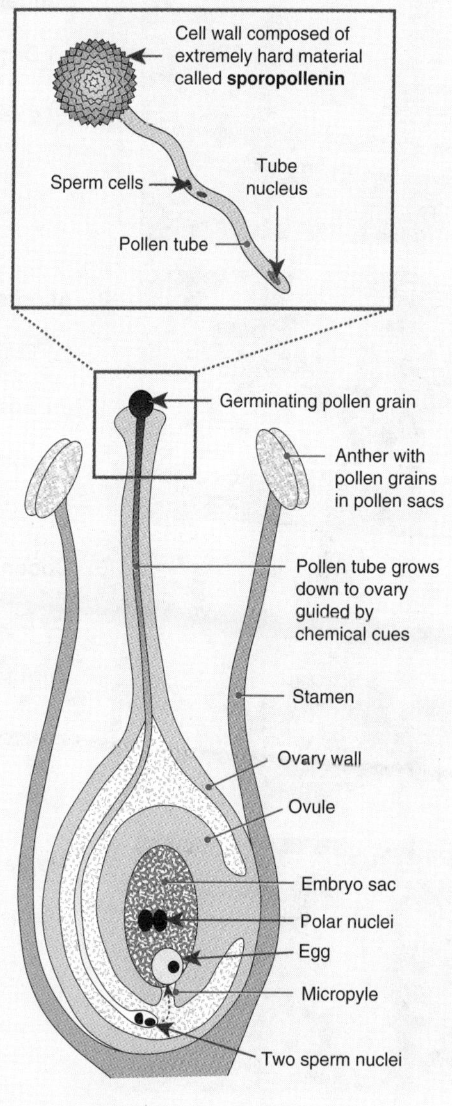
Cell wall composed of extremely hard material called **sporopollenin**
Sperm cells
Tube nucleus
Pollen tube
Germinating pollen grain
Anther with pollen grains in pollen sacs
Pollen tube grows down to ovary guided by chemical cues
Stamen
Ovary wall
Ovule
Embryo sac
Polar nuclei
Egg
Micropyle
Two sperm nuclei

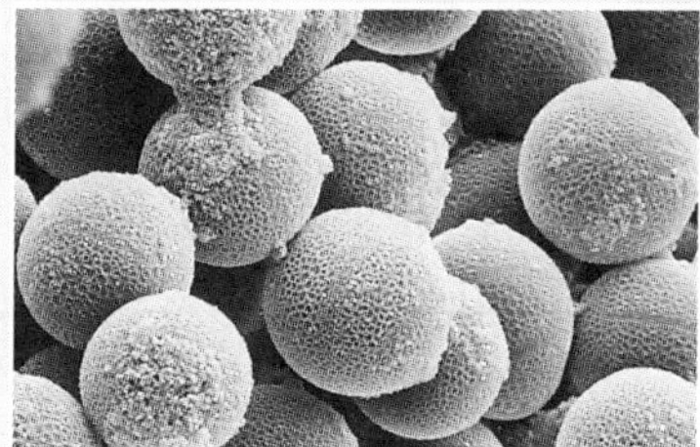

SEM: *Primula* (primrose) pollen

1. Distinguish clearly between **pollination** and **fertilization**: ______

2. Describe the role of the double fertilization in angiosperm reproduction: ______

3. Name the main chemical responsible for pollen tube growth: ______

ISBN: 978-1-927173-16-9

Seed Dispersal

Flowering plants have evolved many ways to ensure that their seeds are dispersed (transported from the parent plant), providing opportunities to expand their range. If a seed is carried into an area suitable for its germination, it will become established there. In some cases the seed itself is the agent of dispersal, but often it is the fruit. The chief agents of seed dispersal are wind, water, and animals. Many seeds are readily dispersed by water, even when they lack special buoyancy mechanisms. Wind also spreads the seeds of many plants. Such seeds have wing-like or feathery structures that catch the air currents and carry the seeds long distances. Plants that rely on animals to spread their seeds may have hooks or barbs that catch the animal hair, sticky secretions that adhere to the skin or hair, or fleshy fruits that are eaten leaving the seed to be deposited in feces some distance from the parent plant. Other dispersal mechanisms rely on explosive discharge or shaking from pods or capsules (e.g. poppy).

For each of the examples below, describe the method of dispersal and the adaptive features associated with the method:

1. **Dandelion** seeds are held in a puff-like cluster:

   (a) Dispersal mechanism:

   (b) Adaptive features:

2. **Acorns** are heavy fruits in which the fleshy seeds are encased in a resistant husk:

   (a) Dispersal mechanism:

   (b) Adaptive features:

3. **Coconuts** are heavy buoyant fruits with a thick husk:

   (a) Dispersal mechanism:

   (b) Adaptive features:

4. **Maple** fruits are winged, two-seeded samaras:

   (a) Dispersal mechanism:

   (b) Adaptive features:

5. **Wattle** (*Acacia* spp.) seeds are enclosed in pods. A fleshy strip surrounds each seed:

   (a) Dispersal mechanism:

   (b) Adaptive features:

6. **New Zealand flax** (*Phormium* spp.) produces seeds in pods:

   (a) Dispersal mechanism:

   (b) Adaptive features:

Related activities: *A Most Accomplished Traveller*
Weblinks: *Seed Dispersal*

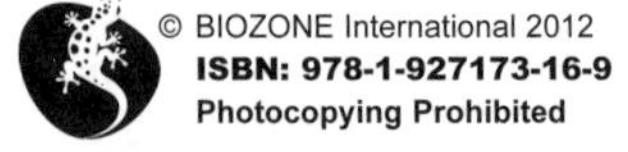

ISBN: 978-1-927173-16-9

A Most Accomplished Traveller

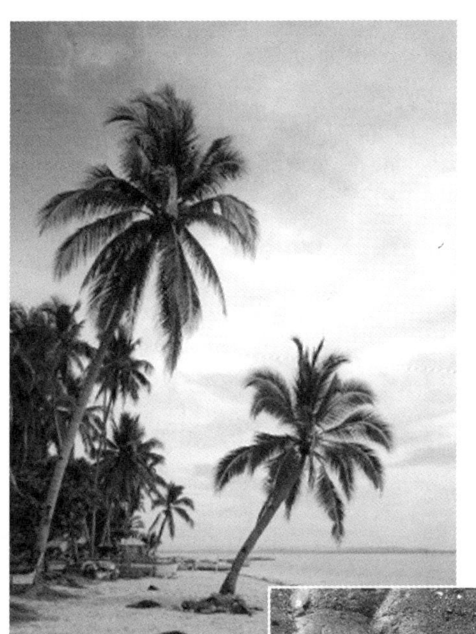

Above: *Coconut palms fringing a beach, Thailand*

Right: *Coconut germinating on black sand on the island of Hawaii*

The origin of the coconut (*Cocos nucifera*) is one of botany's mysteries. It is so extensively cultivated and so widespread in the wild, determining its origin and dispersal around the globe is extremely difficult. Suggestions have been made that the coconut originated on the coastline of the Gondwanan continent, and spread to volcanic islands where competition had been eliminated by volcanic activity. It has also been suggested that the coconut originated in South Asia or South America. Fossils show that it has been wide spread for some time, with the oldest known fossils of coconut-like palm trees found in Bangladesh and fossils from New Zealand showing it was established there some 15 million years ago.

Coconuts are a single seeded fruit (a type known as a drupe and not, in fact, a nut at all), most commonly seen as the seed with the fibrous husk removed. The coconut fruit possesses a number of features that have allowed it to spread throughout the tropics. The fibrous husk allows it to float and keeps out the seawater. Its oval shape is very stable and allows it to ride high in the water. The seed is the largest of any plant except the coco-de-mer (*Lodoicea maldivica*) and it has a thin but tough shell. The endosperm takes up only a small lining inside the seed, leaving a hollow that is filled with liquid. As the seed matures on its voyage across the sea, the liquid is absorbed. This adds to the buoyancy of the fruit. Additionally, the seed takes a long time to germinate, from 30 to 220 days. The seed is never dormant, as this is not necessary in an equable tropical climate, so the long germination period is an adaptation to extended periods of ocean-going travel between islands. Coconuts can still be viable after travelling for as long as 200 days and covering up to 4000 km.

Before the arrival of humans in the Pacific, coconuts were already widespread, but because of its valuable features, it has been even more widely dispersed by humans, both in prehistory and in modern times. Not only is it a portable, storable food and water source conveniently sealed in a hard shell, but the husk fibers can be used for making ropes, bedding, and many other products. Coconut oils, flesh, and fibers are still extensively used today.

The only tropical coastlines the coconut failed to reach were those of the Atlantic and Caribbean as these bodies of water do not mix with the Pacific or Indian Oceans except in polar regions. However, the arrival of Europeans around 1500 AD, soon saw the coconut cross the isthmus of Central America to colonize new coastlines. It is now common in every tropical and subtropical region on Earth.

1. Explain why the origin of the coconut palm is difficult to determine: ______

2. Describe the features of the coconut that allowed it to spread across the Pacific and Indian oceans: ______

3. Explain why humans found the coconut so useful: ______

4. Explain why the coconut never established in the Atlantic before humans introduced it: ______

ISBN: 978-1-927173-16-9

Related activities: *Seed Dispersal*

Seed Structure and Germination

After fertilization has occurred, the ovary develops into the fruit and the ovules within the ovary become the **seeds**. Recall that there is a double fertilization in plants; one sperm fertilizes the egg to form the embryo, while another sperm combines with the diploid endosperm nucleus to give rise to the triploid endosperm. The development of the endosperm is important and begins before embryonic development in order to produce a nutrient store for the young plant. A seed is an entire reproductive unit, housing the embryonic plant in a state of dormancy. During the last stages of maturing, the seed dehydrates until its water content is only 5-15% of its weight. The embryo stops growing and remains dormant until the seed germinates. At germination, the seed takes up water and the food store is mobilized to provide the nutrients for plant growth and development.

Dicot seeds: soy (above) **cashew** (below) There are two fleshy cotyledons. These store food absorbed from the endosperm.

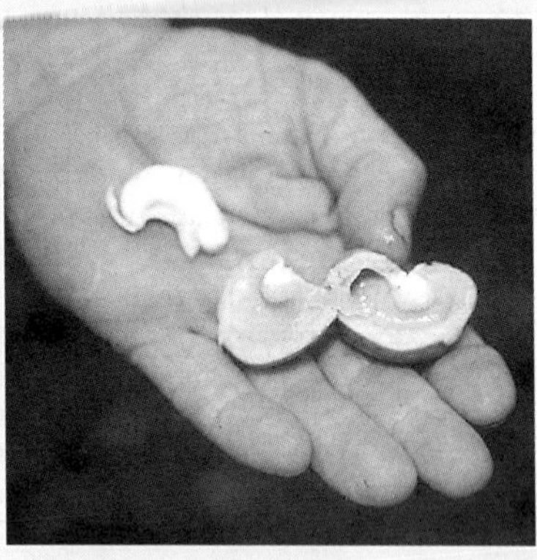

Germination requires rehydration of the seed and reactivation of its metabolism. The seed absorbs water through the seed coat (testa) and micropyle. As the dry substances in the seed tissue take up water, the cells expand, metabolism is reactivated, and embryonic growth begins. Activation begins with the release of gibberellin (GA) from the embryo. GA enhances cell elongation, making it possible for the root to penetrate the testa. It also stimulates the synthesis of enzymes, which hydrolyze the starch to produce sugars. The mobilized food stores are delivered to the developing roots and shoots.

Seed Structure and Formation

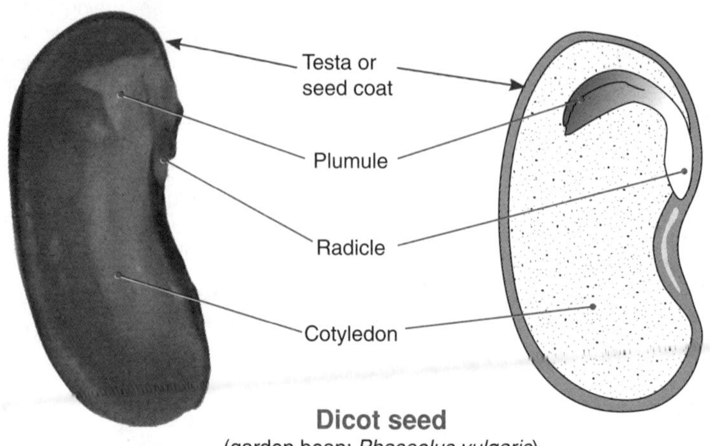

Testa or seed coat
Plumule
Radicle
Cotyledon

Dicot seed
(garden bean: *Phaseolus vulgaris*)

Every seed contains an embryo comprising a rudimentary shoot (plumule), root (radicle), and one or two cotyledons (seed leaves). The embryo and its food supply are encased in a tough, protective seed coat or **testa**. In monocots, the endosperm provides the food supply, whereas in most dicot seeds, the nutrients from the endosperm are transferred to the large, fleshy cotyledons.

Germination in a Dicot Seed

(garden bean: *Phaseolus vulgaris*)

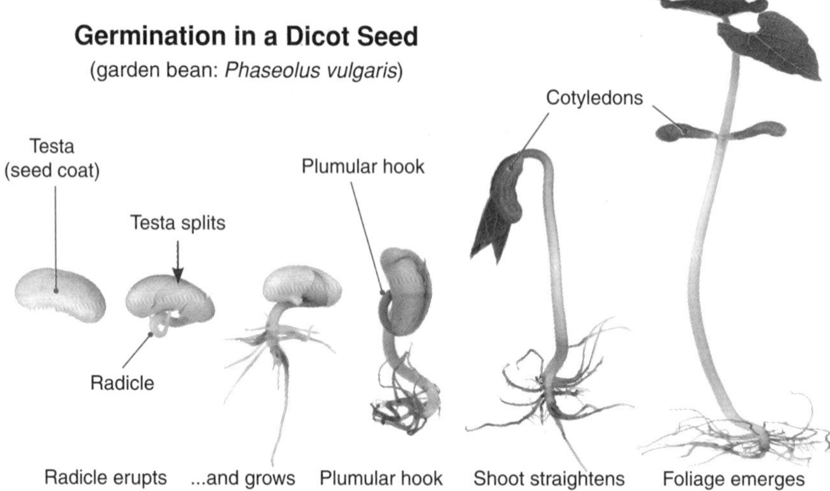

Testa (seed coat)
Testa splits
Radicle
Plumular hook
Cotyledons
Radicle erupts from the seed....
...and grows rapidly downwards
Plumular hook protects the emerging stem.
Shoot straightens after emerging from the soil and lateral roots develop.
Foliage emerges and secondary roots emerge from the lateral roots.

1. What is the purpose of a **seed**? ______________________________

____

2. (a) State the function of the endosperm in angiosperms: ______________________________

   (b) State how the endosperm is derived: ______________________________

3. What is the role of the testa? ______________________________

4. Explain why the seed requires a food store: ______________________________

____

5. Why must stored seeds be kept dry? ______________________________

Related activities: *Pollination and Fertilization, Events in Germination*

ISBN: 978-1-927173-16-9

Events in Germination

Seed germination refers to the beginning of seed growth. It involves a process of rehydration of the seed and reactivation of normal metabolism. The development of the **endosperm** begins before embryo development in order to produce a nutrient-rich store for the young plant. In dicots (e.g. beans), the endosperm is transferred to the two fleshy cotyledons during seed development and may disappear altogether. However, in moncots (grasses and grains) the endosperm remains as the main food source for the embryo. The single cotyledon is called the **scutellum** and the shoot (plumule) is sheathed in a **coleoptile**.

Metabolic Events in Germination

Germination begins when a mature seed begins to take up water through the micropyle and testa (**imbibition**). Imbibition causes the seed to swell and the testa to split. The food stored in the seed is hydrolyzed to produce substrates for respiration (e.g. glucose). Germinating seeds have a high oxygen requirement and respire rapidly. Water is essential to the germination process: it enables expansion of the growing cells and activates the enzymes needed for germination. Some of these enzymes are produced in response to the release of gibberellic acid (GA) from the growing embryo. Water is also required for the hydrolysis of stored starch and for the translocation of the mobilized food to the sites of growth.

The series **A** - **E** below shows the conditions within a **monocot seed** at the times that appear directly above on the graph. (**A**) GA is released from the embryo. (**B**) Digestive enzymes (filled black dots) produced by aleurone cells. (**C**) Enzymes mobilize food in the endosperm, releasing soluble nutrients (open dots). (**D**) Embryo's cotyledon absorbs food and delivers it to the shoot and root. (**E**) The first leaves are photosynthesizing by the time food reserves have been depleted.

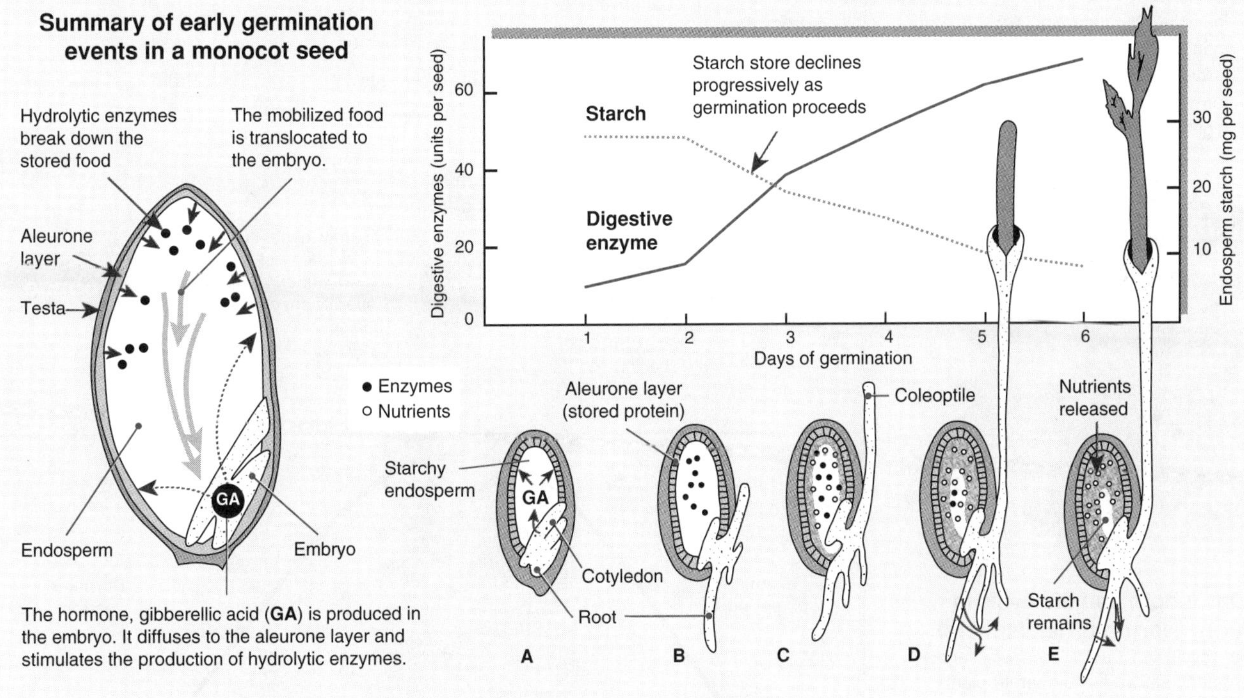

Summary of early germination events in a monocot seed
Hydrolytic enzymes break down the stored food
The mobilized food is translocated to the embryo.
Aleurone layer
Testa
GA
Endosperm
Embryo
The hormone, gibberellic acid (GA) is produced in the embryo. It diffuses to the aleurone layer and stimulates the production of hydrolytic enzymes.
Digestive enzymes (units per seed)
0
20
40
60
Starch
Starch store declines progressively as germination proceeds
Digestive enzyme
Endosperm starch (mg per seed)
10
20
30
1
2
3
4
5
6
Days of germination
• Enzymes
○ Nutrients
Starchy endosperm
Aleurone layer (stored protein)
Cotyledon
Root
Coleoptile
Nutrients released
Starch remains
A
B
C
D
E

1. Identify the two processes involved in seed germination and describe the purpose of each:

(a) ___

(b) ___

2. Explain the role of the following in germination of a typical monocot seed:

(a) Water: ___

(b) Oxygen: ___

(c) Gibberellic acid: ___

ISBN: 978-1-927173-16-9

Photoperiodism in Plants

Photoperiodism is the response of a plant to the relative lengths of daylight and darkness. Flowering is a photoperiodic activity; individuals of a single species will all flower at much the same time, even though their germination and maturation dates may vary. The exact onset of flowering varies depending on whether the plant is a short-day or long-day type (see next page). Photoperiodic activities are controlled through the action of a pigment called **phytochrome**. Phytochrome acts as a signal for some biological clocks in plants and is also involved in other light initiated responses, such as germination, shoot growth, and chlorophyll synthesis. Plants do not grow at the same rate all of the time. In temperate regions, many perennial and biennial plants begin to shut down growth as autumn approaches. During unfavorable seasons, they limit their growth or cease to grow altogether. This condition of arrested growth is called **dormancy**, and it enables plants to survive periods of water scarcity or low temperature. The plant's buds will not resume growth until there is a convergence of precise environmental cues in early spring. Short days and long, cold nights (as well as dry, nitrogen deficient soils) are strong cues for dormancy. Temperature and daylength change seasonally in most parts of the world, so changes in these variables also influence many plant responses, including germination and flowering. In many plants, flowering is triggered only after a specific period of exposure to low winter temperatures. As described in the previous activity, this low-temperature stimulation of flowering is called **vernalization**.

Photoperiodism

Photoperiodism is based on a system that monitors the day/night cycle. The photoreceptor involved in this, and a number of other light-initiated plant responses, is a blue-green pigment called **phytochrome**. Phytochrome is universal in vascular plants and has two forms: active and inactive. On absorbing light, it readily converts from the inactive form (**P_r**) to the active form (**P_{fr}**). **P_{fr}** predominates in daylight, but reverts spontaneously back to the inactive form in the dark. The plant measures daylength (or rather night length) by the amount of phytochrome in each form.

Summary of phytochrome related activities in plants

Process	Effect of daylight	Effect of darkness
Conversion of phytochrome	Promotes **$P_r \rightarrow P_{fr}$**	Promotes **$P_{fr} \rightarrow P_r$**
Seed germination	Promotes	Inhibits
Leaf growth	Promotes	Inhibits
Flowering: long day plants	Promotes	Inhibits
Flowering: short day plants	Inhibits	Promotes
Chlorophyll synthesis	Promotes	Inhibits

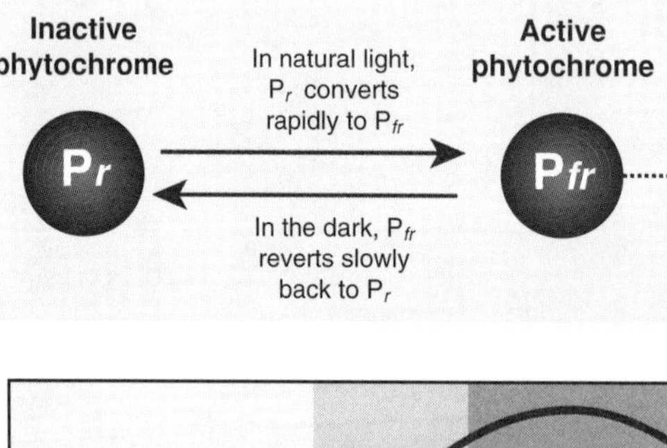
Inactive phytochrome
In natural light, P_r converts rapidly to P_{fr}
Pr
In the dark, P_{fr} reverts slowly back to P_r
Active phytochrome
Pfr

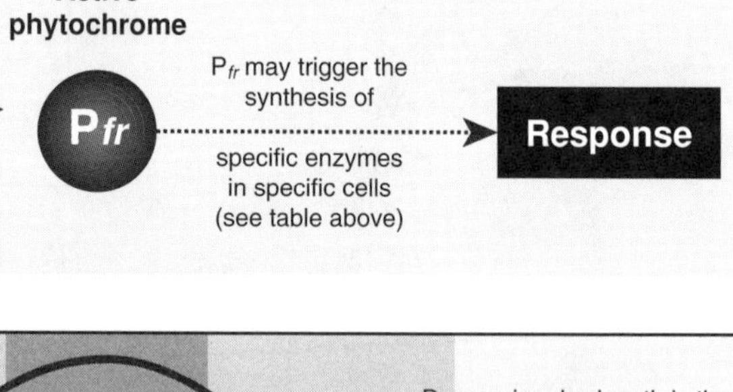
P_{fr} may trigger the synthesis of
specific enzymes in specific cells (see table above)
Response

Day length and life cycle in plants (Northern Hemisphere)

The cycle of active growth and dormancy shown by temperate plants is correlated with the number of daylight hours each day (right). In the southern hemisphere, the pattern is similar, but is six months out of phase. The duration of the periods may also vary on islands and in coastal regions because of the moderating effect of nearby oceans.

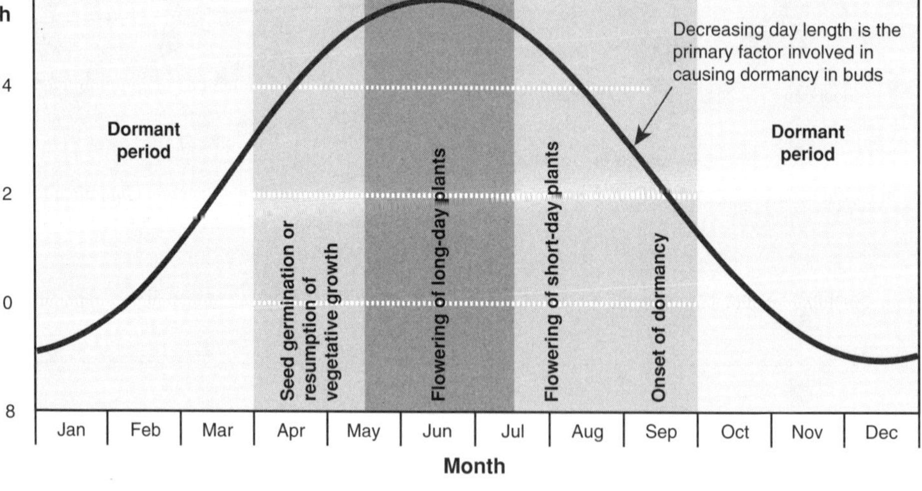
Daylength (hours)
16
14
12
10
8
Dormant period
Seed germination or resumption of vegetative growth
Flowering of long-day plants
Flowering of short-day plants
Onset of dormancy
Decreasing day length is the primary factor involved in causing dormancy in buds
Dormant period
Jan
Feb
Mar
Apr
May
Jun
Jul
Aug
Sep
Oct
Nov
Dec
Month

1. Describe two plant responses, initiated by exposure to light, which are thought to involve the action of phytochrome:

   (a) ____________________

   (b) ____________________

2. Discuss the role of phytochrome in a plant's ability to measure daylength: ____________________

Related activities: Tropisms and Growth Responses
Weblinks: Photomorphogenesis

Periodicals: How plants know their place

ISBN: 978-1-927173-16-9

Long-day plants

When subjected to the light regimes on the right, the 'long-day' plants below flowered as indicated:

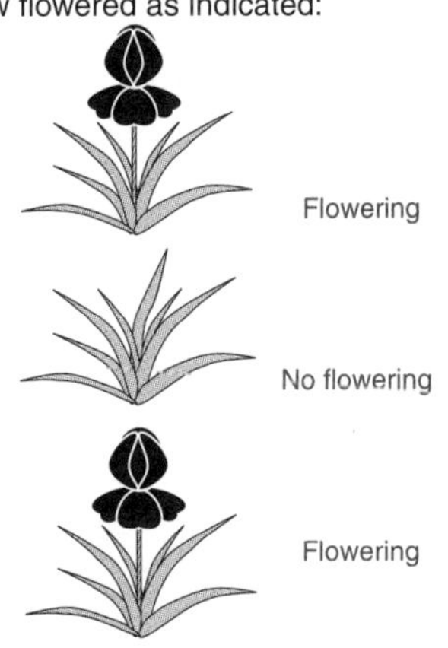
Flowering
No flowering
Flowering

Examples: *lettuce, clover, delphinium, gladiolus, beetscorn, coreopsis*

Photoperiodism in Plants

An experiment was carried out to determine the environmental cue that triggers flowering in 'long-day' and 'short-day' plants. The diagram below shows 3 different light regimes to which a variety of long-day and short-day plants were exposed.

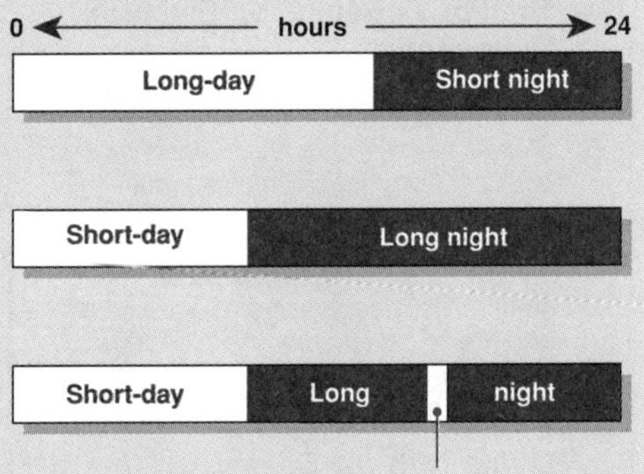

0 ← hours → 24

Long-day	Short night

Short-day	Long night

Short-day	Long	night

Long night interrupted by a short period exposed to light

Short-day plants

When subjected to the light regimes on the left, the 'short-day' plants below flowered as indicated:

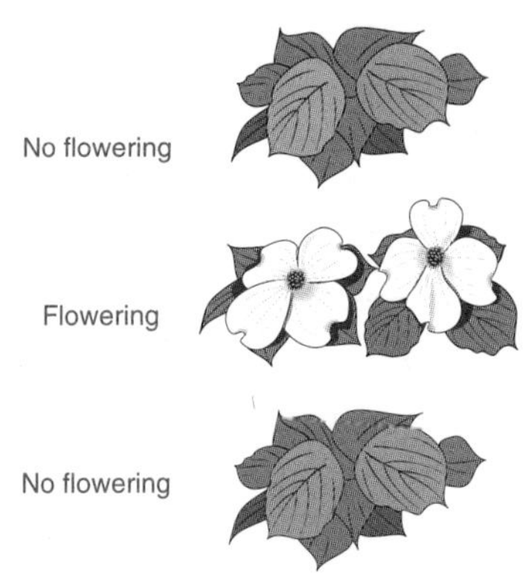
No flowering
Flowering
No flowering

Examples: *potatoes, asters, dahlias, cosmos, chrysanthemums, pointsettias*

3. (a) What is the environmental cue that synchronizes flowering in plants? ______

   (b) Describe one biological advantage of this synchronization to the plants: ______

4. Discuss the role of environmental cues in triggering and breaking **dormancy** in plants: ______

5. Discuss the adaptive value of **dormancy** and **vernalization** in temperate climates: ______

6. Study the three light regimes above and the responses of short-day and long-day flowering plants to that light. From this observation, describe the most important factor controlling the onset of flowering in:

   (a) Short-day plants: ______

   (b) Long-day plants: ______

7. Using information from the experiment described above, discuss the evidence for the statement "*Short-day plants are really better described as long-night plants.*"

ISBN: 978-1-927173-16-9

Appendix

PERIODICAL REFERENCES

SCIENCE PRACTICES AND STATISTICAL ANALYSIS

- **The Truth Is Out There**
  New Scientist, 26 February 2000 (Inside Science). *The philosophy of scientific method: starting with an idea, formulating a hypothesis, and following the process to theory.*

- **Descriptive Statistics**
  Biol. Sci. Rev., 13(5) May 2001, pp. 36-37. *An account of descriptive statistics using text, tables and graphs.*

- **Percentages**
  Biol. Sci. Rev., 17(2) November 2004, pp. 28-29. *The calculation of percentage and the appropriate uses of this important transformation.*

- **Experiments**
  Biol. Sci. Rev., 14(3) February 2002, pp. 11-13. *The basics of experimental design and execution: determining variables, measuring them, and establishing a control.*

- **Be Confident with Calculations**
  Biol. Sci. Rev., 23(2) Nov. 2010, pp. 13-15. *How to use some commonly used calculations in biology. Includes percentages, rates, and magnification.*

- **Dealing with Data**
  Biol. Sci. Rev., 12 (4) March 2000, pp. 6-8. *An account of the best ways in which to deal with interpreting graphically presented data in exams.*

- **Drawing Graphs**
  Biol. Sci. Rev., 19(3) Feb. 2007, pp. 10-13. *A guide to creating graphs. The use of different graphs for different tasks is explained and there are a number of pertinent examples described to illustrate points.*

- **Estimating the Mean and Standard Deviation**
  Biol. Sci. Rev., 13(3) January 2001, pp. 40-41. *Simple statistical analysis. Includes formulae for calculating sample mean and standard deviation.*

- **Describing the Normal Distribution**
  Biol. Sci. Rev., 13(2) November 2000, pp. 40-41. *The normal distribution: data spread, mean, median, variance, and standard deviation.*

CELL BIOLOGY

- **Size Does Matter**
  Biol. Sci. Rev., 17 (3) Feb. 2005, pp. 10-13. *Measuring the size of organisms and calculating magnification and scale.*

- **Getting in and Out**
  Biol. Sci. Rev., 20(3), Feb. 2008, pp. 14-16. *Diffusion: some adaptations and some common misunderstandings.*

- **What is a Stem Cell?**
  Biol. Sci. Rev., 16(2) Nov. 2003, pp. 22-23. *The nature of stem cells and their therapeutic applications.*

- **The Power Behind an Electron Microscopist**
  Biol. Sci. Rev., 18(1) Sept. 2005, pp. 16-20. *The use of TEMs to obtain greater resolution of finer details than is possible from optical microscopes.*

- **What is Endocytosis?**
  Biol. Sci. Rev., 22(3), Feb. 2010, pp. 38-41. *The mechanisms of endocytosis and the role of membrane receptors in concentrating important molecules before ingestion.*

- **Living with the Enemy**
  New Scientist, 25 Oct. 2008, pp. 26-33. *The sheer diversity of mutations that can turn cells cancerous and drive tumor growth gives endless opportunities to outwit our defence. How can we protect ourselves effectively?*

- **The Cell Cycle and Mitosis**
  Biol. Sci. Rev., 14(4) April 2002, pp. 37-41. *Cell growth and division, stages in the cell cycle, and the complex control over different stages of mitosis.*

- **What is Cell Suicide?**
  Biol. Sci. Rev., 20(1) Sept. 2007, pp. 17-20. *An account of the mechanisms behind cell suicide and its role in normal growth and development.*

THE CHEMISTRY OF CELLS

- **Water, Life, and Hydrogen Bonding**
  Biol. Sci. Rev., 21(2) Nov. 2008, pp. 18-20. *The molecules of life and the important role of hydrogen bonding.*

- **Glucose & Glucose-Containing Carbohydrates**
  Biol. Sci. Rev., 19(1) Sept. 2006, pp. 12-15. *The structure of glucose and its polymers.*

- **Designer Starches**
  Biol. Sci. Rev., 19(3) Feb. 2007, pp. 18-20. *The composition of starch, and an account of its properties and roles.*

- **What is Tertiary Structure?**
  Biol. Sci. Rev., 21(1) Sept. 2008, pp. 10-13. *How amino acid chains fold into the functional shape of a protein.*

THE STRUCTURE AND FUNCTION OF DNA

- **DNA: 50 Years of the Double Helix**
  New Scientist, 15 March 2003, pp. 35-51. *A special issue on DNA: structure and function, repair, the new-found role of histones, and the functional significance of chromosome position in the nucleus.*

- **DNA Polymerase**
  Biol. Sci. Rev., 22(4) April 2010, pp. 38-41. *How DNA polymerase operates during DNA replication in the cell and how the enzyme is used in PCR.*

- **What is a Gene?**
  Biol. Sci. Rev., 15(2) Nov. 2002, pp. 9-11. *A synopsis of genes, mutations, and transcriptional control of gene expression.*

- **Transfer RNA**
  Biol. Sci. Rev., 15(3) Feb. 2003, pp. 26-29. *A good account of the structure and role of tRNA in protein synthesis.*

ENZYMES AND METABOLISM

- **Enzymes: Nature's Catalytic Machines**
  Biol. Sci. Rev., 22(2) Nov. 2009, pp. 22-25. *Enzymes as catalysts: a very up-to-date description of enzyme specificity and binding, how enzymes work, and how they overcome the energy barriers for a reaction. Some well known enzymes are described.*

- **Enzymes: Fast and Flexible**
  Biol. Sci. Rev., 19(1) Sept. 2006, pp. 2-5. *The structure of enzymes and how they work so efficiently at relatively low temperatures.*

- **Enzymes**
  Biol. Sci. Rev., 23(3) Feb. 2011, pp. 20-21. *A good account of enzymes. Graphs showing factors affecting the rate of enzyme reactions. Cofactors, inhibitors, synoptic possibilities.*

- **The Double Life of ATP**
  Scientific American, Dec. 2009, pp. 60-67. *ATP the fuel inside living cells, also serves as a molecular messenger that affects cell behavior.*

- **AcetylCoA: A Central Metabolite**
  Biol. Sci. Rev., 20(4) April 2008, pp.38-40. *The role of acetyl coenzyme A in metabolizing fat and carbohydrate.*

- **Lactic Acid: Who Needs It?**
  Biol. Sci. Rev., 18(2) Nov. 2005, pp. 6-90. *An account of the biological roles of lactic acid, including its production in anaerobic metabolism in muscle.*

- **Experiments**
  Biol. Sci. Rev., 14(3) February 2002, pp. 11-13. *The basics of experimental design and execution: determining variables, measuring them, and establishing a control.*

- **Photosynthesis**
  Biol. Sci. Rev., 23(2) Sept. 2010, pp. 20-21. *Photosynthetic processes including the absorption spectrum and chloroplast structure and function.*

- **Chloroplasts: Biosynthetic Powerhouses**
  Biol. Sci. Rev., 21(4) April 2009, pp. 25-27. *Informative account of the structure and role of chloroplasts.*

- **Photosynthesis...Most Hated Topic?**
  Biol. Sci. Rev., 20(1) Sept. 2007, pp. 13-16. *A useful account documenting key points about photosynthesis.*

Appendix

PERIODICAL REFERENCES

CHROMOSOMES AND MEIOSIS

▶ **Mechanisms of Meiosis**
Biol. Sci. Rev., 15(4), April 2003, pp. 20-24. *A clear and thorough account of the events and mechanisms of meiosis.*

▶ **What is a Mutation?**
Biol. Sci. Rev., 20(3) Feb. 2008, pp. 6-9. *The nature of mutations: causes, timing, and effects. Sickle cell disease is the case study described.*

▶ **Genetics of Sickle Cell Anaemia**
Biol. Sci. Rev., 20(4) April 2008, pp. 14-17. *The molecular and physiological basis of sickle cell disease.*

HEREDITY

▶ **What is Variation?**
Biol. Sci. Rev., 13(1) Sept. 2000, pp. 30-31. *The nature of continuous and discontinuous variation. The distribution pattern of traits that show continuous variation is discussed.*

▶ **Mendel's Legacy**
Biol. Sci. Rev., 18(4), April 2006, pp. 34-37. *Explores the accuracy of Mendel's laws in light of today's knowledge.*

▶ **The Y Chromosome: It's a Man Thing**
Biol. Sci. Rev., 20(4) April 2008, pp. 2-6. *The Y chromosome is at the root of sex determination. This account discusses the nature of the Y chromosome, non-disjunction and Y chromosome disorders and the inheritance of Y linked diseases.*

▶ **The Color Code**
New Scientist, 10 March 2002, pp. 34-37. *Researchers are uncovering the five to ten genes responsible for skin pigmentation.*

GENETIC ENGINEERING AND BIOTECHNOLOGY

▶ **DNA Polymerase**
Biol. Sci. Rev., 22(4) April 2010, pp. 38-41. *How DNA polymerase operates during DNA replication in the cell and how the enzyme is used in PCR.*

▶ **Bioinfomatics: What Use Is It?**
Biol. Sci. Rev., 15(4) April 2003, pp. 2-5. *A look at how bioinformatics can help us to better understand the human genome and those of other species*

▶ **What is...Genomics?**
Biol. Sci. Rev., 20(2) Nov. 2007, pp. 38-41. *The nature of genome projects.*

▶ **Revolution Postponed**
Scientific American, Oct. 2010, pp. 42-49. *The HGP has failed so far to produce the medical miracles that were promised. Biologists must decide where to go from here.*

▶ **How We Are Evolving**
Scientific American, Oct. 2010, pp. 23-29. *New genetic analyses provide evidence that natural selection is acting quickly on human populations to promote helpful alleles.*

▶ **Birds, Bees, and Superweeds**
Biol. Sci. Rev., 17(2) Nov. 2004, pp. 24-27. *GM crops: their advantages and applications, as well as the risks and concerns associated with their use.*

▶ **Tailor-Made Proteins**
Biol. Sci. Rev., 13(4) March 2001, pp. 2-6. *Recombinant proteins and their uses in industry and medicine.*

▶ **Rice, Risk and Regulations**
Biol. Sci. Rev., 20(2) Nov. 2007, pp. 17-20. *The genetic engineering of one of the world's most important cereal crops is an ethical concern for many.*

▶ **The Engineering of Crop Plants**
Biol. Sci. Rev., 20(4) April 2008, pp. 30-36. *Crop plants can be engineered to increase the nutritional value of foods and to improve non-food crops as sources of raw materials for industry.*

▶ **Food / How Altered?**
National Geographic, May 2002, pp. 32-50. *Biotech foods: what are they, how are they made, and are they safe?*

▶ **Embryonic Stem Cells**
Biol. Sci. Rev., 22(1) Sept. 2009, pp. 28-31. *The future of embryonic stem cell research. Problems and solutions.*

ECOLOGY

▶ **Getting to Grips with Ecology**
Biol. Sci. Rev., 22(3), Feb. 2010, pp. 14-16. *A good overview on ecological principles using ladybirds and aphids as example organisms.*

▶ **All Life is Here**
New Scientist, 24 April 2010, pp. 31-35. *A look at the variation of biodiversity from the tropics to the poles. Biodiversity hotspots are also covered.*

▶ **The Lake Ecosystem**
Biol. Sci. Rev., 20(3) Feb. 2008, pp. 21-25. *An account of the components and functioning of lake ecosystems.*

▶ **Microbes and Nutrient Cycling**
Biol. Sci. Rev., 19(1) Sept. 2006, pp. 16-20. *The roles of microorganisms in nutrient cycling.*

▶ **The Case of the Missing Carbon**
National Geographic, 205(2), Feb. 2004, pp. 88-117. *The role of carbon sinks in the Earth's carbon cycling.*

▶ **Global Warming**
Time, special issue, 9 April 2007. *A special issue on global warming: the causes, perils, solutions, and actions. Comprehensive and engaging.*

▶ **Think or Swim**
New Scientist, 18 September 2010, pp. 40-43. *Sea levels are predicted to rise as a result of global warming. This article looks at the implications of sea level rise on human populations.*

EVOLUTION

▶ **A Fin is a Limb is a Wing**
National Geographic, 210(5) Nov. 2006, pp. 110-135. *An excellent account of the role of developmental genes in the evolution of complex organs and structures in animals. Beautifully illustrated, compelling evidence for the mechanisms of evolutionary change.*

▶ **Was Darwin Wrong?**
National Geographic, 206(5) Nov. 2004, pp. 2-35. *An account of the scientific evidence for evolution. A good starting point for reminding students that the scientific debate around evolutionary theory is associated with the mechanisms by which evolution occurs, not the fact of evolution itself.*

▶ **The Enemy Within**
Scientific American, April 2011, pp. 26-33. *Antibiotic resistance is spreading in the transfer of genes that confer resistance in a new pattern globally. New medications are not being developed quickly enough to treat gram-negative bacteria.*

▶ **Polymorphism**
Biol. Sci. Rev., 14(1) Sept. 2001, pp. 19-21. *A good account of genetic polymorphism. Examples include the carbonaria gene (Biston), the sickle cell gene, and aphids.*

▶ **The Moths of War**
New Scientist, 8 Dec. 2007, pp 46-49. *New research into the melanism of the peppered moth reaffirms it as an example of evolution, reclaiming it back from Creationists.*

CLASSIFICATION

▶ **A Passion for Order**
National Geographic, 211(6) June 2007, pp. 73-87. *The history of Linnaeus and plant classification.*

▶ **World Flowers Bloom after Recount**
New Scientist, 29 June 2002, p. 11. *A systematic study of flowering plants indicates more species than expected, especially in regions of high biodiversity such as South American and Asia.*

▶ **The Family Line - The Human-Cat Connection**
National Geographic, 191(6) June 1997, pp. 77-85. *An examination of the genetic diversity and lineages within the felidae. A good context within which to study classification.*

▶ **The Loves of the Plants**
Scientific American, Feb. 1996, pp. 98-103. *The classification of plants and the development of keys to plant identification.*

Appendix

PERIODICAL REFERENCES

DIGESTION

▶ **The Anatomy of Digestion**
Biol. Sci. Rev., 23 (3) Feb. 2010, pp. 18-21. *The role of each of the components of the human digestive system is described, and explanations are provided about what happens when things go wrong with digestion.*

▶ **The Liver in Health and Disease**
Biol. Sci. Rev., 14(2) Nov. 2001, pp. 14-20. *The various roles of the liver, a major homeostatic organ.*

THE TRANSPORT SYSTEM

▶ **The Heart**
Bio. Sci. Rev. 18(2) Nov. 2005, pp. 34-37. *The structure and physiology of the heart.*

▶ **Keeping Pace - Cardiac Muscle & Heartbeat**
Biol. Sci. Rev., 19(3), Feb. 2007, pp. 21-24. *The structure and properties of cardiac muscle.*

▶ **A Fair Exchange**
Biol. Sci. Rev., 13(1), Sept. 2000, pp. 2-5. *The role of tissue fluid in the body and how it is produced and reabsorbed.*

▶ **Red Blood Cells**
Bio. Sci. Rev. 11(2) Nov. 1998, pp. 2-4. *The structure and function of red blood cells and details of oxygen transport.*

DEFENSE AGAINST DISEASE

▶ **Finding and Improving Antibiotics**
Biol. Sci. Rev. 12(1) Sept. 1999, pp. 36-38. *Antibiotics, their production & testing, and the search for new drugs.*

▶ **Are Viruses Alive?**
Scientific American, Dec. 2004, pp. 77-81. *The nature of viruses, including viral replication and an evaluation of the status of viruses in the world.*

▶ **Skin, Scabs and Scars**
Biol. Sci. Rev., 17(3) Feb. 2005, pp. 2-6. *The roles of skin, including its role in wound healing and the processes involved in its repair when damaged.*

▶ **Fight for Your Life!**
Biol. Sci. Rev., 18(1) September 2005, pp. 2-6. *The mechanisms by which we recognise pathogens and defend ourselves against them (overview).*

▶ **Looking Out for Danger: How White Blood Cells Protect Us**
Biol. Sci. Rev., 19 (4) April 2007, pp. 34-37. *The various types of leucocytes and how they work together to protect the body against infection.*

▶ **Lymphocytes - The Heart of the Immune System**
Biol. Sci. Rev., 12 (1) Sept. 1999 pp. 32-35. *An account of the role of lymphocytes (includes the types and actions of different lymphocytes).*

▶ **Immunology**
Biol. Sci. Rev., 22(4) April 2010, pp. 20-21. *A pictorial but information-packed review of the basic of internal defence functions.*

▶ **Hard to Swallow**
New Scientist, 26 Jan. 2008, pp. 37-39. *Many people fear that vaccines are unsafe and cause health problems. Particular reference to the polio and measles vaccines.*

▶ **Boosting Vaccine Power**
Scientific American, October 2009, pp. 56-59. *This article looks at vaccines and immunization, and how researchers are boosting the effectiveness of vaccines to be even more specific.*

▶ **Monoclonals as Medicines**
Biol. Sci. Rev., 18(4) April 2006, pp. 38-40. *The use and efficacy of monoclonal antibodies in therapeutic and diagnostic medicine.*

▶ **AIDS**
Biol. Sci. Rev., 20(1) Sept. 2007, pp. 30-12. *The HIV virus can evade the immune system and acquire drug resistance. This has prevented effective cures from being developed.*

GAS EXCHANGE

▶ **Gas Exchange in the Lungs**
Bio. Sci. Rev. 16(1) Sept. 2003, pp. 36-38. *The structure and function of the alveoli of the lungs, with an account of respiratory problems and diseases.*

▶ **Humans with Altitude**
New Scientist, 2 Nov. 2002, pp. 36-39. *The short term adjustments and long term adaptations to life at altitude.*

NERVES, HORMONES AND HOMEOSTASIS

▶ **Refractory Period**
Biol. Sci. Rev., 20(4) April 2008, pp. 7-9. *The nature and purpose of the refractory period in response stimuli. The biological principles involved are discussed with in the context of the refractory period of the human heart.*

▶ **Bridging the Gap**
Biol. Sci. Rev., 21(2) Nov. 2008, pp. 2-6. *Communication between nerve cells across synapses; the biology of chemical transmission.*

▶ **Homeostasis**
Biol. Sci. Rev., 12(5) May 2000, pp. 2-5. *Homeostasis: what it is, the role of negative feedback and the autonomic nervous system, and the adaptations of organisms for homeostasis in extreme environments (excellent).*

▶ **Food for Thought**
Biol. Sci. Rev., 22(4), April 2010, pp. 22-25. *A clear, thorough account of how the body maintains its supply of glucose long after the nutrients absorbed from a meal have been exhausted.*

MUSCLES & MOVEMENT

▶ **Human Muscle: Structure and Function**
Biol. Sci. Rev., 19(4) April 2007, pp. 25-29. *The structure and function of muscle in humans: contraction and the mechanics of locomotion.*

▶ **How Skeletal Muscles Work**
Biol. Sci. Rev., 22(4) April 2010, pp. 10-15. *The structure and function of muscle in humans: contraction, sliding filament theory, and types of muscle.*

REPRODUCTION

▶ **Spermatogenesis**
Biol. Sci. Rev., 15(4) April 2003, pp. 10-14. *The process and control of sperm production in humans, with a discussion of the possible reasons for male infertility.*

▶ **Measuring Female Hormones in Saliva**
Biol. Sci. Rev., 13(3) Jan. 2001, pp. 37-39. *The female reproductive system, and the complex hormonal control of the female menstrual cycle.*

▶ **The Placenta**
Biol. Sci. Rev., 12 (4) March 2000, pp. 2-5. *Placental function and the use of the placenta for prenatal diagnosis and gene therapy.*

PLANT SCIENCE

▶ **Cell Differentiation**
Biol. Sci. Rev., 20(4), April 2008, pp. 10-13. *How tissues arise through the control of cellular differentiation during development. The example provided is the differentiation of blood cells.*

▶ **Sending Plants Around the Bend**
Biol. Sci. Rev., 12(4) March 2000, pp. 14-17. *An account of how plants perceive and respond to stimuli around them. Tropisms are fully covered.*

▶ **How Trees Lift Water**
Biol. Sci. Rev., 18(1), Sept. 2005, pp. 33-37. *Cohesion-tension theory and others on how trees lift water.*

▶ **High Tension**
Biol. Sci. Rev., 13(1), Sept. 2000, pp. 14-18. *Cell specialization and transport in plants: the mechanisms by which plants transport water and solutes.*

▶ **Cacti**
Biol. Sci. Rev., 20(1), Sept. 2007, pp. 26-30. *The growth forms and structural and physiological adaptations of cacti.*

▶ **How Plants Know Their Place**
Biol. Sci. Rev., 17(3) Feb. 2005, pp. 33-36. *Plant responses to light and the action of phytochromes.*

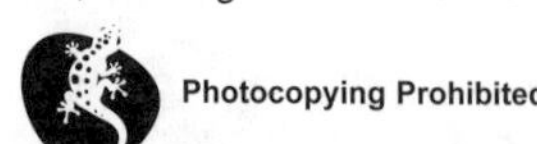

Index

Abiotic factor 214
ABO blood group 159
Absorption spectra 124
Absorption, of nutrients 275
Acclimatization, to altitude 319
Accuracy, defined 13
Acetylcholine 282, 327
Acquired immunity 301
Actin 363
Action potential 326
Action spectrum 124
Activation energy, of enzymes 108
Active immunity 301-302
Active transport 64-66
Activity, of enzymes 107-108
Adaptation, to altitude 319
Adaptations, plant 387-388
Adaptive radiation 241
ADH (antidiuretic hormone) 339
Adrenaline, and heart control 282
Aerobic respiration 116, 118
AIDS 307-309
Alcoholic fermentation 120
Allele 134
Allosteric enzyme regulation 112
Altitude, adaptations to 319
Alveoli 313-14
Amino acid 83-84, 94
Amylase 269, 271, 274
Amyloplast 47
Anaerobic pathways 120
Aneuploidy 136
Animal cell, features of 49-50, 54
Animals, features of 257-259, 261
Annealing 195
Antagonistic muscles 359
Antibiotic resistance 197, 247
Antibiotics, actions of 291
Antibodies 293, 298-299, 305-306
Anticodon 105
Antidiuretic hormone (ADH) 339
Antigen 293, 299, 305
Antigenic variability, pathogen 250, 251
Apical meristem 374
Apoptosis 68, 71
Arterial system 278
Artery 284-285
Assumptions 11
ATP synthase 119
ATP 89, 114-15
Atria, of heart 279
Atrioventricular node (AVN) 282
Autosome 143
Auxins, role 380
Axon 323-324

B-cell 287, 293, 298 300
Back cross 155
Bacteria cell, features of 43
Bacteria, as pathogens 291
Bar chart 24
Base pairing rule 95
Beneficial mutation 141
Bilateral symmetry 255
Bile, digestive 273
Binomial nomenclature 253
Biochemistry, cellular 74
Biogeochemical cycle 216
Biosphere 214
Biotic factor 214
Birth 352
Blastocyst 349
Blood cell count 288
Blood clotting 295
Blood glucose, regulation 328, 332
Blood pressure 280
Blood vessels 284-285
Blood, components 287-288
Blunt end 193
Bowman's capsule 337-338
Breathing 315
Bronchi 313
Bronchioles 313
Budding, in yeast cells 70
Bundle of His 282

Calvin cycle 126, 128
Cancer 68
Capillary 284
Capillary network 286
Carbohydrates 74, 77, 79-80
Carbon cycle 224-225
Cardiac cycle 281
Cardiac muscle 282
Carnivore 215-216
Carotenoids 124
Carrier, genetic 179
Carrying capacity 232, 234
Cartilage 358
Catalase activity, measuring 18, 25
Catalyst, defined 107
Cattle, selective breeding 240
Cell cycle 69-70
Cell cytoskeleton 53
Cell division 67
Cell junction 53
Cell respiration 114-118, 312
Cell sizes 38
Cell structure 51-53
Cell theory 37
Cell types 43, 47-50, 54-55
Cell wall 43, 47, 53
Cellulose, structure of 79-80
Central nervous system (CNS) 321-322
Centrioles 49, 52
Chemical synapse 327
Chemiosmosis 117, 119
Chemoreceptors 331
Chlorophyll 123-124
Chloroplast 47, 51, 123-124
Cholinergic synapse 327
Chromatin 91
Chromosome 92, 134, 143, 145
Chymosin, production of 199
Circulatory system, human 278
Classification keys 263-265
Classification system 253
Clonal selection theory 300
Cloning 209-211
Clotting, role in defense 295
Coding strand, of DNA 104
Codominance 158-160
Codon 103
Coenzymes 110
Cofactors, of enzymes 110-111
Cohesion-tension 383
Colon 272
Companion cell 378
Competitive inhibition 111
Complement system 293
Concentration gradient 61
Condensation reaction 78, 82, 83
Consumer 215
Continuous data 14
Continuous variation 177
Contraction, of muscle 362
Control, experimental 17
Corpus luteum 344, 351
Cortical reaction 348
Cotransport 64
Coupled transport 64
Crossing over 137-138, 165-166
Cyclic phosphorylation 126
Cytokinesis 69
Cytoplasm 47, 49

Darwin's finches, evolution 247
Darwin's theory 241
Data, transforming 15, 22
Data, types 14
Decomposer 215-216
Defense systems 293-298
Denaturation 85,108
Density dependent factor 233
Density independent factor 233
Density, of populations 232
Dependent variable 17
Depolarization 326
Descriptive statistics 21, 27-28
Detritivore 215-216
Diabetes mellitus 332-333
Dichotomous key 263-265
Dicot, features 367, 369-370
Differentiation, of cells 45
Diffusion 41, 61, 312
Digestive system, human 269-272
Dihybrid cross 155, 162
Dihybrid inheritance 165-167, 172-173
Directional selection 243-244
Disaccharides 77
Discontinuous data 14
Disease
- HIV/AIDS 307-309
- Huntington's 150
- viral 292
- Whooping cough 304
Disruptive selection 243, 247
Distribution (of data) 27-28
Distribution, of populations 232
DNA 89-93
DNA ligase 195, 197, 199
DNA model 95-97
DNA profiling 186-188
DNA replication 99, 102, 137
DNA sequence, changes to 141
DNA technology, applications of 189
Dormancy, plant 396
Down syndrome 136
Duodenum 271

Ecological efficiency 219
Ecological pyramids 221-222
Ecosystem, components of 214
Effectors 321-322
Egg (ovum) 346-347
Electrocardiogram (ECG) 281
Electron micrographs 54-56
Electron transport chain 116-117, 119
Embryo 349
Embryonic stem cell 211
Emergent properties 44
Endocytosis 65
Endoplasmic reticulum 47, 49, 51
Energy, in an ecosystem 216, 219
Enzyme applications 113
Enzymes 86, 107-113
- digestive 269, 271, 274
- in DNA replication 101
Epinephrine, and heart control 282
Error bars 24
Erythrocytes 287
Essential amino acid 83
Estrogen 351
Ethics
- of HGP 191
- of IVF 354
Eukaryotes 39, 255-259
Evolution, evidence for 239-251
Excretion, of waste 334
Exocytosis 65
Exon 93
Exothermic reaction 108, 117
Experiment
- catalase activity 25
- plant growth 19-21
Experimental control 17
Expiration (exhalation) 315
Exponential growth 234

Facilitated diffusion 61
Fallopian tube 343
Fatty acids 81
Feedback inhibition 112
Female reproductive system 343
Fermentation 118, 120-121
Fertilization 348-349
- *in vitro* 353-354
- plant 391
Fever 330
Fibrous protein 86
Fick's law 61, 312
Fimbriae 43
Fitness 249
Flagella 43, 52
Fluid mosaic model 59
Follicle stimulating hormone 344-345
Food chain 215, 217
Food production, GM 205
Food web 216-218
Forensic profiling 188
Formula, of molecules 75
Fossil record 239
Fungi, features of 260

Gametes 347
Gametogenesis 67, 345-346
Gas exchange 312-314, 368
Gas transport 318
Gel electrophoresis 185
Gene cloning 197
Gene mapping 190
Gene markers 197
Genetic code 94
Genetic counseling 179
Genetic crosses 155, 157
Genetic gain 240
Genetic modification 192, 197, 201-202
- ethics 207
Genome project, human 190-191
Genome, definition 133
Genotype 151, 238
Germination, of seeds 394-395
Global warming 226-230
Globular protein 86
Glomerulus 337
Glucagon 332
Glucose catabolism, control of 112
Glucose, blood levels 332
Glycocalyx 43
Glycogen, structure of 79
Glycolysis 116-118
GMOs 192, 196, 201
Golden rice 201
Golgi apparatus 47, 49
Graafian follicle 344, 346
Grana 123
Graphs 24
Greenhouse effect 226-227
Greenhouse gas 226
Gross primary production 219
Growth curves, populations 234
Growth responses, plant 379
Guard cell 368
Gut, human 270-272

Halophytes 387-388
Harmful mutation 141
Heart 279, 281, 283
-control of activity 282
Heartbeat, generation of 282
Hemoglobin 318
- mutations to 142
Hemostasis 295
Hepatic portal system 278
Hepatocyte, structure 49
Herbivore 215-216
Herd immunity 302
Heterozygous, defined 134
Histogram 24
Histone protein 91
HIV 307-309
Homeostasis 329, 331
Homologous chromosomes 134, 138
Homologous pair 134
Homologous structure 241
Homologues 134
Homozygous, defined 134
Hormone, auxin 380
Hormones, functions of 328, 344, 351
Horse evolution 239
Human chorionic gonadotropin 351
Human genome project (HGP) 190-191
Humoral immune response 298
Huntington's disease, diagnosing 150
Hydrolysis reaction 78, 83
Hyperglycemia 333
Hypothalamus 330
Hypothesis 10-11

Immune system 298
Immunity 301-302
Immunoglobulin 299
Immunological memory 300
Immunological response 302
Implantation 343, 349
In vitro fertilization 353-354
In vivo gene cloning 197
Independent assortment 137, 154
Independent variable 17
Induced fit, model of enzyme activity 108
Infectious disease 291
Infertility 353-354
Inflammation 297
Influenzavirus 251
Inheritance patterns, examples 171

Index

Inheritance, laws of 154
Inhibition, of enzymes 111
Inorganic ions 74
Insecticide resistance 248
Insectivorous plants 372
Inspiration (inhalation) 315
Insulin 332, 328
- recombinant production 203-204
Intron 93
Ion pumps 64
Isomer, carbohydrate 78

Joints 357, 359

Karyotype 143-144
- exercise 145-147
Kidney 334, 336-338
Krebs cycle 116-117

Labor 352
Lactation 351-352
Lactic acid fermentation 120
Lactose intolerance 113
Lagging strand, of DNA 101
Large intestine 272
Leading strand, of DNA 101
Leaf, structure 366, 368, 373
Leukocytes 287, 293
Ligament 357
Ligation, of DNA 195
Linear magnification, calculating 40
Link reaction 116-117
Linkage, of genes 163-164, 167
Lipase 269, 271
Lipids 74, 81-82
Liver cell, structure 49
Liver, role in digestion 273
Lock and key model, of enzyme activity 107
Logistic growth curve 234
Long-day plant 396
Loop of Henle 337-338
Lungs 313-317
Luteinizing hormone 344-345
Lymphocytes 287, 293, 298, 300
Lysosome 49-50

Magnification 40
Male reproductive system 342
Mean (of data) 21
Median (of data) 21
Meiosis 67, 135, 137-138
- Meiosis vs. mitosis 135
- exercise in modeling 139-140
- non-disjunction 136
Membrane, plasma 58
Memory cells 300
Mendel, Gregor 153-154
Menstrual cycle 344, 351
Meristems 374
Metabolic pathways, control of 112
Metabolism, defined 112
Microsatellites 186-187
Microvilli 271, 275
Middle lamella 47
Migration, population 232
Mitochondrion 47, 49, 51, 114
Mitosis 69-70
Mitosis vs. meiosis 135
Mode (of data) 21
Monoclonal antibodies 305-306
Monocot, features 367, 369-370
Monohybrid cross 155, 157-162
Monosaccharide 77
Morula 349
Motor neuron 323
Movement, mechanics of 357-360
mRNA 103, 104
mRNA-amino acid table 94
Multicellularity 44
Multiple allele systems 159-160
Muscle structure 359, 361-362
Mutation 137, 141-142, 151
Myelinated neuron 323-324
Myofibril 362
Myogenic nature, of the heart 282
Myoglobin 318
Myosin 363

Natural selection 241-251
Necrosis 71

Negative feedback 329, 332
Nephron 337-338
Nerve impulses 326
Nervous system 321-322
Net primary production 219
Neuromuscular junction 361
Neuron 321, 323-324
Neurotransmitters 327
Neutral mutation 141
Node of Ranvier 323-324
Non-competitive inhibition 111
Non-cyclic phosphorylation 126
Non-disjunction 136
Non-myelinated neuron 323
Non-specific resistance 293
Nuclear membrane 47
Nuclear pore 47, 49
Nucleic acids 74, 89
Nucleolus 47
Nucleotide 74, 89-90, 103
Nucleus 47, 49, 52
Null hypothesis 11
Nutrient cycles 223

Oocyte 346
Oogenesis 67, 346
Organelles 47-48, 51 50
Organic molecules 74-75
Osmosis 61-62, 382
Ovarian cycle 344
Ovary 343
Ovulation 343
Ovum 346-347
Oxidative phosphorylation 117
Oxytoxin 351

Pacemaker, of heart 282
Pancreas 271, 332-333
Particulate inheritance 154
Passive immunity 301
Passive transport 61-63, 66
Pasteur, Louis 291
Pathogens 291
- resistance in 250-251
PCR 182-183
Pea plant experiments 153
Pedigree analysis 174-176
Pentadactyl limb 241
Peppered moth, and evolution 244
Peptide bond 83
Peripheral nervous system 321-322
Peroxisome 49
pH, maintaining 331
Phagocyte 287, 296
Phagocytosis 65
Phenotype 151, 238
Phloem 369, 378, 389
Phospholipid, structure of 82
Photoperiodism 396-397
Photosynthesis 123, 125
Photosystem 126
Phototropism 381
Phytochrome 396
Pigments, photosynthetic 123-124
Pinocytosis 65
Placenta 350-351
Plant adaptations 387-388
Plant cells, features of 47-48, 55
Plant growth 379
- experiment 19-21
Plant hormones 380
Plant structure 366
Plant tissues 48
Plant,
- modifications 371-372
- support structure 375-376
Plants, features of 256, 260
Plasma cells 300
Plasma membrane 47, 49, 51, 58-59
Plasmodesmata 47
Plasmolysis 63
Plateau phase, of growth 234
Platelets 287
Pluripotent cell 211
Pneumatophores 371
Podocyte 337
Pollen 390
Pollination 390-391
Polygenes 177
Polymerase chain reaction 182-183
Polymorphism 241

Polypeptide bond 83
Polysaccharides 79-80
Polytene chromosome 92
Populations 232-233
Positive feedback 329
Potency, of cells 45
Potometer 384-385
Precautionary principle 231
Precision, defined 13
Prediction 10-11
Pregnancy testing 306
Primary growth, plants 374
Primary response, immunological 302
Producer 215-217
Progesterone 351
Programmed cell death 71
Prokaryotic cell, features of 43
Prolactin 351
Protease 269, 271
Protein synthesis 103, 106
Proteins 74, 85-86
Protista 39, 255
Proton pump 64
Pulmonary system 278
Purine 89
Purkyne tissue 282
Pyrimidine 89

Qualitative data 14
Quantitative data 14
Quantitative investigation 17-18

Radial symmetry 255
Random sampling, effect on data 29
Range (of data) 21
Receptors 321-322
Recognition sites, for restriction enzymes 193-194
Recombinant bacteria 199
Recombinant DNA 197
Recombination, of genes 165-166
Red blood cells 287
Reflexes 323, 325
Refractory period 326
Regulation, of populations 233
Replication fork 99, 101
Repolarization 326
Reproductive system, human 342-33
Resistance, development of 248-251
Respiratory membrane 313-314
Respiratory pigment 318
Respiratory system, human 313-314
Resting potential 326
Restriction enzymes 193-194, 197 199
Rhizomes 371
Ribosome 47, 49, 51
Ribosome, role in translation 105
RNA Polymerase 101,104
RNA, structure 89
Root hairs 369-370
Root, uptake at 382
Roots 366, 382, 368, 371
RuBisCo enzyme 128
Saltatory conduction 324, 326
Sample variability 29-30
Sarcomere 361
Scientific method 10
Secondary growth, plants 374
Secondary production 219
Secondary response, immunological 302
Seeds 392-394
Segregation, law of 154
Selection 238, 243
Selective breeding 239-240
Self renewal, of cells 45
Seminiferous tubule 345
Sensory neuron 323
Sensory receptor 322
Sex chromosomes 143, 168
Sex linkage 169-170, 175-176
Sex, determination of 168
Sexual reproduction 151
Short tandem repeats 186-187
Short-day plant 396
Sickle cell mutation 142
Sieve tube 378
Silent mutation 141
Sinoatrial node 282
Sinusoid 284
Skeletal muscle 361-362

Skin color, human 177-178, 245-246
Skin, roles of 294, 330
Sliding filament theory 363
Sodium-potassium pump 64
Somatic cell nuclear transfer 209-210
Spearman Rank correlation 31
Specific resistance 293
Sperm 345, 347
Spermatogenesis 67, 345, 347
Spirometry 316-317
Spread (of data) 27-28
Stabilizing selection 243
Standard deviation 28-29
Starch 79-80, 274
Statistics, descriptive 21
Stem cell 44-46, 211
Stems 366, 368
Sticky end 193
Stimuli 321
Stomach 271
Stomata 368
Stroma 123
Student's *t* test 32-34
Substrate-level phosphorylation 117
Substrate, enzyme 107 100
Sugars 77
Surface area to volume ratio 41-42
Synapse 327
Synovial joint 357-358

T-cell 287, 293, 298
Tables 23
Taxonomic groups, features of 254-259
Taxonomy 255
Temperature, control of 329-330
Template strand, of DNA 99, 104
Tendons 359
Test cross 155
Testosterone 342
Thermoregulation 329-330
Threshold potential 326
Thylakoid membrane 123
Tight junction 49
Tissue engineering 46
Tissue fluid 284
Transcription 103-104
Transition state 108
Transitional phase, of growth 234
Translation 103, 105
Translocation 389
Transpiration 383-386
Transport, cellular 61 66
Triglyceride 81
Triplet code 94, 103
Trisomy 21 136
tRNA 105
Trophic structure 217
Tropisms 379
Tumor 68
Turgor 63, 375

Ultrafiltration, at kidney 338
Urine 337-339
Uterus 343

Vaccination 303-304
Vaccines 301-304
Variables, types 14, 17
Variance 29
Variation, sources of 151-152
Vein 284-285
Venous system 278
Ventilation system, human 313-315
Ventricle, of heart 279
Vessel elements 377
Villi 271, 275
Viral diseases 292
Virus, HIV 307-309
Viruses, as pathogens 291-292

Waste products, human 334
Water budget 335
Water, role of 76
White blood cells 287, 293

X-chromosome 168
Xerophytes 387-388
Xylem 369, 377

Y-chromosome 168
Yeast cell, budding 70